MODERN CONCEPTS AND ADVANCED PRINCIPLES IN CROP PRODUCTION

Dr. S. C. Panda, Ph.D.

Consultant, World Bank, Sutra Consulting,
Emeritus Scientist, ICAR,
Former Dean, College of Agriculture,
Dean of Research, Dean, PGF-cum-DRI, OUAT,
Professor & Head, Dept. of Agronomy, ADR,
Chief Scientist (Water Management),
Regional Co-ordinator, NDP (Eastern Region of India),
Program Associate (Multiple Cropping Res.), Ford Foundation

TITLE: Modern Concepts and Advanced Principles in Crop Production
AUTHOR: Dr. S. C. Panda

Published by:

AGROBIOS

AGROBIOS (INDIA)
Behind Nasrani Cinema
Chopasani Road, Jodhpur - 342003
E. Mail: agrobiosinsia@gmail.com
website: www.agroacademics.com

This book contains information obtained from authentic and highly regarded sources. Every effort has been made to trace copyright holders and to obtain their permission for the use of copyright material. Reprinted material is quoted with permission, and sources are indicated. A wide variety of references are listed. Reasonable efforts have been made to publish reliable data and information, but the author and the publisher cannot assume responsibility for the validity of all materials or for the consequences of their use.

First Published: 2014
© (2026) All Rights Reserved

All rights reserved. No part of this book or part thereof, including title of the book may be reproduced or used in any format in any form or by any means—graphic, electronic, or mechanical, including photocopying, recording, taping, or information storage and retrieval systems—without the written permission of the author and publisher. The copyists shall be prosecuted.

ISBNs: 978-81-7754-435-0 (Hardcover)

AGROBIOS (INDIA)
Managing Director: Dr. Updesh Purohit
Founder & Editor: Dr. S. S. Purohit

Services
Book Design & Layout: Yashee Computers
Illustrator & Printers: Agrobios Digitals
Cover Illustration: Reena

PRINTED IN THE INDIA

AT THE FEET OF LORD SHREE JAGANNATH, PURI, ORISSA

DEDICATED TO MY FATHER, SHYAM SUNDAR PANDA AND MOTHER, SHNEHALATA PANDA

FOREWORD

DR. B. PANDA
B.Sc. (Vet.) Hons. (Mumbai), M.S., Ph.D. (Maryland, USA)
Founder Director (Retd.), Central Avian Research Institute (ICAR)
Former Member of the Board of Management, OUAT, Bhubaneswar.
Fellow: International Poultry Hall of Fame
(World's Poultry Science Association)
National Academy Of Veterinary Sciences (India)

Residence:
C 6, Palaspalli
Bhubaneswar-751020
Orissa,(India).
Tel: 0674- 590376
Fax: 91-674-418142
E-mail: (Attn. Dr. B. Panda)
bpanda-29 @ hotmail.com

"Success is the ability to go from failure to failure without losing your enthusiasm"

Global warming as a consequence of climate change has become an issue of major concern throughout the world, more so, for developing countries like India, where the multifarious demands of the large population are putting an enormous strain on the natural resources. Various GCMs generally predict a rise in temperature, increased frequency of floods, droughts and increase in pests and disease infestation in the country, when are likely to adversely affect agriculture, livestock production and fisheries. Some of the key elements of the coping mechanism include use of crop varieties having multiple tolerances to various abiotic and biotic stresses; adoption of resource conservation technologies that use less water and energy and integrated farming systems, which are risk resilient. To face this challenge of climate change, we need to marshal all our resources within the country and evolve location specific mitigation technologies which can help in coping with extreme weather conditions and sustain agriculture and livelihood security of the poor.

The pressure of an ever increasing population and periodic famine due to unexpected flood and drought have forced and awakened the agricultural scientists of India to evolve new plant types and to develop a suitable production technology for such high input responsive crop plants. The last two decades had been an era of real revolution in agriculture, especially in the field of crop production which has brought a spectacular change in the existing cropping system in various agro-climatic zones of the country. The breakthrough in crop yields has made our nation self-sufficient in food grains and even some quantity of these grains is being exported to other needy countries. Thus, the modern technology has changed us from net importer to exporter of the food grains. The Indian farmers have shown a surprising dynamism, adaptability and progressiveness in accepting the recent crop production technology and predominant subsistence farming is being changed into an enthusiastic enterprise however, it lacks perfection. Moreover, even today the national average yield of various crops is miserably poor and incomparable to that obtained in other counties. **Transfer of technology** from research level to the farmers had been probably the major hurdle in achieving the targeted yield. Illiteracy and ignorance of our poor farmers are creating hindrance for new plant types to express their full yield potential when grown under faulty management in the field. Higher input levels, faulty planning coupled with poor marketing facilities often result in a marginal profit from farming due to which majority of marginal and sub-marginal farmers are obviously discouraged.

The technology of crop production is a vastly developing subject. In recent years, there has been such rapid progress in the research efforts in this branch in many of the developing countries of the world that has become difficult to keep pace with the subject. This is particularly true in the tropical regions of the world and in the Indian sub-continent. The students of Agronomy in the Agricultural Colleges need to be taught the latest technology so as to make them effective as extension workers as also to train them for becoming competent teachers and researchers. A comprehensive treatise on crop production would not only help the students but also the teachers in the Agricultural Colleges. In modern agricultural technology 'management' forms an essential part. If agriculture were to compete with other small and medium scale industries in rural areas, there should be proper management of the natural resources and the various inputs which go with it. Unfortunately, this aspect is mostly neglected in Indian agriculture.

Besides, **the modern production technologies** with high input requirement offers on tangible solution to the small and medium farmers' problem of low income and low productivity from cropping. To meet the growing demands, to solve small and medium farmers' problems and to achieve the targeted production levels with stability, ecological sustainability and equitability in the 21st century, a viable technology called 'Integrated Farming System Approach' suited to varied farm situations and agro-climatic conditions seem to be the answer. The farming system approach incorporated in this book has taken due importance through recycling of no cost/low cost material of one enterprise as input on the other enterprise thus achieving great reduction in the cost of cultivation. Agronomists have made efforts to develop low-cost farming systems based on the principles of productive utilization of farm wastes and available resources with restricted use of purchased inputs.

There are 115 million operation holdings in the country and about 80 % are marginal and small farmers. To fulfil the basic needs of house hold including food (cereal, pulses, oilseeds, milk, fruit, honey, meat, etc.), feed, fodder, fibre, etc. warrant an attention about **Integrated Farming System (IFS).** Undoubtedly, majority of the farmers are doing farming since long back but their main focus was individual components but not in a integrated manner. At the ICAR and State Agricultural Universities level, lot of efforts have been made aiming at increasing the productivity of different components of farming system like crop, dairy, livestock, poultry, piggery, goat keeping, duckery, apiculture, sericulture, horticulture, mushroom cultivation, etc. individually but lacking in their integration by following farming system approach. The integration is made in such a way that product of one component should be the input for other enterprises with high degree of complementary effect on each other. The fodder fed to the cattle produces milk. The dung, urine and litter produce farm yard manure and energy used for crops and fish pond. The situation of fish pond is utilized as manure to crops. The farm yard manure can substitute about 25 ^ of recommended N P and K for crops, besides, improving the physical and biological properties of soil. The fish pond water can be used by gravity method while there is breakdown in electricity supply. Oilseeds provide nectar for honeybees, edible oils for human and oilseed cake for animal feed. Integrated nutrient management can enhance the productivity of cereals by 0.5 to 1.0 t/ha. Processing of different products enhances the value addition to the extent of 25 to 50 % besides generating

50-75-man days/family/year of employment. The fish pond embankment comprising 20-30 % can be used for growing cucurbits and fruit trees which provide effective soil cover to check the soil erosion and also make the system economically viable. The apportionment of farm land involving different enterprises, viz., cereal (40 %), pulses (10 %), oilseed (10 %), horticulture (15 %), fishery (10 %), livestock (10 %), poultry/piggery/goatery (2 %), storage, threshing floor, implement shed, vermicompost, straw storage and farm building etc. (3 %) for the efficient use of resources available and to enhance the input use efficiency. The preliminary research investigations advocated the benefits of productivity improvement by 30-50 % depending upon the number and kind of enterprises and their management. The information on farming system in a systematic way is presented in this book. The methodology is explained keeping in mind the work done so far to realize better productivity, profitability and sustainable production systems that would help to solve the fuel, feed and energy crisis, create more employment avenues, ensure regular income and encourage agricultural oriented industry.

Dr. M.S. Swaminathan has highlighted some of the challenges pertaining to food and nutrition security that lie ahead. According to him, sustainable food and nutrition security involves improving current needs in an agricultural production without sacrificing the prospects from meeting the needs of future generations. Sustainability is a dynamic concept. For achieving quantitative and qualitative sustainability, attention has to be paid to ecological, equity and economic factors. For the scientists, he has given seven priority areas. The emphasis has to be laid on sustainable development, biological diversity, hazardous waste disposal, air and water pollution, global climate change etc.

It is now fully realized that, except climate, all other resources in crop production can be exploited to a limited extent. A fuller exploitation of the weather resources, therefore, is the major hope in greater production to meet the demands of the staggering increase in population. It has been demonstrated that, apart from weather forecasting, which solves a number of our daily problems, meteorology, has very useful application to various activities, such as the selection of crop-production sites, irrigation control, soil and water conservation, the amelioration of field climate and the adoption of the best-fitted agronomic and cultural practices in crop production.

In India, agriculture not only provides food for all, but also employment to 70 % of the population, generates 40 % of the national income and consumes about 10 % of the commercial forms of **energy.** Crop cultivation requires application of both animate (bullock, human power) and inanimate (tractors, tillers etc.) **forms of energy** at different stages. Nutrients are provided through farm yard manure, chemical fertilizer or both. Pesticides are required to check or prevent pest attack. Irrigation is done either manually (manually and animal operated) or through diesel/electric pump sets (to lift ground water). To meet the basic food needs of our expending human population, a productive sustainable agricultural system must become a major priority. The depletable resources are fossil fuels, which are non-renewable since the rate of their utilization far exceeds the rate at which they are formed.

The concept of **Integrated Farm Management (IFM)** is more feasible for small farm-holdings and it serves as a tool for linking environmental safety with weed control besides offering scope for **conservation of agro-biodiversity.** The component elements of fish culture and poultry rearing in transplanted rice have shown to compliment weed control by 30 % besides imparting sustainability as improved soil health and economic returns. The technology has been up scaled for adoption by farmers through on-farm demonstrations and farmers training. Under rainfed upland conditions, integrating goat rearing, especially for grazing during the off-season complimented weed control by 47.17 % and addition to goat manure to fields grazed upon by goats contributed for 31.34 % weed control during the cropping season. Technology for integrating a medicinal herb Coleus amboinicus with insect agents *Neochetina eichhorniae/bruchi* for the bio-control of water Hyacinth has also been evolved and demonstrated in aquatic systems. Naturally regenerated stands of *Prosopis juliflora* in social sites are shown to be better managed by slashing and digging out tools followed by rehabilitative cropping with sunflower + Cowpea. Another invasive alien weed *Parthenium hysterophorus* is shown to be resisted by avenue plantations with *Tamarindus indica.*

Most of our agriculture was rainfed. Agronomic research therefore, was revolving around increasing yield under this scenario available at that part of time. Simple fertilizer trial, tillage practices to conserve moisture and control weeds and mixed cropping were the main aspects of research. Farming was done for subsistence. Yields were sustained at low levels. Then, there was a breakthrough. Plant architecture was changed. Short duration dwarf varieties came into existence. Agronomic research also shifted from simple fertilizer trials to complex fertilizer experiments, mono-cropping to multiple cropping and scheduling irrigation. Use of herbicide increased to control weeds. With the expansion of irrigation, scheduling irrigation became main focus and legumes went out of the cultivation and cereal-cereal cropping became predominant. Spread of rice-wheat cropping system in Indo-Gangetic Plain region is the glaring example of this. Due to this cereal-based crop rotation, natural resources started degrading and weeds developed resistance to herbicides. Factor productivity started declining. Presently these are our cultivation practices i.e., our conservation, which is chemical based and input intensive, this we wish to change to bring bio-intensive agronomic practices, which will include massive use of biomanures, biofertilizers, and biopesticides. The research on rhizospheric engineering and carbon sequestration has to be strengthened to improve the soil health and nutrient use efficiencies. Rhizospheric engineering refers to bringing changes in root architecture by modifying planting methods and crop geometry and also improving root zone soil profile through rhizodeposition. In irrigated eco-system, where legumes could not be introduce due to their sensitivity to water, sugarcane is to be brought in the system, because several beneficial microorganisms are associated with its roots, which mobilizes soil materials. Also, large quantity of root mass of sugarcane upon decomposition adds substantial quantity of organic matter to the soil.

System of Rice Intensification (SRI) developed in Madagascar in 1980's is gaining wider acceptance in many countries including India due to its advantages and input saving over conventional method of rice cultivation (Uphoff, 2003). The components of SRI include the use of young seedlings, careful transplanting of single seedling per hill, wider spacing, saturation of field with reduced irrigation, aerated soil conditions by frequent soil disturbance using cono weeder for weed management and the use of organic manures. In India, the rapid increase in area under high yielding varieties of rice accompanied by increased usage of fertilizers has led to increased incidence of pests and diseases. Farmers rely mostly on pesticides for reducing the losses caused by the pests. Rice receives about 20 % of total pesticides in the country. The rice plants grown under SRI method are less susceptible to insect pests and diseases due to their healthy growth (Ngo, 2007). There are some evidences of significant impact of climate changes in India. There have been some changes with respect to moisture and thermal regimes, which have significant impact on agriculture and food security. Instability in rice production will affect food security. Adhering to the System of Rice Intensification (SRI) can provide considerable stability in rice production and food security along with increased resource use efficiency under changing climate.

The nano-technology is likely to revolutionize the agricultural production system. Nutrition with desired elements and flavor stored in the capsules burst open in the stomach and deliver the nutrients in quantities needed by the organ. Similarly they release the toxins in the stomach of the harmful insects of crop plants. **Nano-sensors** will alert the farmers before the occurrence of the pest, disease and nutrient deficiency and damage to the crops. The resource conservation technologies for anaerobic crops, after the harvest of aerobic rice save expenses and time for land preparation.

Some developing countries have embarked upon systematic **biotechnology research** programmes, but many others are unable to do so. International cooperation will be necessary for transfer of technologies to developing countries In the foreseeable future biotechnology will support conventional plant breeding producers but not replace them.

To achieve agricultural sustainability, a special programme is necessary for collection, evaluation and enhancement of genetic resources for the promotion of sustainable advances in crop productivity. Scientific management of water, better use of degraded soil, adoption of inland fisheries, aquaculture, livestock, setting up small scale agro-processing industry in rural areas to sustain rural economy, women's education etc. are important points by which sustainability in agricultural productivity can be fulfilled. Attention to crop, livestock farming and agro-forestry will be helpful in generating more jobs and income.

India is endowed with **a rich and vast diversity of natural resources**, particularly soil, water, weather, multipurpose trees and bio-diversity. To realize the **potential of production system** on a sustained basis, efficient management of the natural resources is very crucial. With the advent of high yielding crop varieties and intensive cultivation, the food grain production has increased from 51 million tons (m t) in 1950-51 to a record figure of 210 m t during 2007-2008. This impressive achievement pulled the country in to self-sufficiency for food demand. With adoption of intensive agriculture to meet the varied growing demands for food, fuel, fiber, feed, fertilizer and products in the recent year, the natural resources are, however, put under intense strain resulting in fast degradation and lowering of their production efficiency.

Agricultural Marketing is a broader sense is concerned with the marketing of farm products produced by farmers and of farm inputs required by them in the production of these farm products. Thus, the subject of Agricultural Marketing includes product marketing as well as input marketing.

It is quite clear that employment policy in India cannot be successful unless it takes into account these various types of unemployment prevailing in India. For every different type of unemployment different type of remedy is needed. It is quite possible that while we may be reducing unemployment in the industrial sector, disguised unemployment may remain as such as before or while disguised unemployment is being cured seasonal unemployment persists. The complexity of the unemployment problem has been recognized by our planning authorities and accordingly we find that unemployment problem being dealt with by breaking it first into these various categories and then devising remedies.

The text-cum-reference book to meet precisely the felt need is an outcome of the author's active involvement in teaching, research and extension guidance in the field of agronomy for over thirty years. The author presented the book entitled, **"Modern Concepts and Advanced Principles in Crop Production"** in a scientific and systematic manner to understand the fundamentals clearly and easily which is the beauty of this book. I believe that this book will be very useful to the undergraduate and postgraduate students in agriculture, teachers in agricultural institutions and those who are interested in the subject. The book is divided into twenty five chapters and covers comprehensively the content of the courses of Modern Concepts and Advance Principles in Crop Production for undergraduate and post graduate students of Agricultural Universities in the country. The chapters have been arranged in such a manner as to lead the students through the entire gamut of Agronomy, Soil Science, Extension Veterinary Science and Agricultural Economics.

I am confident that this book will serve as a text book for agronomy, Soil Science, Extension, agricultural economics and veterinary students, a reference for research scientists and teachers in the areas of agro-meteorology for potential crop productivity, for food and nutrition security in Indian economy, farm management and agricultural marketing on crop production, integrated farming systems, dry land agriculture, cropping systems, production technology management under different situations, soil fertility management, water management, weed management, nutrient management, and avian and animal sciences, along with product and input marketing. This book will also serve as a guide to the extension officials of the department of agriculture. I congratulate Dr. S. C. Panda for his pains taking effort in bringing out this book covering comprehensively the content of all courses in agricultural economics and agronomy offered for U.G. and P.G. students of the agricultural universities in the country and the latest technologies for crop production associated with integrated enterprises in farming systems to meet the growing interest in sustainable agriculture. I am confident that this book will be widely accepted among the students. I extend my best wishes to Dr. Sharat Chandra Panda for the success of this book.

Bhubaneswar. **Dr. Bhagabat Panda**

PREFACE

"Butterfly keeps flying from one flower to another in search of nectar without losing enthusiasm. Similarly, one should never lose enthusiasm as it is the only force, which drives people to perform at their best and allows them to achieve the worthiest of causes"

We increasingly face **ecological** and **environmental** problems as a result of injudicious use of fertilizers and other chemicals (pesticides, herbicides etc.) in intensive agriculture. The deteriorating soil and water quality, and the raising agrochemical toxicity in farming are serious concerns. All these factors jeopardize efforts to sustain growth in food production. The big questions before us are "Can we sustain high productivity with deterioration of soil and water environment?

Most of the additional food grain production must come from irrigated and potential rainfed lands. We need to use more complex technologies and management practices to further intensify crop production systems and to conserve resource base from which all food is produced. The immense task before us is **"How to orient our research to generate innovative technologies".**

Plants like, human being, animals, birds and other living organisms need energy for their survival and proper functioning. Unlike other living organisms, they use to manufacture their food through conversion of solar energy into chemical energy vide process called as photosynthesis. The photo-synthates, thus synthesized, are partly consumed by plants for their growth and development till their active vegetative growth and rest is accumulated which we harvest. The entire process is regulated by various elements which are known as plant food elements. The list of such elements, is gradually enlarged with the **advancement of physiological and biochemical knowledge.** These elements, though present in soil yet the quantity being so meager need to be supplemented through their respective carriers called as fertilizers.

India witnessed self sufficiency in food production by shifting from the traditional organic to chemical industrial agriculture in the second half of the past century. The production levels plateau and several challenges of soil health, crop production and environment have become critical in the new millennium. The need paradigm shifts in agricultural practices and research priorities. The **System of Rice Intensification (SRI)** conserves water but it is difficult to practice. **Organic Farming** is eco-friendly but, organic source of nutrients are not sufficient to meet the nutrient requirement of crops and the yield reduction is substantial. It has little scope for farmers to export the produce to developed countries after meeting the standards of trade and cumbersome procedures. The integrated supply of nutrients is potential to sustain the crop yields and restore the soil-health. The transgenic crops opened new vistas in agricultural production. But, such genetically modified crops are frequently unstable and inactivated. They lose the newly acquired traits due to gene silencing. **The cell and tissue culture techniques** may substitute the traditional crops for the production of food like starch, protein and fats in fermentation tanks. **The nano-**

technology is likely revolutionizing the agricultural production system. Nutrition with desired elements and flavor stored in the capsule burst open in the stomach and deliver the nutrients in quantities needed by the organ. Similarly, they release the toxins in the stomach of the harmful insects of crop plants. **Nano-sensors** will alert the farmers before the occurrence of the pest, disease or nutrient deficiency and damage to crops. The **resource conservation technologies** for anaerobic crops, after the harvest of aerobic rice save expenses and time for land preparation. This will facilitate early sowing of the crops after the harvest of rice on the carry over soil moisture or a shallow irrigation. The use of heavy machinery for sowing may compact the soil in the long run. The residue on the soil surface adds organic matter but, it is also likely that it may hinder proper seedling growth. The rice stubbles may harbor the harmful pests and disease inoculums, refuge the rats and invite the snake problem.

Water is an important life giving natural resource. Yet, human kind not done enough to replenish, conserve and safeguard sources of water supply. On contrary, we face the real prospect of reduced supply of water because of **climate change and global warming**. Irrigation is the major source of water and it consumes about 80 % of the current water utilization in India. It is estimated that with increasing demand from other competing sectors, the share of water is likely to reduce by 10 to 15 in the next two decades. In addition, shrinking water resources due to falling ground water tables, silting of reservoirs and pollution of water resources from untreated domestic waste, industrial effluents and agro-chemicals are threatening the sustenance of existing levels of agricultural production. At the national level, the **ground water contributes** to about 50 % in the irrigated areas and its intensive development has been one of the key-driving elements in ushering the green revolution but over exploiting of ground water has led to several other problems affecting the sustainability of the production base. Its unscientific exploitation is creating problems and leading to many areas in the country being classified as dark and grey blocks. The per caput water availability is declining continuously and likely to reach the stress and even scarcity levels in some regions in the next few years. There are variety of **irrigation water saving technologies** that can improve the water use efficiency and productivity of water. These include *in-situ* water conservation, water auditing, increasing in irrigation efficiency adoption of improved irrigation methods and conjunctive use. Even a marginal improvement in the **water use efficiency** in irrigation sector will result in saving of substantial quantity of water which can be utilized either for extending the irrigated area or for diverting the saving to other sectors of water use. It is now widely recognized that water, especially irrigation water is finite and a vulnerable resource. There is also a wide consensus that water development and management should be based on a participatory approach, involving all stakeholders.

Energy is a critical aspect of national development process. It is expended in agricultural operation in food processing and transportation, in the production of fertilizers, pesticides, and farm equipments. It is necessary for industrial operations that provide jobs. It is required for cooking, for house hold lighting and heating, and for the construction and operation of the infrastructure needed for schools, health centres and water supply.

Precision agriculture is a management strategy that employs detailed site specific information to precisely manage production inputs. **Precision farming** can contribute in any ways to long-term sustainability of agriculture. The idea is to know the soil and crop characteristics unique to each part of the field, and to optimize the production inputs within small portions of the field. Of all the production inputs, nutrients occupy the top position and nutrient management is crucial to the success of any farming system. The usual practice is to apply nutrients at one rate throughout the farming area. Such practice could lead to wastage of resources and maximum yields could not be achieved since, spatial variability is altogether ignored in the management option. **Precision nutrient management system** offers improved land stewardship, optimizes resource usage, since every part of a field receives precise amount of fertilizer required to maximize crop yields. Various strategies of precision nutrient management system are being developed among which management zone technique is gaining importance. The management zone strategy reduced the average N applied from 6.3 % to 46.1 % compared to uniform N management. The variable rate N application using management zones increased net returns from 11.75 to $ 39.17/ha over uniform N management.

Farming Systems represent integration of farm enterprises as cropping systems, animal husbandry, fisheries, poultry farming, etc. for optimal utilization of resources bringing prosperity to the farmers. A judicious mix of cropping systems with associated enterprises like dairy, poultry, piggery, fishery, sericulture, etc. suited to the given agro-climatic conditions and socio-economic status of farmers would bring prosperity to the farmer.

Worldwide, **conservation farming systems** play a major role in sustainable agricultural production. Even after realizing the full potential of irrigation, a major part of the farming areas of most countries will remain rainfed. Land holdings of the poor, small and marginal farmers are mostly dependent upon rainfall. Common property resources like grazing land, tree lots and village tanks which are the mainstay of the livelihood of the rural landless people are again entirely dependent upon the rainfall of the area. Rainfed areas have therefore, assumed great importance for adequate and equitable food and livelihood security of the increasing population of the developing countries. Land degradation is another threat to our food and environmental security. Almost 150 million ha land is degraded due to erosion, water-logging and salinization. Tillage practice in modern agriculture of developed countries had most concern to sustain production in conservation of natural resources. Recycling of crop residues, growing crops of high volume biomass, change in cropping pattern, agroforestry, silvipasture are the some of the modules in conservation agriculture (CA). Annually, 400 m ha –m water is available through rainfall, out of which 75 m ha-m water lossed in erosion and courtly facing problem for drinking water, industrial use and very limited water (30 %) is available for irrigation. Harvesting the rain water *in-situ* and collecting through various mechanical measures and use of it for at least one protective irrigation during dry spell needs almost attention in conservation agriculture. Crop management in changing weather and efficient utilization of bio-diversity to develop **ideal plant genotypes** for biotic and abiotic stress tolerance is another challenge in CA. Conservation agriculture is environmentally sustainable and generates considerable net social gains to society. Caution must be taken to avoid blanket adoption of just everywhere; it should be site-specific and need based.

Organic Farming is developing rapidly and is now practiced in more than 120 countries of the world. Globally, 30.4 m ha of the agricultural land, are currently managed organically. India ranks 33rd position in total area under organic cultivation. India has now become a leading supplier of organic herbs, organic spices, organic basmati rice, etc. to the developed nations. The export realization during 2007-08 was around 78 million US $ registering a 200 % growth over the previous year. Though organic farming is an inevitable phenomenon, it alone cannot feed the billion people of India. To ensure the food and nutritional security, rather than promoting organic farming enmasse and universally, it would be desirable to carefully delineate areas for organic farming. In India, there is apple scope for pure organic farming in the rainfed areas, where there is little or no use of fertilizers and other agro-chemicals.

The Government of India has considered efforts to promote the organic farming without compromising the food grain production to feed the ever growing population in India. Launching of "National Programme for Organic Production"(NPOP) and the dedication of "National Institute of Organic Farming" INOP) were the landmark achievements in the arena of organic farming in India.

It is applied clear that agronomy is not a static subject of research. It changes with market forces, plant architecture and prevailing edaphic environment and is greatly influenced by the climatic change. One has to have the desired flexibility and orientation to change according to the need. Currently, Indian agriculture has major concerns in global warming, gradual drying of perennial rivers, changing rainfall pattern and loss of organic carbon content of the soil. **Rhizospheric engineering,** which encompasses modification of root architecture and modulation of soil-microbial activities, should become the major area of research for young Agronomists. Research on carbon sequestration and modification of root architecture through planting methods and crop geometry for increasing foraging volume of roots is needed to sustain higher productivity and to improve the nutrient-use efficiency. Mobilization of soil nutrients using micro-organism holds the promise. Agronomists must take this opportunity to lead multi-disciplinary team for developing bio-farming practices.

The study of **human recourses** is vital from the point of view of economic welfare. It is particularly important because human beings are not only instrument of production but also ends in themselves. It is necessary to know in quantitative terms the number of people living in a country at a particular time, the rate of which they are growing and the composition and distribution of population, etc No country, whatsoever is its socio-political setup, can bring about quick results in the sphere of population control and family limitation. It is, therefore, of imperative necessary that sustained efforts be made over a considerable period of time to achieve positive results.

The agricultural situation in India has undergone a rapid change in last two decades. Investment in agriculture sector, both in public and private sectors, has risen. Agricultural production, in general, has achieved reasonable growth rate. But the growth rate has not only to be maintained, but accelerated and fluctuations in agricultural production are to be minimized. The efforts are already under way to evolve location specific technologies, transfer them to farmer's fields and assure input supply to the farmers in right time, place and quality. The rate at which new technology, and yield increasing inputs are adopted by the farmers is affected the prices of inputs and output. Simultaneously, consumers also expect the availability of goods at reasonable prices. For achieving these conflicting objectives, **marketing system** for agricultural commodities and inputs has to play a very crucial role. This necessitates an understanding of the marketing system and market structure for agricultural products and inputs by all the sections of the population viz., farmers, traders, consumers, extension workers, scientists, sociologists, administrators, planners and politicians.

Integrated farming system will **generate appreciable employment potentialities** providing more man-days in a year. Even for the educated person's particularly agricultural graduates, such farming systems are likely to be more profitable than white collar jobs thus particularly reducing the job demands by the graduates. As a whole, such system will increase production of agriculture in the state, particularly, fruits, vegetables and produce from diary, duckery and fishery which are of much shortage in the state.

At present, there is no comprehensive text book on **"Modern Concepts and Advanced Principles in Crop Production"** and applied aspects suitable for farmers.

This book will provide comprehensive information on the subject matter and fulfil the needs of students and other professionals. This book makes an attempt to present the available information on **the modern concepts of agricultural research and advanced principles in crop production** easily understandable manner.

This is a book containing all sorts of chapters on modern concepts and advanced principles in crop production. Though this book primarily written to serve as a text book/reference for the students of agriculture in under graduate and post graduate levels and technologists in developing organizations, it is hoped that this book will be valuable for similar groups in the third world countries of Asia and Africa. This book also serves as a valuable reference for the candidates preparing Agricultural Research Services and other competitive examinations. Professional Institutions in Soil Conservation, Krishi Vigyna Kendras and Rural Institutions and similar other Institutions would find this book very much helpful. The farmers may refer this book to practice organic farming, integrated farming and farming systems as the considerable emphasis is placed for obtaining maximum, profitable production per unit area per unit time. Water, being the most important resources and input its management is essential for achieving the target. So, considerable emphasis is, therefore, placed for obtaining maximum, profitable production per unit of water from a unit of land in unit time. This book is very helpful to the engineers for field operations to increase the water use efficiency for crop production. This book is the most helpful to the students of economics for the under-graduate and post graduate levels, teachers, extension specialists, economists, farm managers and other professionals.

This book will serve as text book for students of Agronomy, Agro-meteorology, Agriculture Engineering, Economics, Veterinary, Social Science, a reference for research scientists and teachers in the areas of crop production, integrated farming systems, production technology management under different situations, changing climate and food and nutrition security under Indian Economy.

This book contains 25 chapters which include Introduction; Growth and development of crops; Crop growth in relation to environment; Growth and factors affecting growth; Crop growth and yield maximization; Factors affecting crop production; Crop ecology and Agro-ecological zones of India; Modern concepts in tillage; Soil fertility and productivity; Integrated nutrient supply for crop production; Organic farming; Precision farming; Dry land farming; Farming systems; Crop and land management practices; Bio-conservation of the environment; Systems of crop production; Concept of ideal plant type; Production and package of practices; Natural disaster management; Role of sustainable agriculture; Soil amendment; Bio-fertilizers; Land utilization and production pattern; and Modern concepts of agricultural research and Bio-technology: A science of securing future with adequate references. Each chapter is brief, comprehensive and full-fledged with respect to technical information.

It is hoped that the book will serve a good reference source for those interested to study the modern concepts of agricultural research and the principles of crop production. The author welcomes suggestions from the readers for the improvement of the book in future editions.

The author acknowledges his indebtedness to authors of books from which most of the material in the text has been drawn. In several cases, it has not been possible to obtain permission for reproduction for which the author and publishers offer their sincere apologies.

The author is deeply indebted to ICAR for its assistance provided at various levels for preparing the manuscript. Special mention is made for the valuable help received from Sri K.C. Sahoo, Research Fellow of the Emeritus Scientist Project, OUAT, Bhubaneswar.

The author would particularly like to thank Dr. Bhagabat Panda, Founder Director (Retd.), Central Avian Research Institute (ICAR) for his encouragement, invaluable help and providing excellent foreword. Not only has he checked the contents with meticulous care, but also he has made numerous suggestions for improvements and additions. His help and advice have added greatly to the value and reliability of a work such as this.

In preparing this book, I have received helps, suggestions and encouragements from Dr. D.P. Ray, Vice Chancellor, OUAT; Dr. N.K. Dhal, Dean, College of Agriculture; Dr. J. M.L. Gulati, Professor and Head, Agronomy; Dr. B. Behera, Professor, Agronomy; Dr. T. Barik, Professor, Agronomy; Dr. L. M. Garnayak, Professor, Agronomy (CSR); Dr. D.K.Bastia, Associate Professor, Agronomy and other teaching staff members and scientists of the Department of Agronomy, College of Agriculture, OUAT, Bhubaneswar, also Dr. A. Acharya, Director, and Dr. S. Sasmal, Director-cum-Coordinator, OCTMP, Sutra Consulting P. Ltd., Bhubaneswar. I am grateful to all of them who communicated the encouraging comments regarding the text and constructing suggestions for the improvement of the book.

I express my profound love and affection to my wife Mrs Kalpana Panda, son Ar. Mrutyunjay Panda, Architect and daughter-in-law Mrs. Tanuja Panda, Asst. Professor, Dept. of Computer Science & Application (C.S.A.), OUAT for their immense help in preparing this manuscript. It is my heartfelt pleasure to express my love and affection to my daughter Mrs. Bijaylaxmi Panigrahi, son-in-law Dr. Shasanka Panigrahi, daughter Mrs. Beenapani Nanda, and son-in-law Er. Shibasis Nanda, who always co-operated and encouraged me for this hard work,.

Suggestions and comments are also invited from experienced teachers, scholars and students for improving the quality of the book as there is always ample scope for improvement.

I also wish to thank the Agrobios (India), Jodhpur for the publication of this book.

2, Dharma Vihar, Khandagiri,
Bhubaneswar-751030 (Orissa)

Dr. S. C. Panda

CONTENTS

1 Introduction

"A country which wants to progress, wants to progress in a hundred ways We have, therefore, to take into consideration the order of preference what is first, what is second and what is third. There are so many things we want to do in India, and we want to do them quickly and passionately. The question of finding the proper way to reach a certain goal becomes important".

Jawaharlal Nehru

SUSTAINABLE AGRICULTURAL DEVELOPMENT

Sustainable agricultural development implies that the environment capital stocks should be held constant over time. Maintaining the natural capital stocks leads to the protection of both environmental functions useful to humans and the habitats and ecosystems essential for biological diversity. Sustainable development is the development that meets the needs of the present without compromising the ability of future generations to meet their own needs. Thus, sustainable development may mean everlasting development. It is a situation in which the elements of development do not decrease over time.

Economic growth refers to a process of increase in national income or output of a country over a period of time. In absolute terms, economic growth is measured by the increase in output or income from one period to another. The rapid growth of population stands as an obstacle to the smooth development of our economy and is reflected in the economic condition of our people. Poverty, illiteracy and diseases are the outcome of this population growth. Hence, a thorough examination of different aspects of this becomes essential.

In recent years, one of the important goals of the planners and policy makers of every country is to achieve **'sustainable development'.** The term first came into prominence in the 'World Conservation Strategy' presented in 1980 by the International Union for the conservation of Nature and Natural Resources. It was popularized by the World Commission on the Environment and Development's study "Our Common Future" (1987), which is also known as the Brundtland Report. Economic development mainly takes place through industrialization and agricultural development. In course of industrialization and agricultural operations, natural resources or natural capital such as oil in the ground, mineral resources, ground water, forests, river, natural environment, etc. undergo depletion. Unless stock of natural capital is kept constant, future development will be at stake. Thus, for sustainability of development, natural capital stock should at least be kept intact.

According to Brundtland Report (1987) "If needs are to be met on sustainable basis, the earth's natural resource base must be conserved and enhanced". According to Repetto (1986) "sustainable development is a development strategy that manages all assets, natural resources, and human resources as well as financial and physical assets, for increasing long-term wealth and well being".

Seventy per cent of the Indian population solely depends on agriculture those feed nearly 16% of the global population. About 26% of the gross national product comes from agricultural sector. The burgeoning population of India with a growth rate of 2.2 % compounded with the effects of green revolution coming to a standstill has threatened our food security. Nearly 40% of the total Indian population live below the poverty line those cannot afford two squares of meals a day. If it is focused towards the middle of this century, the food and feed requirement will be nearly doubled that of the present situation. Out of 328.8 m ha of total geographical area, 18% is under forest, 13.5% is not available for cultivation. Only 144 m ha is available for cultivation which has to sustain the one billion of population.

Agriculture is practiced by 105 million farm families who live in 6 lakh villages. About 76.2 % of the total farming community belongs to marginal and small farmers, who possess only 29% of total operated area (Agril. Census 1985-86). On the other hand the per capita availability of land from the time of independence of the country has declined from 0.48 ha to 0.15 ha by 2000. The income from seasonal field crops alone from small and marginal farm is hardly sufficient now to sustain the farmers' family. With the gradual decline in farm size, it would be increasingly difficult to produce enough food for the farm family in coming years. Under such circumstances, the farmers will have to under take some land based enterprises which will compliment their existing farming activity to get more income and employment, leading to a better standard of living. Such enterprises include livestock, fishery, poultry, duckery, apiary, mushroom production and tree farming etc. A judicious combination of any one or more of such enterprises with cropping and effective recycling of residues/ wastes would ensure better utilization of available resource and employment to the family labour during off season. Eastern India is characterized by the highest population density, low or stagnant growth in agriculture, per capita income and investment in infrastructure.

Agriculture in Orissa is largely rice based. Orissa has total geographical area of 155.7 lakh ha out of which 56.06 lakh ha is under forest, 4.45 lakh ha of land is cultivable waste, 60.48 lakh ha land is late sown area. Besides 8.38 lakh ha land is put to non agricultural uses (Orissa Agril. Statistics, 1998-99). This has only 30% of the net area sown under irrigation and fertilizer consumption of the state is only 36 kg per ha. Average rainfall during the cropping season is 1151.1 mm. Besides it has 480 km coast line, 0.49 m ha area under salinity, 0.34 m ha area under water logging and ill-drainage. Out of 33.5 million population of Orissa, 76% belong to small and marginal farmers and they occupy 62% of the total land holdings. Moreover, about 55% of the population mostly consisting of small and marginal farmers is below the poverty line. The average holding units for marginal farmers are less than one hectare. With the gradual decline in farm size, it would be increasingly difficult on the part of a marginal farmer to produce enough food for his family in coming years.

The benefits of modern technology have, however, been restricted to favourable farming situations. Only 25 to 20 % of the modern technology have reached the farmers in the ecologically favourable areas but much less (< 5%) in the handicapped ecologies. This is often because the technology has not been consistent with the existing farming situations. In the horticulture and animal husbandry sectors, not even 1 % of the technology has reached the small and marginal farmers. Only 44%, out of over 500 districts, are contributing to the half of total food grain basket of the country. This clearly suggests that the technology supposed to be scale and resource neutral has been confined to the districts with favourable farming conditions. Since, there is no further scope for horizontal expansion of land for cultivation and because of fragmentation and subdivision of land holdings and decline in the per capita availability of land day by day, there is no alternative left than to go in for vertical expansion and diversification of farm enterprises which have less demand on space and time particularly for small and marginal farmers with very limited resources, especially in rainfed areas (70% of total cultivated land). The new farming system research strategy, should therefore, necessarily concentrate on developing technology with participatory approach within the biophysical and socio-economic environments in which the farmers operate.

The most horrifying feature of many developing countries' demographic picture can be revealed from their burgeoning population over years. India with 2.2% (328 million ha) of the world geographical area supports about 16% of the global population increasing at a pace of 1.9% each year, which is equal to the population of Australia. The direct or/and indirect pressure has been exerted on the demand of food and land. There being no scope for horizontal expansion of cultivated land with dwindling man to land ratio, ever increasing problematic areas and with much touted "Green revolution" confined to a potentially endowed pockets benefiting well off farmers, the resource poor and socio-economically under privileged farmers' lot is becoming deplorable (Though there are 90 million small and marginal farmers out of 105 million farm families residing in 6 lakh villages). So, the only way out for maintaining sustainability in food production is through vertical expansion with diversification of farm enterprises particularly for resource poor small and marginal farmers by focusing on farmers' participatory research and generation of low cost location specific relevant and technically feasible technology and transfer of such technology through an effective and rejuvenated extension net work. Embracing all these purposes Farming Systems Research (FSR) has been developed for sustained increase in food production through vertical expansion.

World population has been increasing by leaps and bounds. The population of India is expected to reach 1370 and 1660 m in 2030 and 2050 A.D. respectively. The food production of the country has reached an all time high of 204 m t during 2000 A.D. A food production level of 289 and 349 m t is needed to satisfy the projected population in 2030 and 2050 A.D., respectively. The net cultivable area is 142.8 m ha. Unlike the population spurt and corresponding food need for 2050, there is every chance that the land area under cultivation will decrease due to diversion of some of the cultivable area to buildings and industrial purposes. It is anticipated that the land area available for cultivation in 2050 A.D. would be 130 m ha. So, it becomes necessary to increase the productivity almost to double of what we are producing today. This could be made possible by putting the land, both irrigated and rainfed, under intensive cultivation. Fortunately, most of our country lies in tropics and so is blessed with abundant solar energy, thus making cropping round the year possible. The only way to increase agricultural production is to increase the productivity per unit area per unit time. In the scientific cropping systems, genotypes, geometry of planting and management practices are designed to increase the productivity, simultaneously making efficient use of available resources and stabilizing yields (Palaniappan and Annadurai, 1999).

Modern agriculture emphasizes two more dimensions viz., **time** and **space**. Time concept relates to increasing crop intensification in a situation where there is no constraint for inputs including irrigation. In other words increasing the cropping intensity in areas where the production potential viz., land is underutilized even with full resource potential. It is a time bound programme where, for most of the field crops, it is considered for 365 days or 12 months. In the case of long duration and perennial crops, the duration of each rotation will vary from two or three years depending upon the duration of the constituent crops. The areas, where only one crop (100 per cent) two crops (200 per cent) three crops (300 per cent), are raised in a year, leaving the land fallow for two to eight months, the cropping intensity has to be increased to 200, 300 and 400 per cent, respectively within one year. This will ultimately help to enhance the productivity. In the case of rainfed areas where there is no possibility of increasing the intensity of cropping, the other concept viz., **'space concept'** can be applied. By way of building up of crops/other agricultural allied components in a vertical dimension (otherwise called multi-tier system), the land equivalent area can be increased. Thus by making use of these time and space dimensions either in irrigated or in rainfed areas within specified time span, say a year, and unit area of land, a hectare, productivity is sought to be increased by repeated and/or intensified cropping. This calls for urgent action on the part of Agronomists and Scientists of related disciplines to devote their attention on the design and testing of cropping systems for different regions. Income through arable farming alone is insufficient for bulk of the marginal farmers. Activities such as dairying, poultry, fish culture, sericulture, biogas production, edible mushroom cultivation, agro-forestry, agro-horticulture, etc., critical importance in supplementing their farm income. It fits well with farm level infrastructure and ensures fuller utilization of by-products.

Integrated Farming System (IFS) is the answer to the problem of increasing food production and farm income and for improving nutrition of the small-scale farmers with limited resources without any adverse effect on environment and agro-ecosystem. In a cropping system, the amount of by-products can be as high as or higher than marketable produce. This may go to waste if not utilized in an animal enterprise. Hence, integration of different agricultural and allied enterprises with crop activity as base will provide a way to recycle low cost produces at farm level/ no cost waste materials of farm from one enterprises to another and thus reduce the cost of production of economic produce and finally to enhance the net income of the farm as a whole (Annadurai et al., 1994)

The future of Indian agriculture depends on the development of appropriate farming system as applicable to resource poor farm families and as suited to different agro ecological zones. Restricting the use of purchased inputs in farming could be achieved through multiple cropping and diversified farming including animal husbandry, forestry, sericulture etc., The demand must be met from limited resources: land, water, energy and labour. Increasing food supplies with limited natural resources is a daunting challenge to the scientific community. Under this situation, Integrated Farming System (IFS) seems to be the answer, considering the current scenario in agriculture in India. Besides facilitating cash income and increased employment opportunities, IFS minimizes the quantum of purchased inputs in farming by effective recycling of products and by-products among the component enterprises; reduce the ill effects of inorganic fertilizers and chemicals (pesticides, herbicides etc.,)/or pollution hazards.

> ***'Nature does the business and man is merely the manager'. 'Corruption is the mother of vegetation'.***

Marketing is as critical to better performance in agriculture as farming itself. Therefore, market reform ought to be an integral part of any policy for agricultural development. Although a considerable progress has been achieved in technological improvement in agriculture by the use of high yielding variety seeds and chemical fertilizers, and by the adoption of plant protection measures, the rate of growth in farming in developing countries has not attained the expected levels. This has been largely attributed to the fact that not enough attention has been devoted to the facilities and services which must be available to farmers if agriculture is to develop.

Agricultural Marketing

Agricultural Marketing was, till recently, not fully accepted as an essential element in agricultural development in the countries of Asia and the Far East. Although opinions differ as to the extent and precedence, there was general agreement till 1970 that the question of markets for agricultural commodities had been neglected. Agricultural marketing occupies a fairly low place in agricultural development policies of developing countries.

The National Commission on Agriculture has emphasized that it is not enough to produce a crop or an animal product; it must be satisfactorily marketed.

THE PAST AND PRESENT FORMS OF AGRICULTURE

It appears from excavation, legends and remote sensing tests that about ten thousand years ago man invented agriculture. It is believed that women were the pioneers for cultivating useful plants from the wide flora, while men went out for hunting and gathering food. Some nomadic aborigines used fire and weapons. They did not have tools except for digging sticks and moll use shells. During the drier season they wandered in bands. The wetter and colder seasons restricted their movements. They would depend on their food gathering. They would dig out edible roots and rhizomes having buried the small ones therefore, subsequent harvests. They would hunt wild animals and use their flesh as food and skin as clothing. Thereafter, they entered upon a more secure stage of living. Probably woman by her intrinsic insight first observed that plants come up from seeds. Man might have consciously planted seeds and nurtured the sprouts to harvest. They used fire and digging sticks to prepare garden plots in which they would grow small grains. Once the seeds were planted the band would move on and return later for the harvest.

A primitive form of agriculture was shifting cultivation, in which people working with the crudest of tools, cut down a part of the forest, burnt the undergrowth and started new garden sites. After a few years when these plots lost fertility, choked with grass or bush or became heavily infested with soil-borne pests, cultivators would shift to a new site. This system would support a sparse population where abundant land was available.

Man invented the use of five animals as beasts of burden. They started domesticating wild animals facultative in nature. People with more members in a group selected somewhat better areas (near a stream or river) and settled down in permanent village sites and started cultivating the same land more continuously. However, the tools, crops and cropping methods might have been primitive. This system was probably the rudimentary system f settled farming. Since then agricultural activity became subsidiary to gathering and hunting.

Thereafter, an advanced form of primitive agriculture was the subsistence form of agriculture. This was based on the principle "grow it and eat it" instead of on a commercial basis. In some cases it was based on the crop production and in other cases on a more intimate blend of crops and livestock. Hunting and gathering were subsidiary occupations. Men in meaningful members began to settle down to sedentary life.

A sparse population of nomadic tribesmen followed their flocks and herds in the seasonal pursuit of water, forage and shelter. Commercial ranching gradually blended with subsistence farming. Under shifting cultivation land abandoned became degraded. Subsequent subsistence farming followed naked fallow rotation. Thereafter, legume rotation was adopted. Under mixed farming comprising field crops and animals, field grass husbandry was adopted and the same fields were used for cropping and grazing.

The concept of property in land evolved and the individuals held the privilege of autonomous action. The area over which a band wandered and carried on their food-gathering was considered to be for the joint and mutual exploitation of the group. They held a territory assuming they could defend it from other bands by the right of prior occupance. Any individual property in land was right to a certain garden plot and domestic site, earned by virtue of having cleaned, improved and used it. During their movement from one location to another they would come in contact with a large number of plant and animal types. They selected the most useful ones for domestication and improved them.

Battles, fights on rivalries among different groups and individuals for property or passion helped to blend newer ideas for the components, commodities and comfort of life. Mutual exchanges, gifts, offerings etc., also blended the style of living, food habits and nature of earning. This right to use the land with the progress of time was recognized on a continuing basis year after year. With still further evolution land was held to be inheritable and the ownership title was granted and defended by group authority. These progressive stages of security in mobile and immobile properties, contributed proportionately to the greater stability and well being of the society. They concentrated on the welfare of their individual families, and the produce of their land and animals. They devoted themselves more intensively to utilize their land to the best of their ability. However, they held together to defend any unwanted enemy of their society. This was still the transitional stage from food gathering to found growing.

The self sufficient agricultural village was the least common denominator of a permanent human society. It developed and perpetuated itself without many tools or other capital and without dependence on the outside world. Several crafts also developed for and around agricultural production. Agriculture then experienced extreme adversity as long as the minimum of food, feed and fibre were raised. The standard of living was low. With the span of time there was the periodic and geographic rise of different civilizations. Agricultural production expanded and towns and cities developed, trade and arts flourished.

Downtrodden people, slaves and bonded labour were the principal sources of human labour. Draft animals were the plodding oxen. Farming equipment was animal drawn and consisted largely of hand tools. It was organized on a commercial basis and management was developed to a fine art. The skills developed were inherited through the generations of cultivators who became a major section of the then civilized society. Advanced farming practices included seed selection, green manuring with legumes, rotation of crops, use of animal and crop refuse as manure, irrigation, pasture management and rearing of birds.

In subsequent stages the feudal manor emerged and was essentially an agricultural village. It consisted of a nucleated village built around the manor-house and surrounded by the village land. The best lands were laid out in large arable fields, the medium lands were used as grasslands for mowing and grazing and the least productive areas were used for grazing, wood lands, funeral grounds and other common activities. Being a self-contained the village included a worship alter and a meeting place.

Arable lands were divided into different plots according to topography and use. During the favourable season crops were grown in the fields and livestock was kept away from the crops but after the harvest cattle would graze in these areas under the supervision of herdsmen. During the cropping season livestock were allowed to graze in common grazing lands and in adjoining woodlands and forests. By-products of crops were used to feed the animals during stall feeding. Quack remedies based on locally available plants with therapeutic values and minerals were the major method of treatment for common ailments of animals. Infestation of insect-pests was also combated by indigenous methods.

There were ponds and tanks in the village. Some of these water reservoirs were reserved for drinking water, washing purposes, fisheries, or small-scale irrigation. The banks of water systems were used for arboriculture and fruit growing.

Emperors, monarchs, kings, zamindars, governors, heads of states, etc., would claim authority over their respective geographical land areas and rule over them. Cultivable areas would be distributed to lease holders, tenants or share croppers directly or through intermediaries. Such cultivators or tillers would enjoy little benefit except for mere subsistence.

The enclosure movement began to be consolidated in some countries during the 14th century. Open and scattered fields of a tenant were compacted into a farm unit under an individual owner. These consolidated, fenced and individually controlled farm units were managed by the farmers with commercial intent. Land as an inheritable and transferable property was subdivided and fragmented to an indefinite extent without the consideration of its economic viability to the tenants who depended on subsidiary occupations to supplement their earnings subsequently. In long and medium term cycles, natural calamities affected agricultural production and the farming community and the kingdom or nation as a whole would fall in distress. Facing such famine was a futile effort except by large scale emigration. The toll of human and animal lives at such times was considerable. No other occupation could compensate for these disasters.

Land, though an immobile property was capable of producing an enormous quaintly of mobile short to long –term crops meant for the livelihood of a large number of people. The helping hands of the peasantry and ploughmen were usually family members. There were also different types of human labourers and work animals. The then tools and implements were a slightly improved form of primitive ones. Bullock or horse drawn tools were limited to the iron banded wooden plough, bullock cart, etc. Therefore, wooden and iron implements and equipment were developed. All of them were manually operated or drawn by animal power.

There were many ups and downs, shortage and disasters in this expended (both spatial and temporal) development of agriculture in site of the diligent endeavours of human for a greater dependence on agricultural production. Modern agriculture started during the early eighteenth century. Crops were grown in sequence without allowing the field to remain fallow. Crop rotation was introduced. Organic recycling was intensified. A number of exotic crop plants and animals were introduced. The crops selected were clean, stable in yield and well adapted to the local agro-ecological systems. Land improvement was introduced. Farm implements and equipment were improved. The care and management of crops were advanced and specialized, crop protection practices were introduced and preservation and processing improved. Gradually livestock and flock breeding and rearing and crop farming became diversified. Sheep meant for wool were bred for the production of better and more efficient meat animals. Dairying became a profession. Fish rearing and fishing became an occupation. Greater emphasis was laid on feeding, breeding, caring for and weeding out of different livestock and birds meant to meet various requirements more efficiently and economically. Crop plants also drew due attention and they were grown in different categories as field crops, plantation crops and garden crops.

In the nineteenth century the real science of agriculture started. Nutrition, physiology, ecology, etc., developed. Research and development in fundamental and basic sciences were brought under applied aspects in agriculture. It looks the shape of a teaching science. The most neglected occupation of a downtrodden and dull community drew the attention of aristocrats, scientists, researchers and others in almost all walks of life. Research on specialized branches of several disciplines was developed and scientific findings were translated into production practices. Laboratories, farms, stations, centres and institutes for research, teaching, training and demonstration were developed. Books, journals and scientific and popular literature were introduced. News media and audio-visual aids were developed to disseminate new findings and information to the rural masses engaged in agriculture.

At present a lot of development on hydrological, mechanical, chemical, genetical and technological aspects of agriculture is in progress. Government is apportioning a greater share of development for agricultural improvement. Central and State Departments of Agriculture, Agricultural Universities and Institutions are flourishing. International Research Institutes are being set up. Farmers specially small and marginal ones of the underdeveloped and developing countries are supplied with improved seeds, fertilizers, irrigation, equipment, electrical energy, pesticides, improved genetic stocks of cows, chicks, kids, pigs, fingerlings, silk worms and their improved feed. Even funds for both recurring and non-recurring expenditure with reasonable cost or rate of interest, or free of cost, as well as technology services are being provided freely. Preserving, processing, pricing, marketing, insuring, distributing, consuming and exporting importing policies are strengthening. Agro-based industries and crafts are developing. Need based planning; programming and executing are in progress. Improved use-efficiency of non-monetary, less-monetary inputs is in the process of adoption. From specialized farming, diversified and integrated farming is emerging. Land reforms, development and diversion from degradation are going on. As the size of operational holdings is rapidly deceasing to marginal and sub-marginal levels, farming by individuals is going to be recast to cooperative, community or contractual farming.

Agriculture now is not merely production oriented but is becoming an agribusiness consisting of enterprises such as livestock, poultry, fishery, silviculture, sericulture, apiculture and lac culture along with field and forage crops and plantation crops. These are the stepping stones of the green, white and yellow revolutions of today and simultaneously of the multicolour revolution with manifold quantitative, qualitative and generative yields.

It is surprising that even in the era of nuclear energy in industrial development and expeditions to space; man retains his animality and primitiveness in food habits and is a competitor with his animals for food grains.

Even though agriculture was, is and will have tremendous potential for life and longevity in this luminous world, yet it is far behind the technologies developed in other sectors such as industry, space expeditions and defense.

The science of Agronomy has made it possible to generate suitable technology for varied agro-climatic regions and integrate the results of other allied sciences, so as to deliver these appropriate innovative and tested practices to the farmers at large.

MODERN CONCEPTS OF AGRICULTURAL RESEARCH

India witnessed self sufficiency in food production by shifting from the traditional organic to chemical industrial agriculture in the second half of the past century. The production levels plateau and several challenges of soil health, crop production and environment have become critical in the new millennium. This need paradigm shifts in agricultural practices and research priorities. **Systems Rice Intensification** conserves water, but it is difficult to practice. **Organic farming** is eco-friendly but, organic source of nutrients are not sufficient to meet the nutrient requirement of crops and the yield reduction is substantial. It has little scope for farmers to export the produce to developed countries after meeting the standards of trade and cumbersome procedures. **The integrated supply of nutrients** is potential to sustain the crop yields and restore the soil health. The transgenic crops opened new vistas in agricultural production. But, such genetically modified crops are frequently unstable and inactivated. They lose the newly acquired traits due to gene silencing. **The cell and tissue culture techniques** may substitute the traditional crops for the production of food like starch, protein and fats in fermentation tanks. **The nanotechnology is likely to revolutionize the agricultural production system.** Nutrition with desired elements and flavor stored in the capsules burst open in the stomach and deliver the nutrients in quantities needed by the organ. Similarly they release the toxins in the stomach of the harmful insects of crop plants. **Nano-sensors** will alert the farmers before the occurrence of the pest, disease or nutrient deficiency and damage to the crops. The **resource conservation** technologies for anaerobic crops after the harvest of aerobic rice, save expenses and time for land preparation. This will facilitate early sowing of the crops after the harvest of rice on the carry over soil moisture or a shallow irrigation. The use of heavy machinery for sowing may compact the soil in the long run. The residue and organic matter but, it is not likely that it may hinder proper seedling growth. The rice stubbles may harbor the harmful pests and diseases inoculums, refuge the rats and invite the snake problem.

INDIAN AGRICULTURAL PLANNING

Dr. M.S. Swaminathan, eminent agricultural scientist, stresses the need for a new vision on Indian Agricultural planning, with the accent on ensuring the prosperity of the farmers and the betterment of the lot of the rural poor.

The progress in Indian Agriculture during the last 50 years can be broadly classified under three areas. Fist, considerable progress has been made in developing research and educational infrastructure, essential for the development and testing of technologies suitable for the different agro-ecological regions. There is at least one agricultural university in every State and there are many Central institutions, national bureaus, and all India coordinated research projects, dealing with crop husbandry, animal husbandry, fisheries and forestry.

Secondly, a reasonably efficient input production and delivery system has been developed. Thus, there are fairly effective instruments for knowledge and skill transfer, credit supply and production and distribution of seeds, fertilizers and other inputs.

Thirdly, policies essential for stimulating higher production by small farmers and increased consumption by the rural and urban poor have been gradually evolving. Thus, progress is being made in agrarian reform, rural infrastructure development, input and output pricing policies and marketing arrangements. At the same time, steps have been taken to strengthen policies to promote greater consumption by the rural and urban poor, through the National Rural Employment Programme, the Employment Guarantee and other projects.

Food Production

Food production has on the whole remained above the rate of population growth. There is, however, widespread hunger in the country owing to inadequate purchasing power. There are also considerable variations in the pattern of growth, with regard to regions as well as crops. Several parts of the country are yet to benefit from new technologies. Many groups of crops, particularly pulses and oilseeds are yet to see the growth rates essential to improve the per capita production and consumption. The total food grains production has also tended to stagnate partly due to climatic factors and partly due to inherent weaknesses in agricultural planning and management.

Five basic principles which should shape the future development programmes in agriculture can be set out.

1. The pattern of production advance should be "land saving crop husbandry and grain saving animal husbandry".
2. Ecological sustainability and equity in the use of national resources should constitute the basic norm of all agricultural and rural development programmes. Protecting the livelihood of the poor is essential for enduring ecological security.
3. Explicit attention should be given to promoting beneficial growth linkages among the primary, secondary and tertiary sectors of economic activity. At least 200 million persons in the rural areas should be found skilled jobs in the off-farm sectors by the year 2000.
4. Agricultural technology should be intellectually satisfying and economically rewarding to attract and retain youth farming and related activities in villages. Over 50 per cent of the population is below the age of 21 and the agricultural future is in its hands.
5. The comparative strengths in agriculture with regard to exports should be carefully identified and nurtured. For this purpose, coastal and mountain area planning and post harvest technologies require particular attention.

These ground rules should constitute the fundamental basis of the development of packages of technologies, services and policies. Progress in agrarian reform has been patchy. Unless the tiller has a long term stake in the land, there will be little interest in making the investments on land essential for sustained high yields. A few of these aspects are worth discussing an length along with indications of what can be done.

Ecological Sustainability

Nearly 90 per cent of the food requirements will have to come from land-based farming. Land is a shrinking resource for agriculture owing to competing demands for its use. Population growth also is leading to increasing pressure on the carrying capacity of land. The water resources too are not being used in a sustainable manner. With deforestation, there is erosion of biological diversity and frequent flash floods in the downstream areas.

How can a dimension of ecological sustainability be imparted to technology development and transfer? What steps are needed to each agro-ecological area to promote scientific management practices in the following three categories of land?

a) Conservation areas which ought to be preserved in their pristine purity;
b) Restoration areas where the biological potential of land needs to be upgraded through regeneration and rehabilitation, and
c) Sustainable intensification areas where production and productivity can improve without harm to the long term production potential of land.

For incorporating sustainability considerations in agricultural research and development, appropriate definitions and tools measurement as indicated below are needed.

- **Production:** Crop output per year.
- **Productivity:** The ratio of the value of the output divided by the value of the inputs, plus changes in stocks of capital. The change in capital stocks should include the measurement of soil health over time.
- **Sustainability:** A non-negative trend in the productivity of a given plot over time.

The quality of produce, consistency of supply and competitiveness of cost would determine the ability to increase export earnings from land and water based occupations. The areas where there is considerable comparative advantage in international markets are horticulture (fruits, vegetables and flowers), aquaculture (both inland and coastal), sericulture, plantation crops, animal products, agricultural biomass-based commodities and skilled services. There is a gradual change in flood habits the world over. The health foods of the future will depend heavily on horticulture. A wide range of vegetables, fruits and flowers can be grown in the country because of the wide range of climatic and soil conditions. Therefore, the National Horticulture Board should be made really effective so that it can provide timely and relevant services to the small farmers, particularly in post-harvest technology.

A plan is also needed for getting the full advantage of the vast coastal areas from where goods can be exported at low cost. The coastal areas provide not only opportunities for conventional fisheries and mariculture, but also for energy plantations, particularly with casuarinas, and for production of a wide range of horticultural products.

Through an agro-forestry system of land management near the coast, appropriate silvi-horticultural and silvi-pastoral systems for the production of a wide range of fruits and food grains as well as animal products can be promoted. Integrated coastal area planning involving agriculture and industry is essential if India's position in international trade in agricultural products is to be substantially increased.

Knowledge of labour-intensive technologies will be another area where a conscious attempt can be made to establish a commanding position in the world of agriculture. Even now there are small initiative in the production of hand-pollinated hybrid seeds in flowers and vegetables for export. Biotechnology industry and software development for computer-aided agriculture are areas where a major push forward has to be made. In the coming years, over 70 per cent of computer costs will relate to software. The vast university student population could be orientated to the acquisition of new skills to increase the share of income from intellectual property. Young people are to be trained specially for providing consultancy services within the country and outside and for increasing the income from agricultural know-how.

The year 1988 marks the 20th anniversary of the coining of the term, "Green Revolution" by Dr. William Gadd of the U.S. Department of Agriculture. In 1968, Indian farmers produced about 16.6 million tonnes of wheat. The highest production earlier was about 12.3 million tonnes in 1965. This quantum jump in wheat production as well as the promise of a similar leap in rice opened up by the semi-dwarf rice varieties promoted to Dr. Gadd to say that what is happening is a revolution of farmers' fields in the tropics and sub-tropics.

Wheat production in the country has now reached 47 million tonnes. The average yield which was 850 kg per hectare in 1960 is now 2046 kg. Punjab alone now produces over 11 million tonnes of wheat, which nearly equals the production of the entire countryin 1964-65. The average yield of wheat in Punjab now is over 3.5 tonnes per hectare. The average yield of rice grown often in the same field during the *kharif* season is nearly 3.2 tonnes per hectare. Thus, the average yield from a rice-wheat rotation in a year is about 6.7 tonnes per hectare, which is one of the highest in the world. Punjab has also made spectacular progress in many other areas of agricultural production like potato, cotton, sugarcane, milk and tree crops.

It is not often realized that equally impressive progress has been made in several areas of animal husbandry. Egg production rose from 1,832 million in 1950-51 to 17,000 millions in 1987-88, milk output from 17 million tonnes to 46 million tonnes and wool production from 27.5 million kg to 41 million kg. There has been good progress in inland and coastal fisheries.

The Green Revolution in Punjab is not a miracle. It happened only because the following pre-requisites for the success existed in the middle-sixties:

1. Land consolidation and leveling,
2. Owner cultivation resulting in a long term stake in land,
3. Rural communication,
4. Rural electrification, and
5. A dynamic Agricultural University.

Had these not existed, the Punjab farmers would have found it difficult to convert the yield potential of the management-responsive varieties of wheat rice into actual production. Ultimately, yield is a product of interaction between the genetic efficiency of the plant and the management efficiency of the farmer. It is the dedication, hard work, innovativeness and enthusiasm of the farmers which have been te catalyst of change in India's agricultural destiny.

Nutrition Security

This implies physical and economic access to balanced diet and safe drinking water to all people at all times. While a degree of food self-sufficiency at current levels of consumption capacity has been achieved, there is a long way to go in the matter of developing a national nutrition security system. What are the dimensions of challenge for a human population for nearly a billion by the year 2000?

Cattle Fodder, Fuel Wood Needs

In statistical terms, 230 to 250 million tonnes of food grains have to be produced. Food production has increased by about 75 million (from 80 to 155 million tonnes) during the last 30 years. This performance has to be repeated during the next 12 years. The present livestock population includes 192 million cattle, 69 million buffaloes, 48 million sheep, 96 million goats, 10 million pigs and 193 million poultry, leading to a total of 608 million animals. In addition, there are two million horses, a million camels and many other animals. The animal population suffers from chronic under nutrition. An animal fodder and feed security system is yet to be developed. Except for Punjab where around 10 per cent of the irrigated cultivated area is under fodder crops, fodder cultivation is yet to receive the attention it deserves. About 700 million tonnes of livestock fodder has to be produced annually by the year 2000.

Over 200 million tonnes of fuel wood per year by 2000 have to be produced as it is still the most important source of cooking fuel. These and the numerous other commodities needed for industrial and domestic consumption and export will have to be produced either from the same or even less land.

JOBS IN OFF-FARM SECTORS

Economic access to food implies opportunities for acquiring adequate purchasing power. This, in turn, involves opportunities for gainful employment. The most deprived sections of the community are those who have neither productive assets nor regular employment. Unless about 200 million additional jobs are found in the off-farm sectors in the rural areas in the next 12 years, the earning capacity of rural families will be low and millions of landless men and women will continue to perform unskilled jobs and will have to seek work in rural areas. Technologies for women specific occupations need particular attention. The poorer the household, the greater is the importance of independent access to income for women. An integrated rural development strategy which can ensure the livelihood security of the poor in their native habitat needs to be devised.

How can these tasks be accomplished on an ecologically sustainable basis. The national commons consisting of land, water, flora, fauna and the atmosphere are being damaged. The world Commission on Environment and Development chaired by Mrs. Gro Harlem Brundtland, Prime Minister of Norway, has given suggestions on managing the global commons on an ecologically sustainable basis. An action plan involving the public, professionals and political leaders for conserving and utilizing the national commons on sustainable and equitable basis is needed. All the basic agricultural assets are in different stages of degradation, but it is not too late to arrest and reverse this process.

For this, a Policy for sustainable management of the National Commons consisting of the following components is required:

1. **Land Use:** Block level land management practices with appropriate attention to conservation, restoration of degraded land and sustainable intensification for agriculture purposes have to be developed.
2. **Water:** Rain water and aquifer (ground water) management, wasteland management, conservation of coastal mangroves and the sustainable use of surface water – all require attention not only from the Government, but also from the entire community. The vast river and ocean resources need to be kept free of pollution and siltation.
3. **Flora and fauna:** A national action plan for the conservation and effective use of the biological diversity is needed. With the rapid destruction of forest eco-systems, particularly in the Himalayas, Western Ghats, Vindhyas and Eastern Ghats, considerable biological wealth is likely to be lost.
4. **Atmosphere:** Besides atmospheric pollution, the long term consequences of deforestation and increased consumption of fossil fuels are receiving serious global attention. A recent global conference held in Toronto in Canada considered the implication of the changing atmosphere for global security. The participants concluded that the future of agriculture will be greatly influenced by (a) climate warming, rising sea level and altered precipitation patterns induced by the 'heat trap' effects of green house gases, (b) depletion of the ozone layer and (c) increased air and water pollution.

There can be a time lag of several decades between the emission of gases into the atmosphere and their full biological and social impact. Therefore, emissions released so far have made some degree of global warming and a rise in sea levels inevitable.

The earth has remained a comfortable place for life began despite a 25 per cent increase in the heat output from the sun. Will this legacy be maintained? If ecological economics (that is, economic policies dealing with the interests of generations yet to be born) and economic ecology (that is, ecological guidelines which can help accelerate economic growth on a sustainable basis) are not promoted actively, ecological access to food may become the most important food and nutrition security challenge in the next century.

The rapid growth in population must be arrested if a harmonious balance between the human population and natural resources is to be achieved. The animal population must also be held in check. Sustainable development and a small family norm are the twin foundations on which a better quality life for all can be ensured.

Until the 18th century, most human communities were predominantly agricultural. The developed nations gradually become industrial societies. Now the developed world is growing into an information society where knowledge and information are sources of great economic power and prosperity. The growth of a conservation society where the best in traditional and modern technologies in agriculture, industry and information sciences could be integrated in a meaningful manner has to be fostered whereby attention will be paid to minimizing demand or national resources and just to maximizing output.

AREAS FOR ANALYSIS

The areas which need in-depth analysis followed by appropriate action can be briefly indicated.

a) **Conservation of basic agricultural assets:** The growth of village level associations for sustainable agriculture has to be fostered. Unless the people of every village or watershed or the command area of an irrigation project are intimately involved in the sustainable utilization of natural resources, particularly land and water, it will be difficult to arrest environmental degradation. Village school children should be involved in the preparation of a "Charter for Nature" for sustainable use of natural resources.

b) **Land saving crop husbandry:** It is feasible to undertake intensive crop production on a sustainable basis, every effort should be made to increase yield per crop and the number of crops per year. If this is not done, land which ought not to be under annual cropping either because it is degraded or erodable, will continue to come under annual cropping. For example, if 120 million tonnes of rice from 30 million hectares are produced by the end of this century, about 12 million hectares can be released for agro-forestry and other sustainable land management systems. Growing rice in areas where water is very scarce can also be avoided.

c) **Grain saving animal husbandry:** Large quantities of grains cannot be diverted for the feeding animals. A massive programme if research and development on fodder crops multi-purpose trees, grass and legume pastures, microbiological up-gradation of all cellulosic wastes and organic recycling is needed. The new Ministry of Food Processing should also deal with animal food and fodder. It should consider the scientific utilization of the entire biomass and not just grains. Whole plant utilization and better animal and fish biomass utilization need urgent attention if a dynamic animal husbandry programme is to be promoted and mixed farming is to be made more employment-intensive. Grains like maize will still be needed for feeding poultry to sustain and enlarge the impressive gains in egg production.

SCOPE FOR NEW TECHNOLOGY

a) **Youth orientation to farm technology:** The future of agriculture is in the hands of youth. Unless farming becomes intellectually satisfying in addition to being economically rewarding, it will be difficult to attract or retain youth in rural professions. Ecologically sound technologies are knowledge-intensive. There are also opportunities for integrating frontier technologies like biotechnology, bio-processing, computer technology, satellite imagery and micro-electronics with traditional technologies combining the strength of both. Knowledge-intensive technologies also help to save on capital and land.

b) **Employment orientation to farming systems research and development:** Farming systems research involves a holistic approach and extensive collaboration with farm people. Generation of better growth linkages between the primary, secondary and tertiary sectors of economic activity has to be included as an explicit goal of farming systems research. Unless

At least three jobs are generated in the secondary and tertiary sectors for every job in the routine farming operations, it will be difficult to overcome the job famine in rural areas. Farming will be overburdened with labour and the income of landless labour families will continue to remain low and uncertain.

An employment and income orientation to research can be given only if all the links in the production and post-harvest phases of agriculture are attended to in coordination. For this purpose, at the level of every block, a technology integration experiment station is needed on the lines of the Prefecture Experiment Station in Japan. These centres, to be operated by the agricultural universities jointly with farmers' associations, banks and the State Departments concerned should develop methods by which more food, jobs and income can be generated on a sustainable basis from the available land, water, labour and credit resources in the block. While scientific research can be done in the agricultural universities, technology development can be done only in farmers' fields with the active participation of farming families. Many of the problems arise from the failure to recognize the distinction between science and technology in agriculture.

c) **Increasing the efficiency of small farm management:** The average size of an operational holding is getting smaller and smaller. Nearly 80 million of the 90 million holdings in 1980-81 belonged to small and marginal farmers. A small farm is ideal for knowledge-intensive agriculture, but a small farmer suffers from serious handicaps arising from the cost, risk and return structure of farming. To help the small farmers to improve the efficiency of management of farms;

d) Research on knowledge and input delivery systems which can help all farmers irrespective of their innate input mobilizing and risk taking capacity is needed to derive economic benefit from new technologies. Very little attention has been paid so far to research on delivery systems;

e) Group endeavour should be promoted in areas like land and water management, pest management, nutrient supply and post-harvest technology. Unless individual initiative, group endeavour and Government support become mutually reinforcing, the efficiency of small farm management will continue to be low; particularly in rainfed areas where water harvesting and equitable distribution of the concerned rain water are extremely important for higher and more stable production.

Indian agriculture needs a new vision if further progress is to be made. One can not all the times be talking about the millions of tonnes of food grains the country needs to feed its growing population. This is, of course, important for the urban population, but farmers will get enthused to making greater efforts only if their toil in the sun and rain could lead to a better life for them and their children. Agriculture should, therefore, be viewed as an instrument for promoting agrarian and rural prosperity and better quality of life in villages. The fallout of such an approach will be not only increased production, but also more agricultural exports. The challenges are great, considering the extent of rural poverty and illiteracy. Women are yet to be helped to play their pivotal role in the production and post-harvest phases of agriculture. Fortunately, the minimum essential infrastructure and information tools which can help to accelerate progress are already there. The task is not easy, but can be accomplished, if political will and professional humility can be combined. Farmers should occupy the pride of place in agricultural planning.

SAVING THE ECOSYSTEMS

There is particular urgency in restoring the damaged ecosystems of the Himalayas and Western and Eastern Ghats. Damage to these ecosystems is not only eroding the livelihood security of the people living in these regions, but also the crop and livelihood security of millions of rural people living downstream. The Indo-Gangetic region has been described as one of the great bread baskets of the future. If deforestation and soil erosion in the Himalayas continue at the current rate, the Indo-Gangetic area may become within 50 years a sad example of man-made desertification. This is why the youth should take greater interest in all areas of agriculture.

Finally, the present information age has provided a unique opportunity to mobilize the mass media for stimulating awareness, analysis and action on matters of national importance. Software development for media use needs more support. The media have helped prevent starvation deaths by serving as an early warning system. Mass media support will be essential for the spread of knowledge-intensive and ecologically sustainable agriculture.

UNEMPLOYMENT AND ITS SOLUTION

Unemployment in an undeveloped country is different from unemployment in a developed country. In a developed country, the mobility of labour is very high. It is trained and skilled there is always unutilized productive capacity in the industries. As a result, unemployed labour can be easily employed in the existing industries. The cause of unemployment normally happens to be just some defect in the working of the economic system. Because of a deficiency in effective demand the business people feel that increased production is not profitable. However, if demand rises, the production becomes profitable and as a result, unemployed labour is quickly absorbed.

In an undeveloped country, like India, the problem is entirely different. Here unemployment is basically structural. That is to say that labour is too much in supply as compared with land and capital resources with which to employ it gainfully. For this reason, unemployment here is of chronic nature and can be cured in the long run only through increased resources with which to employ the labour. Further, labour in India is unskilled and as such cannot be employed readily in modern industries unless trained first. Also labour mobility in India is very low because of the customs, attachment to soil, unfamiliarity with the conditions prevailing in different regions and so on. As a result, it may be that while in some parts of the country labour is unemployed in the others there may be actual scarcity!

To put the nature of unemployment problem in India into concrete terms, there are basically two sections of the unemployed: (1) educated and (2) non-educated. And there are further sub-divisions of these. The educated unemployed are mainly in the urban areas and further fall into the categories of clerical type and skilled type. Non-educated unemployed persons in the urban areas cannot get employment in the cities and have a tendency to go back to their villages whence they came. Therefore, non-educated employed are mainly in the villages and these also mainly in the agricultural field.

In addition to this, we have seasonal unemployment. Our economy being mainly agricultural and our agriculture being confined mainly to one or two crops a year the people engages in it usually do not find enough work throughout the year. There is the crop-season and there is the lean season. This seasonal employment could have been remedied if people could engage themselves in cottage industries. However, this is rarely practiced.

Then there is a peculiar type of under employment where more workers are doing the same job which a less number of them could do. A field is cultivated by five persons while three could easily do the same thing in as good a manner, so two out of the five are really unemployed. But their unemployment is not open-even to themselves it may not be known. All of them are busy, but every one of them is doing less work than he can do. This disguised unemployment is there because of the joint system and because of the fact that any number of the agricultural family who cannot find employment outside is asked to share the fortune of the family. It is, therefore, nothing but a sort of unemployment insurance in such a way that nobody appears to be unemployed.

Thus, it is quite clear that employment policy in India cannot be successful unless it takes into account these various types of unemployment prevailing in India. For every different type of unemployment different type of remedy is needed. It is quite possible that while we may be reducing unemployment in the industrial sector, disguised unemployment may remain as much as before or while disguised unemployment is being cured seasonal unemployment persists. The complexity of the unemployment problem has been recognized by our planning authorities and accordingly we find that unemployment problem being dealt with by breaking if first into these various categories and then devising remedies.

The extent of unemployment in India must be assessed before we resort to remedies. It has been one of the greatest difficulties to ascertain the extent of unemployment in India. Adequate data have not been available for the purpose. Employment exchanges are unreliable for the purposes. So many people, especially the rural people, do not get themselves registered with the employment exchanges and these exchanges do not cover all the cases of the unemployed and the disguised unemployed nor are they able to tell us effectively the handicaps by labour mobility. The census which could have given us a correct picture of unemployment failed to do so and there are other standard sources of information. The Census gives us a three-fold economic classification of population: (1) self-supporting, (2) earning dependents and (3) non-earning dependents. The first two categories are the persons, who are employed but the Census included in them even those people who received income but were not really employed, e.g., beggars, rent receivers, etc.

The **real solution lies** in long-term steps which should check the population growth of the country. But the problem cannot wait its solution for such a long period. Certain steps should be taken to reduce its magnitude immediately and they should be such that the persons employed are able to make an adequate living and still not be a burden on the society. Therefore, the steps to solve the unemployment problem should be divided into long-term and short-term categories and should be undertaken simultaneously. **Creation of industrial estates and small industrial units** and renting out to promising young men at reasonable rates, to produce spare parts, plastic goods, and products ancillary to large-scale industries, it will help directly in the industrialization of the country, and reduce unemployment to some extent.

Efforts should be made to reduce the disinclination of the educated people for manual jobs. The importance of formal education for securing employment should be reduced. The Government should try **to develop small-scale and consumer goods industries in rural areas** to reduce pressure on agriculture and cut seasonal unemployment. Such a programme needs organization for marketing, provision of raw materials, and guidance of high standard. Otherwise, all the government help would be going down the drain.

Another field in which the government can help reduce unemployment is to **encourage techniques which are labour intensive** but not costly. In the case of River Valley Projects, Major and Minor Irrigation Projects, road building, digging of irrigation works and many other fields, there is the simple choice to opt for labour instead of heavy machinery. The Government should think of making use of more labour in these projects.

Along with these, the Government must popularize the family planning programmes and put the same within the reach of every one.

In this way, with labour intensive industrialization and public works, and family planning ensuring near zero population growth, the problem of unemployment may be successfully solved.

REDUCE POPULATION EXPLOSION

By far the most serious obstacle to the growth of the Indian economy appears to be the terrific rate of growth of population.- an addition of between 1.2 and 1.5 crore of people per year. Now the rate of population growth has gone still further, to about 2.5% per annum. The population problem has assumed serious proportions due to a drop in the death rate in the face of a persistently growing **birth rate which leaves a big rate of survival.** The **death rate has come down** subsequent to the eradication of epidemics like plague and cholera and endemic diseases like malaria. Penicillin and other antibiotics also have been greatly responsible for a drop in the death rate. A drop in the death rate should normally be welcome, particularly when working hands are saved and progeny is not wasted after a good deal of investment. A drop in the death rate should be particularly welcome when there is a corresponding fall in the birth rate as well. The fall in the birth rate lags invariably behind the drop in the death rate, leading to the phenomenon of a sudden spurt in the rate of growth of population. The phenomenon is observed to occur in the first phase of industrialization in almost every economy and there is a reversal only when there is a substantial improvement in the standard of living of the people. A part from the drop in the death rate, there is also likely to be a sort of stimulus being given to the birth rate due to increasing opportunities of employment and the spread of the temptation to enjoy economic betterment in the form of marriage and children. There is a considerable time-lag between the drop in the death rate and the full in the birth rate, resulting in a tremendous growth of population which in more ways than one comes in the way of rapid economic development.

The immediate problem that comes up in an economy like ours is one – most vital of all – of finding adequate food to feed the growing millions. As it is, there is an acute shortage of food in term of both quantity and quality – which has resulted in a viable deterioration in the economic structure and in the health of the bulk of our population. Most of them are reduced to unhealthy living conditions with a deplorable vulnerability to a variety of debilitating diseases, incapacitating the workers to put in the normal quota of work. The weak are inexorably caught up in the vicious circle of poverty – the incapacity to work, resulting in low income and giving a further onward push down hill towards the valley of virtual living extinction. The situation becomes all the more tragic when there is a continual addition of hapless children to the families to those already caught up inextricably in the quagmire of poverty. The plight grows from bad to worse with every addition that is made to the size of the family and very much like a recklessly loaded boat; the whole family sinks into misery and eventually, leads to famine which take a heavy toll of life.

The people of India suffer misery under the unbearable weight of over population and the biological litter of human species is thrown to find its way in an atmosphere of gloom and misery.

Land per head, capital per head, transportation, power and water, special amenities like houses, schools and hospitals must necessarily keep declining when income does not keep pace with an over growing population. The enormous crowds and queues that we see in residential buildings, bus stops, market-places, employment exchanges and petty political meetings are a sharp pointer to the enormous number of people that keep on pouring. Man being eaten by his own creation, is the sorry spectacle of over population.

We cannot get away from the fact that half the population of India today has to live on less than fifty rupees per day to cover the expenses of food, shelter, medical assistance, education, transportation and what not. About 90% of the people live under conditions of inhuman poverty and yet they breed profusely. **Poverty, famine, unemployment, disease and a hopeless deterioration of the physical mental and moral structure of the human personality is the price that we are paying for unrestrained population explosion.**

Universality of marriage and unrestrained procreation is the greatest obstacle to India's economic development. In the vigorous tug of war between the rate of growth of income and the rate of growth of population, the strength exercised by the population factor is a terrible force to reckon with. The **use of contraceptives** is a way out to save the Indian economy from the ever increasing swarm of unwanted children, without sacrificing valuable human vitality so much required for economic development and it may eventually lead to planned families of optimum strength and satisfaction.

Late marriage, to certain population growth, reminds us of Malthus. It is difficult to implement as early marriage is still an economic compulsion for many poor population, apart from being a taboo. Hence, only economic uplift can work wonders in limiting population.

Human resources development is vital. In fact, environment development programme and human resource development programme should go together. An educational policy which emphasizes on conservation and management of natural resources and family planning will go a long way to achieving sustainability.

"Some people dream of success while others wake up and work hard at it."

Ants work hard around the clock without sleeping. It makes us release us realize that we can achieve whatever we want only if we have the will to do it and work hard towards it.

2 Growth and Development of Crops

"Scientific discovery and scientific knowledge have been achieved only by those who have gone in pursuit of it without any practical purpose whatsoever in view"
Max Plank

GROWTH

The term growth is applied to quantitative changes occurring during development and it may be defined as an irreversible change in the size of a cell, organ or whole organism. External form of an organ is primarily the result of differential growth along certain axes.

DEVELOPMENT

Development is applied in its broadest sense to the whole series of changes which an organism goes through during its life cycle, but it may equally be applied to individual organs, to tissues or even to cells. Plant development involves both growth and differentiation.

DIFFERENTIATION

During development there are appear not only quantitative differences in number and arrangement of cells within different organs, but also qualitative differences between cells, tissues and organs to which the term differentiation is applied. For example, differentiation of plant body into shoots and roots. Thus, growth and differentiation are the two major developmental processes.

THE DYNAMICS OF GROWTH

For individual cells or organs, growth is potentially unlimited and begins as an exponential pattern. However, mutual interaction within an individual impose limitations on growth, the actual growth curve becomes sigmoid in nature. The overall growth of an organism includes an early exponential phase. Growth of a higher plant during its exponential phase is analogous in accumulation of capital at continuous compound interest. If the plant's initial weight is W_0, and the rate of compound interact is r, the total weight after a certain time't' will be:

$$Wt = W_0 (1: r) t$$

Growth

Growth is irreversible increase in size or weight. Growth is attained mainly by **photosynthesis** less what is lost through respiration.

PHOTOSYNTHESIS

Photosynthesis links inorganic and organic world. Inorganic compounds like water and carbon dioxide are combined to form organic compounds like sucrose, starch etc. The main principle in agronomy is harvesting as much solar energy as possible and converting solar energy into chemical energy by photosynthesis. Almost all the agronomic practices are aimed at intercepting maximum possible solar radiation. The primary sugars produced during photosynthesis are the raw materials for subsequent production of more complex sugars, starches, cellulose and other plant constituents. Photosynthesis provides the raw material for plant growth and for the development of storage organs such as leaves, stems, tubers and fruits. Photosynthesis can be studied under **three processes:** entry of carbon dioxide (CO_2), photolysis and carbon dioxide reduction.

Entry of Carbon Dioxide

Carbon dioxide entry into the plant can be easily understood by the analogy of flow of electricity. The flow of electricity is reduced by the resistance offered by the conducting material. Similarly, carbon dioxide entry into leaf faces resistances such as boundary layer resistance, stomatal resistance and mesophyll resistance. On the leaf surface, a thin layer of undisturbed air is present which is called boundary layer. It offers resistance to the flow of carbon dioxide which is called boundary layer resistance. The magnitude of this resistance ranges from near zero in very turbulent air to 400 s/m in still air. It means that carbon dioxide takes 400 seconds to move a distance of one metre in still air while in turbulent air carbon dioxide moves quickly as there is negligible boundary layer.

After passing through the boundary layer, carbon dioxide has to go through stomata which may be fully opened, partially opened or fully closed. The resistance offered by stomata is stomatal resistance which ranges between 50 to 100 s/m in the case of fully opened stomata and very large values when it is tightly closed. Subsequently carbon dioxide has to move from stomatal cavity to mesophyll cells which are the actual site of action for photosynthesis. The resistance offered by mesophyll cells is mesophyll resistance which ranges between 100 s/m and 1000 s/m. The flow of carbon dioxide also depends on concentration gradient of carbon dioxide between air and mesophyll cells. If the concentration of carbon dioxide is very high in the air, it moves quickly into plant cells.

Photolysis

In plants, the molecule absorbing radiant energy in the visible range is referred to as a pigment. To the human eye, these pigments are coloured; chlorophyll is green, carotenoids red or yellow and phytochrome blue. Plants also contain a variety of pigments such as anthocyanins and flavones. Only the chlorophylls and carotenoids, however, function in photosynthesis by absorbing radiant energy which is then coupled to the reduction of carbon dioxide and formation of organic substances.

Solar radiation contains small units called photons. When the photons strike pigment molecules, the electron of the pigment from lower orbit moves to the upper orbit. In this condition, the pigment molecule is said to be excited and it gains energy. This energy is spent **in three ways:** released as heat or as fluorescence or utilized in chemical reaction in the presence of enzymes like ferredoxin, plastocynin, etc. After the energy is dissipated by either of the three ways, the pigment molecules come to normal state.

When the pigment molecule is in close proximity to reacting molecules, the energy is transmitted to reacting molecules and these reacting molecules use this energy for breaking down water i.e. **photolysis**. By photolysis, water is broken down into hydrogen ion and nascent oxygen. The hydrogen ion molecules are utilized for reducing NADP to NADPH. The energy released in this process is utilized for oxidative phosphorylation i.e., conversion of ADP to ATP. Phosphorylation occurs in two ways: cyclic and non-cyclic reactions.

A. Cyclic Phosphorylation

In the cyclic reaction, the excited electron is transferred first to an iron containing protein, ferredoxin. From ferredoxin, the electron is transferred by means of carrier molecules to plastocyanin, a copper containing protein. The electron completes its cycle when it is transferred from plastocyanin to its original position in the chlorophyll molecule. It arrives at this starting point in its original non-excited state (Fig. 2.1).

The energy is not lost but it is used in the conversion of ADP to ATP.

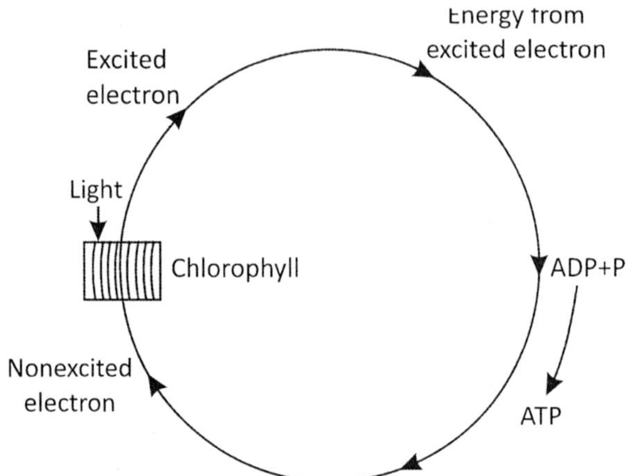

FIG. 2.1: Cyclic Phosphorylation.

B. Non-Cyclic Phosphorylation

This reaction occurs at two sites, photosystem F (PS 1) and Photosystem II PS II) which are connected by electron © transport system. In PS I, light energy excites an electron which is passed along through a series of enzymes until it ultimately binds hydrogen ions (produced in PS II) to a special hydrogen carrier, nicotinamide adenine dinucleotide phosphate (NADP).

$$\text{NADP} + \text{H}^+ \longrightarrow \text{NADPH}$$

The high energy electron is temporarily lost from the chlorophyll molecule of PS I, but it is replaced by an electron that comes from water as a result of the reactions associated with PS II. In PS II, the end results are:

$$4\text{H}_2\text{O} \longrightarrow 4\text{H}^+ + \text{OH}^+$$

$$4OH^- \longrightarrow 4OH^+ - 4\,e^-$$
$$4OH \longrightarrow 2H_2O + O_2$$

Reduction of Carbon Dioxide

The reduction of carbon dioxide is also called dark reaction as light is not necessary. Dark reaction does not mean that the this reaction occurs only in dark. With the energy supplied by ATP (which is formed during light reactions), carbon dioxide combines with hydrogen (supplied by NADPH) and forms carbohydrates. Based on the method of reduction of carbon dioxide, plants are classified into C_3, C_4 and CAM plants.

C_3 Plants: The initial product of carbon assimilation is the three carbon compound. The enzyme involved in the primary carboxylation is ribulose-1, 5-biposphate carboxylase. Photorespiration is high in these plants. Rice, wheat, cotton, soybean etc. belong to this group.

C_4 Plants: In these plants, the primary product of carbon fixation is 4-carbon compound which may be malic acid or aspartic acid. The enzyme responsible for carboxylation is phosphoenol pyruvic acid (PEP) carboxylase. This enzyme has high affinity for carbon dioxide and capable of assimilating carbon dioxide even at a lower concentration. Photo respiration is negligible in this plant type of plants and photosynthetic rates are higher. Crops with C_4 type of photosynthesis are maize, sorghum, pearl millet and other minor millets. C_4 minor millets, even under good management do not out yield C_3 rice. The advantage of C_4 photosynthesis at caboxylation level seems to be lost at crop level. The great photosynthetic advantage of C_4 plants compared with C_3 plants at the level of carboxylation is attenuated or reduced at the whole leaf level. It is due to combination of stomatal and mesophyll resistance at the leaf level and by mutual shading and periods of low light at plant community level. As a result, no consistent advantage of the C_4 pathway is evident in crop growth rates and crop yields of C_4 plants. The real value of the C_4 pathway probably lies elsewhere, in its better adaptation to higher temperature and higher insolation conditions whereas C_3 plants perform better under cool conditions with only moderate insolation. Since the conditions for optimum growth of C_4 and C_3 plants are quite different, valid comparisons of their performance can hardly be made.

Crassulacean Acid Metabolism (CAM): In these plants, the stomata are open at night and large amount of carbon dioxide is fixed as malic acid which is stored in vacuoles. During day, as the stomata are closed, there is no possibility of carbon dioxide entry. Carbon dioxide which is stored as malic acid is broken down and released as carbon dioxide. This carbon dioxide is reduced as in the case of C_3 plants. Pine apple and sisal are the known crops with CAM type of photosynthesis (Table 2.1).

TABLE 2.1: List of crops with different types of photosynthesis

C_3	C_4	CAM
Rice, soybean, wheat, groundnut, Barley, cotton, oats, sunflower, rye, potato, tomato, sugar beet.	Sorghum, maize, napier grass, pearl millet, sugarcane, Panicum sp., finger millet, amaranthus.	Pineapple, sisal

Among these three groups, C_3 which is stored as malic acid, is broken down and released which is stored as malic acid, is broken down and released plants have lower water-use efficiency. In case of CAM plants, stomata are closed during the day. Transpiration is, therefore, negligible. For limited amount of water lost in transpiration, more carbon dioxide is fixed. In C_4 plants, as PEP enzyme has high affinity and activity with carbon dioxide, for known size of stomatal operture, more carbon dioxide is fixed for the same amount of water loss. C_4 and CAM plants have, therefore, high water-use efficiency. Plants with C_4 dicarboxylic pathway of photosynthesis use considerably less water per unit of additional dry weight than do calvin cycle (C_3) plants. CAM plants have the lowest transpiration rates and much less water. C_4 plants are able to utilize low concentration of carbon dioxide and can thrive in situations where carbon dioxide supply is less.

RESPIRATION

Respiration is almost a reverse process of photosynthesis. The summary reaction of respiration is:

$$C_6H_{12}O_6 \longrightarrow 6CO_2 + 6H_2O$$

Respiration supplies the necessary energy for the execution of various biological and chemical reactions in plants. It also provides intermediate products that are necessary for the development of complex compounds.

Based on function, respiration in plants can be divided into growth and maintenance respiration. Growth respiration provides energy and products for the synthesis of structural and storage compounds. It depends on the nature of compounds being synthesized and rate of photosynthesis. It is not directly affected by temperature. Maximum respiration provides energy for the entire plant for routine works such as maintenance of membranes, proteins, cellular organization and for ion uptake. It depends on body weight and ambient temperature. Growth respiration is necessary for the normal growth whereas maintenance respiration is considered a wasteful process especially under the high temperature. As the rate of maintenance respiration depends on temperature, carbohydrates are lost without being utilized for growth. Low light and high temperature during grain filling period is considered as unfanourable for high yields.

Dark and Photo Respiration

The respiratory mechanism common to all types of plants (C_3 and C_4) is called dark respiration since it occurs regardless of light. The additional respiratory mechanism of C_3 plants is called photo respiration and occurs only during high light and oxygen availability. Under high light intensity, high level of energy and oxygen is available in the leaves. The energy absorbed by the chlorophyll pigments has to be disseminated either to convert ADP to ATP or NADP to NADPH or by some other means. In the absence of electron acceptor, hydrogen peroxide is produced which is toxic to plants. Under this situation, energy is dissipated by photo respiration. The enzyme in photo respiration is the same ribulose-1, 5-bisphosphate carboxylase which fixes carbon dioxide. As this enzyme also accepts oxygen, it is also called as ribulose-1, 5-bisphosphate oxygenase. Photorespiration is high in C_3 plants which are considered as wasteful.

NET PHOTOSYNTHESIS

The total amount of photosynthesis produced due to the process of photosynthesis is called gross photosynthesis. Plants spend some amount of carbohydrates through respiration. Net photosynthesis is the amount of carbohydrates added in the plant by photosynthesis less the amount of carbohydrates spent in respiration. The amount of net photosynthesis depends on light intensity and carbon dioxide concentration.

When the light intensity is low, the photosynthates manufactured due to the process of photosynthesis may be just equal to the amount of carbohydrates spent in respiration. The light intensity at which photosynthesis and respiration are equal is called light compensation point. Leaves of plants normally show increasing rates of photosynthesis with increasing solar radiation above the light compensation point. However, increase in the rate of photosynthesis stops at a particular light intensity beyond which there is no further increase in photosynthesis. The light intensity at which there is no further increase in the rate of photosynthesis is called light saturation point. Light saturation is high in C_4 plants as compared to C_3 plants (Fig. 2.2).

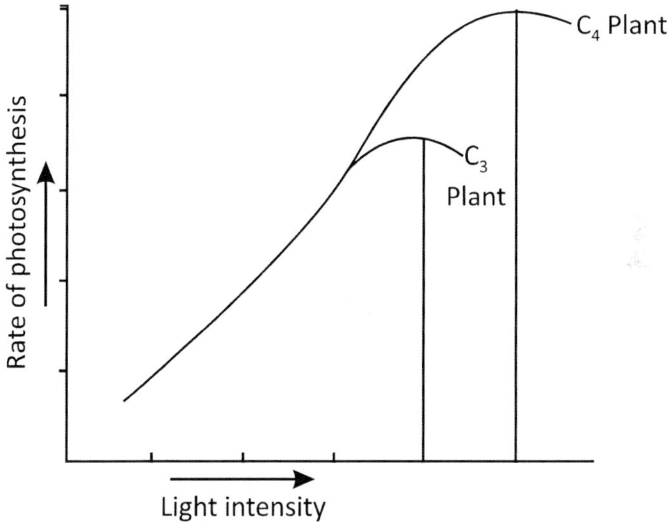

FIG. 2.2: Light saturation point of C_3 and C_4 plants.

Photosynthesis is limited by low carbon dioxide concentration. The concentration of carbon dioxide at which respiration and photosynthesis are equal is called carbon dioxide compensation point. It is lower in C_4 plants than in C_3 plants.

TRANSLOCATION OF ASSIMILATES

Much of the carbon assimilated moves from the sites of fixation to other organs where it is used for their maintenance and growth. This movement takes place through sieve elements of the phloem. The carbohydrates are usually translocated as sucrose and to some extent as starch. The rate of translocation is related to the rate at which carbon enters the phloem but this is not always a simple relationship with the rate of photosynthesis. In most circumstances, the rate of photosynthesis is greater than the rate of translocation and there is a net accumulation of carbon in leaves, usually as starch. The major long term sinks for assimilates are the storage organs like grains, tubers and fruits. Some of the photosynthates thus translocated are consumed in maintenance respiration and some are used in growth respiration.

The products formed during respiration are immediately used for building complex materials like cellulose, protein, fats, pectins etc. which are used as building material for the production of leaves, stems, flowers etc. The photosynthates translocated to storage organs may be stored as carbohydrates, oils, proteins etc.

GROWTH OF PLANT

Growth in plants has various phases which start with activation of the embryo and ends with maturation of the seed. The processes involved in growth are very complex and are vitally influenced by environmental factors which themselves form a complex. Growth of any organism increases in size and also in complexity. The quantitative increase in matter is accompanied by changes in the nature of their plant parts which are of qualitative character. These qualitative changes are associated with the development phase as distinct from growth phase. These phases run their course during the life cycle of the plant with modifications engendered by the environment.

Germination

Seeds when sown in moist soil, germinate under favourable temperature and sufficient aeration. Seeds of some plant species do not germinate even under favourable conditions and these seeds are said to be dormant. In simpler terms, seed dormancy is blocking the growth of embryo. The period of dormancy may be for a few days, weeks, or months. Dormancy may be due to several reasons: improper development of embryo, high concentration of growth inhibiting substances, hard and impermeable seed coat. Seed dormancy in crops is less significant, as dormancy is for a very short period or is broken during the period between harvesting and next sowing. However, it is important with seeds of some pasture and weed species. Based on the light requirement for germination, seeds are grouped as (1) positive photoblastic, those that require light for inducing germination, (2) negative photoblastic, those that do not germinate when exposed to light, and (3) non-photoblastic, and those that germinate either in light or dark. Seeds of many cultivated crops belong to non-polyblastic group.

Flowering plants can be broadly grouped into monocotyledons and dicotyledons. Monocotyledons have single cotyledon in their seeds. Cereals and millets belong to this group. The important parts of monocotyledon seed are embryo, endosperm and aloeurone layer (Fig. 2.3).

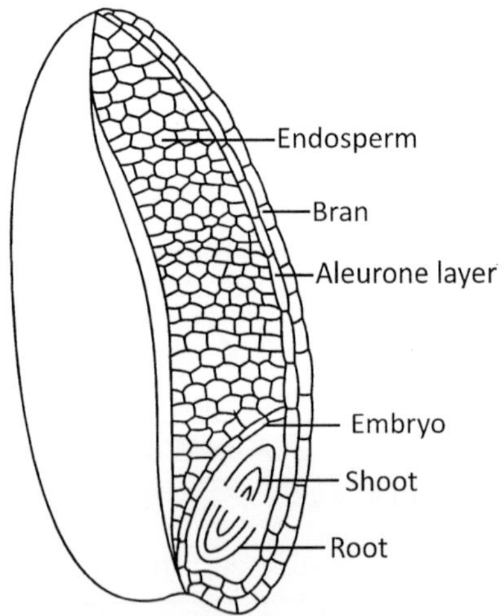

FIG. 2.3: Longitudinal section of wheat seed.

Carbohydrates are stored in endosperm whereas embryo contains miniature root and shoot. In other words, seed is more or less partially formed plant enclosed in protective casing reserve of essential food. Under appropriate conditions, this plant has the ability to utilize these reserves and develop into an autotrophic plant. In dicotyledonous plants, the seed contains two cotyledons which provide food material during the **processes of germination.**

Biochemical Process of Germination: Germination can be grouped into biochemical and morphological processes. When the seed is sown into moist soil, it absorbs water twice its weight. As a result, the seed bulges and breaks the seed coat. Subsequently, embryo secretes gibberellins which in turn induce the secretion of hydrolytic enzymes by aleurone layer in cereals. Hydrolytic enzymes break down complex food materials like starches and proteins into simple sugars and amino acids respectively. Fats which occur chiefly in the cotyledons contain oil bearing seeds and in the embryos of cereal seeds are split by enzymes called lipages into fatty acids and glycerol. These in turn undergo chemical changes and form sugars. The simpler products thus formed are soluble and can be translocated to growing points until the plant becomes autotrophic. The energy required for the chemical and biological process, 25 to 50 per cent of stored food is used up in respiration.

Morphological Process of Germination: The radical is the first organ to emerge from the seed in nearly all cases. The radical is soon followed by the plumule of young shoot. In many dicotyledonous plants, such as groundnut, beans etc. the cotyledons emerge from the soil and function as first leaves. The three first leaves are brought to soil surface without damage because they are present inside the thick cotyledons. The cotyledons are brought out of the soil by covered elongation of hypocotyls. This type of germination where cotyledons emerge out of the soil is called epigeal germination (Fig. 2.4a).

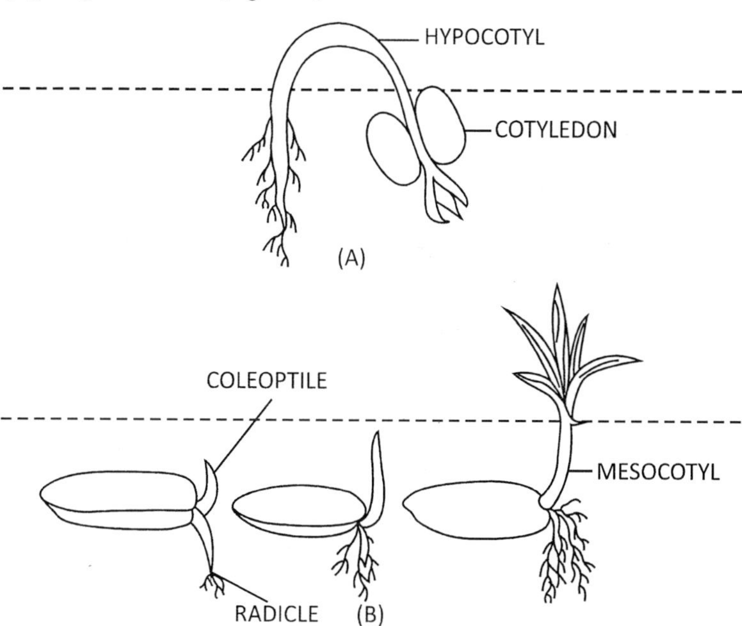

FIG. 2.4: (a) Epigeal and (b) hypogeal germination.

In monocotyledonous plants, the cotyledon remains in the soil. The plumule grows or is pushed upward by the elongation of epicotyls. This type of germination is called hypogeal germination since the cotyledon is below or inside the soil (Fig. 2.4b). Dicotyledons also exhibit hypogeal germination when sown deep. For example, groundnut when sown deep the reserves in cotyledons are spent for the elongation of hypocotyls and only true leaves come out of the soil. The coleoptyl of grasses emerges from the soil as pale tube like structure that encloses the first true leaf. A slit develops at the vent on the tip of the coleoptiles and leaf emerges through it. The radical develops seminal roots. Seminal roots emerge utilizing seed reserves. The average number of root primordial in the embryo shows general variation in the region of 4-6 in wheat and about 9 in barley. The number actually developing varies widely and depends on nutrient and water availability.

Emergence

The emergence of crops is critical for good stand establishment. Generally it is hindered by soil crust, high salt concentration, water logging etc. Plants with bigger seeds and with epigeal germination can overcome the soil crust problem and emerge out. After emergence, the cotyledons become green and support the seedlings in dicotyledonous plants. In the case of monocotyledonous plant like wheat, the emerged seedling becomes independent when the second leaf has fully expanded and the third just appears. Seedling is able to take up nitrogen, phosphorous, zinc and possibly most minerals from the soil. The reveres present in the seed usually have more of carbohydrates or lipids which are sufficient till plant becomes autotrophic, while minerals present in the seed are not sufficient up to that stage.

Root Growth

In monocotyledononous plants, radical elongates and constitutes primary root. Primary roots usually grow vertically down wards while secondary roots that appear during the seedling stage grow horizontally for a few centimeters before turning down wards. The higher order laterals are spread randomly. These roots are called **seminal roots.** At the end of the seedling stage, fresh roots develop from lower nodes of stem. These roots are called **nodal roots** or **adventitious roots**. The seminal roots penetrate downwards to the greatest depth. The bulk of adventitious roots may grow first horizontally and then vertically while the last formed nodal roots may grow upwards also. The adventious roots constitute a major portion of total root mass and become important for the rest of the life of the plants, while the seminal roots decay gradually. When the seedlings are transplanted, they establish by producing nodal roots.

In dicotyledonous plant, the root system is tap root type compared to fibrous root system of most of the monocotyledonous plants. The main root elongates at rates of 1 to 20 mm per day and eventually may reach a depth of 2 m. However, most of the roots are present in top layers of the soil only. The rate of elongation is related to temperature, soil moisture, photosynthates and nutrient supply. The appearance and death of the root hairs occur continuously.

Inflorescence

The growth of inflorescence in length usually shows two phases: an initial period of slow elongation during the time of formation of new spikelets, followed by the period of rapid elongation of the internodes of the rachis. The latter appears to be associated with increased production of gibberellins. The upper three internodes of the culm and the inflorescence grow simultaneously compacting for photosynthates and mineral nutrients. Both the internodes and inflorescence get equal amounts of assimilates. Top leaves contribute more to ear growth and almost half of the pre-anthesis weight of the ear is from the flag leaf.

Potential yield is determined during the early stage of differentiation of inflorescence by the number of tillers that form inflorescence and number of spikelets formed on each spike. Spikelet differentiation appears to stop when stamens are differentiated in the most advanced floret. Low temperature prolongs the duration of inflorescence growth but length of inflorescence and number of spikelets is more than those at optimal temperature. High light intensities increase the number of spikelets and also number of effective tillers. Mineral nutrients influence growth of inflorescence through their influence on leaf area and number of effective tillers, but not directly on inflorescence. Moisture stress at early stage of growth of inflorescence delays the formation of spikelets but recovers after relief of the stress. Prolonged moisture stress at later stage of inflorescence growth reduces both number and size of spikelets. In panicles of cereals, moisture stress at panicle emergence causes death of spikelets and at anthesis germination and growth of pollen tube is affected.

Seed and Fruit Growth

Growth of ovary may continue at slower rates up to anthesis and picks up after fertilization. If fertilization does not occur, the ovary then gradually shrinks and the flower may drop except in the case of parthenocarpic species. The growth of seed and fruit may develop differently or separately. In certain cases, seed growth continues for long time even after the stoppage of fruit growth. For example, seed of groundnut grows at a faster rate after the stoppage of fruit growth. Later, some of the constituents of fruit are retranslocated to the seed. Carbohydrates and nitrogen are retrabslocated to grain from the cob in maize.

There are usually large differences between sizes of ovaries at different positions and in the rates at which these develop. This results in an appreciable variation in fruit and seed size at maturity. However, the average seed size of any particular variety is remarkably constant over a wide range of environments. Seed size is more influenced by genetic factors than environment except moisture stress.

Photosynthates are applied to seed and fruit from leaves, fruit itself and retranslocated from the other parts. The relative contribution of these sources differs depending on the species and environment. Generally, in wheat and barley, most of the contribution to grain is from flag leaves.

In rice and wheat, more leaves are green at the time of grain filling and therefore, current photosynthesis is more important. Another source of carbohydrates to grain filling is the stored carbohydrate from stem which is about 10 to 20 per cent. During moisture stress, when the normal photosynthesis is affected, stem contribution may exceed 20 per cent of final grain weight.

Supply of nitrogen, phosphorus and other mineral nutrients to the fruit come either directly via the roots from the soil or by re-export from other tissues. The amount absorbed from the soil depends on the root growth and the amount of available nutrients remaining in the soil. However, as root growth is reduced after flowering, most of the mineral nutrients are from older leaves. The flag leaf, the leaf just below the ear, is more important suppliers of carbohydrates to grain and older leaves are important suppler of nitrogen.

Storage of water during fruit growth leads to reduced size and hence less total yield, probably through its effect on rate of photosynthesis and also by hastening leaf senescence. If moisture stress is for a short period, the reduction may be compensated by subsequent rapid growth rate.

The yield achieved by a crop is the result of environmental factors acting on physiological processes. When nutrients are not limiting, the important environmental factors are solar radiation, temperature and water. After seed geminates, primordial are initiated. Vegetative primordial expand to form a canopy and root system, which by intercepting radiation and extracting water and nutrients from the soil, produce dry matter (Fig. 2.5). Some of this is used to germinate a larger canopy or more extensive root system, the rest is allocated to expanding reproductive primordial.

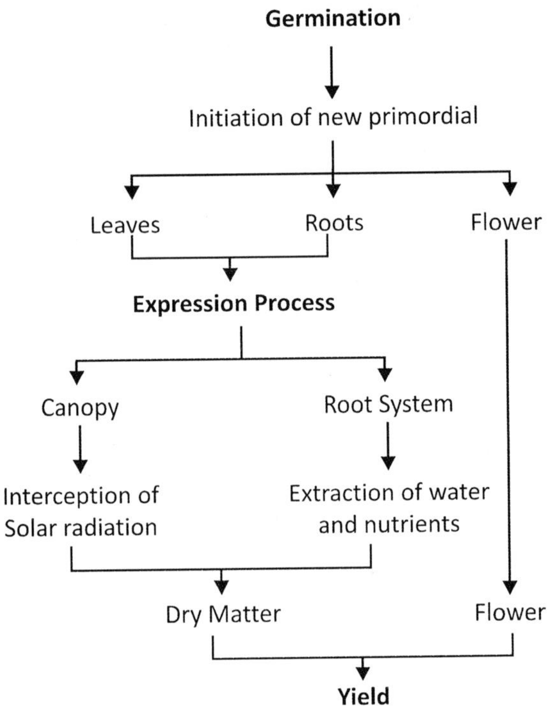

FIG. 2.5: The main process of growth a crop.

DEVELOPMENT

Growth is an irreversible increase in mass or weight whereas development denotes phasic change. The development of a plant from germination to maturity can be considered as a series of discrete periods, each identified by an accompanying process of change in the structure, size or weight of specific organs. Developmental processes relate to varying degrees of expansion or growth in weight of an organ. Some, such a leaf initiation, are associated with only very small changes in size or weight, others such as flowering in millets which follows expansion of the stem and panicle can only proceed after considerable increase in size and weight.

A characteristic of developmental processes is that they are discrete. Seeds either germinate or do not germinate; and a leaf primordium is invisible or visible. Therefore, they can only be defined in terms of time, and not in length, area, volume or weight. The duration of a developmental process is usually measured between events that are detected visually. Such durations are those between wetting a seed and the subsequent elongation of the radical to a defined length and between fertilization of flower and the maximum weight of grain or pod. The rate of development is simply the reciprocal of this duration.

Developmental Stages

Age of the crop is expressed as crop growth stages. However, the term growth stages is misleading as these stages do not indicate increase in dry weight or mass. They can be called as developmental stages as they indicate age or progress towards maturity.

The life cycle of a crop is divided into different stages based on morphological characters, developmental stages etc. The demarcation of crop into different developmental stages helps in better communication among farmers, extension workers and scientists. The expression of crop age or stage by calander days is misleading because the duration of certain physiological stages are altered by climate. Panicle initiation in cereals, which is an important stage for application of fertilizers and irrigation, may be delayed or advanced by photoperiod and temperature. In rice, low temperature delays panicle initiation. The developmental stages are helpful in identifying the time for scheduling irrigation, application of fertilizers, herbicides etc. Certain stages of crop growth are very sensitive to moisture stress. For example, moisture stress at crown root initiation stage reduces the grain yield of wheat considerably and it can not be compensated by subsequent better management practices. Fertilizer application is made in split doses to avoid leaching losses. But sufficient quantities are to be applied each time to meet the crop demand. As nutrient requirement at panicle initiation stage is high, sufficient dose is applied at this stage. Similarly, for effective control of weeds and to reduce adverse effect of herbicides on crop, they have to be applied at proper stage of the crop.

The crop developmental stages applicable to most of the crops are: **germination, emergence, seedling stage, maximum vegetative growth stage, primordial differentiation, flowering, fruit growth, fruit maturity, physiological maturity and harvest maturity.** These crop stages are named differently to suit the peculiarities of individual crops.

Germination and Emergence

After sowing of the crop, the seeds absorbed moisture and put forth radical and plumule and it is known as **germination**. On subsequent growth, the plumule is pushed through the soil and comes out of soil and this is known as **emergence**.

Seedling Growth

Subsequent to emergence, the plants put forth leaves and no longer depend on seed reserves. During the stage, the plants establish by developing sufficient number of seminal roots and leaves. This stage ends when the plants put forth tillers or branches.

Maximum Vegetative Growth Stage

During this stage, crops grow at a faster rate. Plants start producing branches or tillers and the crop covers the ground, as much as possible, to intercept more radiation. The loss of the plant population that occurs during early stages can be overcome in this stage by producing more number of tillers or branches. The number of ears per unit area is decided during this stage. This stage is variously called as tillering stage, active tillering stage, branching stage etc. depending on the crop.

Primordial Differentiation

This stage is called as panicle initiation stage in cereals and millets, squaring in cotton, flower bud initiation in sunflower etc. With starting of this stage, the plants enter reproductive phase. The number of fruits or grains is decided during this stage. This stage is more sensitive to moisture and solar radiation in cereals (Fig. 2.6).

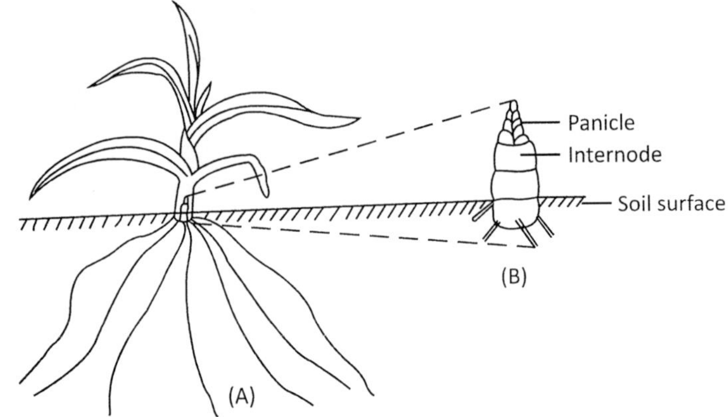

FIG. 2.6: Panicle initiation stage (a) concealed in leaf sheets (b) exposed and enlarged.

Flowering Stage

This stage also called as 50 per cent flowering, anthesis, blooming, etc. The crop is said to be 50 per cent flowering when 50 per cent of plants put forth flowers. Opening of flowers and shedding of pollen is called **anthesis**. Blooming also indicates opening of flowers. This stage is very sensitive to moisture stress. Fruit set is development of flowers into fruits and this is a very important stage.

Fruit Growth

During this stage, photosynthates are accumulated in the fruits. Fruit growth is also called as **grain growth** in cereals, **pod growth** in groundnut and pulses, **boll growth** in cotton etc.

Fruit Maturity

During this stage, biochemical reaction takes place in the fruits for the synthesis of proteins, oil and starches from simple carbohydrate. These complex products are synthesized in fruit itself or synthesized in leaves and translocated to fruits.

Physiological Maturity

When fruit growth is complete and photosynthates are no longer translocated to fruits, it is known as physiological maturity. An abscission layer forms between fruit and panicle. Crops, if necessary, can be harvested at this stage.

Harvest Maturity

At physiological maturity, the moisture percentage of grains is about 20 per cent. In addition, all the grains might not be at physiological maturity. Therefore, it is a general practice, to allow the crop for 10 or more days after physiological maturity to reduce the moisture percentage and also to provide time for the development of late formed grains.

PHYSIOLOGICAL GROWTH COMPONENTS

Internal as well as applied auxins and chemicals play a profound role in the growth and metabolism of the rice plant. These aspects have been reviewed by Sircar (1965), 1971) who emphasized the role of IAA, as a natural growth substance in rice.

Maximum Dry Matter Production

The first pre-requisite for high yields is a high production of total dry matter per unit area. The amount of dry matter produced depends on photosynthesis which is in turn depends on large and efficient assimilating area, adequate supply of solar radiation and carbon dioxide and favourable environmental conditions. Some amount of photosynthates is used in transpiration which is essential for maintenance and development. The total yield of dry matter is, therefore the amount of dry matter produced less the photosynthates used for respiration. Finally, the amount of economic yield depends on the manner in which the net dry matter produced is distributed among the different parts of the plant.

Assimilating Area

The leaves of the plant are normally its main organs of photosynthesis and the total area of leaves per unit area of the land surface is leaf area index (LAI). Variation in total rice area of a plant may result from changes in leaf number and size. Leaf number of growing points, the length of time during which leaves are produced, the rate of leaf production during this period and the length of life of the leaves. Leaf size is determined by the number and size of the cells of which leaf is built and is influenced by light, moisture regime and available plant nutrients in the soil.

Optimum LAI

Leaf Area Index (LAI) increases slowly in the early stage of crop growth and rapidly after seedling stage. As LAI increases, light interception is more resulting in higher dry matter production. However, at high LAI, mutual shading of leaves occur and as shaded leaves respire similar to that of unshaded leaves; they contribute less to the total dry matter. There is, therefore, an optimum LAI for maximum dry matter production which is reached when the largest number of leaves receive just sufficient sunlight for photosynthesis to balance respiration. Below optimum LAI, light is not being fully intercepted and above optimum LAI, the leaf area is not being fully utilized at maximum efficiency. At higher than optimum LAI, lower leaves become parasitic even under conditions of full light intensity. Optimum LAI differs with the crop and their leaf orientation. Optimum LAI is between 3 to 4 for crops with horizontally oriented leaves and 6 to 9 for crops with upright leaves.

Extinction Coefficient

The proportion of incident light that is intercepted by the crop canopy does not depend on LAI alone but also on the architecture of the plant community. The arrangement of leaves on stem of the plant and the arrangement of plants in the field decide the crop architecture. The extinction coefficient (K) is a measure of the light intercepting efficiency of the leaf area.

FACTORS INFLUENCING GROWTH AND DEVELOPMENT

Growth and development of a plant are influenced mainly by solar radiation, temperature, soil moisture, soil aeration and mineral nutrients. In addition, other factors also influence crop growth at different stages.

Germination

Germination is influenced by temperature, soil moisture, light, aeration and dormancy of seeds while soil physical conditions and depth of sowing influence emergence.

Temperature

The rate of germination increases linearly with temperature over the range from the minimum to 2-3°C below optimum temperature. The decrease in rate at high temperature and progressive downward shift of this rate with time is an usual response. The maximum temperature above which no germination occurs is usually within the range of 35-45°C.

Soil Moisture

Most of the crop seeds germinate well within the moisture regime of field capacity to 50 per cent available soil moisture. The effect of change in matric potential of the soil is attributed to the associated increase in mechanical strength of the soil rather than availability of water or to changes in area of contact with liquid water. For better contact with the seed, the mean diameter of soil aggregates should be less than about one fifth of the seed.

Depth of Sowing

After the crop is sown, soil gradually dries out if there is no further addition of moisture to the soil, and if the seeds are sown shallow, they may not germinate. If the seeds are sown deep, seeds may not emerge as the seed reserves may not be sufficient to put forth sufficient growth. Even if, they just manage to come to the soil surface, the seedling may be too weak to survive. Soil crust is the main hurdle for the emergence of crops like foxtail millet, pearl millet etc. as the seeds are small. Oxygen content, light and dormancy are also some of the factors that influence germination and emergence. However, their influence under field conditions is not of much practical value except under certain exceptional conditions. For example, germination may be affected due to lack of oxygen by water logging if rains follow immediately after sowing. Light quality and dormancy are important for seed weeds and fodder crops. For example, seeds of stylo and boo babul need treatment prior to soaking to break dormancy.

Seedling Growth

The early stage of vegetative growth can be called as **seedling stage.** The seedling growth period starts with autotrophic life of the emerged seedlings and ends with initiation of tillering or branching. During seedling stage or juvenile stage of the crop, there is no competition for light, nutrients and moisture among them. Leaf and root are the components that are growing at this stage.

During emergence, the developing growing point is near ground level by elongation of internodes above coleoptile. The next 2 to 3 internodes in accropetal succession remain arrested and thickened to form crown of the plant. Subsequently, from this crown, both tillers and adventitious roots arise. During seedling growth, seminal roots penetrate downwards and branch at lower layers of the soil. Seminal roots support the plant during the juvenile stage. As there are no adventitious roots during this stage, the seedlings are uprooted easily with high winds. If the seminal roots cannot succeed in coming in contact with receding soil moisture, the seedling may die. The final plant population is decided during germination, emergence, and the seedling stage. Once, the crop passes through the seedling stage, the possibility of reduction in number of plants per unit area is far less.

Leaf Growth

In cereals and grasses, the tip of the leaf (distal region) matures earlier while basal region is still growing. The photosynthates prepared in the distal region are utilized in the basal portion and part of it may be exported to other plant parts. A similar sequence of events is also seen in the leaves of dicotyledonous plants. The rate of cell division and expansion are more in earlier formed leaves. Therefore, relative growth rates are less at later stages of crop growth.

Solar radiation, temperature, mineral nutrients and water status are important factors that decide the size of the leaf. The size that can be reached i.e., potential leaf size, is decided by solar radiation and temperature but its realization depends on nutrient supply. Leaf expansion depends on nitrogen supply and high nitrogen application leads to larger leaves. The weight and volume of leaves increase with solar radiation but leaf area is reduced. Leaf expansion is normal if the relative water content (water content of leaves compared to water content at saturation) is about 90-100 per cent. If it falls below 70-75 per cent, leaf expansion stops. The relative water content of leaves is more in young leaves compared to old leaves. Cell expansion is more affected by moisture stress than cell division.

The changes in physical structure of leaf during its development are also accompanied by changes in chemical composition and physiological activity. Nitrogen, phosphorus, chlorophyll, cell wall constituents increase until it acquires maximum area. The total nitrogen and phosphorus contents and their compounds gradually decrease with the age of the leaf. The activities of enzymes also follow the same pattern except with abscisic acid. These changes are reflected by changes in physiological process such as photo-synthesis and respiration. The rate of photosynthesis per unit area increases up to full leaf expansion and subsequently it decreases. Respiration rate is higher in very young and developing leaves and decreases with age of the leaves.

The leaf of a dicotyledonous plant depends entirely on the supply of carbohydrates from older leaves till it unfolds. Its peak period of export of assimilates is when it reaches full size. Net important of nitrogen, phosphorus and potash continues until the leaf reaches its full size. In the earliest stages, most of the nutrients seem to be received from older leaves, but as the leaf grows, higher proportion of the nutrients is obtained directly from the roots. The leaf often becomes a net exporter of these major nutrients at full expansion. The subsequent decrease in physiological activity is no doubt associated with these losses and denotes the beginning of senescence. The wheat lamina imports all of its carbohydrates until emergence and even at full expansion it imports one fourth of its dry weight from lower leaves. About half of the dry weight of sheath comes from sources other than the attached lamina.

The plant often produces more carbohydrates then is used immediately in growth. This surplus is stored in various tissues and is utilized when the current supplies become inadequate to support the requirements of growth and respiration. At later stages, senescence sets in and less of chlorophyll occurs. Subsequently, organelles like plastids, endoplasmic reticulum, mitochondria etc. and all membranes are disrupted. Gradual loss of water occurs and finally the leaf dies.

Tillering and Branching

The growth of auxillary buds into shoots is called **branching**. The branches that arise from basal nodes of the stem or crown are called tillers, especially in cereals and grasses. Initially, tillers appear at a rapid rate and after maximum number of tillers is produced, some of the late formed tillers die. The rate of production and number of tillers per plant depend on variety, availability of water, mineral nutrients and photosynthesis. At the time of initiation and early growth, tillers depend on older leaves for photosynthates. However, with the emergence of their own leaves, they become independent. Death of the tiller occurs due to insufficient supply of photosynthates, nutrients etc. to terminal bud and primordial leave. If the tiller survives until the shoot apex has been induced to flower, death does not occur and normal sequence of flowering and grain filling proceeds. Generally, tiller production stops at panicle initiation stage and tiller death continue up to emergence of ear on the main stem. Successive tillers are smaller and yield small amounts of grain. The late formed tillers are induced to flower early and the inflorescence develops quickly so that all tillers may come to maturity at about the same time.

PLANT GROWTH SUBSTANCES

Internal as well as applied auxins and chemicals play a profound role in the growth and metabolism of the rice plant. These aspects have been reviewed by Sircar (1965, 1971) who emphasized the role of IAA, as a natural growth substance in rice.

Plant growth substances are biochemical produced in plants (endogenous) or synthetic substances applied to plants externally (exogenous) which cause modifications in plant growth and development. Plant growth substances produced by the plant are referred to as phytohormones. Growth substances initiate biochemical processes which help in increased production of roots, shoots, flowers etc.

Plant growth substances are mainly grouped into growth promoters (which enhance growth) and growth retardants (which reduce growth). There are five major groups of endogenous growth substances present in plants viz., auxins, cytokinins, gibberellins acid and ethylene. Auxins, cytokinins and gibberellins are growth promoters while abscisic acid and ethylene are growth retardants. Growth promoters exist in multiple forms while abscisic acid and ethylene are in single form. The mechanism of growth substances is not well understood because of multiplicity of active molecules. Some important growth regulators and their use are given in Table 2.2.

TABLE 2.2: Some important plant growth regulators and their use

Abbreviation or Common name	Chemical name	Trade name	Major use
ABA	Abscisic acid	Abscisin	Defoliant
Acifluorfen retardant	5-[2-chloro-4-(Trifluoromethyl) Phenoxy]-3-nitrobenzoic acid	Blazer	Growth
Chlormeequat (CCC)	2-chloroethyl-trimethylammonium chloride	Cycocel	Lodging preventor
4-CPA	2-(3-chloropheneoxy) propionic acid	Fruitone	Fruit thinner
2,4-D	(2,4-dichlorophenoxy) acetic acid	Plantgard	-
Diaminozide (SADH)	Succinic acid-2, 2-dimethyl hydrazine	Alar	Growth retardant
Ethephon	2-chloroethyl phosponic acid	Cepha	Sugarcane ripener

Abbreviation or Common name	Chemical name	Trade name	Major use
Glyphosate	N-(-Phosphonomethyl) glycine	Round up	Sugarcane ripener
GA_3	Gibberellic acid	Berelex Pro-gibb	Grape enlarger
Kinetin	6-furfurylaminopurine	-	Dormancy breaker
IAA	Indol-3-acetic acid	Rhizopon-A	Plant cell enlarger
Maleic hydrzide (MH)	1, 2-dihydro-3, 6-pyidazinedione	Sproutstop	Growth retardant
Mepiquar chloride	N, N-dimethyl peperidinium chloride	DPC	Growth retardant
NAA	1-naphthaleneacetic acid	Fruitone-N	Fruit thnner
TIBA	2,3, 5-triodo-benzoic acid	Regim-8	Growth retardant

Auxins

Auxins stimulate elongation of cells of stems and coleoptiles. The endogenous auxins are Indoleacetic acid and Phenolacetic acid. Some of the synthetic auxins are Naphthaleneacetic acid (NAA), Indole-3 acetic acid, Indole-3-propionic acid and 2, 4-D. Auxins promote cell enlargement resulting in elongation of coleoptile, stem etc. The presence of auxins in apical bud suppresses the auxiliary buds and these results in apical dominance. Cell division and root formation are some of the functions of auxins.

Plant organ development in tissue culture is strongly affected by the balance of auxins to cytokinins. With high concentration of both auxins and cytokinins, cell often grow amorphously without differentiation. A high ratio of auxin to cytokinin causes induction of roots in callus cells, while high cytokinin to auxin ratio enhances the formation of shoots.

Auxins are used as herbicides when applied at higher doses. 2, 4-D and 2, 4, 5-T are used as herbicides. Other uses have been to enhance fruit set and fruit ripening. NAA induces flowering in pineapple. IBA is used for inducing roots of cutting.

Cytokinins

Cytokinin is stimulating cell division. Some of the endogenous cytokinins are: Kinetin, Zeatin, and Isopentenyladenosine. One of the potent synthetic cytokinins is 6-Benzyladenine. Cytokinins promote orderly development of embryos of seed, break dormancy of seeds and buds. They delay senescence.

Gibberellins

Gibberellin is first isolated from the fungus *Gibberella fuzikuror.* Subsequently gibberellins were found in plant tissues also. There are 50 different forms of gibberellins designated as GA_1, GA_2, GA_3, GA_7, etc. Gibberellins generally increase cell division and cell elongation. The gibberellin commonly available is GA_3 and is known as gibberellic acid. They promote cell elongation and increase in size of leaf, flower and fruit. Dormancy is broken and flowering is induced by gibberellins.

Abscisic Acid

Abscisic acid (ABA) induces leaf and fruit abscission. Dormancy is induced due to accumulation of abscisic acid. Moisture stressed plants produce abscisic acid which facilitates stomatal closure and help in maintaining cell turgidity.

Ethylene

Ethylene stimulates the swelling or isodiametric growth of stems and roots. Regulation of ethylene can trigger ripening of fruits or delay in ripening process.

Other Substances

Phenolics which are derivatives of phenol also show growth regulating activity. Phenolics inhibit cell division, cell enlargement and germination of seed. Triacontanol is a long and straight chained primary alcohol (1-Hydroxytriacontane). Lucerne leaves contain considerable quantities of triacontanol. When applied as a spray to the foliage, it increases plant growth. It is reported to have increased the growth and yield of soybean, maize and wheat.

Cycocel (CCC), phosfon-D, maleic hydrazide (MH) are growth retardant. CCC and MH retard stem elongation and the leaves of the treated plants become thick and dark green. They help in overcoming drought to some extent by reducing transpiration rate.

The well known regulatory activities of five major growth substances are:

Rooting:..Auxins, gibberellins, cytokinins, ethylene
Growth:.....................Auxins, gibberellins, cytokinins, abscisic acid, ethylene
Flower initiation: ....All the five
Fruit set: .... All except abscisic acid
Fruit growth:...All the five
Fruit ripening:.. All except abscisic acid
Abscission:...All the five
Senescence:...All the five

The regulatory activity of each growth substance differs with concentration used or combination of growth substances applied. They have widely overlapping functions. The most extensive use of plant growth substances in field crop production is for weed control with 2,4-D. Plant growth substances are extensively used in horticulture for rooting cuttings (IAA), regulating ripening of fruits (ethylene etc. (Table 2.3)

TABLE 2.3: Plant growth substances used in agriculture and horticulture

Plant growth substance	Use
2,4-D	Herbicide
Maleic hydrazide	Growth retardant, sucker control in tobacco
Glyphosine	Sugarcane ripener
Ethepon	Sugarcane ripener
Abscisic acid	Cotton defoliant
Triacontanol	General Growth stimulant
Cycocel	Dwarfing agent in wheat, reducing excessive Vegetative growth in cotton
NAA	Increasing number of flowers
Gebberellins	Production of seedless grape, increase in size of fruits
Ethylene	Regulating ripening of fruits

GROWTH EXPRESSION

Growth of a plant can be expressed in different equations or models. Scientists tried to express growth versus time of other growth factors.

Sigmoid Growth Curve

All plants pass through various stages of growth. Growth is the best expressed by means of a curve plotted against time. Growth curves are helpful in understanding the general pattern of growth. The S-shaped or sigmoid curve is typical of growth pattern of individual organs, of a whole plant and of population of plants (Fig. 6.10). It consists of five distinct phases:

1. Initial lag periods during which internal changes occur that are preparatory to growth. The increase in size or weight is very slow or negligible during this period.
2. It is followed by log phase or log period of growth or the grand period of growth during which growth is very fast. The logarithm of growth when plotted against time is a straight line during this phase.
3. Subsequently, a phase in which growth rate gradually diminishes.
4. A point at which the organism reaches maturity and growth ceases.
5. Later, senescence and death of organism sets in, giving rise to another component of the growth curve.

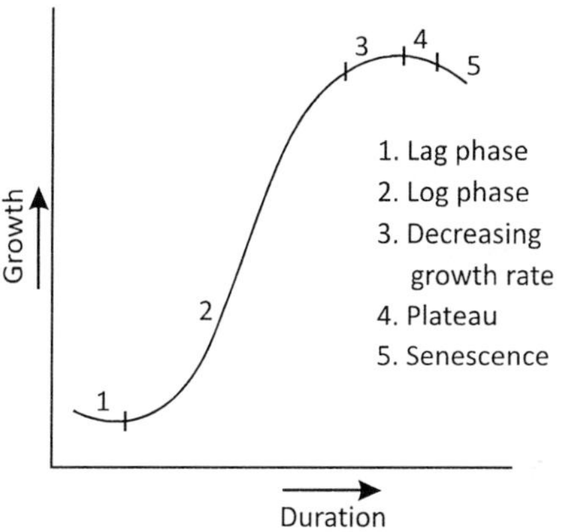

FIG. 2.7: Sigmoid growth curve.

Limiting Factors

Plant growth is a function of various environmental or growth factors. Several concepts related to their influence on crop growth are developed.

Liebig Law of Minimum: Justus Von Liebig proposed law of minimum in 1940 according to which the growth of the plants is limited by the plant nutrient present in smaller quantity, all others being in adequate amounts. This has been re-stated as barrel concept. A barrel with staves of different lengths can not contain anything above the height of the shortest stave. In like manner, growth can be no greater than allowed by the factor lowest in availability.

Blackman's Law of Optima and Limiting Factors: Blackman in 1905 proposed law of optima and limiting factors. It states that when a process is conditioned as to its rapidity by a number of separate factors, the rate of the process is limited by the pace of the slowest factor.

Mitscherlich's Law of Diminishing Returns: Mitscherlich in 1909 developed an equation relating growth with the supply of plant nutrients. When plants are supplied with adequate amounts of all but one limiting element, their growth is proportional to the amount of this one limiting element. Plant growth increases as more of this element is applied, but not in direct proportion to the amount of the growth factor added. Increase in growing with each successive addition of the limiting element is progressively smaller. Mitscherlich expressed this mathematically as:

$$Dy/dx = (A\text{-}y)\ C$$

When dy/dx is growth rate, A the maximum possible yield with sufficient level of all growth factors, y the yield obtained at a given level of factor, x and C are proportionality constants.

Inverse Yield-Nitrogen Law: Wilcox (1929) proposed inverse yield-Nitrogen law which states that the power of growth or yielding ability of any crop plant is inversely proportional to the mean nitrogen content in the dry matter. A crop plant with high mean percentage of nitrogen in dry matter has less dry matter production potential than a crop plant with low percentage of nitrogen.

Macy-Poverty Adjustment: Macy in 1936 proposed that a relationship exists between sufficiency of a nutrient and its percentage content in plant. According to him there is a critical percentage of each nutrient in each kind of plant. Above that point, there is luxury consumption and below that point there is poverty adjustment. This poverty adjustment is proportional to the deficiency until a minimum percentage is reached.

ANALYSIS OF GROWTH, DEVELOPMENT AND YIELD

The analysis of growth and development helps is devising methods for attaining maximum growth of crops in the field. This analysis based on responses to various factors of environment, provides to improve yields of crops by modifying the controllable factors of the environment. When all such factors can be controlled, better growth and development of the crop is achieved. Growth analysis, therefore, points to the limiting factors in the environment and indicates maximum potentialities of crop plants. The basic principle behind this concept of analysis is the estimation of crop grow that various stages and finally reasoning for yield variation. This would give an insight not only on the performance of a particular genotype, but also on the impact of superimposed agronomic practices on the crop at any particular stage of growth as well as on the final yield. This technique would involve estimating growth components and yield components. The growth analysis technique aids in identifying and defining an ideal plant type. This would also identify the factors responsible for high yield and thereby provide the basis to tailor agronomic practices.

Growth

Crop physiology, with the aim of understanding the dynamics of yield development of crops, began with the work of W.L. Balls on cotton in 1917. Gregory (1917) suggested the rate of increase in dry weight per unit leaf area, the net assimilation rate, could be used as measure of the act photosynthetic efficiency of the leaves. Blackman (1919) described plant growth by what he called an efficiency index of relative growth rate. Briggs *et al.* (1920) preferred the term unit leaf rate to net assimilation rate. Later, dry matter accumulation was described in term of net assimilation rate on the one hand and of growth in leaf area on the other. Watson (1947) developed the concept of leaf area index.(LAI) and it provided meaningful way of analyzing growth in crops. In 1953, Monsi and Saeki of Japan combined the concept of LAI with that of extinction coefficient of light penetrating a crop canopy to quantity profiles of light and photosynthesis within a crop. Large differences were shown to exist between plant communities in their light extinction coefficient as determined by the inclination, size and other attributes of leaves. The work of Monsi and Saeki led to attempts in rice breeders in Japan to develop a more comprehensive definition of an ideal plant type and also on the modeling of crop photosynthesis. Crop geometry has been analyzed almost entirely in relation to light penetration with little concern for effects on carbon dioxide renewal. Yield has been assumed to be limited primarily by photosynthesis and the supply of assimilates by crop physiologists are now beginning to give more attention to other processes which may also limit yield capacity.

The most common parameters used in growth analysis are LAI, crop growth rate, relative growth rate, net assimilation rate and leaf area duration.

Leaf Area Index (LAI)

Leaf area is important for photosynthesis. Its estimation indicates both assimilating area and growth. For crop production leaf area per unit land area is more important than leaf area of individual plants. Leaf area index is the ratio between leaf areas to ground area.

$$\text{LAI} = \frac{\text{Leaf Area}}{\text{Ground Area}}$$

Crop Growth Rate (CGR)

It indicates at what rate the crop is growing i.e., whether the crop is growing at a faster rate or slower rate than normal. It is expressed as g of dry matter produced per day.

$$\text{CGR} = \frac{W_2 - W_1}{t_2 - t_1}$$

Where W_1 and W_2 are dry weights of plants at times t_1 and t_2 respectively.

Relative Growth Rate (RGR)

This parameter indicates rate of growth per unit dry matter. It is similar to compound interest, wherein interest is also added to the principal to calculate interest. It is expressed as g of dry matter produced by a g of existing dry matter in a day.

$$\text{RGR} = \frac{\text{Log}_e\ W_2 - \text{Log}_e\ W_1}{t_2 - t_1}$$

Where W_1 and W_2 are dry matter at times t_1 and t_2 respectively.

Net Assimilation Rate (NAR)

It indirectly indicates the rate of net photosynthesis. It is expressed as g of dry matter produced per dm^2 of leaf area in a day. For calculating NAR, leaf area of individual plants has to be used but not leaf area index.

$$\text{NAR} = \frac{(W_2 - W_1)\ (\text{Log}_e\ L_2 - \text{Log}_e\ L_1)}{(t_2 - t_1)\ (L_2 - L_1)}$$

Where L_1 and W_1 are leaf area and dry weight of plants at time t_1, and L_2 and W_2 are leaf area and dry weight of plants at time t_2.

Leaf Area Duration (LAD)

Yield of dry matter is a function of leaf area, net assimilation rate and duration of leaf area. Leaf area duration of a crop is a measure of its ability to produce leaf area on unit area of land throughout its life. LAD is calculated by the formula:

$$\text{LAD (Days)} = \frac{L_1 + L_2}{2} \text{ X } (t_2 - 1) + \ldots\ldots + \frac{L_{n-1} + L_n}{2} \text{ X } (t_n - t_{n-1})$$

Where L_2 is leaf area at the time t_1; L_2 leaf area at time t_2; L_n leaf area at the time t_n and L_{n-1} leaf area at the time t_{n-1}.

Developmental Analysis

The developmental rates can also be estimated in a similar way. Some of the developmental rates that are useful are leaf production rate, panicle emergence rate, rate of flowering, days to 50 per cent flowering and days to maturity.

Leaf Production Rate (LPR)

Leaf production rate can be estimated by counting the number of leaves on tagged plants at periodical intervals. It is expressed as number of leaves per day.

$$\text{LPR} = \frac{L_{n2} - L_{n1}}{t_2 - t_1}$$

Where L_{n1} and L_{n2} are number of leaves at times t_1 and t_2 respectively. Similarly, tiller production rate can also be calculated.

Panicle Emergence Rate (PER)

Panicle emergence rate indicates the rate at which panicle emerges from the leaf sheath. As panicle emerges due to the force of internodal elongation, this parameter is helpful in moisture stress studies.

$$\text{PER} = \frac{Pl_1 - Pl_2}{t_2 - t_1}$$

Where Pl_1 and Pl_2 are length of emerged panicle at time t_1 and t_2 respectively

Rate of Flowering (FR)

This parameter indicates whether most of the tillers or branches flower in quick succession or not. In other words, it indicates synchronous flowering. It is expressed as number of flowers that appear per day.

$$\text{FR} = \frac{F_{r2} - F_{r1}}{t_2 - t_1}$$

Where F_{r1} and F_{r2} are number of flowers that appear at times t_1 and t_2 respectively

Days to 50 Per Cent Flowering

It is the number of days in which 50 per cent of the plants flower.

Days to Maturity

It is the number of days in which plants attain maturity.

Yield Analysis

Biological and Economic Yield

The total dry matter produced by a crop is known as **biological yield** and a fraction of the biological yield which is useful for man is known as **economic yield.** The relationship between biological yield and economic yield is expressed as:

Biological yield x K = Economic yield

Where K is a constant and is expressed as the coefficient of effectiveness or harvest index. In other words, harvest index is the ratio of economic yield to biological yield. Harvest index can also be expressed as percentage.

$$\text{Harvest Index} = \text{Biological yield/Economic yield}$$

A balance between the productive parts of the plant and reserves which form economic yield is essential. A considerable increase in the yield of the economic product is usually dependent on an increase in total dry matter produced, though this is not an absolute relationship and always true. In many cereals, most of the dry matter in the grain is produced by photosynthesis after the ears emerge. Grain yield, therefore, depends to a large extent on the photosynthetic activity of the parts of the plant that are still green after anthesis. Most of the dry matter of the grain is produced by the part of the root above the flag leaf node. Flag-leaf contributes nearly 40 per cent of the total dry matter in the grain in many cereals. Green parts of the ear also contribute considerable quantity of dry matter to the grain. Some amount of carbohydrates formed before flowering are stored in culms and leaf sheath and later re-translocated to the grain.

Yield Attributes

Economic yield is expressed as a function of factors which contribute to yield. And these yield contributing factors are known as yield attributes.

$$Y = a \times b \times c \times d$$

Where Y is economic yield, and in case of cereals a, b, c, d is final plant population (number of plants at the time of harvest), number of effective tillers (ear producing shoots), number of filled grains/ear and the weight of the individual grain respectively. Instead of individual grain weight, 100 grain weight (in case of bold grains) and 1000-grain weight (in case of small sized grain) are taken as yield attribute and this is also known as test weight. The yield attributes for different crops vary but basically yield attributes consist of final population, number of fruits or grains per plant and weight of individual fruit.

Limiting Processes—Source of Sink

Economic yield of a crop may be limited by source (photosynthesizing organs) or sink (storage organs) or the capacity to translocate assimilates form one part to another or other limiting processes such as water or nutrient uptake or transport. The seasonal sequence of conditions plays a major role in determining whether source or sink is more limiting. In cereals, grain filling is largely dependent on photosynthesis and environmental conditions after flowering, but the capacity for storage is determined by conditions before flowering and these may have dominant influence on yield. In plants where source and sink are developed simultaneously, like groundnut, red gram, cotton etc., the adjustment of the two components may be more readily achieved.

The rates of both photosynthesis and leaf senescence are controlled by the demand for assimilates. Photosynthesis and storage may appear to be in balance although storage process may actually be limiting. Consequently, parallel ranking for photosynthetic rate and yield need not imply that yield is limited by photosynthetic rate and attention is paid to selection of plants with more erect leaves or higher photosynthetic rate or calvin cycle plants lacking photorespiration.

Evans (1972) emphasized the fact that photosynthesis, translocation and storage were so closely interdependent that it was difficult to determine which limited the yield most. On the basis of researchers on other crops, Asana (1968) and Mehrotra

(1973) indicated that still higher yielding varieties could possibly be evolved through breeding for higher photosynthetic efficiency and low photorespiration, for more favourable partitioning of assimilates provided and for greater sink capacity (Asana, 1968; Mehrotra 1973). However, for such breeding programmes to be successful, it is essential to screen thoroughly the available genetic material for desirable physiological traits and develop simple techniques for the identification of cultivars or segregating material for such traits. Breeding for physiologically efficient genotypes for the wet monsoon season is a challenging problem as the bulk of the rice crop is raised during this season, during which the present average yield is only about one tonne per hectare. The constraints for higher yield during wet season are mostly low panicle density, the high spikelet sterility and the low harvest index. The available evidence shows that sterility may be due to disturbed nitrogen metabolism in the panicle and possible accumulation of toxic nitrogenous compounds in the panicle at flowering. The low panicle density is often attributed to a high mortality of tillers during the cloudy monsoon season. The partitioning of dry matter to the panicle and an improvement in the harvest index are found to be controlled by hormones. A clear understanding of these mechanisms in different genotypes is necessary for the further improvement in the yield potential over the existing high yielding cultivars during the wet monsoon season.

"Perfection comes from practice"

3 Crop Growth in Relation to Environment

"What are the new strategy can be expected to achieve is not so much a sharp acceleration in the overall rates of growth of agricultural (or, even of food grains) outputs as in giving considerable stimulus to private investment in the bigger sized farms in regions which have the necessary climate and soil conditions, as well as sources of water supply, for adopting the new technology".

K. N. Raj

Plant growth is influenced by **aerial** and **soil environment**. Suitable environment is necessary for better germination, growth and yield of crops. **Aerial environment** is less amenable for modification. However, by adjusting plant population and weed control undue competition for light can be reduced. **Soil environment can** be altered substantially by tillage, fertilizer application and irrigation facilitating free aeration, reducing mechanical resistance and providing adequate nutrients and water.

Soil environment can be studied under **physical, chemical and biological environments.** The favourable physical environment for proper crop growth is deep soil without hard layers or high water table so that crop can exploit larger soil volume. The soil has to contain higher available water holding capacity, proper aeration and less soil strength or mechanical resistance. The favourable soil chemical environment is higher nutrient availability. The biological environment consists of favourable organisms in the soil. Thorough understanding of soil properties is essential for its management for higher productivity.

SOIL FERTILITY AND PRODUCTIVITY

Soil fertility is the status or the inherent capacity of the soil to supply nutrients to plants in adequate amounts and suitable proportion. **Soil productivity** is the capacity of the soil to produce crops with specific systems of management and is expressed in terms of yields. All productive soils are fertile, but all fertile soils need not be productive. It may be due to some problems like water logging, saline or alkaline condition, adverse climate etc. Under these conditions, crop growth is restricted though the soil has sufficient amounts of nutrients.

According to modern usage, soil fertility is the capacity of the soil to produce crops of economic value and to maintain health of the soil without deterioration.

TABLE 3.1: Difference between soil fertility and productivity

Soil Fertility	Soil Productivity
– It is considered as an index of available nutrients to plants.	– It is a broader term used to indicate yields of crops.
– It is one of the factors for crop production. The other factors are water supply, slope of the land, depth of the water table etc.	– It is the interaction of all the factors that determine the magnitude of yields
– It can be analysed in the laboratory	– It can be assessed in the field under particular climatic conditions.
– It is the potential status of the soil to produce crops.	– It is the resultant of various factors Influencing soil management.

FERTILITY LOSSES AND MAINTENANCE

Soil fertility is depleted due to the removal of nutrients by crops and weeds, leaching of nutrients, soil erosion and gaseous loss of nutrients either by volatilization or denitrification. Restoration of the soil fertility is a constant problem. Maintenance of the soil fertility should be considered both on a temporary and long term basis. The temporary measures include suitable cultural practices such as addition of organic manures, green manures, bio-fertilizers and fertilizers. The long term measures are reclamation of the soils by amendments and adopting suitable soil conservation practices.

PHYSICAL ENVIRONMENT

Soil Texture

Primary Particles

Soil texture refers to the relative proportions of sand, silt and clay. Generally, rock fragments larger than 2 cm in diameter are stones, materials between 2 cm to 2 mm diameter are gravel, and the soil mineral matter smaller than 2 mm in diameter is the fine earth (Table 3.2). Sand, silt and clay together constitute fine earth.

TABLE 3.2: Classification of soil particles (International system)

Particle	Diameter (mm)
Stone	> 20
Gravel	2 – 20
Fine earth	< 2
Coarse sand	0.2 – 2
Fine sand	0.2 – 0.02
Silt	0.02 – 0.002
Clay	< 0.002

Sand: The sand particles of 0.2 to 2 mm in diameter are classified as coarse sand and of 0.02 to 0.2 mm in diameter as fine sand. Sand particles are small pieces of unweathered rock fragments. Unless these particles are coated with clay or silt, they do not exhibit properties such as plasticity, cohesion, stickyness, moisture and nutrient retention etc. Because of the large size of sand particles macrospores exist between them which facilitate free movement of air and water.

Silt: The size of the silt particles are in the range of 0.02 to 0.002 mm in diameter. Because of an adhering film of clay, they exhibit some plasticity, cohesion, adhesion and adsorption. These silt particles can hold more amount of water than sand but less than clay. Both silt and sand particles are approximately spherical and cubical in shape.

Clay: Clay fraction is the most important constituent of soil that controls most of the soil physical and chemical properties. These particles are of size less than 0.002 mm in diameter. They have the highest surface area since surface area is inversely related to size. They can adsorb and retain water and nutrients. Some clays swell on wetting and shrink o n drying. They exhibit properties like flocculation (grouping or clustering), deflocculation and plasticity. Clay behaves like a weak acid which is neutralized by bases such as calcium and magnesium ions, thus serving as store house for several nutrients.

Textural Classes

Soil texture is classified based on the proportion of predominant size fraction of sand, silt and clay. If the soil contains more than 80 per cent of silt fraction, the soil is designated as silty soil. Soil with more than 85 per cent of sand is called as sandy soil and soil with 40 per cent clay as clay soil. If all the three fractions are in sizable proportion, the soil is called as loamy soil. Silty clay soil means a soil in which the clay characteristics are outstanding and which also contains sufficient silt. Ranges of sand, silt and clay percentages for each textural class are presented in Fig 3.2 and Table 3.3.

Textural Classes of Soils

The properties of predominant size of fractions of sand silt and clay is the basis for classification of soil texture. If the soil contains more than 85 % sand, soil is called sandy soil. When > 80 % silt then silty soil, and when > 40 % clay then clay soil (Fig. 3.2).

TABLE 3.3: Textural classes of soils

Textural classes	Ranges (4) of		
	Sand	Silt	Clay
Sand	85–100	0–15	0–10
Loamy sand	70–90	0–30	0–15
Sandy loam	43–80	0–50	0–20
Loam	23–52	28–50	7–27
Silt loam	0–20	88–100	0–12
Sandy clay loam	45–80	0–28	20–55
Clay loam	20–45	15–53	27–40
Silty clay loam	0–20	40–73	27–40
Sandy clay	40–65	0–20	35–45
Silty clay	0–20	40–60	40–60
Clay	0–40	0–40	40–60

Three broad and fundamental groups of soil textural classes are sand, loams and clays. An ideal loam is defined as the mixture of sand, silt and clay particles that exhibits light and heavy properties in about equal proportions. Roughly it is a half-and-half mixture on the basis of properties. Loamy soils are best for agricultural production because they retain more water and nutrients than sandy soils and have better drainage, aeration and tillage properties than clay soils.

Soil texture influences soil physical and chemical properties like water holding capacity, nutrient retention, nutrient fixation, nutrient availability, drainage, strength, compressibility and thermal regime.

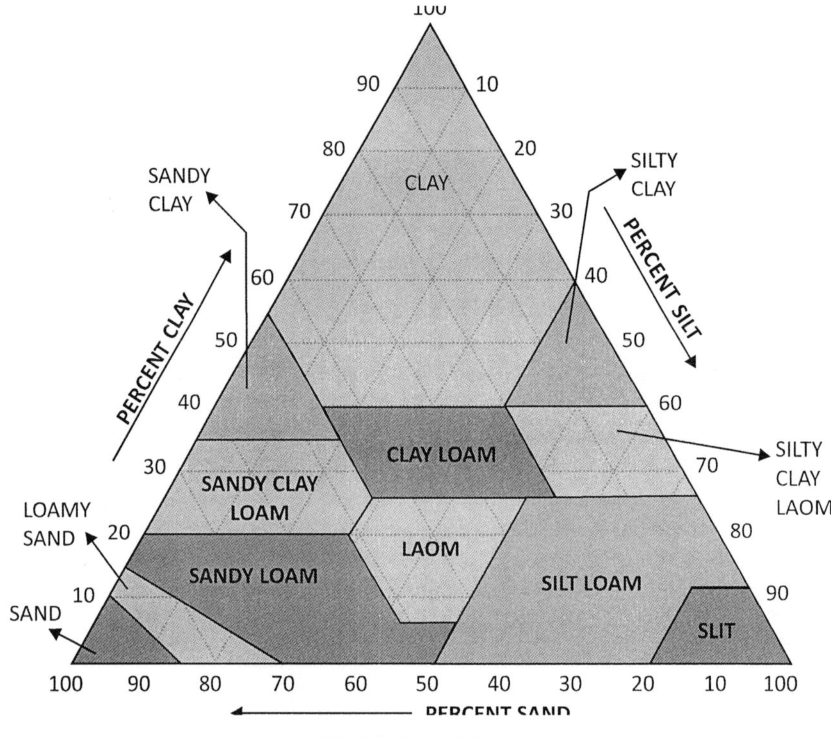

FIG. 3.2: Textural Class.

The clay imparts cohesion and stickyness and fine texture to the soil and induces resistance to the passage of implements or in other words it makes the soil **'heavy'.** Comparatively sandy soils offer little resistance to implements and are called **'light soil'.** The term heavy and light refer to the degree of resistance offered by the soil to the passage of implements, but not to its weight. In fact the heavy clay soil weighs less than an equal volume of the light sandy soil.

Rice, cotton, sorghum, coriander are grown on **heavy textured soils** which include clay loam, silt clay loam, silty clay and clay.

Medium textured soils like **loam, silt loams, silts** and **sandy loams** are suitable **foe most** of the crops.

Sandy, loamy sand, sandy loam and **sandy clay** are **light textured soils** and these are suitable for groundnut, potato, tobacco, pearl millet and leguminous fodder crops.

Significance of Soil Texture

Soil texture is a permanent feature of soil and its change over years is negligible. It can be changed by adding sand or silt or clay as an amendment to improve physical condition. For example, tank silt is added to sandy soil to improve the water holding and nutrient retention capacities. Sand is added to heavy clay soil to improve internal drainage. But these operations are expensive.

Soil texture influences soil physical and chemical properties like water holding capacity, nutrient retention, nutrition fixation, nutrient availability, drainage, strength, compressibility and thermal regime. Sandy soils are of open character facilitating good drainage and encourage circulation of air and water. They are usually loose, friable and hence easy to handle in tillage operations. Clay soils have a high capacity to adsorb and retain nutrients and moisture. They usually have fine pores and a moderate to poor drainage and aeration. They are difficult to handle in tillage operations. Loamy soils exhibit properties intermediate between sand and clay soils. They are considered the best for the agricultural production, because they retain more water and nutrients than sandy soils and have better drainage, aeration and tillage properties than clay soils.

The suitability of a soil to a particular crop depends on texture in addition to soil depth, depth of water table, salinity and alkalinity. Though a particular crop can be grown on a variety of soils, the most suitable soil for each crop differs. Rice, cotton, sorghum, coriander are grown as heavy textured soils which include clay loam, silty clay and clay. Medium textured soils like loams, silt loams, silts and sandy loams are suitable for most of the crops. Sandy, loamy sand, sandy loam and sandy clay are light textured soils and these are suitable for groundnut, potato, tobacco, pearl millet and leguminous fodder crops.

Soil Structure

The primary soil particles viz. sand, silt and clay are usually grouped together in the form of aggregates. The arrangement of primary particles and their aggregates into certain defined patterns is called **soil structure.** Natural aggregates are called fragments while the word clod is used for an artificially form soil mass. Stable aggregates are those that resist break down by disruptive forces such as water and wind. Individual aggregates are known as peds or **secondary units**. Soils that do not have aggregates with naturally preserved boundaries (peds) are considered to be structure less. Two forms of structure less condition are recognized.

1. **Single grain:** Particles are easily distinguished. Particles are unattached to each other and each particle acts individually. Movement of the air and capillary water is the highest. Such struxture is usually found in sandy soil and upper surface of podsols.
2. **Massive:** Individual particles adhere closely to each other but the mass lacks planes of weakness. When clayey soils are ploughed under wet condition, its individual particles adhere closely to each other and soil becomes massive or puddled. Such puddle soil retains water due to lack of pore spaces and thus is suitable for rice.

Soil conditions and characteristics such as water movement, heat transfer, aeration, bulk density, amount and size of pore space (porosity), availability of plant nutrients and growth of micro organisms are much influenced by soil structure. In fact, the important physical changes imposed by the farmer i.e. ploughing, cultivating, drainage, liming and manuring his land are structural rather than textural.

Soil Structure Formation

The aggregate formation is influenced by the amount and nature of colloidal clay. The nature of adsorbed cations also influences it. Ca^{++} and H^+ bring about better aggregation than Mg^{2+} and K^+. Wetting of clay particles with a liquid like water is required to form aggregates because water molecules show dipole movement. Sesquioxides i.e. iron and aluminium oxides, act as cementing agents for binding sand and silt particles to form aggregates.

Aggregates formed with sesquioxides are more stable than those formed by silicate clay. Colloidal organic matter is more effective in forming aggregates than clay.

Types of Soil Structure

The shape and arrangements of aggregates is to be referred. Soil particles differ in size, shape and orientation. In general three broad categories of soil structure are recognized viz. Single grained Massive and Aggregated. Between single grain and massive two extremes, an intermediate condition in which the soil particles are associated in quasi-stable small clods are known as aggregates or peds.

Shapes of Soil Structure

The shapes of aggregates observable in the field:

1. **Platy:** Peds are arranged in relatively thin horizontal layers, plates, leaflets or lenses. It occurs in recently deposited clay soils, surface layers of virgin soils but also in subsoil horizons i.e. **'B'** (Fig. 3.3)

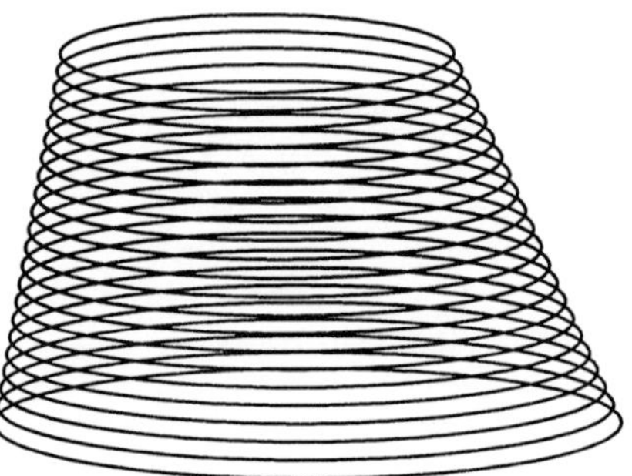

FIG. 3.3: Platy structure.

2. **Prism like:** Vertically oriented aggregates or pillars occur in 'B' horizon of clayey soils in arid and semi-arid regions. When the tops are flat, these vertical aggregates are called **Prismatic** and when rounded they are known as **columnar**. The size of prism like structure is up to 15 cm in diameter (Fig. 3.4).

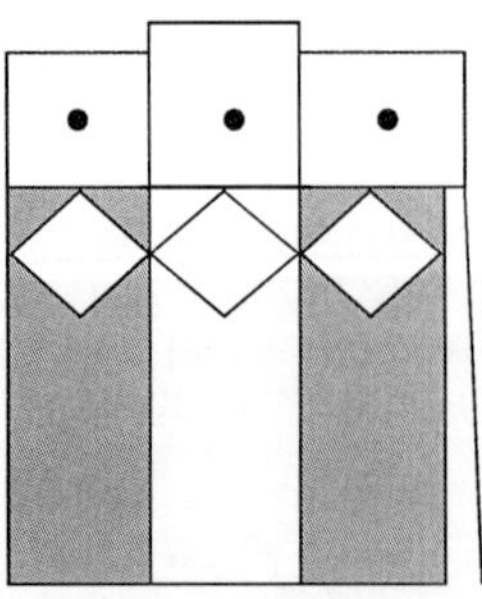

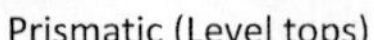

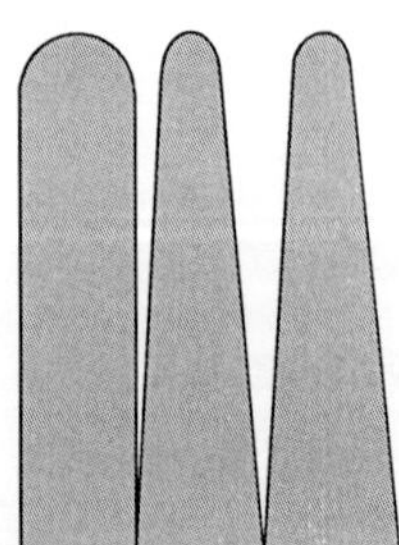

FIG. 3.4: Prismatic (Level tops) and Columnar (Rounded tops).

3. **Block like:** Original aggregates have been reduced to block, irregularly six faced, cube like blocks of soil, 2-8 cm in size, common in heavy sub-soils particularly of humid regions and in upper part of 'B' horizon. In cube like block structure, edges are sharp and rectangular faces distinct (Fig. 3.5).

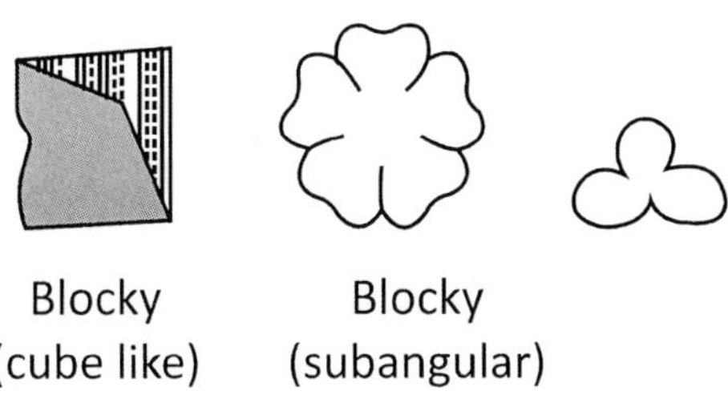

FIG. 3.5: Block like Structure.

4. **Spheroidal:** Rounded aggregates or peds not more than 2 cm in diameter (generally) often found in a loose condition in the 'A' horizon. Such units are ordinarily called granules and when granules are especially porous the term 'Crumb' **is** used. It is the characteristics of furrow slice and subject to wide and rapid change. For the farmers, crumb and granular structure is favourable for plant growth. Formation of crumbs is more pronounced in clayey than in sandy soils. Clay humus and lime in the soil promote the formation of crumbs (Fig. 3.6).

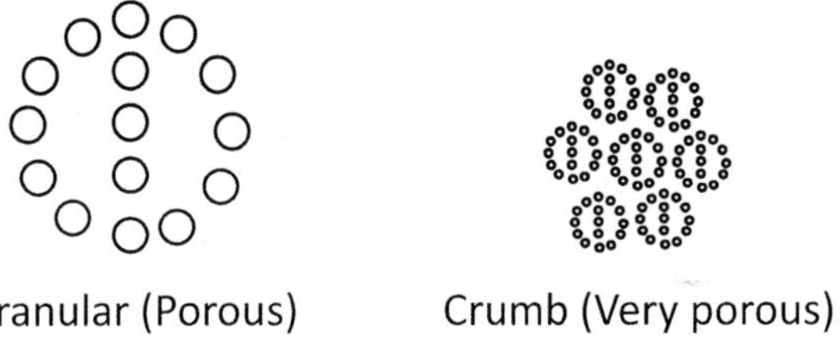

FIG. 3.6: Spheroidal structure.

The shapes, sizes and densities of aggregates generally vary within the profile. The aggregates of B horizon are bigger due to weight of the top layers and less shrinkage and expansion activity as there are less fluctuations of soil moisture. The profile in semi-arid region contain granulated A horizon with a prismatic B horizon. In humid temperate regions, the A horizon contains granulated aggregates but B horizons have platy or blocky aggregates.

Other Classification of Soil Structure

According to the soil aggregates, soil is grouped into two structural heads:

A. **Soil having Definite Cleavage**
   i) Structure having Vertical cleavage:
      i) Prismatic and
      ii) Columnar structure.
   ii) Soil Structure of horizontal cleavage:
      i) Platy
      ii) Laminar (very thin plates).
   iii) Soil Structure of both vertical and horizontal cleavage in mixed form
      i) Blocky structure

ii) Nut like structure: When particle aggregates pf blocky structure are somewhat smaller, such soil structure is known as nut-like structure and found in B horizon.
iii) Granular
iv) Crumb structure.

B. **Soils having Indefinite Cleavage**
i) Single –grain structure.
ii) Massive Structure.

Large masses of soil aggregates are called 'Clods"

Soil Structure Classes: According to the size of the aggregates, five size classes are very fine, fine, medium, coarse and very coarse.

Soil Structure Grades: Refer to degree of aggregation i.e. degree of inter aggregation adhesion and to aggregate stability. Four grades are:

1) **Structure less**: Single grain and Massive.
2) **Weak:** Poorly formed peds.
3) **Moderate:** Moderately durable peds and well formed.
4) **Strong:** Durable peds and quite evident in undisturbed soil.

Factors Affecting Soil Structure

1. **Soil Management:** A good soil management (proper tillage) with a proper crop rotation system has the effect of maintaining the soil in a good state of aggregation.
2. **Adsorbed cations:** Na + and K+ ions disperse the soil, while Ca^{2+}, Ba^{2+} have favourable effects on the aggregation. Similarly the presence of soluble salts favour flocculation.
3. **Micro-**organisms: Filamentous fungi and microbial decomposition products of organic matter have a binding effect on soil particles; burrowing activities of insects, excrements of earthworms change the soil structure.
4. **Variation of soil moisture:** Drying of soil forms cracks and big clods, poorly drained soils have unfavourable structure.
5. **Influence of Organic Matter:** Binding effect, water holding, porosity are increased by adding organic matter to soil. Cementing action of humus is equivalent to certain inorganic compounds like oxides of Fe which provide the most of the long-term aggregate stability. Polyvalent inorganic cation e.g. Ca^{2+}, Mg^{2+}, Fe^{2+} and Al^{3+} cause flocculation which act as bridges between organic matter and soil clays. There is only one practical way to improve the soil structure is to add organic matter to the soil.

The shapes, sizes and densities of aggregates generally vary within the profile. The aggregates of B horizon are bigger due to weight of the top layers and less shrinkage and expansion activity as there is less fluctuation of soil moisture. The profiles in semi-arid region contain granulated A horizon with a prismatic B horizon. In humid temperate regions, the A horizon contains granulated aggregates while B horizons have platy or blocky aggregates.

Aggregate Stability

Aggregate stability is a measure of resistance of aggregates to break down when subjected to potentially disruptive processes. Aggregates form, disintegrate and reform in course of time. For instance, a newly cultivated field may, for a time, exhibit a nearly optional array of aggregate sizes with large pores in between the aggregates favouring high infiltration rate and unrestricted aeration. The soil structure may begin to deteriorate due to beating action of rainfall, repeated traffic, frequent cultivation etc. They form again when there is decomposition of organic matter.

Dispersion of clay particles from micro-aggregates is promoted by adsorption of complexing organic acids which increase the negative charge on clays. These acids are produced by plant roots, bacteria and fungi. However, the dispersability of clay in micro-aggregates is offset by the binding action of polysaccharides, mainly mucilages produced by bacteria, but also produced by plant roots and fungal hyphae. The stability is also enhanced by multivalent cations which act as bridges between organic colloids and clay. The aggregate stability also depends on the presence of carbohydrates predominantly from microbial sources.

Aggregate stability is a term used in relation to the destructive action of water. On wetting, the aggregates collapse due to weakening or solubility of bonds or due to swelling. If wetting is non-uniform, one part of the aggregate will swell more than another resulting in friction. Sudden swelling causes trapping of air inside the aggregate resulting in explosion of the aggregate is the air is compressed beyond a point. Rain drops and flowing water detach the particles from aggregates. Thus, both rainfall and irrigation reduce the stability of aggregates. Water stable aggregates are necessary for prolonged good physical condition of the soil.

Management of Soil Structure

The main aim of improving aggregate stability and its management are soil conservation and better growth of crops. The compact soil mass is to be broken to favourable size aggregates and these aggregates are to be maintained. The improvement and stability of aggregates can be achieved by growing suitable crops, adopting suitable cropping systems, applying crop residues and farm yard manure.

Significance of Soil Structure

Soil structure is an import soil property which influences the **soil environment** through its effect on the amount and size of pore space, water holding capacity, availability of plant nutrients and growth of microorganisms. The altered soil environment affects germination and root growth of crops.

Fine clay particles i.e., colloids, flock or group together due to cohesion and form a cluster. Sand and silt particles with a coating of colloidal clay are bound together into aggregates due to sticky nature of clay particles.

The amount and nature of colloidal clay influences aggregate formation. The nature of adsorbed cations also influences aggregate formation. Calcium and hydrogen ions bring about better aggregation than magnesium and potassium ions. Wetting of clay particles with a liquid like water is required to form aggregates because water molecules show dipole movement. Sesquioxides i.e., iron and aluminium oxides, act as cementing agents for binding sand and silt particles to form aggregates. A part of the iron in solution acts as a flocculating agent and remaining acts as a cementing agent. Aggregates formed with sesquioxides are more stable than those formed by silicate clay.

Organic components such as cellulose substances, on decomposition release sticky materials resembling mucus or mucilage. The humic and fulvic acids produced during decomposition of organic matter are also sticky in nature and help in aggregate formation. Colloidal organic matter is more effective in forming aggregates than clay. Clay is adsorbed on humus forming clay humus complex and it helps is forming stable aggregates, Fungi and actinomycetes produce sticky materials which are helpful in forming aggregates. Prominent among the many microbial products capable of bounding soil aggregates are polysaccharides, hemicelluloses and several other polymers. Such materials are attached to clay surfaces by means of cation bridges, hydrogen bonding, vander Waal forces and anion adsorption mechanism.

Pore Space

The volume of soil not occupied by soil particles is known as **pore space.** Pore space μoccupied by air and water or both. Plant roots exist and grow in pore space. It directly controls the amount of water and air in the soil and thus indirectly controls plant growth and crop production. The pores are classified into micro- and macro-pores depending on the diameter of the pore. Pores with a diameter of 2 to 20 μm are called **micro-pores** and pores with a diameter of 200 μm to 0.3 mm are called **macro-pores.**

The amount of pore space depends on particle size, texture, structure and biological activity. Clay soils have less micropores compared to sandy soils. However, the total pore space is 50 to 60 per cent in clay soils. Sandy soils content 25,000 pores/m^2 and clay soils have 25 million pores/m^2. Addition of organic matter increases the pore space. Borrowing animals increase pore space. Pore space can be modified by tillage and irrigation.

Other Properties

The root system of plants grows into the soil exploring water and nutrients. They encounter various soil layers with different physical and chemical properties. The suitability of the soil to grow particular crop depends on soil profile, soil depth, water table and soil strength.

Soil Profile

Soils have distinct layers at varying depths below the surface. A vertical section of the soil to expose the layers is called **soil profile**. The upper layer is usually higher in organic matter and darker in colour than the layers below. This upper layer is called **A horizon** or **top soil**.

The middle part of the soil profile usually contains more clay. This layer is called **B horizon or subsoil.** The A and B horizons together are referred to as the **solum or true soil.** Crops feed mostly in the top soil and to a very less extent in subsoil. The C horizon commonly referred to as soil parent material occurs beneath the solum and extends downward up to bed rock.

Soil Depth

The depth of the soil is the last boundary up to which the roots of plants can exploit. It differs with the soils ranging from a few centimeters to a few metres. The profile depth of the soils can be grouped into deep, medium deep and shallow based on root depth up to which roots can enter.

Water Table Depth

The depth at which water table is present also limits the lower boundary for root growth. The depth of water table is near the soil surface in water logged soils to several metres below in other soils. All crops suffer due to shallow water table except rice. Water table fluctuates with season and is shallow during rainy season and deeper in summer season.

Soil Strength

Soil strength or soil mechanical resistance indicates the resistance offered by the soil to root penetration. It is measured with cone penetrometer and expressed as kg/cm^2 or bars. It ranges from 0.2 bars in puddle submerged rice fields to 22 bars in dry black soil. Soil strength mainly depends on soil moisture. It is less with high moisture content and increases with decrease in soil moisture content. The other factors that influence soil strength are **bulk density** and soil compaction.

Particle Density of Mineral Soil

It is the mass (or weight) of a unit volume of soil solids. It is denoted as Particle Density (PD) or Dp was previously known as Real specific gravity. PD of the most mineral soils: 2.60-2.75 g/cm^3 or Mg/cm^3 (mega gram per cubic metre. 1 Mg = 10^6 gram). For general calculation Dp of average arable mineral surface soil is **about 2.65 g/cm^3.**

Dp of organic matter is 1.1-1.4 g/cm^3.

Particle density depends on the chemical composition and crystal structure of the mineral particle. The size of the particles of a given mineral and the arrangement of the soil solids have no effect on Dp.

Bulk Density (Apparent Specific Gravity)

It is the mass (or weight) of a unit volume of dry soil including both solids and spaces (i.e. pores). Fine textured soils like clay, silt loam etc. have lower Bulk Density (BD), than sandy soil due to high pore spaces. Soils with high proportion of pore space to solids have lower BD (or Db). Increase in pore space, decreases the BD.

Bulk Density (BD) of general soil calculation is **1,33 g/cm^3** (just half of PD) Deeper the profile, higher the BD, the reasons are low content of Organic Matter, less aggregation, root penetration and compaction caused by overlying layers weight. Cropping increases the BD of the top soils in all cases while pore space decreases proportionately.

$$BD = \frac{\text{Weight of soil}}{\text{Volume of solids and pores}}$$

or i.e. Mass volume

Pore Space / Solid Space

Let

$$\begin{aligned} Ws &= \text{Wt. of oven dry soil (Solids)} \\ Vs &= \text{Volume of oven dry soil (Solids)} \\ Vp &= \text{Volume of Pores.} \\ Vs + Vp &= \text{Total soil volume.} \end{aligned}$$

By definition:

$$PD = \frac{Ws}{Vs} \text{ or i.e. } Ws = PD \times Vs \qquad \text{(i)}$$

$$BD = \frac{Ws}{Vs + Vp} \text{ or i.e. } Ws = BD \times (Vs + Vp)$$

From the (i) and (ii) equations

$$PD \times Vs = BD \times (Vs + Vp)$$

i.e.

$$\frac{Vs}{(Vs + Vp)} = \frac{BD}{PD}$$

Since

$$\frac{Vs}{(Vs + Vp)} \times 100 = \text{per cent solid space.}$$

So,

$$\% \text{ solid space} = \frac{BD}{PD} \times 100$$

% Pore Space = 100 - % slid space
% Pore Space = 100-(BD/PD x 100)

Soil type	Total pore space in %
Clay	50-60 %
Loamy	30=50 %
Sandy	20-30 %

- In sandy soils, there are 25000 pores/m^2 whereas in clay, 25 x 10^6 pores/square metre.
- Pores less than about 0.06 mm in diameter is known as **Micropores** where as more than 0.06 mm in diameter is known as **Macropores.**

Cropping reduces macropores space for ready movement of air by about one half.

Soil Consistence

It is to be referred to the resistance of soil materials to deformation or rupture. It depends upon the degree and kind of forces which attract one molecule to another. It varies with size, shape, and arrangement of soil particles and the nature of the water films around it, nature and amount of inorganic and organic colloids, when water dries from soil particles, soils develop greater cohesive force and mechanical strength.

Soil Colour

It is due to either mineral matter or organic matter and mostly to both. Organic Matter imparts black to dark gray tinges, iron compounds for red, brown and yellow tinges, and silica lime and other salts give light, white and grey tinges. Red colour means relatively old and intensively weathered, well drained such as laterites. Red colour is related to unhydrated ferric oxide although manganese dioxide and partially hydrated iron oxide may also contribute to it. The yellow colour is due to oxides of iron ($2Fe_2O_3.3H_2O$). Yellow colour indicates more moist soils than red colour. Soil colour directly influences soil temperature.

Soils have various shades of black, yellow, red and grey colours. It is an easily observable character which is useful in classification. Some of the soil groups of India have been named after their dominant colour. The colour of surface soil may be different from the colour of its lower horizons.

These different shades of colour are due to various reasons. One of the factors responsible is the parent material e.g., red sand stone gives rise to red soil. The second factor is the soil organic matter in surface soil which imparts dark brown to black colour. Thirdly, the presence of certain materials in soils like titamin compounds impart darker colour. Iron compounds like haematite, give red and limonite yellow colour. Preponderance of silica or lime results in whitish or grayish tinge to the soil. Accumulation of salts makes soils white or black, depending on the type of salts. If the soil is well drained like the lateritic soil, the ferric compounds of iron are commonly formed and they give red colour to the soil, but if the drainage is poor, the soil colour is more greenish or bluish in water logged soils. Sometimes, in the subsoil layers, there is a colour variation or mottling of soil indicating alternate oxidizing and reducing conditions at that particular depth mainly due to fluctuations in water table.

Soil colour, besides being useful for classifying soils, is indirectly helpful in indicating many other properties of soils. For example, dark brown soils, indicate high organic matter and fertility, red colour shows good aeration, white colour accumulation of salts etc.

Soil Air

Gaseous phase of the soil: Under moist field conditions macropores generally constitute the air space. Ordinarily, the occupation of nearly one-third of the pore space in the soil by air and two-third by water constitutes the most favourable condition for plant growth. Plant growth suffers when oxygen diffusion rate (ODR) below 40 x 10^{-8} g per cm^2 per minute.

The growth of root is ceased at 20 x 10^{-8} g/cm^2/minute ODR.

Percentage by volume

	O_2	CO_2	N_2
Soil Air	20.6	0.25-1.0	79.2
Atmosphere	20.96	0.03	79.0

Field air capacity is the fractional volume of air in a soil at field capacity. It depends upon soil texture. Soil air constitutes in sandy soil, more than or equal to 25 % (i.e. > 25 %), loamy: 15-20 % and in clay less than 10 % of the total soil volume. Clay soils retain more water hence is the lower air capacity.

Volume Fraction of Soil Air

The volume ratios of the three constituent phases of soil, namely, solids, water and air are continually changing as the soil undergoes wetting or drying, swelling or shrinking, tillage or compaction and aggregation or dispersion. Specifically, water and compete for the same pore space, and their volume fractions are also related that an increase of one generally decreases the other.

Field air capacity is the fractional volume of air in a soil at field capacity. The air capacity depends on soil texture. In sandy soils, it is of the order of 25 per cent or more. In loamy soils, it is generally between 15 and 20 per cent and in clay soils which tend to retain more water, it is likely to fall below 10 per cent of the total soil volume. Fine textured soils with macro aggregates of the order of 5 mm or more in diameter generally have a considerable volume of micropares (inter aggregated) which drain very quickly and remain air filled practically all the time. Hence, such soil exhibits an air capacity of 20 to 30 per cent. As the aggregates are dispersed or broken down by mechanical forces, the micropores tend to disappear so that a strongly compacted soil may contain less than 5 per cent air by volume at its characteristic field capacity volume of soil moisture.

Composition of Soil Air

The composition of soil air differs from atmospheric air. The carbon dioxide content and relative humidity are higher and oxygen is less compared to atmospheric air (Table 3.4).

TABLE 3.4: Composition of soil air

Particulars	Percentage by volume		
	Nitrogen	Oxygen	Carbon dioxide
Atmospheric air	79.00	20.95	0.03
Soil air (surface)	79.20	20.60	0.25
Sandy soil air	79.20	19.95	0.30
Loamy soil air	79.20	19.20	0.62
Clay soil air	79.20	19.69	0.66

Particulars	Percentage by volume		
	Nitrogen	Oxygen	Carbon dioxide
Manured soil air	79.20	18.23	1.85

Movement of Soil Air

Soil air is virtually in continuation of atmospheric air and there is continuous exchange of gasses between atmosphere and soil air. The circulation of air in the soil mass is known as **aeration**. The movement of air is influenced by temperature, water and diffusion. As soil air gets heated during the day, it expands and moves into the atmosphere. When the soil begins to cool, it contracts and atmospheric air is drawn into the soil. When water is added to soil, soil air is displaced by the incoming water and air reenters when soil dries. Diffusion is the main process in exchange of air between soil and atmosphere. Nitrogen, oxygen and carbon dioxide are the important gases in both atmosphere and soil air. Each gas exerts its own partial pressure in proportion to its concentration. It is the partial pressure that regulates its movement. The direction of movement is from regions of higher partial pressure to region of lower partial pressure.

Influence on Crop Growth

When the oxygen content of the soil is limited, the growth of the most of crops is affected and stops when oxygen content reaches a level below two per cent. Under field condition, exclusion of oxygen takes place due to excess amount of water in the soil. Root growth is decreased due to decrease in oxygen and increase in concentration of carbon dioxide. Germination is inhibited in the absence of oxygen. Decomposition of organic matter is reduced and change in microflora occurs. Potato, tobacco, cotton, linseed, tea and legumes need higher level of oxygen in soil air compared to other crops. Cereals except rice are intermediate in oxygen requirement and rice can tolerate very low level or even complete absence of oxygen in the soil. Nutrient and water uptake are reduced when the oxygen content of soil is less as it affects activity and permeability pf roots.

Control of Soil Air

Soil air can be controlled mainly by providing good drainage facilities. The selection of suitable method of irrigation is necessary to avoid application of excess water.

Soil Temperature

- For optimum root development = 10-27°C
- For Opt. crop root development = 15-45°C.

Black cotton soils absorb about 86 %, alluvial soils 40 % and grass covered soils about 60 % of solar radiation. On an average, the specific heat of soils ranges from 0.20 to 0.23.

Soil temperature is one of the most important soil properties that affects crop growth. The major source of heat is sun and heat generated by the chemical and biological activity of the soil is negligible.

Factors Affecting Soil Temperature

The amount of solar energy received on unit area of the soil decides the soil temperature. The energy received from the sun or absorbed by the soil which is converted into heat. It is partly used for increasing the temperature and partly for evaporating soil moisture.

Specific heat is the heat in calories required to raise the temperature of one gram of a substance to one degree C. The specific heat of water is 1.0 cal/g that of soil is 0.2 and organic matter 0.5. It means one calories of heat raises the temperature of water by 1°C, that of soil by 5°C and that of organic matter by 2°C. The specific heat of the soil increases with increase in moisture content. The specific heat is 0.33 cal/g, when the soil moisture content, is 20 per cent and 0.38 cal/g with 30 per cent moisture. Thus, wet soil is cold while lighter soils which hold less water warm up much more quickly.

When the soil moisture content is 20 per cent, 20 g of water is present in 100 g of soil.

- Calories of temperature content for 20 g of water = 20 cal.
- Calories of temperature required for 100 g of soil = 100 x 0.2 = 20 cal.
- Total calories required for 100 g of soil = 20 = 20 = 40 cal with 20 per cent moisture
- Specific heat = 40/100 = 0.33 cal/g.

The flow of temperature in the soil is expressed as thermal conductivity. It increases with water content and decreases with porosity. In moisture soils, with higher thermal conductivity, heat flows into lower layers and surface temperatures are, therefore, low. Tillage improves porosity and decreases thermal conductivity resulting in higher surface temperature.

The latent heat of vaporization for water is 580 calories. Evaporation from the soil surface reduces its temperature. Soil under vegetation is cooler in summer and warmer in winter. A good vegetative cover intercepts considerable portion of solar radiation and the soil beneath it does not warm up. Bare soil loses its temperature by re-radiation in winter. Vegetation acts as an insulting material and reduces the heat loss. Dark coloured soils get warmed up quicker than light coloured soils, since the former reflect back less radiation to atmosphere and thereby absorb more heat. Sandy soils warm up quicker than clay soils, since the latter have more Wtewr holding capacity.

Effect of Crop Growth

As about half of the plant is in the soil, the temperature of the soil has considerable influence on aspect of growth, especially in early stages of crop. Germination entirely depends on soil temperature as the seed is inside the soil unlike a grown up plant. Too high or too low temperature injures or kills the seed. Optimum soil temperature for germination for different crops is: winter cereals 15 to 18°C, maize 32°C, sorghum, sugarcane and cotton 30°C, and mustard 18 to 23°C.

Germination and emergence is hastened with increase in temperature up to a limit. Sorghum requires 51 heat units for emergence with a base temperature of 10.4°C. Wheat seeds put forth radicle 6 days after sowing at 5°C, 3 days after sowing at 10°C and within 2 days after sowing at 18°C. The hypocotyls length of cotton grows at a faster rate up to 30°C and later the rate of growth is reduced.

Low temperature can be harmful at seedling stage. Crops such as maize or sorghum need a high soil temperature for active seedling growth. Their growth is slow while parasitic fungi grow actively and weaken or kill the seedlings. No microbial activity is noticed below 5°C or above 50°C with optimum between 25° and 35°C. Mineralization, nitrogen fixation in legumes and other microbial activity are all temperature dependent. Nutrient availability is higher at optimum temperature.

Control of Soil Temperature

Control of soil temperature is difficult and expensive. However, some management practices slightly alter the temperature. Moisture content of the soil controls soil temperature to some extent. A soil that is over wet due to water logging and poor drainage has low temperature. In such soils, temperature can be raised by draining excess water. Water heats and cools slowly compared to soil. This principles is used to moderate soil temperature in lowland rice. About 10 to 30 cm of water is maintained in Japan during vegetative stage of rice to avoid adverse effect of low temperature. Plant canopy reduces soil temperature in warmer region. The seedlings of tea, coffee, tobacco etc., are raised under shade to avoid damage by high soil temperature. Mulching materials like straw of cereals reflect back some of the radiation and reduce soil temperature. Black materials like farm yard manure, charcoal, polythene film etc., absorb most of the radiation and increase soil temperature. Tillage influences soil temperature through its influence on aeration and bulk density.

Soil Water

(A) Physical Classification of Soil Water

Hygroscopic Water

1. Such water held at hygroscopic coefficient.
2. Tension varies from 31-10,000 atmospheres.
3. Held mostly by soil colloids.
4. Mostly non-liquid and therefore, biological inactive.
5. Moves mostly in vapour form.
6. Film thickness not more than 15-20 layers of water molecules and maximum thickness of 4-5 mμ
7. Related to texture and organic matter content of soil.

Capillary Water

1. Such water present (or held) in capillaries of soil.
2. Held between **field capacity and hygroscopic coefficient in micropores.**
3. Soil-water tension or tension of film. Varies 1/3 atm. (or 0.1) – 31 atm.
4. Functions as soil solution.
5. Moves by film adjustment from thick to thin film.
6. Related to organic matter content and texture of soil.

Gravitational Water

1. Such water held by a negative tension of 0.3 atm. Or less, loosely held between less than 0.1 to 0.3 atm.
2. Free water which drains out, also called drainage water.
3. Usually present in macropores.
4. Undesirable and nutrient leaching takes place.
5. Downward movement of water due to gravitational force.

(B) Biological Classification of Soil Water

Superfluous Water

1. Moisture tension $\leq$ 1/3 atm. (less than or equal to).
2. Such water held at field capacity but not beneficial for higher plants.
3. Harmful for crops if too much.
4. Also called gravitational water.

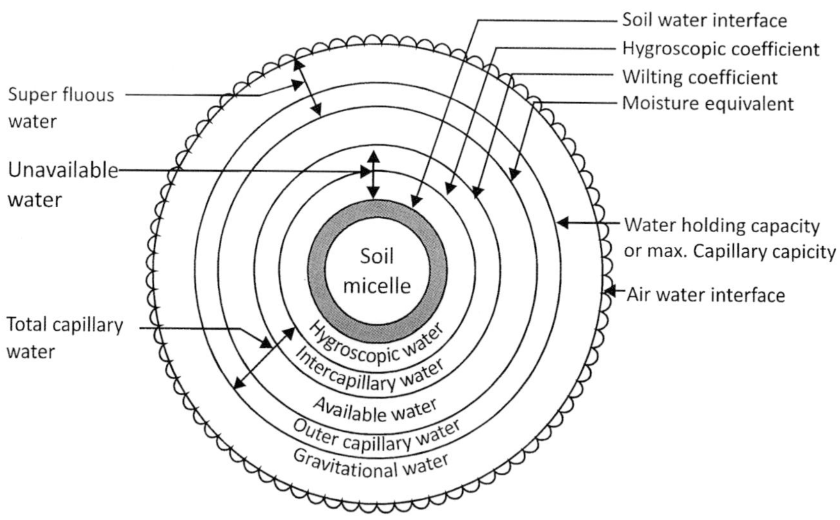

FIG. 3.7: Different layers of soil water.

Available Water

1. Portion of capillary water lying between field capacity (1/3 atm.) and wilting coefficient (15 atm.)
2. Availability of moisture depends upon crops when wilting coefficient is measured at 15 atm. Pressure; it is a measure of soil condition and not that of plant because ability to absorb soil moisture differs according to plant types.

Unavailable Water

Such water held in soil at the permanent wilting point. It includes hygroscopic water which is removed too slowly by plants to prevent wilting.

Optimum growth of plants occurs when soil moisture content near field capacity with a moisture tension of 1 bar or less.

Field Capacity

When macropores drain their water and are filled by air but micropores or capillary pores are still filled with water then soil is called at field capacity. The matric tension is around 0.1-0.3 atm. The pF range of capillary moisture is 6.0-4.2. The pF range of available moisture is 2.54-4.2. Moisture with about pF 2.54 is very favourable for plant growth.

Permanent Wilting Percentage

Concept by Briggs and Shantz: Although plants not dead, but they are now in a permanently wilted condition and will die if water is not added. The tension is 15 atm. For most of the crops the water remaining in the soil is found in the smallest of the micropores and around individual soil particles. The pF at 15 atm. is about 4.2.

Hygroscopic Coefficient

After the removal of liquid water completely from micropores, the remaining water is associated with the surface of soil particles. The tension is about **31 atm**. The pF is about 6.0. The moisture held at this point is called hygroscopic Coefficient.

Moisture Equivalent

The weight percentage of water retained by a previously saturated sample of soil 1 cm in thickness after it has been subjected to a centrifugal force of 1000 times gravity for 30 minutes. Moisture equivalent was introduced by Briggs and Mc. Lane (1907), Briggs and Shantz (1912) used the moisture equivalent as an indirect measure of wilting point (WP).

$$WP = \frac{\text{Moisture equivalent}}{1.84}$$

However, this factor i.e. 1.84 can not be used for all soils. Moisture equivalent is assumed to represent moisture held under field conditions or the field capacity. The pF at Moisture Equivalent (1/3 atm.) is about 2.54.

- Maximum capillary capacity (MCC)
- MCC = Water holding capacity-Hygroscopic Coefficient.

Water Retention

Water is held around soil particles in thin layers due to adhesion and cohesion. It is held with greater force (i.e., 10,000 bars) near the particle surface. As the thickness of water layer increases, the force with which it is held is reduced. When the soil particles are holding water with less than 0.3 to 0.1 bar force, it is lost into deeper layers due to gravitational pull. When pure water is held in hypothetical place where no force is acting on it, its capacity to do work is zero. It can move in any direction depending on the external force applied. When water is held by soil particles or if it contains salts or due to gravitational force, its ability to do work or its potential is restricted. Therefore, water potential in soils is always negative. Water potential of soil is expressed in bars or mega pascals. The soil water, held between - 0.3 to - 15 bars, is considered as available water. Soil water moves from higher potential to lower potential. Movement of water is also an important factor in water availability. As the water near the root surface is exhausted, it has to be replenished by the movement of water from higher potential to lower potential.

Function of Soil Water

Soil water has major influence on modifying the soil environment. The soil strength is reduced due to its presence. It moderates the soil temperature from extremes. Nutrient availability and mobility increase with water. Biological activity including mineralization increases. Soil water is controlled by rainfall, irrigation and drainage. Rainfall received is effectively used by following soil and water conservation measures.

Soil Taxonomy and Survey

Soil profile characteristics such as depth, clay and organic matter content and the degree of horizon development vary greatly from location to location. To study and understand the soil at different locations and to communicate the information, systems of soil classification have been developed. Examination and study of soil in the field should be focused on a three dimensional unit, sufficiently large (1 to 10 m^2) so that the nature of its horizons can be ascertained and the range of its properties identified. This smallest volume of soil that can be called as 'a soil' is called a **pedon.** A pedon, because of its very small size can not be used as the basics unit for a workable field soil classification system. However, two or more pedons, termed a **polypedon,** all of which are within the defined limits of a single soil series (soil individual) can serve as a basic classification unit.

Soil Taxonomy is an orderly arrangement of soils in the field for the purpose of more efficiently studying their properties, with the ultimate aim of increasing crop productivity. Three soil scientists deserve the primary credit for organizing world soil taxonomy and for bridging the language barrier. V, V. Docuchaev (1846-1903), a Russian, is credited with the first publication on soil taxonomy. He noted the relationship among climate, vegetation and soil characteristics. Glinke (1867-1929) translated some of Dokuchaev's work to German, adding some of his own ideas in the translation. Marbut (1863-1935) then translated it into English. Both Glinka and Marbut were renowned soil taxonomists in their own right and each added to world knowledge by their respective translations. The soil survey staff of US Department of Agriculture, in corporation with soil scientists in other countries, developed a new comprehensive system of soil classification based on soil properties. This system has been in use in the united States since 1965, and it is used, atleast to some degree, by scientists in 45 other countries.

Basis of Classification

Soil taxonomy maintains the natural body concept along with two other major features that make it more useful. First, the system is based on soil properties that are easily verified by others. Secondly, the unique nomenclature employed gives a definite connotation of the major characteristics of the soil in question.

Soil taxonomy is based on the properties of soils as they are found today. The specific criteria used in grouping the soils are soil properties although soil genesis is one of the basic objectives of the system. The diagnostic surface horizons, called epipedons, include upper part darkened by organic matter, the upper eluvial horizons or both. Part of the B horizon may also include, if darkened by organic matter. Six epipedons are recognized, though four are of any prominence.

Modern System of Classification

The soil survey staff of United States Department of Agriculture (1975) proposed a new system of Soil Classification called the **Seventh Approximation.** This is based on present soil properties which are the consequence of the soil genesis. There are six categories of classification in soil taxonomy: **order, suborder, great group, subgroup, family and series**. These categories are similar to those used for the classification of plants.

Order: This category is largely on the basis of soil forming processes as indicated by the presence or absence of specific diagnostic horizons. There are eleven soil orders in soil taxonomy. Nine of eleven orders are mineral soils. The eleventh order (Andisols) is only recently added.

	Orders	Major characteristics
1.	Entisols	– Young soils with little profile development.
2.	Inceptisols	– Embrionic soils with few diagnostic features, no horizon of illuvial clays, moisture adequate for crops in most years.
3.	Mollisols	– Dark coloured surface horizons, high base saturation, relatively high organic matter.
4.	Alfisols	– Relatively low in organic matter, relatively high in base saturation, an illuvial horizon of silicate clays, moisture adequate for crop growth.
5.	Ultisols	– Low in base saturation, highly weathered, an illuvial horizon of silicate clays, moisture adequate for crop growth in most years.
6.	Oxisols	– No illuvial horizon of silicate clays, no weatherable minerals, clay very inactive.
7.	Vertisols	– High in swelling clays, deep, cracks when soil is dry.
8.	Aridsols	– Relatively low in organic matter, few horizons, inadequate moisture for crop growth.

	Orders	Major characteristics
9.	Spodosols	– An illuvial horizon of amorphous aluminium and organic matter with or without amorphous iron.
10.	Histosols	– Peat or bog, more than 30 % organic matter.
11.	Andisols	– From volcanic ejecta, dominated by allophone or aluminium humic complex.

Suborder: These are subdivisions or orders that suggest genetic homogeneity. Moisture content, climatic environment and vegetation which helps determine the nature of genetic processes, also helps determine in suborders in which a given soil is found. There are 47 soil suborders in the US system of soil taxonomy.

Great groups: There are about 230 known great groups in the world. There are classified primarily on the basis of:

- Similarities in kind, arrangement and degree of expansion of horizons,
- Similarities in soil moisture and soil temperature, and
- Similarities in degree of base saturation.

Subgroups: Great groups have been divided into three subgroups:

- First subgroup possesses only the properties of the concerned great group,
- Second subgroup possesses some of the properties of other subgroups, great group, suborders and orders in addition to properties of concerned great group, and
- Third subgroup possesses some properties that have not been found in any other great groups, suborders and orders in addition to properties of the concerned great group.

Family: Subgroups have been divided into families on the basis of similarities in physical and chemical properties affecting their response to management practices, especially to the penetration of plant roots. Important properties considered are soil texture, clay minerals, soil reaction, annual mean soil temperature and rainfall. About 6600 soil families are recognized in US.

Series: Families are subdivided into lowest category of the system called series. One series may be differentiated from another in a family through differences in one or more but not necessarily all of the soil properties. Series are established mainly on the basis of profile characteristics. The soil series have been divided into **phases** which describe soil texture, percentage slope, stoniness, wetness, erosion, soluble salts, etc.

Soils of India

Soils of India are broadly divided into five groups:

1. **Alluvial: Entisol, Inceptisol and Alfisol.**
   Such soils include deltaic alluvium, calcarius alluvium, coastal alluvial soils and coastal sands, formed by transportation in streams and rivers and deposited inflood plains or along coastal belts, generally deep soils, geologically it is of two types:
   a) Khadar or newer alluvium which is sandy, light coloured and less kankary.
   b) Bhangar or older alluvium, more clayey, dark and full of kanker.
2. **Black soils: Vertisol, Inceptisol and Entisol.**
   The typical soil derived from deccan trap is called regur or black cotton soils: the sedimentary soils derived from granite and gneiss mostly, clayey in nature, practically uniform throughout profile, parent rocks are generally hornblendic

type, rich in lime and magnesia and limesoda-type feldspar, varying quantities of gypsum deposited in the lower layers in 45 % of the black soils. It is highly argillaceous, very fine-grained and dark and contains a high proportion of Ca and Mg-carbonates. It is rich in lime, magnesia and alumina, moderate potash and poor in P_2O_5, N and organic matter, rich in montmorillonite and beidelitic group of clay minerals, high CEC – 40-60 me/100 g of soil.
Occurrence: Maharashtra, West M.P., parts of Andhra Pradesh, Gujrat and some parts of Tamil Nadu.

3. **Red Soils: Alfisol, Utisol and Inceptisol.**
Redsoil in Tami Nadu occupy the largest area. Occurrence vast areas of Tamil Nadu, Karnataka, Goa, Daman and Diu, South East Maharashtra, East Andhra Pradesh, M.P., Orissa and Chhota Nagpur, Santhal Praganas, Mirzapur, Jhansi, Hamirpur. Such soils formed from ancient crystalline and metamorphic rocks, colour due to wide diffusion of iron rather than to a high content, poor in N, P_2O_5 and humus, potash. Some red soils are of lateritc origin and of a quite different nature, rich in kaolinitic type of minerals. Light soils, silica-sesquioxide ratio is 2.0,. Soils generally are shallow and low in soluble salts. pH of soil 7.0-7.5, the soils under lift irrigation are commonly called garden lands and hungry and thirsty, opposite of black soils in almost all physical and chemical properties.
In the light and frequent rains of south west monsoon, red soils permit sowings being done earlier than other soils hence called early soils. The black soils are not much benefited by light showers in the early season due to its lower permeability. Thus, such black soils are moisture sufficiently in the later period of season-called 'Late soil'. Aeration and warming up of soil are hampered in heavy soils.

4. **Laterite Soils: Alfisol, Ultisol and Oxisol.**
Occurrence: Hills of Karnatak, Kerala, M.P., Eastern Ghat of Orissa, Maharashtra, West Bengal, Tamil Nadu, and Assam; poor in lime and magnesia and N. Occasionally P_2O_5 but K_2O deficient, Occasionally higher humus formed under heavy rainfall and high temperature alternating. The heavy rains leach the salts and left it a porous matrix with a characteristic vesicular structure, facilitating easy drainage. Mixture of hydrated oxides of Al and Fe with small amount of manganese oxide, titania and others are present

5. **Desert Soils: Entisol, Aridsol**
Occurrence: West Rajasthan, Haryana, Punjab, lying between Indus river and Aravalli range. The desert sand is composed of quartz but feldspar and hornblende grains also occur with a fair proportion of calcareous grains.

Genetic System of Classification

The basis of Russian pedologists genetic classification is genesis or origin of the soil and its subsequent development. The soil is looked upon as the result of five main soil forming factors: parent material, climate, organisms, topography and time. The soils inherit their properties from the parent material in the initial stages of soil formation. However, the inherited characters gradually disappear and the soils are dominated by acquired character from climate and vegetation.

According to genetic systems, soils of the world have been classified into three orders:

- **Zonal soils:** Climate and vegetation largely determine the soil characters. Soils developed from different parent material under similar climate and vegetation possess same general properties.

- **Intrazonal soils:** Local conditions like drainage, soluble salts, etc. determine the general characteristics of this order. The characteristic due to climate and vegetation are usually masked to a great extent by the local conditions.
- **Azonal soils:** Soils of this order does not posses well developed profile as the climate has no significant role in their formation.

Soil Survey

Soil surveys consist of soil maps and reports that describe the natural resources of an area, usually a country. Included in the soil survey report is description of climate, principal agricultural enterprises, generalized soil map and from 20 to 50 detailed folded soil maps on a photographic base. Soil survey report also explains limitations for major land use, interpretations for each soil map unit of productive potential for major crops and engineering use for enhancing human environment.

Soil survey is the systematic description of soils in the field, their grouping into well defined mapping units to identify their best use and show their location on the map. The purpose of soil survey is to:

- Classify soils into well defined mapping units (soil series, phases, etc.),
- Show their distribution in the field on the map,
- To find out the best use to which a soil can be put, and
- To predict their response to management practices.
- Soil survey is carried out in three stages:

1. Pre-field operation consisting of studying existing data with regard to geographical, agronomic and climatic aspects of area to be surveyed,
2. Preliminary study of soil, legend preparation, soils identification, plotting of boundaries and collection of soil samples for laboratory analysis, and
3. Revision of aerial photographs, interpretations, laboratory analysis of soil samples, analysis and evaluation of data and presentation in a report.

There are three main types of survey: reconnaissance and detailed and the third one is detailed reconnaissance.

1. **Reconnaissance survey** use 1:250,000 to 1:5,000,000 scale. Soil profiles are located at intervals of 3 to 6 km depending on soil heterogeneity. Photographs of profiles and landscape are taken.
2. **Detailed survey** aims at bringing out differences in respect of soil and terrain features to enable interpretation and classification of soils into land use classes, subclasses and capability units. Base maps used are in the scale of 1:8000 or aerial photographs in the scale of 1:20,000 or larger. Auger samples of soil are examined at every quarter to half km intervals depending on soil heterogeneity.
3. **Detailed reconnaissance survey** consists of carrying out reconnaissance survey first, focused by detailed survey of about 15 per cent of the area where research projects are to be established.

The final soil survey report and maps are prepared in six sections:

A. Introduction,
   i) General description of the area with regard to location and extent, physiography, rivers, major streams, drainage, geology, climate and vegetation,
   ii) Socio-economic conditions such as transport, marketing, educational and medical facilities.

B. Present land use and agriculture including cropping and farming systems adopted in the area,
C. Soils giving detailed methods of survey, soil mapping units, physical and chemical properties and land capability classes.
D. General problems of the area with broad suggestions and recommendations, and
E. Summary and conclusion.

Details of the methods and procedures for soil survey in India are found in the manuals prepared by National Bureau of Soil and Land Use Planning, Nagpur, India.

Soil Surveys in India

The need and importance of soil survey in agricultural development in India have been recognized for many years. The scientific study of soils of the country was taken up as early as 1898 when four major soil groups occurring extensively were differentiated (National Commission on Agriculture, 1972). With the initiation of agricultural research in the country the emphasis was on soil fertility. Soil profile studies started receiving attention only during the 1980s when soil genesis and development came to be recognized as significant factors for the purpose of soil classification. The standard comprehensive soil survey was started in 1956 when the All India Soil Survey and Land Use Organization was established. The modern system developed by the United States Department of Agriculture has been in use since 1969. A task of carrying our and coordinating the soil survey work in the country is done by the National Bureau of Soil Survey and Land Use Planning, Nagpur. The survey in progress is designated to furnish data about all important soil characteristics and associated features for the largest variety of users.

Land Capability Classification

Soil survey maps and reports are the basis for systems of land capability classification. **Land capability classification** is grouping of soils into different classes according to their capability for intensive use and treatments required for sustained use. It emphasizes the need for using the land only for what it is suited best to realize optimum returns without land degradation. Land capability classification system developed by USDA is useful for agriculture. **Eight land capability classes** are recognized in this system of classification.

- **Class I soil** has few limitations on their use. They are deep, well drained and nearly level. They are fertile or responsive to fertilizer application. A variety of crops can be grown intensively with recommended management practices.
- **Class II soils** have few limitations such as gentle slope, moderate erosion problem, inadequate depth, slight saline and alkaline problem and relatively restricted drainage. Less intensive cropping system must be followed. Simple management practices such as contour cultivation can maintain the soil for crop production.
- **Class III soils** can be used for crop production with special conservation practices like terracing. Soil erosion, shallow water permeability, low moisture retentivity, moderate salinity and low fertility are the limitations in their use. Smothering crops such as legumes are more ideal than row crops.
- **Class IV soil** can be used for cultivation with severe limitations on the choice of crops. Steep slopes, severe erosion, shallow soil depth, low moisture retentivity, salinity or alkalinity restricts profitable crop production. These lands should be used for close growing crops or grasses with soil conservation practices.

- **Class V soil** is generally not suitable for grain crops due to limitations such as rocky soil, ponded areas with no drainage facility, etc. Pastures can be improved on this class of land.
- **Class VI soil** is not suitable even for growing grasses. Limitations are the same as those for class V land, but they are more rigid. Their use may be restricted to woodland or wild life.
- **Class VII soils** are severely limited even for growing grasses and forest trees. They are steep rough stony soils of extremely shallow depth.
- **Class VIII lands** are not suitable even for forest trees as they are steep rough stony mountains. Land use is restricted to recreation, wild life, etc.

In each of the land capability classes there are subclasses that have some kind of dominant limitations for agricultural use. The four kinds of limitations in the subclasses are risks of erosion (e), wetness (w), soil depth (s) and climate ©. Thus soil class III (e) indicates risk of erosion in Class III.

CHEMICAL ENVIRONMENT

The soil chemical environment is dynamic and reactions that maintain dilute solution of nutrient elements are indispensible for continual plant growth. The nutrient transformation and its availability in soils depends on pH, clay minerals, cation and anion exchange capacity.

Soil pH

The pH is defined as the negative logarithm of hydrogen ion activity. The pH of a soil indicates its acidity or alkalinity.

Influence of pH

The pH influences nutrient availability, soil physical condition and plant growth.

Nutrient Availability: The pH influences rate of nutrient release through the influence on decomposition, cation exchange capacity and solubility of materials. The important source of nitrogen and sulphur is organic matter. Decomposition of organic matter is slowed at pH below 6 and fasted between 6 and 8. Availability of Fe, Al, Mn, Ca, Zn are influenced by pH through its effect on solubility of their nutrient compounds. Availability of phosphorous is high within a pH range of 6.5 to 7.5 (Table 3.5) due to higher solution of phosphorous compounds. At low pH, phosphorous is precipitated as Fe and Al phosphates while at high pH, calcium phosphate is formed which is less soluble. Metallic cations such as Fe, Mn, Cu, Zn and also B precipitate at high pH, hence their availability is less in alkaline soils. The availability of K and B are influenced by pH through its effect on cation exchange capacity. At lower pH, H^+ ions replace K and leaching if K occurs. At high pH, potassium compounds are converted into non-exchangeable form and thus availability is reduced, Boron is leached out at low pH.

Soil physical conditions: Soil physical condition is related to pH and high sodium content which deflocculates soil colloids. It results in destruction of soil structure and movement of water and air are impaired in the soil.

Plant Growth: High or low pH of nutrient medium has less influence on plant growth when provided with sufficient nutrients. However, soil pH influences plant growth through its effect on nutrient availability and soil physical condition.

TABLE 3.5: Optimum pH range for availability of different nutrients

Nutrient	Optimum pH range
N	6.0 – 8.0
P	6.5 – 7.5
K	6.0 – 7.5
S	6.0 and above
Ca and Mg	7.0 – 8.5
Fe	6.0 and below
Mn	5.0 – 6.5
Bo, Cu, Zn	5.0 -7.0
Mo	7.0 and above

Suitable pH for Different Crops

Soil reaction preference differs with crop. Rice and tea prefer acidic soil reaction and most of the other crops prefer neutral pH (Table 3.6).

TABLE 3.6: Soil reaction preference of crop plants

Crop	pH	Crop	pH
Rice	4.0 – 6.0	Soybean	5.5 – 7.0
Wheat	6.0 - 7.5	Pea	5.5 – 7.0
Sorghum	6.0 – 7.5	Sugarcane	6.0 –7.5
Maize	6.0 – 7.5	Cotton	5.0 – 6.5
Barley	6.0 – 7.5	Potato	5.0 – 5.5
Millets	5.0 – 6.5	Tea	4.0 – 6.0
Oats	5.0 – 7.5		

Causes for Acidity and Remedies

Within a pH range of 5 to 7, the acidity of soils is due to exchangeable hydrogen ions. At pH < 5.0, H^+ ions replace Al^{++} from the lattice of clay mineral. Aluminium thus released is hydrolyzed with liberation of H^+ ions. Iron behaves similarly and contributes to acidity.

$$Al^{3+} + H_2O \longrightarrow Al\,(OH)^{2+} + H+$$
$$Al\,(OH)^{2+} + H_2O \longrightarrow Al\,(OH)^{2+} + H^+$$
$$Al\,(OH)^{2+} + H_2O \longrightarrow Al\,(OH)^{2+} + H^+$$

On soil with low cation exchange capacity (CEC) and in high rainfall regions, leaching of bases occurs resulting in reduction of pH. Addition of acid forming fertilizers increases acidity. When potassium chloride is applied to the soil, it dissociates into K^+ and Cl^- in soil solution. Potassium replaces hydrogen from exchange complex which reacts with chloride to form hydrochloric acid.

$$KCl \longrightarrow K^+ + Cl^-$$
$$K + H+ \longrightarrow H^+ + Cl \rightarrow HCL$$

During the decomposition of organic matter, several acids are formed which increase the acidity. In acid sulphate soils, pH is low due to formation of sulphuric acid by oxidation of sulphur and sulphides.

Acid soils can be converted into neural soils of lime. The liming materials are oxides, hydroxides and carbonates of calcium and magnesium. Calcium carbonate, the common liming material, hydrolyses into calcium, bicarbonate and hydroxyl ion in the soil. The hydroxyl ions neutralize the activity of hydrogen ions.

$$CaCO_3 + H_2O \longrightarrow Ca^{++} + HCO_3^- + OH^-$$

Similarly, calcium oxide releases hydroxyl ions and increases pH,

$$CaO + H_2O \longrightarrow Ca^{++} + OH$$

Causes for Alkalinity and Remedies

Soils with pH higher than 7, are called alkaline soils. The reasons for alkalinity and presence of free carbonates of Ca and Na are the main reasons for alkalinity. Exchangeable sodium has higher effect on pH than of Ca and Mg since sodium hydroxide is a stronger base. Hydrolysis of carbonates of Ca and Mg releases hydroxyl ions which increase pH. The presence of excess sals also increases pH. Alkaline soils are classified based on the electrical conductivity, exchangeable sodium percentage (ESP) and pH (Table 3.7).

TABLE 3.7: Classification of alkaline soils

Class of the soil	Limiting value		
	EC (dSm)	ESP	pH
Saline	> 4	< 15	< 8.5
Saline-alkali	> 4	> 15	< 8.5
Alkali	< 4	> 15	8.5 – 10

Electrical conductivity (EC) of soil is defined as the reciprocal of the electrical resistance of the extract of the soil which is one centimeter long and a cross sectional area of one square centimeter. It is generally expressed as dSm at 25°C and used to express the salinity of the soil. Exchangeable sodium percentage (ESP) is defined as the degree of saturation of the soil exchange complex with sodium.

The classification of alkaline soils was modified and sodium adsorption ratio (SAR) was taken into consideration (Table 3.8).

TABLE 3.8: Modified classification of alkaline soils

Class of the soil	EC (dSm)	SAR	pH
Saline	> 2	-	-
Saline-sodic	> 2	> 15	> 8.5
Sodic	-	> 15	-

In saline soils, excess salt is the root cause of high pH of the soil. This effect can be removed by leaching with rain or or irrigation water. Free drainage is provided to leach the salts beyond the root zone of crops.

Reclamation of alkali soils requires removal of a part or most of the exchangeable sodium and its replacement by the more favourable calcium ion in the root zone. Chemical amendment for reclaiming alkali soils have been broadly grouped into three categories: (1) soluble salts like gypsum ($CaSO_4$ $2H_2O$), Calcium chloride ($CaCl_2$), phosphor-gypsum, (2) sparingly soluble calcium salts like calcite, calcium carbonate, and (3) acids or acid formers like sulphuric acid, iron sulphate, aluminium sulphate, lime sulphur, pyrites etc.

The quantity of an amendment required for improvement of an alkali soil depends on the amount of exchangeable sodium which in turn depends on soil pH. The gypsum requirement of soils is usually determined by equilibrating gypsum solution of known strength with the sodic soils and estimating the calcium deficit. Graphical relationship between soil pH and approximate gypsum requirement is presented below (Fig. 3.8). The required gypsum is spread on the soil surface and incorporated to the required depth. Reclamation of the surface 10-15 cm soil through incorporation of gypsum has proved quite efficient for the rice-wheat sequence.

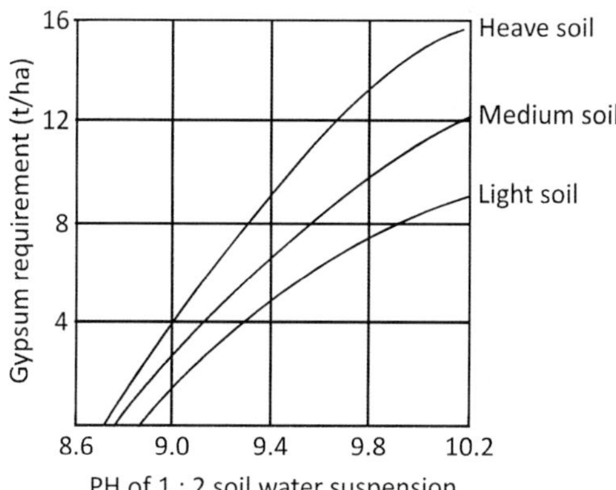

FIG. 3.8: Approximate gypsum requirement of soils.

Calcium present in the gypsum replaces sodium from the exchange complex. Sodium combines with sulphate to form sodium sulphate resulting in reducing exchangeable sodium. When exchange complex has sufficient calcium, flocculation occurs resulting in aggregation which in turn improves movement of water and air.

Clay Minerals

A mineral is a naturally occurring inorganic substance with a definite chemical composition and distinct physical characters. A mineral that forms the original component of a rock is known as **primary mineral.** It is generally, formed anhydrous (without water molecule in its composition) and is originally formed by cooling and solidification of molten mass. Feldspar, homblende, mica etc. are some of the primary minerals. A mineral that has been formed as a result of subsequent changes in rocks is known as a **secondary mineral.** It is hydrous and is formed due to weathering or metamorphosis of primary mineral. Some of the secondary minerals are kaolinite, montmorillonite, illite, limonite, gibbsite, dolomite, calcite, apatite, gypsum, and pyrite.

Hydrated aluminosilicate secondary minerals with particle size less than 0.002 mm in diameter are clay minerals. The clay minerals together with organic matter constitute the colloidal fraction of the soil which is the active seat of physical and chemical properties of soil. Colloids are particles of size less than 0.0002 mm in diameter. The individual particle of a colloid is called **micelle**. Most pf the clay minerals that are less than 0.0002 mm in diameter and also humus exhibit colloidal properties. Soil colloids are of two types viz., **inorganic** and **organic colloids**. The

inorganic colloids may be either silicate or non-silicate clay minerals. **Humus** is an **organic colloid.**

Silicate Clay Minerals

The three important silicate clay minerals are kaolinite, montmorillonite and illite. Depending on the number of silica and alumina layers, they are classified as 1:1 and 2:1 clay minerals.

Kaolinite Group

Each crystal unit of kaolinite group consists of one each of silica and alumina. Hence, these are 1:1 type of clay minerals. These two sheets are held together by oxygen atoms which are mutually shared by Si and Al atoms of their respective sheets. These units are in turn held together rigidly by hydron bonding. Because of the strong bonding between crystal units, the distance between two units is fixed and no expansion ordinarily occurs between the units when wetted. Cations and water, therefore, do not enter between the structural units of the micelle or colloids. The effective surface area of kaolinite is thus restricted to its outer surface. Cation exchange capacity is less due to its low surface area. The size of kaolinite units range from 0.1 to 5.0 μm in width, Kaolinite group of clay minerals are bigger than those of other group.

Montmorillonite Group

The units of montmorillonite micelle belong to 2:1 type with two sheets of silica and one sheet of alumina. These sheets of a unit are held loosely by weak van der Waal's forces and the units in turn are loosely held by very weak oxygen to oxygen linkages. Hence, those are easily expandable allowing water molecules and cations in between these units. In addition to external surfaces, the movement of water and cations between crystal units exposes a very large internal surface area which greatly exceeds the external surface area. The size of montmorillonite crystals ranges from 0.1 to 1.0 μm in width and are much smaller than kaolinite micelle. Montmorillonite clays have higher CEC, plasticity, shrinkage and swelling properties.

Illite Group

Illite group of clay minerals belong to 2:1 type. These micelles are held by K+ ions more tightly than oxygen linkages as in montmorillonite. Thus, these micelles are less expandable than montomorillonite.

Non Silicate Clay Minerals

Hydrous Oxides of Iron and Aluminium

These colloids are present in red and yellow soils of tropical and sub-tropical regions. These are oxides of iron and aluminium with different number of water molecules, depending on the clay mineral. For example, gagacthite (Fe_2O $3H_2O$ is an iron oxide with one water molecule and gibbsite (Al_2O_3 $3H_2O$) is a hydrous aluminium oxide with three water molecules in its structure. These micelles also carry negative charges, but less than those of kaolinite. Most hydrous oxides are not sticky and lesser in plasticity and cohesiveness than silicate clay minerals. A soil dominated by hydrous oxides has much better physical condition.

Allophanes

Amorphous mineral matter in soils is called allophones which are a combination of silica and sesquioxides. They have an approximate composition, Al_2O_3 $2SiO_2$, which is found in many soils. Allophones have high cation and anion exchange capacities.

Organic Soil Colloids (Humus)

Humus is a product of decomposition of plant animal residues which is fairly stable, amorphous and brown to black in colour. On equal weight basis, organic colloids exhibit five to seven times higher absorption of water and cations than inorganic colloids. The charges of humus colloids are pH dependent. Under strongly acidic conditions, hydrogen is tightly bound and not easily replaceable by other cations. Therefore, these colloids exhibit low negative charge. With increase in pH, hydrogen from carboxyl group followed by hydrogen from phenolic group, ionize and are replaced by Ca, Mg and other cations.

Cation Exchange Capacity (CEC)

CEC is a very important property of the soil from plant nutrition point of view. Colloids or micelles have large surface area per unit weight. Most clay micelles have a crystalline structure while humus micelles are amorphous. These micelles contain negative charges on their surfaces. The negative charges on clay micelles are due to isomorphic substitution, ionization of hydroxyl groups and exposed carboxyl and hydroxyl groups. Isomorphic substitution of silica or alumina atom by an atom of similar geometry, but of low charge results in higher negative charge. Silica hydroxides on ionization get extra negative charge.

$$SiOH + H_2O \Leftrightarrow SiO^- + H_3O^+$$

Negative charges on humus micelles are due to exposed carboxyl (COOH) and hydroxyl (OH) groups. Hydrogen in these groups is exchangeable with cations.

Due to the presence of negative charges, cations are adsorbed on surface of the micelles (Fig. 3.9). Divalent cations are more strongly adsorbed than monovalent cations. Cations that are adsorbed on the surface of micelles are capable of exchanging with those in solution. This process of exchange of cations between solid and liquid phases is called cation exchange. These cations are held over the micelles in two layers and it is known as electrical double layer (Fig. 3.10). The quantity of cations is expressed in milli equivalents.

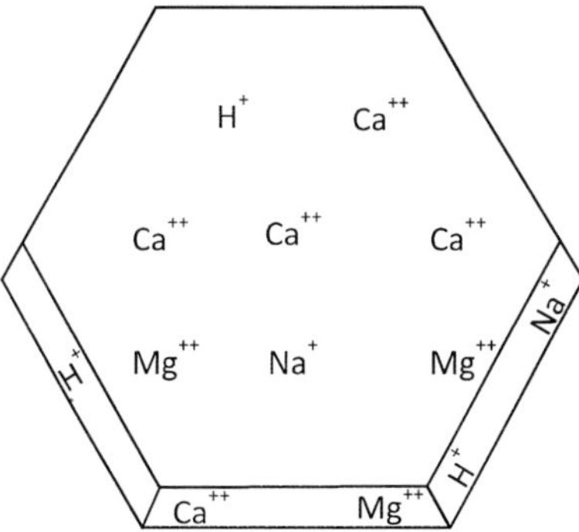

FIG. 3.9: Diagrammatic representation of clay micelle with cations.

One gram of hydrogen is one equivalent weight because the atomic weight of hydrogen is considered as one. Thus one milli equivalent of hydrogen is one-thousandth of a gram of hydrogen or one milligram equivalent weight, which is shortened to milli equivalent. Each cation element has its own distinct equivalent weight based on how much of it will react with or displace hydrogen from the soil exchange sites. The equivalent weight of an element is its atomic weight divided by its charge. The atomic weight of potassium (K^+) is 39 and its charge is one i.e. 39 grams of it will displace one gram of H^+. Thus, the equivalent weight of K^+ is 39/1 or 39. Calcium (Ca^{++}) has an atomic weight of 40 and charge of 2; the equivalent weight of Ca^{++} is 40/2 or 20.

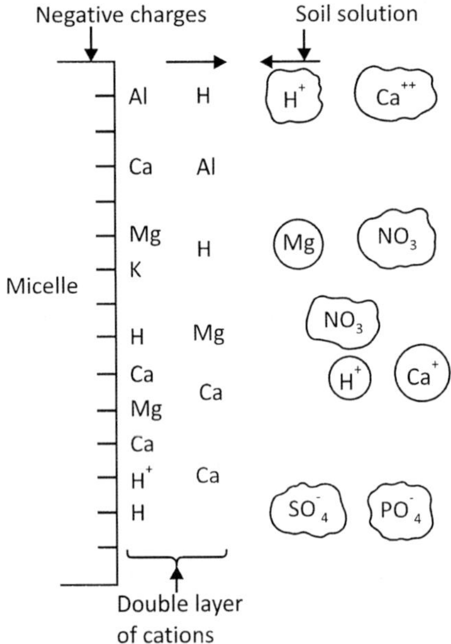

FIG. 3.10: Diagrammatic representation of relationship Between CEC and soil solution.

CEC of soils influences the capacity of the soil to hold nutrients such as Ca, Mg, NH_4^+ etc. The percentage of bases in the total cations (except hydrogen) present on the exchange complex is called **base saturation**. Hydrogen is a cation, but it is acidic and not basic like Ca, Mg, K, Na. Per cent base saturation indicates the proportion of basic cations in CEC. For example, let CEC of a soil be 20 m eq. and CEC of Ca, Mg, K and Na as 15 m eq. The per cent base saturation is {(15/20) x 100}, i.e., 75 per cent. Availability of these nutrients increase with their saturation per cent.

Anion Exchange Capacity (AEC)

AEC is less in magnitude compared to CEC. The sites for anions for exchange to and from soil solution and also from soil constituents are present at the amine groups in humus and the edge of a silicate clay mineral when the bonding between clay mineral and cation terminates. In addition, release of OH^- from Fe and Al cause positive charge on clay minerals. Similar to exchange of cations from clay mineral and soil solution, anions are also exchanged.

Soil Solution

The solubility of a material determines its ability to move in soils. Soluble salts are chlorides, sulphates, nitrates and some carbonates and bicarbonates of Na, K, Mg, Ca and NH_4. When a material is dissolved in water, they dissociate into cations and anions. Water molecules rearrange around these ions as they are dipole and form shells around the ions. These ions especially cations, are easily available to plants. These are absorbed easily by plants from soil solution than from exchangeable sites.

The concentration of soil solution and its pH are changing continuously. Soil solution and clay minerals exchange cations depending on addition or depletion from soil solution. When fertilizers are added, soil solution acquires more cations and some of them reach clay complex. When nutrients are absorbed by plants, cations from clay complex move to the soil solution. Irrigation or rain water dilutes the soil solution while dry period increases its concentration. The pH of the soil solution varies during crop growth. When plant absorbs, cations, roots release H^+ ions in exchange resulting in decrease in soil pH. Similarly, if anions are absorbed, OH^- ions are released into the soil solution increasing pH. The small increase in H^+ or OH^- concentration does not reflect on soil pH due to buffering capacity of soil. The CEC is the principal buffering mechanism in soil. Usually, some of the exchangeable cations are basic (ca^{++}, Mg^{++} etc.) and others are acidic (H^+, Al^{+++}) but their acidic or basic properties are expressed only when they are in solution. Most soils have over 100 times as many cations adsorbed on the micelles as in the solution at any one time. Only active position of the H^+ ions is measured in the pH, but there is a continual exchange or equilibrium between exchangeable adsorbed ions and ions in solution.

Nutrients

Source

Carbon, hydrogen and oxygen are mainly obtained from water and air. A part of the carbon is obtained from carbonates present in the soils. The important sources for the rest of the plant nutrients are organic matter and mineral matter. Nitrogen is obtained by the plants mainly from organic matter. Some plants are capable of fixing nitrogen. For the rest of the nutrients, minerals are main source and to some extent organic matter (Table 3.9).

Plant nutrients present in minerals and organic matter are in non-available form. Weathering of minerals and mineralization of organic matter make the nutrients available to plants. The available nutrients applied in the form of fertilizers may become non-available by precipitation in the soil or by immobilization by organisms.

TABLE 3.9: Source of plant nutrients in soil

Nutrient	Source
C	Carbonate
N	Organic matter
P	Apatite (Calcium phosphate minerals), Fe and Al phosphates, organic matter
K	Micas, Feldspars
Ca	Dolamite, Calcite, Apatite, Calcium carbonate, Gypsum, Augite
Mg	Dolamite, Muscovivite, Biotite, Olivine, Augite, Homblende
S	Pyrite, Gypsum, Organic matter
Fe	Pyrite, Magnetite
B	Tourmaline
Cu	Chalcopyrite, Olivine, Homblende, Augite, Biotite
Mn	Magnanite, Pyrolusite, Olivine, Homblende, Augite
Mo	Olivine
Zn	Sphalerite, Olivine, Homblende, Augite, Biotite
Cl	Apetite

Transformation

The nutrient present in the soil are subjected to physical, chemical and biological changes or transformations. Several substances are produced during these transformations which are not directly connected to plant nutrition. Some may be toxic and others are harmless. These transformations may either release or fix nutrients.

Nitrogen Transformation

On mineralization, organic matter releases ammonium and nitrate and these are available to plants. Nitrate may be further transformed into molecular nitrogen and escape into the atmosphere. The optimum soil moisture for mineralization is 50 to 70 per cent water holding capacity of soils. Mineralization is temperature dependent process and rate of mineralization is more at high temperatures.

Fertilizer materials added to the soils are transformed and several losses occur during this process. Immediately after application urea, it is hydrolyzed in the presence of urease and forms ammonium carbonate. Most of the arable soils have enough urease to cause rapid hydrolysis.

$$\begin{matrix} H_3N \\ | \\ CO \\ | \\ NH_2 \end{matrix} + H_2O \xrightarrow{\text{Urease}} (NH_4)_2\,CO_3$$

Ammonium carbonate is an unstable compound and decomposes into ammonium and carbon dioxide. Ammonium is adsorbed on the cay complex, a portion is absorbed by the crop and around 11 per cent is lost as volatilization. Since urea is soluble, leaching losses occur especially under lowland condition. Therefore, mixing urea with soil and incubating for 24 to 48 hours before application is suggested. During this period, urea is converted into ammoniacal ion and absorbed on the complex. This reduces teaching losses. This technique is applicable only to acid and neutral rice soil. In alkaline soils, this method increases volatilization.

The applied and native forms of nitrogen are subjected to several losses. They are volatilization, leaching, denitrification, ammonium fixation and immobilization.

Change of material from solid to vapour form is called **volatilization.** Urea and ammoniacal fertilizers are subjected to volatilization. Urea hydrolyses and ammonium carbonate is formed which dissociates and increases the pH around urea granules. The increase in pH initiates volatilization losses.

$$NH_4 + H_2O \longrightarrow NH_3$$

Volatilization of ammonia is maintained by soil alkalinity. Otherwise, volatilization of ammonia decreases pH. In the case of ammonium fertilizers, external resource of alkalinity i.e., soil alkalinity, is necessary for initiating volatilization. Ammonium ions react with water and hydroxyl lions and form ammonia and water. Ammonia gas escapes into the atmosphere.

$$NH_4^+ + OH^- + NH_3\uparrow + 2H_2O$$

Volatilization losses increase (1) with increase in pH due to OH ions, (2) decrease in CEC due to less adsorption, (3) increase in temperature due to high rate of reaction, (4) increase in $CaCO_3$ content, and (5) increase in dose of nitrogen. Volatilization losses range from 0.1 to 20.0 per cent and can be reduced by incorporation of fertilizers or by placement.

Placement of nutrients beyond root zone along with water is known as **leaching**. Soluble fertilizers move along with water into deeper layers. Nitrate forms of nitrogenous fertilizers are subjected to higher leaching losses compared to other forms. The leaching losses with urea are around 7.5 per cent of applied nitrogen where urease activity is less. They are less with ammoniacal fertilizers as ammonium ions are adsorbed on clay complex. Leaching losses of these fertilizers occur in submerged soils and in soils with less CEC. The leaching losses that occur in a 12 week period from different fertilizer sources, applied in 3 splits in sandy soils are high with ammonium nitrate (19.5 per cent) and lowest with urea (7.4 per cent). The leaching losses during this period are around 9.7 per cent with ammonium sulphate nitrate, 9.4 per cent with calcium ammonium nitrate and 7.7 per cent with ammonium sulphate.

Nitrates present in the soil are converted into elemental nitrogen and are lost into the atmosphere. Denitrification losses are more in submerged soils.

Fixation of ammonium is due to trapping of these ions within crystal lattice of montmorillonite, illite, and vermiculite minerals. They fix more ammonium when the soil is dry due to concentration of these minerals. On expansion, NH_4^+ is released by exchanging with Ca^{++}, Mg^{++}, Na^+ and H^+ ions. Fixed ammonium is available in microorganisms which convert them to nitrate and also to some plants which absorb them. Therefore, there is very little difference between fixed NH_4^+ and exchangeable NH_4^+. The amount of nitrogen fixed from 4 to 54 per cent of total N content.

Mineralization is faster in aerobic soils and is carried out by fungi, actinomycetes and bacteria. When large quantities of organic matter are added to the soils, the microorganisms utilize available nitrogen in the soil for their multiplication. The temporary locking up of nitrogen in microorganisms is called **immobilization**. It is revere process of mineralization.

Phosphorous Transformation

Phosphorous is present in the soils in organic and inorganic forms. Inorganic phosphorous is more than organic phosphorous. Mineralization of organic matter releases phosphorous in available form, the inorganic forms of soil phosphorous are Ca-P, Fe-P and Al-P. The dominant form of phosphorous in vertisols is Ca-P and alfisols Fe-P. In Indian soils, Ca-P is around 40 to 50 per cent of total phosphorous in neutral and alkaline soils and more than 50 per cent in alkaline soils and more than 50 per cent in alkaline soils. Fe-P and Al-P are less than 10 per cent in these soils. Phosphorous in soil solution is in the form of primary and secondary orthophosphates ($H_2PO_4^-$ and HPO_4^{2}). Primary orthophosphate ($H_2PO_4^-$) is more in acid soils and secondary orthophosphate (HPO_4^{2}) more in alkaline soils. Phosphorous in soil solution depends on rate of decomposition of organic matter and rate of reaction with inorganic fraction. Organic and inorganic phosphorous is in equilibrium with phosphorous in soil solution.

$$\text{Organic P} \Leftrightarrow \text{Solution P} \Leftrightarrow \text{Insoluble P}$$

Phosphorous equilibrium is temporarily altered by addition of soluble fertilizers or by rapid mineralization due to cultivation. When fertilizers, containing soluble phosphates are added to the soil, phosphorous is retained or fixed by different mechanisms. Fixation phosphorous is by adsorption, isomorphous substitution and double decomposition. Insoluble phosphates are weathered and phosphates are released into the soil solution. Thus, equilibrium is maintained.

Fixation is influenced by pH, nature and amount of clay, free oxides of Fe and Al, Calcium carbonate and organic matter. At pH 2 to 5, fixation is mainly by formation of insoluble Fe and Al phosphates as Fe and Al are more active in acid soils. Within the pH range of 4.5 to 7.5, phosphate is fixed on clay particles. It occurs in two ways: (1) by placing two hydroxyl ions to aluminium in clay with one ion of primary orthophosphate, and (2) by forming clay phosphate linkage.

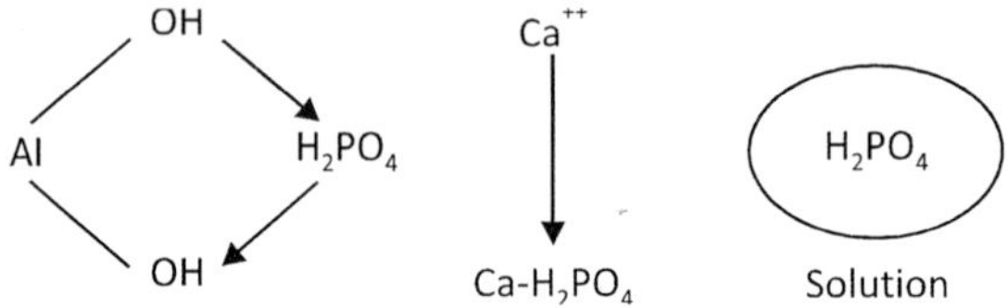

Within the pH range of 6 to 10, phosphate is precipitated as phosphate of calcium or magnesium as calcium and magnesium are more in alkaline range. Phosphorous fixation is more when the clay content is high. Vermiculite and smectite clay fix more phosphates than kaolinite. Black soils have higher fixing capacity than red soils. Presence of oxides of Fe an Al results in formation of sparingly soluble phosphate. Aluminium oxides are more important than iron oxides in fixation. Phosphate fixation is high in soils with high calcium carbonate. Soluble phosphates are first converted into tricalcium phosphates and subsequently to carbonate apatite. The exchangeable calcium forms clay-calcium phosphate linkage or calcium enters soil solution and combines with phosphate.

Potassium Transformation

Water soluble potassium is present in the soil solution as K^+ ion. It is in equilibrium with exchangeable K. Exchangeable K is present on soil clay complex which is adsorbed. It is only one per cent of K present on soil clay complex which is adsorbed. It is only one per cent of K present in mineral soil. Fixed potassium is bound between the units of clay minerals. These three forms viz., fixed, exchangeable and water soluble K are in dynamic equilibrium in the soil.

$$\text{Fixed K} \Leftrightarrow \text{Exchangeable K} \Leftrightarrow \text{Water soluble K}$$

If one form is depleted, it is replenished from the other forms. Unavailable K or fixed K in mica and feldspars are made available slowly due to the weathering process.

$$K\,Al\,Si_3\,O_8 \longrightarrow KOH + H\,Al\,Si_3\,O_3$$

On an average, four-fifths of the K removed by the cropping system of wheat pearl millet is from non-exchangeable form in sandy loam soils containing mainly illite clay mineral. Exchangeable K adsorbed on clay complex is released into soil solution when its concentration is decreased due to plant absorption. When fertilizer is added, K in the soil solution is increased. A fraction of it goes to exchangeable form and in due course even to fixed form.

Calcium Transformation

The important sources of calcium are dolomite, calcite, apatite and calcium feldspars. It may be taken up by plants, lost in drainage, re-adsorbed by clay or a small fraction may be re-precipitated as secondary calcium compound. Calcium is not fixed or made unavailable in the soil. Calcium in soil solution and in exchangeable complex are in a state of dynamic equilibrium. Higher concentration of calcium in solution is adsorbed on clay complex. Exchangeable calcium is released into soil solution when its concentration is low in soil solution.

Magnesium Transformation

Magnesium is available to plants due to weathering of biotite, dolomite, chlorate, serpentine and olivine. The released magnesium may be absorbed by plants and microorganisms, lost in drainage or precipitated as secondary minerals. Magnesium in soil solution and exchangeable form are in dynamic equilibrium, similar to calcium.

Sulphur Transformation

Earth's crust contains about 0.06 per cent of sulphur. It is present as sulphides, sulphates and in organic combination with carbon and nitrogen. The original source of sulphur in soil is sulphides of metals in plutonic rocks. During weathering process, these sulphides are oxidized to sulphates. These sulphates are present as insoluble sulphate salts in arid and semiarid regions.

The nature of applied sulphur is in the form of sulphates of Ca, Mg, K, Na or NH_4 in soil solution in arable soils. These sulphates may be absorbed on 1:1 clay or hydrous oxides of Fe and Al. It may be adsorbed by plants and microorganisms. Sulphates are reduced to sulphides in water logged soils and form H_2S, FeS etc. Elemental sulphur is oxidized to sulphates by microorganisms in aerated soils.

Transformation of Iron

Iron is the most abundant metallic element in earth's crust. It occurs in soil as oxides, hydroxides and phosphates in primary and secondary minerals. A small amount of iron is released due to weathering of primary and secondary minerals Iron deficiency occurs in calcareous soils and soils with high phosphorous. The available and exchangeable iron decreases with increase in pH as it is inversely related to solubility of iron. Higher concentration of phosphorous causes deposition of iron on the root surfaces or just inside the root, possibly as iron phosphate. Iron is oxidized in the rhizosphere. Grasses have more oxidizing power than legumes. Iron availability is reduced with oxidation. Its availability is higher in submerged soils due to reduction of insoluble ferric compounds to soluble ferrous compounds. Organic matter, native or added, improves the availability of iron.

Manganese Transformation

The primary minerals of manganese are present chiefly as oxides and to some extent as carbonates and silicates. The predominant secondary minerals are pyrolusite and manganite. Manganese in the soil is generally considered to exist in divalent (Mn^{++}), trivalent (Mn_2O_3) and tetravalent (MnO_2) forms in equilibrium.

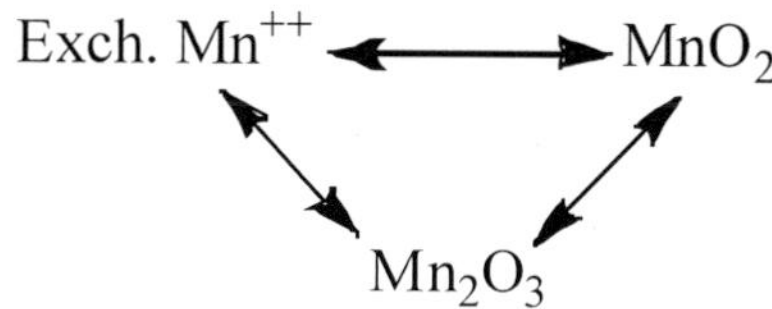

Availability of manganese increases in acid soils as in the case of iron. In acid soils, the available manganese is up to 16 per cent of total manganese content of soils, while in alkaline soils it seldom exceeds one per cent. Its availability is more in submerged soil due to reduction of insoluble manganese compounds to soluble manganous compounds. Tri- and tetra-valent forms are present under high pH and oxidizing conditions. Calcium carbonate decreases manganese availability due to strong binding of Mn^{++} on calcium carbonate. Organic matter decreases the availability of manganese.

Transformation of Copper

Copper in earth's crust occurs mainly as sulphides and most abundant mineral of copper is chalcopyrite (Cu Fe S_2). It is also present in secondary minerals and organic complex. Weathering process releases copper. The solubility of applied copper fertilizer decreases with a rise in soil pH. It is due to adsorption of Cu^{++} on soil colloids and due to precipitation as copper hydroxides. Copper is fixed strongly on organic matter.

Transformation of Zinc

Zinc in earth's surface is found chiefly as sphalerite, a sulphate mineral. It also exists a constituent of organic matter. When soluble zinc salt is added to the soil, it undergoes several transformations by adsorption, precipitation, entry into clay lattice and by microbial consumption. The transformations are mainly influenced by soil pH. Within 24 hours after application of zinc fertilizers, only two per cent soluble zinc is present in soil solution in soils with pH above 7.0. The adsorption by clay complex and organic matter is high at higher pH. Generally, zinc availability is more in acid soils. Its availability is reduced with higher level of phosphates due to low solubility of zinc phosphate and also due to high organic matter as zinc is strongly adsorption of zinc on carbonates of calcium and magnesium and formation of insoluble carbonates.

Transformation of Boron

Tourmaline, a fluorine bromosilicate is the important bromine containing mineral. As it is resistant to weathering, only a small fraction of boron is available to the plants from this source. Most of the boron in cultivated fields is from organic matter. Mineralization of organic matter releases boron, some of it leached and some is held on clay. Irrigation water is another source of boron for crops. Boron added to the soil remains mostly soluble in acid soils and is subjected to leaching especially in light textured soils. Boron adsorption increases with pH. More than 72 per cent of applied water soluble boron is converted into water insoluble at pH above 7.5. The availability of applied boron is, therefore, reduced in alkaline soils.

Transformation of Molybdenum

Molybdenum is present largely in the crystal lattice of primary and secondary minerals. Organic matter is another source of molybdenum. It is present in the soil as MoO_4^{+}. It is either adsorbed on clay complex or present in the soil solution. Unlike other micronutrients, its availability is high at high pH. MoO_4^{2-} is exchanged for two hydroxyl ions from the clay. Oxides of Fe and Al increase the adsorption of molybdenum. Higher level of P fertilizers increases availability of molybdenum, but S reduces it. The ions of MoO_4 and SO_4 are similar in size and charge and they compete at root surface for entry into the plant. Higher soil moisture increases molybdenum availability due to decrease in ferric iron which is responsible for adsorption.

Transformation of Chlorine

Chlorine is present in soils as a chloride and it is highly mobile in soil. It is not adsorbed on clay complex.

Factors Influencing Micronutrient Availability

Several factors influence the transformation and in turn availability of micronutrients. The important factors are **soil pH, organic matter, calcium carbonate content,**

amount of soil phosphorous and submergence. The influence of these factors on micronutrient availability are summarized in Table 3.10.

TABLE 3.10: Factors influencing micronutrient availability in soils

Particulars	Availability of Micronutrients	
	Increases	Decreases
Organic matter	Fe, B	Mn, Zn, Cu
Acidity	**Fe, Mn, Cu**	
Alkalinity	Mo	
Calcareous soils	-	Zn, Mo
Phosphorus	Mo	Fe, Zn
Submergence	Fe, Mn, Mo	

Modification of Chemical Environment

Chemical environment of the soil is modified by addition of organic matter and fertilizers and by irrigation and tillage.

SOIL BIOLOGICAL ENVIRONMENT

The soil biological environment is composed of several organisms belonging to both plant and animal kingdom. Because of their number and activity, several changes take place which are generally helpful to the plants while a few of them are harmful. The activity and number of soil organisms can be modified by management practices so as to provide suitable environment for plant growth.

Classification of Soil Organisms

Soil organisms belong to both plant kingdom (flora) and animal kingdom (fauna). In soil flora and fauna, microscopic and highly developed organisms exist which are called micro- and macro-organisms respectively. The macro-flora of soil consists of roots of higher plants while bacteria, fungi, action-mycetes and algae constitute the soil micro-flora. Protozoa and nematodes are important soil micro-fauna and earth worms, moles, ants etc., are the macro-fauna (Fig. 3.11).

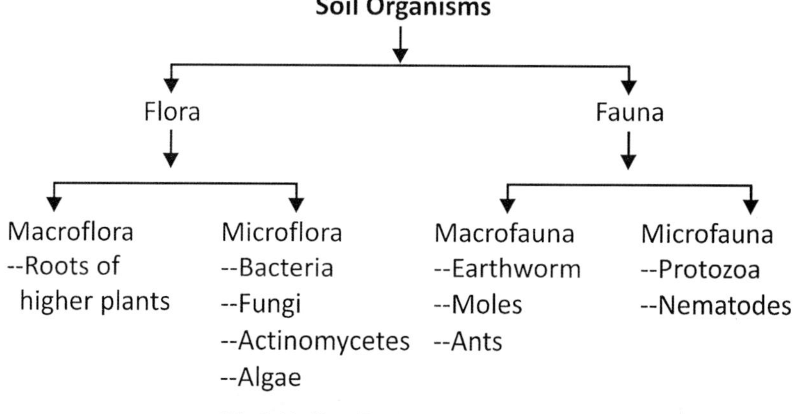

FIG. 3.11: Classification of soil organisms.

Roots of Higher Plants

Space around the root surface of higher plants is called **rhizosphere.** Rhizosphere provides a special environment for rapid proliferation of microorganisms due to exudation of energy rich compounds. Bacteria mostly colonise on the rhizosphere compared to the rest of the soil. About 7 billion bacteria are present in a gram of rhizosphere soil compared to 25 million in non-rhizosphere soil.

Microflora

Based on the oxygen requirement, microbes are divided into three groups: **aerobes**, which survive and multiply under sufficient level of free oxygen; **anerobes**, which survive without free oxygen, but obtain from other components; and **facultative microbes,** which grow better with sufficient level of oxygen, but can adopt even to a very low level of oxygen.

Based on the temperature requirement, microorganisms are grouped into **psychrophiles** (<10°C), **mesophile** (20-40°C) and **Thermophiles** (>40°C). Autotrophs can fix carbon dioxide whereas heterotrophs depend on organic matter for their carbon.

Population and activity of micro-flora depend on soil physical and chemical environment. Soil moisture is the most important factor in deciding the type of the microorganisms present in the soil. When soil moisture is near or below field capacity, aerobic bacteria flourish. When the soil is saturated or submerged only anaerobic bacteria exist. Actinomycetes dies under anaerobic conditions. The optimum moisture for aerobic bacteria is 50 to 75 per cent water holding capacity while actinomycetes predominate in dry regions.

Most micro-flora of the soil are mesophiles and grow well at 155 to 45°C with optimum being 37°C. Higher carbon dioxide concentration is toxic to microbes. Direct sun light is injurious to most micro-flora. Optimum pH differs with type of microflora. Bacteria grow well within pH range of 6.5 to 8.0. Fungi prefer acidic range (4.5 to 6.5 pH) while actinomycetes prefer slightly alkaline conditions. Heterotropic microflora requires organic matter for their carbon form. They proliferate when soil organic matter is high.

Soil Bacteria

Bacteria are the smallest and simplest forms of plant life. They are single called organisms, most of which are either spherical or rod shaped. The bacteria weigh about 0.01 to 0.4 per cent of soil weight. They are the most abundant (4.6 to 8.4 million per g of soil) among the microbes present in the soil. The most abundant species of soil bacteria are *Pseudomonas, Azatobacter, Bacillus, Arthobactor, Aerobacter, and agrobacterium.* The important species of bacteria from plant nutrition point of view are: *Rhizobium, Closttridium, Beijerinckia, Nitrobacter and Nitrosomonas.* As these bacteria are heterotropic, their population depends on soil organic matter. Bacteria are involved in several processes and transformation of nutrients. Nitrogen fixation is an important process carried out by bacteria and this has considerable influence on agriculture. Several steps in mineralization of organic matter are carried out by bacteria which release several nutrients like N, P, K, S and Fe. Bacteria are also involved in decomposition of cellulose and oxidizing sulphur and manganese.

Soil Fungi

The number of soil fungi per unit of soil is less than the number of soil bacteria (0.9 to 1.4 million per g of soil) but their weight is approximately equal because of their bigger size than bacteria. The fungi dominate in acid soils, since bacteria and actinomycetes are inactive. The optimum pH range is 4.5 to 6.5. The important genera of fungi are *Aspergillus, Fusarium, Trichoderma, Mucor, Pencillum etc.* They perform two important functions: (1) act as scavengers and decompose several, types of materials (Table 3.11). Bacteria cannot decompose lignin while fungi can. (2) certain fungi feed on protozoa and nematodes and thus maintain micro-flora equilibrium in the soil.

TABLE 3.11: Materials decomposed by fungi

Material	Organism responsible
Cellulose	Penicillium, Trichoderma, Aspergillus, Fusarium,
Lignin	Basidiomycetes
Protein	Fusaira, Aspergillus, Mucor, Citromycetes

Mycorrhiza another soil fungi, solubilizes phosphates in soils and makes available to plants on which they live. The mycorrhizae that live inside roots (endotrophic) are *Phoma* and *Rhizoctonia.* The *ectotrophic mycorrhizae,* which develop on root surfaces, are *Amanita,* and *Boletus.*

Actinomycetes

Actinomycetes are thread like organisms and are intermediate in size between bacteria and fungi. These are 1.7 to 4 million per g of soil and next to bacteria in abundance. *Streptomyces* is the most common genera of actinomycetes. They are widely distributed in soils, composts, etc., and their population is more in top layer of the soil of tropical dry regions and grass lands. They are aerobic and heterotrophic and their activity is less in acid soils. The optimum temperature for mesophiles of actinomycetes is in the range of 28°C to 37°C. Thermopiles are present in manure and composts. They decompose all types of materials and are also capable of decomposing more resistant organic matter. However, the rate of decomposition is slow compared to bacteria and fungi.

Algae

Algae are autotrophic and they contain pigments to fix carbon dioxide. Both micro and macroscopic organisms are present in this group. Depending on the colour of the pigment, algae are grouped into: (1) **Cyanophyta** – blue green algae which occur mainly in tropical regions; (2) **Chlorophyta** – grass green algae which exist in grasslands of temperate regions; (3) **Xanthophyta** – yellow green algae; (4) **Bacillariophyta** – golden green algae. The last two groups are less abundant. The population of blue green algae in rice fields is about 43,000/g of soil. The most common genera are *Anabaena* and *Nostoc.* They fix atmospheric nitrogen and are used as bio-fertilizers.

Soil Fauna

The macro-fauna consists of earthworm, moles, ants, etc. These organisms consume plant residues and help in decomposition of organic matter. Protozoa and nematodes constitute the micro-fauna. Protozoa feed on bacteria and actinomycetes. Some nematodes are parasitic on crop plants and some live on organic matter.

Influence on Soil Environment

Enzymes

Microorganisms secrete several enzymes such as urease, sulphatases, phosphates, cellulose, amylase and proteinase. These enzymes are utilized for the breakdown of complex compounds present in plant residues. However, some of these enzymes are not used immediately and are adsorbed on the clay complex and are used at subsequent stages.

Nitrogen Fixation

Atmosphere contains 78 per cent nitrogen by volume. Some micro- and macro-flora are capable of fixing nitrogen. Nitrogen fixation has greater significance in agriculture and plant ecology. Nitrogen fixation by microorganisms is estimated as 190 mt per annum while industrial fixation is around 50 mt. Some of them fix nitrogen in association with higher plants and are called symbiotic nitrogen fixers. Some free living organisms fix nitrogen and are called asymbiotic nitrogen fixers.

Symbiotic Nitrogen Fixation

Bacteria belonging to the genus *Rhizobium* infect roots of leguminous crops. The bacteria obtain water and carbohydrates from leguminous plants, fix atmospheric nitrogen and supply to the host plant. Thus, both leguminous plants and bacteria are benefited by their association.

Rhizobium

Rhizobium is a heterotrophic, aerobic bacteria occurring in soil. The rhizosphere of legumes stimulates the growth of *Rhizobium* which infects the roots and forms nodules. The nodules of legumes are small (2-3 mm in diameter), spherical to club shaped and are formed on roots. The *rhizobium* present in the root nodules are known as **bacteroids** and each host cell contains several bacteroids. Though bacteria is aerobic, the enzyme (nitrogenase) functions only in the absence of oxygen. The host cells that contain bacteroids produce a reddish protein called **teghaemoglobin.** Its functions are similar to that of haemoglobin in mammals. It has high affinity for oxygen and even the low level of oxygen present around the host cell is absorbed. It releases oxygen to bacteroids slowly and in just sufficient quantities. Bacteroids contain oxidase enzymes which have high affinity to oxygen and utilize effectively and completely even at low levels. Thus, the enzymes that reduce nitrogen are provided with required anaerobic condition. The activity of nodules can be seen by opening them. The active nodules contain leghaemoglobin when exposed to oxygen, and turn pinkish in colour.

In the bacteroids nitrogen is reduced to ammonium (NH_4^+) in the presence of nitrogenase. Ammonium ion moves out of bacteroids into host cells where it is transformed into aspargine. It moves through cortical cells, then through xylem vessels and reaches leaves where it is assimilated or redistributed. A part of the nitrogen fixed by legumes is exuded into soil as amino acids. Thus, in legume grass intercropping system, grasses are benefitted by the nitrogen supply.

Nitrogen fixation by *Rhizobium* is influenced by soil chemical environment. Nitrogen fixation is less or negligible if the soil pH is less than 5 (five). High nitrogen content and low phosphorous content reduce nitrogen fixation. *Rhizobium* is an aerobic bacteria and water logging makes the nodules ineffective. Leaves of legumes become yellow after temporary water logging as nitrogen supply is reduced due to death of *Rhizobium.* However, recent reports indicate that soybean when continuously frown at saturation can overcome the initial set back and adopt to anaerobic conditions. Fixation is normal subsequently. The amount of nitrogen fixed by *Rhizobium* varies depending on the crop and environmental conditions. It ranges from 20 kg/ha with green gram to more than 250 kg/ha with Lucerne.

There are some actinomycetes which fix nitrogen on non-leguminous plants. *Causurina alder lives* in association with Causurina and fixes atmospheric nitrogen.

Non-Symbiotic Nitrogen Fixation

Some free living bacteria are capable of fixing atmospheric nitrogen. Some of them are *Azotobacterchroccum, A. vinefundi,* and *Clostridium pastourianum, Azoto bacter* is a heterotrophic bacteria and sensitive to acidity. The optimum pH range for *Azotobacter* is 6.5 to 8.0. *Clostridium* is anaerobic bacteria and can survive in wide range of soil pH (5.0 to 9.0). It fixes less nitrogen than *Azotobacter. Azospirillum* present in the *rhizosphere* of grasses fixes atmospheric nitrogen. The genus *Rhizospirillum* is photosynthetic anaerobic bacteria capable of nitrogen fixation. Several species (around 40) of the genera of *Nostoc* and *Anabaena* are the important nitrogen fixing organisms in blue green algae.

Transformation of Nutrients

Mineralization of Organic Matter

A variety of bacteria are involved in breaking down complex organic matter to mineral form of nutrients. During the process, N, P, K, S, and several micronutrients are made available to plants.

Nitrogen Transformations

The ultimate source of nitrogen is atmosphere and immediate and important source of nitrogen is soil organic matter. Plant materials and microorganisms are the source of soil organic matter. Plant materials contain nitrogenous and non-nitrogenous compounds. The nitrogenous compounds are amino acids, amino sugars, nucleoproteins, nucleic acids, amides, purins, pyrimedines, alkaloids etc. The non-nitrogenous compounds are cellulose, hemicelluloses, starch, sugar, gum, lignins, tannins, fats, oils, waxes, resins, and pigments. Among them, cellulose and hemicelluloses are more abundant and constitute 30 to 60 per cent of dry matter (Table 3.12).

TABLE 3.12: Plant Composition

Compound	Range (%)
Cellulose	20 – 30
Hemicellulose	10 – 30
Lignin	10 – 30
Proteins	1 – 15
Fats and waxes	1 – 8

They are decomposed by cellulose decomposing bacteria viz. *Cytophaga, Sporocytophaga, Angiococus, Polyangium.* These bacteria secrete celluloses which convert cellulose into mono- and disaccharides. Anaerobic bacteria belonging to the genus *Clostridium* decompose cellulose.

The proteins present in the organic matter are hydrolysed into polypeptides which are subsequently broken down into amino acids by the activity of *Bacillus* and *Pseudomonas* (Fig. 3.12). The amino acids are further broken down into ammonium by the process known as **ammonification.** The ammonium thus released is first attacked by *Nitrobacter* and converted into nitrite, Nitrite is attacked by *Nitrosomonas* and the end product is nitrate.

The heterotrophic organisms feed on organic carbon for their energy. While mineralizing the proteins, part of mineral nitrogen is utilized by these microorganisms for their tissue build up. Materials such as cereal straw, stubbles of sorghum and other millets contain high C : N ratio. As sufficient quantity of carbon material is available, bacteria multiply using the mineralized nitrogen. The nitrogen mineralized from the plant residues ends up as protein in microorganisms and available form of nitrogen is not released into the soil. Thus, there is a temporary lock up of nitrogen which is made available after decomposition of dead microorganisms.

Aeration is very important for nitrification as these bacteria are aerobic. Nitrification ceases below pH 5.0, the optimum being 6.5 to 7.5. The optimum temperature for nitrification is 30°C to 35°C. Nitrates are further attacked by denitrifying bacteria and reduced to elemental nitrogen. It is a wasteful practice from the agricultural point of view, but it is an essential step in nitrogen cycle to maintain constant level of atmospheric nitrogen. The important denitrifying bacteria are *Pseudomonus* and *Bacillus*. These bacteria are aerobic, but denitrification is more under anaerobic conditions.

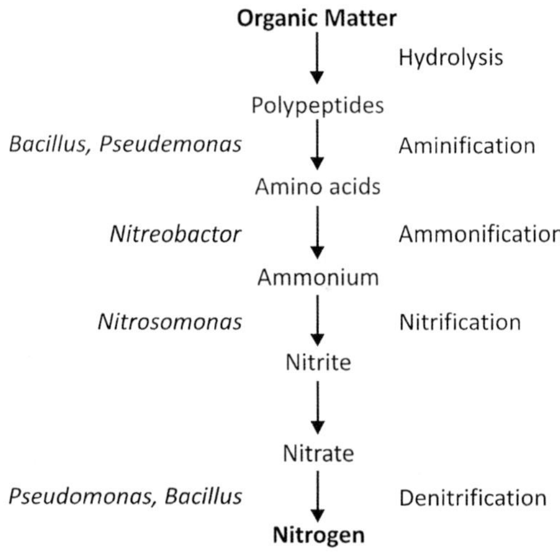

FIG. 3.12: Flow chart showing Organisms involved in Mineralization of Nitrogen.

Phosphorus Transformation

The organophosphorus compounds of organic matter are phytin, phospholipids, nucleoproteins, nucleic acids, phosphorylated sugars, coenzymes etc. Bacteria, fungi and actinomycetes act on these materials by secreting phosphates and break them down into simpler compounds. The optimum C : N : P ratio is 100:10:1 and if the C : P ratio is more than 100:1, immobilization of phosphorus occurs.

Sulphur Transformation

Plant residues contain sulphur in the form of proteins, amino acids and vitamins. Mineralization of these compounds releases sulphates. The C: S ratio of 50:1 is critical. Decomposition of plant materials with wider C: S ratio causes immobilization of sulphur. In anaerobic soils, sulphates are reduced to hydrogen sulphide by the action of *Desulfovibro spp.* of bacteria. Hydrogen sulphide is toxic to crops. Elemental sulphur when applied to the soil is converted to sulphate by *Thriobacillus spp.*

Transformation of Other Nutrients

Mineralization of organic matter releases potassium from organic matter. Some bacteria and fungi are capable of acting on alumino silicates and release potassium.

Iron present in organic matter is released during mineralization by bacteria. Manganese oxidizing bacteria make manganese unavailable to plants.

Degradation of Toxic Chemicals

Several toxic chemicals such as insecticides, fungicides, bactericides and herbicides sprayed on crops or applied reach the soil. Plants also secrete and release allelopathic chemicals into the soils. Accumulation of these chemicals causes considerable damage to the crops. Microorganisms act on these chemicals and degrade them into simpler products.

Modification of Biological Environment

Soil biological environment can be modified by cultural practices to provide suitable environment for crop growth. Tillage loosen the soils and incorporates plant residues thus increasing aeration and organic matter. As a result, microbial population proliferates and nutrients are released due to faster mineralization. Toxic chemicals like herbicides, and allelopathic chemicals are also broken down faster due to better aeration. Irrigation provides sufficient soil moisture and increases the microbial population. Application of amendments brings soil pH to near neutral. The favourable pH increases microbial population. Application of organic manures increases heterotrophic microflora. Oil cakes of neem, karanj, and groundnut increase the population of nematodes trapping fungi. Neem cake is also toxic to nematodes. Thus, these materials reduce nematode attack on the crop. Application of nitrogenous fertilizer decreases the population of bacteria and increases actinomycetes and fungi. Phasphatic fertilizer increases the population of nitrogen fixing bacteria, while nitrogenous fertilizer reduces their population.

ENVIRONMENT OF SUBMERGED SOILS

Rice is grown in submerged soils. Paragrass, a fodder crop, is capable of growing in water logged soils. Several other field crops are subjected to periodic submergence due to wet spells, cyclones or due to over irrigation. The physical, chemical and biological properties of soils are altered due to submergence.

Physical Environment

When soils submerged, changes occur in soil water, soil aeration, soil structure and soil strength Most of these changes are favourable for the growth of rice.

Soil Water

Compared to upland soils, amount of water is plenty in submerged soils. All the micro- and macro-pores are filled with water. In addition, there is standing water above the ground. The energy status of soil water in upland soils generally ranges from -0.33 bar to -15 bars. In submerged soils, it is always zero. Therefore, the availability of water to plants is more in submerged soils.

Soil Air

Immediately after flooding, most of the air escapes from the soil pores. Air may be present in a few pockets in subsoil. Oxygen entry is very slow through water. A thin layer of 2 mm to 12 mm surface layer of soil is oxidized due to diffused air. Since rice plant has the capacity to transport air from leaves of roots, the rhizosphere of rice roots are oxidized. Carbon dioxide released during decomposition of organic matter is dissolved in water to form carbonic acid.

Soil Structure

Soil structure is purposely destroyed in rice cultivation by puddling to avoid percolation losses of water. Even in unpuddled soils also, soil structure is destroyed if there is continuous standing water.

Soil Strength

Soil strength decreases with the amount of water. When there is continuous standing water, soil strength or soil mechanical resistance is around 0.2 bars to 0.3 bars, while upland soils, it may range from 3 bars to 21 bars (Table 3.13).

TABLE 3.13: Effect of soil moisture on soil strength (Just before irrigation)

Moisture regime	Soil strength (bars)
Submergence	0.2
Saturation	1.2
50 per cent depletion of Available soil moisture	15.2
Rainfed	21.4

Chemical Environment

Several chemical changes take place when a soil is submerged and most of them are favourable for the growth of the rice crop.

Change in Redox Potential

Transfer of electrons from one substance to the other is known as oxidation-reduction reactions. The substance which loses electron is oxidized and that which gains is reduced. The tendency of a given system to oxidize or reduce the susceptible substance is known as redox potential. In case of flooded soil, sulphates, nitrates etc. are susceptible substances. Eh measures the intensity with which oxidation reduction is proceeding and is expressed in volts. It is positive in aerobic soils and negative in highly reduced soils. Within hours after flooding, free oxygen escapes. The oxygen present in the soil constituents like NO_3, MnO_4, FeO_3, etc. is consumed by anaerobic bacteria. The reduction of compounds proceeds in the following sequence, O_2, NO_3^-, MnO_4^{4-}, FeO_3^{2-}, CO_2, N_2 and H^+. Redox potential of water and top layers of submerged soils is +3 to +5 volts. The lower layers have a stable Eh of 0.2 to -0.3 volts.

Electrochemical Changes in Flooded Soil

1. A decrease in redox potential,
2. An increase in pH,
3. An increase in specific conductance.

Decrease in Redox Potential: Eh falls sharply upon flooding, reaches a maximum within a few days, rises rapidly to a maximum and then decreases asymptotically with time.

The course of Eh changes is determined by the initial aerobic potential, the temperature, organic matter content and nature and content of electron accepters present in the soil.

Temperature below 30°C retard fall in Eh above 35°C, no significant effect up to 60°C. Green manure causes a steeper decline in Eh than straw.

1) Soils low in organic matter (<1.5%) or high in active Mn (>0.2%) maintained positive Eh values for almost 6 months.

2) Active ferric iron at the concentration of 4-5% iron did not retard the decline of Eh until the potential had fallen to about -50 mv.
3) Soils low in active iron and Mn (coarse textured soils) with more than 3% organic matter showed the steepest declines in Eh and attained Eh values lower than –200 mv within 2 weeks of submergence.

pF is the logarithm of the height in cm of a column of water which represents the total stress with which water is held by a soil. Redox potential is the measure of the intensity of reduction without changing the concentration of reduction products.

Changes in pH Values in Flooded Soil

The flooding causes a decrease of soil pH during first day or two, followed by increase to a maximum content value of 6.5 to 7.5 in 2 to 3 weeks.

Changes in Specific Conductance

The specific conductance of a soil (not un duly high in nitrate or sulphate) increases in the early phases of submergence, reaches a maximum roughly coincident with peak reduction.

Neutral and slightly alkaline soil starting with high specific conductance attain values exceeding 2,000 micromhos/cm (provided sufficient organic matter is present) and decline slowly. Strongly acid soils have low initial specific conductance, slow steep increases during the first 30 days of flooding and decline sharply thereafter in striking similarly to the conductance time changes for soluble Fe^{+2} and Mn^{+2}.

Change in Soil pH

The pH of most acid and alkaline soils converges between 6 and 7 within 2 to 3 weeks after flooding. Alkaline soils contain mainly sodium carbonate and to some extent calcium carbonate. Under submergence, carbonates release carbon dioxide which dissolves in water and forms a weak acid known as **carbonic acid.** Due to its influence, the pH of the alkaline soil is reduced. Acid soils contain mainly ferric hydroxide which is insoluble. Under submergence, it is reduced to ferrous hydroxides which are highly soluble. It ionizes in soil solution into ferrous and hydroxide ions. The hydroxides release hydroxyl ions which increases the pH of the acid soils. Due to the favourable pH attained with submergence, nutrient availability is increased.

Reduction of Iron

Hydrates of ferric oxides are reduced to ferrous compounds within two to three weeks after submergence. As a result, the soil colour changes from brown to gray. Large amount of ferrous ions enter soil solution. Concentration of water soluble iron from 0.1 ppm shortly after submergence increases to 600 ppm within two weeks. Later, concentration of iron decreases, but stabilizes at a higher level than that present just before submergence.

In strongly acid soils with high organic matter, the reduction of iron compounds results in high concentration of iron which is toxic to rice.

Reduction of Manganese

Manganese is more rapidly reduced and rendered soluble than iron compounds. Manganous ions increase steeply within one week after submergence, and availability is higher under submerges conditions.

Supply and Availability of Nitrogen

The supply and availability of nitrogen is more in submerged soils. Higher supply of nitrogen is due to fixation of atmospheric nitrogen by rhizosphere bacteria. The availability of nitrogen is more as organisms involved in mineralization are different from those that occur in upland soil. The decomposition is slow under submerged conditions, but immobilization is less. Actinomycetes and fungi convert 15 to 40 per cent of carbon in aerated soils and require higher nitrogen for their tissue development. These organisms are absent under flooded conditions and most of the decomposition takes place by microorganisms which require less nitrogen. Mineralization stops at ammonification stage since aerobic nitrifying bacteria are absent resulting in accumulation of ammonia. The end products of organic matter decomposition in well drained soils are carbon dioxide, nitrates, sulphates, and humus while in submerged soils the end products are carbon dioxide, hydrogen, methane, acetic acid, butyric acid, ammonia, amines, hydrogen sulphide and partially humified residues. Among them, hydrogen sulphide, acetic acid, butyric acid and carbon dioxide are accumulated to near toxic levels when large quantities of organic matter is added.

With regard to losses, nitrogen is lost mainly by volatilization, surface runoff, denitrification and immobilization. Volatilization of ammonia and surface runoff occur from soil surface.

Ammonium present in oxidized zone diffuses into reduced zone where it is taken up either by plants or immobilized by microorganisms. Major portion of the NH_4 – N in oxidized layer is converted into NH_3 – N by nitrification. Because of its solubility, NO_3 – N, quickly reaches the reduced zone. It is denitrified and N_2 escapes as gas (Fig. 3.13).

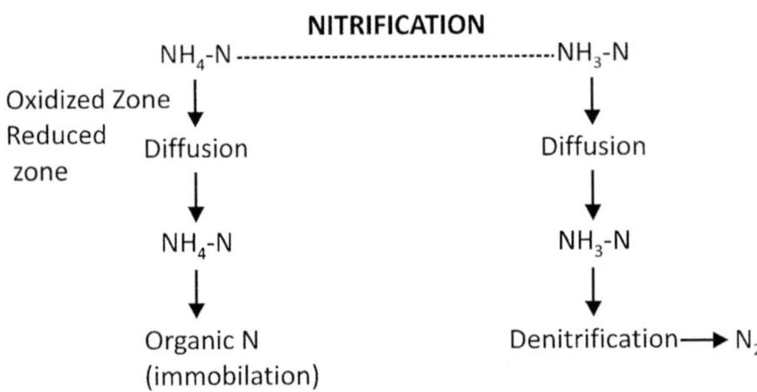

FIG. 3.13: Nitrification-denitrification reaction In submerged soils.

Availability of Phosphorus

Phosphorus in soils exists in different forms viz. (1) iron and aluminum phosphate, (2) phosphates adsorbed or precipitated with iron and alluminium hydrous oxides, (3) phosphates held by anion exchange on clay and hydrous oxides, (4) calcium phosphate, and (5) organic phosphates. Iron and alluminium phosphates are predominant in acid soils and calcium phosphate in alkaline soils. Availability of phosphate increases in submerged soils due to hydrolysis of Fe and Al phosphates due to increase in pH. It is also due to reduction of insoluble ferric phosphates to soluble ferrous phosphates. The solubility of hydroxyl apatite increases due to decrease in pH in alkaline soils. Desorption of phosphates also takes place from clay and oxides of Al and Fe. Hence, phosphate availability in submerged soils increases considerably within two weeks after submergence.

Availability of Potassium

Potassium availability marginally increases in submerged soils. Flooding increases ferrous and manganous ions which displace potassium from exchange complexes. A portion of displaced potassium is lost due to leaching

Availability of Zinc and Copper

Availability of zinc is reduced under flooded conditions due to precipitation as zinc hydroxide {$Zn(OH_2)$} or as zinc carbonate ($ZnCO_3$) due to high concentration of carbon dioxide. It is also precipitated as zinc supplied (ZnS) under reduced conditions. Availability of copper is also reduced due to formation of insoluble copper sulphide.

Availability of Sulphur

Sulphates are reduced to hydrogen sulphide under reduced conditions. Hydrogen sulphide is toxic to plants and microorganisms. Hydrogen sulphide reacts with heavy metals viz. Fe, Zn and Cu and forms insoluble sulphide. However, due to presence of oxidized zone in the rhizosphere of rice roots, sulphur is available to rice.

Biological Environment

The biological environment of submerged soils is quite different from that of aerated soils. The important organisms of submerged soils are *Azolla,* Blue green algae, denitrifying bacteria and rhizosphere bacteria. Actinomycetes and fungi are absent in submerged soils. Non-symbiotic bacteria present in anaerobic soils and those in the rhizosphere of rice roots fix atmospheric nitrogen. Nitrogen fixation in rice rhizosphere starts a month after transplantation and reaches a peak at heading. These bacteria can fix up to 63 kg N/ha.

"Your aspirations are your possibilities."

4 Growth and the Factors Affecting Growth

> *"The important thing in science is not so much to obtain new facts as to discover new ways of thinking about them"*
> ***William Bragg***

The success of the farming operation depends largely on the growth of the crop. If plant growth and yield of the harvest are good, the farmer will have been successful and barring an unfavourable economy, he will receive a return on his investment in labour and capital in keeping with the quality and quantity of the crop. On the other hand, if plant growth has been poor and the yield of the harvest is low, the farmer will receive a correspondingly lower return, if any at all.

From the present stand point of a profitable agriculture, growth and the factors affecting it occupy a place of importance second to none in the farming enterprise. Because of this, some of these factors, and the effect they may have on limiting plant growth responses to an adequate supply of plant nutrients, are discussed briefly.

GROWTH EXPRESSIONS

Growth is defined by Webster as ***the progressive development of an organism,*** but there are several ways in which the development of a plant can be expressed. Growth may refer to the development of some specific organ or organs or to the plant as a whole, and it may be stated in terms of dry weight, length, height, or diameter.

Growth as Related to Time

Whether growth is given as the increase in dry weight or in height of the plant, there is a fairly constant relationship between the measure of growth employed and time. The general pattern is one of initially small increases in size, followed by rapid growth, and then by a period during which the size of the plant increases slowly or not at all. The pattern is illustrated in Fig. 4.1).

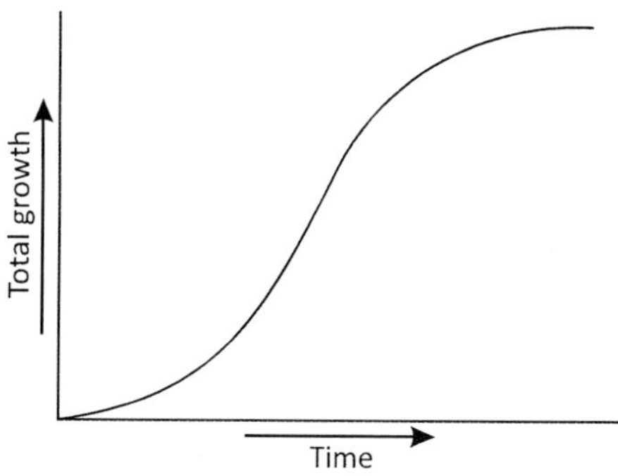

FIG. 4.1: A generalized curve illustrates the growth pattern of an annual plant.

Growth Related to the Factors Affecting it

Growth curves are, of course, helpful to an understanding of the general pattern of plant development. They indicate nothing, however, of the factors affecting growth, such as the supply of mineral nutrient elements, light, carbon dioxide, and water. The plant is a product both of its genetic constitution and its environment. The genetic pattern is a fixed quantity for a give plant and determines, so to speak, its potential for maximum growth under an environment favourable for development; from the stand point of that particular plant, attainment of this maximum development is dependent on a favourable environment. In other words, plant growth is a function of various environmental or growth factors, which may be considered as variables whose magnitude and combination determine the amount of growth that will be made. Symbolically, this may be expressed as:

$$G = f(x_1, x_2, x_3, \ldots x_n)$$

Where G = some measure of plant growth

and $x_1, x_2, x_3, \ldots x_n$ = the various growth factors

Further, if all but one of the growth factor is present in adequate amounts, an increase in the quantity of this limiting factor will generally result in increase in plant growth. Symbolically,

$$G = f(x_1)\, x_2\, x_3 \ldots\ldots x_n$$

This, however, is not a simple linear relationship. It has been shown repeatedly that the addition of each successive increment of a growth factor results in an increase in growth that is less than that obtained from addition of the proceeding increment.

Mitscherlich's Equation

The growth equation discussed in the preceding paragraph is a generalized expression relating growth to all of the factors involved. In 1909, E.A. Mitscherlich, a German, developed an equation which related growth to the supply of plant nutrients. He observed that when plants were supplied with adequate amounts of all but one nutrient their growth was proportional to the amount of this one limiting element which was supplied to the soil. Plant growth increased as more of this element was added, but not in direct properties to the amount of the growth factor added. The increase in growth with each successive addition of the element in question was progressively smaller. Mitsherlich expressed this mathematically as:

$$dy/dx = (A - y)C \quad (1)$$

Where dy is the increase in yield from an increment of the growth factor dx, dx is an increment of the growth factor x. A is the maximum possible yield obtained by supplying all growth factors in optimum amounts, y is the yield obtained after any given quantity of the factor x has been applied, and C is a proportionately constant which depends on the nature of growth factor.

The value of C, as determined by Mitscherlich, was 0.122 for nitrogen, 0.60 for P_2O_5, and 0.40 for K_2O. Numerous workers since Mitscherlich have found that C in fact is not a constant value, but instead a value that varies rather widely for different crops grown under different climatic conditions. It is the lack of constancy of C that has caused much of the criticism of the Mitscherlich growth equation. The same criticism may be applied to the **Spillman equation**, for it reduces itself to the same form.

Spillman Equation

As is so often the case in scientific investigations, this same principle was independently enunciated several years later by an American, W. J. Spillman. Spillman expressed the relation as:

$$Y = M(1 - R^x) \quad (2)$$

Where y is the amount of growth produced by a given quantity of the growth factor x, x is the quantity of the growth factor, M is the maximum yield possible when all growth factors are present in optimum amounts, and R is a constant.

Further work by Spillman showed that, though of a different form, his equation and that of Mitscherlich could both be reduced to

$$Y = A(1 - 10^{-cx}) \quad (3)$$

where y is the yield produced by a given quantity of x, x is the growth factor, A is the maximum yield possible, and c is a constant dependent on the nature of the growth factor. None of these expressions is conveniently handled as written, but they may be stated in the equivalent form

$$\log (A - y) = \log A - 0.301 (x) \quad (4)$$

The symbols used are the same as those in equation 3.

If the function is graphed, the curve obtained appears as shown in Fig. 4.2.

Calculation of Relative Yields from Addition of Increasing Amounts of a Growth Factor

The value 0.301 replaces the constant c when yields are expressed on a relative basis of A = 100. When conventional units of yield are employed, the value of c, as already mentioned, varies with the growth factor.

If A, the maximum yield, is considered to be 100 per cent, Equation 4 reduces to

$$\text{Log} (100 - y) = \log 100 - 0.301 (x) \quad (5)$$

It is possible to determine the relative yield expected from the addition of a given number of units of x. It will be helpful if the student observes how these calculations are made.

If none of the growth factor is available, that is x = 0, then y = 0; but suppose one unit of x is present. Then

$$\text{Log} (100 - y) = \log 100 - 0.301 (1)$$
$$\text{Log} (100 - y) = 2 - 0.301$$
$$\text{Log} (100 - y) = 1.699$$

$$100 - y = 50$$
$$y = 50$$

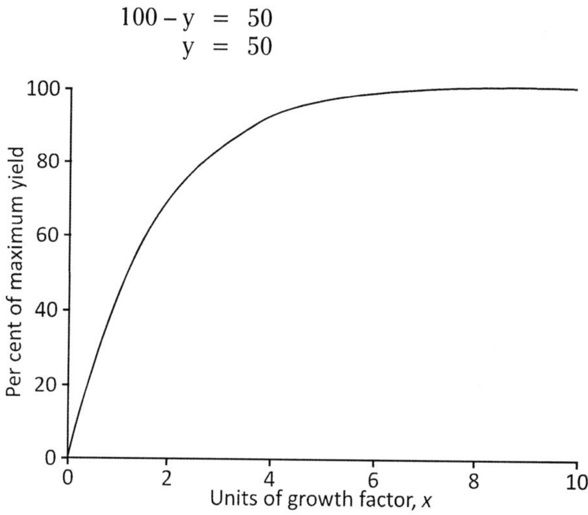

FIG. 4.2: The percentage of maximum yield as a function of Increasing additions of a growth factor x.

The addition of one unit of the growth factor x results in a yield that is 50 per cent of the maximum.

Assume, however, that two units of the growth factor were present. In this instance

$$\text{Log}\,(100 - y) = \log 100 - 0.301\,(2)$$
$$\text{Log}\,(100 - y) = 2.000 - 0.602$$
$$\text{Log}\,(100 - y) = 1.398$$
$$100 - y = 25$$
$$y = 75$$

The same operation may be repeated until 10 units of the growth factor have been added. The result of such a series of calculation is given in tabular form.

Unit of growth factor (x)	Yield (%)	Increase in yield (%)
0	0	--
1	50	50
2	75	25
3	87.5	12.5
4	93.75	6.25
5	96.88	3.125
6	98.44	1.562
7	99.22	0.781
8	99.61	0.390
9	99.80	0.195
10	99.90	0.098

It is obvious that the successive increases of a growth factor result in a yield increase that is 50 per cent of that resulting from addition of the preceding unit until a point is reached at which further increases, for all intents and purposes, are of no consequence.

The Baule Unit

Reference has been made so far only to units of a growth factor. Certainly this is a rather meaningless term, and it is necessary to understand its relation to familiar terms before proceeding with discussion. This unit has been designated as a baule after the German mathematician who collaborated with Mitscherlich. Baule suggested that the unit of fertilizer, or any other growth factor, be taken as that amount necessary to produce a yield that is 50 per cent of the maximum possible. As almost everyone knows, plants require different absolute amount of nitrogen, phosphorus, and potassium, but the amount in pounds of each required to produce a yield that is 50 per cent of the maximum possible is termed one baule unit. Obviously, according to this concept, one baule of a growth factor is equivalent to one baule of any other growth factor in terms of growth-promoting ability. The values of the baule unit in pounds per acre of N, P_2O_5 and K_2O are 223, 45, and 76, respectively, as calculated from the results of Mitscherlich's work.

Yield Increases in Response to More than One Growth Factor

An extension of the baule unit concept has been developed for cases in which two or more growth factors are limiting. It has been shown that when all growth factors but one, x, were present in optimum amounts the addition of one unit of this factor x would produce a yield that was 50 per cent of the maximum possible. Suppose, however, that all except two growth factors, x_1 and x_2 were present in optimum amounts. If one baule of each is added, the yield obtained will not be 50 per cent but 50 per cent times 50 per cent, or 25 per cent of the maximum. If all growth factors but three were present in optimum amounts, the simultaneous addition of one unit of each of the three would result in a yield that was 50 x 50 x 50, or 12.5 per cent of the maximum. The relationship is expressed by the general equation.

$$y = A(1 - 10^{-0.301x1})(1 - 10^{-301x3}) \quad (6)$$

in which x_1, x_2, and x_3 are quantities of the growth factors to be added. The equation is very simply handled by determining the percentage yields that correspond to given values for x_1, x_2, and x_3 and then multiplying these figures together.

It is of interest to note that the rule of decreasing increments applies even though all growth factors are not present in optimum amounts. This makes possible a calculation of the maximum possible yield under a given set of climatic and genetic conditions when fertilizers are added in increasing amounts.

It also makes possible the calculation of the greatest possible crop production from a limited amount of fertilizer, for the greatest increases in yield result from the addition of the first unit increment. The German people made extensive use of this concept during World War II. It is probably; fair to say that the practical application of Mitscherlich's ideas was largely responsible for preventing a greater degree of starvation in Germany than actually occurred during the war years.

The Agrobiology of O. W. Wilcox

O.M. Wilcox, an American, has employed the Mitscherlich equation to extend a branch of science that he calls quantitative agrobiology. This science, he states, "Comprises the general and specific ***quantitative*** relations between plants and the outer factors of their growth and yield". Wilcox assumes that the Mitscherlich equation is valid, an opinion that is by no means shared by many agricultural scientists. He further develops from the Mitscherlich equation the inverse nitrogen:

yield concept which says in effect that the yield of a crop is inversely proportional to as nitrogen content. Symbolically,

$$y = \frac{k}{n} \tag{7}$$

Where Y is the yield, n is the percentage of nitrogen in the crop, and k is a constant. To support this contention, Wilcox cites a number of examples of crop yields and nitrogen percentages that follow this general rule.

From the Mitscherlich equation, Wilcox further extends the inverse yield: nitrogen concept and evaluate the constant k in Equation 7. He finds that the value of k is 318 lb/A, and states that this is the maximum amount of nitrogen that can be absorbed in one season by an annual crop growing on an acre of land. This presumably would occur with the production of a maximum yield which could happen only if none of the various growth factors was limiting. It is obvious that the crop yields would increase of infinitum if any the nitrogen content of the crop on a percentage basis could be made to decrease.

It is the inverse yield: nitrogen law and the criticism directed toward Wilcox's ideas. For a thorough treatment of this tropic attention is invited to the references listed at the end of this chapter which deal with the Witscherlich concept and Wilcox's agrobiology. These papers are lively and well presented and should prove to be interesting reading for both graduate and undergraduate students of the plant and soil science.

Although there are numerous examples reported in which the Mitscherlich-Baule-Wilcox equations fall to describe plant growth adequately, as a function of plant nutrition inputs, it is felt that these concepts should be considered, for they represent an original attempt to develop a unified theory of plant growth as a function of mineral nutrition. The ideas of Mitscherlich, Baule, and Wilcox are not completely without foundation. Plant growth as a function of nutrient inputs is logarithmic and generally follows a pattern of diminishing increases, as expressed in the mitscherlich equation. The growth of annual plants *does* tend to reach a maximum with increasing inputs of nutrients under a particular set of environmental conditions, and often the plants that produce the highest yield of dry matter have the lowest percentage of nitrogen in their tissues. However, it remains for posterity to determine whether a single expression may be developed that will universally predict the amount of growth that can be produced from the input of a given quantity of plant nutrients when environmental and genetic growth factors are adequately described.

Factorial Experiments and Regression Equations

Much fertilizer research has made use of what is known as the factorial experiment. In such studies the effect on crop yield of several levels of different input factors is evaluated simultaneously in one experiment. For example, the effect of several levels of nitrogen and phosphorus on the yield of a crop can be studies simultaneously in one experimental layout. If three levels of nitrogen and three levels of phosphorus are examined the experiment is said to be a 3^3 factorial and requires nine treatments to evaluate the effect on yield of all combinations of nitrogen and phosphorus. The results are usually subjected to various statistical treatments. An equation, normally termed a regression equation, is developed in which the yield is functionally related to the inputs of the fertilizer variables.

Research workers at the Iowa and North Carolina Experiment Stations were among the first to make extensive use of this type of experiment, which is now popular in America. When climate, soil type, plant population, fertilizer placement, and other factors are uniform and content, such studies help to predict fertilizer requirements, but the equations developed from them are not universally applicable. An example of this type of equation is given later in the chapter in the discussion dealing with the effect of moisture on plant growth. Such factorial studies are an outgrowth of the Mitscherlich concept, and their equation are generally similar to form to that of the Mitscherlich equation.

Growth Expressed as a Rate-of-Reaction Equation

A slightly different approach to the yield nutrient input equation has been suggested by Hagin, who was compared the pattern of plant growth to the rate of a chemical reaction which can be expressed by

$$v = V - \frac{S + K_m}{VK_m} \quad (8)$$

Where v is the rate of reaction at the substrate concentration S, V is the maximum rate of reaction, and K_m is a constant. Hagin has modified this expression to

$$y = A - \frac{Ac}{X + b + c} \quad (9)$$

Where Y is the yield obtained when x units of a plant nutrient are added to a soil containing b units of the same nutrient, A is the maximum yield obtainable under the given conditions, and c is a constant which Hagin states may be termed *a growth rate factor.*

Using yield data collected from some Danish pot experiments, Hagin devised equations of the Mitscherlich and the rate-of-reaction types. Both could be fitted to the date, but, of course, the equations differed, from soil to soil.

Bray's Nutrient Mobility Concept

A modification of the Mitscherlich-Baule-Spilllman concept has been proposed by Bray and his co-workers at the university of Illinois. In brief, it is claimed that crop yields obey the percentage sufficiency concept of Mitscherlich for such elements as phosphorus and potassium, which are relatively immobile in the soil. This concept, in turn, is based on Bray's nutrient mobility concept which states that as the mobility of a nutrient in the soil decreases, the amount of that nutrient needed in the soil to produce a maximum yield (the soil nutrient requirement) increases from a variable net value, determined principally by the magnitude of the yield and the optimum percentage composition of the crop, to an amount whose value tends to be a constant.

The magnitude of this constant is independent of the magnitude of the yield of the crop, provided that the kind of plant, planting, pattern and rate, and fertility pattern remain constant and that similar soil and seasonal conditions prevail. He further states that for a mobile element such as nitrate nitrogen Liebig's law of the minimum best expresses the growth of a crop. Bray has modified the Mitscherlich equation to

$$\log (A - Y) = \log A - C_{1b} - C_x \quad (10)$$

where A, Y and x have the connotation already given, C_1 is a constant representing the efficiency of b for yields in which b represents the amount of an immobile but available form of nutrient, such as phosphorus or potassium, measurable by some suitable soil test, and C represents the efficiency factor for x, which is the added fertilizer form of the nutrient b.

On the basis of work done in Illinois, Bray has shown that the values for C_1 and C are specific and fairly constant over a wide area in the state, regardless of yield and season for each of the following crops: corn, wheat, and soybean. The factors that will after the values, however, are wide differences in soil type, plant population and planting patterns, and the form and distribution in the soil of the immobile nutrient element under study. Bray's concepts form a useful basis for making fertilizer recommendations in those situations in which the previously noted conditions are uniform. An excellent discussion of the use of these concepts may be found in the paper by Vavra and Bray in the references listed at the end of this chapter.

Method of Systematic Variations of M. V. Homes

A somewhat different approach to the problem of plant nutrition and its relation to plant growth has been suggested by Professor M. Homes of Brussels. Unfortunately, Homes's ideas have not received much attention in the United States, and it is not intended here to do more than introduce the reader very briefly to them. The discussion appertaining thereto is correspondingly short.

Factorial fertilizer experiments carried out in the United States, and in many other areas for that matter, have been devoted largely to a determination of those rates and ratios of N-P-K that should be applied to various crops to produce maximum yields under different soil and climatic conditions. Homes, however, believes that maximum plant growth is determined by a unique ratio of potassium to magnesium to calcium, a unique ratio of nitrogen to phosphorus to sulphur, a unique cation-anion ratio, and a unique total salt concentration. To determine the unique ratios of K : Ca : Mg and NO_3:: PO_4:: SO_4 he employs his *methods des variante ssystematique,* which can best be explained by illustrating with the elements potassium and calcium. Using a sand or solution culture technique, Homes varies the amounts within a fixed concentration of Ca + K – from K, 100 per cent; Ca, 0 per cent to K, 0 per cent; Ca, 100 per cent. Yields in quantities of dry matter produced are measured; and the maximum yield corresponds to the most favourable ratio of K: Ca. This is illustrated by the graph in Fig. 4.3.

From the simple system of two nutrients (a binary system) Homes develops his idea for the three cations, (potassium, calcium and magnesium) and for the anions (nitrogen, phosphorus, and sulphur).

Using cotton as the test crop in a sand culture and anion: cation ratio of 1:1 at a total ionic concentration of 1,000 meqs per plant, Homes obtained the data given in Table 4.1. He then sums up the yield for the N-S-P treatments, which is 152, and for the potassium, calcium, and magnesium treatments, which is 221. To obtain the optimum balance for the anions, he makes the following calculations:

NO_3	SO_4	PO_4
$\frac{86}{152}$ x 100 = 57	$\frac{33}{152}$ x 100 = 22	$\frac{33}{152}$ x 100 = 22

For the cations

K	Ca	Mg
$\frac{73}{221}$ x 100 = 33	$\frac{79}{221}$ x 100 = 36	$\frac{70}{221}$ x 100 = 32

The optimum ratio of anions is thus

$$57\ NO_3 : 22\ SO_4 : 22\ PO_4$$

And for cations this ratio is

$$33K : 36\ Ca : 32\ Mg$$

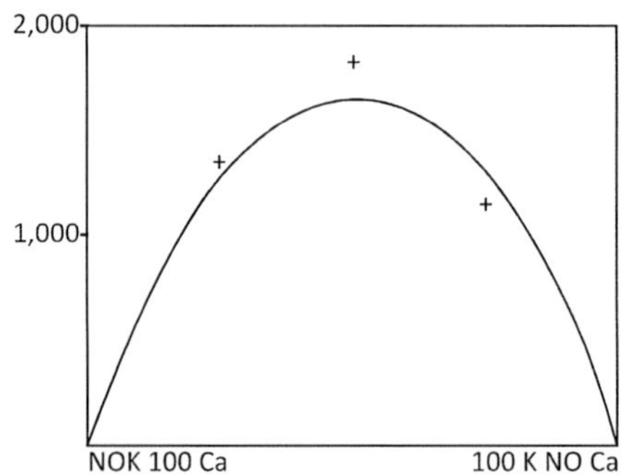

FIG. 4.3: Yield as a function of varying concentrations of Potassium and Calcium.

The ***optimum cation : anion ratio*** is established in the same way by using the comparison of two treatments in which the sum of one anion (e.g., NO_3) and one cation (e.g., Mg) is kept constant, and their relative proportions inside that sum are varied according to the same system.

The optimum total ionic concentration can also be deduced from the relation between yield and total concentration. It requires, in addition to the already mentioned treatment, one or two other treatments with the same ratios and varying total amount of nutrients.

To adapt this method to the determination of fertilizer treatments for the highest yield under field conditions Homes has modified his experimental technique to account for the amounts of nutrients present in the untreated soil. He estimates these amounts from yield measurements made on untreated soils.

Homes has been careful to point out that in the determination of these ratios the level of all nutrients, other than those under study, as well as other environmental conditions should be kept at levels optimum for the growth of plants. He has tested his theory on several crops in what was formerly the Belgian Congo and claims that the method has given predictable and satisfying results and that it affords a mean of determining the type of fertilizer that is to be used for maximum crop production on various soil types.

TABLE 4.1: Yield of Cotton Grown at Various Concentrations of Cations and anions

Treatment	Anion balance			Cation balance			Yield (Lint + seed) per cent
	NO_3	SO_4	PO_4	K	Ca	Mg	(g)
N	60	20	20	29	45	26	86
S	20	60	20	29	45	26	33
P	20	20	60	29	45	26	33
K	50	25	25	60	20	20	72
Ca	50	25	25	20	60	20	79
Mg	50	25	25	20	20	60	70

Source: From Homes Soils Fertilizers, 18:1 (1959).

As already indicated, Homes's views have not been widely studied in the United States. However, they have been criticized by European workers, among whom was Professor P. Prevot of the IRHO (Institut de Recherches pour les Huiles et Oleagineux) in Paris. He points out the fundamental difference between the factorial method and the *methods des variants systematique* by stating that former permits an evaluation of the effect of increasing levels of one or more experimental variables, all other factors being kept constant. The latter method, on the other hand, studies the effect of the simultaneous and inverse variations of two or three factors, the sum of which is kept constant, with the intention of determining a ratio that is associated with a maximum yield of plant material. Prevot has developed his criticism of Homes's work in an article written with Ollagnier, which is cited in the references at the end of this chapter.

Limited Applications of Growth Expressions

It is obvious that there is a lack of agreement among the various concepts which purport to describe the relation between plant growth and nutrient element input. Steenbjerg and Jakobsen of Demmark have recently addressed themselves to this question. In commenting on the variability among growth response curves, they point out that: the constants in formulas are not constants because the variables in the formulas are not independent variables". Factors other than nutrient interactions obviously affect the shape of yield curves. They include other environmental factors which are discussed in the next section. The change in shape and position of yield-plant nutrient input curves with changes in environmental conditions is of the greatest importance to the practical agriculturist. The understanding of the interplay among these factors and their successful manipulation by the farm operator to bring about maximum crop production at a minimum production cost determines in the final analysis the degree of success of the farm operation.

The term ***limiting growth factor*** or more simply ***limiting factor,*** used frequently throughout this book, has been clearly illustrated by the variable nature of the response curves and surfaces previously discussed. If, for example, a crop has inadequate moisture, the application of given amount of fertilizer will provide a lower yield than if moisture were adequate. Another example, and an important one, is the application of fertilizer to a crop growing on a soil that is too acid for maximum growth, regardless of the amount of fertilizer added. If lime is not applied, acidity becomes the limiting factor that keeps yield responses to the added fertilizer low and reduces the farmer's return on his investment. The importance to practical farm operations of the concept of a growth pattern and how it may be altered by various "limiting factors" cannot be overstated. The remainder of this chapter is devoted to those factors, both naturally occurring and artificially induced, that have an impact on crop responses to added fertilizer nutrients.

FACTORS AFFECTING PLANT GROWTH

Most people intuitively feel that plants can grow only so large. But how large is that? As already pointed out, maximum growth, according to the Mitscherlich-Spillman concept, is that attained by a plant when none of the growth factors is present in less than the amount required – or a relative yield of 100 per cent. This is a circular type of definition, and unless these optimum conditions can be defined in other terms than maximum yields, how can they be established? Further still, because the maximum possible yield is not known in absolute terms, how can it be Known when such a condition has been attained?

These are intriguing questions, but it remains for posterity to supply the answers. Even though some absolute maximum yield value for different crops can not now be determined, agriculturists are becoming more aware of high yields and the factors that may limit further increases in production. The importance of understanding the relation between growth and the factors affecting it cannot be too strongly emphasized. All growth factors probably have not been identified. Those that have been defined, however, can be classified as genetic or environmental. The remainder of the chapter deals with the relation of these factors to plant growth.

Genetics

The importance of genetics in the growth of agricultural crops is shown by the large increases in yields which have come about through the use of hybrid corn and other improved crop plants. The high yielding potential and other characteristics (such as quality, disease resistance and drought hardiness) are related to the genetic constitution of the plant. Just why one variety or hybrid outperforms another is not known, but research in recent years has demonstrated the influence that the genes exert on physiological processes by a controlling mechanism on the synthesis of enzymes. Research in this and related fields will doubtless provide the answers to many of the puzzling questions concerning plant behavior.

Variety and Plant Nutrition Needs

It is obvious that a hybrid corn producing 150 bu/A. will require more plant nutrients than one producing eighty bushels. This important fact has sometimes been overlooked when a shift to the higher yielding variety is made. Under low fertility conditions a given variety may not be allowed to develop the full potential of its yielding capacity. In fertile soils the new variety will deplete the soil more rapidly and eventually yields will be depressed if nutrient supplements are not provided.

A significant amount of plant improvement research including breeding has been carried out under average fertility conditions because it was felt that such conditions would most nearly approximate those found on most of the farms. However, the fallacy of this approach has become evident. It has led to the selection of hybrids or varieties which were the most efficient extractors of plant nutrients from the soil and frequently obviated the selection of superior lines which were discarded because of an unsuitable environment, in this case low soil fertility. In a much more fruitful approach new genetic material is tested at several levels of soil fertility, which permits it to express its yielding capacity as a function of plant nutrient levels. The more promising lines can then be selected for release, additional research, and possible further improvement.

In research leading to the development of new hybrids or varieties, as well as in the use of these new lines in the farming enterprise, pest control measures must be adopted whenever necessary. Frequently, greater vegetative growth results from the use of these newer materials under higher fertility conditions, and insects, disease, and weeds may be encouraged. If such pests are not controlled, many otherwise promising high yielding materials will be discussed in a research programme or fail to perform adequately on the farm. This is another example of the limiting factor concept.

Variety-Fertility Interactions

The differential response of various crops to applied plant nutrients has been demonstrated on numerous occasions. In general varieties that have a small range of adaptation tend to show significant variety-fertilizer interactions, whereas those with a wide range of adaption do not. As early as 1922 the Tennessee Experiment Station was recommending varieties of corn on the basis of the fertility level of the soil. The varieties selected for poor soils were entirely different from those suggested for the more fertile soils.

In United States the present trend is to supply adequate plant nutrients on the low fertility soils. The recommendations for varieties or hybrids thus do not need to be made on the basis of the fertility level of the soil but rather on their ability to withstand insects, diseases, or unfavourable conditions of moisture or temperature. Soil fertility need no longer be a limiting factor.

Importance of Progressive Research in Plant Genetics

The genetic constitution of a given plant species limits the extent to which that plant may develop. No environmental conditions, no matter how favourable, can further extend these limits. It is imperative that there be a forward moving plant-breeding programme that will produce new varieties or hybrids capable of achieving maximum growth under specified environmental conditions. What has been done in the area of plant breeding and genetics is amply attested to by the crop production records kept by the better growers in this and other countries. What can be done will be determined by tomorrow's needs and the imagination and skill of geneticists and plant breeders the world over.

Environmental Factors

Environment is defined as the ***aggregate of all the external conditions and influences affecting the life and development of an organism***. Among the environmental factors known to influence plant growth, the following are probably the most important:

1. Temperature,
2. Moisture supply,
3. Radiant energy,
4. Composition of atmosphere,
5. Gas content of the soil,
6. Soil reaction,
7. Biotic factors
8. Supply of mineral nutrient elements.

Many environmental factors do not behave independently. An example is the inverse relation that exists between soil air and soil moisture or between the constant of oxygen and carbon dioxide in the soil atmosphere. As the soil moisture increases, the soil air decreases, and as the carbon dioxide content of the soil air increases, the oxygen content decreases.

Another example is the relation between the diffusion rate of oxygen in the soil and soil temperature. The partial pressure of oxygen in the root environments is extremely important to the growth of the plants. The maintenance of this pressure is related to the diffusion rate of oxygen to the root surface which in tern is influenced by the soil temperature.

"Success is where preparation and opportunity meet."

5 Crop Growth and Yield Maximization

> *"Pure science, undertaken without regard to applications to human needs, is usually ultimately of direct benefit to mankind"*
> *Irving Langmuir*

Plant growth is affected by the biological (biotic) and physical (abiotic) factors. The physical factors mainly consist of climate which accounts for greater variation in crop yields. Hence, it is important to remove all the physical factors limiting plant growth resulting in substantial increase in the yield of crops per unit area per unit time. The crop growth generally lies in the range of 15-20 g/m^2/day for crops without stress. Phytosynthetic efficiency of crop plants should be stimulated for full utilization of all biological and physiological reserves in the plants. Higher crop production is possible by increasing the biological solar energy conversion.

Crop growth and development are primarily governed by environmental conditions, i.e., **climate** and **soil.** The success and failure of crops is intimately related to the weather during the crop periods. The modification of weather, except to a limited extent, is difficult and uneconomical. However, it is possible to get high crop yields by adjusting cropping patterns and by following suitable agronomic practices to mitigate the adverse effects of weather. Weather has significant influence on every phase of agricultural activity from preparatory tillage to harvesting and storage. A sound knowledge of the climatic factors and its interactions with crop is essential for successful agriculture.

CROP GROWTH

Growth in general, may be defined as a vital process which brings about a permanent change in any plant or its part in respect to its size, form, weight and volume. Growth occurs in meristematic region, the activity of each meristem influences the activity of other meristems. The main epical shoot meristem is active and is retards the activity of more recently initiated lateral bud meristem. This phenomenon is usually referred to as **'apical dominance.'**

The higher plants, growth includes cell division, enlargement and cell differentiation. Cell differentiation is the transformation of genetically identical cells of common origin (e.g., zygote) into different tissues. This is accomplished by various structural, functional, and physiological specializations. Growth and differentiation result in development leading to characteristic form and chemical composition of plants.

The growth may be grouped into two types in higher plants: (1) **limited** and (2) **unlimited**. In the limited type (e.g., annual plant), the plant grows up to a predetermined size and stops growing any further. On the other hand, woody perennials exhibit a pattern of unlimited growth where the plant adds to its mass by growth every growing season of the year.

Growth Curve

Under favourable conditions, there is characteristic course of increase in the plant growth. If a graph is drawn (growth rate vs time), it results in a sigmoid or S shaped curve. There are three well marked regions in a standard growth curve.

1. The initial slow growing phase called the lag phase,
2. A rapid growing phase called log phase, the growth is an logarithmic phase,
3. The third phase is the period of declining growth and the organisms maintains the size it has already achieved. This is called as steady stage phase (Fig. 5.1).

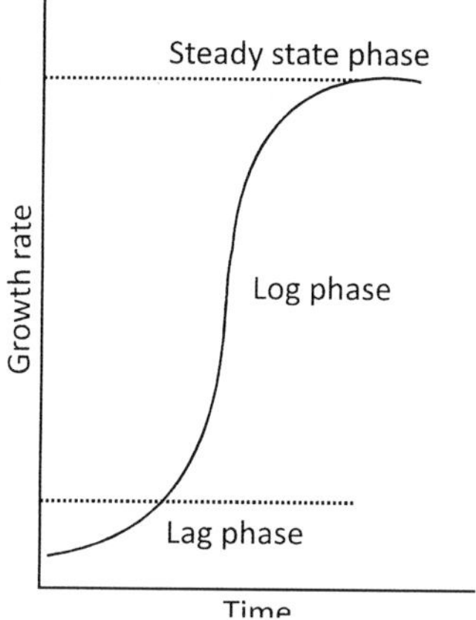

FIG. 5.1: Sigmoid curve.

YIELD MAXIMIZATION

With the best management practices under limited environment, it is possible to achieve the highest crop yield. Maximum yield achievable under a production system includes the best of all controllable factors needed to produce, the highest possible yield. The yield maximum generally varies with climate, soils and growing season. In a place where all the parameters are optimum during the a perfect growing season, maximum yield should be approximate to the maximum potential yield. In this situation, all the production factors will be interrelated.

Research in U.K. has shown that yields of 12-15 tonnes of wheat and barley, 85-100 tonnes of potatoes and 75-85 tonnes of sugar beet per hectare are possible when all the growth factors are optimized. Johnson (1982) estimated the theoretical maximum yield of maize and soybean in the United States at over 30 and 11 tonnes/ha respectively. However, the maximum yield that is achievable is much less than the theoretical maximum. The potential of a crop can frequently be increased by changes in management practices which may involve the change of a single or multiple interactions.

All the vegetation depends on solar energy, CO_2, water and nutrients and solar energy is considered as the basis for determining productivity potential. If all the factors are non-limiting, the radiation use efficiency determines the photosynthetic rate, dry matter production and yield. The highest crop yields reported in most of the field experiments in India are given in Table 5.1. However, the National average yield is much lower than the potential and the highest experimental yield.

TABLE 5.1: Highest yield of crops in India

Crop	Yield (t/ha)
Rice	14.0
Wheat	10.0
Maize	11.5
Sorghum	8.5
Pigeon pea	3.5
Bengal gram	4.5
Mustard	3.8

Source: Sinha, 1989

Maximum Economic Yield (MEY)

Maximum Economic Yield is the yield level where the net profits are maximized in response to increasing use of inputs and management. Maximum Economic Yield is slightly lower than the maximum yield and is the point where the last increment of an input just pays for itself. Maximum Economic Yield also varies with the location, economic and soil conditions. This also represents the farmer's greater potential for economic survival. It is possible to get maximum net profit by selecting cost-effective inputs and management practices to produce MEY. Each must be cost-effective when used singly but may provide the most profit from the an interaction effect. The interaction should occur without limitations, changes in fertility and cultural practices play an important part.

An integrated multiple production strategy should be worked out to achieve MEY. Choice of crop cultivars, timeliness of planting or sowing, treatment of seeds, maintenance of adequate plant population, amount and method of application of manures and fertilizers, optimum irrigation and appropriate plant protection measures are important to reap the maximum yield potential.

PLANT POPULATION

Yield of crop is the result of final plant population which depends on the number of viable seeds, germination per cent and survival rates. Establishment of optimum plant population is essential to get maximum yield. In crops grown on stored soil moisture under rainfed conditions, population should not be high to deplete most of the moisture before crop matures and not low to leave moisture unutilized. Under conditions of sufficient soil moisture and nutrients, higher population is necessary to utilize other growth factors efficiently. Once soil moisture and nutrients are not limiting, yield of crop is limited by solar radiation. The level of plant population should be such that maximum solar radiation is intercepted.

Yield of Individual Plant and Community

The full yield potential of individual plant is achieved when at wider spacing. When sown densely, competition among plants is more for growth factors resulting in reduction in size and yield of the plant. Yield per plant decreases gradually as plant population per unit area is increased (Fig. 5.2).

However, the per unit area is increased due to efficient utilization of growth factors. Maximum yield per unit area can, therefore, be obtained when the individual plants are subjected to severe competition.

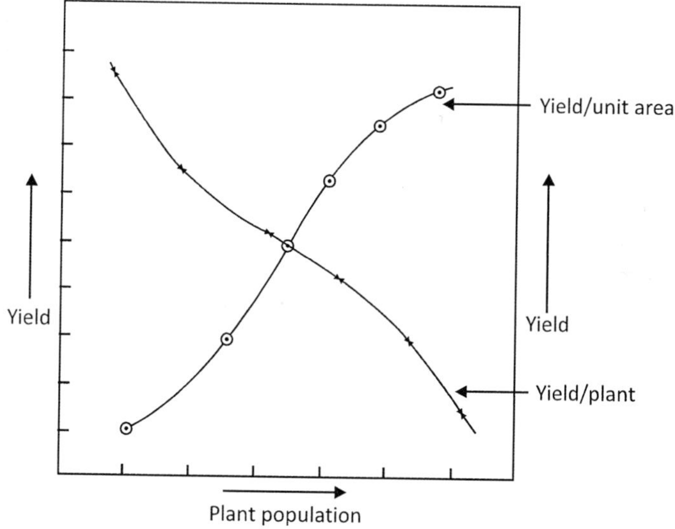

FIG. 5.2: Yield of individual plant and community as Influenced by plant population.

Plant Population and Growth

High plant density brings out certain modifications in the growth of plants. Plant height increases with increase in plant population due to competition for light. Sometimes it may happen that moderate increase in plant population may not increase but decrease plant height due to competition for water and nutrients but not for light. Increase in plant height due to higher plant population is advantageous for better light interception due to exposure of individual leaf at wider vertical interval. Another adaptation of dense plant stands is reduction of leaf thickness. Leaf orientation is also altered due to population pressure. The leaves are erect, narrow and are arranged at longer vertical intervals under high plant densities. This is a desirable architecture to intercept more light. Dry matter production per unit land area increases with the increase in plant population up to a limit when the reduction in the growth of a plant, is more than compensated by increase in the number of plants per unit area.

Plant Population and Yield

Decrease in the yield of individual plant at higher plant density is due to reduction in the number of ears in indeterminate plants. In determinate plants, wherein the terminal bud ends in a flower or inflorescence, the reduction in yield is mainly due to the reduced size of ears or panicles. Highly branching or tillering plants behave as indeterminate plants and yield reduction is due to reduction in the number of seeds. Red gram produces about 20 pods per plant at 3.33 lakhs/ha while it produces more than 100 pods per plant at 50,000 plants/ha. Conversely, non-tillering or non-branching plants produce fewer yields due to reduction in size of ears as in the case of maize and sorghum. Among all the yield attributes, test weight (seed size) is the stable character under wide range of plant population. Under very high population levels, plants become barren.

Efforts are made to quantify the relationship between plant population and yield. Hollidar (1960) suggested two types of response curves, asymptotic and parabolic.

Asymptotic Response

In a situation where entire dry matter is the economic product as in the case of fodder crops or most of dry matter as in tobacco, the response to increasing plant density is asymptotic (Fig. 5.3)

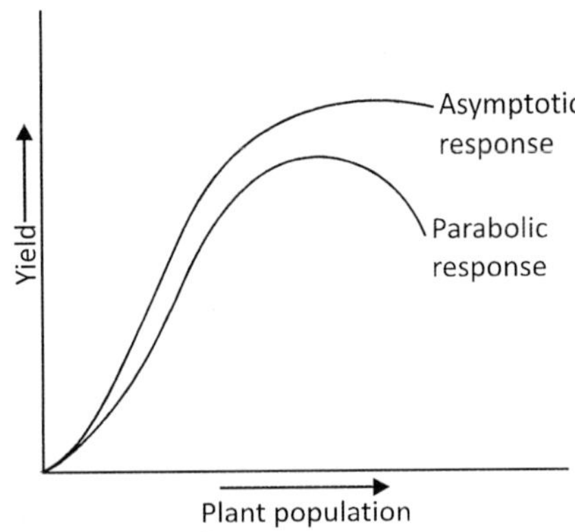

FIG. 5.3: Diagrammatic representation of asymptotic and Parabolic response curves.

When the plants are widely spaced, the dry matter of individual plant increases with increase in plant density. This indicates lack of appreciable competition between neighbouring plants. Further, increase in density. Further increase in density, increases the dry matter of individual plant at a diminishing rate, signifying that competition between plants is felt, and that the dry matter production of the individual plant is reduced. However, this reduction is more than compensated by the increase in the number of plants. A further increase in plant population results in a plateau i.e., with increase in plant population there is no increase in dry matter per unit area. It shows that increase in production due to increased plant population compensates almost equally for the reduction in the production of the individual plant. The plateau continues for large increases in plant population. This is why, for fodder crops, dense stands are recommended to get maximum yield. The extra expenditure is the seed cost only. In addition, dense stands provide for lean stems and more leafy fodder compared to sparse population where stem dry matter is high. This type of response gives the asymptotic curve which is expressed as follows:

$$Y = Ap + \frac{1}{1 + Abp}$$

Where Y is yield of dry matter/unit area; A apparent maximum yield per plant; p number of plants/unit area and b linear regression coefficient.

The term 1/(1 + Abp) represents the maxima of which the maximum plant yield (A) is reduced by increasing competition resulting from greater plant density. It is termed as competition factor. Maximum yield that is practically no competition is denoted as A or apparent maximum yield perplant.

Parabolic Response

Parabolic curve is used to describe plant population and yield relationship when the economic yield is a fraction of the total dry matter. In this case also, yield increases with increase in plant population, then reaches a maximum. However, unlike in asymptotic curve, yield decreases with further increase in population (Fig. 5.3). Holliday (1960) suggested that the curve can be fitted to the quadratic equation.

$$Y = a + bp + cp^2$$

Where Y is yield/unit area; p plant population and a, b, and c are regression coefficients.

There is a drawback in representing plant population-yield response as a quadratic function. With increase in plant population, after reaching a maximum, yield does not fall suddenly. There is a plateau foe some range of plant population. The extent of plateau depends on elasticity of plants. This aspect is not considered in the quadratic equation.

Square Root Function

The disadvantages of quadratic function can be overcome by square root function.

$$Y = a + bp + c\sqrt{p}$$

It is almost similar to quadratic function except that root of p is taken instead of square (Fig. 5.4).

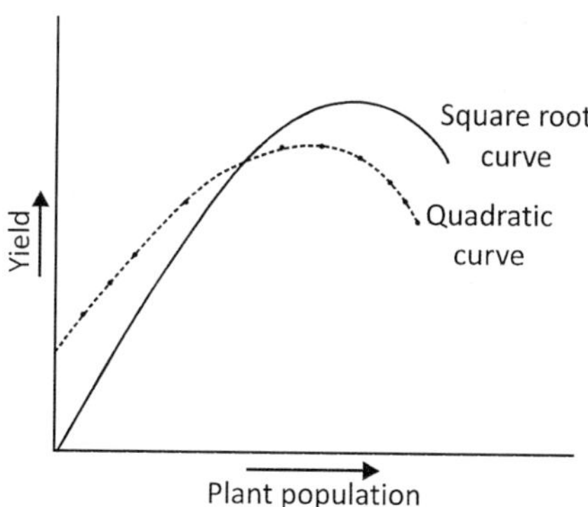

FIG. 5.4: Diagrammatic representation of square root and quadratic curve.

Reciprocal Functions

Reciprocal functions can also be used to express relation between population and yield. The increase of average yield of individual plant is directly related plant density. In case of biological yield it is expressed as follows:

$$\frac{1}{w} = a + bp$$

Where w is weight of individual plant, p population/unit area, and a and b are constants. To describe economic yield and plant density, the reciprocal equation is:

$$1/w = a + bp + cp^2$$

Where a, b and c are constants.

BIOLOGICAL AND ECONOMIC YIELD

The economic and biological yields are interrelated. When the data of these two parameters are plotted on the same graph, it can be seen that the level of population at which biological yield reaches a plateau is the level of plant population at which maximum economic yield is obtained.

In young plants, most of assimilates are used for the production of stem and leaves. As the plants enter reproductive stage, assimilates are partitioned to stem, leaves and inflorescence. Once the seeds are set, most of assimilates move to the seed. There exists an internal competition within the individual plant between vegetative and reproductive parts. This internal competition is severe under high plant population. As plant density increases, changes may occur in the allocation of assimilates to different parts of the plant due to which greater proportion of plants may become barren. Grain production then shows a decline, whereas the total dry matter production may remain constant.

Donald (1963) explained the relationship between yield and plant density with the help of inter-plant competition, intra-plant competition and the time at which competition sets in. The competition between plants is the inter-plant competition and competition between different plant components within the same plant is the intra-plant competition. These competitions set in at different periods with different levels of plant population. In the widely spaced plants or with less population density, competition is absent during early stages of growth. More number of flowers are initiated per plant and as growth proceeds, inter-plant competition sets in during fruit set and fruit development. The load of inflorescences is more which leads to competition among inflorescences of the plant. This loss of efficiency at widest spacing is evident in fewer seeds per inflorescence and reduced seed size compared with somewhat more dense stands. Thus, intra-plant competition may be intense at lower densities.

In moderately dense stands, inter-plant competition becomes operative at the time of flower initiation. The number of flowers produced is reduced and the plant is capable of filling all the seeds that set in. The seeds per inflorescence and size of seeds per unit area are more in this condition.

In extremely dense stands, the competition at the time of flower initiation is instance and plants in such community suffer both from inter-plant and intra-plant competition resulting in lower yield.

PLANT POPULATION AND GEOMETRY

It will be useful to know optimum or critical population densities to exclude population as limiting factor for crop yields.

Plant population is defined as the **number of plants per unit area**. **Geometry** is the **pattern of distribution** of plants over the ground or the shape of the area available to the individual plants. When plants are widely spaced dry matter production may first tend to increase linearly with increase in plant density due to no or minimum competition between the adjoining plants. Further increase in density may reduce the dry matter production probably due to competition between plants and the yield of the individual plants may be reduced. However, this is compensated by the increase in the number of plants per unit area. Any further increase in plant density results in a plateau of practically constant dry matter production. This indicates that equilibrium has been reached at which the increase in production due to increased plant number compensates the reduction in the production of the individual plants.

Optimum number of plants is required per unit area to utilize efficiently the available production factors such as water, nutrients, light and CO_2. Maximum exploitation of these factors is achieved when the plant population puts forth maximum pressure on all the factors of production. As a result, individual plants are put under severe stress because of inter-plant competition. Normally maximum yields are obtained from plant population which does not allow plants to achieve their individual maximum potential. Thus, the entire community of plants is considered for higher production rather than the individual plant performance.

RESPONSE OF CROP PLANTS TO POPULATION AND GEOMETRY

Modern varieties are much more responsive to closer plant spacing than traditional varieties. Though the yield per plant is lower at closer spacing, greater number of plants per ha compensates the loss in yield of individual plants (Table 5.2)

TABLE 5.2: Grain yield of modern and traditional varieties of rice

Spacing (cm)	Grain yield (t/ha)		
	Modern varieties		Traditional tall varieties
	Low tillering	High tillering	
15 x 15	6.2	7.5	4.1
25 x 25	5.7	8.2	4.5
35 x 35	5.3	7.4	4.4
45 x 45	4.2	6.9	4.3

Source De Datta, 1981.

Maize hybrid grown under irrigated condition require 8.0 – 10.0 x 10^4 plants per ha for higher production. Flannery and Woodard (1981) designed an experiment to measure the effect of maize hybrids, plant population, irrigation, macro and micro-nutrients application and their interaction on maize silage and grain yield on sandy loam soils. The maize hybrid produced 19.6 t grains/ha and 100 t silage when grown at 10 x 10^4 plants/ha. At Coimbatore, a population of 11.0 x 104 plants/ha produced the highest yield of 10.3 t/ha during *kharif* season and further increase in population did not give any beneficial yield increment.

There is also negative effect of very high population density, particularly in branching crops. In soybean, there is elongation of internodes and plant height due to competition for sunlight under dense situation, because of the etiolation. There is also lodging with increased plant density in soybean, particularly in determinate types. Increase in plant population also results in decrease in side branches and number of pods in soybean. There has also been yield reduction per plant from 1.2 g to 8,2 g when plant density increased from 8 plants to 40 plants/m^2 and this is mainly due to reduction in pod number. To maximize seed yield in determinate type of soybean varieties, 5 x 10^5 plants/ha is found optimum.

Maintenance of optimum plant population is the biggest contributor to narrow down the yield gap in high value crops like cotton. An extra yield of 5 q/ha could be achieved with the adoption of closer spacing of 75 x 60 cm in Varalaxmi hybrid cotton. However, cotton cultivars like MCU 5 and MCU 7 registered the highest yield

(24 and 25 q/ha respectively) when grown at a spacing of 75 x 30 cm. The wider spacings have decreased the plant population per ha to the extent that it adversely affected the yield.

Under dry land conditions, maintenance of adequate plant population enables insurance against yield reduction. Farmers from vertisol areas generally adopt low seed rates and plant population primarily as an insurance against rainfall failure. While such practice is well suited to non-responsive traditional varieties, it results in below-optimum yields in the case of newly available improved crop varieties, particularly at high levels of inputs and stored moisture. The highest yield with improved *rabi* sorghum genotype was obtained at 1.35 lakh plants/ha under favourable moisture conditions in Bijaphur. In the year of low rainfall and poor soil moisture, however, plant stands higher than 50,000 or 60,000/ha exerted negative influence on sorghum yield even in the case of improved genotypes. In contrast to sorghum, safflower showed no response to plant density over a range of 3 to 5 plants/m^2 owing to its multiple branching habit.

Calculation of Plant Population

Plant population can be easily calculated for a given crop if the spacing is known. For example, maize is planted with a spacing of 60 x 20 cm. The plant population is calculated as follows:

Area occupied by a single plant	=	0.60 x 0.20 m = 0.12 m^2
Population for 0.12 m^2	=	1 plant
For one hectare or 10000 m^2	=	10000/0.1112 = 83333 plants/ha

OPTIMUM PLANT POPULATION

Moderately dense stands or optimum plant population are necessary to obtain maximum yield. Optimum plant population depends on size of plant, elasticity, foraging area, nature of the plant, capacity to reach optimum leaf area at an early date and seed rate used.

Size of Plant

The spread or the volume occupied by the plant at the time of flowering has influence on the spacing to be adopted for these crops. Plants of red gram, cotton, sugarcane etc., occupy larger volume of space in the field compared to plants of wheat, rice, finger millet etc. Even the varieties of the same crop differ in the size of the plant. In red gram, the cultivar LRG-30 grows to a height of 1.5 to 2.0 m with a horizontal spread of 1.0 to 1.5 m while the average height and spread of ICPL 87 is 70 and 30 cm respectively.

Elasticity of the Plant

Variation in size of plant between the minimum size of the plant can reach under unlimited space and resources is the elasticity of the plant. LRG-30 red gram can produce a few pods when the plant attains a minimum size of 20 g dry weight but it can attain a size to produce a dry weight of 2,000 g per plant. Instead of the weight of the plant, it is more meaningful to consider elasticity in the number of branches and pods per plant. The elasticity of the plant for these characters can be estimated by co-efficient of variation (CV). The higher the CV more is the elasticity of the plants. The elasticity of red gram for branching and number of pods are 30 and 80 per cent respectively. Elasticity of growth and yield characters of plants are higher in indeterminate and long duration crops. The optimum plant population range is quite high for indeterminate crops. For indeterminate red gram varieties, the optimum population ranges from 55 to 133 thousand plants/ha. The elasticity of plants is due to branching or tillering. For determinate plants, the elasticity is less and the optimum plant population range is small as in maize, sorghum etc. In castor, closer spacing is followed in combination with nipping of the auxiliary buds to get uniform and early maturity. The removal of auxiliary buds reduces the elasticity of the plants.

Soil Cover

The crop should cover the soil as early as to intercept maximum sunlight. The intercepted solar radiation and dry matter production are directly related. Light energy is instantaneously available and it has to be instantaneously intercepted. Closely spaced soybean crop which attained 95 per cent light interception within 54-55 days gave 26-32 per cent higher yield compared to wider spaced crop.

Dry Matter Partitioning

Dry matter production is related to the amount of solar radiation intercepted by the canopy. As plant density increases, the canopy extends more rapidly, more radiation is intercepted and more dry matter produced, especially during the early stages before the canopy closes (Fig. 5.5)

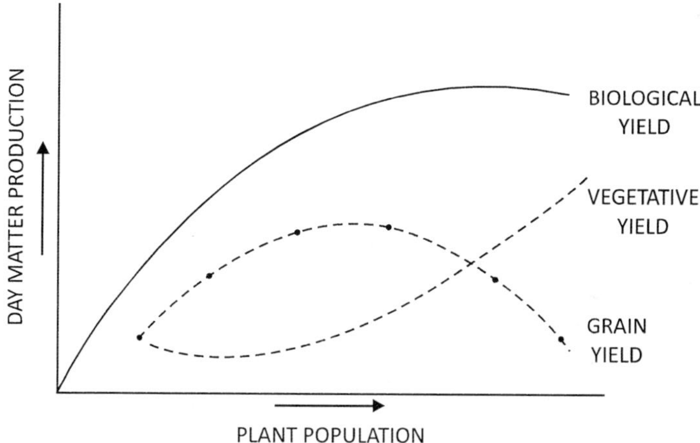

FIG. 5.5: Diagrammatic representation of the relationship between plant population and dry matter production.

Each plant must have certain amount of dry matter in vegetative tissue before any assimilate is allocated to fruiting structures. As density increases, the amount of dry matter in vegetative parts also increases. With regard to the biological yield, it increases with increase in plant population up to a point and with further increase in plant population no additional biological yield can be obtained. Economic yield increase in plant population up to a point and subsequently decreases with increase in population.

OPTIMUM PLANT POPULATION AND ENVIRONMENT

Optimum plant population for any crop varies considerably due to environment under which it is grown. It is not possible to recommend a generalized optimum plant population since the crop is grown in different seasons with different management practices.

Time of Sowing

The crop is subjected to different weather conditions, when sown at different periods. Among the weather factors, the most important factors that influence optimum plant population are day length and temperature. Photo-sensitive varieties responds to day length resulting in change of size of the plant. Red gram plants sown as winter crop will have half the size of those grown in monsoon season. Optimum plant population is 55,000 plants/ha for monsoon season. Crop of re gram and this is increased to 3.33 lakh plants/ha for winter crop. As low temperature retards the rate of growth, higher population is established for quicker ground cover. In sorghum, when the climate is favourable during pre-an thesis period, the optimum population is two lakh plants/ha and when it is not congenial for growth during pre-an thesis, it is four lakh plants/ha.

Irrigation

It is well known that plant population has to be less under rainfed conditions compared to irrigated conditions. However, this is applicable mostly to crops grown on residual soil moisture and to some extent in seasons where prolonged dry spells between two rains occur. Under these conditions, evaporation is negligible as the surface soil is dry and transpiration is the main component in evapo-transpiration (ET) and with higher plant population, transpiration is more resulting in stress in the last phase of the crop. Under adequate soil moisture, ET is similar with different plant densities. Under higher plant densities, more water is lost through transpiration than through evaporation as soil temperature is less due to shading of plants. Under adequate irrigation or under evenly distributed rainfall conditions, higher plant population is recommended.

Fertilizer Application

Dense plant stand is necessary to fully utilize higher level of nutrients in the soil to release potential yields. Nutrient uptake increases with increase in plant population. Higher plant population under low fertility conditions leads to development of nutrient deficiency symptoms. For example, rice does not respond to plant population without N application. With application of a moderate dose of N (50 kg/ ha)., higher plant population gives more yield than lower plant population. However, by the application of higher dose of N (100 kg/ha), even moderate plant densityalso gives good yield due to its tillering nature. In non-tillering plants like maize, higher the fertility, more should be the plant population to get higher yield. Generally, close planting of the crop reduces weed infestation. But it creates problem for subsequent weeding operation with implements. Wider row spacing facilitates the use of inter cultivation implements. Increasing plant population increases the proportion of ears or fruits in the upper layer of the canopy which facilitates case in harvesting. Cotton sown with a close spacing, produces mostly sympodial branches and most of the bolls appear in the top layer of canopy. Similarly, with higher population, rice panicles appear in the upper layer of the canopy.

PLANTING PATTERN

Planting pattern influences crop yield through its influence on light interception, rooting pattern and moisture extraction pattern. Different planting patterns are followed to suit different weed control practices and cropping systems. Plant geometry refers to the shape of plant while crop geometry refers to the shape of space available for individual plants. Crop geometry is altered by changing inter- and intra-row spacing.

Square Planting

It is responsible to expect that square arrangement of plants will be more efficient in the utilization of light, water and nutrients available to the individual plant than in a rectangular arrangement. In wheat, decreasing inter-row spacing below the standard 15-12 cm i.e., reducing rectangularity, generally increases yield slightly. In crops like tobacco, inter cultivation in both directions is possible in square planting and helps in effective control of weeds. However, square planting is not advantageous in all crops. Groundnut sown with a spacing of 30 x 10 cm (3.33 lakhs/ha) gave higher pod yield than with some amount of population in square planting. Pod yield is reduced either by increasing rectangularity or approaching towards square planting.

Rectangular Planting

Sowing the crop with seed drill is the standard practice. Wider inter-row spacing and closer inter-row spacing is very common for most of the crops, thus attaining rectangularity. The rectangular arrangement is adopted mainly to facilitate inter cultivation. Sometimes only inter-row spacing is maintained and intra-row spacing is not followed strictly and seeds are sown closely as solid rows.

Miscellaneous Planting Arrangement

Crops are sown with seed drill in two directions to accommodate more number of plants and mainly to reduce weed population. Crops like rice finger millet are transplanted at the rate of 2-3 seedlings per hill. Transplanting is done either in narrow or randomly. Shipping of every alternate row is known as **skip row planting.** When one row is skipped, and the population is adjusted by decreasing intra-row spacing, it is known as **paired-row planting.** It is generally resorted to introduce an intercrop.

GAP FILLING AND RESOWING

It is common that gaps exist in rows of the crop due to several reasons like poor quality seed, soil crustation, very shallow or very deep placement seed etc. Gape filling is done to maintain population. However, gap filling is not advantageous for short duration crops. Groundnut emerges from the soil 5-7 days after sowing and gaps are known only them. Though the gap filled plants are delayed by seven days at sowing, they are delayed by more than 20 days at maturity due to severe competition they face by the adjacent plants. They do not contribute to yield. The plants around the gaps grow better if gap filling is not done and compensate for the loss. If the gaps are more than five at a stretch, or when the entire row is missing, then there is possibility of some advantage by gap filling. However, gap filling is advantageous in long duration and widely spaced crops. In most environments, only stand reduction of over 50 per cent needs to be replanted. Weed control in the fields with over 40 per cent random stand reduction may be difficult.

THINNING

In certain situations where germination and emergence is a problem, high seed rate is used. During favourable season, this cause very high population and left as it is, results in weak seedlings due to overcrowding. Low yields are obtained due to severe competition. Barren plants are also observed under excess population.

Thinning is needed to obtain optimum population by removing excess population. However, thinning is expensive and a time consuming operation.

SEED RATE

The required number of plants/unit area is decided by calculating the seed rate. The seed rate depends on spacing or plant population, test weight and germination percentages. It is calculated by the formula:

Seed rate (kg/ha) = Spacing x test weight x germination (%) x 1,00,000

The seed rate for a particular crop would depend not only on its seed size/ test weight, but also on its desired population, germination percentage and purity percentage of seed. It is calculated as follows:

Area to be sown:

$$\text{Seed rate (kg/ha)} = \frac{(1000\text{m}^2) \text{ x Test weight x } 100 \text{ x } 100}{1000 \text{ x } 1000 \text{ x Germination (\%) x Purity (\%) x Spacing}}$$

LINEAR PROGRAMMING

Linear programming is a budgeting technique that is more refined and systematic than the conventional budgeting method. It is a technique that uses computational procedure.

Linear Programming *vs*. Budgeting

Similarities: Both in linear programming and budgeting it is assumed that input-output coefficients, prices and restrictions are certain. Both use the concept of linearity and take into consideration risk and certainty in alternative plans. Both are concerned with determination of the optimal solutions to problems.

Differences in Use: Budgeting technique can gainfully be employed when simple decision problems are involved. For example, it may be taken the solution of a single problem of determining the optimum use of fertilizer, choice of competing enterprises or organizing a farm for profitable production. Linear programming, a mathematical counterpart of budgeting, has the capacity to process large quantities of data and is to be preferred to the determination of representative optimal farm plans for regional comparison. Further, if the number of restricting resources and possible enterprises are large, linear programming will provide precise answers. In linear programming we get the marginal value products of resources and the range could expand or contract the level of product.

Linear Programming: The term linear connotes **'straight line'** relationships employed in linear programming. It is an interesting mathematical discipline that can be substituted for some of the portions of arithmetic. Its theory can be developed without using calculus.

Definition: Dorfman et al, 1958 define linear programming as "the analysis of the problems in which a linear function of a number of variables is to be maximized (or minimized) when those variables are subject to a number of restraints in the form of linear inequalities".

Requisites of Linear Programming

The following quantitative components are needed for the application of linear programming to the solution of a problem.

1. **An Objective:** Profit maximization or cost minimization or any other objective.
2. **Resource and Non-negativity Restrictions:** Imitational factors and resources must exist. None of the activities can be carried out at a negative level. Both are logical restrictions.
3. **Production Process:** Production processes or activities must be defined that conform or nearly conform to the assumptions of independence, linearity and divisibility. Further, they must be subject to input-output measurements and pricing.

Basic Assumptions

The following assumptions are inherent in linear programming:

(1) Linearity, (2) Divisibility, (3) Additivity, (4) Fineteness and (5) Single value expectations.

Programming Techniques

There are different techniques used in order to prepare the optimal farm plans.

Linear Programming

It is more systematic and accurate method of determining mathematically the optimum combination of enterprises or inputs so as to maximize the income or minimize the cost within the limits of available farm resources. Linear programming (L.P.) was developed by G.B. Danzig to enhance efficiency in military planning during World War II. There are two areas in which L.P. is commonly used in agricultural decision making.

1. Whole Farm Budgeting
2. Cost Minimization.

Elements of Linear Programming

Maximization/minimization of a linear function of variables, that is subject to

i) linear inequalities
ii) Must assume non-negative levels.

Our objective function must be maximized or minimized, Say:

$$\text{Max. } z = \sum_{J=1}^{n} p_i x_j$$

Where

z = Profit function

X_j = level of j th activity say acreage under wheat, rice, dairy etc ≥ 0

p_j = net return from one unit of j th activity.

The linear inequalities determine the limits within which we have to maximize/minimize the objective function. Thus, the necessary condition is that total resource requirements of different activities should not exceed its availability and the resource requirements equation should be linear.

$$\text{e.g. } \sum_{i=1}^{n} a_{ij} x_j \leq b_i$$

where a_{ij} = Requirement of i th resources for each unit (say an acre) of jth activity

x_j = level of jth activity

b_i = total availability of ith resource

Foe example land available = 10 acres

$lx_1 + 1x_2 + 1x_3 + \ldots\ldots \leq 10$

Use of land for each crop per acre is one acre and the total requirement should be less than total available land i.e. 10 acres.

All the activities should appear at non-negative level i.e. $x_j \geq 0$

Assumptions of Linear Programming

1. **Linearity:** It means absence of interaction between resources and activities i.e. every unit increment of input adds equal increment to output.
2. **Divisibility/Continuity:** It is possible (i) to produce fractions of output, (ii) to use fraction of resources but in some cases it is not possible to use resources in fraction e.g. 6.4 buffaloes can be understood but can not be used. Similarly, it is practically not possible to put 1.03 acres under maize.
3. **Additively/Independence**: Enterprises or activities are assumed to be independent of each other. For instance resources required by one enterprise are independent of whether other enterprise is included or irrespective of the sources allocated to other enterprises. But, in actual farm situations, however, there are many dependent activities, e.g. in peculiar crop rotation the input requirements of crops are dependent upon each other.
4. **Finiteness**: It means that resource constraints should be limited i.e., opportunities open to the farmers should not be unlimited.
5. **Single Value Expectations:** It means that all the coefficients in linear programming are known with certainty.
6. **Short Term Static Model:** Linear Programming is a static model. It provides the optimum values under given conditions at a point of time.

Matrix Construction

There are two basic parts of the matrix:

i) Resources or resource constraints: constituting rows.
ii) Alternative or activities: constituting columns.

The nature of various resource constraints i.e. mutually exclusive, manually inclusive needs to be very carefully considered. The activities are generally four types:

1. **Real or Production Activities:** Real or production activities are those which constitute the principal alternatives of planning framework. In column of real activity its per unit (acre) requirements against various constraints are placed, e.g. land required is 1 acre.
2. **Disposal Activities:** These activities are introduced to make equality.
3. **Hiring/Borrowing Activities:** These are introduced for those resources which can be supplemented. For example, a farmer is hiring labour or borrowing capital (which is scarce) till its MVP is greater than the acquisition cost.

4. **Transfer Activities:** These are introduced in model to take care of certain aspects such as transfer of stock resources from one period to another e.g. transfer of surplus *kharif* capital in *rabi* season.

   We write + 1 for what is transferred and -1 against which is not transferred.

Post-optimality Analysis: It is the analysis made after getting the optimal solution. This will give information regarding:

a) **Value of additional resources:** It is very important in making decisions. The value of an additional unit of resource is also called as shadow price or marginal value product.
b) The amount of change in net income of each enterprises needed to make current optimal plan no loner optimal e.g. under what price conditions will it pay him to grow more wheat and less gram. It depends on relative profits. How large a change in net revenue is needed to make a non-optimal enterprise part of the optimal plan can also be examined.

Problems of Linear Programming

1. It does not take into account economics of size or diminishing marginal returns e.g. cost of growing per acre maize is same for one acre and 20 acres.
2. It does not account for risk because solution is often based on price information of a given year without considering the variations in the expected return.
3. The technique is designed to maximize income and may leave out other objectives or goals of the farmers unless they are carefully built in.
4. One must differentiate the various resources according to their productive characteristics otherwise results may be unrealistic.
5. It is costly method in terms of both computer time and trained manpower, when compared with budgeting.

Integer Programming

One of the assumptions of linear programming problem is that the variables are continuous i.e. variables can attain fractional as well as integer values. The assumption is unrealistic for certain type of problems e.g. livestock. So in Integer Linear Programming the value of the variables are constrained to be discrete and entirely integer non-negative values.

Dynamic Programming

The method was developed in 1950 through work of Richard Bellman. It is a multi-variable optimization problem decomposed into a series of stages, optimization being done at each stage w.r.t. one variable only. It may also be called as recursive optimization. It can be used for both discrete and continuous variables and deterministic as well as stochastic models.

Linear Relationship in Farming

It may be pointed out that farming is mostly subject to the law of diminishing returns, where linear relationships are represented by constant returns. However, constant returns can be seen in certain spheres of farming. For example, field operations – ploughing, sowing or harvesting involve linear relationship between inputs and outputs over all the hectares of a farm. Each added hectare operated would require equal number of labourers, bullocks and field operations and add equal increments to output. Similarly, the technique can be used in poultry production or mixing cattle feed rations to obtain maximum income or minimum cost.

Example: An extremely Simplified Problem:

Problem: A farmer has 4 hectares of land and 375 human labour days available with him. with these two restrictions, there is an additional restriction that he can not grow maize on more than 2 hectares because of the physical condition of the soil. The other crop considered is jowar. His objective is to maximize net income.

The following information is available from experimental farms with regard to the resource requirements per unit of crop land and net income per hectare.

Resource Requirement/Ha

Resource	Jowar	Maize	
Land (in ha)	1	1	Let X_1 be the area under jowar and X_2 be the area under maize
Labour days	100	112	
Net income (in Rs.)	200	400	

Algebraic Procedure

$$\text{Land - } X_1 + X_2 \leq 4 \text{ hectares}$$
$$\text{Labour} - 100\,X_1 + 112\,X_2 \leq 375 \text{ days}$$
$$\text{Profit - } 200\,X_1 + 400\,X_2 :$$

Profit Maximization

Profit will be maximum if are under X_2 (maize) is maximum but we cannot put more than 2 hectares under it as the physical condition of the soil does not permit.

Since, X_2 = 2 hectares, we can have,
X_2 = 2 hectares only

- 2 hectare of maize will utilize 2x112 days = 224 days. Now the days
- available for jowar are 375-224+151 days

∴ 100 days required for 1 hectare of jowar
∴ 151 would be needed for 1.51 hectare of jowar

- Now, the area under X_1 and X_2 is (1.51+2.00) =3.51 hectares
- So, the area under fallow would be = 0.49 hectare
- Total = 4.00 hectares

Therefore, maximum profit = Rs. 200 x 1.51 + 2 x 400 = Rs. 1102/-

Programming for Economic Use of Fertilizers

In commercial production of crops it is always essential for the cultivators to know the cost-benefit ratio of the fertilizer application; however, fertilizer use efficiency depends upon several factors. But all the factors being constant, the additional yield with every added unit of fertilizer to the soil makes a basis to determine the most economic or optimum dose of fertilizer which a farmer should apply for growing a crop. Application of a unit of fertilizer is economical, if the value of the increase in the crop yield due to the quantity of fertilizer added is greater than the cost of fertilizer used. If a unit of fertilizer does not increase the yield enough to pay for its cost, its application will not be economical and will not add the net profit even after a constant increase in yield. While deciding a balanced fertilizer scheduling one should consider following points:

How much Fertilizer to Apply

Roughly it depends upon the soil test values but it is always better to consider the response pattern and the price of added dose and the value of the produce so obtained will make a basis to decide the most optimum dose of fertilizer. From the following example given in the Table 5.3 the optimum dose of nitrogen, phosphorus and potash can be worked out:

TABLE 5.3: Determining Optimum Level of Fertilizer Application in Dwarf Rice

Dose	Yield (q/ha)	Marginal yield (q/ha)	Marginal Return (Rs.)	Total Return (Rs.)	Total Cost (Rs.)	Marginal Cost (Rs.)	Return per ha (Rs.)
A. Nitrogenous Fertilizers:							
0	10.0	0	0	2500	0	0	2500
30	22.0	12	3000	5500	192	192	5308
60	30.0	20	5000	7500	384	192	7116
90	60.0	30	7500	15000	576	192	14424
120	55.0	-5	-1250	13750	768	192	12982
150	40.0	-15	-3750	10000	960	192	9040
B. Phosphatic Fertilizers:							
0	10.0	0	0	2500	0	0	2500
30	25.0	15	3750	6750	300	300	3450
60	35.0	25	6250	8750	600	300	8150
C. Potassic Fertilizers:							
0	10.0	0	0	2500	0	0	2500
20	26.0	16	4000	6500	40	40	6460
40	36.0	10	2500	9000	80	40	8920

The above table indicates that the farmer can use N_{90} P_{60} K_{40} because nitrogen beyond this dose shows decline in yield and similar is the case of P and K which show a reducing trend in additional yield obtained from added dose.

Optimum Time of Fertilizer Application

Time of fertilizer application plays a vital role in profitability of the farmer because the fertilizer use efficiency is greatly influenced by it. The Table 5.4 given below indicates the appropriate time of fertilizer application.

TABLE 5.4: Optimum Time of Fertilizer Application in Hybrid Maize

Time of application	Grain yield in q/ha	
	N_{150} P_{80} K_{100}	N_{120} P_{60} K_{80}
1. Entire dose as basal	38	32
2. Half N + P and K as basal, remaining half N at knee high stage of crop	55	48
3. 1/3 N + P and K basal, 1/3 N at knee high stage and 1/3 N at tasseling stage	60	58

The grain yield mentioned in the above table shows that the time of fertilizer application has its great influence even within the same dose. It further, indicates that in the case of light textured soils lower may give better results when applied in three splits than higher doses applied in two splits. Thus, by applying fertilizers at an appropriate time one can increase fertilizer use efficiency and lower doses can be equally effective.

Right Method of Fertilizer Application

Fertilizer use efficiency and thereby the profitability is found to be greatly influenced by various methods of fertilizer application which the farmers adopt in growing crops. In general, it has been found that the farmers prefer broadcasting over placement because the broadcasting is quick, cheaper, easy and less expensive. However, the nutrient recovery is found to be very poor. Contrary to this, the placement method gives better results even at lower level of fertilizer application which is due to higher recovery of nutrients. Thus, by placement the rate of fertilizer application can be reduced.

Besides these, there are some other methods also by which the most economic dose of fertilizers can be calculated. These methods are explained as under:

Production Function

Production function is a measure of the relationship between the yield of crops and input used to produce that yield. Thus, it helps in establishing the levels of output for corresponding levels of inputs. For producing crops we use more than one input viz., land, labour, seed, water, machines, farm animals, pesticides, fertilizers etc. Therefore, the production function is obviously calculated by using following equation:

$$Y + f(x_1, x_2 \ldots\ldots x_8)$$

Here Y is the crop yield, x_1, to x_8 are various inputs used to get Y yield and f is the relationship between inputs and outputs. The production function may be calculated by the following methods:

(a) Cobb-Douglas Function: The Cobb-Douglas function is very frequently used to determine the production function by the help of following equation:

$$Y = AX_2^{b1} X_2^{b2}$$

Where Y is output, x_1 and x_2 are the two inputs, A is a constant, and b_1 and b_2 are the elasticities of the production.

According to **Cobb-Douglas** the sum of the elasticities ($b_1 + b_2$ in two variable cases) will indicate the nature of returns to the scale which makes it clear that:

i) $b_1 + b_2 = 1$ shows constant returns to the scale i.e., an equal proportionate increase in the return with the application of both the inputs. Thus, x_1 and x_2 will result in the same proportionate change in Y (return).
ii) $b_1 + b_2 = 1$ indicates increasing returns to the scale, whereas,
iii) $b_1 + b_2 = 1$ shows decreasing returns.

(b) Mitscherlich Spillman's Production Function: According to Mitscherlich Spillman the production function may be calculated by using following equation:

$$Y = A(1\text{-}B^{x1})(1\text{-}\mu^{x2})$$

Where Y = output, A is a constant factor showing the maximum response from increasing both factors, B and μ are ratios by which the marginal products of x_1 x_2 decline.

(c) The Production Function may be calculated by using **Quadratic Production Function** for analysis of data and drawing conclusions:

$$Y = a + b_1x_1 + b_2x_2 - b_3x_1^2 - b_4x_2^2 + b_5x_3x_2$$

Marginal Product

This refers to the additional unit of output by the use of an additional unit of input where all other inputs remain unchanged. Under this measure, the marginal product of fertilizer will be the number of extra-kilograms of food grains produced by the application of an extra-kg of fertilizer nutrient (N, P, K). The marginal product is also referred as the incremental response ratio of fertilizer use. Mathematically this may be easily calculated by partially determining the function of a relevant input viz., in the case of the quadratic production function as explained earlier:

$$Y = a + b_1x_1 + b_2x_2 - b_3x_1^2 - b_4x_2^2 + b_5x_1x_2$$

The marginal products of x_1 and x_2 inputs can be estimated separately by using following equation:

$$\text{MP of } x_1 = \frac{Y}{x_1} b_1 - 2b_3x_1 + b_5x_2$$

and marginal product of

$$x_2 = \frac{Y}{x_2} b_2 - 2b_4x_2 + b_5x_1$$

Marginal Cost and Marginal Revenue

This refers to the additional cost incurred and the additional income earned respectively by producing an additional unit of output. The farmer gets maximum profit when the marginal revenue becomes equal to that of marginal cost as explained below:

$$P(Y) = R(Y) - C(Y)$$

Where P is profit, Y is output, R is revenue and C is cost.

Therefore, we have to set $\frac{dp}{dY} = 0$ for maximizing profit P which may be calculated as under:

$$\frac{dP}{dY} = \frac{dR}{dY} - \frac{dC}{dY} = 0$$

$$\text{Or } \frac{dR}{dY} = \frac{dC}{dY}$$

where $\frac{dR}{dY}$ is marginal revenue and $\frac{dC}{dY}$ is the marginal cost.

Suppose a farmer applies 100 kg N and gets 50 quintals of wheat. If by using an additional kg of N through urea which costs Rs. 6.35 and he gets as extra yield of 10 kg wheat valued Rs. 36.50 then obviously farmer can use more nitrogen. He will go on applying until has marginal revenue balances his marginal cost of fertilizer.

According to other school of thought there are three major concepts for making fertilizer recommendations, however, there is little difference in conceptual belief. These concepts are maintenance, cation saturation ratio and sufficiency level.

Maintenance Concept

Crop plants keep on removing plant nutrients during their life span irrespective of soil fertility status. This necessitates application of supplemental dose of nutrients to replace that amount removed by the crop plants so that there is no depletion in soil fertility. This concept is known as maintenance concept. It has little practical utility hence is not in practice.

Cation Saturation Ratio

Nutritional experiments reveal that there should be ideal cation ratio for achieving best results hence nutrients are recommended to maintain the required cation. According to this thought an ideal soil should have following composition of exchangeable cations:

Calcium	65%
Magnesium	10%
Potassium	5%
Hydrogen	20%

The ideal ratios of these cations should be as follows:

Ca: Mg-	6.5: 1
Ca: K	13.0: 1
Mg: K	2: 1

Thus, any nutrient falling short of the above ratios needs to be applied and the recommendation is made accordingly. However, wide variations in these ratios have no adverse effect, therefore, it has no practical utility.

Sufficiency Level Concept

This concept is based on nutrient use efficiency or crop response to applied doses of nutrients. Presently this concept is popularly adopted while making recommendation of fertilizers. There are several approaches in this concept with their relative merits and demerits fertilizer recommendations.

Blanket Recommendation

It is not scientific because in this method we do not consider the status of inherent soil fertility. We normally consider the nutrient requirement of any crop and the same is recommended to be supplied through fertilizers without taking residual values into account. In this method, sometimes, there is gradual build up in P and K status into the soil whereas N is usually lost from the soil system. Thus, the fertilizer use efficiency is reduced and farmers can not get better profit.

Response Equation

Different crops and their improved varieties are tested at various nutrient levels in all possible climatic zones and the response of these plant types is compared with local standard practices having blanket application by farmers. Mathematical equation of nutrient doses and their corresponding yield increases in the form of a curve to find out most economic dose of fertilizers to be recommended for including into package of agronomic practices.

Curve Fitting: Usually crop response to fertilizer doses is quadratic means yield increases with increase in the rate of fertilizer application up to a certain level beyond which the rate of increase in the yield with every additional dose of fertilizer is reduced. Thus, the yield reaches to a peak or plateau at a particular level of fertilizer and further increase in fertilizer dose results in a reduced yield which is considered to be toxic level (Fig. 5.6):

Figure given makes clear that nearly one ton yield of wheat grains may be obtained with inherent soil fertility. There is linear response becomes curvilinear (from B to C) which indicates that wheat yield is in decreasing trend with every increase in fertilizer dose. Beyond 90 kg N/ha, the grain yield does not increase in N beyond 110 kg/ha is considered to be toxic level.

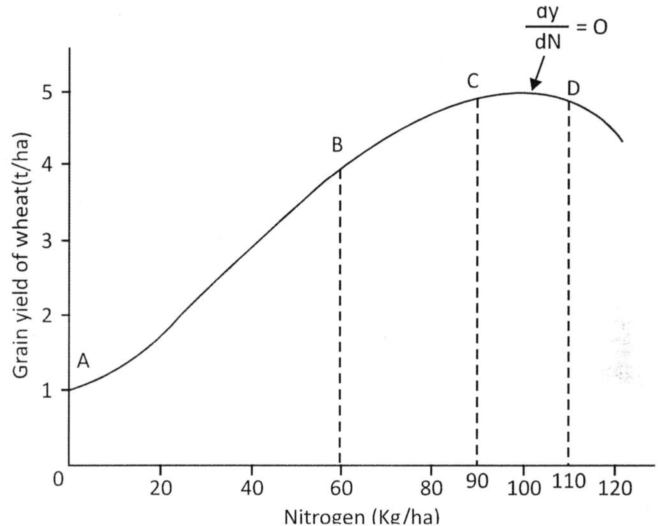

FIG. 5.6: Response curve of wheat to nitrogen doses.

Mathematically this curve may be expressed into following equation:

$$Y = a + bN + cN^2$$

Where Y is grain yield (kg/ha), N is nitrogen dose (kg/ha) and a, b and c are constants. Of these a is intercept i.e., yield without application of N. Constant b (slope) indicates response rate (kg grain/kg of applied N per hectare) where as constant c represents the curvature of the response line showing an adverse effect of excess dose of N which has usually a negative sign. These constants can be found out by using following equations.

On summation, $Y = a + b\,N + cN^2$ becomes

$$Y = n\,a + b \sum N + c\,N^2 \qquad (1)$$

By multiplying the equation with N

$$YN = a\,N + b\,N^2 + c\,N^2 \qquad (2)$$

By multiplying this with N again

$$Y\,N^2 = a\,N^2 + b\,N^2 + c\,N^2 \qquad (3)$$

...Where n stands for number of observations recorded.

DETERMINATION OF PHYSICAL OPTIMUM

The response rate d Y/ d N (as mentioned in the curve) is found to be higher at lower doses of applied nitrogen (N). Gradually, this rate decreases and becomes zero at some level of applied nitrogen beyond which if N level is in creased d Y/d N becomes negative. Therefore, the maximum grain yield can be obtained by increasing N doses up to the level where response rate becomes zero as expressed below:

$\frac{dY}{dN} = 0$ (at this the grain yield will be maximum).

DETERMINATION OF ECONOMIC OPTIMUM

It depends on the current market price of the grain as well as fertilizers. Based on this approach the farmer can go on applying nitrogen till he gets the highest return per rupee investment in terms of fertilizers. This may be calculated by using following equation:

$\frac{PN}{PY}$ Where P is price, N is nitrogen and Y is physical yield.

This may be made clear by assuming the price of wheat grain as Rs. 4/kg and price of N as Rs. 8/kg then the price ratio will be worked out as:

P N/P Y or 8/4 which shows that the yield of wheat should be at least 2 kg/kg of applied nitrogen (N). Thus, the farmer can go on applying nitrogen at the till this value becomes zero i.e., P N/P Y = 0, beyond which he will be in loss. Therefore, it may be concluded that narrower the ratio, greater will be the profitability and vice-versa.

SOIL TEST CROP RESPONSE (STCR) APPROACH

This is also called as Rationalized Fertilizer Prescription approach in which inherent soil fertility and yield level of the crop are taken into account while recommending the fertilizer doses. Fertilizer recommendations are made for different crops and different zones separately. Fertilizer recommendations based on such equation has been given as under for rice crop:

$$
\begin{aligned}
FN &= 4.39\ T - 0.6723\ SN \\
FP_2O_5 &= 2.83\ T - 6.110\ SP \\
FK_2O &= 1.41\ T - 0.329\ SK
\end{aligned}
$$

The soil phosphorus and potassium in the equation are considered in elemental form while fertilizer P and K are in oxidized forms as P_2O_5 and K_2O.

In the equation FN stands foe fertilizer nitrogen, FP_2O_5 stands for fertilizer P_2O_5, FK_2O stands for fertilizer K_2O, SN = soil nitrogen, SP = soil phosphorus, SK = for soil potassium and T is targeted yield of rice.

Example: Calculate fertilizer doses to be recommended for the one hectare rice crop by using following STCR values.

Targeted yield (T) .... 6 tones or 60 quintals
Soil nitrogen (available) .... 250 kg/ha
Soil P_2O_5 (available) .... 35 kg/ha
Soil K_2O (available) .... 350 kg/ha

Solution: Since the P and K are considered in elemental forms in the equation, therefore, the P_2O_5 and K_2O are to be converted into elemental forms as given below:

$$\begin{aligned}
\text{Soil P} &= 35 \times 0.43 = 15.05 \\
\text{Soil K} &= 350 \times 0.83 = 290.5 \\
\text{F N} &= 4.39\ \text{T} - 0.6723\ \text{SN} \\
&= (4.39 \times 60) - (0.6723 \times 250) \\
&= 263.40 - 168.07 \\
&= 95.33\ \text{kg N/ha} \\
\text{F } P_2O_5 &= 2.83\ \text{T} - 6.11\ \text{SP} \\
&= (2.83 \times 60) - (6.11 \times 15.05) \\
&= 169.80 - 91.95 \\
&= 77.85\ \text{kg } P_2O_5\text{/ha} \\
\text{F } K_2O &= 1.41\ \text{T} - 0.329\ \text{SK} \\
&= (1.41 \times 60) - (0.329 \times 290.5) \\
&= 84.60 - 95.57 \\
&= -10.97\ \text{kg } K_2O\text{/ha}
\end{aligned}$$

The fertilizer recommendation, therefore, will be application of 95.33 kg N/ha, 77.8 kg P_2O_5 but K_2O having surplus availability in the soil, should not be applied for achieving targeted yield of rice (6 tones/ha as given).

Thus the blanket recommendation of 120:60:60 the farmer will apply excess of N where as P will remain in short supply which will result into imbalanced fertility and the grain yield will be poor, therefore, the farmer will get lesser profit.

This approach needs separate equations for different crops and for different regions which are not available; therefore, the approach has limited applicability.

TABLE 5.6: Soil Test Crop Response Equation for Rice under Different Regions

Variety and soil type	STCR equations		Suitable Region
IR 20 (Black soil)	FN	= 2.54 T – 0.54 SN	Tamil Nadu
	F P_2O_5	= 2.27 T -3.00 SP	
	F K_2O	= 2.64 T –1.32 SK	
IR 20 (Black soil)	FN	= 7.94 T – 0.46 SN	Madurai and parts of
Kharif	F P_2O_5	= 1.46 T – 3.73 SP	Tiruneleli
	F K_2O	= 3.01 T – 0.11 SK	
IR 20 (Alluvial soil)	FN	= 7.36 T – 0.66 SN	Thanjavur
Kharif	F P_2O_5	= 2.15 T – 6.46 SP	
	F K_2O	= 3.07 T – 0.18 SK	
IR 8 (Alluvial soil)	FN	= 3.45 T – 0.29 SN	Parts of Bellary,
Kharif	F P_2O_5	= 2.80 T – 3.88 SP	Raichur, Chirtradurga,
	F K_2O	= 2.00 T – 0.11 SK	Dharwar.
(Black soil)	FN	= 6.28 T – 0.40 SN	Karnool, Cuddipah
	F P_2O_5	= 3.47 T – 5.09 SP	(A.P.)
	F K_2O	= 2.06 T – 0.08 SK	
IR 8 (Red soil)	FN	= 2.78 T – 0.27 SN	Eastern M.P.
Kharif	F P_2O_5	= 5.45 T – 11.03 SP	
	F K_2O	= 1.88 T – 0.23 SK	
IR 8 (Chesnut)	FN	= 3.02 T – 0.63 SN	Punjab
Kharif	F P_2O_5	= 1.77 T – 5.99 SP	
	F K_2O	= 2.75 T – 1.39 SK	
Ratna	FN	= 3.28 T – 0.18 SN	West Bengal

Variety and soil type	STCR equations		Suitable Region
(*Kharif*)	F P_2O_5	= 4.80 T – 5.02 SP	
	F K_2O	= 2.82 T – 0.54 SK	
IR 8 (Alluvial soil)	FN	= 3.44 T – 0.31 SN	Delhi
Kharif	F P_2O_5	= 2.76 T – 4.18 SP	
	F K_2O	= 2.02 T – 0.24 SK	

Source: Mineral nutrition, manures and fertilizers (Principles of Agronomy) p. 229 written by Reddy and Reddi (1992), Kalyani Publishers, New Delhi.

DIAGNOSIS RECOMMENDATION INTEGRATION SYSTEM (DRIS) APPROACH

This is recently suggested approach in which plant samples are collected, analyzed for different nutrient contents and they are expressed as ratios of nutrients with others. Using various observations, thus recorded, suitable ratios of nutrients are established for obtaining higher crop yields. The nutrient ratios of test crops are compared and whose ratios are not optimum for higher yields are supplemented by top–dressing. The method offers greater promise for longer duration crops, however, it is being tested for short duration crops like soybean, wheat etc.

MODELLING APPROACH

The approach is more suited to temperate regions, where the rate of oxidation being very slow, the soils are rich in organic matter and rate of mineralization is slow. It is used to recommend nitrogen fertilization by estimating soil nitrogen depending upon rate of its mineralization based on soil temperature by using following equation:

$$Y = K[(T-15) \times D]^n$$

Where Y is amount of mineralized N; T is soil temperature (°C); K is co-efficient related to the potential of mineralized nitrogen; 15 is threshold temperature and D is incubation period.

> ***"If you go as far as you can see, you will then see enough to go even further."***

6 Factors Affecting Crop Production

"If you are late in doing one thing in agriculture, you are late in all things".

Cato, 2nd Century B.C.

Field crops are classified into different groups based on various criteria. The most useful classification is from the agronomic standpoint based on their usage. Field crops are classified as as cereals, pulses, oilseeds, fibre crops, sugars, forage crops, narcotics, tuber crops, plantation crops, spices, etc. Roots and tubers, plantation crops, flowers, spices, fruit trees and vegetables are generally referred to as horticultural crops. Among the field crops, cereals and pulses are called as **food crops**; the others as **cash or commercial crops**. At present only 200 species are used as food and 80% of the food supply come from eight species only.

The genetic constitution of a given plant species limits the extent to which that plant may develop. No environmental conditions, no matter how favourable, can further extend these limits. It is imperative that there be a forward moving plant-breeding programme that will produce new varieties or hybrids capable of achieving maximum growth under specified environmental conditions. What has been done in the area of plant breeding and genetics is amply attested to by the crop production records kept by the better growers in this and other countries. What can be done will be determined by tomorrow's needs and the imagination and skill of geneticists and plant breeders the world over.

Environment is defined as the aggregate of all the external conditions and influences affecting the life and development of an organism.

The life of a plant is so intimately related with the **environmental factors** of a place that every species of plant seems to be fitted to a particular set of environmental conditions within the limits of which it is able to grow and reproduce its own kind. The preferred environment of a plant is called its **habitat.**

Though the environmental factors have to be studied individually, they do not act in isolation from one another. All these factors interact with one another to influence the plant (Fig. 6.1).

Every living organism is a product of its biological heredity and its environment. To survive and reproduce with which it regularly comes in contact. The crop may not attain the maximum efficiency in any one of the environments occupied; but to achieve the maximum efficiency the environmental conditions should be ideal. The factors can be broadly classified under the following types:

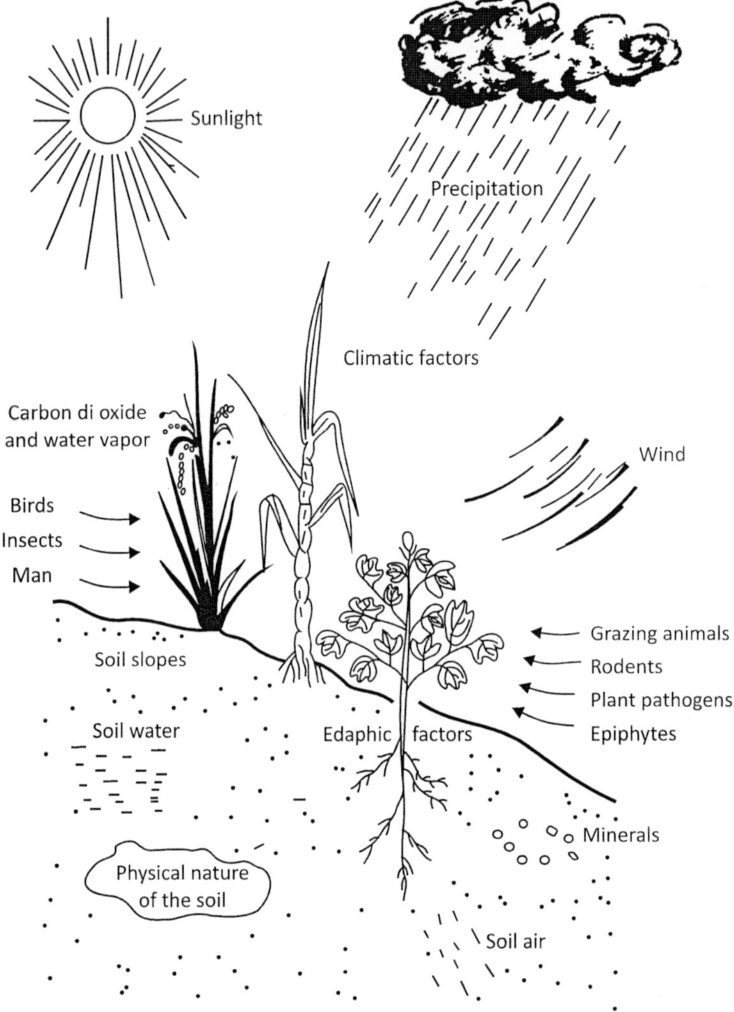

FIG. 6.1: Environmental factors affecting a plant.

CROP GROWTH

I. Internal Factors (Genetic or Hereditary)
II. External Factors (Environmental)
   a) Climatic
   b) Edaphic
   c) Biotic
   d) Physiographic
   e) Man (Anthropic)

GENETIC FACTORS

The importance of genetics in the growth of agricultural crops is clearly shown by the increase in yields that have been obtained by the use of improved seeds. Plant Breeders have incorporated many desirable characteristics into their hybrids and varieties. The following outline gives a list of useful characters which are considered as desirable breeding aims although it may not be possible to combine all these useful properties in a single variety.

1. High yields under given environmental conditions.
2. Early maturity (in some cases, later maturity).
3. Better resistance to lodging.
4. Drought, flood and salinity tolerance.
5. Greater tolerance to insects and diseases.
6. Chemical composition of grains (high percentage of oil, increase in protein quality or quality, etc.).
7. Quality of grains (fineness, coarseness, etc.).
8. Quality of straw (sweetness, juiciness, etc.).

The increased yields and other desirable characters are related to the genetics make up of the plant. These characters are inherent in each individual and are transmitted from one generation to another by genes.

ENVIRONMENTAL FACTORS

Climate

Variations in the climate can take place in relatively small areas due to the topographical or other factors, though the climate of a particular region is fairly uniform over a wide area. These local variations are limited to small areas in which they can be arranged both vertically and horizontally. Some of the examples are the microclimates of rice fields in garden lands, groves of coconut and areca-nut and sugarcane plantations. The atmospheric factors which affect the crop plant are called climatic factors.

1. Precipitation,
2. Temperature,
3. Atmospheric humidity,
4. Solar radiation
5. Wind velocity, and
6. Atmospheric gases.

Precipitation

World's supply of fresh water is obtained almost entirely as precipitation resulting from evaporation of sea water and land surfaces. The processes involved in the transfer of moisture from the sea to the land and back to the sea again form what is known as the **hydrologic cycle**. The first stage in the hydrologic cycle is the evaporation of water from the oceans. This vapour is carried over the continents by moving air masses. If the vapour is cooled to its dew point, it condenses into visible water droplets which form cloud or fog. Under favourable meteorological condition, the tiny droplets grow large enough to fall to earth as precipitation.

Precipitation includes all water which falls from the atmosphere to the earth's surface. Precipitation occurs in a variety of forms such as rainfall, snow, hail, etc. Besides, there may be fog drip and dew that may contribute to moisture. Fog consists of water droplets so small that their fall velocities are negligible. Fog particles which contact vegetation may adhere; coalesce with other droplets and eventually from a drop large enough to fall to the ground. On clear nights, the loss of heat by radiation from the soil causes cooling of the ground surface and of the air immediately above it. Condensation of the water vapour present in the air results in a deposit of dew. Crops grown in winter such as wheat and cabbage are efficient dew collectors and can be grown under dry condition with little irrigation. The yearly precipitation, both in total amount and seasonal distribution greatly affects the choice of cultivated crops of a place. Low and ill-distributed rainfall are common features of dry land farming wherein drought resistant crops like sorghum, pearl millet (*Pennisetum typhoides*), Italian millet (*Setaria italica*) and other minor millets are commonly grown. On the other hand in places of heavy and regular rainfall areas such as in the Western Ghats of India crops like rice are grown in flat areas while tea, coffee, rubber, etc., are grown in the slopes.

Rainfall is one of the most important factors influencing the vegetation of a place. Most of the crops receive their water supply from rain water. Rain water the source of soil moisture so essential for the life of a plant. The interaction of rainfall with temperature has a very profound effect on the vegetation and soil of a place. Very heavy rainfall and high temperature in the equatorial and tropical zones cause the formation of most highly developed vegetation of the world. The kind of vegetation and soil formed due to climatic conditions are indicated below.

Relationship Between Climate, Natural Vegetation And Soil

Dry cold					**Wet cold**
	Perpetual snow and ice			Perpetual snow and ice	
	Tundra			Tundra soils	
	Taiga			Podzol soils	
	Arid	Semi arid steppe land	Subhumid grass forests	Humid rain forests	
	Desert			Podzol	
	Grasses & Shrubs			Grey brown podzol soils	
	Sierozems and desert soils	Chestnut and brown soils	Prairee soils and chernozems	Red yellow podzol soils	
Dry hot				**Laterite soils**	**Wet hot**

It may thus be seen that while desert grasses and shrubs are found in desert soils of dry hot climate, rain forests are seen in laterite soils of wet hot climate.

Temperature

Temperature may be considered as a measure of intensity of heat energy. The range of maximum growth for most agricultural plants is between 15 and 40°C (59 to 140°F). The temperature of a place is largely determined by its distance from the equator (latitude) and the altitude. Based on the above, the vegetation are classified as tropical, temperate, taiga, tundra and polar as indicated in Fig. 6.2).

Some investigators have classified the vegetation of the world into four classes on the basis of prevailing temperature conditions as stated in Table 6.1.

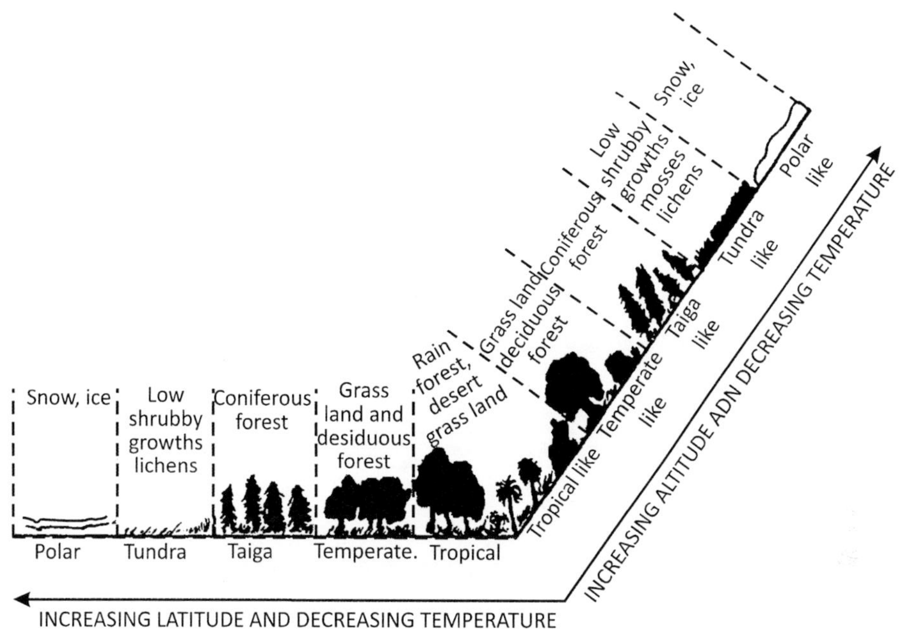

FIG. 6.2: Relationship of latitude and altitude to vegetation.

TABLE 6.1: Classification of vegetation based on temperature

Class	Region	Temperature Conditions	Type of vegetation	Common crops cultivated
Megatherms	Equatorial and tropical	High temperature throughout the year	Tropical rain forest	Tropical crops like rubber, cassava, varieties of rice, etc.
Mesotherm	Tropical and Sub-tropical	High temperature alternating with low temperature of winter	Tropical deciduous forests	Sub-tropical crops like maize, sorghum, etc.
Microtherms	Temperate and high altitude plants (upto 4000 m of tropical and sub-tropical)	Low temperature	Mixed coniferous forests	Temperate crops like wheat, oat, potato, etc.
Hekistotherms	Arctic and Alpine regions (above 5000 m in tropcs and 4000 m in temperate regions)	Very low low temperature	Alpine vegetation	Pines, spruce, etc.

Every plant community has its own minimum, optimum and maximum temperature known as their cardinal points. The cardinal points of some of the common crops are indicated in Table 6.2.

TABLE 6.2: Cardinal temperature of certain crops

Crops	Minimum (°F)	Optimum (°F)	Maximum (°F)
Wheat	39-40	77	86-90
Barley	39-40	68	82-86
Oats	39-41	77	86
Maize	46-50	89	104-111
Sorghum	46-50	89	104
Rice	50-53	90	97-101
Tobacco	55-57	82	95

The effect of temperature in affecting the yield of a crop is indicated below for potatoes (Table 6.3.).

TABLE 6.3: Yield of potatoes at varying temperature

Place	Mean temperature in growing Period in °F	Average yield in q/ha
Nilgiris (Tamil Nadu)	60	125
Delhi	64	118
Patna (Bihar)	66	83
Bangalore (Karnataka)	70	52

A part from the reduction in yield many injuries on the plants and adverse effect on soil conditions occur under both extremes of temperature.

***Cold injury*:** Plants are injured due to cold temperature in the following ways:

1. **Chilling injury:** Plants growing in hot climate, if exposed to low temperature (which is above the freezing point) for some time are found to be killed or injured severely. Some effects of chilling are chlorotic condition or bands on leaves of sugarcane, sorghum and maize when exposed for 60 hours at 2 to 4°C.
2. **Freezing injury:** This is generally, caused in plants growing in temperate region. When the plants are exposed to very low temperature, water is frozen into ice crystals in the intercellular spaces. With further fall in temperature, water is withdrawn from the cells, resulting in increase in the size of ice crystals in the intercellular spaces. The protoplasm of the cell is dehydrated and mechanical distoration takes place resulting in killing of the cells. Frost damage to potatoes, tea etc., in winter in the hilly areas like Nilgiris is a typical example of the freezing injury.
3. **Suffocation:** During winter the ice or snow form a thick cover over the ground and the crop suffers for want of oxygen. Ice in contact with roots inhibits diffusion of carbon dioxide and the respiratory products may become harmful to plants.
4. **Heaving:** Injury to plants is caused by a lifting upward of the plant along with the soil from its normal position in temperate regions where snowfall is a common phenomenon.

***Heat injury*:** Very high temperature often stops growth. The plant faces incipient starvation due to high respiration rates. The plant is stunted and if such a condition persists for a long period the plant is killed. Direct temperature effects are noticeable in young seedlings and transplanted crops. High temperature causes sterility in flowers. The general effects of excessive heat are defoliation, pre-mature dropping of fruits and in extreme cases death of plants.

Effect of Temperature on Crop Production

Plants can grow only within certain limits of temperature. For each species and variety there are not only optimal temperature limits, but also optimal temperatures for different growth stages and functions, as well as lower and upper lethal limits.

While light is involved in only one aspect of photosynthetic process namely as a source of energy for the conversion of carbon dioxide and water to carbohydrates, there are certain biochemical processes preceding and following the reduction of carbon dioxide, which are affected mainly by temperature. As long as light is limiting, temperature has little effect on the rate of photosynthesis. However, when light is not limiting, the biochemical processes associated with photosynthesis become the limiting factor, so that the effect of favourable temperatures on the rate of photosynthesis is measured.

In general, high temperature associates growth processes. Rarely are high temperatures per se the direct cause of death of plants, provided the water supply is adequate. However, beyond a certain limit, which depends on the crop, the stage of development and physiological processes involved, high temperatures may have detrimental effects on crop production. The photosynthetic process appears to become heat inactivated at excessively high temperatures which however, do not inhibit respiration, so that apparent photosynthesis declines rapidly under these conditions.

Reduction of growth and difficulties in fertilization, even in heat loving crops such as sorghum, occur at temperatures that are often well below the lethal limit. The harmful effects of excessive temperatures are usually aggravated by lack of available moisture. Hot, dry winds will further increase the damage.

In the Great Plains of U.S.A., it is found high summer temperatures may even override the effects of summer rainfall on the yield of crops. Thus, the variability in yields of spring wheat was clearly associated with variability in average maximum temperatures in June, and variability in maize yields was most closely associated with variability in average maximum temperatures in July. With increasing temperatures, evapo-transpiration increases. Hence, a moisture supply that is adequate in a cooler climate may be deficient in a warmer region – a situation which may be aggravated by hot winds. The daily alternations of high and low temperature may affect grain production.

Atmospheric Humidity

Water is always present in the atmosphere in the form of invisible water vapour. This is normally known as **humidity** of the air. When the atmosphere contains the maximum possible amount of water vapour it is said to be saturated at the particular temperature and pressure. Any increase in temperature, water remaining constant, will make the air unsaturated. In unsaturated condition, the water vapour content of air is usually expressed as **relative humidity** which is the ratio between the actual humidity per cent and the saturated humidity possible at that temperature.

The relative humidity of a place, besides being affected by temperature and pressure, is also affected by wind, exposure, vegetation and water content of soil. The evaporation of water from plants or a body of water is directly dependent on the relative humidity of the atmosphere. To be more precise, the rate of transpiration is directly proportional to the difference between the amount of water vapour actually present and the amount that would be needed to saturate the atmosphere. The difference is termed **vapour pressure deficit** and as compared to relative humidity is much more inactive of the potential rate of evaporation. Evapo-transpiration of crop plants increases with temperature but decreases with high relative humidity affecting the quantity of irrigation water. Moist air favours the growth of many fungi and bacteria and these affect seriously the crop. The blight diseases of potato and tea are common examples of diseases spread under moist weather. Similarly, many kinds of insect parasites such as aphids and jassids thrive well under moist conditions.

Solar Radiation

Solar energy provides two essential needs of plants – (i) **light**, required for photosynthesis and for many other functions of the plant – including seed germination, leaf expansion, growth of stem and shoot, flowering, fruiting and even dormancy and (ii) **thermal conditions** required for the normal physiological functions of the plant.

***Light*:** Light is one of the most important climatic factors for many vital processes of the plant. It is indispensable for the synthesis of the most important pigment of the plant, i.e., the green chlorophyll. The chlorophyll is capable of absorbing radiant energy and converting it into potential chemical energy of carbohydrates. The carbohydrates manufactured by the plants during photosynthesis are the only link between the solar energy and the living world. It regulates the rate of transpiration by controlling the opening and closing of stomata.

Light affects the plants in four ways:

1. Light Intensity,
2. Quality (wave length) of Light,
3. Duration (photoperiod) of Light, and
4. Direction of Light.

1. Light Intensity

The variations in light intensity are always accompanied with change in temperature and relative humidity and therefore, it is difficult to evaluate light effects alone. Generally speaking light intensity falling at a particular place is normally enough for the plants and their physiological phenomena viz. photosynthesis.

In photosynthesis about one per cent of the light energy is converted into potential chemical energy. Very low light intensity reduces the rate of photosynthesis and may even result in the closing of the stomata. This results in reduced vegetative growth of the plants. Very high light intensities are detrimental to plants in many ways. It increases the rate of respiration and thus disturbs the photosynthesis-respiration balance. It causes rapid loss of water resulting in the closure of stomata. The most harmful effect of light intensity is the phenomenon of solarization in which all the cell contents are oxidized by atmospheric oxygen. This oxidation is different from respiration and is termed photo-oxidation.

Light intensity varies greatly under natural conditions. Most of the crops show marked response to changes of light intensities. During cloudy weather the production of dry matter is limited. Depending upon the nature of the crops such as *heliophytes* (sun loving) and *sciophytes* (shade loving), the dry matter production is affected. Many species produce maximum dry matter under high light intensity if water is available in plenty. However, crops like betel, sunflower, buckwheat and tobacco grown during summer produce greater dry matter if slightly shaded. Except under glass house or shade conditions, intensity of light can not be controlled in the production of most agricultural crops.

Light Intensity and Crop Yield

Light is the most important factor influencing the photosynthetic activity of a crop, as it is the energy source for the whole process. Light utilization by the crop has two limitations: (a) The maximum quantum yield at low light intensities and (b) Light saturation at high light intensities. The variation in yield of crops particularly the photo insensitive varieties such as dwarf rice in tropics under same input and management conditions is attributed to the effect of sunlight and the connected thermal conditions.

2. Quality of Light

When white light is passed through a prism it is dispersed into wavelengths of different colours; violet 400-435 millimicrons (mm); blue 435-490; green 490-574; yellow 574-595; orange 595-626 and red 626-750. The principal wavelengths absorbed in photosynthesis are in the violet-blue and the orange-red regions. Among these short rays beyond the violet such as X-rays and gamma rays and longer rays beyond red such as infrared are detrimental to growth. Red light seems to be the most favourable light for growth followed by violet-blue.

Ultraviolet and shorter wavelengths of the visible light are scattered and infrared and longer wavelengths are absorbed by moisture of the atmosphere. On clear days such scattering and absorption mat be 10-15 per cent of the total light, whereas on cloudy days it may be upto 100 per cent. Such a light is called **diffuse light** or **sky light.** Ultraviolet and shorter wavelengths kill bacteria and many fungi.

3. Duration of Light

This is considerable importance from the farmers' point of view in selecting the variety of a crop. The length of the day has greater influence than the intensity. The response of plants to the relative length of day and night is known as '**photo-periodism**'. Plants which develop and produce normally when the photo-periodism is greater than a critical minimum (more than 12 hours of illumination) are called '**long day plants**' and those develop normally when the photoperiod is less than a critical maximum (less than 12 hours illumination) are called '**short day plants**'. Some plants are found to be unaffected by photoperiod and are called '**day neutral**' plants.

Plant characters like floral development, floral initiation, bulb formation, rhizome production etc. are all influenced by photo-periodism. If a long day plant is subjected to short day periods the internodes may be shortened to give a rosette appearance and flowering will not take place. In the same way when a short day plant is subjected to long day period, the growth of the plant becomes abnormal and there is no floral initiation.

Among crop plants, soybean, maize and millets are examples of short day flowering plants while sugar beet, wheat, barley are long day plants and others are intermediate (day neutral) in that they flower at any day lengths. Among the horticultural crops while straw berries, chrysanthemum, etc., are short day plants, crop like lettuce, spinach and cruciferous plants are long day plants. Tomato, asparagus, etc. are not sensitive to day length.

5. Direction of Light

Shoots roots and leaves show different orientation to the direction of light. In temperate regions, the southern slopes show better growth of crops than the northern slopes due to the direction of light contributing more sunlight towards the southern side.

Wind Velocity

The velocity of wind at a place depends on various factors such as geographical situation, topography, altitude, distance from sea shore, flat plains, vegetation etc. Its action on plants may be direct or indirect. As a direct effect, wind causes mechanical lacerations and bruises on the tissues of crop plants. High winds tear the leaves and the plants are in danger of being uprooted and blown away. Lodging of crop plants in wheat, maize, sugarcane, rice, etc. are caused by violent winds.

A very high velocity of wind (gale, blizzard, hurricane, cyclone etc.) breaks dead and living branches of trees and even uproots them completely. The effect is however, less when there is a windbreak. In bare deserts high velocity of wind causes constant soil erosion and this makes it difficult for plants to grow. Shelter belts of vegetation are raised to arrest the movement of soil by erosion. Wind is also responsible for causing rainfall to a very large extent. In India, the monsoon type of rainfall is largely determined by particular patterns of wind movement. Wind has a powerful effect on the humidity of atmosphere.

An important function of wind is to help in pollution of flowers and dispersal of seeds, fruits and micro-organisms. Indirectly the velocity of wind affects the rate of transpiration in plants. Much damage is caused to crops by hot dry winds at the time of flowering. The internal water balance of plants is affected resulting in poor seed setting. Another for of damage is blossom injury caused by evaporation of secretions from the stigmas.

Effects of Wind on Crop Production

Winds affect growth mechanically and physiologically. The sand and dust particles carried by the wind may damage plant tissues. Emerging seedlings may be completely covered or alternately the roots of young plants may be exposed by strong winds. Winds may also cause considerable losses by inducing lodging, the breaking of stalks and shedding of grain.

The physiological effects of winds consist mainly in increasing transpiration as well as evaporation from the soil. Hot dry winds, may however, adversely affect photosynthesis, and hence productivity, by causing closure of the stomata even when soil moisture is adequate.

The hot dry winds may also occasionally have beneficial effects. The main reason for the low incidence of the dangerous yellow rust of wheat (*Puccina glumarum*) is the susceptibility of the pathogen to high temperature that occurs when seasonal winds blow. In rare years when no winds blow in the spring season, the rust may cause devastating reductions in yield.

Moderate winds have a beneficial effect on photosynthesis by continuously replacing the carbon dioxide absorbed by the leaf surfaces. The favourable effect of wind on photosynthesis is most marked under low turbulence of wind at the lower leaf-layers, increasing wind strength permits even deeper leaf layers to approach the maximum rate of photosynthesis under the amounts of radiation received.

Atmospheric Gases

The atmosphere surrounding the earth contains a mixture of gases viz., carbon dioxide (0.03%), oxygen (20.95%), nitrogen (78.09%), argon (0.93%) and miscellaneous gases (0.02%) in a constant proportion. Variations from this may be found over industrial plants which release large quantities of gaseous fumes such as SO_2, CO_2 and CO as well as water vapour and mineral particles.

Carbon dioxide: The CO_2 is the main source of carbon for the various types of organic compounds in the body of the plants. It is the main raw material for the manufacture of carbohydrates by the photosynthetic process of the green plants. Photosynthesis is approximately proportional to the concentration of CO_2 in the air surrounding the foliage of the crop. The CO_2 which gets incorporated into the organic compounds of the plants returns to the atmosphere by respiratory break down of these compounds and by death, decay and combustion of plants.

Increased growth and greater yield of vegetables are possible under green house conditions by increasing the CO_2 content. Certain gases like SO_2, CO and HF when released into the air in sufficient quantities are toxic to plants. Injuries to crops have also been reported in the vicinity of factories producing alumina and phosphates. In both cases fluorine is released and this causes the damage. Near about cement factories, the fall out of cement particles also causes damage to the crops.

The CO_2 content dissolved in water is fairly high. So the aquatic plants get CO_2 from water. In soil, excess CO_2 causes poor root respiration. This causes adverse effect on the metabolic activities of roots in plants.

***Oxygen*:** Every living organism whether plant or animal, sustains its life because of oxygen. Atmospheric oxygen is vital to its life and normal growth. The amount of oxygen is normally constant in the air because plants give off oxygen during photosynthesis.

***Nitrogen*:** The nitrogen of the atmosphere is made available to the soil by lighting, rainfall and nitrogen fixing micro-organisms. Free living bacteria like *azotobacter*, symbiotic bacteria like *rhizobium*, blue green algae, etc. fix a good amount of nitrogen in the soil. The decomposition of the protein content of the dead plants and animals by micro-organisms also adds to the nitrogen pool of the soil. The nitrogen in soil is made available to the crops by the activity of nitrifying bacteria. The plant tissues are largely consumed by animals as food. Part of the nitrogen in the tissues is again returned to the soil in the form of animal excreta or dead animal tissues.

Edaphic Factors

The land plants are completely dependent on the soils in which they grow for anchorage, water and mineral nutrients. The nature of the soil is normally so complex and dynamic and its effects on plants so pronounced that an agronomist has to attach great importance to the study of soil and its management. The factors which affect the crop growth through the soil are (1) soil moisture, (2) soil air, (3) soil temperature, (4) soil mineral matter, (5) soil organic matter, (6) soil organisms and (7) soil reaction.

Soil Moisture

Plant tissues constitute about 90 per cent of water. The moisture lost through transpiration is made up only by absorbing water from the soil. The important functions of water in the plant system are:

1. It gives turgidity to plants and keeps them erect.
2. It regulates the temperature in the plant body.
3. It acts as a solvent and nutrient carrier.

Indirectly water present in the soil helps the plants in many ways:

1. It supplies the essential raw materials for production of carbohydrates by photosynthesis.
2. It promotes physical, chemical and biological activities in the soil.
3. The physical properties of soil such as proper structure formation, friability, penetrability plasticity, shrinkage, cohesion etc. Are altered by the soil moisture content. Besides the thermal properties of soil such as thermal conductivity, absorptive etc. is affected by moisture content.
4. The forms and contents of different elements in the minerals, the different chemicals changes such as hydration, hydrolysis etc. and the concentration of different salts affecting plant growth are affected by moisture content.

5. The forms and types of different microbial organisms which may affect plant growth both beneficially and adversely are controlled by the moisture content of soil.
6. Gaseous diffusion in soil for proper aeration depends on the moisture content.

Soil moisture is in a highly dynamic condition and varies from time to time. The moisture is held within the soil by the attractive forces of the soil particles.

Soil Air

Aeration of the soil is absolutely essential for the absorption of water by roots. Absorption of water by the roots takes place at a rapid rate in well aerated soils while at a greatly retarded rate in soils deficient in oxygen supply.

Oxygen is needed for the respiration of roots. In well aerated soils, carbon dioxide released in respiration of roots and micro-organisms is freely exchanged with the air above the soil. In poorly aerated soil the carbon dioxide gets accumulated and is detrimental for the water absorption process of the plant.

Sol air is also useful in increasing the nutrient availability of the soil, (1) by breaking down the insoluble minerals into soluble salts, (2) by decomposing plant and animal remains and (3) by bringing about nitrifying and nitrogen fixing processes of bacteria.

Soil Temperature

The prevailing temperature of the soil besides affecting the physical and chemical processes going on the soil also influences to a great extent the rate of absorption of water and solutes, the germination of seeds and the rate of growth of the underground portions of the plant body. The maximum metabolic activities of the plants and the maximum absorption of water by the roots take place generally between 20 and 30°C. Lowering of temperature below 20°C causes an appreciable reduction in the rate of absorption of water. Cold soils are therefore, not conducive for rapid growth of most agricultural crops.

Soil temperature is one of the factors that control the micro-biological activity and the process involved in the availability of nutrients to plants. Nitrification does not begin until the soil reaches a temperature of about 5°C (40°F).

Soil Mineral Matter

The mineral content of the soil is derived from the weathering of the rocks and minerals and consists of particles of different sizes. The principal minerals occurring in the earth's crust are indicated in Table 6.4.

TABLE 6.4: Principal minerals in earth's crust

Minerals	Per cent
Feldspar	48
Quartz	36
Mica	10
Limestone and magnesium limestone	2
Hornblend and Augite	1
Olivine and Serpentine	1
Clays	1
Other minerals	1

1. Feldspars are aluminium silicates with varying amounts of silicates of K, Na and Ca.
2. Quartz is silicon dioxide and is found in all crystalline rocks; also occurs as sand and is the principal component of granite and sand stones.
3. Mica is silicate of Al and K with silicates of Fe, Mg and Na.
4. Limestone contains carbonate of Ca. With magnesium carbonate it occurs in dolomite.
5. Hornblend is composed of silicate of Ca, Mg and Fe with sodium silicate.
6. Augite is made up of silicates of Ca, Mg, Fe and Na in varying proportions.
7. Olivine is a ferro-magnesium silicate, and serpentine is hydrated magnesium silicate.
8. Clays are hydrated aluminium silicates and may contain varying amounts of Ca, Mg, Na and Fe. They usually contain iron oxide and carbonaceous matter together with varying amounts of quartz, feldspar and mica.
9. Other minerals occurring in the soil include Tourmaline (borosilicate of Al with alkali metals and Fe or Mg), Rutile (Titanium oxide), Zircon (Zirconium silicate), Glauconite (hydrated silicate of Fe and K), Magnetite and Hematite (oxides of Fe).

In course of weathering of rocks, the readily soluble products are carried away in drainage waters whereas comparatively insoluble products accumulate to form soil. This is particularly so in humid climates and in consequence, silica, alumina and iron oxide, as such or in silicate combinations form the major portions of soils in heavy rainfall areas. Silicon in silicate form also dissolves, leaving laterites, which consist chiefly of aluminium iron oxides. In arid regions, leaching plays an insignificant part and therefore, the chemical composition of soils is very similar to that of the rocks from which they originate.

Inorganic Compounds

Compounds of Si, Ca, Mg, Fe, K, Na and Al form the principal chemical constituents of soils. Besides these, the sol contains small quantities of a large number of other mineral elements like B, Mn, Mo. Zn, Cu, Co, I and F which are known to be essential for the well being of plants in very small quantities. Such elements are called **minor-** or **trace-elements.**

The total amounts of elements contained in soils depend partly on the nature of the rocks from which they are formed and partly on their age and the extent to which soluble products have been leached away. The chemical composition of different horizons of a soil also shows a good deal of variation. Generally, speaking the A horizon is rich in insoluble compounds as compared with the B horizon. Usually the elements that are commonly leached out are the ones required in largest quantities by plants.

Table 6.5 give the percentage inorganic composition of eight representative Indian soils and Table 6.6 indicate the successive horizons of a lateritic soil. The different components of the soil may be classified as in Table 6.7.

The mineral matter of soil may be classified texturally based on the size of particles (Table 6.8).

Based on the proportion of different size groups of soil particles the soil is classified for better understanding on their characters (Table 6.9).

TABLE 6.5: Percentage inorganic composition of representative Indian soils

Type	State	SiO_2	Al_2O_3	Fe_2O_3	MnO	CaO	MgO	K_2O	Na_2O	P_2O_5	SO_3
Sandy	Rajasthan	91.72	2.92	2.36	-	0.35	0.78	0.33	0.08	0.08	0.04
Sandy loam	Uttar Pradesh	84.84	5.30	4.52	-	0.91	0.52	0.16	0.03	0.01	-
Sandy loam	Punjab	-	10.89	-	-	0.06	0.29	0.71	-	0.34	-
Loam	Tamil Nadu	82.11	5.19	5.70	2.12	1.10	1.27	0.24	0.86	0.06	3.04
Loam	Uttar Pradesh	86.98	4.10	3.52	-	0.42	1.29	0.56	-	0.06	-
Silt loam	Assam	86.49	9.16	7.00	0.21	0.14	0.90	1.85	-	0.24	-
Clay	Tamil Nadu	65.16	13.76	9.27	0.25	2.18	2.47	0.14	0.01	Tr.	–
Clay	Maharashtra	63.80	Not available	-	1.09	1.60	1.60	-	0.23	-	-

Source: Hand Book of Indian Agriculture, I.C.A.R., New Delhi.

TABLE 6.6: Percentage chemical composition of different horizons of red Lateritic soil from Nilgiris, Tamil Nadu

Horizons	Depth in cm	SiO_2 (soluble)	$Fe_2O_3 + Al_2O_3$	CaO	MgO	K_2O	P_2O_5	N
1	0-22.5	0.53	32.60	0.140	0.03	0.656	0.138	0.154
2	22.5-75	0.58	30.22	0.084	0.02	0.667	0.138	0.109
3	75-112.5	0.57	37.15	0.056	0.03	0.568	0,149	0.050
4	112.5-117.5	0.55	26.15	0.098	0.04	0.480	0.245	0.030
5	117.5-162.5	0.50	35.73	0.084	0.02	0.580	0.222	0.030

Source: Hand book of Indian Agriculture, ICAR, New Delhi.

TABLE 6.7: Composition of soil

Content	Per cent
Mineral matter	30
Soil water	30
Soil air	30
Soil organic matter	5 – 10

TABLE 6.8: Textural classification of soil

Name of particles	Particle size (mm)
Clay	0.0001-0.005
Silt	0.005-0.05
Very fine sand	0.05-0.10
Fine sand	0.10-0.25
Medium sand	0.25-0.50
Coarse sand	0.50_2.00

Sands tend to menace the size of pore spaces in soils and this aid in the movement of air and water. In general, fine textured soils hold much of water and make them available to crop growth over longer periods. Root penetration is retarded by high silt and clay content of the soil.

Loam soils are the best for the growth of plants since they contain coarse sand as well as fine silt and clay particles. Loam soil has good aeration, good water infiltration and water movement, easy root penetration and also good water holding capacity and fertility.

TABLE 6.9: Mechanical composition and textural class of some typical Indian soils (surface soils)

Soil group	Per cent composition of different size fraction				Textural classes
	Coarse sand	Fine sand	silt	Clay	
Desert soil	73.6	18.5	3.8	4.4	Coarse sand
(Ajmer)	79.4	13.4	3.7	4.8	Coarse sand
	89.4	8.7	1.4	3.4	Coarse sand
Alluvial soil	0.2	57.8	25.7	14.1	Sandy loam
(Kanpur)	1.2	36.3	27.9	27.8	Loam
	2.0	72.1	14.5	8.4	Loamy sand
Black cotton soil	0.7	11.4	21.8	45.2	Clay
(South India)	18.9	13.5	16.1	47.4	Clay loam
	41.2	18.2	11.4	23.9	Coarse loam
Laterites and	10.0	43.0	20.0	27.2	Sandy loam
Red soil	32.0	19.3	18.0	32.3	Loam
(Bengal)	24.0	29.0	31.2	18.0	Sandy loam

Source: Hand book of Indian Agriculture, ICAR, New Delhi.

Soil Organic Matter

Besides inorganic substances, soils contain organic matter in amounts ranging from less than 1 per cent in arid sandy soils to as much as 90 per cent in peaty soils. The presence of organic matter distinguishes soil from disintegrated rock material.

The organic matter is added to the mineral matter of the soil every year. Though the percentage of organic matter is less than 5 % of dry weight of well developed soil, it has a marked influence on the soil properties and growth of crops. The organic matter of the soil is largely derived from (1) dead and decaying roots of plants and living organisms present in the soil and (2) dry leaves, twigs, dead plants and animals.

The organic matter of the soil is a source of mineral nutrients essential for the crop growth. During decomposition, the humus is mineralized. It has increased water holding capacity due to its content of organic colloids. It can hold by adsorption a great amount of minerals particularly the cations in the ionic form, thus increasing the exchange capacity of soils. It is a source of food for most of the soil organisms. As a source of plant nutrient it contains 95 % of the total nitrogen, 50 to 60% of total phosphorus and 10 to 20% of total sulphur.

Soil Organisms

Raw organic matter in the soil is not directly used by plants as food. It must be broken down first into humus and then into simpler products before it can be utilized. This work is done by different kinds of micro-organisms which inhibit the soil by the billions. Decomposition of organic matter primarily forms part of the feeding and growth processes of these microscopic plants and animals. Sugars, starches and proteins are broken down first, then cellulose and fatty substances (lipids), and lastly lignin or woody substances.

The principal kinds of living organisms found in the soil are as follows:

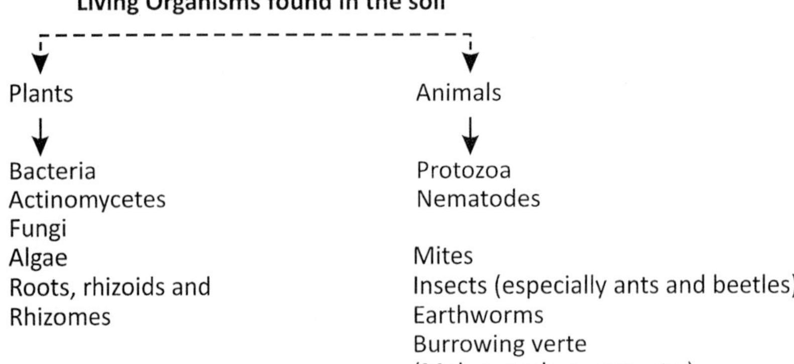

A large number of bacteria and fungi bring about the decay of the organic substances. Their mineralization process releases a variety of nutrients to be made available to the plants. The ammonifying bacteria convert proteins into ammonia. The nitrifying bacteria oxidize ammonia to nitrites and nitrates. A number of bacteria and blue green algae fix atmospheric nitrogen in the soil under anaerobic conditions of water logged soils; certain bacteria cause denitrification, releasing free nitrogen which gets lost into the air. Many bacteria and fungi are pathogenic causing diseases of plants. Some of the fungi form mycorrhizal association with root system of higher plants for helping them in the absorption of water and minerals.

The larger organisms like earthworms; rodents etc. improve the aeration of the soil due to their burrowing habit. They play a key role in the weathering of rock and minerals and in the formation of soil.

Soil Reaction

Soils may be neutral, acidic or alkaline depending upon their content of the basic salts and acidic components. Neutral soils are the best for the growth of most crops.

Soil acidity beyond a particular limit is injurious to plant growth due to the following reasons:

1. High acidity (particularly high alumina content) is toxic to plants.
2. High acidity interferes with the absorption of several nutrients particularly the cations such as K, Ca and Mg due to their low content in soil. Nutrients like P are fixed in acidic soils.
3. Decomposition of organic matter by micro-organisms may be reduced.
4. The activities of nitrifying and nitrogen fixing bacteria may be checked.
5. Certain types of fungoid diseases like potato scab may be favoured.

Similarly, high alkalinity of the soil also adversely affects crop growth. The soil alkalinity is due to the excess of cations like Na, K, Ca and Mg found in the sol. Among these the presence of high amount of sodium interferes with nutrient and water absorption. This may interferes with water movement and infiltration in the soil as well as the microbial activity.

Biotic Factors

Beneficial or harmful effects are caused by other plants and animals on the crop plants. These are called as biotic factors.

Plants

Competitive and complementary nature among field crops: Competition between plants occur when there is demand for nutrients, moisture and sunlight. Different crops do have optimum leaf area indices for the maximum utilization of the available resources of sunlight and nutrients for their maximum production. While closer spacing reduces the yield of individual plants, wider spacing will reduce the total yield because of lesser population of plants per unit area. Hence, optimum spacing of crops is an important agronomical practice. But in the case of different crops when grown together as in the case of mixed cropping, there is mutual benefit resulting in better yield. Ceral-legume mixture is a common practice throughout the world.

Competition between weed and crop: Weeds are plants growing where they are not wanted. In nature, certain plants, the seeds of which are not sown come up at all times under various conditions of soil, climate and environment. Weeds reduce crop yields on account of their competition with crops for water, soil nutrients and light. They increase the cost of labor and equipment, render harvesting difficult, reduce the quality and marketability of agricultural produce, harbour insect, fungal, virus and bacterial organisms and some are poisonous to human beings and livestock.

In non-irrigated areas the competition between weeds and crops is largely for water. The transpiration coefficient for Bermunda grass (*Cynondon dactylon*) is 813, while it is only 430 for sorghum. A saving of 750 to 1250 tons of water in a hectare of soil, 180 cm deep, is possible by keeping it free of weeds. In irrigated tracts, the competition is severe for nutrients. The removal of nutrients by the weeds in the unweeded control is much higher as compared to the weeded plots.

Weeds in fallow land deplete the soil of both moisture and nutrients. Weeds block drainage and impede the flow of water in rivers and canals. Control of weeds is a must for the well being of crops, soil and water.

Plants as parasites: A plant parasite is dependent on its host plant for its existence. Under favourable conditions these parasites exert a profound influence on the plant community. Common examples of this are the parasitic fungi, bacteria, virus etc. These cause different kinds of diseases on agricultural crops. These micro-organisms obtain their food either by breaking down dead plant and animal remains (saprophytes) or by attacking living plants and animals (parasites). In order to obtain their nutrients, the parasitic organisms kill the tissues and cells of the host plant as a result of which either the whole plant or a part of it is damaged and killed, or considerable disturbance takes place in the normal metabolic processes of the living plant. In some cases they also produce toxic substance

Controlling the diseases caused by these parasitic organisms has become a routine practice in crop production. They may be controlled by evolving disease resistant varieties, use of chemicals and cultural practices and field sanitation.

Symbiosis

The apparent independent existence of individual plants, on close scrutiny, is found to be not so. Certain organisms are normally found associated with others to form a community. In the community, the different organisms have manual relationship with each other and with the prevailing environment of the place. This biological relationship among the organisms is termed as '**symbioses**'.

The symbiotic relationship between legumes and rhizobia which results in nitrogen fixation is of great significance to crop production. Plants by themselves are unable to utilize elemental nitrogen present in the atmosphere. Two groups of bacteria participate in the capture of gaseous nitrogen and help plants in utilizing it, the nodule organisms (*Rhizobium sp.*) of the legumes and the free living bacteria (*Azotobacter*) of several kinds.

The legume bacteria use the carbohydrates of their host as energy and fixes up atmospheric nitrogen. A part of it is passed to the infected host. The free living bacteria acquire their energy from the soil organic matter, fix the free nitrogen and make it a part of their own tissue. When they die, the nitrogen available by their body tissues is used by the crop plants.

Animals

***Soil animals*:** The soil fauna including protozoa, nematodes, rotifers, snails and insects constitute a highly important part of the environment of plant roots. All these organisms contribute to organic decomposition, when using the organic matter for their living.

Among these, insects and nematodes cause considerable damage as crop pests. Crops are subjected to insect attack throughout their period of growth. Even after harvest, the grains may be damaged by insects. The average loss due to insects is reported at about 20% throughout the world.

***Beneficial organisms*:** Many crop plants are pollinated primarily by insects. Bees and wasps are probably the pollinators of the greatest importance. Moths, butterflies and beetles do considerable pollination. Earthworms are influential in changing soil conditions. Burrowing facilities aeration and drainage and the ingestion of organic matter and mineral matter result in a constant mixing of these materials in the soil. This probably tends to make better plant growth.

***Small animals*:** Many small animals like rabbits, squirrels and field rats constitute an important factor. They effect changes by consuming green plants, burrowing and soil mixing. These cause extensive damage to field and garden crops.

***Large animals*:** Domesticated animals like cattle, goat, sheep and horses as well as wild animals like deer, elephants, pigs and porcupines do considerable damage to crop plants by their grazing and browsing habits. Extensive areas nearer their habitat suffer greater damage if no protection is made.

Physiographic Factors

The geological strata of a place and topography affect the crop considerably.

Geological strata: The nature of the geological strata accounts not only for the kind of parent material utilized in soil formation, but also on the nature of crops grown in these soils for proper utilization.

Topography: The nature of the surface of the earth is known as topographic factors. They include:

1. Altitude of the place.
2. Steepness of slope.
3. Exposure of the slope to light and wind.
4. Direction of mountain chains.

The topographic factors affect the crops indirectly by modifying climatic and edaphic factors of a place.

Altitude of the Place

The continuous increase in altitude and its effect on the vegetation can be best observed on the mountains. Increase in altitude is usually associated with:

1. Decrease in temperature.
2. Increase in precipitation,

3. Increase in wind velocity.

It has been found that the temperature decreases at the rate of 6 to 7°C per 1000 metres except in the valleys and lowlands due to the downward descent of cold air. Due to increased rate of precipitation and strong winds, the type and nature of soil and vegetation are affected. The organic matter of the soil increases and this in turn increases the nitrogen content and the acidity of the soil.

As we ascend the mountains, changes in temperature etc. give a definite pattern of vegetational zonation, similar to the sequence of vegetation met with the moving from equator to the poles.

Steepness of Slope

Steep slope causes swift run-off of water after a rainfall. This decreases the moisture content of soil. Besides the soil becomes unstable and may even be completely eroded. The humus is also not allowed to accumulate. The bare rocks are exposed. These areas may not be useful for raising field crops. To counteract the harmful effect of steep slopes, soil conservation measures are employed to reduce the flow of water and to conserve soil.

Exposure of the Slope to Light and Wind

A mountain slope exposed to weak intensity of light and strong dry winds as the case of northern slopes of temperate regions and the Himalayas may have poor crops due to want of moisture and sunlight. Similarly, the slopes exposed to high velocity of wind as in the case of western slopes in the hilly areas of Tamil Nadu have poor crops due to the damage caused by winds.

Direction of Mountain Chains

The distribution of rainfall throughout the country during the monsoon seasons is largely governed by the direction of mountain chains. This pattern of rainfall largely governs the type of agricultural crops grown under the dry conditions under different soil conditions.

Anthropic Factor

Man or woman has produced profound changes in plant environments. Improvements effected by plant breeders towards increasing crop yields, introduction of exotic plants etc. have greatly influenced crop growth.

Successful farming does not depend merely on a good knowledge of the physical, chemical and biological properties of a soil. It is a matter of both soil and its management. The most important consideration in the latter is the correct application of the relationship between the soil and the crop to be grown on it. Although problems of soil management differ widely with the nature of the soil, the climatic conditions and the types of crops to be grown, yet, there are certain fundamental factors that should govern the choice of soil management practices everywhere. Good soil management must be based on the following guiding principles.

1. Selection of the right crops for a given soil or the right soil for a given crop.
2. Maintenance of soil so that it provides suitable conditions for the successful growing of selected crops.
3. Improvement of the productive capacity of soil.
4. Recommendation of only economically profitable husbandry methods.

In carrying out these, one has to collect and carefully weigh accurate information on the different points mentioned above. Field surveys, enquiries from experienced farmers, actual experiments in farms, and the results of investigations in laboratory and glass house must all be pressed into service.

In doing so, the scientific soil and crop management practices have greatly influenced the crops by increasing their yield manifold.

WEATHER AND ITS SIGNIFICANCE IN AGRICULTURE

The life of man depends fundamentally upon agriculture. In spite of the fact that he has greatly advanced in the technological field, the foundation of agriculture from the standpoint of feeding the population are still quite insecure. This is because climate, one of the great factors in crop production, still remains uncontrolled by man. The year to year variations in climate and sudden departures from normal values have deleterious effects on crop plants and keep the food supply highly fluctuating. Excessive and untimely rains, scanty rains, hot and cold waves dust-storms, thunder-storms, hailstorms and high winds are a few of the weather abnormalities which have a tremendous negative influence on crop production. In this study, the various elements of climate highlight the various detrimental effects of their abnormalities on the normal growth of the plants. This point is of particular interest to the workers in the field of agro-meteorology, because one of the aims of the subject is to analyze the weather abnormalities and the associated biological responses for mitigating them for better crops and more food.

The most important variables of climate, commonly known as the elements of climate, are solar radiation, atmospheric temperature, humidity, winds, rainfall, the precipitation of sleet, snow or hail. All these have got their own significance for the life of crop plants. The healthy growth of the plants, their high yield and good production depend upon certain optimum values of these elements. The role of these elements in crop life is analyzed and is given here in a summarized form. This analysis, of course, is highly arbitrary because of the fact that the elements influence the growth and development of the plants not in their individual capacities but in conjunction with one another. The importance of an individual element varies to a certain extent on the intensity of other elements.

Crop growth and development are primarily governed by environmental conditions, i.e., **climate** and **soil**. The success or failure of crops is intimately related to the weather during the crop periods. The modification of weather, except to a limited extent, is difficult and uneconomical. However, it is possible to get high crop yields by adjusting cropping patterns and by following suitable agronomic practices to mitigate the adverse effects of weather. Weather has significant influence on every phase of agricultural activity from preparatory tillage to harvesting and storage. A sound knowledge of the climatic factors and its interactions with crop is essential for successful agriculture.

Weather is a state or condition of atmosphere at a given place and at a given time. It is daily variations or conditions of lower layers of the atmosphere. Weather pertains to smaller area like village, city or even a district, smaller duration of time, i.e., part of a day or complete day, and is expressed by numerical values of meteorological parameters. Some examples are hot day, rainy day, cloudy weather, dry weather etc. Climate is a generalized weather or summation of weather conditions over a given period region during a comparatively longer period. Climate is weather conditions related to large areas like zone, state, country, part of continent or whole of country; longer duration of time like month, season or year and best described by the normal and averages, e.g., cold season, tropical climate etc.

Meteorology is the science that deals with laws and principles as they apply to atmospheric phenomenon. It is a science of atmosphere and its attendant activities.

Meteorology is also defined as a branch of physics that deals with physical process in the atmosphere that produce weather. Agricultural meteorology is the branch of applied meteorology which deals with the response of crops to the physical environment. It is the study of physical process of the atmosphere that produces weather in relation to agricultural production, and also deals agro-climatology, instrumentation and weather forecasting. Climatology is the science dealing with the factors which determine and control the distribution of climate over the earth's surface. Agro-climatology deals with the relationships of climatic regimes and agricultural production. It also includes instrumentation and weather forecasting.

Weather Elements

When it is desired to give the state of weather at any particular time and at particular place, the following elements are considered: (1) Solar radiation, (2) Air temperature, (3) Atmospheric pressure, (4) Wind movement, (5) Relative humidity, (6) Precipitation and (7) Clouds.

Solar Radiation

The radiation from the sun is the main source of energy which sustains the organic life on the earth's surface. All the three broad divisions of the solar spectrum are significant for plant life. The shorter wave lengths of the spectrum, known as ultraviolet rays, are chemically very active, and if the plants are excessively exposed to this segment of the solar spectrum, the effects are detrimental. The atmosphere, however, acts as a regulator of this type of radiation and a very small amount of ultraviolet rays reaches the earth's surface, and is normally tolerated by the plants. The part of the spectrum on the opposite side of ultraviolet rays is known as infra-red and is the long wave radiation. Infra-red of water vapours, this radiation does not cause any harm to the plants; rather, it supplies the necessary thermal energy to the plant environment. The middle part of the spectrum, which is visible and is known as light, is most important for the plants. Virtually, all plant processes are directly or indirectly influenced by this part of the spectrum. The light of correct intensity, quality and duration is essential to normal plant development. Light relations are frequently responsible for plant abnormalities and disorders.

The intensity of light is indispensable to photosynthesis. It governs for the distribution of photosynthates among the different organs of the pants. It also affects the production of tillers, the stability, the strength and the length of culms, the yield and the total weight of the plant structures, and the size of the leaves and the roots.

The length of the day, i.e., the duration of the light period, has profound effects on the contents of soluble carbohydrates and on the form of the carbohydrates present. The duration of light period accelerates the flowering of the cryophytes (cold-loving plants) and delays that of the non-cryophytes.

The duration of light period suppresses the growth of the shoots and the leaves and favours that of the roots, tubers and fruits. When other environmental factor is not a limiting one, the duration of the light period increases the photosynthesis.

Temperature

The heat energy of a plant comes to it mainly as radiant energy from its environment. Plants also receive additional energy through conductance and convection from the surrounding air, when it is warmer than the plants. The overall source of this energy is the solar radiation, a component of which a transformed into the sensible heat. Temperature is the index of the availability of this energy. The atmospheric temperature is the most important climatic variable which affects plant life. It provides the energy for several processes, like respiration. The growth of the higher plants is restricted to a temperature between 0 and 60°C, and the crop plants are further restricted to a narrower range of 10° to 40°C. However, each species and variety of plants and each age-group of plants has its own upper and lower temperature limits. When a specific range, a plant will be considerably damaged, it may be even killed. In an order range, its growth may stop and it may become dormant, but it may remain alive. Yet there is a third range of temperature in which the growth of the plant at a particular stage of the development is optimum. Thus, it is the amplitude of variations in temperature rather than the mean values of it, which is more important to plant growth.

The High Maximum (Mid-Day) Temperature

The mid-day temperature increases the saturation deficit of the plants. It accelerates photosynthesis and the ripening of the fruits. The maximum production of dry matter occurs when the temperature ranges between 20° and 30°C, provided the moisture is not a limiting factor. The high temperature can devernalize the cryophytes, especially the buds of the sun-exposed deciduous trees. When high temperature occurs in combination with high humidity, it favours the development of many plant diseases. High temperature also affects plant metabolism.

The High Maximum (Night) Temperature

High temperature during night increases respiration. It favours the growth of the shoot and leaves at the cost of roots, stolons, cambium and fruits. It governs the distribution of photosynthates among the different organs of the plants, favouring those which are generally not useful. The high night temperature affects plant metabolism. It accelerates the development of non-cryophytes.

Low Minimum (Night) Temperature and Frost

Most of the crop plants are injured and many are killed when the night temperature is very low. Tender leaves and flowers are very sensitive to low temperature and frost. Plants which are rapidly growing and flowering are easily killed by it. Low temperature interferes with the respiration of the plants. If it occurs when the soils are wet, it results in the accumulation of harmful products in the plant cells. Frost also interferes with plant metabolism.

Humidity, Rainfall and Precipitation

The importance of water to the plants is paramount in a number of ways. It is a raw material in photosynthesis. It is a universal solvent and dissolves all plant nutrients present in the soil. It is the medium by which the solute enters the plant and move about in the tissues. It increases the chemical activity of the compounds in the plants. Its presence in the plants is essential to the normal functioning of the cells and to the maintaining of their turgidity. The presence of water within the plants and the environment saves the plants from lethal temperatures.

Vapours in the Atmosphere or Humidity of the Air

The humidity of the air is of considerable importance to the life of the plants. It affects the intensity and quality of solar radiation, evaporation and transpiration. There are certain plants which directly utilize the water vapours from the atmosphere without demanding a change in its state into the liquid form. It is also a source of soil moisture.

High air humidity reduces the saturation deficit, but favours many plant diseases and pests. It increases the growth of the shoots and leaves at the expenses of fruits, stolons, bulbs and cambium. On the other hand, low atmospheric humidity increases the saturation deficit and accelerates transpiration. When the transpiration is high, the water that reaches the leaves is not sufficient for them even when there is no dearth of moisture in the soil. Because of this excessive transpiration, the stomata close and photosynthesis is interrupted. When the saturation deficit is extreme, most of the plants wilt and some are even killed because of physiological drought.

Rain

Although rain is of tremendous importance of the plants, it is of little direct significance, in general. It provides water for transpiration. A rainy weather is always associated with reduced evapo-transpiration. It favours plant diseases and pests. It also interferes with many farming operations, such as the preparation of seedbed, sowing, harvesting and the threshing of crops. Rain increases soil moisture, but it causes a fall in actual soil fertility. The deficiency or absence of rainfall causes the soil horizons to dry up, hampers the supply of moisture to the plants, damages the plants and reduces the yield.

Snow

Snow can be beneficial or injurious to the plants in a number of ways. Certain trees in cold climates have a strong performance to a particular amount of snow-fall. The Snow which melts in spring is a big source of moisture for the crops. The heavy snow-fall on the other hand, especially when followed by strong winds, breaks off the leaves, twigs and branches of the plants and can reduce the growth for a number of years. Broken trees are often attacked by insects, fungi and other pathogens. The glazing or coating of the leaves and stems causes a lot of damage to the plants by way of suffocation, accumulation of toxic materials and oxygen deficiency. Ice, in contact with the crowns and the roots, inhibits carbon dioxide diffusion.

Hail

Hail affects the crop plants suddenly and little can be done to prevent it. A hailstorm can decimate foliage, break branches, and shatters the fruit. Cereal crops are more seriously damaged. When the hailstorm is severe, the leaves are knocked off, the stalk and the heads are broken. When the crop is ripe a hailstorm shatters the grain. Lisions are caused on the plants owing to the mechanical rapture of the mesophyll cells. On many occasions, a hailstorm virtually beats a crop to the ground and the damage to it may be as high as hundred per cent.

Wind

Wind is an important weather element and both its direction and velocity are significant. The influence of wind is both local and regional. It influences the configuration and the distribution of plants in a region. It influences the plant life, both mechanically and physiologically. The influence is more pronounced on crop plants on flat lands near the sea coast and on the higher slopes of the mountains. Wind affects the plants directly by increasing transpirations and the intake of carbon dioxide and by causing several types of mechanical damages. Less significant effects are numerous, including the transporting of cold and heat waves, the moving clouds and fogs and the changing of water, light and temperature conditions.

Under natural conditions, wind increases transpiration. However, this increase is only up to a certain point, beyond which either it becomes constant or begins to fall. With the increasing wind velocity, there is a greater increase in cuticular transpiration than in stomatal transpiration. Wind increases the turbulence in the atmosphere, thus increasing the supply of carbon dioxide to the plants and resulting in greater photosynthetic rates. However, the increase in photosynthesis is again up to a certain wind speed, beyond which its rate becomes constant.

When the wind is hot, it accelerates the desiccation of the plants by replacing the humid air by dry air in the intercellular spaces. When the hot and dry winds blow at the time the cells are expanding and maturing, it results in the dwarfing of plants. This is because the cells can not attain full turgidity in the absence of optimal hydration and become fixed at subnormal sizes. When the developing shoots come under the influence of a strong wind pressure from a fixed direction, the normal form and position of the shoots are permanently deformed. Another severe injury to the plants caused by strong winds is lodging. This injury is commonest in crop plants, such as maize, wheat, and sugarcane. Strong winds break the twigs and shed the fruits from many plants. Further, the crops and trees with shallow roots are often uprooted. Many trees which bear comparatively big fruits have preference for light winds. Crops grown on sandy soils in areas where strong winds prevail are damaged because of abrasion. When the plant cover is not thick, strong winds remove the dry soil, so as to expose the roots of the plants and kill them. The eroded material from one place becomes a hazard to the existence of small plants in places where it is deposited. This is because the deposited material sharply reduces the aeration around the roots of the plants. Winds which blow from closed seas and lakes do a lot of salt spraying on the windward coastal areas, making it impossible to grow crops which are sensitive to excessive salt.

Considering the detrimental effects of strong winds and their influence on transpiration and photosynthesis, it seems that only a moderate wind speed is optimal for plant growth. There are, however, slight variations with plant species and other environmental conditions.

ENERGY MANAGEMENT IN CROP PRODUCTION

Energy is a critical aspect of a national development process. It is expended in agricultural operation in food processing and transportation, in the production of fertilizers, pesticides, and farm equipments. It is necessary for industrial operations that provide jobs. It is required for cooking, for household lighting and heating, and for the construction and operation of the infrastructure needed for schools, health centres and water supply.

To obtain an understanding of energy matters, it is essential to convert energy rapidly to a common equivalent. The tones of coal equivalent (t.c.e.) are such a unit and have been adopted by the United Nation (UN) bodies. Since, the chemical composition of coal is not a fixed quantity, it is necessary to assume a representative figure. A commonly used value for the t.c.e. is 2.9 x 10°J or 29 GJ. Electricity is normally quoted in terms of kilowatt hours or units. It is helpful to know that a tone of coal per year is approximately equivalent to an average power of 1 kW or 1 t.c.e. per year – 1 kW.

Energy Source

The energy sources can be classified in a number of ways based on the nature of their transaction, as commercial and non-commercial sources of energy. All energy sources, particularly the commercial ones, are natural. Coal, oil and nuclear sources constitute commercial sources, while firewood, biomass and animal dung constitute non-commercial sources. Also, the energy sources are classified based on animate and inanimate characteristics. Energy sources could also be classified as exhaustible/ non-renewable or non-depletable/renewable resources. The distinguishing feature of an exhaustible resource is that, it gets exhausted when used as an input of a production process, and at the same time, its undisturbed role of growth is nil. That is, the temporal services provided by a given stock of an exhaustible source are finite.

Further, based on conventionality in derived energy, energy sources could be classified as environmental (coal, oil, hydro, nuclear, etc.) and non-conventional (solar, wind, tidal, geothermal, biogas, etc.) sources. They are also classified a primary types (coal, firewood) or secondary type (electricity). Energy in its primary form can be of different kinds. The main types are chemical (fossil fuels-coal, oil, natural gas, peat; biomass – wood, agricultural residues, etc.) potential (water at a certain height), Kinetic (wind, waves), radiation (sun), heat (geothermal reservoirs, ocean thermal reservoirs) and nuclear (uranium). The primary form of energy must generally be converted into secondary or final forms of energy before it can be used. For instance, the potential energy of a waterfall (primary energy) is converted into electricity (secondary energy), which is transmitted and transformed to supply (final) energy, where it is converted into mechanical energy (useful energy) for productive operations.

Important types of secondary energy are electricity and mechanical energy. But chemical energy is also important as a secondary energy, for instance, refined oil products. Final energy is the energy that reaches the consumer. This can be electricity at a suitable voltage, or chemical energy in kerosene or batteries. Most of the energy sources are substitutable to each other due to the fact that some form of energy can be converted to other – such as coal to electricity, use of photo-electricity to drive a chemical reaction, wind energy to pump and store water that could be used to produce electricity when required, or solid biomass to produce liquid or gaseous fuels of higher calorific value. All forms are ultimately converted into heat. This gives rise to the inter-fuel substitution process with which an economy can substitute its abundantly available resources to the scarcely endowed one.

Classification of Energy

On the basis of source, the energy is classified and direct and indirect energy.

Direct Source of Energy

The **direct sources of energy** are those that release the energy directly like man power, bullocks, stationary and mobile mechanical or electric power units, viz., diesel engines, electric motor, power tiller and tractors. The direct energy may be further classified as **renewable** and **non-renewable** sources of energy depending upon their replenishment.

Renewable direct sources of energy: In this category the energy sources, which are direct in natural but cab be subsequently replenished, are grouped. The energies which may fall in this group are human beings, animals, solar and wind energy, fuel wood, agricultural wastes, etc.

Non-renewable direct source of energy: In this category, direct energy sources which are not renewable (at least in near further say next 100 years) are classified. Coal and fossil fuels exemplify non-renewable direct sources of energy.

Indirect Source of Energy

The indirect sources of energy are those which do not release energy directly but release it by conversion process. Some energy is invested in producing indirect source of energy. Seeds, manures (farm yard and poultry), chemicals, fertilizers and machinery can be classified under indirect sources of energy. Again, on the basis of their replenishment, these can be further classified into renewable and non-renewable indirect source of energy.

Renewable indirect source of energy: Seeds and manures can be termed as renewable indirect source of energy as they can be replenished in due course of time.

Non-renewable indirect source of energy: The energy sources that are not replenished come under non-renewable indirect source of energy. Chemicals fertilizers and machinery manufacturing are the non-renewable indirect source of energy.

On the basis of comparative economic value the energy may be classified as commercial and non-commercial.

Commercial Energy

The energy sources like petroleum products (diesel, petrol and kerosene oil) and electricity, which are capital intensive, exemplify commercial sources of energy. Considering the fact that most of the commercial sources are also non-renewable and to some extent are imported to India, efforts are made to conserve such sources of energy.

Non-Commercial Energy

Each and every energy source has some economic value. Some energy sources are available comparatively at low cost whereas others are capital intensive. The energy sources which are available cheaply are called non-commercial sources of energy whereas the ones which are capital intensive are called commercial energy resources.

Human labour and bullocks exemplify the category of non-commercial source of energy. Human labour and animals are readily available and can be used as a source of power directly. The commonly available and less expensive materials like fuel wood, twigs, leaves, agro-wastes and animal dung, etc., are also classified as noncommercial sources of energy.

Energetics and System Modelling in Cropping Systems

System evaluation constitutes rationalization of cropping within a defined frame work to examine the intent of various cropping systems. The transformation from a unit set of resource input to a unit set of product-output implies a system of integrating production resources to generate composite output. Attempts have been made to examine the various indices and methodologies developed to evaluate the land use efficiency, biological potential stating production efficiency and interference in association and succession, technical feasibility, economic viability, energetic and system modeling in cropping system research.

Cropping System Research (CSR) involves a critical study of externally applied, naturally available and environmentally exploitable resources integrated to harness temporal and spatial benefits in crop production. The interaction between the various production functions is both intimate and intricate. However, quantitative evaluation of a system's performance is imperative to accept or to reject and tailor it to needs commiserating with national production priorities. It calls for developing workable yardsticks for the quantification of benefits accruing from a particular system paying due regards to a large number of factors involved in its composition and performance. Efforts have successfully been made to develop and test various measurement indices for quantitative determination of Land Use Efficiency, Biological Potential (realized and realizable), Technical Feasibility, Economy Viability and Systems Modeling. An effort has also been made to put together the indices or procedures or models developed by various workers to evaluate cropping systems (Refer Chapter 7-Assessment of Agronomic Interactions). Further, energetic and system modeling in cropping system research is given below.

Energetic Approach

The energetic approach utilizes a system of calorific quantification of both the input materials and the product outputs. It has been widely used, but the recent fossil fuel crisis has renewed great deal of interest in this approach.

It claims that what really matters is the intrinsic value of the things given up or obtained or both; intrinsic value is hardly quantified by *market place pricing systems*. Calorific quantification is an attempt to indicate the intrinsic values of goods and efforts through a process that is not significantly influenced by the dictates of market systems.

In both, monetary and energy evaluation, input and output is weighed in common units, the latter having an excluding weakness of the former with only one limitation of not under stable by the actual user (farmer). This can be achieved by translating the evaluation yardstick into monetary values of final stage or by appropriately blending both the systems.

To determine energy use and energy yields, total resource accounting of total energy use, however, the process has to be extended. For that purpose, four basic energy indicators have been proposed by Liberero (1977).

1. Pro-Modernity Index (PMI)

It measures the extent to which energy input in the farm production process comes from the goods developed through the use of modern technology or manufacturing process, such as, machinery, manufactured fertilizers, insecticides, etc. The index is comparable to monetization.

2. Pre-Industry Index (PII)

Its corollary, Pro-subsistence Index (PSI) measures the magnitude of output used as raw materials for industrial processing or to support industrial workers (or used for direct consumption). It is comparable to the index of commercialization.

3. Energy Efficiency Index (EEI)

It measures the rate at which the use of energy inputs generates energy outputs. It is comparable to the Economic Efficiency Index (EEI).

4. Output Parity Index (OPI)

It measures the parity of energy output to energy input. It combines both the economic and energetic approaches. Operationally, it is the ratio of price/unit of energy output to the price/unit of energy input.

Burnett (1982) gave following indices for energetic of cropping systems.

5. Energy Intensiveness (M Cal / Rs. I)

It can be computed as indicated below.

$$I = \frac{\sum_{i=1}^{n} (E_i \; L_i)}{\sum_{i=1}^{n} (P_i \; Y_i \; L)}$$

Where, Li = Proportion of land under crop i, Ei = Energy input for crop I (M Cal/ ha)$_2$, Pi = Price of crop I (Rs./unit) and Yi = Yield of crop I, (unit/ha).

6. Renewable Natural Energy (RNE)

Renewable natural energy has been defined as the annual solar radiation plus the annual rainfall in excess (or minus the annual rainfall in deficit) of the global average, all measured in CEK Cal. (Coal Equivalent K Cal).

$$N = (\text{Area in sq m}) \frac{(\text{Solar K Cal})}{\text{Sq m/yr}} \times \frac{0.0005 \text{ CE K Cal}}{\text{Solar K Cal}} + \frac{\text{Rain } 81.1 \text{ cm}}{\text{Year}} \times \frac{11.28 \text{ CE K Cal}}{\text{cm sq m}}$$

n energy-quality factor for rain was calculated from the data on solar constant, the mean planetary albedo, the earth's radius, the global average annual rainfall, the earth's surface area and the energy–quality factor, relative sunlight to CEK Cal. One cm of rain/sq m embodies solar energy equivalent to 11.28 CEK Cal. The energy quality factor for rain is 11.28 CEK Cal/cm, sq m.

At the local level, rainfall in excess of global average (81.1 cm/yr) constitutes an embodied solar energy subsidy from outside the region, while a rainfall deficit in comparison to the global average constitutes an embodied solar energy export.

7. Fossil Fuel Based Feed Back (FFBF)

It represents the purchased inputs viz., fuel, electricity, equipment, fertilizer, seed, feed and stock fuels.

8. Nutritional Energy Equivalent (NEE)

An Energy Quality Factor (EOF) relating to protein and non-protein energy (sugar, starches, fats, and oils) was computed using the standard human nutritional requirements of 40.0 K Cal (Odum, 1971 and 0.55 protein/kg of body weight (United Nations, 1973). Since, protein has an energy content of 5 K Cal/g, 2.75 K Cal of the total are supplied by protein leaving a gap of 37.25 K Cal to be met by non-protein sources. The EQF was 13.55 K Cal non-protein energy/ K Cal of protein energy.

To determine the nutritional equivalent content of a food item, it was first apportioned into protein and non-protein parts. The protein energy was then converted to the quality of the non-protein energy using the EQF, and then multiplied by the Net Protein Utilization factor (NPU) ti account for protein quality. The two quantities, thus obtained summed up and divided by two (since it was a two factor analysis).

$$\text{NE content} = \frac{\text{Non-protein} + (\text{Protein energy} \times \text{EQF} \times \text{NPU})}{2}$$

NEE factors were defined as the ratios of NE content to actual heat kilocalorie content ranged theoretically from 0.50 for pure sugar to 6.775 NE/K Cal for pure protein of NPU 1. The NEE factors for some representative foods used in computation are given in Table 6.10.

TABLE 6.10: NEE factors for various food items

Food item	NEE/K cal	Food item	NEE/K cal
Apple	0.56	Pork	1.52
Beef	1.29	Potato	0.99
Cabbage	1.47	Rice	0.94
Chick pea	1.99	Soybean	2.04
Corn grain	0.88	Sweet potato	0.77
Eggs	2.82	Tomato	1.39

Food item	NEE/K cal	Food item	NEE/K cal
Fish	4.70	Wheat	1.16
Honey	0.52		
Milk	1.86		

9. Energy Ratios and Indices (FRI)

Energy ratios and indices based on inputs, (renewable natural energy + fossil-based feedback. F). F includes energy consumed both in food production and supporting the agro-system.

10. Investment Ratio (F/N)

The investment ratio is defined as the ratio of fossil-fuel-based feed back (F) to the renewable natural energy base (N). It is a measure of the relative work contribution by fossil fuel-base and renewable natural inputs. The greater the investment ratio, the greater is the fossil-fuel intensity and dependence of the system.

11. Energy Productivity Ratio (Y/F)

Ratios showing food yield/unit of energy input are most often used in evaluating the energetic of agricultural systems. The energy productivity ratio defined for this study was food production in nutritional equivalents (Y) to fossil-fuel-based feed back (F).

12. Life-Style Support Energy (F/P)

The life style support energy is defined as the ratio of fossil fuel-based feedback (F), to the number of agro-ecosystem inhabitants supported (P). It is a measure of the per capita amount of feedback from the main economy. One twenty-fourth of the number of people whose daily nutritional equivalent requirements could be met by exported food are included as agro-ecosystem inhabitants for the purpose of this index as one sixth of total US energy budget is consumed in the food system and one-fourth of this quantity is consumed in agricultural production.

While adopting of either economic or energy approach may be convincingly argued, the complex problems of cropping system not favour any single approach. It has become necessary to work in terms of the whole biophysical system of agriculture and release this to socio-economic system of a human community. The key approaches to the problem might be to use the methodology of economics with the calorie units of energetic.

SYSTEM MODELLING

Krezymuski (1980) in Poland attempted to design a model based on crop productivity and labour inputs to evaluate crop rotations. He converted crop yields to cereal units and weighted to take account of soil conditions and preceding crops in rotation. Total work requirement in terms of men and machinery was also taken into consideration. He then, developed calculations and evaluation indices for a series of model crop rotations in identical environmental conditions. However, the method requires further improvement for its adoption.

ENERGY

Energy is the most important component of economic infrastructure. There is a high degree of correlation between per capita energy consumption and the level of economic development. It is power, which drives the wheel of economic development. India is underdeveloped because its per capita energy consumption is much less compared to developed countries. There are two sources of energy (a) non-renewable energy (Commercial energy) like coal, petro-products and electricity and (b) renewable energy (non-commercial energy) like wind foams and solar energy. These are two conventional sources of energy.

Non-Commercial Energy

Sources of non-commercial energy include firewood, agricultural wastes and dried dung. These are non-commercial because these are available generally without payment in rural areas. Non-commercial sources of energy are renewable. The commonly available and less expensive materials like fuel wood, twigs, leaves, agro-wastes and animal dung, etc. are also classified as non-commercial sources of energy.

Commercial Energy

Source of commercial energy consists of coal, oil and electricity. These sources are commercial in the sense that they command a price in the market. Sources of commercial energy except hydroelectric power are non-renewable (or exhaustible). The energy sources like petroleum products (diesel, patrol and kerosene oil) and electricity which are capital intensive exemplify commercial sources of energy. Considering the fact that most of the commercial sources are also non-renewable and to some extent are imported in India, efforts are made to conserve such sources of energy.

ENERGY SOURCES GROUPED UNDER DIFFERENT CATEGORIES OF ENERGY

The various energy sources grouped under different categories are given in Table 6.11

TABLE 6.11: Energy sources grouped under different categories of energy

Category of energy	Sources of energy
Direct energy	Human, Animal, Fuel wood, Agricultural waste, Diesel, Electricity, Kerosene, etc.
Indirect energy	Seeds, Farmyard manure, Chemicals, Fertilizer, etc.
Renewable energy	Human, Animal, Fuel wood, Agricultural waste, Seeds, Farmyard manure, etc.
Non-renewable energy	Petrol, Diesel, Electricity, Chemicals, Fertilizers, Machinery, etc.
Commercial energy	Petrol, Diesel, Electricity, Chemicals, Fertilizers, Machinery, Seeds, etc.
Non-commercial energy	Human, Animal, Fuel wood, Agricultural wastes, Farmyard manure, etc.

ENERGY EQUIVALENTS FOR VARIOUS SOURCES

The energy equivalents for various inputs and outputs are given in Table 6.12

TABLE 6.12: Equivalents for Various Sources of Energy (Energy, MJ)

Particulars	Units	Equivalent	Remarks
A. INPUTS			
1. Human Labour			
Adult man	Man-hour	1.96	1 Adult Woman = 0.8 Adult man
Women	Woman-hour	1.57	1 Child=0.5 Adult man
Child	Child-hour	0.98	

Particulars	Units	Equivalent	Remarks
2. Animals			
Bullocks	Pair-hour	14.05	Body weight above 450 kg
Large	Pair-hour	10.10	Body weight 350-450 kg
Medium	Pair-hour	8.07	Body weight less than 350 kg
He-buffalos	Pair-hour	15.15	He-buffalo=1.5 medium bullock
Camel or Horse	Animal-hr	10.10	Camel or Horse= Medium bullock pair
Mule and other	Animal-hr	4.04	Small animal=0.4 medium bullock pair
3. Diesel	litre	58.31	
4. Petrol	litre	48.23	
5. Electricity	KWh	11.93	
6. Machinery			
a. Electric motor	Kg	64.30	Distribute the weight of the machine
b. Primemovers	Kg	68.40	very equally over the total life span of the
other than electric motors (including self-propelled Machines)			machinery (in hours). Find the use of machinery (hours) for the particular operation or crop.
c. Farm machinery excluding self-propelled machines	Kg	62.70	
7. Chemical fertilizers			
Nitrogen	Kg	6.60	Estimate the quantity of nitrogen, P_2O_5
P_2O_5	Kg	11.1	and K_2O in the chemical fertilizer. Then
K_2O	Kg	6.7	compute the amount of energy input from chemical fertilizer.
8. Farmyard manure	Kg	0.3	Dry matter
9. Chemicals Super chemicals	Kg	12.0	Chemical requiring dilution at the time of application
Zinc sulphate	Kg	20.9	
Inferior Chemicals	Kg	16.0	DDT, gypsum, or any other chemicals not Requiring dilution at the time of application
10. Seed			
a. Output of crop production system and processed			Same as that of output of crop production system
b. Output of crop Production system and is processed before using for Seed (e.g., potato, groundnut, cotton etc.)			Add 1.5,1.0 and 0.5 MJ/Kg for potato, groundnut and other seed respectively to the equivalent energy of the product of crop production system
B. OUTPUTS			
1. Main Product			
1. Cereal crop such as Wheat, Maize, (dry mass) Sorghum, Bajra, Barley, Oats, Rice (not shelled rice)	Kg	14.7	The main output is grains

Particulars	Units	Equivalent	Remarks
2. **Pulses such as** Mash, Mung, Lentil, (dry mass) Arhar, Soybean, Peas, Beans, etc.	Kg	14.7	The main product is grains
3. **Oilseeds such as** Cotton seed, Groundnut pods(not shelled), Sesame, Rape seed, Mustard, Linseed, Sunflower Seeds, etc.	Kg	25.0	The main output is seed except for groundnut, where it is pods
4. **Sugarcane** (harvested mass)	Kg	5.3	The main output is cleaned canes
5. **Vegetables**			
a. **Root or Tuber** vegetables			
High food values: Sweet potato, Topioca, etc.	Kg	5.6	
Medium Food Values: Colocasia, Potato etc.	Kg	3.6	
Low Food Values: Carrot, Radish, Onion, Beelroot, etc.	Kg	1.6	
b. **Fruit or Seed Vegetables**			
Broad beans, Cluster Beans, Ladyfinger, Musk melon, Water melon, etc.	Kg	1.9	
Gourd family Vegetables	Kg	0.8	
Drumstick Giant, Green papaya, Pumpkin, Tomato, Chillies, etc.			
c. **Leaf Vegetables**			
Cabbage, Spinach, Green Mustard, Leaves, etc.	Kg	1.2	
6. **Fruits**			
a. **Higher Food Values**	Kg	11.8	
Tamarind, Grapes etc.			
b. **Low Food Values**	Kg	1.9	
Fruit, Guava, Mango, Amla, Apple, Peach, Ber, Chiku, Cashew Fruit, Citrus, etc.			
7. **Fibre Crops**	Kg	11.8	Main product is fibre
Cotton, Sunhemp, (Dry mass) Sunflex, Jute, etc.			
8. **Fodder Crops**	Kg	18.0	The main product is dry or green fodder
Berseam, Lucerne, Senji, Oats, Maize, Bajra, Sorgham, Cowpea, Guara, Napier, etc.			
9. **Green Manuring Crops**			
Dhaincha, Cowpea, Sunhemp, etc.	Kg	18.0	Nutrients added to the soil through
10. **Fuel Crops**	Kg	18.0	The main product is fuel wood.
Sunhemp, Dhaincha, etc.	(dry mass)		
11. **Crops for Fodder, Fuel and Green Manuring Crops**			

Particulars	Units	Equivalent	Remarks
a. **Fodder from cereal crops:**	Kg	14.7	The main output is seed.
Maize, Oat, (dry mass) Bajra, Sorghum, etc.			
b. **Other fodder crops**	Kg	10.0	The main output is seed.
(Berseam, Lucerene, Senji, Guanara, ect.)	(Dry mass)		
Fuel Crops and Geen Manuring Crops			
11. By Products			
Straw, Vines,	Kg	12.5	
Stalk, Cobs, Plant wood	(dry mass)		
Etc.	Kg	10.0	
Fruits, vines, Straw (dry mass) Plant wood, etc.			
Leaves, Vines and	Kg	10.0	
Straw From vegetables	(dry mass)		
Cotton seed	Kg	25.0	
Fibre Crop seed other (dry mass) than Cotton			
Fuel crop seed	Kg	10.0	
Sugarcane leaves and tops	(dry mass)		
12. Animal Production System			
Milk			
Buffalo	Kg	4.9	
Cow, goat, etc	Kg	2.8	
Eggs (whole)			
Duck	Kg	7.58	
Hen	Kg	7.41	
Meat (cleaned)			
Poultry	Kg	4.56	
Mutton	Kg	4.98	
Fish	Kg	4.26	
Dung (dry)	Kg	18.0	

Source: Singh *et al.,* 1988)

ENERGY USE

1. Energy output to input ratio

$$\text{Energy output} : \text{Energy input} = \frac{\text{Energy output (GJ/ha)}}{\text{Energy input (GJ/ha)}}$$

2. **Energy productivity:** Energy productivity of a crop or system is the rice equivalent yield per unit energy input

$$\text{Energy productivity (kg REY/GJ)} = \frac{\text{Rice equivalent yield (kg/ha)}}{\text{Total energy input (GJ/ha)}}$$

3. **Energy output efficiency:** Energy output efficiency is the per day energy output of a crop or system.

$$\text{Energy output efficiency (GJ/ha/day)} = \frac{\text{Total energy output (GJ/ha)}}{\text{Duration (day)}}$$

4. **Energy intensiveness:** It is the energy output of the system per rupee invested.

$$\text{Energy intensiveness (MJ/Re)} = \frac{\text{Energy output of a system}}{\text{Cost of cultivation of the system (Rs/ha)}}$$

Where, Gj = Gega Joule, MJ = Mega Joule, REY = Rice Equivalent Yield

"Strong determination can out do anything."

7 Crop Ecology and Agro-Ecological Zones of India

> *"We must also initiate positive measures to reduce regional imbalances as otherwise the tension caused by such balances will inhibit the very process of development".*
> ***Indira Gandhi***

The term 'Ecology' is derived from 'Oikos' meaning home. Crop ecology means the relationship of a crop or crop communities with its surroundings say environment. A local climatic condition that differs from surrounding areas because of differences in relief, exposure or cover is called microclimate. From the agricultural point of view the microclimate is studied at two levels i.e. at plant level or at soil level.

MICRO-CLIMATE AT PLANT LEVEL

The eco-climate of the plant i.e. the climate just above and beneath the plant's canopy, is different from the climate of the surrounding environment. The microclimate at the plant level i.e. the eco-climate varies with the type of plants, morphological features of plant i.e. height, condensed canopy, shape and size of leaf and stem and soil colour. Soil colour modifies the microclimate by affecting the sunlight temperature, humidity, CO_2: O_2 level, air velocity and organisms. The plants having large morphological features reduce the air movement and the penetration of sun rays to the ground surface. Even emerging seedling alters the climate near the soil surface by reducing air movement and by shading the ground. In the vegetation covered areas, the extreme variations in the temperatures of the soil surface are reduced and the eco-climate of the plant is more humid and cooler than the surroundings which favours the spread of insect pest and disease. The plant canopy changes the intensity and quality of light when the light rays pass through it. In the day time when sunrays directly fall on the leaves, the leaf temperature may be higher than the air temperature and at night may be lower than air temperature.

MICRO-CLIMATE AT SOIL LEVEL

Soil colour is one of the factors which affect microclimate. Due to soil colour, parts of incoming radiations are reflected. The percentage of the incoming radiation that is reflected is called **albedo.** Albedo varies according to soil type, plant cover, moisture content of soil, soil organic matters, etc. Due to differences in albedo of the soil surfaces, different thermal regimes are set up even when incoming radiations are indenticals. For example a mulch on the soil surface becomes very hot by absorbing most of the incoming radiations and keeps the soil cool during day time and vice-versa during night hours.

Wet soil absorbs more incoming radiations than dry soil. Most of the absorbed energy is used to evaporate the soil water due to which moist soil takes longer time to heat up or to cool down. The heat gain or heat loss by the soil is five times quicker than by the water because the specific gravity of the soil is 0.2 and the sp. gr. of the water is 1.0. High soil temperature restricts the root growth. In the arid region warm (20°C-30°C) moist soil is the favourable medium for most of the cultivated crops.

Harvest Index

The effectiveness of photosynthesis depends on:

i) a large and efficient assimilating area.
ii) an adequate supply of solar energy and CO_2.
iii) favourable environmental conditions.

The total product of photosynthesis throughout the lifetime of the crop growing in the given circumstances depends upon the following factors:

a) the size of the assimilating area
b) the efficiency of assimilation and CO_2 fixation.
c) the period of efficient assimilation and CO_2 fixation.

In the Agronomy, the efficient utilisation of assimilation and CO_2 fixation i.e., Photosynthesis is expressed in terms of Harvest Index. H.I. is also called 'Coefficient of Effectiveness'.

$$\text{H.I.} = \frac{\text{Economic yield (e.g. grain)}}{\text{Biological yield (e.g. grain + straw)}} \times 100$$

$$= \frac{\text{Sink}}{\text{Source}} \times 100$$

Grain yield can be calculated by the formula:

$$Y = a \times b \times c \times d$$

Where,
y = grain yield per unit area
a = no. of plants per unit area
b = no. of fertile tillers per plant
c = no. of grains per ear
d = weight of the individual grain

In wheat, barley and similar cereals, the photosynthetic activity of the ear, which is situated at the top of the stalk, makes a considerable contribution to grain formation. Practically all the dry matter of the grain is produced by the part of the shoot above the flag leaf node. Of this dry matter, the ear contribute about 50% in wheat and upto 70% in barley.

Sink- Source Ratio

Source: Such part of the plant where photosynthesis occurs; generally leaves are the source.

Sink: Where photosynthates accumulate is called sink. Generally grain is the sink but stem (e.g. Potato), root (e.g. Sugar beet) or leaf (e.g. Palak) may be the sink.

For higher HI, development of sink is the most concerned. But the development of sink depends on three factors:

a) Good Source
b) Partitioning of Photosynthates, and
c) Translocation of Assimilates

In spite of good source, if partitioning of photosynthates is not well developed, sink will be poorly formed and thus is the lower H.I. Therefore. for higher sink-source ratio, the followings are necessary:

a) Source should be good.

i) Partitioning of source i.e. photosynthates should be well developed.
ii) Translocation of Photosynthates must be good.
iii) And lastly photosynthates do accumulate as sink.

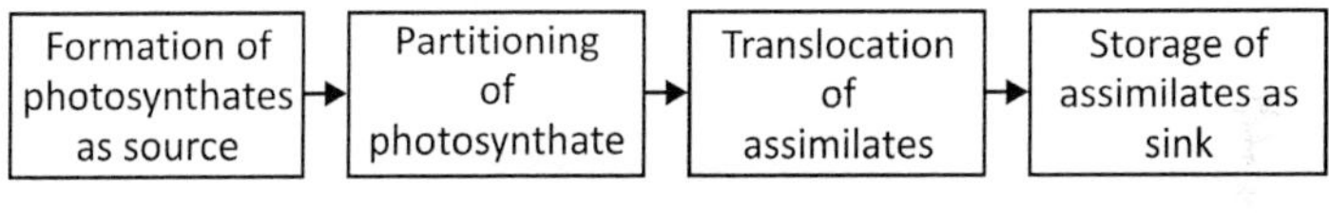

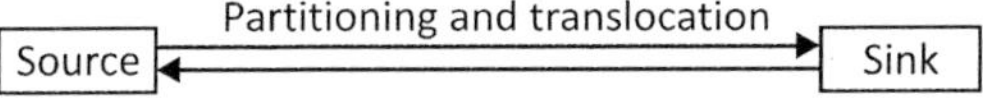

The formation of sink from source is a reversible process but source to sink is more pronounced.

With the appearance of inflorescence, competition for assimilates between the leaves and the inflorescence takes place. When the seed formation starts, most of the assimilates move to the grain. The internal competition is more severe, as competition between plants increases i.e. plant population pressure, which is illustrated by the figure.

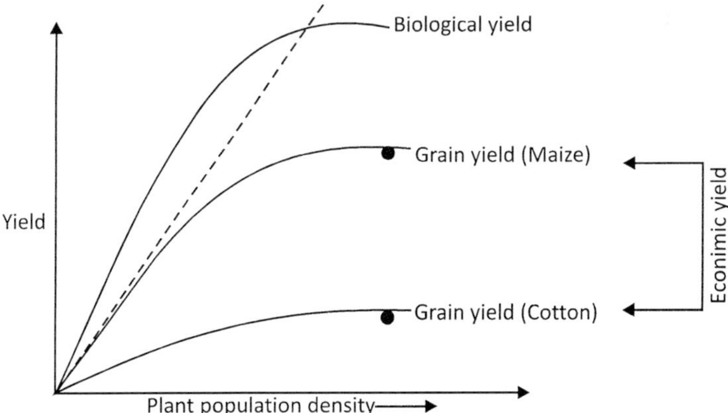

IDEOTYPES (MODEL PLANTS)

The production of every plant is affected by the plant's type, climatic condition, soil type and management factors. According to Army and Greiver (1967), the production of the plant might be increased by changing plant's type and increasing the period of grain-filling in a certain climatic condition where management and soil factors are not liming. There is a direct relationship between the plant's type and the crop production because the orientation and number of leaves play the important role in the CO_2 assimilation (photosynthesis).

The concept of ideotype was given by **Donald** in 1968. He illustrated that there should be minimum competition between the crops and the crop must be competent one to compete with weeds. The single plant would give the better result in a group when the crop has at least competition with the same type of the crop. Ideotype is the model plant which may be defined as " a biological model which is expected to perform or behave in a predictable manner within a defined environment ". On the basis of environment Donald & Hamblin (1976) identified two forms of ideotypes i.e. isolation ideotypes and competition ideotypes. Competition ideotypes are suitable for mixed cultivation.

IDEOTYPES FOR SOME CROPS

Wheat

According to Donald (1968), thc ideotype for wheat crop has following features:

i) To avoid lodging, there should be short strong stem.
ii) Few small erect leaves to allow the sunshine into its canopy.
iii) A large erect ear.
iv) The no. of fertile florets per unit area should be more so that HI is more.
v) There should be awns to increase the photosynthetic area.
vi) A single culm to avoid wasteful vegetation growth.
vii) Resistant to insect pests & diseases.
viii) There must be proper partitioning and translocation of assimilates.

Maize

The plants have erect upper leaves and the lower leaves gradually become horizontal to allow the sunshine into its canopy and for proper movement of air into the field. The height of the plant is to be 1.5 metre in which 2-3 cobs may be produced on the nodes near the tassel.

Gram

Pande and Saxena (1973) proposed the ideotypes for gram having following features-

i) The vegetative growth must be stopped before the starting of reproductive stage.
ii) The plant is to have erect branching. In the prevalent varieties of gram, the spreading and branching of its canopy is just like umbrella which interferes to penetrate the sunshine into its canopy causing humid conditions favourable for insect pests & diseases.
iii) To harness the long photoperiod and favourable temperature at the time of flowering, there should be 2-3 longer pods in the leaf axis and 2-3 seeds in each pod.

Arhar

The growth of arhar varieties in the beginning is too less i.e. in the first two months only one or two branches are come out. Therefore Arhar is unable to harness the solar energy properly in the first two month. The flower's drop is also a major problem. The filling of pods according to Hydro-dynamic model sets up the competition between vegetative and reproductive phases. Considering all the views Pande and Saxena suggested ideotypes having following features:

i) The fast growth of plant's canopy at least in the beginning.
ii) The reproductive phase starts after the closure of vegetative growth.
iii) Long floral axis having 2-3 flowers in each trifoliate axis.
iv) Synchronised flowering i.e. all the flowers come out at the same time.
v) Active root nodules for the long time.
vi) Resistant to insect pests & diseases.

Rainfed Upland Rice

i) Short growth duration (85-100 days)
ii) Effective deep root system
iii) Dwarf (less than 100 cm) plant having erect leaves and thick stem.
iv) Early strong fertile tillering.
v) Synchronised flowering.
vi) Good number of panicles at higher density about 400 panicles per square meter.
vii) Highest number of grains per panicle.
viii) Moderate seed dormancy.
ix) Resistant to insect pests and diseases e.g. resistant to *Helminthosporium sp.*

Rainfed Wheat

An ideotype of rainfed wheat was conceived and proposed by Dr. Asana:

i) Large no. of spike lets,
ii) Large peduncle,
iii) Strong and deep root system,
iv) Flat leaves parallel to soil and
v) A grain development period coinciding with mean max: temp. of 25°C.

Ideotypes for Dry Land Farming

i) Short growth duration
ii) Effective root system
iii) Drought tolerance
iv) High yield potentiality with altered morphology viz:
   a) Plant with few leaves just sufficient to maintain photosynthetic output and growth (to minimise the use of water)
   b) Leaves horizontally disposed for better light interception contrary to vertically disposed most effective under irrigated conditions.

CLIMATE AND PLANT RESPONSES

The intensity and duration of light, the temperature, the atmospheric conditions, including the humidity and wind velocity, the quantity and pattern of precipitation etc. constitute the climate of any place.

1. Light – General Description

Light is a form of radiant energy. It is an essential factor for the primary production of plant materials upon which all other living organisms depend directly or indirectly. Solar energy is received in the form of electromagnetic waves.

Visible range .... 380–720 nanometers (nm)
Chloroplast pigments absorb........................ 330–740 nanometers wavelength

Light is form of energy and hence it can be converted into other forms of energy such as heat.

Quantity of light is expressed as g cal/m^2/year

Intensity of light is measured as candle power.

Foot candle. The light intensity from a standard candle at one foot distance is called foot candle (F.C.)

Lux. The light intensity from a standard candle at one metre distance, a metre candle or lux.

1 Foot candle = 10.76 lux.

2. Compensation Point

It is that intensity of light at which the rate of photosynthesis is just sufficient to meet the requirements of respiration. At this point the dry weight of plant does not increase. This point differs from species to species and in the same species in different individuals of different age.

PLANT CLASSIFICATION BASED ON PREFERENCE TO LIGHT

- *Heliophytes.* These plants grow in open sun.
- *Sciophytes.* These plants grow in shade.
- *Facultative sciophytes.* Heliophytes, which can grow in shade also but not so well.
- *Obligate heliophytes.* Heliophytes which fail to grow in shade.
- *Facultative heliophytes.* Sciophytes which may also grow in light.
- *Obligate sciophytes:* Sciophytes that fail to grow in the sun.

Photoperiodism

The length of day and night periods in the twenty four hour cycle also regulate the leaf fall, flowering and other phonological processes in plants. This phenomenon is called photoperiodism.

Critical Periods

Critical day length (CDL). The critical day length is that continuous duration of light which must not be exceeded in short day plants and should always be exceeded in long day plants in order to bring them to flower.

CDL: For short day plants .... 15.6 hrs
For long day plants .... 9 hrs

Critical dark period:. It is that continuous dark period which must not exceed in long day plants and should always be exceeded in short day plants. Critical dark period is more important in short day plants.

CLASSIFICATION OF PLANTS ON THE BASIS OF PHOTOPERIODISM

- *Short day plants.* These plants flower below a critical day length or photoperiod e.g. tobacco, chrysanthemum. Also are called as the long night plants.

- *Long day plants.* They generally require long photoperiods which must be above a critical length of light hours e.g. henbane, spinach, radish, oats, barley.
- *Day-neutral plants.* Those species in which phenological events are indifferent to length of light or dark periods e.g. cotton and maize.
- *Short-long day plants.* These plants require short photoperiods for floral initiation and long photoperiods for blossoming e.g. *Trifolium repens and Secale cereale.*
- *Long-short day plants.* These plants come to flower only when they experience long photoperiods followed by short photoperiods e.g. *Bryophyllum* and Castrum.
- *Intermediate plants.* These plants can flower only within a definite range of light hours, above and below which no flowering occurs e.g. some varieties of sugarcane.
- *Amphiphotoperiodic plants.* They behave just opposite to the intermediate plants, flowering both above and below an intermediate day length. The plants remain vegetative in intermediate day length e.g. *Media elegans.*

Processes Involved in the Production of Flowers

Photoreceptor

Light wavelengths effective in producing response generally lie between 580 nm and 780 nm. Photoreceptor for this region is phytochrome which functions above 550 nm and is found in all photosensitive organs. Phytochrome can be extracted from leaves.

Flowering Hormone

It is believed that the flower inducing stimulus is in the form of a chemical. It has been named as florigen.

VERNALIZATION

Vernalization has been defined as a process which promotes flower initiation by a previous cold treatment. It has been successfully applied in a number of plants mostly belonging to the families gramineae and leguminoseae.

Examples are: wheat, oat, rye, pea, clovers, etc.

1. Site of Vernalization

The perception of stimulus of cold treatment is normally received by the shoot apex in contrast to leaves in photoperiodism. In seeds the stimulus is perceived by the embryo and in the embryo the site of vernalization is the plumule.

2. Requirements of Vernalization

Low Temperature

Usually between:........0-10°C
Maximum:........17°C
Minimum:........6°C

- *Active embryo or shoot.* Vernalization does not occur in dry seeds. Seeds should be semi-germinated and should have an active embryo.
- *Aerobic respiration.* Vernalization is an aerobic process. It also requires an active metabolism. The aerobic process has, however, a high temperature coefficient.
- *Water.* For activation of embryo, the seeds are soaked in water for 22-24 hours to the moisture content to 30-40 per cent.

3. Devernalization

The effect of vernalization can be prevented by high temperature (35-40°C), reduced water supply and unfavourable photoperiods. In general, devernalization can be carried out immediately after vernalization treatment.

TEMPERATURE

Temperature is a measurement of the degree of hotness.

1. Cardinal Temperatures

Optimum, maximum and minimum temperatures are called cardinal temperatures.

The cardinal temperatures vary from species to species and within the same species from pant to pant and in the same part at different age.

Optimum temperature. The temperature at which physiological processes are at their maximum efficiency.

2. Temperature Coefficient or Q_{10}

The ratio of rate of reactions at differences of 10°C is called the temperature coefficient.

CLASSIFICATION OF PLANTS BASED ON TEMPERATURE REQUIREMENTS

- *Eurythermic plants.* Most of the land plants thrive in quite a wide range of temperatures from almost freezing to about 55°C and these are called eurythermic plants. But their optimum range is not so broad and show best growth between 20-30°C.
- *Stenothermic plants.* Aquatic environments usually show less fluctuation in temperature and hence many aquatic plants thrive in a narrow temperature range and these are called stenothermic plants.

Lethal Limits of Temperature

The range of temperature beyond which on either lower or upper side the species dies is its lethal limits.

Activity Limit

The range of temperature in which metabolism is active and favourable is called as activity limit.

Thermoperiodism

The temperature of any place changes constantly from hour to hour in a diurnal cycle. Many plants have, in course of time, got adjusted to the rhythmic diurnal cycle in temperature changes for several phenological events. This regulation of phenology to periodic thermal changes is called thermoperiodism e.g. in some tomato varieties alternating temperature of 26°C and 18°C promotes higher fruits production and better vegetative growth.

CLASSIFICATION OF PLANTS BASED ON WATER REQUIREMENTS

1. Hydrophytes

Hydrophytes are those plants which grow partly or fully submerged in water at least for some weeks.

Classification of Hydrophytes

Macrophytes, such as:

- Marginal emergent plants
- Rooted submerged hydrophytes
- Rooted hydrophytes with floating leaves
- Free surface floating hydrophytes

Mesophytes. Mesophytes are plants that normally grow in habitats where water is neither scarce nor abundant. In such habitats the pore space in soil is occupied almost equally by water and soil atmosphere e.g. wheat, maize, barley, peas, gram, sugarcane etc. and species in grasslands, meadows, tropical and temperate forests are all mesophytes.

2. Xerophytes

These are plants that usually grow in extremely dry or arid conditions. There is no hard and fast line between the water status of habitats for meso and xerophytes and the term is only a relative one. Xerophytes do not always show preference for dry habitats for a good growth but they are only adapted to grow in dry habitats and assume characteristic dominance in such habitats e.g. desert etc. due to absence of other types of plants.

Types of Dry Habitats

- Actually dry habitats
  - Wet habitat wherefrom water is not available to plants on account of high osmotic concentration of salts or freezing temperature etc.
  - From the standpoint of habitats, xerophytes may be of the following types:
- *Psammophytes* are plant that grow on sand or small pebbles usually in day habitats like *Acacia Senegal.*
- *Lithophytes* are the plants that grow on rocks where even in good rain, water does not stay. Species of lichens, *Linaria, Selaginella* are the common examples.

- *Phychrophytes* are plants that grow in soils which become very cold and the availability of water is limited on account of frequent freezing. In India, such plants are restricted to the Himalayas.
- *Halophytes* are the plants that grow on highly saline soils where water is poorly available to plant on account of high osmotic concentration e.g. *Brugaria polyrrhiza, Rhizophora spp. etc.*

BIOTIC INTERRELATIONSHIPS

1. Symbiosis

The association of two different species for mutual benefit is called symbiosis.

2. Commensalism

Among animals, there is some slightly different association where the benefit derived by weaker partner is less clear. Such association is called commensalisms.

3. Mutualism

Specialized association of algae and fungi in lichens is called mutualism.

4. Mycorrhizal Association

In mycorrhizal associations, tree roots become infested with fungal hyphae. The fungi derive their food from tree roots and in turn supply water and nutrients that they absorb from the soil, much like root hairs of trees.

5. Parasitic Relationship

Between plants and plants, or animals, and plants are of very widespread occurrence. In fact all plants and animals have some kind of parasite which impoverishes the host. Parasites greatly regulate the population level of plants and animals.

6. Prey-Predator Relationship

This is a kind of relationship of eating and being eaten of one species by another found universally among animals of all kinds of ecosystem. Both the predators and preys regulate the population size of each other and in ultimate effect regulate their own populations through the medium of other.

7. Epiphytism

This is still another type of biotic association where the epiphytes grow on other plants but do not derive their food from the supporting plant. These also grow on a variety of plants occupying different positions, but mostly they are to be found on the stems of higher plants.

8. Amensalism

When chemical exudates or their decomposition products harm or depress one set of organisms while the other remains stable is known as amensalism.

Types: --- Antibiosis --- Allelopathy.

AGRO-CLIMATIC REGIONAL PLANNING

Agro-climatic Regional Planning in India was initiated in **1988** by the planning commission. It means under the seventh five year plan; Regional Planning was come into force on the basis of agro-climatic factors for balanced regional growth. The agro climatic factors are rainfall, temperature, soil, topography, cropping and farming system and water resources.

Agro-Climatic Zones

Agro-climatic zone is a land unit in terms of major climates suitable for a certain range of crops and cultivars (FAO, 1983).

On the basis of criteria of homogeneity in agro-characteristics such as **water surplus and water deficit,** India was divided into 11 agro-climatic zones.

Later on, India was divided into 15 agro-climatic zones on the basis of **water surplus, water deficit** and **cropping system.**

An ecological region is an area of the earth's surface characterised by distinct ecological responses to macro-climates expressed by soils, vegetation fauna and aquatic systems.

Agro-ecological region is the land unit cut out of agro climatic region when superimposed on land form and the kind of soils and soil conditions that act as modifiers to climate and length of growing period.

The classification is on the basis of (i) Growth period (ii) Rainfall/potential evapo-transpiration (iii) Soil (iv) Districts

There are two components of ecological zones:

i) Biotic factors viz Fauna and Flora
ii) Abiotic factors viz water, minerals, light etc.

At present India has been divided into 21 agro ecological zones. The detailed descriptions of 15 agro-climatic zones are as follows:

1. **Western Himalayan Region:** includes Jammu and Kashmir, Himachal Pradesh & Hilly areas of U.P. climatic is cool and humid. Annual rainfall – 1650 – 2000 mm. Important crops are Wheat, Maize, Paddy and Potato. It is divided into 3 sub regions. Cropping Intensity is lowest in J & K and highest in H.P. The productivity level of all the crops is below the national average. To develop this region:
   a) intensify the soil and water conservation planning
   b) Land utility planning should be implemented viz. Agriculture for 30% sloppy land, Horticulture for 30-50% slope, rice for above 50% slope.
   c) Good storage for transportation, Marketing & processing.
   d) Irrigation management like water harvesting technique.
   e) Livestock management etc.
2. **Eastern Himalayan Region:** Sikkim, Hills of Darziling, Arunachal Pradesh, Nagaland, Meghalaya, Manipur, Mizoram, Assam, Jalpaiguri area of W.B. and some parts of Bihar. Climate humid & sub-humid, 1840-3528 mm annual rainfall, 5 sub-regions, main crops: Maize, Paddy, Jute, Rapeseed. Shifting cultivation on 1/3 rd of cultivated area, soil erosion more. For the development of this region:
   a) Soil and water conservation for each watershed.
   b) Educate the people about the harmful effect of land erosion at Panchayat level.
   c) Facilities of good supply of inputs, marketing transportation, processing & storage.

d) Adoption of Integrated Farming system because of different dietary pattern of local people.
e) Good communication &transport facility.

3. **Lower Gangetic Plain:** West Bengal; Productivity of rice is lower then the national average 15 q/ha but this region contributes 12% of the total rice production in India. Crops: Rice, Jute, Rapeseed and Wheat. Rapeseed, Maize & Potato are newly introduced; moist humid and dry humid climate, 4 sub regions. 1302 - 1607 mm annual rainfall. Followings are negative points: (i) lack of good water management; drainage problem. (ii) Use of local variety (iii) Less use of HYV (iv) Poor extension work (v) Inadequate supply of inputs (vi) Poor marketing & processing (vii) Lack of transportation and storage (viii) Dependent on monsoon (ix) soil & water erosion (x) Uneducated & poor farmers (xi) Poor electricity supply.
4. **Middle Gangetic Plain:** Eastern parts of U.P. and Bihar plains. 40% of total cropped area is irrigated cropping intensity 142% (India's national C I 127%) low productivity of rice, 3 sub-regions of eastern U.P. and 9 sub-regions of Bihar, moist sub humid to dry sub humid climate. 1211- 1470 mm annual rainfall, main crops: Sugarcane, Paddy, Maize, Wheat.
5. **Upper Gangetic Plain:** Remaining parts of UP and 3 sub regions, **drysub humid** to **subdry climate,** 721-979 mm annual rainfall. Main Crops: Rice, Wheat, Maize and Sugarcane. Irrigation intensity is 131% and C I is 145%. Irrigation by canal and tube well. Special planning to grow vegetables & fruits scientifically. Milk production and processing units are being established. **Earst while Uttar Pradesh** is divided into **4 agro-climatic zones:**
   a) Western Himalayan Region **(Horticulture)**
   b) Middle Gangetic Plains (Rice, Sugarcane)
   c) Upper Gangetic Plain
   d) Central plateau and Hills
      i) Maximum dry areas: Western dryland region.
      ii) Gujarat Plains and Hills are called oilseed region.
      iii) Plantations and spices are grown in western coast and ghats.
      iv) Coconut cultivation in Island region.
6. **Transgangetic Plains:** Punjab, Haryana, Shiwalik region, Sriganagar, Delhi semi-dry areas of plains, dry areas of Thar (Rajasthan), Maximum net shown area and irrigated area. Cropping Intensity is the highest, low poverty, maximum use of ground water. Followings points for the development of this region:
   a) good water management
   b) management and reclamation of user soil.
   c) Proper use of ground water
   d) Diversification of crops.
   e) Introduction of Leguminous crops in cropping system.
   f) Proper livestock and pasture management
   g) Increasing of area under vegetables, fruits and pastures.
7. **Eastern Plateau & Hills:** Wainganga, Eastern Hilly areas of M.P., inland Orissa, Chotanagpur plateau, plateau and Hilly regions of West Bengal, Chhattisgarh. 5 sub regions, moist sub humid to dry sub humid. 1271-1436 mm annual rainfall, Main crops: Rice, Wheat, Maize, Ragi, 67% small and marginal farmers have 25% of agricultural land only shallow and undulating land. Horticulture development planning is being intensified. Irrigation is through tanks and tubewells.
8. **Central Plateau and Hills:** About 46 districts of M.P. Chhattisgarh, U.P. and Rajasthan. Undulating soil topography, Ravines present, 14 sub regions, 400-1550 mm annual rainfall, semi-arid to dry sub-humid climate, Main crops:

Wheat, Gram, Jowar, Bajra, Paddy. 75% area is rain affected where following programmes are being run:
a) Water shed management
b) Crop diversification
c) Reclamation of Ravine area
d) Other programmes

9. **Western Plateau & Hills:** Main regions of Maharashtra, major parts of M.P. and some parts of Rajasthan. 4 subregions, semi-arid climate, 602-1040 mm annual rainfall. It covers maximum parts of peninsular area. Major crops: Jowar, Bajra, Cotton and Wheat. Net sown area 65%, 11% forest area, irrigated area 12.4%. Irrigation mainly by canal. 50% Jowar of our country is from this region. Best quality of orange, grape and banana. Following suggestion for this zone:
a) Management of minor irrigation
b) Improvement in the conditions of cattle and buffaloes.
c) Cultivation of costly crops.

10. **Southern Plateau and Hills:** About 35 districts of Andhra Pradesh, Karnataka, Tamil Nadu, typically semi arid zones. 81% dry land farming and 111% cropping intensity. 6 sub regions, 677-1000 mm annual rainfall, cultivation of low value crop. Suggestions:
a) Proper utilization of dry land technology in watershed area.
b) Increase fertilizer use efficiency.
c) Minor irrigation programme.

11. **East Coast Plains:** East coast of T.N., Andhra & Orissa. Soils are mainly alluvial and coastal sands. Irrigation through canals and tanks. 6 sub regions semiarid and sub humid climate, 780 – 1287 mm annual rainfall. Main crops: Rice, Groundnut, Ragi, Jowar and Bajra. 20.3% of total rice production in India is from this zone and 17.5% groundnut. Prevalence of alkaline and saline soils. Development of fisheries due to waterlogging. 75% area rainfed. In place of shifting cultivation, integrated Horticulture Agriculture programme should be adopted by tribal people.

12. **West Coast Plains and Ghats:** West coast of T.N., Keral, Karnataka, Maharashtra and Goa. Main occupation is cultivation of spices and plantation crops; 4 sub regions, dry sub-humid to humid, 2226-3640 mm annual rainfall, main crops: Rice, Ragi, Groundnut and Tapioca, suggestions:
a) Storage of rain water.
b) Proper use of ground water.
c) Increase in area under fruits and vegetables.
d) Fisheries management

13. **Gujarat Plains and Hills:** 19 districts of Gujarat, 7 sub regions, arid and low rainfall, 340-1793 mm annual rainfall, 32.5% irrigated land through wells and tube wells. Main crops-Maize, Wheat, Rice, Groundnut, Tobacco, Cotton, Jowar, Bajra. This zone is famous for oilseed crops hence called oilseed region. C I: 114%. About 60% area is drought prone. 78% rainfed; suggestions:
a) Rain water harvesting and its management
b) Dry land farming & management of canal & ground water.
c) Development of Agro-forestry and-Horticulture.

14. **Western Dry Region:** 9 districts of Rajasthan, hot sandy desert, erratic rainfall, high evaporation, scanty vegetation, no perennial rivers, general occurrence of drought, 95 mm annual rainfall, 1.2% forest area, 4.5% pasture area, 6.3% irrigated area, C I: 105%, main crops: Bajra, Gram, Wheat, Rapeseed. Two main plannings:
a) Indira Gandhi Canal and

b) DDP (Drought Development programme).

15. **Island Regions:** Island territories of Andaman & Nicobar and Lakshadeep. 3000 mm annual rainfall spread over 8-9 months. Largely forest zone with undulating land. Coconut cultivation on 50% area of total cropped area, humid climate, main crops: Rice & Coconut. Main emphasis on:
    a) Crop Improvement
    b) Water Management
    c) Fisheries.

Agro-Ecological Zones

India has been divided into 21 agro-ecological zones:

1. **Western Himalaya I:** Cold arid region with shallow sedimentary soils, covering Laddakh and Gilgit. Temp less than 8°C, Rainfall less than 150 mm which is 15% less than annual PET. Growing period less than 90 days. Covers 15.6 million hectare (4.7% of total geographical area).
2. **Western Himalaya II:** Warm sub humid with brown forest & podzolic soils, J & K, H.P., northwest Hilly areas of U.P. (Uttaranchal); 17.7 mha (5.4%), Rainfall 1600-2000 mm, exceeds PET, water surplus, Growing period 150-210 days.
3. **Western Plains:** Hot arid region with desert and alkaline soil. Covers west Rajasthan, Southwest Haryana, Punjab, Kutchha Peninsulsa, north of Kathiawada. 29.6 mha (9%of total geographicl area). Rainfall 300 mm, PET: 1500-1900 mm, Deficit 1200-1600 mm, growing period less than 90 days.
4. **Deccan Plateau I:** Hot arid region, with mixed red & black soil. Part of Deccan plateau including district of Raichur, Bellary of Karnataka & Anantpur of Andhra. 4.7 mha (1.4%), Rainfall 400-500 mm, PET:1800-1900 mm, Growing Period less than 90 days.
5. **North Plain and Central Highland:** Hot semiarid region with alluvial derived soil. Parts of North plain, Central High land and Gujarat plains. 32.9 m ha (10% of total geographical area), Rainfall 400-800 mm, PET: 1400-1800 mm, Growing period 90-150 days.
6. **Central (Malwa) Highland and Kathiawada Peninsula:** Hot, semiarid region with medium deep black soil, west M.P., East Rajsthan, and Gujarat, Kathiawada, 1.84 mha (5.6%), Rainfall 600-900 mm, PET: 1600-2000 mm, Growing period 90-150 days.
7. **Central Highland (Malwa & Bundelkhand):** Hot sub humid region with medium & deep black soil, Central highland including distt. of Raipur, Sagar, Bhopal, Sehore, Shajapur, Hoshangabad. 8.2 mha (2.5%) Rainfall 1000-1500 mm, PET: 1300-1500 mm Growing period 1560-180 days.
8. **Deccan Plateau II:** Hot semarid with shallow & medium black soil, West Maharashtra, North Karnataka, A.P. 33.0 mha (10%), Rainfall 600-1000 mm, PET: 1600-1800 mm, Growing period: 90-150 days.
9. **Deccan Plateau and Eastern Ghat:** Hot semi arid with red and black soil. Major part of Andhra Pradesh, 20.8 mha (6.3%), Rainfall 600-1000 mm, PET: reduction 40%, growing period: 90-150 days.
10. **Eastern Ghat (TN upland) and Deccan Plateau:** Hot semi arid region with red loamy soil, South of Deccan Plateau, T.N. upland, West Karnataka, 22.7 mha (6.9%), Rainfall 600-1000 mm, PET: Deficit 400-700 mm, growing period 120-150 days.
11. **Deccan Plateau and Central Highland (Bundelkhand):** Hot and sub humid region with red and black soils. Bundelkhand part of M.P., North Vidarbha, 13.7

mha (4.2%), Rainfall 1000-1500 mm, PET: 1500-1600 mm, Deficit: 500-700 mm, growing period: 150-180- days.

12. **Eastern Plateau (Chhatishgarh Region):** Hot subhumid with red and yellow soils; parts of eastern plateau covering chhatisgarh region of M.P. and southwest Highland of Bihar. 13.2 mha (4%) Rainfall 1200-1600 mm, PET: 1400-1500 mm, (deficit 500-700 mm) growing period: 150-180 days.
13. **Eastern (Chhotanagpur) Plateau & Eastern Ghat:** Hot sub humid region with red loamy soil, Chhotanagpur plateau of Jharkhand, West of W.B., Eastern Ghat of Orissa, Baster region of M.P., 17.8 mha (8.5%), rainfall 1000-1600 mm, PET: 1400-1700 mm, growing period 150-180 days.
14. **Northern Plains:** Hot sub humid region with alluvial soil, part of Indo Gangetic plain including plain of west Himalaya, Area 12.2 mha (3.7%), Rainfall 1000-1200 mm, PET 1400-1800 mm, growing period 150-180 days.
15. **Eastern Plains:** Hot sub humid region with alluvial soil, North east UP, North Bihar, Foothill of eastern Himalaya, 9.3 mha (2.8%), Rainfall 1400-1600 mm, PET: 1300-1500 mm, growing period: 180-210 days.
16. **Assam & Bengal Plains:** Hot humid with alluvial soil. Plains of the Bramhaputra and Ganges rivers, parts of Assam & W.B., 117 mha (3.6%), rainfall increases from 1400-1600 mm in the Ganges Plain to 1600-2000 mm in the Bramhaputra plain. PET: 1000-1400 mm, growing period more than 270 days.
17. **North Eastern Hills (Purwanchal):** Warm humid region, Red and laterite soil, north east states of Nagaland, Meghalya, Manipur, Mizoram, South Tripura, 10.7 mha (3.3%), Rainfall 1600-2600 mm exceeds PET, Dryspell 100-150 mm in post-monsoon period, growing period more than 270 days.
18. **Eastern Himalaya:** Warm humid, brown-hill soils, Northern Part of W.B., North Assam, Most Part of Arunachal, Sikkim, 8.0 mha (2.4%), Rainfall exceeds 2000 mm occurs throughout year, PET less than precipitation, growing period more than 270 days.
19. **Eastern Coastal Plain:** Hot and sub-humid alluvial derived soil, East Coast Plain, The Cauvery delta to Gangetic delta, 8.1 mha (2.5%), Rainfall 1200-1600 mm, PET: 400-500 mm, deficit 20% less, growing period: 150-210 days.
20. **Western Ghat & Coastal Plain:** Hot humid red laterite, alluvial Sahyadris, West Coast of Maharashtra, Karnataka, Kerala, 10.1 mha (3.0%), Rainfall exceeds 2000 mm, PET: 1400-1600 mm, Deficit 300-400 mm dry spell in Feb-April, growing period more than 270 days.
21. **Islands of Andaman, Nicobar & Lakshdweep:** Hot humid, red loam-sandy loam, Andaman Nicobar & Lakshdweep groups, 0.8 mha (0.3%), rainfall: 1600-3000 mm, exceed PET, except 300-400 mm deficit during December to March, suitable for plantation crops, growing period more than 270 days.

"Innovation comes from deep understanding."

8 Modern Concepts in Tillage

"Corruption is the Mother of Vegetation"

TILLAGE

Tillage refers to mechanical manipulations of the soil that are used to provide necessary soil conditions favourable for the growth of crops. Tillage includes all operations and practices that are used for the purpose of modifying physical character of the soil. Tilling the soil is the most difficult and time consuming operation in field crop production. The word tillage is derived from the Anglo-Sexon words *tilian* and *teolian*, meaning to plough and prepare soil for seed to sow, to cultivate and to rise crops. Manipulation of the soil with tools and implements for obtaining conditions ideal for seed germination, seedling establishment and growth of crops is called **tillage**.

About 30 % of the total cost of cultivation is for tillage operations. Jethro Tull, who lived in the 18th century, is considered the father of tillage. He emphasized the need of the tillage to improve the productivity of the soil as it causes breaking large sized soil particles into finer ones. Soil surface area is thus increased from which the plant roots obtain their food. He had introduced "Horse-hoeing husbandry" in which crops are planted in rows and "horse-hoeing" is done in between.

Tillage was considered as an 'art' and in the recent years, research evidences has focused tillage as 'science'. From time immemorial various methods of ploughing have been tried by trial and error, based on the labour availability and economic status of the farmer. Cultural practices have also undergone quite a lot of changes due to shortage of labour and /or increased cost of labour and availability of machinery. A lot of cultural operations and methods of land preparation vary from place to place and even in the same place, depending upon the climate and crops cultivated.

OBJECTIVES OF TILLAGE

1. To prepare the seed bed which promotes good germination and establishment of the seedlings to a satisfactory level.
2. To control weeds and improve close plant-soil interaction in the rooting zone.
3. To manage the plant residues by incorporating into the soil or to retain on the top layer to reduce erosion.
4. To improve the physical conditions of the soil so that rain water could be absorbed easily and soil erosion can be minimized.
5. To establish specific surface configurations for sowing, irrigation, drainage, etc.

6. To incorporate and mix applied fertilizers and manures into the soil.
7. To destroy the eggs and larvae of insects and their breeding places.
8. To permit root elongation and proliferation due to a low density soil

To achieve the objectives, the soil is disturbed/opened up and turned over. The intensity and depth to which this is affected are usually depended on site and soil type on one hand and on the type of implements used on the other, but this may also be a function of type of crop in same instances.

The aim of tillage is to produce as good a soil condition or tilth as possible for crop establishment and initial shoot and root development. The term **tilth** is used to describe qualitative characteristics of a loose friable (mellow) and crumby condition of the soil favourable for crop production. When the soil is brought to such a condition ideal for crop growth, is called a **seed bed** and is said to be in good tilth. In other words, tilth is physical condition of soil resulting from tillage. The ideal seed bed is a soil in which porosity of the mineral solid matter provides an optimum balance between water holding capacity and freely drained. The optimum balance appears to be the one in which capillary and non-capillary pores are in equal proportion.

TILTH

Tilth is the term used to express the physical condition of soil resulting from tillage. The physical condition is related to the soil structural changes, which promote good seed germination and crop establishment. Tillage operations are aimed at producing and maintaining good tilth. A soil with good tilth will be mellow, friable and adequately aerated, Such soils will have high retentive capacity of rainwater and good aeration with adequate infiltration. Soil tilth is not static but varies with time. The capillary and non-capillary pores are about equal when the soil is in good tilth. Soil aeration, activity of microorganisms and chemical reactions in the soil increases with the increase in non-capillary pores. Total porosity and pore size distribution are important to improve moisture retention and availability of oxygen in the rooting zone. Soils having crumb structure of 1-5 mm size are considered favourable for crop growth. Powdery soil encourages surface crusting when the soil dries after wetting. Soil aggregate should be stable enough to withstand water and wind erosion. There may be deterioration in tilth when the land is continuously cropped by using heavy machinery like tractor drawn implements. Factors such as heavy rainfall and erosion by water and wind may destroy the tilth. The action of wetting and drying or freezing and thawing often regenerates desirable soil tilth.

FACTORS INFLUENCING TILTH

Factors contributing to tilth are soil texture and structure, organic matter content and weather. Texture is an inherent soil property which is difficult to modify, except to a very small scale. Agricultural significance of soil structure is related to its effect on porosity and permeability and on the surface area of the soil fraction. Structure which indicates the way in which the individual mineral particles are arranged to form aggregates is important in determining soil tilth. The structure considered ideal for agriculture is crumb in which the aggregates are small, porous and relatively water stable. The development of soil structure results from the complex interaction between weather and method of cultivation and soil organic matter content. Soils with high organic matter cultivated at optimum soil moisture leads to crumb structure with aggregate stability. Exposure of soil surface to wetting and drying contributes to breakdown of large clods in heavy soils (self tillage) leading to loose surface with aggregate desirable in a good seed bed.

SOIL CONSISTENCE

The term **soil consistence** is used to describe the resistance of a soil at different soil moisture contents to mechanical stress or manipulations. It is a composite expression of these forces of mutual attraction among soil particles that determine the case with which a soil can be worked. It is usually measured by feeling and manipulating the soil by hand or by pulling a tillage instrument through it. Sol consistence is described at three moisture levels: wet, moist and dry. Terms used to describe soil consistence at these moisture levels are given below.

Wet soil		Moist soils	Dry soils
Stickness	Plasticity		
– Non-sticky	– Non-plastic	– Loose	– Loose
– Slightly sticky	– Slightly plastic	– Very friable	– Soft
– Sticky	– Plastic	– Friable	– Slightly hard
– Very sticky	– Very plastic	– Firm	– Hard
		– Very firm	– Very hard
		– Extremely firm	– Extremely hard

Tillage aims at ideal tilth for crop production. Its success depends on its timing to a particular soil condition. Relative ease with which a soil can be worked is usually referred to as soil workability. The term light, medium and heavy are also indicative of the amount of of effort or energy required for tillage. Free (drawbar pull) required to draw a plough through soil with 24 per cent clay is about 580 kg as against 704 kg through a soil with 35 per cent clay.

Soil workability depends as consistency- the combination of soil properties, materials that determines its resistance to externally applied force such as ploughing and traction of animals or vehicles used in cultivation. The amount of movement or deformation that will take place is related to the **share force** which builds up in the soil. This force results from friction and/or cohesion of the constituent particles. Friction built up between the implement and the soil is a function of weight, roughness and shape of the constituent particles. Cohesion is a function of soil moisture content and texture. In a loose dry coarse soil, frictional forces predominate while cohesive force is more important in soils with high clay content. Consistency states are described as cemented or hard, friable, plastic and liquid. Coarse textured soil, with less than 10 per cent clay content, are non-plastic and do not exhibit well defined consistency limits.

The **shrinkage limit** is the state at which the soil passes from moist to dry appearance. Moisture content is just adequate to fill the pores at minimum volume as a result of drying. At **plastic limit,** particle is surrounded by a thin film of water just sufficient to act as lubricant. The film of water is thick enough to reduce cohesion between soil particles at **liquid limit** and results in soil flow under applied force. Between the shrinkage and plastic limits soil is friable and moist, a condition for working the soil with minimum effort and with minimum structural deterioration. Structural deterioration increases at plastic limit.

The difference between shrinkage and plastic limits, in medium and fine textured soils is the **friable index**, which is a measure of optimum moisture range for cultivation. Higher the index, longer will be the period when cultivation is feasible and soil conditions are optimal for tillage. **Plastic index**, the difference between the moisture content at plastic and liquid limits is often used as a general index of workability in terms of soil susceptibility to structural damage. As the plastic index increases, soil consistence increases logarithmically and a proportionally greater effort is required to attain desired tilth. Heavy soils with more than 28 per cent clay are difficult to manage as they become plastic when too moist and hard and cloddy when too dry. Number of days on which they can be easily worked is small compared with that for other soils. Therefore, timeliness and speed of cultivation are of greater importance on these than on any soil. Consequence of tilling heavy soils beyond critical consistency limits are compaction and puddling.

In spite of the fact that tillage is essential for crop production it leads to undesirable soil conditions especially due to over cultivation.

- Frequent ploughing to a constant depth leads to plough pans,
- Intensive soil cultivation deteriorates soil aggregates into dust leading to wind and water erosion, and
- Repeated tillage operations hasten the oxidation of organic matter from the soil.

Size distribution of aggregates and mellowness (friability) of soil indicate the condition of tilth. Higher percentage of smaller aggregates of 1 to 2 mm dia is desirable in dry land agriculture as against aggregates of 5 mm or more for irrigated agriculture. Moisture content at which the soil is worked determines the relative proportion of different sizes of aggregates. Friability is that property of soil by which the clods when dry become more crumbly. A soil with good tilth is quite porous and has free drainage up to water table. Equal proportion of capillary and non-capillary pores permits adequate water retention in the soil, besides adequate soil aeration for crop growth.

TILLAGE MECHANICS

The reaction of soil to forces applied by tillage tools is affected by the resistance of soil to compression, shear, adhesive and friction. Practically, all tillage tools consist of devices for applying pressure to the soil, often by means of inclined planes or wedges. As the tool advances, soil in its path is subjected to compressive stress which is a friable soil results in a shearing action. Shearing of the soil may extend to a considerable distance on either side of the shear plane because of internal friction and cohesive action of moisture films. Commonly used terms in tillage are:

- **Force:** Any action that changes the state of rest or motion of body. It is completely specified by its magnitude and direction and position of its line of action.
- **Pull:** Total force extended upon the implement by a power unit. With tillage implement, it is generally with some angle above horizontal and it may or may not in a vertical plane parallel to the line of motion.
- **Draft:** Horizontal component of pull, parallel to line of motion.
- **Side draft:** Horizontal component pull, perpendicular to the line of motion.
- **Specific draft:** Draft per unit area of tilled cross section. Unit in SI system is N cm^{-2} (N is normal force).
- **Power:** It is rate of doing work. The unit is SI system is KW.
- **Kilowatt-hour:** Work done when one KW is used for one hour.
- **Frictin coefficient of soil (F):** It is the ratio of frictional force and the normal force. The magnitude of the frictional force (T) arising when a soil slides along a metal depends on the magnitude of force (N) normal to the contact surface, mechanical composition and soil moisture content, velocity of sliding and several other factors. It can be expressed as F= T/N.

TYPE OF TILLAGE

Tillage operation can be broadly grouped into two, based on the time during which they are carried out: on-season tillage and off-season tillage.

On-Season Tillage

Tillage for a crop from start of crop season to the crop harvested are known as on-season tillage operations. It includes preparatory tillage and inter-tillage.

Preparatory tillage: Preparatory tillage (preparatory cultivation) refers to tillage operations for raising crop up to sowing or planting as the case may be. It is of two types: primary and secondary tillage. **Primary tillage starts** with onset of the season. Arable soils on an average are ploughed twice, each time in diagonally opposite directions. The objective is to cut, lift, twist and turn the soil while incorporating organic and animal wastes into the ploughed layer. Primary tillage aids in improving the soil moisture regime for subsequent operations even if the rains are delayed. **Secondary tillage** includes operations for obtaining seed-bed for sowing. Since the soil is in loose condition due to primary tillage, there may not be any necessity for ploughing the soil. Blade harrows are worked ones or twice to kill the weeds. Break up clods and to obtain relatively smooth and leveled surface for subsequent sowing. After the crop is sown, the soil may need further tillage to control weeds, break up crusting or create soil mulch. All such tillage operations up to crop harvest are known as inter tillage or intercultural operations.

Low land rice grown under land submergence is prepared by puddling. Manipulation of soil under water with tools and implements to break down the soil aggregates is called **puddling** and the soil is called puddled **soil.** It disrupts the continuity of the pore spaces, reduces pore space and increases bulk density. Dispersed clay breaks transmission of pores and forms a thin layer over the surface. As a consequence of all these changes, deep percolation is greatly reduced/ The three main objectives of puddling are: ideal soil condition for planting seedlings, reducing deep percolation loses of water and weed control. Standing water in the field due to puddle soil, however, has several advantages in rice production.

Off-Season Tillage

Tillage operations during uncropped season for special purposes other than that for immediately raising the crop in the season are said to be **off-season tillage.** Taking advantage of summer rains, the field is ploughed deep (deep ploughing) to eradicate perennial weeds and to destroy soil borne pathogens, pests and parasites and to check erosion. This tillage is known as **summer tillage.** To break open the hard plough pans, reduce soil compaction and to improve drainage, the field is ploughed very deep and subsoil plough is used. This tillage is usually known as **subsoiling.** Vertisols are often left uncropped during a part of rainy season due to unfavourable weather or soil conditions or both. The soil is tilled during this period to control weeds and to create soil mulch for conserving moisture to subsequent crops grown during post rainy season. This tillage is called **fallow, mulch tillage** or **clean tillage.** In dry land agriculture, tillage initiated with the onset of summer showers is continued periodically until sowing, during crop period and even after the crop harvest to avoid weed growth, hard pans and soil water erosion besides conserving soil moisture. Such tillage is usually referred to as **year-round tillage.**

FACTORS INFLUENCING PREPARATORY TILLAGE

Several factors influence the preparatory tillage. Prominent among them are:

- **Crop:** It dictates the type and extent of preparatory cultivation. Smaller crop seeds like finger millet, tobacco, etc. require a fine seed bed which can provide intimate soil seed contact as against relatively coarser seed bed for large size seeds such as sorghum, maize, maize, pulses, etc. Delicate seeds such as tomato, chillies, cabbage, cauliflower require finer tilth with shallow cover over seeds. Root or tuber crops require deep tillage, Rice needs puddling.
- **Soil type:** It dictates the range of soil moisture and the time (duration) during which the soil can be tilled. Light soils require early and rapid land preparation due to free drainage and low retentive capacity as against heavy soils which require considerable time to start tillage. The energy requirement for light soils is less than that for heavy soils. Loamy soils can be brought to good tilth with little expenditure of energy.
- **Climate:** It influences soils moisture content, drought required for tilling and the type of cultivation. Low rainfall and low water retentive capacity of shallow sdil do not permit deep ploughing at the start of the season/ Contour tillage is necessary for crops on sloppy lands. Heavy soils developing cracks during summer (self tilled) need only harrowing. Light soils of arid regions need coarse tilth to minimize wind erosion.
- **Type of farming:** It influences the intensity of land preparation. In dry land agriculture where only one crop is taken per year, deep ploughing is necessary to eradicate perennial weeds and to conserve soil moisture. Land is tilled for each crop although the year under intensive irrigated farming. Repeated shallow tilling is adequate under such intensive cropping.
- **Cropping system:** Cropping system involving different crops need different types of tillage. In general, crop following rice needs repeated preparatory cultivation for obtaining an ideal seed bed. Crops following tuber crops like potato require minimum tillage. Similarly, crops following pulses need lesser tillage than that of following sorghum, maize or sugarcane.

TILLAGE IMPLEMENTS

Digging sticks and simple hoes were the earliest tools of cultivation used by men. Still widely followed in less developed parts of Africa, they do little more than scratching the surface of the ground. The plough which digs deeper, developed in the history of man has remained the most effective and widely used implement with modifications and alterations. Wooden plough drawn by bullocks is still used in many parts of India. Mould board plough came into use in the middle ages. The scientific design of the iron mould board plough was worked out towards the end of eighteenth century. Invention of the design of the mould board plough has been claimed by three people: William Amos of England, James Small of Scotland and Thomas Jefferson of USA. The invention was by James Small and Thomas Jefferson independently and they will have equal priority. The design by William Amos appeared much later. Among other innovations in Britain in eighteenth century were the seed drill and the horse hoe introduced by Jethro Tull.

In India, traditional wooden plough is a multipurpose implement used for land preparation, sowing the seed, inter-cultivation, placement of fertilizer, forming bunds and irrigation channels and even for harvesting tuber crops. Tillage implements can be broadly grouped according to the work.

- **Primary tillage (ploughing):** Traditional wooden plough, bose plough, disc plough, turnwrest plough, wetland puddler, cage wheels.

- **Secondary tillage (harrowing):** Tractor drawn cultivators, tractor drawn disc harrow, blade harrow, spike tooth harrow, spring line harrow and chain harrow, traditional blade harrows like *pedda guntaka* and *guntaka*.
- **Intercultivation:** Traditional wooden plough, small blade harrows like *danthies*, *rekkala guntaka*, *metla guntaka*, *meesala guntaka*, improved HM *guntaka*, star weeder and rotary weeder (for rice).
- **Off**-season tillage: Traditional wooden plough (big), mole plough, subsoil plough, chisel plough, ridge plough, blade harrows, buck scrapper.
- **Implements for specific purposes:** Ridge plough, bund former, traditional wooden markers (making intercepts for planting seed), wooden plank for covering seeds after sowing and for leveling the soil.

MODERN CONCEPT OF TILLAGE

Conventional tillage involves primary tillage to break open and turn the soil followed by secondary tillage to obtain seed bed for sowing or planting. With the introduction of herbicides in intensive farming systems, the concept of tillage has undergone radical change. The depth of tillage has been reduced and cultivator replaced the plough as primary tillage implements, particularly on heavy soil. In addition, there has been an increasing trend to reduce the number of cultivations in order to avoid undesirable results of over cultivation. Reduction or elimination of tillage has been accompanied by direct drilling of seeds. In countries like USA and UK, demand for labour and energy costs in crop production forced the farmers to restrict the tillage operations to the minimum during 1950s.

The greatest success and potential for restricted tillage appear to be on heavy soils, where timeliness of cultivation is so critical. The main advantages are that direct drilling favours perennial weeds and certain pests and diseases. Nutrient leaching may be greater than on ploughed land. Response of direct drilled crops to fertilizers is lesser because of greater weed competition or slower crop establishment. Soils are variably more compacted and proportion of large pores decreases.

Minimum Tillage

It aims at reducing tillage operations to the minimum necessity for a reasonably good seed bed to establish adequate crop stand and favorable edaphic environment for growth and development of crops. It has certain advantages over no tillage in that it improves soil physical condition due to incorporation of organic residues leading to higher water infiltration into the subsoil, lesser resistance to root growth, lesser surface runoff of rain water and satisfactory crop stand establishment. Different minimum tillage practices are **row zone tillage, plough-pan tillage** and **wheel track tillage.** In all these systems, primary tillage is as usual. However, secondary tillage is replaced by direct sowing in which the sown seed is covered through in the row zone with the equipment used for sowing.

Zero Tillage

It refers to crop production with least soil disturbance by controlling unwanted vegetation by any other means. It is an extreme form of minimum tillage. However, seeding requires some tillage which may not be possible to eliminate. It is a system in which mechanical soil manipulation is reduced to traffic and sowing only. Herbicides are used to destroy the vegetation present for taking up sowing. Zero or no tillage system is applicable for soils with- a coarse textured surface, good internal drainage, high biological activity of fauna, favourable initial soil structure and adequate quantity of crop residue as mulch.

Conservation Tillage

The main objective is to conserve soil and soil moisture. It is a system of tillage in which organic residues are not inverted into the soil such that they remain on surface as protective cover against erosion and evaporational losses of soil moisture. If stubbles from the protective cover on surface it is usually referred to as **stubble mulch tillage**. The residues left on soil surface interfere with seed bed preparation and sowing operations.

The concept of minimum or zero tillage may not work under Indian conditions, except on vertisols of certain areas and to certain extent under irrigation. The limiting factor of dryland crop production on vast areas of alfisols is soil moisture. Minimum tillage may not permit storage of as much rain water as possible in soil profile for subsequent use by the crop.

IMPLEMENTS AND THEIR USES

A large number of implements are used for different tillage operations (Fig. 4.1-13). Based on operations for which they are designed, implements are classified into different groups.

I. Implements used for ploughing.
   a) Country plough.
   b) Mould board plough.
   c) Disc plough.

II. Cultivators and harrows.
   a) Junior hoe.
   b) Disc harrow.
   c) Spring harrow.
   d) Spike tooth harrow.
   e) Indigenous blade harrow.

III. Implements for wetland operations.
   a) Wetland puddler.
   b) Helical blade puddler.
   c) Green manure trampler.
   d) Cage wheel.

IV. Leveling implements.
   a) Wooden leveling board.
   b) Buck scraper.
   c) A-frame ridger.
   d) Ditcher.
   e) Float.

V. Implements used for soil compaction and breaking clods.
   a) Iron roller.
   b) Wooden or stone roller.

VI. Implements for special operations.
   a) Bund former.
   b) Ridge plough.
   c) Bed furrow former.
   d) Rotovator.
   e) Chisel plough.

f) Sub-soiler.

VII. Sowing/planting implements.

a) Seed drills.

b) Paddy transplanter.

VIII. Weeders.

a) Dryland weeders.

b) Rotary weeders.

EFFECT OF TILLAGE

Soil is the medium in which crops are grown. To ensure proper growth of plants, soil has to be maintained in proper physical, chemical and biological condition. Tillage of the soil has been considered as one of the numerous soil management practices which are adopted for ensuring proper soil health. Tillage means manipulating or stirring of soil. The word is derived from the Anglo-Saxon words *tilian* and *teolian,* meaning to plough and prepare soil for seed, to sow, to cultivate and to raise crops. The objectives are to obtain proper tilth. Tilth is the physical condition of the soil resulting from tillage. Soil is good tilth when it is mellow, friable and adequately aerated. Soils with such good tilth are supposed to permit rapid infiltration of rain water without being too open to lose excessive water by evaporation and to offer very little resistance to root penetration.

Tillage is an oldest art associated with development of agriculture. Despite considerable improvement in various aspects of art and science of raising crops, tillage has continued to be the most difficult and labour consuming item of work although the new minimum or no tillage concept is becoming a practical proposition in some parts of the world.

OBJECTIVES OF TILLAGE

Tillage is one of the most expensive of the operations in raising crops. The purpose of tillage should, therefore, be studied to try to reduce this cost. The world over, there are eight reasons for tillage for upland crops (Donahue, Follett, Tulloch, 1983).

1. To prepare a mellow but firm seedbed that permits the incorporation of organic matter, lime, and fertilizers and a uniform planting depth for uniform seedling emergence.
2. To control weeds that robs crop plants of water, nutrients, and sunlight.
3. To improve the physical condition of the soil, to increase aeration and to warm up soil.
4. To conserve moisture by breaking the surface crust and thereby increase infiltration of rain water and irrigation water.
5. To remove stubbles of previous crops with a view to preparing land for the next crop.
6. To cover crop residues so that seeding and cultivating equipment can operate more efficiently.
7. To control certain insects and diseases by ploughing under of certain crop residues.
8. To control wind erosion by establishing cloddy ridges at right angles to erosive winds.

Tillage can, however, be a hazard to soil productivity, as explained here:

1. In preparing a mellow but firm weed free seedbed, there is a danger that natural and desirable peds will be fractured, lose their water-stability, and develop

into hard clods. This usually results from excessive tillage on fine-textured soils when they are tilled too wet. Furthermore, herbicides are becoming increasingly available around the world to control weeds.

2. Proper tillage at the right soil moisture content does immediately increase aeration and water infiltration. However, most long term studies indicate that crops with extensive root systems (forage crops for instance) results in the best soil physical condition.
3. Although tilling does break the surface crust and increases infiltration, over the years there are fewer crusts and greater water infiltration when some crops with good root system are grown with *no tillage*.
4. Covering crop residues does help to bury crop residues so that they will not interfere with subsequent seeding and tillage. However, equipment is now made to permit satisfactory seeding through surface crop residues.
5. Hard pans are man-made compacted layers at plough depth that restrict water movement and root development. They develop when the soilis ploughed too wet, with heavy machinery, and at the same depth each year.

EFFECT OF TILLAGE ON SOIL CHARACTERISTICS

Tillage affects many soil characteristics particularly tilth, moisture, aeration, temperature and strength.

Soil Tilth

Soil tilth is a collective term describing the whole of the physiological conditions of the soil. The farmer will strain the soil through his fingers to look the at the aggregate and clod size. The roughness of the surface and the case with which the clods break down will be observed.

Plants respond to variations of soil water, soil air, soil temperature and soil strength. The purpose of tillage is to alter the tilth or fabric of soil so that water, air, temperature and strength conditions in the soil are improved for plant growth and for long time productivity of the soil.

Soil tilth is not static but varies dynamically with time. Mechanical forces may change the roughness of the soil surface, the total porosity and density of the tilled layer and the aggregate or clod size. In spite of dynamic nature, measurements of tilth can be made and the effects of the soil tilth on soil management and plant growth studied.

The roughness of a surface is an index of the amount of water that can be stored in the soil depressions and may also be related to the resistance of the soil surface to sealing. Roughness may also be related to rates of evaporation and transfers of heat and air between the soil and the atmosphere.

Total porosity and pore size distribution are important parameters in a tilled soil. Porosity in the tilled layer is important not only because of its relationship to moisture and oxygen availability but also because it serves as a reservoir for water during intense rain.

The primary purpose of tillage is often to reduce aggregate or clod sizes. Aggregates must be small enough around the seed and seedling root to prevent undue drying of the soil, to provide sufficient soil solution, and to provide adequate aeration besides root contact. Yet, the aggregates should act be so finely divided as to encourage severe surface crusting when dry. Aggregation of soil particles and arrangements of the aggregates within the soil have a large influence on consistency and moisture relationship of the soil.

Tilth may improve or degenerate because of the forces of water and wind. Tilth may be destroyed by heavy rainfall and erosion by water and wind. The action of drying, wetting, freezing and thawing of soil often regenerates desirable soil tilth. Early ploughing of fine textured soil helps in the seed bed preparation later.

Soil Moisture

Tillage may have a marked influence on soil moisture through its effect on infiltration, runoff, temporary surface storage, internal storage and availability to plants.

When rain strikes a soil it either infiltrates or is retained on the surface, or flows off. If soil material has been finely divided by tillage, a surface crust often will form quickly and the infiltration rate will be reduced. During rainfall, a soil surface with uneven micro relief can store considerable water in the small depressions for later infiltration. Once water enters the soil its rate of movement will depend on the internal transmission characteristics of the profile. Increased porosity from tillage may act as an important reservoir for temporary storage of water during rains.

Crop residues ploughed under to form a mat at the bottom of the plough layer can reduce the rate of downward flow of water and restrict the air content of the plough layer for considerable periods of time following water application.

The amount of water retained by a soil may be influenced by the soil density and aggregate size. Ploughing or loosening a dense soil particle pores which can store water and may increase the storage capacity, but the soil structure may be too coarse.

The question of whether a loose soil will lose more or less water than a compact soil by evaporation is a complicated one. At relatively high moisture content, water conductance to an evaporating surface will be greater in packed soil. Actually, conductance to a planted seed is partly assured by firming the soil. When the soil surface is dry, evaporation is principally as vapour through pore spaces and hence likely to be greater in looser soils. But there will be the benefit of higher infiltration in loose soils when there is rainfall. Hence, the question of leaving the surface compact or loose depends on the season whether to conserve moisture or to obtain rapid drainage for tillage.

Soil Aeration

Respiration by plant roots and micro-organisms depletes oxygen and releases carbon dioxide and minute quantities of other gases into the soil atmosphere. Hence, a constant inflow of oxygen into and an outflow of respiratory gases from the soil is necessary for plant growth.

A productive well drained soil has approximately 50 % total porosity. Air-filled pores commonly fluctuate from 15 to 30 % of the total volume depending on the water content, soil density and structure.

The volume composition of the atmosphere is 78.09 % nitrogen, 20.95 % oxygen and 0.3 % carbon dioxide with traces of other rare gases. Soil air is usually higher in carbon dioxide and lower in oxygen, though in the one meter surface the oxygen percentage is usually above 17.5 %. The amount of carbon dioxide varies widely depending on temperature, organic matter supply and microbial respiration.

Both oxygen deficiency and carbon dioxide excess may be detrimental to plant roots, but the relative importance of the two depends on the plant species and the soil temperature.

Soil Temperature

Tillage influences soil thermal properties. Loosening a soil decreases the heat capacity because of a larger volume fraction of soil air. Thermal conductivity is decreased with decreasing volume fraction of solids and increased with increasing water contents. Loosening the soil often results in a larger difference between maximum and minimum temperatures but this may be comparatively less than air temperature.

Soil temperature affects germination, nutrient availability, plant development, grain yield and nutrient content of grain.

Soil Strength

Plant roots can exert tremendous forces in penetrating the soil, in packed soils pore size may be too small for root tips to enter. Critical pore size varies with kind of plant, but there are pore diameters below which roots may not penetrate. Pore rigidity, as well as will affect root penetration.

Soil crusting can reduce seedling emergence. Even where seedlings rupture and lift a small block of crushed soil, final population is apparently reduced.

There is evidence that wheat and sorghum seedlings transfer water frm their roots to the shoot tip thus softening the surface crust.

Soil crusts form when a puddle soil surface dries. Crusting is most severe on soils with low aggregate stability and where the surface has been smoothed and packed by tillage. High silt is conductive to high strength on drying.

Crust strengths are reduced by water either due to dew or further rainfall, and are likely to be weakened by succession wetting and drying. Careful soil management such as taking care to prevent puddling will help reduce the problem of crusting.

EFFECT OF TILLAGE ON NUTRITION AVAILABILITY

Tillage operations influence the nutrient availability in soil through their modifying influence on soil water, soil air, soil temperature and root impedance. As compared to ploughed soil, nitrogen deficiency was seen in maize crop when the field was listed, disked and sub-surface tilled. When nitrogen fertilizer was applied, maize yield was increased in the case of all treatments except ploughing. Similarly, potassium was also more available on ploughed soils than the listed, disked or sub-surface tilled soil.

Uptake of phosphorus and potassium was found low in soils of high bulk densities (unploughed field) as compared to the ploughed field with low bulk densities. Both phosphorus and potassium uptake were reduced by the higher degree of packing. The greatest decline was seen when air-filled pore space was lowered from 18 to 14 %. When moisture stress was allowed to increase, the uptake decreased in all soil bulk density levels.

Under mulch cover, the availability of nitrogen is reduced. The lower availability particularly in the early growing season is attributed to a number of reasons including microbial immobilization over a longer period of time than with base soil, cooler soil temperatures and hence, less mineralization of nitrogen and possibilities of leaching out of the row zone because of water conservation in the soil furrows. But available phosphorus tends to accumulate near the surface on stubble mulch soils.

Maize growth retardation accompanied by chlorosis similar to that resulting from deficiency of nitrogen or of some microelement has been observed on stubble mulched plots. It has been suggested that phytotoxic substances produced during the crop residue decomposition may be a contributing factor to the growth retardation.

When sufficient amount of fertilizers are added in no-tillage fields similar to conventional tilled area, the crop yield is found equal and sometimes better.

In doing so, no-tillage practices are needed in meeting the requirements. As already indicated, the objectives of tillage include:

1. Preparation of seed bed with proper tilth.
2. Incorporation of residues and organic matter.
3. Availability of plant nutrients.
4. Control of weeds and to a certain extent pest and diseases.
5. Control of soil erosion.

EFFECT OF TILLAGE ON ORGANIC MATTER

Much emphasis was once placed on the organic matter content of a soil as an indication of its productivity. After the availability of chemical nitrogen, by supplementing it, the yield of crop has increased tremendously. It is now felt that it is not wise to rely on the organic matter for high yields.

Bradfield (1963) sums up the subject as follows:

For most farmers, the only economical way to get more organic matter in their soils is to grow more organic matter on their own farms. Larger crops will mean more roots, more stalks and stubble, more food for livestock, and hence, more manure to return to the soil. The cheapest way to grow these larger crops is by more liberal fertilization and by use of good soil building rotations in which soil is so handled that maximum is obtained from the fertilizers. Organic farming with chemical fertilizer will resulting higher yields per hectare and even more organic matter in the soil.

Maize crop fertilized with nitrogen may contain as much as 1.5 % in the stalks. In contrast, maize receiving small amount of nitrogen may have less than 0.5 % in the stalks. Because of this and the much larger amount of organic matter produced, the stalks from the well fertilized maize will be more effective in maintaining organic matter than that from poorly fertilized maize.

In areas of high mean annual temperatures such as in tropical regions, the decomposition of organic matter continues over a considerable part of the year. It is difficult to increase the organic matter content under such conditions. In contrast, in the temperate regions, the soil organic matter content if depleted can be increased more readily by cropping systems.

EFFECT OF ADDED PLANT NUTRIENTS ON ORGANIC MATTER

The amount of fertilizer added in a cropping system affects not only the yield of harvested crop and its composition, but also the amounts of crop residues produced. This increased quantity of crop residues brought about by greater amounts of plant nutrients is important in organic matter maintenance. In addition, high yield means more extensive root systems which distribute more organic matter in the soil. The addition of plant nutrients increases not only the grain yield but also the straw and roots.

EFFECT OF ADDED ORGANIC MATTER ON AVAILABILITY OF NUTRIENTS IN THE SOIL

The larger quantity of carbon dioxide evolved during organic matter decomposition is thought to be important to the release of certain nutrients, inorganic phosphorus in particular. The carbon dioxide dissolves in water and forms carbonic acid. The result is a decrease of soil pH. This effect would be of greater importance on neutral or alkaline soil. Under such conditions the temporary reductions in pH would increase the rate of release of other elements such as boron, zinc, manganese and iron as well as phosphorus.

Certain of the intermediate products of organic matter decomposition are believed to form complexing or chelating ions. Phosphorus and some of the micronutrients attached to these ions are maintained in a weakly ionized state. The ions are retained against fixation by the soil but remain in a form that can be utilized by plants.

Should Organic Matter be Maintained or Increased?

To answer this question it is necessary to review some of the functions of organic matter:

1. It acts as a store house of nutrients – Nitrogen, Phosphorus, Sulphur, and so on.
2. It increases exchange capacity.
3. It provides energy for micro-organisms activity.
4. It releases carbon dioxide
5. It stabilizes structure and improves tilth.
6. It provides surface protection and thus increases infiltration.

It is interesting to note that all of these functions (except surface protection) depend on decomposition. Hence, the production of large quantities of residues, and their subsequent decay, is necessary to good crop and soil management. It is significant that one of the plant nutrient problems of the Arctic region is the resistance to decay of organic matter under low temperature and organic matter accumulations even on gravel ridges. In contrast, in the sub-tropics and tropics, although much organic matter is produced, it decays very rapidly.

Maintenance of organic matter for sake of maintenance alone is not a practical approach to farming. It is more realistic to use a management system that will give sustained top profitable production. The greatest source of soil organic matter is the residue contributed by current crops. Consequently, the selection of the cropping system and the method of handling the residues are equally important. Proper management and fertilization will produce higher yields, the principal part in which farmers are interested. A by-product of these high yields is the organic residue. Therefore, those soils that are being managed to produce large yields are improved at the same time.

TILLAGE AND THE ENVIRONMENT

Reduced tillage will reduce soil sediments that move into the air (dust) and water. Thus, the reduced tillage seems desirable. However, reduced tillage also increases the amounts of pesticides applied. There is a tendency to use the more persistent for better long term weed control, these are more likely to be persistent long enough to move into waters. It is typical to 15-40 percent more pesticides in reduced tillage compared to needs in conventional tillage. Higher pesticide additions increase solubilized pesticides. Any eroded soil sediment has higher amounts of both pesticides and phosphorus adsorbed to it. The reduction in fossil fuels burned while using reduced tillage will probably be insignificant in its effect on the environmental air pollution. Crop rotations would reduce the amounts of pesticides required and would be a logical consideration to help reduce environmental pollution from crop lands.

BEST TILLAGE

To suggest that there is a **best tillage** is to ignore that there are special needs of each climate, each soil, each crop, each rotation, each social and economic society, and each farmer's preferences. Each system will be tailored to its own needs and peculiarities. The following items are a particular summary of the considerations to ponder when planning of a tillage system for a field or a farm.

1. If soil erosion is not a problem, serious thought should be given to the weed, insect, and disease problems before converting to reduced tillage or no-till. These items are easier to control using conventional tillage.
2. Use of reduced tillage if soil erosion control is needed. Reduced tillage is much more adoptable to more systems than is no-till. It allows for some tillage, such as an overall cultivation or cultivation of narrow strips.
3. Reduced tillage requires better management skills and planning than does conventional tillage. Reduced tillage will reduce soil preparation and planting time, but it may not reduce total costs. If you are using reduced tillage primarily to reduce costs, do some careful bookkeeping to be sure you are ahead economically.
4. Different types and/or toughness of planters and tillers are needed in reduced tillage. Will the changeover be too great a financial burden?
5. The least adaptable soils to reduced tillage are clayey soils in cold climates where crop residues cause slow drying and warming. In such areas, the least amount of residue still capable of controlling erosion should be used.
6. Preparation of good seedbed will be needed for some small seeded crops, such as tomatoes, lettuce, carrots, sugar beets, alfalfa, and clovers.
7. Fertilizer and lime additions as strips on the soil surface ("horizontal bands") are nearly as effective as are the normal banding and broadcast-incorporation. In cold periods (<15°C), both potassium and phosphorus uptake may be reduced. The reduction may be temporary and may not alter yields appreciably. The greatest nutrient deficiency will probably be in rainfed crops having a prolonged drought, because most nutrients and roots are in the top 5-8 cm of soil. On basic soils, urea applications may have large losses as ammonia, unless the urea is watered in with irrigation. Injection of ammonia if slits do not close adequately.
8. Crop rotations, although they may appear to produce less income, may be as good or better financially over the long time. Usually, rotations reduce the weed and insect problems, distribute the work load, and allow more options in tillage, fertility, and crop residue management.
9. The most useful tillage operations depend on the most critical needs. Where weeds are a problem, one or two cultivations are very effective. Where drying and warming are problems, light disking to incorporate residues partially and leave some soil barren of cover is critical. Where surface flow furrow irrigation is planned, ridge furrow construction is the most essential tillage operation.
10. For sustained agriculture on the land, reduced tillage should have these as its priorities, in order of decreasing importance: (1) erosion control, (2) weed control, (3) pest control,(4) preparation for surface irrigation, (5) minimizing environmental pollution, (6) reducing required land preparation time, and (7) reducing fossil-fuel energy needs (future fuel costs will increase rapidly).
11. In the 1983 survey reported earlier, the three most important reasons farmers gave for opposing conservation tillage were (1) inadequate weed control, (2) herbicide costs, and (3) lack of proper equipment. Do you feel competent to solve these three problems, as the beginning problems?. Reduced tillage may be better, but it is not usually easier.

TILLAGE AND CROP PRODUCTION

Introduction

Tillage favours break up of clods, incorporates the organic matter into the soil and kills the weeds to create more favourable seed bed conditions. Physical condition of the soil significantly influences its moisture retention, root spread and penetration. It also influences the supply and uptake of nutrients by the crop plants. Soil aggregation depends on the clay and the humus content in the soil. Soil porosity considerably influences root penetration. Exchange of carbon dioxide from the soil and oxygen from the atmosphere is essential for normal root development and microbial activity. Tillage ensures good soil aeration on heavy clay soils. The critical level of oxygen for better root growth is about 1-2 %. The oxygen supply normally determines the rate of root growth. Due to restricted oxygen supply, the root hair growth is much restricted. High concentration of carbon dioxide in the soil is toxic to the roots. The critical limit of carbon dioxide for most species is between 10-20 %.

Tillage results in better aeration and hence increases the activity of microorganisms. Because of the increased activity of microorganisms, the rate of decomposition of organic matter is rapid, thus enabling the release and availability of plant nutrients.

Weeds are buried into the soil with tillage operations. Deep tillage is always preferred as one of the measures to control the perennial weeds.

Tillage also helps in reducing the insect pest population in certain situations. Insects (pupae) are exposed to the sun's heat and to the attack of birds and are eliminated. The places for living and feeding of the insects are also destroyed. Clean cultivation is generally advocated as one of the methods of controlling insects and diseases. Tillage helps to destroy some of the fungi causing plant diseases, e.g., scab disease causing organism of Gibbrella.

Limitation

Tractors and heavy equipments are used for the land preparation. Hence, there is a possibility of wind and water erosion as the land is left barren after the conventional tillage. There is also soil compaction due to repeated usage of heavy equipments and possibility of development of a dense zone (plough pan) below the plough layer.

TILLAGE AND PLANT GROWTH

There are five interrelated soil physical conditions important to plant growth that are influenced by tillage. These are soil temperature, soil moisture, bulk density, granulation, and porosity **(Donahue, Follett, and Tulloch, 1983).**

Soil Temperature

Soil Temperature when cool is increased by tillage primarily by increasing the evaporation of soil water. Tillage can also help to bury crop residues and thus increase soil temperature by absorption of more heat from the sun. Each species of plant has an optimum soil temperature at which it germinates. For example, planted beans will emerge in 16 days at 15 ° C but in four days at 30 ° C.

Soil Moisture

Soil Moisture and tillage are related to two important ways. Tillage at appropriate time helps the soil to dry out on the surface and planting can be done earlier, and it also increases the amount of infiltration and water storage. Contrastingly, excessive tillage can cause the surface of the soil to become too dry to germinate seeds.

Bulk Density

Bulk Density refers to the weight of a given volume of soil, usually in grams per cubic centimeter. Bulk densities of about 1.1 contain sufficient pore volume to supply adequate oxygen and water to plant roots. Bulk densities above 1.5 usually have too little total pore volume capacity for air and water. Tillage immediately decreases bulk densities by increasing pore volume, however, after a few rains the bulk density may rise again.

Granulation

Granulation is a general term used to characterize soil structure. The soil structural units are groups of soil particles held together by humus, iron, or gypsum. Natural structural units are called *peds* and anthropogenic structural units are *clods*. Peds are usually more water-stable than clods and can withstand heavy rains without disintegration (slaking). Minimum tillage preserves many peds, but excessive tillage turns peds into clods. Organic mulches, animal manures, and sewage sludges increase ped formation. Peds make a better seedbed than clods because peds more uniformly surround each seed and crop root to supply more adequate water, air, and nutrition.

Porosity

Porosity is a term closely related to granulation and bulk density. Porosity means pores or openings between individual particles of sand, silt, clay, and organic matter (humus) as well as opening inside of and among peds and clods. Large pores permit rain water and irrigation water to move into the soil, and if there is enough water, some of it percolates to replenish water tables. Medium size pores hold water against the pull of gravity and is available for absorption by the plant roots for growth and reproduction. The smaller pores hold water so tightly that it is not available to plant roots. Too lightly porosity and a soil is said to be compacted.

TRADITIONAL TILLAGE

Man took the first step towards civilization, when he learned and practiced the art of raising plants of his choice in place of the natural vegetation. The initial stage was the removal of the natural vegetation using available hand tools and implements. In spite of considerable developments that have taken place during the last two centuries with regard to design of tools and implements, control of unwanted vegetation has remained one of the main objectives of tillage. It is only in recent years that potent chemicals are being utilized to achieve this objective in some parts of the world. Tillage operations have been carried out in a routine that has been established as a result of experience through ages. The routine has included primary tillage like ploughing, secondary tillage like clod crushing, leveling, harrowing, smoothening, etc., and inter-tillage or after tillage operations in the standing crops.

Ploughing is the basic tillage operation. Its essential purpose has been considered to be loosening, turn and mix the soil so that any vegetation or manure on the surface is buried. A layer of soil from below is also brought to the surface and exposed to the action of weathering agents and other implements.

A wooden plough is one implement that is available with almost every farmer in India and ploughing operation consumes as much as 70 to 75 percent of total energy expended in carrying out field operations. Attempts made so far to either improved it or replace it with an iron plough has not yielded desired results. Only about 10 percent of the ploughs in the country are of iron type although it is shown that an iron plough is more efficient in many respects than the wooden plough. To get the same soil condition that is obtained with a single pass with an iron plough, the wooden plough has to be worked at least four times.

The results obtained so far indicate that ploughing helps in the reduction of weed population, and in improving aeration and water absorption capacity. Ploughing has proved more beneficial in humid areas where soils are fine-textured and are prone to crusting and packing.

A blade harrow is used as the primary tillage implement in place of the plough in some areas especially where the soils are deep, Vertisols. It breaks clods and smoothens land when worked after ploughing. It is an effective implement for destroying weeds and in creating mulch. Shovel tooth cultivators of various kinds are available for use under special circumstances, with different attachments like sweeps, furrwers, etc. The disc harrow is another implement that is proving useful and popular in some parts of India. It is used instead of the plough for preparing land quickly.

Various kinds of clod crushers are used to break clods. While in some situations simple wooden planks are sufficient to break clods, heavy spiked implements are necessary to deal with large and hard clods raised because of ploughing when soil moisture condition is much below the optimum level.

Leveling and smoothening operations are carried out with various kinds of gadgets to ensure uniform spread of rain water. Under some situations compacting of soil becomes necessary to get better germination of seeds. It is possible to achieve this objectives with the use of simple planks, blade harrow or various kinds of rollers.

In the traditional system, cultivation continues even after sowing until it is almost impossible to do anything because of the spread of the crop. Thereafter cultivation is carried out between rows of crops. This operation is done by means of either hand tools of various kinds or cultivators worked with bullocks or machines.

The main source of power used for farming in India is animal (Arakeri et al., 1959). In recent years mechanical power is being increasingly employed in some parts. Power availability is about 0.4 horse power per hectare as against 0.8 considered optimum the world over. In some parts of the world especially in some African countries human being is the main source of power.

TILLAGE CONFIGURATIONS

Tillage can not be separated from the various practices of laying out land for soil and water conservation and also for planting. At the close of preparatory tillage operations, the land can either be left flat until it is sown or laid into ridges and furrows, flat beds, tie-ridges or raised beds. Ploughing leaves the land cloddy if soil moisture content at the time of ploughing is much less than the optimum. If ploughing is carried out at optimum condition it is likely that the land surface will be just wavy with small ridges and furrows. Both these conditions are favourable for increased penetration of rain water.

As a result of further working of soil with clod crushers, levelers and harrows the land is likely to be left in flat condition. With the receipt of rain, when the land surface is flat and without any plant residue on it, the soil is likely to crust soon and obstruct infiltration of water leading to increased runoff and soil erosion. To get over these problems land can be laid eventually into some sort of configuration that will aid in soil and water conservation.

Laying the land into ridges and furrows, preferably across the slope along the contour, with suitable implements will help in creasing infiltration in initial stages. Once the soil is saturated, water runoff and soil erosion may take place as usual. To improve the situation further, with regard to holding of water uniformly all over the field, checks are provided to each furrow at fixed interval. The system is known as the tie-ridge method.

Scooping, with implements specially designed for the purpose, is suggested with a view to creating pockets or depressions all over the field to hold water. The entire field can also be laid into beds by a running ridger or a bund former criss-cross at a regular interval.

All these methods help in increasing infiltration of water and reduce soil erosion up to a certain limit, beyond which the results have been observed to be the same as if not worst than what is obtained without any configuration. There are other disadvantages too like uneven wetting which hinders other cultural operations at later stages. It is for these reasons that none of the configurations have become popular with farmers so far.

Recently research in Hyderabad, Andhra Pradesh, India on medium deep Vertisols resulted in higher yields and greater market value for maize (corn) and pigeonpea when planted on beds 150 cm wide as compared with planting on 75 cm beds or flat-planting. The increase in Rupee value of maize plus pigeonpea on 150 cm beds was 14 percent greater than on flat planting and 25 percent greater than on 75 cm beds (Table 8.1)(ICRISAT, 1976-77).

TABLE 8.1: Effect of land management on runoff, erosion, and yield on medium deep vertisol at ICRISAT Centre, 1976-77

Land treatment	Runoff (mm)	Erosion (kg/ha)	Yields (kg/ha)		Value (Rs/ Ha)*
			Maize	Pigeonpea	
Flat planting	14	10	2700	530	3780
Narrow ridges (75 cm)	19	20	2470	480	3445
Broad ridge (150 cm)	17	20	3030	620	4310

* Total value of maize and pigeonpea.

SOIL COMPACTION

Soil compaction refers to the fact that soil consists, on a volume basis, of too little pore space and too much solid matter such as sand, silt and clay. A good soil in extremely good physical condition for plant growth may have a bulk density of 1.1. This density may increase by 50 percent to 1.5 if the soil is ploughed too much and especially when ploughed too wet.

Compacted soils may also occur in nature as claypan horizons, silt pan horizons, or indurated horizons cemented by iron, aluminum, silica, calcium carbonate, calcium sulphate (gypsum), or humus. However, most compacted soils are anthropogenic (man-induced). Fine textured soils are often compacted when tilled too wet or by wheels when driven over with heavy loads. Two terms are used for anthropogenic compacted layers: **Induced pans,** which are formed at plough depth, 15 to 20 cm below the surface by tillage or heavy wheel traffic and **Induced crusts,** which are compacted soil at the soil surface about 1 cm thick caused by raindrop splash.

All tillage eventually results in soil compaction. It is true that tillage immediately increases soil aeration, long term results, however, show that after a lapse of some time they lose porosity and productivity.

For example, in a deep, mellow, well-aerated soil, cotton roots can grow nearly two meters deep and spread one meter laterally. In contrast, on a well-tilled soil in Alabama, cotton roots were restricted to a depth of 20 cm and a lateral spread of 33 cm. The root restriction was caused by a tillage pan at 20 cm and tractor wheel compaction between the rows. Furthermore, a soil in California was so compacted by tractor tires that water infiltration was only 1 cm/hour as compared with 4 cm/hour where the tractor did not run. (Donahue, Follett, Tulloch, 1983).

Soil compaction problem (surface crusts and tillage pans) can be over come by one or more changes in cultural practices, as follows:

- Rotate intensively cultivated crops with soil building crops having profuse root system like forage crops wherever possible.
- Practice minimum tillage.
- Reduce the number of times any tractor, implement, or wheeled vehicle passes over the field. This includes motorcycles and recreation vehicles.
- Increase the amount of organic residues, added to the soil and leave it on the surface.
- Use a chisel instead of a turning plough. Use the chisel when the soil is dry to shatter induced pans and crusts.
- Vary the depth of tillage from year to year.
- Never use a field as an access road or as a footpath or cattle trail.

ALTERNATIVES TO CLEAN TILLAGE

Throughout the history of agriculture, tillage has been taken for granted. It has been accepted that some tillage operations are necessary to grow crops. The development of tillage equipment has been done more as an engineering feat rather than from crop needs. Traditional tillage practices have been assumed to be satisfactory and machines designated to do them better. It is only in the middle of 1940s that the wisdom of ploughing the land and clean tillage was questioned. Clean tillage means ploughing and cultivating to kill weeds and bury all crop residues. Such tillage is very expensive, causes soil compaction, induced pans, induced crusts, and results in erosion. By 1960 chemical seed bed preparation and no tillage technique was proposed but the difficulty of placing the seed in a trashy field was posed. To overcome this difficulty new types of seeding implements have been developed and put to use.

Alternatives to clean tillage have already become popular in many countries on upland crops as well as on rice in recent years. 'No Till' protagonists have projected that this technique would be put to practice on a wide scale in the world before the close of the century.

Alternatives to clean tillage on upland crops are as follows (Donahue, Follett, Tulloch, 1983):

No-Tillage (Zero Tillage)

Till-Planting (Strip Tillage)

A wide sweep and trash bars clear a strip over the previous crop row and a narrow planter shoe opens a narrow strip into which seeds are planted. There is no previous tillage and no tillage between rows.

Sweep Tillage

A wide sweep moves through the soil such as after a small grain crop has been harvested, killing all weeds, mellowing the soil, and leaving all plant residues on the surface. The sweep may move through the soil several times to control weeds until the next grain crop is planted. There is no additional tillage.

Chisel Tillage

A narrow chisel 5 to 10 cm wide is run through the soil at a depth of 15 to 30 cm leaving all plant residues on the surface. The chisel shatters tillage pans and surface crusts. Planting is done through the residues with special planters.

Plough-Plant

After the soil is ploughed, a special planter is used and in one pass over the field the seed is planted. There is no secondary tillage after ploughing.

HERBICIDES-CROP RESIDUES

Sometimes suitable herbicides can be used economically to control weeds. When herbicides are used in combination with plant residues left on the surface, soil productivity can be maintained with minimum or no tillage.

Approaches to the problems of tillage in countries like India are showing trends of new directions. Minimum or no tillage technique would probably be most welcome in these countries where power availability for farming is far below the optimum considered necessary for efficient production and where sowing and planting operations are delayed or carried out ina hurried manner on poor seed beds. The technology, if perfected and offered to farmers having no animal or machine power at their command, may prove very advantageous.

Studies carried out so far in India and other tropical countries during the last decade have indicated that minimum tillage may prove to be a dependable and attractive alternative to conventional tillage.(Khan, Roy and Chatterjee, 1981; Sandhu and Brar, 1979; Singh, Bhan and Singh, 1971; Gautaman and Sankaran, 1976; De Datta, Botton and Lin, 1979).

TILLAGE FOR RICE

Contrary to popular opinion, not world rice is grown by wet land system. In India dry and semi-dry systems are also adopted on a large scale. Rice is generally grown either as a wet land or an upland crop, in deepwater, under rainfed conditions, or is direct-seeded and irrigated. Tillage for rice varies with the system of production used.

Tillage for rice grown under the following systems will be discussed separately, flooded, dry system, semi-dry system, deepwater, and alternatives to clean tillage (De Datta, 1981 and Arakeri, Cholam, Satyanarayana and Donahue, 1959).

Flooded Rice Tillage

When rice is grown under the wet land system a bund (levee) of soil is constructed around leveled fields. The height of the bund is normally about 25 cm. The area inside the bund varies with slope of the land. On level land, several hectares may be enclosed by one bund. On steep land, however, bunds enclose only a fraction of a hectare to permit the field within the bund to be leveled. On sloping lands, bunds can be constructed along the contour to minimize leveling. The requirement for a level rice field is to permit a uniform level of water to be maintained, usually at a depth of 5 to 10 cm.

When the season is declared satisfactory for starting tillage within rice fields and irrigation water is available, a few centimeters of water is let into the field to soften the soil. After a few days, the field is ploughed repeatedly and often cross-ploughed to establish a soupy-mud puddle.

Puddling

Puddling of soils in rice fields is done to accomplish these objectives:

1. Less power is required in tillage.
2. Ease of transplanting rice. Plants are pushed into the mud with thumb and forefinger.
3. Weed control. Buried weed seeds and plants do not sprout as readily.
4. Increase in availability of nutrients, especially phosphorus, iron, and manganese.
5. Reduced loss of water by deep percolation because of the formation of a tillage pan.

Disadvantages of puddling wetland rice soils include:

1. A high water requirement (up to 175 percent more, as compared with nonpuddling).
2. Root development is restricted.
3. Rotation of rice with upland crops is difficult because such crops grow best with no tillage pan and good soil structure.
4. Toxic soil accumulation of iron, hydrogen sulfide, and some toxic organic compounds especially on acid sulfate soils.

Dry Systems

In some of the rice growing areas rice soils are ploughed when dry, the rice seed is sown and then the field is flooded with water. This system is also common in other parts of tropical Asia, the United States, Australia, most of Latin America and southern Europe. Weed control is attempted with chemicals or suitable inter cultivating implements. The system is adopted on uplands in good rainfall areas or under irrigation.

Advantages of tillage when the soil is moist (but not with standing water as in puddling) include:

1. Desirable soil structure is maintained or improved.
2. Large power units can be used.
3. Crop rotations with other crops are easier.
4. Plant root development is favoured.
5. Less water is used.

Disadvantages of dry land rice tillage are:

1. Higher fertilizer requirements.
2. Weed control is not as effective.
3. Higher draft requirements.

Semi-Dry System

Rice is produced in medium high rainfall areas in India and in many other countries by adopting a semi-dry system. No irrigation is used. Land is tilled as soon as the first rains soften the soil. Often weeds are allowed to sprout and are killed by a second tillage. Rice seed is then drill sown or broadcast by hand. Weeds are always difficult to control. Inter cultivation is possible in a drill sown crop with suitable implements. In broadcast sown crops hand weeding is attempted. When sufficient water accumulates in the field and when rice seedlings are 20 to 25 cm in height, soil is stirred with implements to create a semi puddle condition. Thus, the crop is grown partly under dry system and partly under wet system in mid lying or low lying fields. Yields are low.

Deep Water Rice Tillage

Along rivers during the dry season, land is prepared for deep water rice varieties by tillage similar to that for upland rice. Seed is sown on dry soil. During the rainy period the rice elongates as the river rises and floods.

Many other tillage variations exist throughout the world, including minimum tillage systems.

ALTERNATIVES TO CLEAN TILLAGE FOR RICE

De Datta, 1981 reported that rice is grown in tropical and subtropical regions where heavy, beating rains cause severe raindrop splash erosion. Furthermore, soil manipulation for rice is time-consuming and costly. In addition, if two or three rice crops are grown on the same field in a 12-month period, the time taken between crops for soil preparation must be shortened.

Dalapon was used in the 1950s in many rice growing areas to control weeds and thus, to reduce the need for some of the time-consuming tillage used to control weeds. Dalapon was replaced by paraquat or Glyphosate in the late 1970s because they killed more kinds of weeds and left fewer toxic residues.

Minimum tillage for rice includes the use of paraquat or Glyphosate as a nonresidual, broad-spectrum, pre-plant herbicide on soil to control volunteer rice and other weeds. Fields are then flooded, and puddling and leveling, are accomplished, followed by transplanting of rice. Substituting herbicides for tillage-to control weeds saves about two weeks of time and uses about 20 percent less irrigation water.

ALTERNATIVES TO CLEAN TILLAGE IN WESTERN AFRICA

Surface applied organic mulches of 4 to 6 tonnes/ha in Nigeria, western Africa, as compared with conventional tillage, increased yields of maize (corn) by 50 percent, sweet potato by 25 percent, cowpea by 23 percent, soybeans by 21 percent, and pigeonpea by 5 percent (Bornemisza and Alvarado,1975. In these studies, besides increasing yields the organic mulch reduced soil temperatures by as much as 8 °C, reduced soil water loss by evaporation, increased soil nitrate nitrogen and phosphorus availability, increased numbers of earthworms, increased rate of infiltration of rainwater, reduced erosion, and resulted in a more desirable soil tilth. The mulch was as effective in soil restoration as 10 to 15 years of bush-fallow in the traditional system of shifting cultivation.

TILLAGE IN EQUATORIAL AFRICA

Tillage systems in relation to mechanization were studied and evaluated in nine countries in Equatorial Africa over an 18-month on the spot study by a team of two agricultural engineers, one agricultural economist, and one soil scientist-agronomist in 1966-1968.(Kline, Green, Donahue, and Stout, 1969). Key points pertaining to tillage systems are enumerated here:

1. In areas with wet and dry seasons as well as in areas where two or more crops are grown each year, timeliness of tillage is often the principal limiting factor in success or failure.
2. Mechanized tillage always requires larger fields and more level land areas than for animal drawn and hand-powered tillage.
3. Mechanized tillage pays best on very fertile, level soils such as on deep Vertisols.
4. Thousands of hectares of Vertisols have lain idle in tall grasses and scattered trees until tractor mechanization became a reality. The reason is because the moisture content during which Vertisols can be tilled is very narrow, thus, demanding rapid completion of all tillage and planting operations. After a rain, proper moisture content for tillage may last for only a few days. Vertisols are very high in clay and are therefore, too "boggy" when wet and too cloddy when dry within a time span of perhaps a week.
5. The size of the farm to be managed is determined largely by the area that can be kept weeded. For example, in Tanzania a farmer with only hand powered tools can keep merely 2.5 hectares of crops weeded. With animal powered tillage, a farmer can till about 10 hectares, and with a tractor, about 50 hectares.
6. For tillage operation, the common Equatorial African method of driving oxen (bullocks) with two men, whip, and voice control could be improved by adopting the Indian technique of using nose ropes, lines, and one man.
7. Chemical herbicides are much too expensive for the average farmer in Equatorial Africa.
8. Animal powered tillage is especially suited to the higher elevations where trypanosomiasis, carried by the tset sefly, will not kill the animals.
9. Oxisols, Ultisols, and many Alfisols in Equatorial Africa are characterized by low structural stability and resultant surface crust formation, very low natural fertility, horizons compacted by crystallization of iron and aluminum, low available water holding capacity, low organic matter, and extreme variability due to land clearing, brush burning, termite mounds, and differential erosion. Minimum tillage with organic residue mulching is a successful soil management technique.

SOIL CAPABILITIES FOR MECHANIZED TILLAGE

Although soils vary greatly, even within a one-hectare field, soil scientists are making soil survey maps that are valuable to every user of soil, including people who must decide what tillage system to use (Donahue, 1970 and Obeng, 1968).

Interpretive use has been made of soil surveys in Ghana, western Africa, to delineate areas on a map suitable for tractor tillage, animal tillage, hand tillage or no tillage. The legend is in the following three categories:

1. Areas where more than 50 percent of the soils are capable of being cultivated *continuously* with any kind of power, hand, animal or tractor.
2. Areas where more than 50 percent of the soils are capable of being cultivated *intermittently* by hand-power or animal-power, in a system of shifting cultivation.
3. Areas where more than 50 percent of the soils are not capable of being cultivated except in small isolated patches because the soil is either too shallow, stony, gravelly, slowly permeable, or is subject to crystallization of plinthite into iron and/or aluminum pans (laterite).

BURNING TO INCREASE TILTH OF VERTISOLS

Donahue, 1972 reported that in central eastern Africa, Vertisols occupy 44.7 million hectares of fairly level, productive, black soils. Hand powered and animal powered tillage are the dominant systems because very few tractors exist. With such primitive power sources, killing weeds and preparing proper tilth for a seedbed is next to impossible. Therefore, farmers have resorted to burning the soil to achieve what tillage can not do.

Traditional agriculture on Vertisols in Ethiopia consists of cropping then until yields decline because of weeds, insects, diseases, and/or poor tilth. Fields are than abandoned for five to fifteen years during which time wild grasses, weeds, and a few woody plants occupy the soils. Then the fields are ploughed with a local, narrow-shovel plough. The sod pieces are picked up by hand and placed in piles perhaps a meter in diameter and 25 to 50 cm high in the middle. The sod pieces are allowed to dry for a few weeks, after which dry cattle manure is put in a depression in the middle of each pile. The cattle manure is lighted and the sod pieces are moved into the hottest part of the fire as burning continuous.

After the fires have cooled, the residues are spread over the field. These residues are free of weed seeds, in good tilth, and have 122 percent more available phosphorus than the unburned soil. Furthermore, the black clay soil has turned reddish brown and has the apparent texture of loamy sand. Burning changes the clay particles into fine water stable aggregates the size of sand grains.

After burning, weeds are more easily controlled and desirable tilth for a satisfactory seedbed is more readily achieved. The soil is cropped for three to five years until yields decline, then the cycle is repeated.

"Mechanization of Agriculture is our Basic Industry."

Dr. P. Deshmukh

9 Soil Fertility and Productivity

"A traveler without knowledge is a bird without wings"
Sadi Gulistan, 1258

Soil fertility is a prerequisite to its **productivity**. From the stand point of crop yields and quality, nutrient supply both from organic and inorganic sources are important. As such it is essential to substitute the inorganic fertilizers with nutrients from organic sources. The **Integrated Nutrient Management (INM)** system envisages use of organic manures, green manures, bio-fertilizers (organics) along with chemical fertilizers (inorganic).

A number of diverse organic sources are available for use in agriculture. These sources can be reduce the mining of soil nutrients and improve overall soil productivity. Recycling of wastes and their conservation into easily transportable and usable forms should be an integral part of soil fertility management programme. Use of legume inoculants and bio-fertilizers along with chemical fertilizers offers an opportunity to achieve long term sustainability in crop production systems without land degradation.

Integrated nutrient management system improves the soil physical conditions in terms of soil structure, aggregate stability, soil moisture retentively and hydraulic conductivity. Such improvement in soil physical conditions contributes to **soil fertility** and **productivity**. Land degradation and environmental pollution can be minimized for sustainable agriculture.

Fertility problem can not be solved merely by supply of plant food elements but their efficient management has also to be given a due thought as fertilizers being one of the costliest inputs require a well balanced scheduling for optimizing dose to get maximum returns. The future fertilizer planning for an intensive farming will be based on nutrient status of soil as assessed by soil testing si that the problems related to residual effect of fertilizers, the compatibility of fertilizers, an appropriate method and time of their application, suitability of fertilizers for various crops and soil types and the fertilizer, pesticide, herbicide interactions may be taken care. Environment, soil-water pollution, formation of waste lands due to high acidity and alkalinity problems. Nutrient deficiency, interaction and antagonism due to continued use of high doses of certain elements would be perhaps more serious consequences rather than that of shortage of food, feed and fodder.

AVAILABLE NUTRIENT ANALYSIS OF SOIL

For analyzing available nutrients, first available nutrients are extracted by using reagent known as extracting reagent e.g.

- For extracting available N, 0.25 % $KMnO_4$ (Potassium permanganate) is used.
- For Phosphorus, 0.5 M Sodium bicarbonate ($NaHCO_3$) pH 8.5 is used.
- For Potash, Neutral Ammonium Acetate (CH_3COONH_4) is used.

After extracting the nutrients actual amount of available nutrient is determined by colorimetric method. On the basis of analysis of available nutrient, soil fertility is classified broadly into 3 groups viz., Low, Medium and High.

Nitrogen (N):	Low:	status of N in soil when less than 250 kg/ha.
	Medium:	250-500 kg/ha.
	High:	> 500 kg/ha.
Phosphorus (P):	Low:	< 20 kg/ha
	Medium:	20-50 kg/ha
	High:	> 50 kg/ha
Potassium (K):	Low:	< 125 kg/ha.
	Medium:	125-300 kg/ha.
	High:	> 300 kg/ha
Zinc (Zn) and Copper (Cu)	Low:	< 0.5 ppm.
	Medium:	0.5-1.0 ppm.
	High:	> 1.0 ppm
Manganese (Mn):	Low:	< 1 ppm.
	Medium:	1-3 ppm.
	High:	> 3.0 ppm.
Boron (B):	Low:	< 0.33 ppm.
	Medium:	0.33-0.67 ppm.
	High:	> 0.67 ppm
Organic carbon (Organic matter)	Low:	< 0.5 %
	Medium:	0.5-0.75 %
	High:	> 0.75 %

Nutrient Determination

Sl. No.	Nutrient Determination	Analysis Process
1.	Total Nitrogen in soil	Kjeldahl Process
2.	Available N	Alkaline permanganate method
3.	Phosphorus in acid condition	Bray No. 1
4.	Phosphorus in neutral and alkaline condition.	Olsen's method
5.	K^+ and Na^+	Flame photometer
6.	Sulphate sulphur (SO_4^{2-})	Terbimatric method
7.	Organic carbon	(a) Walke and Black method
		(b) Rapid oxidation process ($KMnO_4$)

CROP LOGGING: DEFINED BY H. F. CLEMENT

"The crop log is the graphic record of the progress of the crop, contains a series of chemical and physical measurements. These measurements indicate the general condition of the plant and suggest the changes in management that are necessary to produce maximum yield".

Crop Logging is used for **Sugarcane** first time in **Hawai.** Crop Logging is a record of composition of crop, with respect to N. P. K., moisture, sugar and weight of young sheath tissue, and other management at regular intervals during crop growth in the field. From knowledge of nutrient status, additional nutrient requirements of the sugarcane crop are assessed and necessary application of fertilizer is made. So, this is the case with moisture status and irrigation.

SOIL FERTILITY AND PRODUCTIVITY

Soil fertility is the inherent capacity of the soil to supply nutrients to plants in adequate amounts and in suitable proportions. **Soil productivity** is the capacity of the soil to produce crops with specific systems of management and is expressed in term of yields. All productive soils are fertile, but all fertile soils need not be productive due to some problems like water logging, alkaline, saline, and adverse climate etc. According to modern usage, soil fertility is the capacity of soil to produce crops of economic value and to maintain health of the soil without deterioration.

Factors Affecting Soil Productivity

It will be worthwhile to discuss here about the soil productivity so that we may differentiate clearly between soil fertility and soil productivity, though its detailed account is not needed in the text. Soil productivity means the crop producing capacity of a soil which is measured in terms of yield (biomass). The soil productivity depends upon following factors:

A. Internal Factors

Internal factors which may be called as genetic or hereditary factors which can not be manipulated.

B. External Factors

These factors may be regulated to certain extent. They include following sub-factors:.

1. Climatic Factors
   a) **Precipitation:** Temperature which causes cold injury (like chilling injury, freezing injury, suffocation, heaving) and heat injury and atmospheric humidity.
   b) **Solar Radiation:** Quality and duration of light, direction of light.
   c) **Wind Velocity:** It **causes** mechanical and physiological effects.
   d) **Atmospheric Gases:** Extent and nature of CO_2, N and oxygen.
2. **Edaphic or Soil Factors:** like soil moisture, soil air, soil temperature, soil mineral matter, inorganic components (nutrient elements), soil organic matter, soil organisms and soil reaction.
3. Biotic Factors
   a) **Plants:** Competitive and complementary nature. Competition between weeds and crop plants, Plants growing as parasites
   b) **Bacteria**: Symboints, free living.
   c) **Animals:** Earthworms, small animals and large animals.
4. Physiological Factors
   a) Geological strata (parent material)
   b) Topography (Altitude, steepness of slope)
5. Anthropic Factors
   a) **Human Factors** including skill and efficiency.

Thus, the nutrient status or soil fertility is one of the factors responsible for soil fertility.

Soil fertility means the capacity of soil to supply plant nutrients required by a healthy crop. There are various types of losses from the soil system and to have a regulated nutrient supply it is essential to maintain equilibrium between gains and losses. The equilibrium can be maintained only if we know about various factors responsible for availability of plant nutrients.

FACTORS AFFECTING SOIL FERTILITY

1. Physical Conditions of Soil and Nutrient Availability

Soil is a complex polydisperse system composed of solid, liquid and gaseous components which require a proper relationship for an excellent fertility status.

(A). Texture

The soil consists of stone (particles larger than 20 mm in diameter), gravel (particles larger than 2 mm in diameter) and earth comprising of sand, silt and clay with a particle size of below 2 mm in diameter. The particles below 2 mm are grouped as colloidal or clays (0.002 mm size) and non-colloidal (greater than 0.002 mm to 2 mm) particles.

Soils with non-colloidal particles are of open character because of larger sized particles and larger inter-particle spaces, thus their water holding capacity is lower and passage of percolating water is rapid. These soils are well drained, very friable and are poor in soil fertility. The chemical and biological activities of the soil depend upon the specific surface of the soil, that is the properties of the surface to unit, unit volume of particles, expressed as sq. cm/cubic centimeter or sq. m. of the surface/g of soil which is expressed as under:

Specific Surface of Soil Particles

Soil group	Diameter (mm)	Surface area (cm^2/g)
Coarse	2.0 – 0.2	45
Fine sand	0.2 – 0.002	446
Silt	0.02 – 0.002	4,458
Clay	below 0.002	10^7

Thus, greater the clay content in the soil, the greater will be the specific surface area, through it differs from clay to clay viz. montimorillonite has many times greater surface area compared to kaolinite or illite. Greater the surface area, the greater will be the cation exchange capacity and higher will be the soil fertility.

Cation, Exchange Capacity of Some Clay Minerals

Clay minerals	CEC (msl/100 g)
Kaolinite (1:!)	3 – 15
Montmorillonite (2:1)	80 – 120
Vermiculite (2:1)	100 – 200
Illite (2:1)	20 - 50

(B). Structure

The primary particle, after weathering, gets bound together to form larger units formed as secondary particles or crumbs and remain semi-permanent in nature.

The arrangement of these crumbs in a soil mass is called as soil structure. The crumb formation largely depends upon silt and clay contents in the mass, organic compounds present in the soil, concentration of irreversible soil colloids, microbial activity etc. The mechanical forces which help in separating soil mass into aggregates are the action of small animals specially earthworms, tillage processes pressure and differential drying due to freezing, compression caused by roots and localized shrinkage caused by loss of water from the soil, etc. In general there are seven structures which are recognized so far viz., plate like, prism like, block like, columnar, nut like, granular and crumbs. However, light textured soils which are low in soil colloids are structure less as the coarse sand material does not form aggregates.

The stability of the aggregates depends mainly on the cations adsorbed to the soil colloids. Structure becomes poor when Na^+ dominates the exchange complex, as it has a dispersing effect and aggregation of soil particles is prevented. In contrast, Ca plays an important role due to its flocculating ability which helps in formation of stable aggregates with clay minerals. In combination with humic acids and clay minerals Ca also forms very stable organo-mineral complexes.

The cation saturation of soil colloids depends much on soil type and prevailing climatic conditions. Clay minerals are more prevalent in temporate than in tropical and sub-tropical soils. Temperate soils often, therefore, have higher CEC in relation to their clay contents. In addition, the predominance of 1:1 clay minerals in tropical soils into non-specific cation absorption and they have relatively lower K^+ retention capacity,. On the other hand, the presence of high amounts of hydrous oxides, which frequently occurs in tropical soils, can induce very strong phosphate fixation.

(C). Soil Water

The water supply to plants, to a large extent, is related to water retention and water movement in the soil. The water holding capacity raises with the increase in the contents inorganic and organic colloids. Water can be retained by small and medium sized pores, thus water holding capacity is directly affected by soil structure. The amount of water held after excess gravitational water has drained away is expressed in percentage viz., if the soil contains 200 g water per kg dry soil the field capacity will be 29 per cent.

Water supply to plants does not only depend on the field capacity but also on the strength by which the water is absorbed to the soil particles. Water layers directly adjacent of the adsorbing surfaces may be bound by forces as high as 6000 bars to a pF value in the range of 6-7. The strength of water binding, which is generally measured in terms of pF, is called **"water tension"** or **"suction tension".** Soil water remains in capillary tubes formed by collecting various pore spaces. The height of the capillary rise increases as the diameter of the capillary decreases. The capillary rise of water is important in water and nutrient solution supply to the plants. The relationship the diameter of soil pores and water retention is mentioned below:

Diameter of Pore Space and Water Tension Held in Pores

Size of pores	Diameter in micrometer	Water Tension pF
Coarse	greater than 50	Less than 1.8
Medium	50 – 10	1.8 – 2.5
Fine	10 -0.2	2.5 – 4.2
Very fine	Less than 0.2	greater than 4.2

The ability of sandy soils to store available water is thus very poor, whereas it is considerably higher in the clay soil and so is the case with availability of plant nutrients or the soil fertility.

The nutrient uptake studies in rice as reported by Ghildyal (1971) showed that the uptake of four elements (N, P, Fe and Mn) increased with increase in the water content from unsaturated to field capacity. Similarly, availability and response of K increases with increase of soil moisture contents.

(D). Soil Atmosphere

(i). ***Soil Aeration:*** The composition of soil air differs from that of the atmosphere. The CO_2 level of the atmosphere is about 0.03% by volume whereas in the soil it is higher and is in the order of 0.2 to 1% in the surface layers. Soil also contains a correspondingly lower O_2 content and about 20.3% as compared with that of 20.99% in the atmosphere. The composition of soil and atmospheric air are given below:

Composition of Soil and Atmospheric Air

Gas	% by volume in soil air	5 by volume in atmospheric air
Oxygen	20.0	21.0
Nitrogen	78.6	78.03
Argon	0.9	0.94
Carbon-dioxide	0.5	0.03

Higher levels of CO_2 result from respiration of living organisms in which O_2 is consumed and CO_2 is released. This shows that O_2 is essential for the soil atmosphere. The respiration of plant roots depends to a high extent of the O_2 in the soil air. Respiration provides energy for various metabolic processes including active ion uptake by plant roots. The partial pressure of O_2, however, required for root metabolism can be considerably lower than that of the atmosphere.

Lack of O_2 can directly affect the carbohydrate metabolism of roots. Oxidative degradation of sugars is depressed and ethyl alcohol is produced by fermentation. Ethyl alcohol has a detrimental effect on plant growth and can result in considerable yield depression of crops. However, some plant spp. adapted to poor aeration conditions viz., those growing in marsh soils are able to accumulate instead of ethyl alcohol which prevents toxicity. Oxygen deficiency in the roots also impairs the synthesis of phyto-harmones such as cytokiniins and gibberellins. According to Hopkins *et al.*, (1950) the reported effect of O_2 on relative uptake of P and K by tomato plants is mentioned below:

Effect of O_2 partial pressure in the nutrient solution on the relative uptake of P and K by tomato plants

Oxygen partial pressure Bar	Relative uptake	
	P	K
0.21	100	100
0.05	56	75
0.005	30	37

As soon as the supply of O_2 in the soil becomes limited, the growth of most crops is affected and it stops when the oxygen concentration reaches a level below 2%. Some crops like tomatoes, potatoes, beets and peas are very sensitive to oxygen deficiency but others like sorghum, soybean, etc., can survive flooding to certain extent. Most of crop plants can grow well in a soil atmosphere with less than 21% of O_2 present in free air and when the oxygen content in the soil is below 10%, the growth of most of the crops is restricted. Thus, an adequate amount of O_2 must be supplied to the plant roots if normal growth is to be obtained. It has been reported that the concentration of CO_2 results in restricted nutrient uptake by the roots of a large number of plant species. It has been found that the order of reduction in nutrient absorption by corn from silt loam due to restricted aeration was K>Ca>Mg>N>P (Shaw, 1952).

Besides restricted nutrient uptake, under anaerobic condition the end products of anaerobic microorganisms may be accumulated and may reach to toxic levels to growing plants. The anaerobic metabolic end products include substances which are toxic to higher plants such as ethylene, methane, hydrogen sulphide, cyanide, butyric acid and several other fatty acids. Plants affected by these toxins have restricted growth and start showing wilting systems.

Some plant species differ in their ability to cope with reducing conditions in the soil by bringing about changes in root metabolism which prevent accumulation of toxic substances. Barbar *et al.*, (1962) while working with rice and barley reported that rice is more efficient in transferring O_2 from atmosphere to roots through leaves and stems. Rice roots are capable of even excreting O_2 into the surrounding medium and the rice plants may maintain uptake process of nutrients under anaerobic conditions too.

Reducing conditions in the soil medium affect many inorganic and biological processes and hence have an important impact on plant growth. An important feature of water logged soils is their oxidizing/reducing capability as indicated by the redox potential. This redox potential is measured by using an electrode. A positive redox potential is indicative of oxidizing conditions in soil whereas a negative redox potential indicates reducing condition. The redox potential of rice soils ranges between + 0.6 and – 0.2 volts. The submerged condition in rice soils results into an anaerobic microbial reduction of NO_3 to N_2 and N_20. Manganese and iron are also reduced to Mn^{--} and Fe^{--}. The sulphate is reduced to sulphide and at still lower redox potential molecular H_2 and methane are formed along with accumulation of additional toxic substances like butyric acid and fatty acids.

Under water logging conditions following microbial metabolism and reactions take place:

Steps of Microbial Metabolism in Water logged Soils

Steps	Main reaction	Initial redox potential (volts)
First Stage		
First	Oxygen disappearance	+ 0.6 - + 0.5
Second	Nitrate reduction	+ 0.6 - + 0.5
Third	Mn^{--} formation	+ 0.5 - + 0.5
Fourth	Fe^{++} formation	+ 0.5 - + 0.3
Second Stage		
Fifth	Sulphide formation Sulphate reduction	0.0 - - 0.19
Sixth	Hydrogen formation	- 0.15 - - 0.22
Seventh	Methane formation	- 0.15 - - 0.19

Source: Mengel and Kirkby, Principles of Plant Nutrition 1979, pp 49.

Soils to which organic matter has been applied or soils naturally rich in organic matter are prone to low redox potentials because organic matter favours growth and metabolism of anaerobic micro-organisms (Ponnamperuma, 1965); Tanaka and Yoshida (1970) reported that soils rich in active Fe are able to buffer the redox potential to some extent and prevent the second stage (due to high Fe^{3+}/Fe^{2+} ratio).

(b). *Soil Temperature:* Temperature influences metabolic activity and thereby it affects nutrient absorption. It has been observed that the absorption. It has been observed that the absorption of solutes is reduced to low temperature, though the degree of selective absorption of ions may be different at different temperatures. The temperature coefficients for anion absorption are higher than those for cation absorption. Low temperature does not appear to retard the absorption of nitrogen but they affect the capacity of the roots to reduce the absorbed nitrate which is converted into organic nitrogen (assimilation).

Availability of P is also affected by variation in temperature through it does not inhibit root permeability and plant growth. It is found that in many soils rise in temperature increases rate of mineralization of organic matter and P availability to plants is increased.

Chemical analysis of plant parts shows that increase in soil temperature increases K uptake in many plants like soybean, rice and barley. The most optimum temperature had been maximum 32°C and minimum 20°C and the uptake of most of the nutrients drops at a temperature falling below or moving upwards.

(c). Soil Compaction and Tillage: Soil compaction caused by constant surface tilling and cultivation of fibrous rooted crops restricts root penetration and aeration. This has direct effect on the ability of plant roots to absorb both nutrients and moisture from soil. Infiltration of water is also reduced due to which the plants have to face the moisture stress along with nutrient deficiency as the nutrients especially nitrogen are taken by plants in solution form. The lower rate of infiltration results into a rapid runoff loss of water leading splash and sheet erosion due to which a considerable amount of plant nutrients is washed off the field.

Modern high intensity cropping necessitates minimum tillage. An increased traffic of heavy equipment results into a serious compaction problems. This limits P absorption because of lack of root interception. Tillage promotes oxidation of soil or newly incorporated organic residues which facilitate mineralization and sub subsequent release of nitrogen. Lack of tillage results into deficient nitrogen supply, in continuous grass soil or pasture land.

(d). Soil Reaction: Soil reaction as expressed in terms of negative logarithm of the activity of H^+ in solution. In the soil one may distinguish between actual acidity, which is the H^+ ion concentration of the soil solution and potential acidity, which also includes H^+ adsorbed to the soil colloids.

The H^+ concentration of the soil solution has a pronounced effect on a number of soil constituents and especially on the soil minerals, soil micro-organism and on plant roots. High H+ concentration favours the weathering of minerals resulting into release of various ions viz., K^+, Mn^{2+}, Cu^{2+} and Al^{3+}. The solubility of salts like carbonates, phosphates, sulphates is higher in lower pH range.

Aluminium is adsorbed by clay minerals in the form of $Al\,(OH)^{2+}$ which becomes toxic to plant growth therefore, the pH of the soil with high clay content should be maintained in excess of 6.5.

Similarly the low pH conditions in soil favour the reduction of Fe^{3+} and Mn^{2+} to Fe^{2+} and Mn^{3+} respectively. Consequently the level of Fe and Mn ions becomes more in soil solutions of acid soils. Mn like Al can be toxic to plant growth.

The optimum soil pH for nutrient availability and crop growth is related to soil texture as it increases with increase in clay content. In organic soils, in particular, the pH should not be too high because these soils, by nature, are poor in number of plant nutrients, whose availability is suppressed by high pH conditions. This is true especially in case of phosphate, borate, Mn, Cu, and Zn as shown below.

Micro-organisms: The occurrence and activity of soil micro-organisms have their bearing on soil fertility as they help in nutrient mineralization in soil. It has been observed that generally in the low pH range (less than 5.5) fungi dominate in the soil and in the rhizosphere at higher pH range, the bacteria are more abundant. The nitrification of NH_4-N and NO_2-N brought by nitrosomonas and nitrobacter depends considerably on soil pH as they prefer neutral soil conditions. This is why the strongly acidic soils lack nitrate content. The fixation of molecular Nitrogen by free living micro-organisms (*Azotobacter chroococcum*, *Bacillus amylobacter*) and by symbiotic micro-organisms (*Rhizobium*, *Actinomycus alrri*) also prefer neutral pH conditions in the soil.

The pH of the soil sap of many plant species ranges between 5 and 5.5. If the plants respond favourably to higher pH as in the soil medium, this is due to other secondary effects of pH increase. These effects include the alleviation of Al or Mn toxicity or an increase in Mo or P availability.

The soil pH also regulates nutrients uptake pattern e.g. anions including nitrate and phosphate are absorbed at higher rate under weak acid range whereas the uptake rate for cation seams to be higher in more neutral pH range. It has been observed that at very low pH levels (less than 3.0) cell membranes are impaired and become more permeable resulting into a leakage of plant nutrients, especially of K^+ which diffuses out of the root cells into the soil solution.

Most soils developing in temperate climates tend to become acidic. The production of H^+ mainly results from the decomposition of soil organic matter. CO_2 produced by microbial decomposition of organic matter equilibrates with soil water thus forming H^+ and HCO_3^-

$$CO_2 + H_2O \Leftrightarrow H^+ + HCO_3^-$$

The microbiological decomposition of organic matter also produces NH_3 and H_2S which are oxidized in the soil to strong inorganic acids (nitric and sulphuric) and depress soil pH. Atmospheric SO_2 and NO_2 may also induce soil acidification:

$$\text{Organic carbon} \rightarrow CO_2 + H_2O - \Leftrightarrow H^+ + HCO_3^- \Leftrightarrow 2H^+ + CO_3^{2-}$$
$$\text{Organic nitrogen} \rightarrow NH_3 + 2O_2 \rightarrow H^+ + H_2O + NO_3^-$$
$$\text{Organic sulphur} \rightarrow H_2S + 2O_2 \rightarrow 2H^+ + SO_4^{2-}$$

2. Root Growth and Extension

Root growth and development have a considerable influence on nutrient availability which is brought about in two ways- firstly, the root system explores the soil for nutrients to reach the roots to be absorbed through mass flow and diffusion, and secondly, root metabolism creates a nutrient demand which also influences availability. Thus, the more dense and extensive the root system the more will be nutrient availability. Nutrient availability also depends upon the age of root tissues and accordingly young root tissues are particularly more efficient to take up nutrients, whereas more suberized and lignified older root tissues have very poor nutrient uptake behaviour. During dry periods the cortex of the basal part of the root is often desiccated which collapses by forming a mantle of debris.

Roots also affect nutrient availability by excreting organic and inorganic substances. Organic acids and amino-acids released by plant roots can weather minerals by chaleting cations of the lattice and thus during about a collapse of lattice structure, however, the quantities of organic chaleting compounds exuded by plant roots are minute as compared with the quantities of major nutrients required by plants. Moreover, it seems probable that the rioot exudates may influence the availability of micronutrients like Fe, Cu, and Zn which are easily chaleted and thus, remain in soluble stage.

Roots generally take up an excess of anions over cations and excrete HCO_3^- thus have an alkaline effect on the soil. In acid soils particularly this may lead to rise in pH over a few mm from the root surface which may considerably influence the rhizosphere and enhance phosphate availability.

Roots sufficiently supplied with oxygen are found to be more efficient in absorbing nutrients because anaerobic conditions lead to formation of toxic substances which creates a sickly plant growth and thereby nutrients uptake is dropped.

3. Organic Matter Content of the Soil

The importance of organic matter has been well recognized starting from formation of soil (weathering of rocks) to building up of soil fertility for raising a healthy crop with higher productivity. Since, organic matter is the residue of the living organisms which took nutrients from some sources for their survival; therefore, it is obvious that organic substances would liberate plant nutrients in available form during their mineralization. Organic matter, after certain degree of decomposition, forms humus (a storehouse of various nutrients) which improves the soil structure, drainage, aeration, increases water holding, buffering capacity, influences the solubility of the minerals and supplies energy for the growth and development of micro-organisms. All these lead to building up of soil fertility.

In general, the composition of plant tissue (mature and fully dried) ranges around the following values:

Carbohydrates	– Sugars and starches	1 to 5%
	– Hemicellulose	10 to 28%
	– Cellulose	20 to 50%
Fats, Waxes, Tanniins, etc		1 to 8%
Lignins		10 to 30%
Proteins (Simple water soluble and crude proteins)		1 to 15%

(Reported by Buckman and Brady, 1952)

The Nature and Extent of Plant Nutrients Released by Organic Matter in the Soil

Nitrogen

According to Singh and Lal (9176) the furrow slice of majority of cultivated soils contain between 0.02 and 0.4% nitrogen by weight. The organic form of soil nitrogen occurs as consolidated amino-acids or proteins, free amino-acids, amino-sugars and other complex compounds consisting of material which are formed from (i) ammonia + lignin; (ii) polymerization of quinines and nitrogen compounds; (iii) the condensation of sugars and amines. It has been observed that 20-40% of the total surface nitrogen is found in the form of bound amino-acids, 5-10% in the form of combined hexoseamines and about 1% of the total nitrogen is contributed by purine and pyrimidine while other nitrogenous compounds like chlorine, creatinine, and allantoin are probably present in traces.

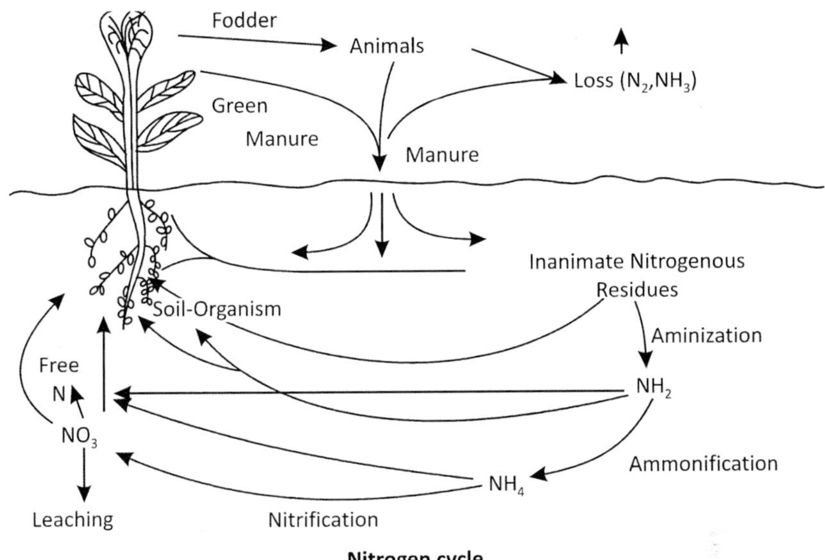

FIG. 9.1: Carbon Cycle.

The organic nitrogen present in soils is not readily available to plants as it has to go under microbial decomposition like ammonification and nitrification and hardly 1 to 3 per cent of total amount present in soil becomes available after mineralization as explained below:

Ammonification: It is the process of combination **of** ammonia by the soil organisms as a product of decomposition of nitrogenous organic matter**.** It takes place in well drained aerated soils having plenty of active basic minerals:

$$\begin{matrix}\text{Protein and}\\ +\\ \text{New protein}\end{matrix} \xrightarrow[\text{digestion}]{\text{Enzymic}} \begin{matrix}\text{Complex amino}\\ \text{compounds}\end{matrix} + CO_2 + \text{Energy} + \text{Other products}$$

$$\begin{matrix}\text{Organic substances with}\\ \text{Amino combinations}\end{matrix} + HOH \xrightarrow[\text{hydrolysis}]{\text{Enzymic}} \begin{matrix}\text{Organic substances}\\ \text{with } OH^-\end{matrix} + NH_3 + \text{Energy}$$

$$2NH_3 + H_2CO_3 \rightarrow (NH_4)_2\,CO_3 \Leftrightarrow 2NH_4 + CO_3^-$$

The ammoniacal nitrogen, thus formed has three facts- firstly ammonifying organisms like Mycorhizal fungi use it in a considerable amount and secondly pass it on to their hosts in some after form readily and thirdly part of ammoniacal nitrogen is used by special purpose bacteria as a course of energy for nitrification process as explained below:

$$2NH_4^+ + 3O_2 \xrightarrow[\text{Oxidation}]{\text{Enzymic}} 2NO_2^- + 2H_2O + 4H^+ + \text{Energy}$$

$$2NO_2^- + O_2 \xrightarrow[\text{Oxidation}]{\text{Enzymic}} 2NO_3^- + \text{Energy}$$

This process is completed by nitric (nitrobacteria like Nitrosomonas and Nitrosococcus) and (Nitrobacter) organism.

Diagrammatically the Nitrogen Cycle may be shown under:

The micro-organisms involved in aforesaid process are the general purpose heterotropic forms that under favourable conditions promote aminisation and ammonification. However, these microbs under inadequate aeration seem to readily abstract oxygen from nitrates and their reduced products (under reduction process). Due to this the nitrogen derived from organic matter is lost in the forms of ammonia and elemental nitrogen as explained below:

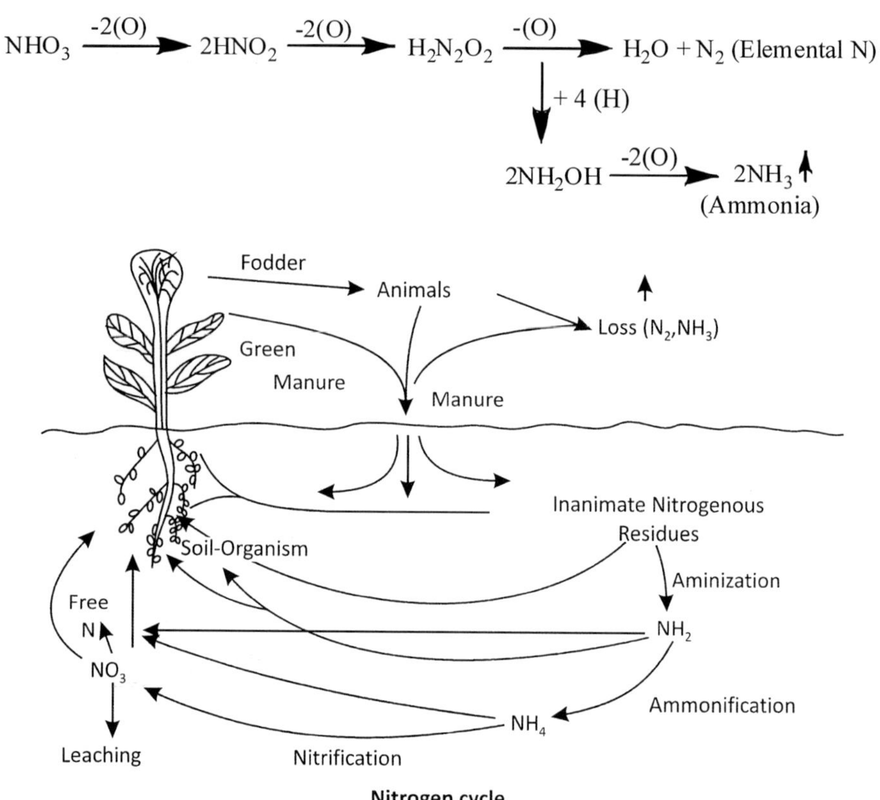

Nitrogen cycle.

FIG. 9.2: Nitrogen Cycle.

Studies have shown volatilization loss of nitrogen from organic soils under anaerobic conditions to the extent of 11 to 80 kg/ha.

If the amount of nitrogen is relatively very low in comparison with carbon then the micro-organisms, involved in decomposition of such organic material, use to utilize the NH_4 or NO_3 for increasing their population to accelerate decomposition process. After completion of microbial decomposition a C : N ratio of 10:1 is commonly observed. It has been found that when organic materials with a C/N ratio of more than 30 are added to the soils then there will be immobilization of soil nitrogen during the initial stages of decomposition. In case of C/N ratio between 15 and 30, there may be neither immobilization nor release of mineral nitrogen for the crop growth. However, the organic materials with a C/N ratio of less than 15 usually start releasing mineral nitrogen early in the decomposition process.

On a world-wide basis, the total nitrogen content of soils ranges from less than 0.02 per cent in subsoil to more than 2.5 per cent in peats, however, the Indian soils have following percentage of nitrogen in them:

Soil Group	Nitrogen percentage
Indus alluvium	0.05
Gangetic alluvium	0.04
Medium	0.05
Black cotton soils Deep	0.06
Red soils	0.03
Laterite soils	0.04

1. Symbiotic Nitrogen Fixation in Soils

Various strains of Rhizobium bacteria present in plant roots of different host legumes use to take energy from the carbohydrates of the plants and in return, give them nitrogen fixed from atmosphere in the quantities given below:

Nitrogen Fixation in the Soil from the Atmosphere

Crop	Nitrogen fixed (kg/ha)
Alfalfa	251
Sweet clover	168
Red clover	151
Alsike clover	141
Soybean	105
Hairy vetch	68
Field beans	58
Field peas	48

The nitrogen fixed by Rhizobium is first utilized by host plants and when the roots and nodules of the host plants are incorporated into the soils almost same quantity becomes readily available to the succeeding crop.

2. Azofication

It is also known as free fixation of nitrogen. Free fixation of N by free living-organisms like blue green algae, some fungi and heterotrophic bacteria such as Azotobacter and Clostridium which are non-symbiotic in nature, is used to fix atmospheric nitrogen into the soil. These organisms take energy required for their growth from decomposing organic materials under neutral soil reaction and fix the atmospheric nitrogen to the extent ranging between 5 and 60 kg/ha per year.

3. Nitrogen Gain through Rain and Snow

These sources, on an average, add about 5 kg nitrogen/ha/year which is more in ammoniacal form of nitrogen in temperate region.

Phosphorus

The organic phosphorus compounds may account for more than half of the total soil phosphorus and a major part of the phosphorus taken up by growing plants is derived from organic forms. According to Bremner (1951) about 25 to 80 per cent of the total phosphorus present in the soil belongs to organic form and it tends to increase or decrease with the organic matter content in the soil. Accordingly surface soil is richer in phosphorus than sub-soils. Thus addition of organic materials increases phosphorus availability. Organic matter produces humus and the humus extracts increase the phosphate solubility. The higher solubility may be accounted for: (i) formation of phosphor-humic complexes that are more easily assimilable by plants, (ii) formation of humate ions which help in anion replacement of phosphate; (iii) formation of a protective covering over sesquioxide particles which reduce the phosphate fixing capacity of the soil. Certain organic anions like citrate, oxalate, tartarate, malate and malonate, etc. produced from decomposing organic matter react with iron and aluminium and form stable compound, thus they are prevented from reacting with phosphorus. Therefore, the phosphorus fixed by Fe and Al is released. There are three main groups of organic phosphorus compounds in the soil like: (a) phytin and phytin derivatives; (b) nucleic acids; and (c) phospholipids. Apparently the first two of these make up most of organic phosphate, with approximately half coming from each other.

Looking at the availability pattern it has been observed that both the phytin and nucleic acidcan be utilized as sources of phosphorus. Apparently the phytin is absorbed directly by the plants, while the nucleic acids are probably broken down by enzymes at the root surfaces and the phosphorus is absorbed in either organic or inorganic form.

Phytin behave like inorganic phosphate by forming iron, aluminium and calcium phytates which become insoluble and thus become unavailable to plants. Under alkaline conditions calcium phytate is precipitated and the phosphoric becomes unavailable.

The nucleic acids get absorbed by clays, especially by montmorillonite under acidic soil reaction and results into phosphate deficiency.

Sulphur

Sulphur behaves like nitrogen in the soil. The soils organic matter goes under microbial decomposition and releases sulphur in organic forms. These organic residues obtained through plant and animal sources and several sulphur containing compounds into the soil like protein containing sulphur amino-acids, cysteine, cystein and methionine, the peptide glutathione, glucosides, alkaloids, thiamine, biotin and other compounds. Major quantity of organic sulphur is released as incompletely oxidized inorganic substances by microbial actions like sulphides, elemental sulphur, thiosulphate, tetrathionate and pentathionate which are later oxidized to sulphate before sulphur is absorbed by plants. According to Barrow (1960) if the C/S ratio of organic matter is below 200, the sulphate gets accumulated in to soil, when the ration is 400; the sulphur is tied up in the soil organic matter. The sulkphur is either released from organic matter or is tied up in the soil organic matter under a C/S ratio between 200 and 400. The process of mineralization completes as per reaction shown below:

$$\text{Org. S} \xrightarrow[\text{Oxidation}]{\text{Enzymic}} SO_4^{--} + H^+ + \text{Energy}$$

Micro-Nutrients

Boron

The availability of B is increased with increase in organic matter content of soils under acidic soils (Berger and Truog, 1945), however, under alkaline soil conditions the availability of Ca is found to be more effective in releasing B than organic matter content. The arid soils of Rajasthan, Gujurat and Punjab have shown a positive relationship between B availability and organic matter content of the soil

(Satyanarayanaan, 1958, Gandhi and Mehta, 1958, Grewal *et al.*, 1969 and Nathan *et al.*, 1969).

Copper

Kanwar (1954) reported that copper fixation in soils was associated with the formation of organo-metallic complexes and Misra and Tiwari (1962) found that only 9.2 to 17.4 per cent of Ca retained by compost is exchangeable by normal neutral salts. Iyer (1959) reported a negative correlation between organic matter and the Ca content of virgin grass land soils. According to Mandal and Jha (1970) the copper availability decreases with increase with increase in organic matter content of soil. However, some other workers have reported a positive correlation between organic matter and exchangeable copper (Rai and Mishra, 1967; Grewal *et al.*, 1969; Lal, 1968).

Iron

The organic matter behaves differently in different soil reaction as in calcareous reaction the organic matter content and availability of iron as expressed by exchangeability and extractability (+0.615 and + 0.448 respectively) but in non-calcareous soils the availability pattern could not be established.

Manganese

There is diverse opinion about the correlation between organic matter content and availability of Mn in the soil. The exchangeable Mn content has been found more at the surface than in the lower horizon (Biswas, 1953; Ranndhawa *et al.*, 1961; Lag and Dev, 1964; Shinde and Datta, 1964; Sharma and Shinde 1968 and Grewal *et al.*, 1969). According to Yadava and Kalra (1964) the organic matter content upto 3 per cent increased Mn availability but a further rise in organic matter content decreased the amount of exchangeable Mn in the forest soils.

Contrary to these findings deficiency of Mn has been reported with increase in organic matter content of the soil due to formation of insoluble complexes (Takker *et al.*, 1969).

Molybdenum

In acid soils presence of organic matter increases the availability of Mo due to protection of Mo from anion adsorption (Kavinandan *et al.*, 1964), however, Iyer (1959) reported an inverse relationship between availability of Mo and organic matter content of soil.

Zinc

The availability of Zn increases with increase in organic matter content of the soil. Positive correlations have been reported by (Eswarappa et al., 1969) between Zn availability and organic matter content of red, laterite and black soils.

4. Cropping System

Requirement of major and minor plant nutrients varies greatly not onlt from one crop to another but also from one variety to another variety of the same crop. Studies have indicated that the depletion of plant nutrient under intensive cropping is several times greater than under single cropping system. It has been found further that selection of new plant types depletes micro-nutrients by four to six times as against growing their traditional varieties.

Antagonistic behaviour of some nutrients to other also creates lot of problem in availability pattern of nutrients as for example a continuous use of high doses of phosphorus makes Fe and Zn deficient which may be due to a reduced mobility or metabolic disturbances but the exact reason is yet to be known.

Therefore, under unthoughtful cropping plan a serious nutrient disorder in the plant body or in soil system may be created.

5. Other Factors

There are several other factors which have the bearing on the fertility status of the soil like sewage irrigation causing into accumulation of heavy metals that may create problem, quantity of irrigation water, frequently and depth of irrigation, drainage, switch over in microbial population caused by use of chemicals like pesticides, herbicides, fungicides, nematicides, etc., secretion of toxins by various plants, fertilizer scheduling, application of soil amendments like lime, sulphur, gypsum, pyrite, etc.

Soil Fertility Vrs Soil Productivity

	Soil Fertility	Soil Productivity
1.	– It is considered as an index of available nutrient to plants.	– It is a broader term used to indicate crop yields.
2.	– One of the factors for crop production, the others are water supply etc.	– It is the interaction of all the factors.
3.	– Can be analyzed in lab.	– Can be assessed in the field under particular climate conditions.
4.	– It is the potential status of the soil to produce crops.	– Resultant of various factors influencing soil management.
5.	– Depends upon physical, chemical and biological factors of soil.	– Depends upon location, fertility, and physical conditions etc.
6.	– Soil fertility is the function of available nutrients of soil. Soil fertility = f (Nutrient status of soil)	– Soil productivity is the function of soil fertility, management and climate. Soil productivity = f (Soil fertility + Management + Climate).
7.	– It is an inherent property of soil.	– It is not the inherent property of soil.
8.	– The fertility of a certain soil is same in all the climates.	– Soil productivity differs according to the variation in climate and location.
9.	– All fertile soils are not productive.	– All productive soils are certainly Fertile.

1. De Datta and Gomez (1975 observed that the N-efficiency was the highest in the first rice crop and declined progressively with successive rice crops.
   K-efficiency however, increased sharply over the years, especially during the dry season.
2. FAO recommended the following practices for adoption in India:
   a) In irrigated rice-wheat system, N should be applied to both the crops, 'P' only to wheat and K and Zn to rice.
   b) In rice-rice-green gram/soybean system, N to both the rice crops, 'P'to dry season rice and, K, S and Zn to the second crop.
   c) In maize + pulse intercropping system, N to maize, P to crops and K, S and Zn to maize.
3. The overall food grain production to fertilizer applied ratio is about 10:1.
4. The Nitrate, form of N is preferable for saline soils because NH^{+}_{4} will be lost as volatilization.
5. Ammoniacal fertilizers should be applied in the reduced zone and Nitrate fertilizers in the oxidized zone.
6. In crops where vegetative part is the economic yield e.g. sugarcane, beet, potato, tobacco, N-supply must be stopped much before maturity otherwise, the quality and maturity may be affected.

7. In seed crops, Last application of N during seed development phase improves the quality and germ-inability of seeds. The application of last doze of N should be foliar.
8. In determinate grain crops like rice, wheat, maize, N-supply at the beginning of the reproductive phase (e.g. panicle initiation) as the last doze increases the number and weight of grains.
9. In determinate plants e.g. rape, sesame. Cotton, N-application at the flowering stage and another at the late flowering stage increases the yield and quality.
10. If there is luxuriant consumption of N, a growth modifier say cycocel may be applied to enhance reproductive growth and yield.
11. Plants require a greater amount of P at their early stages (for root growth) but the plant absorbs K up to harvesting and K-fertilizer becomes available slowly.
12. Rubbing means heat treatment applied to soil by burning refuse placed over it.
13. Teast: Soil containing high proportion of Mo.
14. Mar: Raw humus, a type of forest humus layer of unincorporated organic material.
15. Marling is the application of clay to sandy soils.

FACTORS CONTROLLING SOIL FERTILITY

There are seven main factors for controlling soil fertility given below:

1. Good tilth,
2. Control of weeds,
3. Control of plant diseases, and insect pests,
4. Proper sequence of crops in crop rotation,
5. An adequate supply of available plant nutrients like N, P, K, organic matter and trace elements through FYM, Green manures and fertilizers,
6. Erosion control, and
7. Irrigation and drainage.

SYSTEM OF FERTILITY MAINTENANCE

Successful management of soil fertility is seven in number:

1. Adequate supply of Nitrogen from:	– Crop Residues
	– Farm manure
	– Legumes - Regular Crop, Green Manure
	– Rain and Snow
	– Nitrogenous Fertilizers
2. Addition of Organic Matter by means of:	– Crop Residues
	– Farm Manure
	– Green Manure - Legume, Non-legume
3. Application of Lime If necessary:	– Calcic or Dolomitic
	– Lime stone is most commonly used
4. Addition of Phosphorus:	– Super phosphate or other phosphatic Fertilizers
5. Provision for available Potassium by means of:	– Farm Manure
	– Crop Residues
	– Potash Fertilizers
6. Sulphur deficiencies are met with:	– Flowers of Sulphur or Gypsum or
	– Super phosphate. Sulphur compounds carried in Rain and Snow

7. Application of trace Elements:	– As separate salts or in mixture

CONCEPT OF PRODUCTIVITY

Primary Productivity

It is defined as the rate at which radiant energy is stored by photosynthetic and chemosynthetic activity of producer organisms (chiefly green plants) in the form of organic substances which can be used as food materials.

Gross Primary Productivity

It is the total rate of photosynthesis, including the organic matter used in the respiration during the measurement period. This is also known as total photosynthesis or total assimilation.

Net Primary Productivity

It is rate of storage of organic matter in the plant tissues in excess of the respiratory utilization by the plants during the period of measurement. This is also called apparent photosynthesis or net assimilation.

Secondary Productivity

The rate of energy storage at consumer levels is referred to as secondary productivity.

ROLE OF SOIL IN CROP PRODUCTION

The welfare of mankind depends on the wise use of our soil and water resources. One of the greatest problems facing us today is to use these resources correctly so as to save for the future. Conservation of these soil and water resources depends on an understanding of the forces responsible for their waste or deterioration. Emphasis has to be given to the possibilities of maximum sustained crop production over the range of naturally occurring soil conditions. Sustained production is possible only if irreplaceable resources are not lost. At the present time our national welfare depends in part on our food and fibre crops being produced with minimum expenditures of land, labour and capital. While food and fibre production from other sources such as the sea are important, we are concerned with the primary production from the land.

The growth of the plants on a soil are influenced by the capacity of soil to supply nutrients and water, to offer support to the growing plant, to furnish oxygen to the plant roots and to furnish the plant roots with a suitable range of temperatures. These have to be supplied at optimum quantities and at optimum time during the growth of crops from the time of sowing up to the time of harvest.

Soil Management

Soil management problems frequently consist of determining the most nearly ideal soil condition for the crop in question and determining the most feasible way available for bringing the soil in question closer in properties to that ideal. The other possibility is to determine the best use for a soil with little or no modification.

Over the period of years when soil is used for cropping, the processes of soil deterioration may take place due to erosion, leaching and poor management practices, For continuous maintenance of ideal soil conditions, the management has to be geared for producing good crops continuously in the future also.

It is necessary to keep in mind that one new practice by itself rarely gives a great benefit unless other practices are also adjusted to it. For example, there would be little need to irrigate soils if sufficient fertilizers and improved varieties were not used to obtain optimum yields. The best arable soils have deep rooting zones in relation to the plants to be grown, and have surface soils with good tilth.

In areas subject to wind of high velocity the surface of cultivated soils needs to be cloddy and trashy to help prevent blowing of the fine soil particles. Permanent vegetation must be established and maintained on soils highly susceptible to soil blowing.

Sloping, erosive soils require protection against run off and erosion. Steep, highly erosive soils in regions of high rainfall intensity need nearly full protection by vegetation to intercept the force of rain drops. The following are important in good soil management programmes:

1. Erosion control;
2. Water conservation and use;
3. Drainage;
4. Maintenance of desirable soil reaction (pH);
5. Additions of plant nutrients;
6. Cropping practices;
7. Tillage practices; and
8. Organic matter replenishment.

Soil productivity can be maintained by recognizing the importance of these items and properly using them to meet the particular needs of each farm.

SOURCES OF PLANT NUTRIENTS IN THE SOIL

There are many sources through which the soil gets plant nutrients. These are mentioned below:

1. Addition of Nutrients to the Soil through Crop Residues

New plant types have exhaustive deeper root system and usually the harvesting is also done at a height of about 20-25 cm from the ground which leaves roots and stubbles to the extent of about 3-4 tonnes/crop/hectare in the case of cereals like rice, wheat, barley, etc. Similarly, a huge quantity of leaves or almost entire leaves of oilseeds and pulses are found shaded in the field which comes to around 1.5 to 2 tonnes/crop/hectare. These organic materials in the form of plant refuse or residues add nutrients to the soil after their decomposition.

2. Addition of Nutrients to the Soil through Commercial Fertilizers

The requirement of various crops is also met with application of synthetic/artificial/chemical/commercial fertilizers. These fertilizers include nitrogenous, phosphatic and potassic either in simple (straight) or complex/compound forms. Sometimes these fertilizers also contain micronutrients as subsidiary nutrients like ammonium sulphate contains S along with nitrogen for which it is applied. Similarly single super phosphate contains sulphur along with phosphorus. Thus these nutrient carriers are the chief source to supply plant nutrients into the soil.

3. Addition of Nutrients through Organic Manures

There are some organic resources (plant and animal refuse) which are decomposed or condensed and applied to the soil. The organic manures may be concentrated like oil cakes, blood meal, and bone meal etc. or bulky like compost, farm yard manure, leaf mould, night soil, sewage etc. They supply more of organic matter and small quantities of plant nutrients. These are the prime sources of micro-nutrients alongwith major nutrients,

4. Addition of Nutrients to the Soil through Rains

The elemental nitrogen present in environment combines with oxygen then reacts with rain water. Thus, the nitrogen reaches to the soil in the form of nitrus or nitric acid, The ammonia (which escapes from soil and goes to the atmosphere through volatilization) also reacts with water and comes back to the soil in the form of ammonium hydroxide as explained in the following equations:

$$N_2 + O_2 = 2NO$$
$$2NO + O_2 = 2NO_2$$
$$H_2O + 2NO_2 = HNO_3 + HNO_2$$
$$2HNO_2 + O_2 = 2HNO_3$$

On an average about 5 kg N/ha enters the soil every year through this source.

5. Microbial Fixation of Atmospheric Nitrogen into the Soil

Nitrogen in variable quantities is fixed into the soil through various microbes under favourable soil conditions like good aeration, optimum moisture, good supply of calcium and molybdenum, low initial soil nitrogen level, etc. The nitrogen fixed by these microbes may be classified as under:

(A) Symbiotic Nitrogen Fixation

There is a group of bacteria which in symbiotic relationship (Rhizobium) with host plant (legume) survives in their roots and use to fix nitrogen in the root nodules. These bacteria derive the energy required for this purpose from the carbohydrate of the host plants and in return provide N (through fixation in plant roots of legumes, host). According to Lyon and Brizzel (1934) following quantities of N are fixed by various Rhizobium bacteria associated with different crops (Table 9.1).

TABLE 9.1: Amount of Nitrogen fixed by Various Crops

Crops	N fixed (kg/ha)	Crops	N fixed (kg/ha)
Alfalfa	251	Soybean	105
Sweet clover	168	Hairy vetch	68
Red clover	151	Field beans	58
Alsike clover	141	Field peas	48

The fixed nitrogen is partly used by the host legume plants and rest passed into the soil either by excretion or by falling of roots and nodules which benefits the succeeding crop to be grown in the sequence.

(B) Azofication or Free Fixation

There is a group of free living bacteria, which are heterotrophic in nature like Azotobacter and Clostridium species that plays a prominent role in azofication. These organisms derive the energy required for this process from decomposing organic matter, however, the amount of nitrogen fixed by azofication depends upon the soil pH as below 6 pH of N fixed by this group of bacteria ranges between a few kg to 60 kg N/ha per year.

There are some micro flora like blue green algae like Anabaena, Tolypothrics, Azolla, etc. which may fix the atmospheric nitrogen to the extent of 30-60 kg/ha/ season.

(C) Mycorrhiza

Beneficial association between a plant root and its endophytic fungal associate is known as vascular-arbuscular mycorrhiza (VAM) which helps in increasing P availability and when legumes are inoculated, their symbiotic function is more activated.

(D) Phosphate Solubilizing Micro-Organisms

They secrete organic acids and dissolve fixed P which becomes available to plants.

6. Addition of Nutrients through Green Manuring

Green manuring is a process of turning under decomposable plant materials in green state (other than crop residues) usually from leguminous plants either in place where they are grown (in situ) or brought from a distance. The legumes contain 2.5 to 3 per cent N on oven dry basis and the roots of annual legumes may contain from 5 to 20 per cent of the nitrogen found in entire plants. The approximate amount of nitrogen added by various green manure crops as given by Mirchandani and Khan (1953) is being presented in following Table 9.2.

TABLE 9.2: Amount of N turned under from above Ground Parts of Various Green Manure Crops

Green Manure Crops	Average green Matter (kg/ha)	Moisture (%)	N content in green matter (%)	Amount of N (kg/ ha)
Sunnhemp	19,604	75.0	0.43	83.9
Dhaincha	18,400	78.2	0.43	79.1
Black gram	11,040	83.0	0.41	45.3
Green gram	7,360	75.0	0.53	38.8
Cluster bean	18,400	75.0	0.34	62.6
Cowpea	13,800	86.4	0.49	67.6
Horse gram	9,200	71.5	0.33	30.4
Indigo	8,200	44.7	0.78	71.8

7. Addition of Plant Nutrients through Soil Amendments

There has been a common practice to correct the soil reaction by application of chemical compounds like ground agricultural lime stone, ground dolomite limestone, quick, burnt and unslake lime, hydrated lime and basic slag, etc. as liming material in acidic soils where as gypsum ($CaSO_4 2H_2O$), Sulphuric acid (H_2SO_4), Ferrous sulphate ($FeSO_4 7H_2O$), Aluminium sulphate [$Al(SO_4)_3 18H_2O$)], Sulphur (S), Calcium chloride ($CaCl_2 2H_2O$) and Pyrite FeS_2), etc. are added to the soil for correcting alkalinity or salinity and alkalinity.

These materials add secondary nutrients (Ca, Mg and S) along with micronutrients like Al, Fe, etc. to the soil apart from correcting the soil reaction. Thus, the soil is enriched.

8. Addition of Nutrients through Herbicides and Fungicides

Herbicides like calcium cyanamide etc. and fungicides like blue copper, copper sulphate, Bordeaux mixture, etc. when applied as plant protection materials use to add plant nutrients, though their quantities used are negligible and hardly serve any purpose.

SOIL FERTILITY EVALUATION

Soil fertility is the inherent capacity of soil that enables it to provide essential plant elements in quantities and proportions for the growth of specified plant when other growth factors are favourable. It indicates plant growth in relation to nutrients available in soil. **Soil productivity is** the ability of a soil for producing a specified plant or sequence of plants under a specified system of management. It is usually expressed in term of yield. All fertile soils may not be productive due to adverse climate, water logging, salinity, alkalinity, acidity, etc.

Diagnostic techniques for nutritional disorders are the methods for identifying nutrient deficiencies, toxicities or imbalance in the soil-plant system. Nutrient deficiency can occur when a nutrient is inadequate in the growth medium and/or can not be absorbed and assimilated by plants due to unfavourable environment. Some of the soil conditions associated with nutrient deficiencies are:

Nutrients	Condition inducing deficiency
N	– Low soil organic matter, leaching.
P	– Acidic, organic, leached and calcareous soils, high rate of liming
K	– Sandy, organic, leached and eroded soils, high liming, intensive cropping
Ca	– Acidic, alkali, or sodic soils
Mg	– Similar to calcium
S	– Low soil organic matter, use of high analysis fertilizers with low sulphur
Fe	– Calcareous soils, soil with high P, Mn, Cu or Zn, high rate of liming.
Zn	– Highly leached acidic soils, calcareous soils, and high soil levels of Ca, Mg, and P.
Mn	– Calcareous silt and clays, high organic matter
B	– Sandy soils, acidic leached soils, alkaline soils with free lime.
Mo	– Highly podzolised soils well drained calcareous soils.

Several techniques are used to assess soil fertility status: Biological tests, Visual nutrient, Deficiency symptoms, Plant analysis and Soil testing are the commonly used methods.

BIOLOGICAL TESTS

These methods utilize plants as extracting agents for nutrients from the soil.

Field Tests

These are the most direct way of diagnosing nutrient status of soil. The effect of additional amounts of nutrients in the form of fertilizers applied to the soil on the growth of the crop is measured and subjected to statistical analysis. The quantity of nutrients for optimum crop yield can be broadly determined. The results are location and situation specific and depend on climatic conditions. Field tests are expensive and time consuming.

Laboratory and Green House Tests

These are simpler and most rapid. They still involve higher plants grown in small quantities of soil (pot culture).

Mitscherlich potculture consists of growing oats to maturity in pots containing 6 lb of soil. Yields of N-P and N-K treatments are expressed as percentage of yield from complete N-P-K treatment. Mitscherlich tables are used to read nutrient reserve in the unfertilized soil and to predict percentage increase in yield expected from the addition of a given amount of nutrient.

Neubauer seedling method is based on the uptake of nutrients by a large number of plants grown on a small amount of soil. Available nutrients will be exhausted in a short time. Nutrients removed by plants are determined by chemical analysis. Tables have been prepared to know the minimum values for satisfactory yield of various crops.

Microbiological Methods

These methods use certain microorganisms instead of higher plants.

Sackett and Stewart technique is used to study P and K status of soils. A culture is prepared for each soil. P is added to one portion, K to another and both to a third portion. The cultures are incubated with *Azotobacter* for 72 hrs. The soil is rated very deficit to not deficit in the respective elements, depending on the amount of colony growth.

In **Aspergillus Niger method,** small amounts of soil are incubated for a period of four days in flasks containing appropriate nutrient deficiency. The amount of nutrient absorbed by mycelial is used to measure nutrient deficiency.

Soil plaque method consists of incubating small portion of soil containing starch or sucrose with *Azotobacter*. The intensity of the growth of the colonies indicates the degree of deficiency of phosphorus and potassium.

VISUAL SYMPTOMS

The cheapest diagnostic technique for identifying nutrient disorders in crop plants is visual systems. However, visual symptoms are some times confused with disease, insect or soil moisture stress. There are three steps in identifying nutrient disorders by visual symptoms.

1. Observing plant for its normal growth and development: stunted growth may be due to deficiency or toxicity of all elements, but N and P deficiencies have more influence on growth reduction,
2. Plant part affected: whether foliar symptoms appear on lower or older leaves, or on younger or growing points of the plant, and
3. ecognition of nature of symptoms: chlorotic, necrotic or deformed.

If symptoms appear on lower leaves, they may be due to deficiency of **mobile nutrients** such as N, P, K and Mg. Mobile nutrients are those which can be translocated within plants. Hence, deficiency symptoms occur first on the lower part of the plant. If deficiency symptoms first appear on the upper young leaves, they may be due to deficiency of **immobile nutrients** such as Ca, Fe, Cu, S, B, Mn and Mo. Immobile nutrients are not translocated to the growing region of the plant but remain in older leaves where they were originally deposited after absorption from the soil. Key points in identification of nutrient deficiency or toxicity symptoms in crops are given in Tables 9.3 and 9.4. Practical applicability of this method is rather limited.

- Symptoms are often vitiated by the interaction of elements and also by the intensification of pests and diseases,
- The very fact the crop is showing deficiency symptoms indicates that its growth and development has already been hindered and optimum yield may not be possible even after the remedy,
- Deficiency symptoms may vary from species to species and even from variety to variety, and
- Plants may not exhibit deficiency symptoms due to hidden hunger.

TABLE 9.3: General description of nutrient deficiency symptoms

Nutrients	Symptoms
N	– Stunted growth, chlorosis (yellowing) first on lower leaves, reduced tillering in cereals
P	– Purple orange colour of older leaves, new leaves dark green, poor root system, stunted growth.
K	– Older leaves may show spots or marginal burning starting from tip.
Ca	– Failure of terminal bud and root tips, growing point die and curl, new leaves become white.
Mg	– Interveinal chlorosis of older leaves with veins remaining green, pinkish colour of old leaves.
S	– Chlorosis of younger leaves, severe deficiency leads to chlorosis of entire plant.
Zn	– Characteristic little leaf and resetting or clustering of leaves at the top of the fruit trees. In sorghum, its deficiency is called white bud, cotton leaf and in citrus mottle leaf.
Fe	– Interveinal chlorosis of younger leaves, severe deficiency leads to yellowing of entire leaf first and finally white.
Mn	– Similar to iron deficiency, at advance stages, necrosis develops instead of white colour.
Cu	– Chlorosis of young leaves, rolling and dieback In advanced stages, dead tissue appears along the tips and edges of leaves similar to that of potassium deficiency.
Mo	– Mottled pale appearance in young leaves, bleaching and weathering of leaves.
B	– Thickened or curled leaves, thickened, cracked or water soaked condition of petioles and stems, cracking or rotting of fruits, tubers or roots.

TABLE 9.4: General description of toxicity symptoms

Nutrient element	Symptoms
N	– Abundant foliage with dark green colour, NO_3 toxicity shows marginal burn of older leaves followed by interveinal collapse, NH_4 toxicity leads to blackening around tips of older leaves and necrosis.
P	– Necrosis and tip dieback, interveinal chlorosis in younger leaves, marginal scarch of older leaves
K	– May lead to Mg and possibly Mn, Zn and Fe deficiency.
S	– Reduction in leaf size, sometimes interveinal yellowing or leaf burning
Mg	– May induce K deficiency.
Fe	– Common in lowland rice on acid soils, bronzing of older leaves, induced P, K and Zn deficiency.
B	– Interveinal chlorosis
Mn	– Yellowing starting at the leaf edge of older leaves, uneven chlorophyll distribution, interveinal bronze-yellow chlorosis in leaves.
Cl	– Burning of leaf tips or margins, reduced leaf size.
Zn	– Induced iron chlorosis.

Nutrient element	Symptoms
Cu	– Stunting reduced branching, induced iron chlorosis.
Mo	– Not common
Al	– Yellowing with white interveinal stripe on older leaves

PLANT ANALYSIS

Plant analysis is based on the assumption that the amount of a given element in plant is an indication of supply of that particular nutrient and as such is directly related to quantity in the soil. There are two types of plant analysis: Tissue test on fresh tissue in the field and Total analysis in laboratory.

Tissue Tests

Rapid tests are made in the field on the above ground parts of growing plants to determine nutrient elements in the plant sap of fresh tissue. The sap from ruptured cells is tested for unassimilated nitrogen, phosphorus and potassium. These are semi-quantitative tests for verifying or predicting deficiencies of N, P and K. The sap from the ruptured cells is tested with reagents. The intensity of colour development is compared with standards and used as a measure of the supply of the nutrient in question. In general, latest mature leaf is used for testing. The plant should be tested at the time of bloom or from bloom to early fruiting stage early in the morning or late in the afternoon. Tissue tests have gained considerable importance due to their low cost and simplicity.

Total Analysis

Plant analysis aims at determining the concentration of an element or extractable fraction of an element in a simple from a particular part at a certain stage of crop growth. With the development of instruments such atomic absorption specto-photometers, spark emission spectrometers and inductively coupled plasma (ICP) emission spectrometers, plant analysis has become more sensitive and simplified. Simultaneous analysis of up to 22 elements is possible by ICP. With total analysis, almost all the elements associated with plant nutrition can be determined. Plant parts to be sampled at different growth stages are given in Table 9.5. When sampling, precautions should be taken to avoid soiled, diseased and insect or mechanically damaged plants and to exclude dying or dead tissue.

TABLE 9.5: Plant parts and number of plants to be sampled for field crops at different growth stages (Jones and Steyn 1973)

Crop	Growth stage	Plant part to sample	No. of plants to sample
Wheat	Seedling stage	All the above ground portion	50-100
Rice, barley	Prior to heading	The four uppermost leaves	50-100
Corn	Seedling stage	All the above ground portion	20-30
	Prior to tasseling Below the whorl	The entire leaf fully developed	15-25
	From tasseling and Shooting to silking	The entire leaf at the ear node or immediately below or above it	15-25

Crop	Growth stage	Plant part to sample	No. of plants to sample
Soybean or other beans	Seedling stage Prior to or during initial flowering	All the above ground portion	20-30
		Two or three fully developed leaves at the top of the plant	20-30
Sugar beet	Mid season	Fully extended and mature leaves Midway between the younger centre leaves and the oldest leaf whorl on the outside	30-40
Groundnut	Prior to or at Bloom stage	Mature leaves from the main stems and either cotyledon lateral branch	40-50
Cotton	Prior to or at bloom or when first squares appear	Youngest fully mature leaves on main stem	30-40
Alfalfa	Prior to or one tenth bloom stage	Mature leaf blades taken about one-third of the way down the plant	40-50
Clover	Prior to bloom	Mature leaf blades taken about one-third of the way down from the top of the plant	40-50
Sugarcane	Upto 4 months	Third or fourth fully developed leaf from the top.	15-25
Sorghum	Prior to or at heading	Second leaf from the top of plant	15-25

Interpretation

The basis for plant analysis as a diagnostic technique is the relationship between nutrient concentration in the plant and growth and production response. This relationship should be significant to have complete interpretation in terms of deficient, adequate and excess nutrient concentrations in the plant.

When the nutrients are in **deficiency range,** yields are significantly reduced and there will be sharp increase in yield even with very little change in nutrient concentration in plant. In the **marginal range,** there will be reduction in yield, but plants do noit show deficiency symptoms and both nutrient concentrations and growth increase as more nutrient is absorbed. Within the marginal zone lies **critical nutrient concentration (CNC)** at which growth and yield of crop begins to decrease significantly. The CNC is usually estimated on the basis of 5, 10 or 20 per cent reduction in maximum yield. The third range is **adequate zone,** in which there is no increase in yield, but nutrient concentration increases. In the **excess range,** between adequate and toxic, fertilizer use should be reduced until the nutritional status of plants lies in the adequate range. Growth and yield of crop is reduced but nutrients concentration continues to increase in the **toxic range** leading to toxicity symptoms. Adequate concentrations for essential nutrients in crop plants vary with soil, climate, crop, management practices, etc. and are very difficult to make generalizations. However, such generalized values may give an idea about adequate levels of nutrients in crop plants (Table 9.6).

TABLE 9.6: Average concentrations of essential nutrients in dry matter sufficient for normal growth of crops (Fageria, Baligar and Jones, 1991)

Nutrient	g/kg	Nutrient	mg/kg
Carbon	450.0	Iron	112.0
Oxygen	450.0	Manganese	55.0
Hydrogen	60.0	Zinc	20.0

Nutrient	g/kg	Nutrient	mg/kg
Nitrogen	14.0	Copper	6.0
Phosphorus	1.9	Boron	22.0
Potassium	9.8	Molybdenum	0.1
Calcium	5.0	Chlorine	106.0
Magnesium	1.9		
Sulphur	1.0		

Diagnosis and recommendation integrated system (DRIS) is a comprehensive system which identifies all the nutritional factors limiting crop production for improving crop yields through an ideal fertilizer schedule. Index values which measure how far particular nutrients in the leaf or plant are from the optimum are used in the calibration to classify yield factors in the order of limiting importance. It has several advantages over CNC approach.

- Importance of nutritional balance is taken into account,
- Elemental content in leaf tissue can be universally applied to a particular crop, regardless of where it is grown,
- Diagnosis can be made over a wide range in stages of crop development, irrespective of cultivar, and
- Nutrients limiting yield can be identified and arranged in order of their limiting importance on yield.

Crop logging is a graphic record of progress of crop containing a series of chemical and physical measurements. These measurements indicate the general condition of the crop and suggest changes in management that are necessary to obtain maximum yield. It is developed in Hawaii for sugarcane crop. Critical nutrient concentration approach is used in crop log system. The plant tissue is sampled at regular intervals during the crop season and analyzed for nitrogen, sugar, moisture and weight of the tissue. Analyses are made for P and K at critical times and adjustment practices are introduced as needed.

Isotope dilution technique consisting of radiochemical analysis of plants grown on soils which have been treated with fertilizers containing elements such as radioactive phosphorus may be used to calculate phosphorus supply of original soil. Equations, based on the concept that a plant will absorb a nutrient from two sources in direct proportion to the amount available from each source, reduce essentially to identities in which A = Y.

$$\text{A value: } A = \frac{B\,(1 - y)}{y}$$

Where,

A = available phosphorus in soil,
B = amount of phosphorus applied to soil, and
y = fraction of phosphorus taken by plant from B.

SOIL TESTING

Soil testing is a chemical method for estimating nutrient supplying ability of soil. It may also include any physical measurement depending on the objective of testing. It includes collection of soil samples, sample preparation, laboratory analysis, calibration and interpretation of the tests and recommendations. It has the advantage over deficiency symptoms and plant analysis. In that, one can determine the deficiencies in the soil before the crop is sown. Routine analysis is done for pH, EC, P, K, Ca and Mg. Other determinations, such as organic matter and micronutrients are performed under specific conditions.

Soil Sampling

Since the soil is extremely heterogeneous, the sample collected should be true representative of the area sampled. Most of the immobile nutrients (P, K, Ca, Mg) remain in top layer. As such the results indicate abnormally low values, if top layer is not collected. Sampling depth depends on nutrient mobility and crops to be grown. Mobile nutrients like NO_3^-, SO_4^- and borates should be sampled to a depth of about 60 cm and samples must be taken when biological activity is low. Sampling to tillage depth is adequate for immobile nutrients. For field crops a depth of 15-20 cm is adequate. Soil samples should be taken in zigzag pattern. A representative soil sample is composed of 15-20 sub-samples from a uniform field with no major variation in slope, drainage or past fertilizer history.

Sample Preparation

Samples can be dried in moving air but not in heated air. Samples can also be dried in a cabinet type forced air dryers for 24 to 48 hours. Organic soils are analyzed as received without drying. After drying, samples are ground to pass a 10 mesh (2 mm) screen.

Laboratory Analysis

It involves extraction of nutrients and determination of their concentration in the extract. Nutrient values are expressed as **extractable nutrients rather than available nutrients** because the quantities of nutrients absorbed by plants may be close to those extracted by standard extractants.

Nitrogen: There is no reliable method for testing soil available nitrogen. Soil nitrogen is found in organic forms and its mineralization depends on climatic factors, which vary constantly during the crop season. Various biological and chemical methods have been proposed for estimating organic nitrogen supplying capacities of soils. However, none of these are sufficiently reliable to warrant its routine use in soil testing laboratories. Nitrogen recommendations are, therefore, made on the basis of long term crop response data.

Phosphorus: The most common extractants used in routine analysis of P in various parts of world are:

- **Bray 1:** 0.03 M NH_4F + 0.025 M HCl (pH around 5.5 or less)
- **Olsen:** 0.5 M $NaHCO_3$, ph 8.5
- **Mehlich 1:** 0.05 M HCl + 0.0125 M H_2SO_4

Once the nutrients are extracted, quantitative equipment is used to measure the amounts in the extract.

Potassium, calcium and magnesium: Commonly used extractants of K, Ca, Mg and S are:

- 0.05 M HCl in 0.125 M H_2SO_4 (double acid)
- M NH_4OAc at pH 7.0
- M NaOAc at pH 4.8

Double acid extractant is suitable for sandy soils and acidic soils with low CEC, whereas ammonium acetate extractants are suitable for a wide range of soils.

Micronutrients: Various methods have been developed for determining the availability of micronutrients in the soil. Chelating agents such as DTPA (diethylene-triamine-pentaacetic acid) appears most promising for Zn, Cu, Mn and Fe. Analysis for boron is usually made by the hot water method.

Calibration and Interpretation

The objective of soil calibration is to determine the amount of nutrient to obtain maximum yield. For interpretation of soil test results, crop yields are plotted against soil test values. Absolute yield values should be transformed into relative yield values to eliminate some of the climatic and site influences.

$$\text{Relative yield (\%)} = \frac{\text{Yield of control plot}}{\text{Max. yeld of treated plot}} \times 100$$

Fig 9.3 indicates the relationship between soil test P and relative yield of common bean. Based on relative yield, soil test values are classified as very low (<50% relative yield), low (50-75% relative yield), medium (75-100% relative yield) and high (>100% relative yield). Very low soil test value for a nutrient indicates heavy fertilizer application for soil-building purpose. Relative yield of 50-75 per cent indicates necessity for annual application for high response and soil fertility build up. With medium soil fertility range, normal annual application is used to produce maximum yield. With high fertility range, small applications to maintain soil test level are essential. Table 9.7 indicates typical interpretations of soil tests for P, K and Mg.

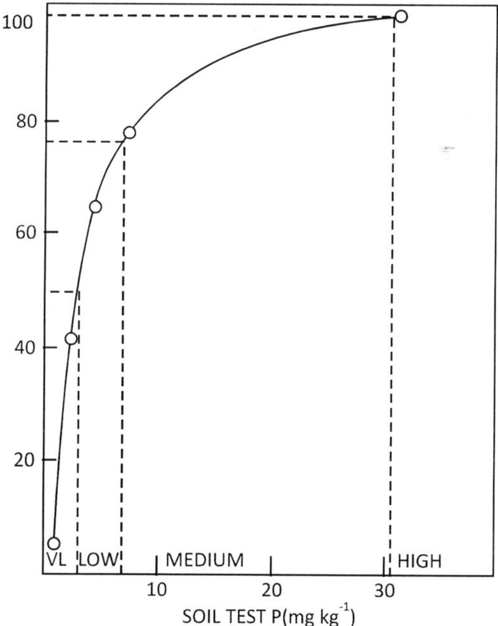

FIG. 9.3: Relationship between soil test P and relative yield.

TABLE 9.7: Interpretation of soil test values for P, K and Mg (Lorenz and Maynard 1980)

Relative rating	Extractable nutrients (mg/kg)				
	Mehlich 1			Bray 1	
	P	K	Mg	P	K
Very low	0-6	0-14	0-17	0-12	0-60
Low	7-13	15-35	18-25	13-25	61-90
Medium	14-23	36-67	26-59	26-37	91-150
High	24-44	68-133	60-126	38-75	151-250

FERTILIZER RECOMMENDATIONS

After calibration and interpretation, soil test data must be associated with appropriate fertilizer recommendations. Each of the four common interpretative categories (very low, low, medium and high) signifies different fertilizer rates. Very low levels require large amounts of fertilizer, low level needs annual application, medium level needs normal applications and high level needs little or no fertilizer. High yield through fertilizer application can be obtained only if all growth factors and cultural practices are optimum.

Cope (1973) describes **index system** which utilizes the concept of percentage of nutrients in the soil. The index expresses the per cent sufficiency of a nutrient based on the amount necessary to produce 100 per cent relative yield. It is usually expressed to the nearest 10 per cent (Table 9.8).

TABLE 9.8: Relationship among fertility index, soil test rating, relative yield and Recommendations (Cope and Rouse 1973)

Fertility Index	Soil test rating	Relative yield of cotton or Legumes (%)	Recommendations
0-50	Very low	<50	– Large applications to build up fertility
60-70	Low	50-75	– Annual applications for higher response and to improve soil fertility
80-100	Medium	75-100	– Normal annual applications for higher yield
110-200	High	100	– Small applications to maintain soil level
210-400	Very high	100	– None until the level drops to high range
410-999	Extremely high	100	– Used for P on lawns, gardens and shrubs. – Further addition may lead to Fe and Zn Deficiency.

SOIL TEST AND CROP RESPONSE (STCR)

Soil test and crop response (STCR) approach based on soil contribution and yield level is used for recommending fertilizer rate. Based on this approach fertilizer recommendations are given for different agro-climatic conditions of India. One such equation for recommending N, P and K for rice is:

$$FN = 4.39\,T - 0.6723\,SN$$
$$FP_2O_5 = 2.83\,T - 6.110\,SP$$
$$FK_2O = 1.41\,T - 0.329\,SK$$

Where FN, FP_2O_5 and FK_2O are fertilizers N, P_2O_5 and K_2O kg/ha to be applied, SN, SP and SK are soil N, P and K kg/ha and T is targeted yield (q/ha).

Soil test correlation approach consists of selecting a group of soils ranging in fertility from high to low in respect of a particular nutrient and testing varying doses of that particular nutrient on a crop. Soil test values are plotted against percentage yield and correlation coefficient between soil test values and per cent yield increase is calculated.

$$\text{Percentage yield} = \frac{\text{Crop yield with adequate nutients - Yield with control}}{\text{Crop yield with adequate nutrients}} \times 100$$

Based on soil test values, soils are grouped into classes. The rating adopted by IARI, New Delhi is:

Nutrient	Low	Medium	High
Organic carbon (%) (For N)	<0.05	0.50-0.75	>0.75
Available P (kg/ha)	<10	10-25	>25
Available K	<120	120-280	>280

The IARI formulated fertilizer recommendations (Table 9.9) based on soil tests for wheat, rice and maize of Delhi.

TABLE 9.9: Soil test fertilizer recommendations (Biswas and Mukherjee 1989)

Soil organic Carbon (%)	N (kg/ha)	Soil available P (kg/ha)	P_2O_5 (kg/ha)	Soil available K (kg/ha)	K_2O (kg/ha)
Low					
<0.20	160	<10	80	<100	60
0.20-0.40	140	11-20	60	101-150	50
Medium					
0.41-0.60	120	21-30	45	151-200	40
0.61-0.80	90	31-40	30	201-250	30
High					
0.81-1.00	75	41-55	15	241-300	20
>1.00	60	>55	10	>300	0

The recommendations are of general nature and are location specific than situation specific, as quantification for an individual situation is not possible.

Permanent Manurial Experiments

Long term field experiments can only give precise information on changes in soil properties and soil productivity. The earliest long term field experiment called **permanent manorial experiment** was started at the Rothamsted Experimental Station, Harpenden, England in 1843. Sir John Lawes was the first to start investigating agricultural problems. Herbert Gilbert later assisted Lawes. In India, such experiments were started at Kanpur in 1885, at Pusa in 1908 and at Coimbatore in 1909. Except at Coimbatore, the experiments have been discontinued. However, new permanent manorial experiments were started at several places during 1930s. The ICAR sponsored Coordinated Project on Long Term Fertilizer Experiments in 1970-71 for critically assessing the effect of modern intensive agricultural systems on soil productivity and crop yield obtained at Bangalore, Barrackpore, Bhubaneswar, Coimbatore, Delhi, Hyderabad, Jabalpur, Ludhiana, Palampur, Pantnagar and Ranchi.

Results so far obtained indicate importance of balanced NPK along with manures, sulphur and zinc in maintaining soil fertility and productivity.

"Hope is, faith holding out its hand in the dark"

10 Integrated Nutrient Supply for Crop Production

"Organic materials were practically the only external source of nutrients to crops before introduction of inorganic fertilizers. The recycling of various forms of residues has the advantage of converting the surplus farm waste into useful product for meeting nutrient requirement of crops, besides maintaining the soil condition and improving the overall ecological balance".

Dr. M. S. Swaminathan

INTRODUCTION

There has been a phenomenal progress in food grain production since middle of the 20th century. In fact from 1950 to 1984, global food production went up to 2.6 fold making possible 40 per cent increase in per capita cereal production despite rapid population growth during that period. But since 1984, per capita production of cereals world-wide has declined each year for a total of 14 per cent over the 4 years period. Global grain reserve is projected to be at the lowest level in the year since immediately after World War II (York Jr., 1989). Now many are questioning our ability to sustain such increases in food production or even maintain current levels.

Thus, **sustainability** is a dynamic concept which goes on changing with change in the needs of steadily growing global population. In the static sense, many traditional agricultural systems were sustainable for centuries in terms of their ability to maintain a continuing stable but low level of production and simultaneously maintaining the ecological balance to a great extent. However, the needs and growing aspirations of the increasing population, have forced changes in production practices that have imposed excessive demands on natural and renewable resources, leading to exploitive agriculture, since we can not afford to go for zero growth in production which only, can keep the environment absolutely sustainable.

Therefore, **"sustainable agriculture** may involve the successful management of natural and renewable resources in crop production to satisfy changing human needs, simultaneously maintaining or enhancing the quality of the environment and conserving natural and man-made resources on long term basis i.e., maintaining an optimum balance between the resources management system and the environmental capital stocks i.e., soil, water, etc."

The strategy for **sustaining satisfactory yield levels** envisages nutrients balances and efficient nutrient cycling. This can be successfully achieved through integrated plant **nutrients supply system.** Integrated nutrients management involves tapping of all the major sources of plant nutrients be it soil nutrient reserve, mineral fertilizers, bulky organic manures, compost, green manures, biological inoculants etc.; in a judicious way and to ensure their efficient use. One important feature of this approach is that the fertilizer recommendation should take into account the cropping system as a whole rather than individual crop in a system.

The nutrient supply from soil can be enhanced by adopting **soil management and conservation practices** to reduce losses of nutrient through leaching, erosion and run off. Amelioration of problem soils and improving soil physical conditions insure maximum possible efficiency of applied nutrients.

There is great need to use available fertilizers most efficiently. This will ensure not only **clean environment** but also higher return by using a somewhat lower fertilizer doses. Fertilizer use efficiency is of high order when nutrients are applied in needed quantities in a balanced proportion at appropriate time in a proper form with suitable method in right way to a crop. Besides, soil moisture, weed control also has a considerable impact on response of fertilizers and efficient use of added plant nutrients. At present, the nutrient use in India is much less compared to that of other countries. Against world average consumption of 95 kg/ha we are using 74 kg/ha with national productivity level of merely 1.1 metric ton per hectare.

Soil is the most important resource for **a sustainable agriculture**, since it contains essential nutrients, stores the water and provides the medium in which plants grow. Thus, an Integrated Nutrient Supply Systems (INSS) may play a vital role in sustaining both soil health and crop production on long term basis.

INTEGRATED NUTRIENT SUPPLY SYSTEM (INSS)

The basic concept underlying the principle of INSS is the maintenance and possibly improvement of soil fertility for sustaining crop productivity on long term basis, which may be achieved through combined use of all possible sources of nutrition and their scientific management for optimum growth, yield and quality of different crops and cropping systems in an integrated manner, in specific agro-ecological situations.

With evolution and availability of dwarf/hybrid varieties of crops, there has been a growing tendency amongst the farmers to harvest immediate gains in term of maximum production per unit area by indiscriminate use of chemical fertilizers and that also in an imbalanced nutrient ratio. In the beginning, there was linear relationship between level of fertilizer use and food grain production but after some time, it is started plateuing and more recently, it has shown a declining trend in spite of using the same high levels of fertilizers per unit area (Gill, 1989). This phenomenon of soil health degradation is much evident in states like Punjab, Haryana and western Uttar Pradesh which had witnessed the green revolution during the late 60's.

The boon in the production of wheat and rice due to availability of high yielding varieties, adequate irrigation facilities led to almost total exclusion of other crops like chickpea, barley, groundnut, sorghum, rapeseed and mustard from these areas. All these created imbalance in the nutrient and water supply systems leading to exploitive agriculture.

To get higher production and replenish the heavy nutrient removal under intensive cereal-cereal sequences, heavy doses of high analysis fertilizer had to be continuously applied. This resulted in wide spread deficiency of zinc in many parts of the country. Other micronutrients like manganese, sulphur and iron are also becoming critical.

Restricted or non-cycling of organic manures and more and more use of inorganic fertilizers have contributed to spread of micronutrient deficient areas.

Integrated nutrient supply system primarily relates to combined application of different sources of plant nutrients (organic and inorganic) for sustainable crop production without degrading the natural resources base the soil and that too, on long term basis.

INTEGRATED USE OF ORGANIC AND INORGANIC SOURCES

An account of combined use of chemical fertilizers with organic manures, viz., farm yard manure (FYM), crop residues and biogas slurry etc. in relation to physical, chemical and biological properties of soil on long term basis, is discussed here. It also relates to raising of mineralization efficiency and nutrient recovery from organic sources.

Addition of organic materials improves the physical properties of the soil. Regular addition of FYM improves the organic carbon, structural status and water retention capacity of silty clay loam soil ((Ranchi) and alluvial sandy loam soil (Sabour). Prasad and Singh (1980) observed the beneficial effect of FYM alone or in combination with fertilizers and lime in increasing the percentage of water stable aggregates of more than 0.5 mm and thereby improving the soil structure. Simultaneously, there was an increase in water holding capacity, decrease in pore space and increase in hydraulic conductivity. A combined application of FYM and chemical fertilizers increased the pore space, water holding capacity and soil structure (Biswas *et al.* 1971). Improvement of soil structure was mainly due to enhanced microbial activity (Gaur *et al.* 1972; Singh *et al.* 1987, Singh and Bhattacharyya 1989).

Different workers have reported a high positive correlation between organic matter content and extractable micronutrients (Misra and Pande, 1976). Application of organic manures in general, improves the availability of micronutrients like zinc, iron, manganese and copper (Mann *et al.* 1978). In most of the studies, organic manures have been primarily valued as source of N, P and K. They are potential source of micronutrients too (Table 10.1).

TABLE 10.1: Micronutrient content (ppm) of some organic manures

Organic manures	Zn	Cu	Mn	Fr
FYM	120	62	410	--
Cow manure	210	61	150	--
Goat manure	2570	1925	6420	--
Sewage sludge	2980	740	150	--
Poultry manure	70	82	191	1280
Piggery manure	198	13	168	1600
Dhaincha leaves	17	--	80	140
Rice straw	20	--	340	280
Sugarcane bagasse	17	--	15	240

Source: Katyal (1985)

The effect of integrated use of organic and inorganic sources of nutrients was more spectacular in acidic soils than normal soils and was equivalent to use of lime as amendment probably due to chelation of Al and Fe ions by organic molecules and reduced levels of these ions (Patwardhan *et al.* 1956 and Nambiar, 1985). Sarkar *et al.* (1973) reported increase in per cent aggregates with increasing levels of FYM in slightly alkaline non-calcareous non-saline sandy loam soil.

Composting is a microbiological process are thus rapid growth and activities of microorganisms resulting in quick and good quality compost depend on various factors like optimum conditions of moisture, C/N ratio of the residues used, temperature and oxygen levels of the composting heaps. At IARI, the uses of composting inoculants like *Aspergillus, Penicillium, Trichoderma, Trichurus,* and *Chaetomium,* has been found to hasten the composting of crop residues with high C/N ratio resulting in improved quality of compost (Gaur *et al.* 1972). In another experiment, amending the compost with low grade rock phosphate followed by inoculation with nitrogen fixing *Azotobacter chroococcum* and phosphate solubilizing *Aspergillus awamori* after about one month of composting has resulted in enhancing nitrogen status of compost by about 4 kg N/tonne of finished compost compared to composting without amendment (Sadasivam *et al.* 1981).

An innovative idea for overcoming zinc deficiency in crop production through integrated nutrient management concept, is the use of zinc enriched organic manures which can ensure better returns than application of zinc sulphate alone (table 10.2).

TABLE 10.2: Influence of Zn-enriched organic wastes in rice-wheat sequence

Treatments	Grain yield (t/ha)		Grain yield (t/ha) in term of response to $ZnSO_4$	
	Rice	Wheat	Rice	Wheat
Control	4.1	2.1	--	--
10 kg Zn/ha as $ZnSO_4$	4.6	2.7	0.5	0.6
Enriched FYM equivalent to 10 kg Zn/ha	4.9	2.6	0.8	0.5
Enriched sewage sludge equivalent to 10 kg Zn/ha	5.3	2.7	1.2	0.5
C.D. at 5%	0.4	0.3		

Source: Sharma and Sinha (1989)

The long term fertilizer experiments conducted all over the country, have shown that neither organic manures alone nor the mineral N, P and K fertilizers can sustain high productivity under modern intensive farming where the nutrient turnover in the soil-plant systems is quite high. Organic manures alone may suffice for lower nutrient demand under low to medium intensity cropping but combination of organic manures and chemical fertilizers becomes imperative to sustain productivity of soil and thereby sustaining a high crop yield.

EFFECT OF INCORPORATION OF CROP WASTES AND RESIDUES

Crop wastes and residues are renewable and readily available, but are scattered organic resources. They are very often directly applied to the field since composting con not be easily done. In addition to improving soil properties, conserving soil moisture and suppressing weeds, application of crop residues results in saving/ substitution of chemical fertilizers to a great extent.

Leguminous plant residues have low C/N ratios and therefore, they degrade faster and help build soil fertility with a good amount of available forms of nutrients. Benefits from incorporation of green manure crops and use of bio-fertilizers have been widely reported. Inclusion of a short duration legume like summer green gram in between wheat and rice, not only increased production of the total system but also helped in enriching the soil by incorporation of its stover. Incorporation of green gram stover as green manure one day before transplanting rice, added 100 kg N/ ha and it indicated a supply of 60 kg N/ha to rice crop alone, besides maintaining a higher availability of N during various growth phases of rice plant.

Increased in grain yield, however, has been attributed to following reasons:

1. Increase in N supply,
2. Favourable N-release pattern,
3. Increase in K supply,
4. Increase in supply of other nutrients, and
5. Improvement of soil structure,

It has been observed that cereal residues have a wider C/N ratio and unless they are supplemented with external nitrogen and phosphorus during decomposition, may cause immobilization of nitrogen and phosphorus for a longer period.

The incorporation of rice straw alone or combined with the higher dose of N was reported to ensure a substantial improvement in the organic matter content of soil as well as satisfactory yield level over a period of 6 years (Finassi, 1976). Better advantage was obtained by chopping of organic materials (Stover/Straw) before incorporation.

The best method of rice straw incorporation (approximately one month before flooding) with supplemental nitrogen and phosphorus, produced the highest average yield in a three years period (6.90 t/ha). Almost double dose of fertilizers were needed to achieve the identical yields from normal fertilizer dose with straw incorporation (Russo, 1976).

Incorporation of straw in sandy loam soil, one week before sowing, was improved nodulation and pod yield of groundnut crop by 95.5 per cent, with 5 tonnes of straw per hectare. Besides, there was residual effect with 10 tonnes of straw per hectare which significantly increased grain yield of the succeeding wheat crop (Table 10.3). Nodulation and yield were enhanced by application of wheat straw which was attributed to CO_2 enrichment inside the crop canopy due to straw decomposition (Hevelka and Hardy, 1976).

TABLE 10.3: Effect of straw incorporation on yield in groundnut-wheat sequence

Straw	Pod yield of groundnut (t/ha)	Grain yield of wheat (t/ha)
Control	1.7	3.6
2 tonnes/ha	2.8	4.0
5 tonnes/ha	3.3	4.5
10 tonnes/ha	2.4	5.2
CD at 5%	0.3	1.0

Source: Gaur and Mukherjee (1979)

Recycling of nutrients and increased grain yield of wheat through incorporation of chopped sugarcane trash one moth before sowing of wheat in medium black soil, has also been demonstrated. Krishnappa and Shinde (1980) reported that field experimented with ^{15} N-labelled rice straw in rice-rice rotation showed marked difference in its mineralization between two soil types. With adequate application of fertilizer N, 30.7 and 10.8 per cent of the applied straw-N was recovered in the first and second crop respectively at Cuttack (sandy loam kaolinite inceptisol, pH 5.5). But in Hyderabad, clayey soil (montmorrilonitic vertisol, pH 8.1), the corresponding recoveries were limited to only 10.7 and 2.2 per cent. The effect of fertilizer N was also more notable in the inceptisol than vertisol.

The decomposition of straw did not appear to be exclusively governed by the C/N ratio or application of fertilizers. A field experiment conducted at Pantnagar for 27 years on rice-wheat rotation with different combinations of N sources, showed that performance of prilled urea + fresh wheat straw was better during the years over urea super granules (USG) + fresh wheat straw at the same rate of applied N and the residual effect of applied fresh wheat straw on wheat yield was more pronounced during second year of experimentation.

When the crop residues can not be incorporated or made into compost, the benefits from the residues can still be derived by using them as organic mulch on soil surface. Mulches also help in moisture conservation and suppression of weeds. Per cent increases in grain yield of wheat, peas, maize and green gram due to organic mulching in different soil types and locations are presented in Table 10.4.

TABLE 10.4: Effect of organic mulching with rice/wheat straw

Crop loam	Red sandy loam	Alluvial	Alluvial	Sierozem Soil	Medium black soil	Acidic red loam
	(Bangalore)	(Barrackpore)	(Delhi)		(Hissar)	(Pune)
Wheat	31.6	67.0	16.1**	66.0	-	
Peas	29.4	9.6	-	-	13.4	
Maize	-	-	95.7	-	-	21.2
Green gram	-	-	31.7	-	-	-

* Mulched material incorporated in soil, previous crop-peas
** Mulched material incorporated in soil, previous crop-maize.
Source: Gaur (1978)

Higher agronomic efficiency and apparent recovery were observed with cowpea residue than green manure or urea-N in upland and lowland rice crops (John *et al.* 1989). The slow mineralization of N from residue and simultaneous uptake and less loss of N increased the N-use efficiency in case of residue.

With the fast expansion of biogas development programme in the country, it is expected that large amount of biogas spent slurry would be available. The digested slurry coming out from biogas plants, has been found to be superior to farm yard manure in respect of nutrient status and capacity to improve soil physical properties. The NPK contents of bio-gas spent slurry (on dry weight basis) are 1.4-1.8 per cent N, 0.4-0.9 per cent P and 0.6-1.0 per cent K. It is a valuable source of humus and micronutrients. However, nutrient content of slurry varies depending upon the initial material value of the dung or other substrates added to bio-gas plants. But due to high moisture content (92-96 per cent), it is difficult to handle the digested slurry. Using of slurry directly with irrigation water has been advocated. Also, it can be composted with crop residues or used as manure after drying and applied in combination with fertilizers.

Efficiency of slurry could be increased by applying a combination of half nitrogen through slurry and half nitrogen through ammonium sulphate (Table 10.5). There was no differential behaviour in decomposition of slurry under flooding (anaerobic) on saturation or moist (semi-aerobic) condition and thus, the only advantage of applying slurry in irrigation water, is for easy handling. Application of the slurry a few weeks in advance of planting and supplementation with chemical fertilizer, will help to overcome its deficiency of slow mineralization and thus, making the nutrients immediately available in early stages of crop growth.

TABLE 10.5: Yield (t/ha) of rice (var. Vani) as influenced by maurial practices

Treatments	Grain yield (t/ha)
Slurry (20 kg N/ha) basal	3.3
Slurry (10 kg N/ha) basal + Ammonium sulphate (10 kg N/ha)	3.6
Ammonium sulphate (20 kg N/ha) --½ basal + ½top dressed	3.7
No nitrogen	3.0

Source: Asthana and Jha (1977)

INTEGRATED USE OF BIOFERTILIZERS WITH OTHER SOURCES OF NUTRIENTS

Biological sources include green manures/green leaf manuring and bio-fertilizers like *Rhizobium* culture, *Azospirillum, Azotobacter,* Blue green algae (BGA), *Azolla,* phosphate solubilizing bacteria and *Mycorrhiza*, and they make up a judicious combination with chemical fertilizers and organic manures. Here, an attempt has been made to critically discuss the effect of integrated use of bio-fertilizers with chemical/or organic sources of nutrition in the context of sustainable crop and soil productivity.

(A). Nitrogen Fixation

There are two aspects of biological nitrogen fixation

i) Recycling of nitrogen fixed by selected leguminous and non-leguminous plants through green manuring and making nutrients available to the succeeding crops, and
ii) Direct fixation of molecular nitrogen into the soil through some symbiotic and non-symbiotic micro-organisms.

Green manuring is an age old practice but it started losing its importance with availability of chemical fertilizers and their fast and immediate effect on crop production from late 60's. In the present era of intensive cropping system, heavy input technology, environmental degradation and other related problems, its inclusion in nutrient supply system is highly valued. Leguminous crops like *Sesbenia spp.*, *Crotolaria juncea*, *Vigna mungo*, *Vigna sinensis* are common. Tree species like *Leucaena leucocephala* also have tremendous potential. Vegetative parts of other trees like *Casuarina, Gliricidia, Pongamia, calotropis, Tephrosia, Indigofera* and *Cassia* can also be used as green manures. Area covered under green manure programmes by different states in 1987-88 was estimated at 5.7 million hectares (Dept. of Agri. and Cooperation, Government of India, 1988-89). Benefits from *Sesbenia* and *Leucaena* green manuring has been equivalent to about 60 kg N per hectare in a rice-wheat system (Singh *et al.* 1987).

A record yield increase of 2.7 t/ha in rice by incorporation of 52 days old crop of *Sesbenia rostrata* before transplanting was reported, whereas application of 60 kg N/ ha through ammonium sulphate increased yield by only 1.7 t/ha over control (Rinando *et al.* 1981). It was estimated that about 33 per cent of the nitrogen accumulated by *S. rostrata* was transferred to the crop. Because of the stem nodules, *S. rostrata* can fix N under water logged conditions. As and when the N content of the medium is high, it gives exceptionally high potential for biological nitrogen fixation.

Green leaf manure of *Sesbenia* and *Ipomea* applied before transplanting released more N than farm yard manure and yielded as much as with 40 kg/ha of fertilizer N (Chatterjee, 1979). According to Tiwari *et al.* (1980), 40 kg N/ha along with green manuring, compared well with the application of 120 kg N/ha without green manuring, with regard to availability of nutrients in the soil, their uptake and rice yield. An increase of 54 per cent in grain yield of wheat succeeding rice with green manure was also observed, indicating notable residual effects of green manure.

Green manures, when buried just before transplanting rice, act as slow-release fertilizers and also create reducing conditions, which help in mobilizing several other nutrient elements. Accumulation of nitrogen in a 60 day old *Sesbenia* crop and a green gram crop was 105 kg and 101 kg per hectare, respectively (Gill,1989). Incorporation of green manure crop just before transplanting showed the best performance, which also allowed more time for green manure crop to grow.

In addition to the role of green manures as a source of organic matter and N, their capacity to mobilize soil P and other nutrients is also more or less universally recognized. A saving of 120 kg N/ha from *Sesbenia* green manuring in a sequence of rice-wheat has been observed, besides other benefits to the soil (IRRI, 1964).

Green manures may also play a key role in the reclamation of problem soils. Decomposition products of *Sesbenia* helped in removal of exchangeable Na from salt-affected soils (Singh, 1969).

Incorporation of green portion of *Ipomea cornea* alone or in combination with inorganic fertilizer (ammonium sulphate) resulted in higher N uptake by rice grain and straw, higher grain and straw yields and residual fertility than ammonium sulphate alone as observed for 6 seasons (Jha *et al.* 1980). This was mainly because of an almost continuous supply of N distributed over the entire crop growth period, the release of more N in submerged soil, increased biological fixation of atmospheric N, release of various beneficial chemical compounds like polysaccharide, glucoside, gums, alkaloids, and growth hormones. Alkaloids acted as slow N releasing materials and preserved native, applied and biologically fixed N. It also increased solubility and availability of many essential nutrients because of release of organic acids during decomposition.

Rhizobium was the first microbial fertilizer to be used on a large scale. There is specific *Rhizobium* for every legume. Inoculation with efficient *Rhizobium* strains is very essential for the nitrogen gains and better crop yields. About 15-30 per cent yield increase in pulses can be achieved by *Rhizobia* inoculation in farmers' fields with respect to pigeon pea, green gram, cowpea, black gram and moth bean as observed in trials conducted under All India Coordinated Pulse Improvement Programme. However, the response varies with varieties in the same crop and with locations. *Rhizobium* legume symbiosis can meet 80 per cent of the nitrogen requirement of the legume crop.

The effect of *Rhizobium* legume symbiosis also results in an increase in the total nitrogen content of the soil. The efficiency increases markedly in a nitrogen poor soil. Growing of groundnut left the soil richer in nutrient status than a more competitive cereal crop like pearl millet as evident from the date presented in Table 10.6.

Numerous trials have confirmed about residual benefits from legumes on the yield increase of the succeeding crop. Such yield increases in succeeding crop grown after different legumes inoculated with *Rhizobium* cultures, were also reported by Subba Rao and Tilak (1977) and Ahlawar *et al.* 1981).

In fixing atmospheric nitrogen, legume as an inter crop as well as component crop in rotation, contributed to N content of soil. The quantity of nitrogen fixed by the legume component in an intercropping system depends on the species, morphology, and density of legume in the system, the type of management and the competitive abilities of the component crops. Evidence in the literature suggests that the nitrogen fixed by the intercrop legume, may be available to the associated crop in the current season and also as residual nitrogen for the benefit of the succeeding and this results in N-economy in various legume based cropping systems. Nair *et al.* (1979) found the residual N transfer from legume based intercropping system and reported grain yield increases in wheat by 30 per cent after maize + soybean intercrop and by 34 per cent after maize + cowpea, compared to wheat after sole crop of maize.

TABLE 10.6: Effect of groundnut and pearl millet on soil fertility

Treatment	Available nutrient (kg/ha)					
	N		P_2O_5		K_2O	
	0-15 cm	15-30 cm	0-15 cm	15-30 cm	0-15 cm	15-30 cm
Initial soil status	172	141	8.4	31.0	148	101
Nutrient added to groundnut	**50**		**100**		**0**	
Fertility status after groundnut	220	204	28.5	10.1	325	269
Deviation	+18	+63	+20.1	-20.9	+177	+168
Nutrient added to pearl millet	**120**		**30**		**40**	
Fertility status after pearl millet	172	173	21.0	16.8	224	297
Deviation	0	+32	+12.5	-14.2	+76	+196

Source: Dwivedi (1981)

The organic matter content of soil is one of the deciding factors which influences the growth and survival of *Rhizobia* in the rhizosphere of legumes. Subba Rao and Tilak (1977) reported that the incorporation of FYM at the rate of 5 tonnes/ha along with *Rhizobia* inoculation to soil, increased the yield of pigeonpea by 48.2 and 38.5 per cent over unamended (inoculated) and amended (uninoculated) soils, respectively. Seed-pelleting with suitable amendments help to protect *Rhizobia* from the effects of acidity/alkalinity of the soil.

Various works have proved that legumes demand high levels of P for subsequent higher amount of N fixation. It has also been shown that legumes are able to extract insoluble forms of soil phosphorus. Thus, inclusion of legumes in intensive cropping systems has many advantages. Other proven sources of biological nitrogen fixation are free living bacteria *Azotobacter* and *Azospirillum*, Blue green algae (BGA) and *Azolla-Anabaena* symbiosis. All of them are situation specific and there is wide variation in their nitrogen fixing potential.

Recent reports have shown positive interaction of *Azobacter* with applied nitrogen and phosphorus. In coordinated trials conducted at IARI and Dharwar on maize crop, combined treatment of *Azotobacter* and application of 40 kg N/ha gave significantly higher yield than application of 80 kg N/ha alone. Similar to *Azotobacter* inoculation, *Azospirillum* has also shown positive interaction with applied N in crops like rice, wheat, oats, barley, sorghum and pearl millet (Subba Rao *et al.* 1979, 1982). The average response to seed inoculation in term of grain and fodder yield is nearly equivalent to 15-20 kg/ha applied N.

The use of BGA and *Azolla* in water logged rice fields, contributes about 20-25 kg N/ha. BGA as well as *Azolla* inoculations show positive interaction with applied N (Venkatraman and Goyal, 1969), however, algae contribution is of greater magnitude at lower levels of N. By use of BGA, about one third of the recommended N fertilizer could be converted without affecting crop yield. But differential crop response to algalization in presence of different sources of combined nitrogen has been observed. Ammonium sulphate and mono ammonium phosphate were found to be better than urea, ammonium nitrate and sodium nitrate. Soil conditions too have great impact on algal performance. In acidic soils, the performance is subdued and can be increased by liming.

(B). Phosphate Solubilisation

Several soil bacteria and fungi possess the ability to bring insoluble phosphates into soluble forms by secreting organic acids. Lowering of soil pH brings about dissolution of immobilized form of phosphates. Some of the hydroxyacids may chelate with calcium and iron resulting in effective solubilisation of soil P and thereby its higher utilization.

Rock phosphate holds potential in acid soils, but there is remote possibility at present for its use in natural and alkali soils. Several strains of phosphorus solubilising bacteria and fungi viz., *Pseudomonas striata, Basillus polymyxa, Aspergillu awamori, Penicillium digitatum* and most recently *Tolypothris celonica, Westiellopsis prolifica, Scytonema cincinatum* and *Celothrix braunii* have been isolated in the Division of Microbiology, IARI, New Delhi. It has been shown that integrated use of the microbial cultures along with a cheap source of low grade rock phosphate can bring about a saving of super phosphate equivalent to about 30-50 kg P_2O_5/ha (Gaur and Singh, 1982, Shinde and Patil, 1985).

Soil inoculation with P-solubilising organism *Pseudomonas striata,* besides increasing grain yield of wheat and pea crops showed residual effect by increasing grain yield of following maize crop in alluvial soil at IARI (Gaur, 1988).

It is now a well established fact that Vesicular arbuscular mycorrhiza (VAM) has beneficial effect on plant growth particularly in P deficient soils. Improved uptake of nutrients mainly phosphorus and minor elements like zinc, copper etc., increased uptake of water, production of plant hormones, increasing activity of *Rhizobium,* are the prime benefits of mycorrhizal inoculation. VAM fungi not only improves plant yield but saves phosphatic fertilizer to the tune of 25-30 per cent. Keeping in view the high phosphorus requirement by legumes and the improvement in root nodulation and nitrogen fixation in legumes in presence of VAM, twin symbiotic associations within a legume plant from *Rhizobium* as well as VAM, hold tremendous promise in integrated nutrient management system.

Recent studies have shown that mycorrhiza responses were much higher in the second and third crop in soils to which rock phosphate was added than in those treated with soluble phosphate. Higher amount of rock phosphate application effect than super phosphate, especially to plants inoculated with mycorrhiza (Goswami and Mishra, 1980).

INTEGRATED NUTRIENT MANAGEMENT IN PROBLEMATIC SOILS

Problematic soils viz., salt affected and acid soils do have various problems in relation to nutrient supply system. Occurrence of deficiencies/toxicities of various elements is a common phenomenon in these soils. The use of soil amendments like lime, gypsum, pyrite, press mud, phosphor-gypsum etc. improved the efficiency of applied fertilizers and also raised the availability of native nutrients in such soils.

P-deficiency can also be corrected by pH by applying soil amendments. In heavy soils of Indo-gangetic plains, addition of gypsum for improvement of alkali soils in terms of both physical as well as chemical properties is important. Mixing limited quantities of gypsum in a shallower depth was more beneficial than mixing in a larger depth (Khosla et al. 1973). Sodic soils are very much deficient in soluble calcium and application of amendments like gypsum, phosphor-gypsum, and press mud in addition to their ameliorating effect on physico-chemical properties of the soil, also supplies Ca and S for plant growth.

Combined application of gypsum (11 t/ha), FYM (25 t/ha) and zinc resulted in yield increase in rice much more than gypsum + FYM contribution but in absence of applied Zn. Similar trend was obtained with other treatments too, whose efficiency was increased by zinc (Table 10.7).

TABLE 10.7: Effect of application of Zn, FYM and gypsum on yield of rice on sodic soils

Treatment	Yield (t/ha)	
	Without Zn	With Zn
Control	5.4	7.2
FYM 25 tonnes/ha	6.6	7.8
Gypsum 11 t/ha	6.7	8.9
Gypsum (11 t/ha) + FYM (25 t/ha)	7.7	9.2

Source: Bhumbla (1972)

Availability of N, P and Mo increases due to liming (Verma and Jha, 1970; Bidappa and Venkat Rao, 1972). Need-based frequent liming of acidic soil not only restores yield stability but also improves productivity of the soil on long term basis. Acid soils need both liming and balanced fertilization for improving productivity of crops and cropping systems. The use of NPK becomes more effective in increasing crop yield in limed acid soil compared to unlimed one (Table 10.8).

TABLE 10.8: Yield (t/ha) of crops due to liming and NPK in acid soils

Crops	Treatments			
	Control	Lime	NPK	NPK + Lime
Wheat	0.6	0.8	1.5	1.8
Barley	0.7	0.7	1.9	2.3
Maize	0.3	0.5	1.1	1.7
Jute	1.7	2.3	2.6	3.3

Source: Mandal *et al.* (1975)

Efficient use of indigenous rock phosphates which are unsuitable for manufacture of phosphatic fertilizers could be applied in acid soils with better economic results than super phosphate (Subrawardy *et al.* 1980). But contrary to that, these rock-phosphates have also been found to be only 40 per cent efficient as super phosphate in terms of direct effect on succeeding crops in acid soils (Kanwar *et al.* 1982).

MANAGEMENT OF INORGANIC SOURCES TO RAISE NUTRIENT USE EFFICIENCY

There are various well proven scientific practices through which the efficiency of applied fertilizers can be raised. This would go a long way to cut down the heavy amount of chemical fertilizers being used, prevent various types of soil and environmental problems, reduce the cost of production without any reduction in the potential yield of the crops/cropping systems. This may be achieved through balanced and optimum nutrition, soil compaction, tillage, proper water management, proper time/method/source of fertilizer application. Thus, raising efficiency of applied inorganic fertilizers in crop production also forms an integral part of the integrated nutrient management system.

Indian agriculture has come to an age when multiple nutrient deficiencies have become prevalent in widespread areas. In such cases, diagnosis may be difficult because of overlapping effects of more than one nutrient and the situation may be aggravated by continued heavy application of one nutrient which can accumulate or induce deficiency of other ones. Some of the important examples of antagonism and synergism are as follows:

Antagonism

1. Excess copper adversely affects iron nutrition.
2. Iron to manganese or zinc to iron is antagonistic.
3. High levels of phosphorus may induce zinc deficiency.
4. High levels of potassium and calcium depress boron absorption.

Synergism

1. Nitrogen usually enhances uptake and utilization of phosphorus and micronutrients.
2. Magnesium may increase phosphorus uptake.
3. High levels of phosphorus may favour molybdenum uptake, which is important for pulses.

Deficiency of secondary and micronutrients in Indian soils has been reported by Tandon (1989). However, balanced fertilizer application has been found to be a dynamic and important component of the integrated approach of nutrient management.

In highly P-deficient soils, application of N or P alone would not result in higher productivity and, therefore, combined application of N and P would be minimum acceptable input package for sustainability of crop production and higher level of fertilizer use efficiency (Meelu *et al.* 1979 and Khanna and Choudhury, 1979). Similarly, higher use efficiency of nitrogen by using K has been observed. Each of the 3 major nutrients N, P and K exert an effect on yields, but best response is obtained when all the three are applied in balanced proportion (International Fertilizer Correspondent, 1989).

A marked depression in yield of maize resulted with application of P alone when zinc was deficient in the soil. Application of zinc resulted in two-fold increase in grain yield of maize. The highest yield was recorded at high levels of both Zn and P (Kanwar and Randhawa, 1974).

In alluvial soils, coarse tilth seed beds produce higher grain yields of wheat than fine tilth seed beds because of better plant growth and reduced leaching of mineral nutrients from the surface (Agarwal *et al.* 1984). Reduced tillage has been a means for both prevention of soil erosion losses and conservation of nutrients and water for efficient crop production. Conservation tillage in maize, reduced soil losses by 92 per cent compared to conventional tillage while losses of N and P were reduced by about 70 and 80 per cent, respectively (McDowell and McGregor, 1984).

Moderate compaction (1.6 g/cc bulk density) of the soil may also lead to major improvements in fertilizer use efficiency (Patnaik *et al.* 1971).

Efficient water management practices are also one of the pre-requisites to obtain higher nutrient use efficiency for sustainable crop production. In rainfed wheat, efficiently usable N rates could vary from 40 to 120 kg N/ha depending on the quality of stored moisture in the soil profile at sowing time.

Use of nitrification inhibitors and slow release fertilizers offers an alternative to split application of N in rice and long duration crops like sugarcane. The recovery of applied phosphorus is normally very low in the first crop and major portion gives residual effect. The magnitude of the residual effect depends upon the rate and the kind of phosphatic fertilizers used, the cropping and management system followed and to a greater extent on the type of soil (Rao and Bhardwaj, 1973). Both direct and residual effects of phosphorus are important for deciding the quality of phosphatic fertilizer for component crops in a cropping system. Placement of phosphatic fertilizers in upland rainfed crops has been reported to be a better method since it reduces the P-fixation.

n seasonal crops, the most common recommendation for potassium has been to apply the recommended dose of potash all as basal at planting time. Over the years, research findings in favour of spit application of K have emerged, for a number of situations, particularly for crop like rice in Assam, Kerala, Meghalaya, Orissa, Uttar Pradesh, where leaching due to high rainfall leads to considerable loss of potassium.

INTEGRATED NUTRIENT MANAGEMENT THROUGH CROPPING SYSTEMS

Nutrients applied to one crop in a cropping system are sometimes not fully utilized by that crop and enough residual and cumulative effects are carried over to second or third subsequent crops in the sequence. During the present era of intercropping systems, the current transfer between legume and non-legumes during the same season and the carry-over effects to subsequent crops, have occupied a prime place in integrated nutrient management system. Researchers done on the aspect have tried to specify compatible legumes for different intercropping and sequential cropping systems, in terms of nutrient contribution.

In the long term fertilizer experiments conducted at 11 centres under major agro-climatic regions of India, attempts have been made to study the depletion of various plant nutrients with their selective application and in different amounts through organic and inorganic sources. Results of these experiments revealed that variety of nutrient imbalances caused by variable nutrient supplies. Such imbalances are due to deficiency or building up of certain nutrients in the soil owing to the residual effect of the applied nutrients (Nambiar and Ghosh, 1964).

In rice-based multiple cropping studies conducted at Coimbatore, the nutrient balance was very much influenced by the fertility treatments (Purushothaman, 1979). There was a possible N balance when full N was applied, but it was negative when half of N dose was applied. While in case of P, the balance was positive for both full and half of the total quantity applied. P application to summer crops like sorghum, finger millet, maize, cotton, only resulted in negative balance of P. Balance of K was negative with all the fertility levels but the depletion was greater when application of K was done only to summer crops.

Results of a long term experiment at Ludhiana have shown that in a soil under the maize-wheat rotation, Olsen-extractable phosphate was built up to 73 kg/ha as a result of continuous P application, it could sustain two maize crops and one wheat crop without P application, after which P application again became limiting for sustaining yields (Singh and Brar, 1985)

Enriching of soil with same nutrients by continuous application beyond a certain stage will lead to reduced or no response to that nutrient and also varied type of nutrient antagonism and deficiencies

In different long term experiments, attempts have been made to see the direct, residual and cumulative effects of organic manures in different cropping systems. Patro and Sahoo, (1977) observed that direct, residual and cumulative effects of 15 tonnes FYM/ha were significant in lateritic soil in rice-rice cropping system. Results of the permanent plot experiment on integrated nutrient supply system (AICARP, 1984-85) also suggest that substation of 35 to 50 per cent of N through FYM/green manuring in kharif and application of 75 to 100 % of the optimum level of NPK in *rabi* season, produced the highest yields in lateritic and red soils. Beneficial effect of organic manures on total N and P status of the lateritic soil has also been reported by Sahu and Nayak (1971). Information on the nutrient balances, as affected by combined application of inorganic fertilizers and organic manures in different cropping systems on long-term basis under specific soil and climatic conditions have been made available.

Inclusion of legumes in sequential or intercropping system has shown benefits to succeeding crop in the sequence and also to the associated non-legume crop by virtue of its N-fixation capacity.

The following conclusions can be broadly generalized from the various short and long term experiments on integrated nutrient management in different cropping systems:

1. In cereal-cereal systems, little or no carry-over benefit of nitrogen is expected. So each crop in such sequence should be fertilized with nitrogen.
2. Residual benefit of P is expected. Normally, it is useful to apply full P dose to *rabi* crops in a system.
3. Soil potassium declines considerably under intensive cropping without K application. Application of N and P is important to sustain productivity of a cropping system.
4. Organic manures do increase the total N and P status of the soil but more benefits are derived when these are applied during *kharif* season. Notable residual and cumulative benefits from application of organic manures, in subsequent crops have been derived.
5. Deficiencies of secondary and micronutrients particularly zinc, manganese, sulphur, and iron have become critical, particularly in areas adopting heavy input-cum-intensive cropping systems.
6. Residual as well as current N transfer from legumes is important for N-economy in cropping systems.
7. Nutrient requirement of a cropping system varies with the nature and types of component crops included and the agro-ecological conditions they are grown.

SUGGESTIONS

1. More extensive use of animal and other organic manures along with mineral fertilizers to provide both organic matter and nutrients to the soil.
2. More and more adoption of longer crop cycles in contrast to soil depleting mono-culture or short term rotations.
3. More and more inclusion of legumes/pulses and green manure crops in cropping systems to add nitrogen and organic matter to the soil.
4. Less use or in some cases no-use of mineral fertilizers and increasing reliance on plant and animal residues.
5. Use of bio-fertilizers partially to substitute commercial fertilizers.
6. Adoption of nutrient management practices to raise fertilizer use efficiency in cropping systems.

In developing countries like India, where most of the farmers who are poor and marginal, may not afford to go for a green manure crop at the cost of some economic crop in sequence. Animal manures in huge quantities may not be available to efficiently contribute to the nutrient needs of the intensive cropping systems, and that also when most of the cow dung is utilized as cakes to meet the domestic needs of fuel. Now it should be decided to which extent such integrated nutrient supply system (INSS) can be substituted for or replaced commercial fertilizers on field scale.

To what extent the use of commercial fertilizers can be reduced through INSS without any reduction in the targeted growth rate of food grain production to accumulate the growing needs of the increasing population of the country. Majority feel that in the event of heavy cut on fertilizer use, it will be difficult to meet the growing needs and no alternative will be left except bringing additional land under cultivation and thereby again damaging or destroying natural ecosystem.

"Better to manure than to plough the land".
Ancient Tamil Proverb

11 Organic Farming

> *"Natural resources determine the course of development and constitute the challenge which may or may not be accepted by the human mind."*
> *W. Arthur Lewis*

Organic Agriculture is one among the broad spectrum of production methods that are supportive of the environment. Organic production systems are based on specific standards precisely formulated for food production and aim at achieving agro-ecosystems, which are socially and ecologically sustainable. It is based on maximizing the use of external inputs through efficient use of on-farm resources compared to industrial agriculture. Thus, the use of synthetic fertilizers and pesticides is avoided.

Pollution free environment is essential for sustaining life of all living beings on the earth. While technology induced farming is required for feeding the surging population, the excessive use of chemical fertilizers, pesticides and weedicides in agriculture has caused environmental imbalance and is causing problems to all living beings on the earth. **Eco-friendly farming** has emerged as the only answer to bring sustainability to agriculture. Eco-friendly farming is a farming of integration of biological, cultural and natural inputs including integrated disease and pest management practices. It is not only advocates for stopping or restricting the use of pesticides, but it emphasizes the need for farming which should create on ecological balance and a micro-environment suitable for health and growth of soil micro-flora, plants, animals, farm workers and finally the vast population which consume the farm produce.

CONCEPT

The concept of **organic agriculture** has been perceived differently by different people. To most of them, it implies the use of organic manures and natural methods of plant protection instead of using synthetic fertilizers and pesticides. It is regarded by some as farming involving the integrated use of fertilizers and organic manures as well as of chemicals and natural inputs for plant protection. In either case the concept has been understood only partially.

Organic agriculture has been defined differently, but the description offered by Lampkin (1990) appears to be most the comprehensive one covering all essential features. As per this description, organic agriculture is a production system which avoids or largely excludes the use of synthetic compounded fertilizers, pesticides, growth regulators and livestock feed additives. To the maximum extent feasible, organic system relies on crop rotations, crop residues, animal manures, legumes, green manures, off-farming organic wastes and aspect of biological pest control to maintain soil productivity and tilth, to supply plant nutrients and to control insects, weeds and other pests. The concept of soil as living system that develops the activities of beneficial organisms is central to this definition.

Organic agriculture does not imply the simple replacement of synthetic fertilizers and other chemical inputs with organic inputs and biologically active formulations. Instead, it envisages a comprehensive management approach to improve the health of underlying productivity of the soil. In a healthy soil, the biotic and abiotic components covering organic matter including soil life, mineral particles, soil and water exist in a stage of dynamic equilibrium and regulate the ecosystem processes in mutual harmony by complementing and supplementing each other. When the soil is in good health, the population of soil fauna and flora multiplies rapidly which, in turn, will sustain the bio-chemical process of dissolution and synthesis of a high rate. The state of soil life and the associated organic transformations will enhance the regenerative capacity of the soil and make it resilient to absorb the effects of climatic vicissitudes and occasional failures in agronomic management.

The success of organic agriculture depends to a great extent on the efficiency of agronomic management adopted to stimulate and augment the underlying productivity of the soil resource. In this context, the concept of agro-ecosystem becomes relevant. A farming system unit is treated as an agro-ecosystem when it attains the semblance of a forest ecosystem in species diversity and multiplicity. The adoption of sequence and mixed cropping models in the presence of compatible species of nitrogen fixing trees with or without the association of livestock components makes the agro-ecosystem benefit from the positive interaction and stimulated cycling mechanisms. As a consequence, the system slowly achieves self-regulation and stability. Agricultural production attained at this stage will be engaging without eroding or deteriorating the natural resource base.

As the **Organic Agricultural System (OAS)** derives if strength from the basic productive capacity of the soil and complimentary interaction among the components of the system, the use of chemical inputs either for soil fertility management or for plant protection is excluded. This renders the system free from the pollution problems usually associated with the use of such inputs. For achieving marked improvement in the soil productivity and for sustaining optimum levels of biological production, OAS lays emphasis on appropriate cropping and farming models, ensuring on-farm diversity and nutrient cycling, conservation and use of organic/biological sources of nutrients, cultural practices conducive to the conservation of soil and water resources and natural and/or biological methods of pest and disease suppression.

With an understanding of the principles of organic agriculture, a straight and simple definition to the concept can be suggested. Organic agriculture is a farming system devoid of chemical inputs, in which the biological potential of the soil and underground water resources are conserved and protected from the natural and human induced degradation or depletion by adopting suitable cropping models including agro-forestry and methods of organic replenishment, besides natural and biological means are used for pest and disease management by which the soil life and beneficial interaction are stimulated and sustained. The system achieves self regulation and stability as well as capacity to produce agricultural outputs at levels which are profitable and enduring over time and, at the same time, consistent with the carrying capacity of the managed agro-ecosystem.

There are also different opinions on nomenclature of **organic farming**. Some call it as **eco-farming** i.e., farming in relation to ecosystem. Others prefer the term **biological farming** (farming in relation to biological diversity), yet others prefer the term **bio-dynamic farming** (biologically dynamic and ecologically sound and sustainable farming) or **macrobiotic agriculture** (agriculture in relation to micro-fauna). Whatever be the name, the basic point is that organic farming is the farming based on natural principles which alone are sustainable. According to Fantilanan (1990), organic farming is a matter of giving back to nature what we take from it. It is safe, inexpensive, profitable and sensible. **Organic farming** is not mere non-chemicalism in agriculture; it is a system of farming based on integral relationship. So, one should know the relationships among soil, water, plants and micro flora and the overall the apex animal. It is the totality of those relationships which is the backbone of organic farming.

Organic farming does not totally exclude the elements of modern agriculture. Varying agro-climatic conditions do need input from the current technological advances. It is basically simple as it abhors excessive ploughing, hoeing, weeding and application of plant protection chemicals and fertilizers. The principal elements to be considered while practicing organic farming are:

1. Maintaining a living soil.
2. Making available all the essential nutrients.
3. Organic mulching for conservation, and
4. Attaining sustainable high yield.

Agricultural practices followed in organic farming are governed by the principles of ecology and are within the ecological means. Limited experience shows that this form of natural farming is the basis for sustainable agriculture and could be highly productive. It should not be misconstrued for reversion to inefficient and less productive farming systems.

ORGANIC VS. NATURAL FARMING

There is a misconception that organic farming is merely to say 'no' to chemicalism. But apart from restricting and to the extent possible eliminating chemicals (pesticides and fertilizers), it has something else also to convey. One who understands the whole concept of organic farming will be certainly inspired by it.

The differences between organic farming and natural farming (based on natural principles) are given below:

Natural farming	Organic farming
– It is not alternative system of farming but part of the philosophy of life involving continuous search to know the true spirit and form of nature.	– In many respects close to natural farming- but does have the philosophical over-tone of natural farming.
– Totally eliminates all the components of modern farming.	– Organic farming does not totally exclude elements of modern farming. It involves limited and essential ploughing, hoeing, weeding and use of chemicals.
– It indicates a 'do nothing' approach.	– It indicates a soil building programme – more Intensive style of natural farming. Application of natural plant protection chemicals (which are not inorganic derivatives), use of organic manures (instead of chemical fertilizers) are permitted.
– The essential principles are:	– Principal elements to be considered in practicing
	– Organic farming is:
– No cultivation	i. Maintaining a living soil.
– No chemical fertilizers.	ii. Making available all essential nutrients

Natural farming	Organic farming
– No weeding, * No plant protection	iii. Organic mulching.

Nonetheless, the principles and practices that lie behind these terms are essentially similar. The objectives of organic agriculture are been concisely expressed in the standard document of the International Federation of Organic Agriculture Movement (IFOAM) as follows:

- To produce food of high nutritional quality in sufficient quantity;
- To work with natural systems rather than seeking to dominate them;
- To encourage and enhance the biological cycles within farming system involving microorganisms, soil flora and fauna, plants and animals;
- To maintain and increase the long term fertility of soils;
- To use, as far as possible, renewable resources in locally organized agricultural systems;
- To work, as much as possible, within a closed system with regard to organic matter and nutrient elements;
- To give all livestock, conditions of life that allow them to perform all aspects of their innate behaviour;
- To avoid all forms of pollution that may result from agricultural techniques;
- To maintain the genetic diversity of the agricultural system and its surroundings, including the protection of plant and wildlife habitats;
- To allow agricultural producers an adequate return and satisfaction from their work including a safe working environment; and
- To consider the wider social and ecological impact of the farming system.

CHARACTERISTICS OF ORGANIC FARMING

The most important characteristics of organic farming are as follows:

1. Maximal but sustainable use of local resources.
2. Minimal use of purchased inputs, only as complementary to local resources.
3. Ensuring the basic biological functions of soil-water-nutrient –humus continuum.
4. Maintaining a diversity of plant and animal species as a basis for ecological balance and economic stability.
5. Creating an attractive overall landscape which gives satisfaction to the local people.
6. Increasing crop and animal diversity in the form of poly-culture, agro-forestry systems, integrated crop/livestock systems, etc. to minimize risk.

Organic agriculture systems are not a repudiation of the assets of modern agriculture technology; neither is they systems of simple elimination of synthetic fertilizers or pesticides. Methods in organic agriculture are less intensive in terms of synthetic and other external inputs compared to the conventional farming methods, but are much more intensive from a biological point of view. Organic agriculture systems include approaches and methods like organic, biodynamic, regenerative, nature farming and premaculture. These were developed during the last 50 years. Although there are some differences among these approaches, the common understanding is that practicing organic agriculture is managing the agro-ecosystem as an autonomous system, based on the primary production capacity of the soil under the given agro-climatic conditions. Agro-ecosystem management implies treating the system, on any scale, as a living organism supporting its own vital potential for biomass and animal production, along with biological mechanisms for mineral balancing, soil improvement and pest control.

PRINCIPLES OF ORGANIC AGRICULTURE SYSTEMS

Organic agriculture systems are based on the three strongly interrelated principles under autonomous ecosystem management: **mixed farming**, **crop rotation** and **organic cycle optimization**. The common understanding of agricultural production in all types of organic agriculture is managing the production capacity of an agro-ecosystem. The process of extreme specialization propagated by the green revolution led to the destruction of mixed and diversified farming and ecological buffer systems. The function of this autonomous ecosystem management is to meet the need for food and fibres on the local ecological carrying capacity.

Mixed Farming

In organic agriculture systems, one strives for appropriate diversification, which ideally means mixed farming, or the integration of crop and livestock production on the farm. In this way, cyclic processes and interactions in the agro=ecosystem can be optimized, like using crop residues in animal husbandry and manure for crop production. Diversification of species biotypes and land use as a means to optimize the stability of the agro-ecosystem is another way to indicate the mixed farming concept. The synergistic concept among plants, animals, soil and bio-sphere support this idea.

Crop Rotation

Within the mixed farm setting, crop rotation takes place as the second principle of organic agriculture. Besides the classical rotation involving one crop per field per season, intercropping, mixed cropping and under sowing are other options to optimize interactions. In addition to plant functions, other important advantages such as weed suppression, reduction in soil-borne insect pests and diseases, complimentary nutrient supply, nutrient catching and soil covering can be mentioned.

Organic Cycle Optimization

Each field, farm, or region contains a given quantity of nutrients. Management should be used in such as way that optimal use is made of this finite amount. This means that nutrients should be recycled and used a number of times in different forms. Second, care should be taken that only a minimum amount of nutrients actually leave the system so that 'import' of nutrients can be restricted. Third, the quantity of nutrients available to plants and animals can be increased within the system by activating the edaphon, resulting in increased weathering of parent material.

OBJECTIVES OF ORGANIC AND CONVENTIONAL FARMING

It can be summarized as follows:

Organic farming	Conventional farming
Organisation	
1. Ecological orientation, second economy, efficient labour input.	Economical orientation, mechanization, Minimizing labour input.

Organic farming	Conventional farming
2. Diversification, balanced combination of enterprises.	Specialization, disproportionate development of enterprises.
3. Stability due to diversification.	Programme based on market.
Production	
4. Cycle of nutrients within the farm, pre-dominantly farm produced materials.	Supplementing nutrients, predominantly through fertilizers.
5. Weed control by crop rotation and cultural practices.	Weed control by herbicides.
6. Pest control based on non-polluting substances.	Pest control by pesticides.
7. Housing of livestock for production and health.	Livestock rarely combined.
Mode of influencing life processes	
8. Production is integrated into environment, building healthy landscapes.	Emancipation of enterprises from their environment by chemical and technical manipul.
9. Balanced conditions for plants and animals; deficiencies need to be Corrected.	Excessive fertilization, necessitating frequent correction of nutrient deficiencies.
Social Values	
10. Optimum input : output ratio	Low input : output ratio.
11. No pollution.	Considerable pollution.
12. Maximum conservation of soils, water quality and wild life.	Using up soil fertility often resulting In erosion and losses.
13. Holistic approach.	Economic motivation.

OPTIONS IN ORGANIC FARMING

There are at least three options available in organic farming. They are:

1. Pure organic farming
2. Integrated green revolution farming
3. Integrated farming system (IFS).

Pure Organic Farming

This excludes the use of inorganics, both fertilizers and pesticides, but advocates the use of organic manures and biological pest control methods. By the year 2000 AD, to meet the demands of the population of a billion people, food production has to reach 230 million tones needing 24 million tones of NPK requirement is to be supplied in the form of organics, either as farm or town compost or green manure, the quantity of organics required will be huge. But, large potential of organic resources remains untapped in the country. Nearly 750 million tones of cow dung, 250 million tones of buffalo manure and 100-115 million tones of crop residues are available. The nutrient value of these organics produced annually is in the order of 2.5, 2.0 and 3 million tones of NPK equivalent respectively. Besides, hundreds of million tones of rural and urban compost could be collected.

Integrated Green Revolution Farming

Under this option, the basic trends of the green revolution such as intensive use of external inputs, increased irrigation, development of high yielding crop varieties and hybrids and mechanization of labour are retained. But much greater efficiency on the use of these inputs is obtained to limit damage to the environment and human health. For the purpose, some organic techniques are developed and combined with the high input technology in order to create integrated systems such as integrated nutrient management (INM), integrated pest management (IPM) and biological control methods which reduce the need for chemicals. Modern biotechnology is also employed to develop higher yielding, pest resistant crop varieties. This option is possible for conditions, including fertile soils, climate and availability of necessary infrastructural facilities like irrigation.

Integrated Farming System

The third option in organic farming is the low input organic farming, in which the farmers have to depend on local resources and ecological processes, recycling agricultural wastes and crop residues. Capital intensive green revolution techniques are simply not a feasible alternative for the poorest of the 1.4 billion farmers who live developmentally less favourable production conditions. In order to cover such risks and to ensure sustainability in their small holdings, the age-old mixed farming systems are prudently integrated with the cropping system.

MANAGEMENT OF ORGANIC FARMING

Management of organic farming system involves:

1. Organization of crop and livestock production and the management of farm resources in such a way that it harmonizes rather than conflicts with natural systems.
2. Achievement of closed cycle to the greatest extent possible between soil, plants, animals and people and an avoidance of environmental pollution.
3. Maintenance of soil fertility for optimum production, relying primarily on renewable resources.
4. Reduction of pest and disease incidence through a carefully designed farm rotation and enterprises structure; use of resistant varieties; the encouragement of beneficial pest predators; and the use of other biological pest control techniques.
5. Use of forms of animal husbandry which respect the welfare and behavioural needs of farm livestock.
6. Use of appropriate farm machinery and cultivation techniques which reduces non-renewable resource consumption.
7. Enhancement of the environment in such a way that wildlife flourishes and it is enjoyable for people both working within the system and viewing it from outside.

These principles will lead to a wider definition of quality than is usually given to food. The following categories have been suggested:

1. **External quality**: Freedom from pest and disease damage, freshness and colour.
2. **Technical quality**: Improved properties of storage and processing.
3. **Nutritional/physiological quality**: Increased content of valuable nutrients such as proteins and vitamins, and the absence of detrimental substances such as nitrates and other agricultural chemical residues.
4. **Environmental quality** of the system of production, with regard to the organization of crop and livestock and management of farm resources, in such a way that they harmonize rather than conflict with natural systems.

ADVANTAGES OF ORGANIC FARMING

1. Organic manures produce optimal conditions in the soil for high yields and good quality crops.
2. They supply all the nutrients required by the plant (NPK, secondary and micronutrients).
3. They improve plant growth and physiological activities of plants.
4. They improve the soil physical properties such as granulation and good tilth, giving good aeration, easy root penetration and improved water holding capacity. The fibrous portion of the organic matter with its high carbon content promotes soil aggregation to improve the permeability and aeration of clay soils while its ability to absorb moisture helps in the granulation of sandy soils and improves their water holding capacity. The carbon in the organic matter is the source of energy for microbes which help in aggregation.
5. They improve the soil chemical properties such as source supply and retention of soil nutrients and promote favourable chemical reactions.
6. They reduce the need for purchased inputs.
7. Most of the organic manures are wastes or byproducts which on accumulation may lead to pollution. By way of utilizing them for organic farming, pollution is minimized.
8. Organic fertilizers are considered as complete plant food. Organic matter restores the pH of the soil which may become acid due to continuous application of chemical fertilizers.
9. Organically grown crops are believed to provide healthier and nutritionally superior food for man and animals than those grown with commercial fertilizers.
10. Organically grown plants are more resistant to disease and insects and hence only a few chemical sprays or other protective treatments are required.
11. There is an increasing consumer demand for agricultural produces which are free for toxic chemical residues. In developed counties consumers are willing to pay more for organic foods.
12. Organic farming helps to avoid chain reaction in the environment from chemical sprays and dusts.
13. Organic farming helps to prevent environmental degradation and can be used to regenerate degraded areas.

BARRIERS TO ORGANIC FARMING

Initially there may be some barriers which inhibit the farmers from adopting organic farming. Land resources can move freely from organic farming to conventional farming; they do not move freely in the reserve direction. In changing over to organic farming an initial crop loss generally occurs, particularly if it is rapid. Biological controls may have been weakened or destroyed by chemicals, which may take three or four years to build up. Organic farmers may be afraid to enter the new market without adequate government support.

COMPONENTS OF ORGANIC FARMING

The components of organic farming are discussed in this section:

Organic Manures

Organic materials such as farm yard manure, biogas slurry, compost, straw or other crop residues, bio-fertilizers, green manures and cover crops can substitute for inorganic fertilizers to maintain the environmental quality. In addition, the organic farmers can also use sea weeds and fish manures and some permitted fertilizers like basic slag and rock phosphate.

The use of organic manures will increase the organic matter content of the soil increasing the water holding capacity. Erosion is reduced by organic manures. Crop rotation with legumes adds to soil fertility. Green manure provides the nutrients and improves the soil.

Non-Chemical Weed Control Measures

Compared to conventional farmers, the organic farmers use more of mechanical cultivation of row crops to reduce the weed menace. No herbicides are applied as they lead to environmental pollution.

Biological Pest Management

The control of insect pests and pathogens is one of the most challenging jobs in tropical and sub-tropical agriculture. Here again non-chemical, biological pest management is encouraged. The conservation of natural enemies of pests is important for minimizing the use of chemical pesticides and for avoiding multiplication of insecticide – resistant pests. Botanical pesticides such as those derived from neem could be used. Selective microbial pesticides offer particular promise, of which strains of *Bracillis thuringiensis* is an example.

NATURAL ENVIRONMENT

The natural resources of a country are of primary importance for the development of the economic life of its inhabitants. As a matter of fact natural resources determine the economic life of nation. For example, **England** is an industrial country but she owes her present position to her coastline, rivers, the proximity of rich coal mines and iron fields, and the temperate climate. The **U.S.A.** is both an agricultural as well as industrial country. It possesses the requisites of industrialization and at the same time its soil at some places is very fertile. **Japan,** too, is an industrial country only because there is no scope for agriculture as the very nature of the soil, is not favourable for agriculture.

Man may grow rich in knowledge and intelligence, however, he may have overcome nature but ultimately he will have to depend on the materials supplied by Mother Nature for the development of his economic life. The **physical factors** like topography, soils, geologic formation, climate and the available flora and fauna, are the basic influences which lead to differences in land use, cropping pattern, settlement and density of population and occurrence of minerals, water and power resources in different parts of the country. In the case of India, a study of agricultural problems should start with an investigation of the physical, edaphic and climatic factors.

In the North lies the Himalaya which separates it from China, in north-west lies Pakistan and in the east it is flanked by Burma, separated from it by Assam hills. In the south, it is bounded by the Arabian Sea in the west and by the Bay of Bengal in the east. Shri Lanka lies on its south-eastern tip. In the Arabian Sea stand the Lakshadeep (Laccadive, Amindivi and Minicoy) and in the Bay of Bengal are the Andaman Nicobar Islands.

The **Republic of India** is a vast country which lies wholly in the northern hemisphere stretching from latitudes 8°4'28" and 37°17'53" north and from longitudes 68°7'33" and 97°25'47" east. It is one of the largest of the three irregular peninsulas of Southern Asia possessing a highly favourable position as regards the rest of the world for purposes of international trade. It measures 3,214 km from north to south and 2,933 km from east to west. It has a land frontier 15,200 km and a costal-line of about 6,100 km. Tropic Cancer divides it into two unequal halves, the northern half lying in the temperate zone and the southern half in the Torrid Zone.

INCREASE IN AGRICULTURAL PRODUCTION

There is no doubt that Indian Land resources and farming efficiency can result in rising per hectare productivity on all farms provided all the factors of production are fully exploited.

The problem of agricultural production is a problem of bringing about proper combination of soil, water, plants and people for production. The agricultural problem is not one of **maintaining,** but of **developing** soil productivity to the highest practicable level and of maintaining it at that level. What is to be sought in agriculture is not any kind of natural balance, but the highest level of cultural balance that modern technology and social organization make possible.

The **physical resources** of soil, water and climate are sufficient to yield at least double, perhaps more than double the current production with full use of machines, chemicals, sufficient water supply and a combination of other good management practices.

Mechanization refers to application of mechanical power to agricultural operation from sowing to harvesting stage through bulldozers, graders, tractors, seed-drills, cultivators, rollers, fertilizer distributors, combine harvesters, and other light farm machineries. **Chemicalisation** means the application of chemical fertilizers. Under **rationalization** are included the application of science to different phases of agricultural production, viz., seed selection, or use of disease resistant, high yielding varieties of seeds, use of insecticides and pesticides, erosion control etc. **Irrigation** gives a larger yield per hectare and an irrigated crop shows a larger response to each unit of fertilizer applied to it than an un-irrigated crop. Besides, availability of irrigation makes it possible to grow more crops where only one crop could be grown depending on rainfall.

It may be pointed out that no one practice is good over a wide range of soils. Nor can we usually expect the adoption of any single practice, by itself, to give an economic result. The big harvest comes from combinations of practices adapted to the particular situation. **"Easy way"** of emphasizing some one or two practices above all others in a "**campaign**" should be avoided because the results are bound to be low for the effort. For example, bunds alone, fertilizer alone, improved village alone, or improved seeds of potentially high yielding varieties alone, may each give a small un-economic increase on good soil under dry culture; but the combination commonly gives a three-six fold increase in the harvest.

Since, in **India**, the basic problem is that of under production, enhancement of all types of farm production is urgently needed. The basic approach, therefore, should be for increasing the output of all crops. **The formulation of incentives for being effective, must suit the particular farms, the farmers, and the crops.** For formulating proper incentives the increase in output should be classified according to the methods such as improvement in yield per hectare, expansion of area, increase in double cropping, improvement in quality and diversification. The nature and character of the method is significant for the suitability and effectiveness or otherwise of any specific incentive.

For better quantitative production, provision of irrigation, high yielding varieties, manures, fertilizers and implements at concession rate, should prove effective.

For better quality production, attractive prices, based on quality differentials accompanied by better seeds, should prove helpful.

For food crops and subsistence farmers, provision of cost subsidies should be made.

For crops grown for the market and for bigger farmers, price guarantees may be better responses.

In case of competing cash crops measures aimed at higher income at low cost may prove more beneficial; and

Lastly, **programme formulation and target fixation** should be done, in view of the local situation, at the village level and mainly by local initiative and should be tailor made to suit local conditions.

Incentives must be formulated in view of the individual circumstances and they should be supplemented by persuasion which is necessary to convince for the producer for taking up any programme.

In fact, **Indian Agriculture has been and become a deficit economy and to a very great extent the Indian cultivator labours neither for profit nor for a net return but for subsistence**. The overcrowding of the people on land, the lack of alternative means of securing a living, the difficulty of finding any avenue of escape and the early stage at which man is burdened with the dependents combine to force the cultivator to grow food whenever he can and on whatever terms he can. The ordinary cultivator on his tiny plot is still a man of small resources, with small means for meeting his small needs. He, therefore, requires all the help which science can afford, and which organization, education and training can bring within his reach.

A variety of factors are responsible for this state of affairs, most important being natural, technical, economic, social, structural, institutional and administrative obstacles in the way of optimal utilization of our land resources. It has rightly been said that "the fundamental problem of agriculture is to transform this occupation from a mode of living into a business proposition for the benefit of the cultivating classes".

If technology, currently known to India, could be applied to Indian Agriculture it would be easy to increase production substantially. However, it should be borne in mind that technology will be of no avail in improving agriculture unless institutional changes (in the form of land reforms) first provide the necessary economic basis for more efficient agriculture. Gradual worsening of debt situation, violent price fluctuations, the middle man's exorbitant share and exploitation and the landlord's apathy are some of the factors which discourage the farmer to adopt new methods of techniques whose failure could ruin him and whose success benefits every one else than him.

Therefore, the measures to trigger individual incentives are:

1. Security of tenure,
2. Security to minimize the risk and uncertainties in agricultural production through price supports,
3. Elimination of middle traders,
4. Provision of marketing facilities, and credit institutions,
5. Strengthening of extension services,
6. Putting farming on a sound economic footing through establishment and improvement in co-operative organizations,
7. Better motivation of farmers,
8. Improvement in physical supplies of superior inputs, and
9. Inculcation of administrative integrity.

Unless the disabilities and hindrances in the way of agriculture are reduced it seems inevitable that much of the teachings of the farm advisory services will fall upon stony ground.

ORGANIC FARMING PRACTICES FOR RICE

Organic manures have many advantages, which chemical fertilizers are lacing. The reasons for using organic manures are partly **agronomic** – a more productive soil and better crops, partly **economic** - making use of the cheap resources in the place of expensive commercial fertilizers and partly **environmental.** Organic manures are a more complete plant food and they provide almost all essential nutrients and trace elements while chemical fertilizers supply only specific nutrients, and thus a soil receiving only chemical fertilizers over a long period is likely to have an unbalanced soil nutrient content. The disadvantages are that some organic materials may contain toxic or pathogenic substances that are dangerous to crop or human health. The most serious problem is heavy metals often present in municipal wastes and sewage sludge, it is difficult to standardize commercial organic manures since the quality of raw materials is so variables; organic materials release their nutrients slowly, as the material decomposes and it meets the needs of crops for a steady nutrient supply over a growing season. But, many farmers feel that chemical fertilizers are more effective as the effect is more dramatic and rapid. The main disadvantage of organic manures is that they are bulky and difficult to handle and transport; some organic manures are messy and with unpleasant odour for farmers to handle (Palaniappan and Annadurai, 2001). If society wishes to reduce environmental problems by recycling organic materials, the government must either subsidies their use or offer some incentive to the farmer.

Organic farming is a productive system which largely avoids or excludes the use of synthetically compounded fertilizers, pesticides, growth regulators and livestock feed additives and to maximum extend possible organic farming systems relays upon crop rotations, crop residues, animal manures, legumes, green manures, off-farm organic wastes, mechanical cultivation, mineral bearing rocks and aspects of biological pest control to maintain soil productivity and tilth, to supply plant nutrients and control insects, weeds and other pests or Organic farming is not merely use of non-chemical fertilizers and pesticides but it is a holistic approach integrating soil, plant, water, soil micro flora and fauna, human, animals and the environment. Prime purpose of organic farming is to build of natural resources such as soil and water for future use. The following are the objectives of organic farming.

1. To develop a sustainable agriculture system for guaranteed adequate food production in the foreseeable future.
2. To develop self-sufficient agriculture system, this would rely as much as possible upon resources from within its own resources.
3. To develop an alternative strategy over chemical farming, this would be a guideline for working of biological processes in natural eco-systems.

More substitution of organic inputs in the place of inorganic inputs for animal raising and agriculture/horticulture does not form organic farming. Understanding the resource capabilities of the farm and using the same in a conservative manner avoiding over exploitation of land, underground water etc. According to Lampkin (1990), the principle elements to be considered in practicing organic farming are: (1) creating a healthy soil, (2) making nutrient and energy flows in the soil ecosystem, (3) keeping the biological life in the cycle and (4) providing sustainable yield.

Because of the indiscriminate use of chemical fertilizers for decades, the organic matter content of soils has come down to less than 1 per cent. In addition, the use of pesticides led to pest resurgence and difficult to control weeds species. The residues of the chemicals cause concern over the safety of food and sustainable production. The addition of chemical fertilizer like nitrogen in plant caused an infant disease like methanoglobinaemia. Hence, the expectation that the organic farming by reverting to the use of manures, green manures, urban waste, rural wastes, etc. can bring sustainability to agriculture and eco-friendliness. Hence, it becomes imperative for the researchers and planners to develop viable strategy to supplement the chemical farming.

CHALLENGES

According to researchers, different farming systems do have implications on the productivity, sustainability, production orientation and ecological situation (Werf and Narayan, 1989 and Pingali *et al*, 1997). While modern commercial (conventional) farming system is unsustainable and market oriented the organic (sustainable) farming do have merits like productivity orientation, ecological sustainability and farm complexity.

Traditional rice cultivation involves mainly direct sowing, inter-culture, weeding, rainfed irrigation and harvest. In such system, rice plants withstood extreme drought, provided more fodder and yield was moderate for subsistence. Research study in the Philippines indicated that in rainfed low land farms, dry seeding of rice the promising technology for improving rice productivity (IRRI, 1992). Subsequently, the introduction of semi and dwarf varieties of high yielding varieties of rice resulted in prone and susceptible to pests and diseases susceptibility required more water, demanded more factory made external inputs (especially agro-chemicals) and yielded less fodder. The propensity to use abundant water and more chemicals increased year after year. In some pockets of rice regions in India, indiscriminate and non-judicial use of chemicals was observed that has also led to health hazards on the labour (Mencher, 1994). Nonetheless, such varieties yielded more, which was essential to feed the growing population. Simultaneously, adverse ecological impacts like water logging, salinity and chemical contamination were reported (Pingali, 2001 and Rajendran, 2002). In addition to this, comprehensive studies on organic farming especially rice cultivation are scanty in India. Thus, the relevance of organic agriculture and research has been realized all over the world.

BIODYNAMIC AGRICULTURE

Biodynamic agriculture is a method of **organic farming** that treats farms as unified and individual organisms, emphasizing balancing the **holistic** development and interrelationship of the soil, plants, animals as a closed, self-nourishing system. Regarded by some proponents as the first modern ecological farming system, biodynamic farming includes organic agriculture's emphasis on manures and composts and exclusion of the use of artificial chemicals on soil and plants. Methods unique to the biodynamic approach include the use of fermented herbal and mineral preparations as compost additives and field sprays and the use of an **astronomical** sowing and planting calendar. Biodynamic agriculture has its basis in a spiritual world-view known as **anthroposophy** as propounded by founder **Rudolf Steiner.**

History

The development of biodynamic agriculture began in **1924** with a series of eight lectures on agriculture given by **Rudolf Steiner** at **Schloss Koberwitz** in what was then **Silesia, Germany,** (now in **Poland** east of **Wroclaw**). The course was held in response to a request by farmers who noticed degraded soil conditions and deterioration in the health and quality of crops and livestock resulting from the use of chemical fertilizers. An agricultural research group was subsequently formed to test the effects of biodynamic methods on the life and health of soil, plants and animals. In the **United States,** the **Biodynamic Farming and Gardening Association** was founded in **1938** as a New York state corporation.

In **Australia** the first biodynamic preparations were made by Ernesto Genoni in **Melbourne** in **1927** and by Bob Williams in **Sydney** in **1939**. Since the 1950s research work has continued at the Biodynamic Research Institute (BDRI) in Powelltown, near Melbourne Australia under the direction of **Alex Podolinsky**.

Today bio-dynamics is practiced in more than 50 countries worldwide.

Biodynamic Method of Farming

Biodynamic agriculture conceives of the farm as an organism, a self-contained entity with its own individuality. "Emphasis is placed on the integration of crops and livestock, recycling of nutrients, maintenance of soil, and the health and well being of crops and animals; the farmer too is part of the whole. Cover crops, green manures and crop rotations are used extensively. The approach also attempts to consider celestial influences on soil and plant development and to revitalize the farm, its products, and its inhabitants.

Biodynamic Preparations

Steiner prescribed nine different **preparations** to aid **fertilization** which are the cornerstone of biodynamic agriculture, and described how these were to be prepared. The prepared substances are numbered 500 through 508, where the first two are used for preparing fields whereas the latter seven are used for making **compost.**

Field Preparations

Field preparations, for stimulating **humus** formation:

- **500**: (horn-manure) a humus mixture prepared by filling the horn of a **cow** with cow manure and burying it in the ground (40–60 cm below the surface) in the autumn. It is left to decompose during the winter and recovered for use the following spring.
- **501**: Crushed powdered **quartz** prepared by stuffing it into a horn of a cow and buried into the ground in spring and taken out in autumn. It can be mixed with 500 but usually prepared on its own (mixture of 1 tablespoon of quartz powder to 250 liters of water). The mixture is sprayed under very low pressure over the crop during the wet season to prevent fungal diseases. It should be sprayed on an overcast day or early in the morning to prevent burning of the leaves.

Both 500 and 501 are used on fields by stirring about one teaspoon of the contents of a horn in 40–60 liters of water for an hour and whirling it in different directions every second minute.

Compost Preparations

Compost preparations, used for preparing compost, employ herbs which are frequently used in medicinal remedies:

- **502**: **Yarrow** blossoms (*Achillea millefolium*) are stuffed into **urinary bladders** from **Red Deer** (*Cervus elaphus*), placed in the sun during summer, buried in earth during winter and retrieved in the spring.
- **503**: **Chamomile** blossoms (*Matricaria recutita*) are stuffed into **small intestines** from **cattle** buried in humus-rich earth in the autumn and retrieved in the spring.
- **504**: **Stinging nettle** (*Urtica dioica*) plants in full bloom are stuffed together underground surrounded on all sides by **peat** for a year.
- **505**: **Oak bark** (*Quercus robur*) is chopped in small pieces, placed inside the skull of a domesticated animal, surrounded by peat and buried in earth in a place where lots of rain water runs past.
- **506**: Dandelion flowers (*Taraxacum officinale*) is stuffed into the peritoneum of cattle and buried in earth during winter and retrieved in the spring.
- **507**: **Valerian** flowers (*Valeriana officinalis*) are extracted into water.
- **508**: **Horsetail** (*Equisetum*)

One to three grams (a teaspoon) of each preparation is added to a dung heap by digging 50 cm deep holes with a distance of 2 meters from each other, except for the 507 preparation, which is stirred into 5 liters of water and sprayed over the entire compost surface. All preparations are thus used in **homeopathic** quantities, and the only intent is to strengthen the life forces of the farm.

Treatment of Pests and Weeds

Biodynamic agriculture sees the basis of pest and disease control arising from a strong healthy balanced farm organism. Where this is not yet achieved it uses techniques analogous to fertilization for pest control and **weed control.** Most of these techniques include using the ashes of a pest or weed that has been trapped or picked from the fields and burnt. Steiner saw weeds and plant vulnerability to pests as a result of imbalances in the soil.

- Pests such as **insects** or **field mice** (*Apodemus*) have more complex processes associated with them, depending on what pest is to be targeted. For example field mice are to be countered by deploying ashes prepared from field mice skin when **Venus** is in the **Scorpius** constellation.
- **Weeds** are combatted (besides the usual mechanical methods) by collecting **seeds** from the weeds and burning them above a wooden flame. The ashes from the seeds are then spread on the fields, which are intended to block the influence from the full moon on the particular weed and make it infertile.

Trademark Protection of Term Biodynamic

The term *Biodynamic* is a trademark held by the **Demeter** association of biodynamic farmers for the purpose of maintaining production standards used both in farming and processing foodstuffs. The trademark is intended to protect both the consumer and the producers of biodynamic produce. **Demeter International** is an organization of member countries; each country has its own Demeter organization which is required to meet international production standards (but can also exceed them). The original Demeter organization was founded in 1928; the U.S. Demeter Association was formed in the 1980's and certified its first farm in **1982.** In **France, Biodivin** certifies **biodynamic wine.** In **Egypt, SEKEM** has created the Egyptian Biodynamic Association (EBDA), an association that provides training for farmers to become certified.

Studies of Efficacy

Studies h ave compared biodynamic farming methods to both other organic methods and to conventional methods. Yields and soil quality have generally been found to differ little from those of other methods of organic farming, significantly from conventional farming methods.

- A study of the effect of biodynamic preparations on compost found that biodynamically treated compost contained 65% more nitrate than untreated compost, as well as significant differences in microbial life, temperature of composition, carbon-dioxide respiration.
- A 1993 study compared soil quality and financial performance of Biodynamic and conventional farms in New Zealand. The study reported that, "The Biodynamic farms proved in most enterprises to have soils of higher biological and physical quality: significantly greater in organic matter, content and microbial activity, more earthworms, better soil structure, lower bulk density, easier penetrability, and thicker topsoil." The biodynamic farms were just as financially viable on a per hectare basis. The study compared biodynamic farms with adjacent conventional farms, but didn't attempt to compare farms of similar size, or with similar crops.
- A further study investigated whether biodynamic preparations had any effect on the yield and growth of **lentil** and **wheat** crops, weed populations and soil fertility in the short term. The study found that "In general, soils and crops treated with biodynamic preparations showed few differences from those not treated". Plots tended with biodynamically treated compost produced results for yield, crop quality and soil fertility that was similar to those tended with non-biodynamic composts and **NPK fertilizers.** Some alteration was observed in the **nitrogenous** chemistry of the soil and grain where biodynamic field sprays were applied; however the study did not ascribe or discern any biological significance to the difference. Among the variables considered by the study, some measured outcomes correlated with biodynamic field spray usage, including a higher per-unit biomass yield ratio for lentils and a lowering of **carbon** and crude **protein** contents in wheat grains. The study's conclusion remarked that "any additional short-term benefits from biodynamic preparations remain questionable.
- A long-term study conducted at a commercial vineyard in California compared vineyard blocks treated with biodynamic preparations alongside those tended with general organic farming methods, to examine effects upon soil and crop quality. "No differences were found in soil quality" during the first six years of the study, and analyses of other indicators including the yield per vine, clusters per vine, cluster and berry weight also showed there were no differences. The study did find a statistically significant (**p-value** < 0.05) difference in the yield-to-pruning weight ratio, indicating an "ideal vine balance for producing high-quality winegrapes" for the biodynamically treated crop, but noted the control vines had been "slightly overcropped". In one particular year of the study the biodynamically treated winegrapes had significantly higher **Brix** and notably higher total **phenols** and **anthocyanins.** In conclusion, the study found that biodynamic preparations "may affect" the vine canopy and chemistry, but showed no effects on the soil and tissue nutrient parameters measured in the study.

- A 21-year study in Switzerland compared the agronomic and ecological performance of biodynamic, organic and two conventional systems. The study found that nutrient input in the biodynamic and organic systems was 34 to 51% lower than in the conventional systems but crop yield was only 20% lower on average, indicating more efficient production. The total energy (for fuel, production of mineral fertilizer and pesticides, etc.) to produce a dry-matter unit of crop was 20 to 56% lower for the biodynamic and organic systems, and pesticide input was reduced by 97% (by 100% for the biodynamic system). In regards to soil aggregate stability, soil pH, humus formation, soil calcium, microbial biomass, and faunal biomass (earthworms and arthropods), the biodynamic system was superior even to the organic system, which in turn had superior results over the conventional systems. With the significant increase in microbial diversity in the biodynamic and organic systems, there was a significant associated decrease in metabolic quotient, indicating a greater ability to use organic material for plant growth.

Related

Ehrenfried Pfeiffer, a biochemist prominent in the early development of biodynamic preparations, developed a process for the bacterial conversion of municipal waste into compost usable in agriculture. The process was first used on a commercial scale in **Oakland, California** in the early 1950s.

Criticism

In a newspaper editorial, Peter Treue argued that similar or equal results can be obtained using standard **organic farming** principles (which he also criticized as unproven in efficacy) and that the biodynamic preparations more resemble **alchemy** or **magic** akin to **geomancy.**

In a 1994 analysis, Holger Kirchmann concluded that Steiner's instructions were occult and dogmatic, and cannot contribute to the development of alternative or sustainable agriculture and that many of Steiner's statements are not provable because scientifically clear hypotheses cannot be made from his descriptions (for example, it is hard to prove that you have harnessed "cosmic forces" in the foods). Kirchmann asserted that when methods of biodynamic agriculture were tested scientifically, the results were unconvincing. Further, in a 2004 overview of biodynamic agriculture, Linda Chalker-Scott also pointed out that many of the research articles comparing biodynamics with conventional agriculture did not separate the use of biodynamic preparations from practices used in organic agriculture. The term "biodynamic" should not be used interchangeably with "organic" agriculture. Chalker-Scott concluded that "scientific testing of biodynamic preparations is limited and no evidence exists that addition of these preparations improves plant or soil quality in organically managed landscapes."

Organic Farming is the agricultural system which aims at cultivation of the land in a way so that the soil is kept dynamic with living activities and in good health, at the same time keeping the environment clean, maintaining the ecological balance and providing stability to the production level without polluting soil, water and air. It is concerned with the best use of natural resources and maintaining their balance. In this system, soil is not only concerned to be physical medium for standing crop but also a shelter of innumerable living entities. In fact, it is only means to achieve the stable/sustainable target of production.

Intensive Agriculture with the use of agrochemicals in large amount has, no doubt, resulted in manifold increase in the productivity of farm commodities but the adverse effects of these chemicals are clearly visible on soil structure, soil microflora, quality of water food, fodder and food materials. Pesticides and nitrates from fertilizer have been detected in ground water in many agricultural regions. High concentration of nitrates in drinking water affects human health, particularly infants and can prove fatal in some cases. Trace of banned DDT and BHC isomers have been found in soil, cereals, pulses, vegetable oils, bovine or human milk, butter fat, fish, meat, eggs, vegetables, fruits, animal feed and drinking water. The World Health Organization has estimated that globally at least three million people are poisoned by pesticides every year, out of whom 20,000 die. A majority of pesticide deaths are reported to be occurring in developing countries, which use only 25 per cent of the global pesticides. Thus, **organic farming** is certainly an answer for making available safe food and clean environment.

Role of Organic Farming

Organic farming is the backbone of sustainable agriculture. The U.S.D.A. report (1980) defined it as, "Organic farming is a production system, which avoids or largely excludes the use of synthetically compounded fertilizers, pesticides, growth regulators, and livestock feed additives". To the maximum extend feasible, farming systems rely upon crop rotations, crop residues, animal manures, off-farm organic wastes, mechanical cultivation, mineral bearing rocks, and aspects of biological pest control to maintain soil productivity and tilth, to supply plant nutrients, and to control insects, weeds, and other pests". The approach is based on the perception that tomorrow's ecology is more important than today's economy. According to the definition of FAO, organic farming should involve successful management of sources for agriculture to satisfy changing human needs while maintaining or enhancing the quality of the environment and consuming natural resources. Organic farming may result in comparable performance as conventional agriculture and crops grown with organic manure application could tolerate the pest and disease attack better. It is an ecologically sound and sustainable way of growing more food.

Organic agriculture aims at to utilize local resources, maintain the long-term fertility of the soil, avoid all forms of pollution, provide foodstuffs of highly nutritional quality, to reduce fossil energy in agricultural practices, to maintain rural environment and also preserve non-agricultural ecological habitats.

Waste Recycling

Organic farming mainly depends on organic recycling. Harnessing the nutrient energy from biological and industrial wastes is of prime importance for maximizing production. When these wastes are recycled as manure for crop production and are subjected to the degradation and assimilative capacity of soil, pollution of streams and/or rivers receiving these wastes is reduced to a large extent as compared to their direct disposal in water bodies. Organic recyclable wastes include crop residues, animal wastes, farm/industrial wastes, municipal and sewage wastes. They are valuable sources of plant nutrients and humus. In tropical and sub-tropical soils found in India, there is a general deficiency of organic carbon and plant nutrients due to rapid loss of these nutrients by biodegradation. To make up for these losses, extensive utilization of organic residues in agriculture is essential. In addition, they also protect the soil from erosion. The manorial value and quality of these wastes could be improved by composting and enriching these organic sources along with inexpensive materials such as rock phosphate. In India, there is a great potential for utilization of crop residues/straw of some of the major cereals and pulses. Approximate availability of straw is to the tune of 141.2 m t which contributes about 0.7, 0.84 and 2.1 m t of N, P_2O_5 and K_2O respectively. Considering that 50% of these residues are utilized as animal feed, the rest could be very well utilized for recycling of nutrients. Crop residues have wide C : N ratio and when applied directly, nutrients may get immobilized leading to initial adverse effect on crop growth. Hence, adequate care is required to use the residues after proper composting with efficient microbial inoculants. Inoculation of these wastes with efficient cellulotic microorganisms produces enough simple sugars for growth and multiplication of beneficial microflora like free living N fixers and phosphate solubilizers in the soil which ultimately increases crop yields.

While the incorporation of crop residues e.g., wheat and rice straw, as such or inoculated with fungal species had beneficial effects on crop yields and improvement in physio-chemical properties of soils. Incorporation of legume residues viz., green gram or soybean stalks in rice, cowpea stalks in pearl millet or wheat saves energy input besides improving the soil physical and chemical properties.

Sugarcane Trash Compost

Sugarcane trash left out after harvest is about 19-38 m t. It is either burnt *in situ* or used as domestic fuel. It can be recycled after composting suitably. Technology has been developed for production of enriched compost from sugarcane trash at Pune (Bhavanase et al., 1995). In the compost pit of 1 x 1 x 1 m^3, chopped sugarcane trash is filled in layers. In the bottom of the pit, trash is filled to 30 cm thickness. A mixture of cellolytic fungi @ 500 g/tonne of sugarcane trash and Azotobacter are added. These bio-fertilizers are mixed with cow dung. Sufficient quantity of water is applied on the surface of trash. A mixture of cellolytic fungi is inoculated: *Trichoderma viride, Trichurus spiralis, Aspergillus awamori, Pencillium spp.* and *Paecilomyces fusiparus.* These inoculants improve the rate of decomposition and mineralization of nutrients. Another layer of sugarcane trash is then added. Above this layer, urea (to supply 8.33 kg N/ha and rock phosphate (5% of trash) are applied dissolving in sufficient quantity pf water. Cow dung with bio-fertilizer mixture should not be applied simultaneously with the fertilizers. Thus, the pit is filled with trash in layers. Water is regularly added on alternate days to maintain 60% moisture in the compost pit. Adequate moisture is important for proper activities of the microbes. Turning of the material is to be done at regular intervals (at least once in a month) to improve aeration and maintain temperature at 50-60°C which is optimum for composting.

Tresh sugarcane trash contains 0.36% N with a wide C: N ratio of 122:1. Small bits of trash decompose earlier than big size trash. Addition of rock phosphate, urea, bio-fertilizer and moisture play significant role in early composting. The composted trash contains higher content of N (1.09%) with reduced C : N ratio (20:1).

Biogas Slurry

Organic manures from animal wastes are very important nutrient sources in building up the soil fertility. In India, the estimated production of dung and urine from bovine population works out to 1002 and 658 million tones respectively. They contribute about 5.71 m t of N, p and K. Proper method of collection and preservation is required to fully utilize the manorial value of these materials. Cow dung is an input in biogas plants. The biogas route has the dual advantage of providing both fuel (gas) and fertilizer (slurry).

The residual slurry that comes out of the digestion tank has a manorial value superior to FYM. The wet slurry should not be used as such. It should be dried for usage as manure. While drying, there is some loss of N and the dry manure contains 1.8%N, 1.10% P_2O_5 and 1.50% K_2O.

Industrial Wastes

Among the industrial by products, spent wash from distilleries and molasses and pressmud from sugar industry have good manorial value. It is important to use only well decomposed pressmud at 10 tonnes per ha. Addition of pressmud improves the soil fertility and enhances the activity of microbes. It is also used as an ameliorating agent in saline and sodic soils.

Coir waste is the byproduct from coir industry. Its product is about 1 million tonne in India. Coir waste can be used as manure after proper decomposition. Composted coir wastes (inoculated with *Pleurotus spp.* and 5 kg urea/tonne) are recommended for irrigated and dry crops. It has been reported that addition of 12.5 t/ha of composted coir waste with recommended fertilizers increased the yield of groundnut and maize.

Municipal and Sewage Wastes

Municipal and sewage wastes form an important component of organic wastes. In India, the total municipal refuse is about 12 m t per annum containing about 0.5%N, 0.3%P and 0.3%K. Sewage sludge is available to an extent of 4 million tonne per annum containing 3%N, 2%P and 0.3%K (Bharadwaj and Gaur, 1985). Sewage sludge, particularly from industrialized cities, is contaminated with heavy metals and thus pose hazard to plants, animals and human beings. Segregation of toxic wastes at the source will minimize the concentration of such elements in the sludge. High N input through excessive hydraulic loading of sewage can lead to ground water contamination. Further, survival of pathogens in soil and on crops causes health risk. This can be minimized by proper treatment, design of hydraulic loading, selection of suitable crops and adoption of modified irrigation system.

Vermicomposting

Vermicompost is organic manure produced by the activity of earthworms. Vermicompost is a mixture of worm casts which are rich in macro and micronutrients, vitamins, growth hormones and immobilized microflora. The average nutrient content of vermicompost is much higher than that of FYM. Vermicompost contains 1.60% N, 5.04% P_2O_5 and 0.80% K_2O with small quantities of micronutrients. The C : N ratio of vermicompost is much lower (1:16) than that of FYM (1:30). Application of vermicompost facilitates easy availability of essential plant nutrients to crops. The activity of earth worms is recognized as beneficial for the improvement of soil physical condition and plant growth.

Numerous species of worms have been identified for vermicomposting. The worm species more commonly used are *Eudrilus eugeniae, Lumbricus rubellus, Eisenia foetida, Perionyx excavatus* and *P. arboricola.* Wide varieties of wastes and byproducts which can not be directly applied to soil as a source of nutrient can be composted by this method. Vermicompost can be used for different crops.

Horticultural crops/Forest trees	100-200 g/tree
Potted plants	50 g/pot
Field crops	5 t/ha

Preparation of Vermicompost

Method of preparation of vermicompost has been described by Purakayastha and Bhatnagar (1997). A tank of 2 x 1 x 1 m is constructed with brick lime in a shaded place. The bottom of the tank (about 5.0 cm) is made up of pieces of brick, stone chips and sand. This will facilitate easy drainage of water. Loamy soil is spread over this to a thickness of 15 cm. About 4-5 kg of diluted dung may also be applied. About

100 number of locally available worm species are placed over the soil bed. Dried leaves or hay may be put to 5 cm thickness on the vermibed. It should be moistened without flooding. A layer (about 19 cm) of straw, leaves of plants, kitchen wastes, etc. are spread over this layer. This can be continued till the bed is completely filled. Liquid cow dung suspension is then applied and waste paper and bigger leaves are used to cover the contents in the tank and finally the tank is fully covered with a gunny cloth.

Watering the tank is very important to keep the contents sufficiently moist (50-60%). After a month time, the contents are turned up and down and covered with gunny cloth. Watering is continued for another month. At maturation, watering is stopped to ensure drying of compost and for inducing the worms to migrate to the vermibed.

Vermicompost can also be prepared in pots and unused packing boxes of 45 x 60 x 25 cm, provided with drainage holes at the bottom. The mature compost is a grey to brown coloured loose granular mass. After removal from the pit, it is sieved, dried and packed. The nature and quality of vermicompost vary with types of waste material and worm culture used.

SCOPE FOR ORGANIC FARMING

Although the scope for organic farming appears limited, there is hope since there is an increasing realization on their international and local fronts about sustainability and organic farming.

After spending almost one third of UNDP assistance on high external input agriculture over the thirty years, the developed countries now decided to change strategy. The FAO/Netherlands conference, the special happening of the Earth summit at Rio de Jeneiro and a follow up international conference entitled "Agenda 21 – Sustaining the Earth Summit" created an incredible wave of activity towards solving the world's environmental and developmental problems.

Another organization, International Federation of Organic Agriculture Movements (IFOAM), which spearheading and coordinating organic agriculture movement, internationally, has 386 full members, 106 corporate associates and 400 individual farmers from 75 countries. Besides, there are various organizations like AGRECOL and SKAT in Switzerland, ILEIA in Holland, Farmer's network worldwide in the U.K. and IASA in USA which provide technical support to interested NGO's.

The specific purpose of the studies is to document the agronomic, ecological and socio-economic sustainability of organic farming.

ORGANIC FARMING IN CALIFORNIA

Organic farmers in California are distributed throughout the State with varying farm sizes. Almost every crop grown in California can be found on organic farms. Although management systems vary according to region, crop, economic and environmental factors, most organic farmers tend to view their farming systems in a more holistic manner. By adopting cover crops, crop rotation, inter-planting, combined crop/livestock systems and general farm diversification schemes, farmers usually achieve several crop management objectives simultaneously and thus require little external supplementation in order to obtain sufficient protection for crops and to sustain soil fertility. The following is a brief description of specific organic farming systems in California that comprise successful examples of sustainable agricultural production.

Organic Rice Production

The best examples of organic rice production in California are the farms of Bundberg brothers. Totally 100 acres are under organic rice in which a 2 year rotation that alternates rice with purple vetch (*Vicia benghalensis*) and fallow is practiced. Dry unsprouted rice seed is drilled directly into the soil until appropriate moisture is available for sprouting, at which point the field is rapidly and briefly irrigated. After the rice plants have become fully established, the fields are keep flooded until they are drained in preparation for harvest. Purple vetch is planted in the fall following rice harvest and again in the fall of fallow year. The vetch supplied about 120-130 kg of nitrogen per ha. In the spring of fallow year vetch is chopped and disked along with largely decomposed rice straw leaving a mulch on the field. The field is shallow tilled with an implement to control weeds. Seeding into mulch followed by intermittent flooding in the early stages of rice growth and development also breaks the life cycles of various weeds, pests and pathogens and eliminating the use of pesticides. It has the disadvantage of significantly lowering the yield but the profits obtained average about US $ 65/acre more than in conventional farms due to premium price they obtain for organic rice.

Organic Grape Production

The organic grape production in California is succeeding very well. The management practices include application of 2000 tons of composted steer manure/year, application of enzyme digested mixture of fish waste materials and no tillage methods with perennial rye grass and cover crops chopped periodically for weed control. Hand weeding is done between grape vines. The primary pest in the vineyard is the grape leaf hopper (*Erytheoreura elegantula*) kept in check by the naturally occurring parasite *Anargus*. The yield obtained is much better than that in the conventional farms. Moreover, 12-25 per cent premium price is obtained. So this system is succeeding financially also.

ORGANIC FARMING – CONCEPT, INITIATIVES, STATUS AND IMPLICATIONS

1. The Concept

Farming is believed to have originated in the river valleys of the subtropical regions of the world which had access to water and fertile soil where food could be grown. Over the period, several agricultural systems developed throughout the world and productivity of the crops increased manifolds. Green Revolution in agriculture which brought spectacular increases in production and productivity in the country during late sixties started showing symptoms of fatigue from the unwanted side effects on natural resources like soil, water and biodiversity and thus human health. As a result:

- The vast areas of soil once fertile have degraded as a result of soil erosion, salinisation or a general loss of soil fertility.
- Water resources have been over exploited and polluted due to excessive requirement of irrigation water and intensive use of agro-chemicals for high yielding varieties.
- Residues of harmful pesticides in food and drinking water have endangered both farmer's and citizen's health and
- Excessive use of external inputs consumes lot of energy from non-renewable resources.

Organic farming is a way of conserving the soil, maintaining its fertility, protecting soil flora and faunal diversity thus preventing pollution of ground water, lakes and rivers. It does not utilize non-renewable external inputs and energy. Since, no chemical pesticides are used for crop production there are low chances of pesticide residues in food. At the same time the organic products are believed to be more healthy and better in quality characteristics like taste, aroma and storage. Thus organic farming has emerged as a dynamic alternate farming system that is believed to produce nutritious food of high quality with good employment opportunities for rural population.

The concept of organic farming is, however, being understood variously. To some it simply implies non-application of any chemical inputs, while a balanced approach suggests identifying inputs based on necessity, nature and way of production and their effect on human health and environment. In addition to this, socio-economic and ethical aspects are also required to be kept into consideration. In simple terms, it is a form of food production system where traditional wisdom and ancient knowledge are amalgamated with modern practices without much dependence on off farm resources.

Organic farming has been defined variously by various organizations. Codex Alimentarius and International Federation of Organic Agriculture Movement (IFOAM) emphasize that only such methods have to be used in organic farming, which minimize pollution of soil, water and air. Its primary goal is to optimize the health and productivity of soil, plant, animal and people. According to Food and Agriculture Organization (FAO):

- "Organic farming is a holistic production management system which promotes and enhances agro-ecosystem health, including biodiversity, biological cycles, and soil biological activity".

International Federation of Organic Agriculture Movement (IFOAM) has defined organic agriculture as:

- "A whole system approach based upon a set of processes resulting in a sustainable ecosystem, safe food, good nutrition, animal welfare and social justice. Organic production, therefore, is more than a system of production that includes or excludes certain inputs".

2. Other Similar Systems

There are a variety of organic farming systems prevalent worldwide. Biodynamic agriculture started in 1924 by Rudolf Steiner embraces holistic and spiral understanding of nature where the farm is a self-contained evolving organism, which keeps external inputs to a minimum. In this, biodynamic preparations are used and requirements include, among others, harmony of cultivation with cosmic rhythms, fair trade and the promotion of associative economic relations between producers, processors, traders and consumers. In this system, energies from cosmos, earth, cow and plants are systemically and synergistically harnessed.

Homa farming involves agnihotra, which meant crop husbandry with the help of burning of whole rice grains mixed with cow-ghee in a copper made pyramid of a specified size, along with chanting of mantras at sunrise and sunset based on the local timing. This practice is based on the tremendous of energy which gathers in atmosphere with the holy smoke while chanting mantra which comes back to the soil and dramatic benefits to the crop growth and productivity.

Panch Gabya consists of five products from cow, i.e., dung, urine, milk, curd, and ghee mixed together with sugarcane juice, coconut water, ripe banana and toddy and then incubated for 15 days and stirred daily. The above mixture is diluted in 1:10 ratio with water and filtered. It is believed to have wonderful effect on vegetative and reproductive growth of plants.

Bill Mollison develop the concept of **"Permaculture"** in 1970s. It is a social design system that works to conserve energy on-farm i.e., fuel from crop, firewood, food calories or to generate more energy than it consumes agriculture.

Nature farming introduced by Mokichi Okada in 1936 rests on the belief in universal life-giving powers that the elements of fire, water, and earth confer on the soil. Utilizing the inherent power of the soil is the underlying principle of this farming.

Ecological Farming is a labour intensive system based on techniques of crop cultivation, involving promotion of renewable source of energy (draught animal, power and electrical energy through garbage and biogas from organic waste). It also involves water use efficiency through conjunctive use of rain, tank and underground well and river water. IPNM and IPM are the essential components of ecological farming.

Finally there is the Traditional agriculture. Owing to its nature, the traditional system does not use synthetic agricultural inputs. Many, though not all traditional systems, meet the production standards for organic agriculture. However, may of these system though reported to be useful in organic crop production by may individuals and organizations are yet to be adequately researched to establish their quantitative and qualitative effects on production crops and environment.

3. Global Initiatives

In spite of being an age-old practice, systematic efforts to develop organic farming are only three and a half decade old. In 1972 the International Federation of Organic Agriculture Movement (IFOAM) was started by bringing together all the global organizations involved in organic production, certification and promotion. The concept of organic farming took off with increased consumer interest in countries like USA, Canada, Australia and Japan. Their strong concern and interest to be supplied with wholesome, environmental friendly products has been largely responsible for the growth of the sector in the West. In 1998, IFOAM basic standards for organic farming and processing were formulated which summarize methods of production and processing of organic products. The IFOAM also set up a European Union regional group for undertaking a constant dialogue with the European Commission on this issue.

In 1999, Codex Alimentarius Commission of FAO formulated and adopted Guidelines for the Production, Processing, Labeling and Marketing of Organically Produced Foods. These guidelines set out the principles of organic production from the farming stage through the preparation, storage and transport, labeling and marketing of crop products. They are intended to enable member counties (including India) to draw up their own rules, on the basis of these principles, while taking account of specific national features. The same year it also embarked on an organic farming work programme, mainly concerned with promoting organic farming in the developing countries. India is a beneficiary of this programme, while Switzerland, USA, and Germany can be considered as pioneers in organic farming. In 2003, an International Task Force on Harmonization and Equivalence in Agriculture was launched in Nuremburg, Germany. This is a joint initiative of FAO, UNCTAD, and IFOAM. In 2005, EU's revised draft was circulated to member counties.

4. Indian Initiatives

In traditional India the entire agriculture was practiced using organic techniques. Even presently much of the forest produce of economic importance like herbs, medicinal plants, etc., by default come under this category. However, there has been consistent increase in the number of certified farmers in the country every year. Many Organic Farmer's Association, have taken the lead in adopting and spreading technology.

Institutional support for organic production was created by launching a National Programme for Organic Production (NPOP) by Agriculture and Processed Food Export Development Authority (APEDA), Ministry of Commerce. The NPOP supports promotional initiatives, accreditation of inspection and certification agencies, and offers support to agri-business enterprises to facilitate export. APEDA has been interacting with European Union (EU), United States Department of Agriculture (USDA), Japan and IFOAM for recognition of equivalence of Indian quality assurance system. While the country follows IFOAM basic standards, it has been provided equivalence recognition by EU.

In 2004, Ministry of Agriculture, Government of India, launched a nation wide initiative, " National Project on Organic Farming (NPOF) with a budget allocation of Rs. 57 crores with the objective to promote organic farming in India. The project aims at promoting organic farming through facilitating access to organic inputs, streamlining production, facilitating certification and developing domestic markets for organic commodities.

In 2004-05 a National Centre for Organic Farming was established under the Ministry of Agriculture at Ghaziabad to provide institutional support and to facilitate farmers to move into organic crop production, by providing suitable logistics of knowledge and materials inputs like bio-fertilizers. The National Horticulture Mission launched by DOAC in 2005 offers assistance for taking up organic farming of horticultural crops.

5. Current Status

Organic Farming is currently being practiced in almost all countries of the world and its share both of agricultural land and in terms of number of farms is growing very fast. Accordingly to SOEL-FIBL Survey (2007), the total organically managed area is more than 30 million hectares worldwide. In addition to this, the area of certified "wild harvested plants" is an additional 19.7 million hectares. Presently, a major part of this area is located in Australia, China and Argentina. The share of global organic land is maximum in Oceania (39%), followed by Europe (23%), Latin America (19%), Asia (9%), North America (7%) and least in Africa (3%). However, the percentage of total agricultural land under organic cultivation is maximum in Europe. As far as the organic farms are considered these numbers globally are maximum in Europe (187,697) and least in Oceania (3,689) (Table. 11.1 and Fig. 11.1). The number one country for organic production is Italy; Mexico for Coffee; and Dominican Republic for cocoa. The leaders in organic grape production are Italy, Spain and France. Italy, Spain and Tunisia have the largest areas of organic olives.

According to NCOF, 2008, the land area under organic management in India was 4, 13,850 in 2006-07. Twenty nine states of India are already having area under organic production ranging from 176.44 ha in Goa to 112233.41 ha in Rajasthan (Table 11.2). Eleven states in the country have more than 5000 ha area under organic cultivation while 5 of these have more than 10,000 ha area under organic cultivation.

TABLE 11.1: Organic land and farms by continent

Continent	Organic land area	Organic farms
Africa	890,504	124,805
Asia	2,893,372	120,805
Europe	6,920,462	187,697
Latin America	5,809,320	176,710
North America	2,199,225	12,063
Oceania	11,845,100	2,689
Total	30,558,183	633,891

Source: SOEL-FIBL Survey 2007.

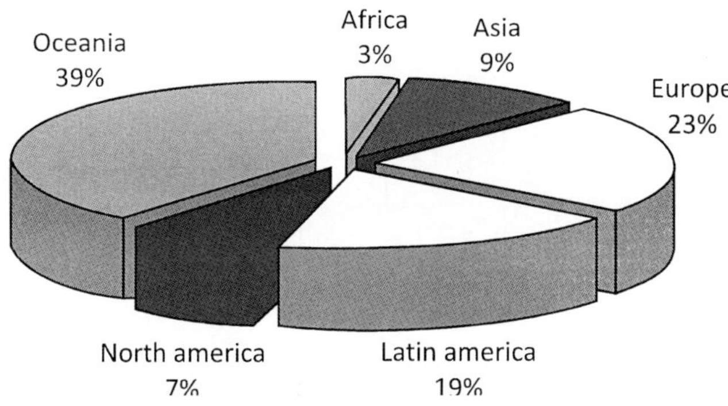

FIG. 11.1: Percentage share of organic land area by continents.

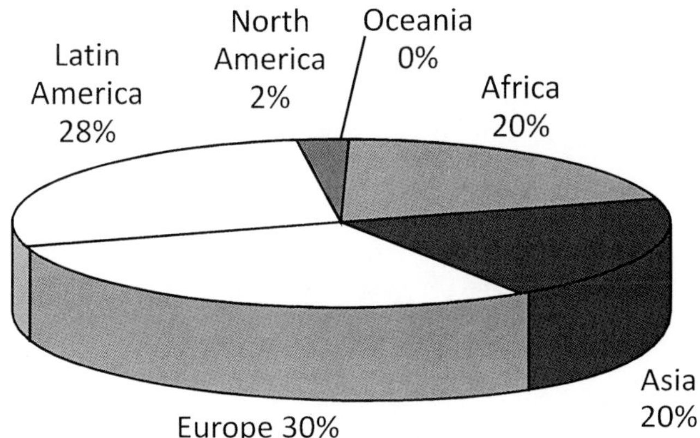

FIG. 11.2: Percentage share of organic farms by continents.

TABLE 11.2: Total Certified Organic Areas

Sl No.	State	Year		Sl. No.	State	Year.	
		2005-06	2006-07			2005-06	2006-07
1.	Andhra Pradesh	1661.42	6286.97	15.	Manipur	347.65	5644.33
2.	Arunchal Pradesh	557.76	830.24	16.	Maharashtra	18786.69	96879.06
3.	Assam	1817.504	1913.71	17.	Madhya Pradesh	16581.37	112233.41
4.	Bihar	0	0	18.	Mizoram	300.40	333.40
5.	Chattisgarh	293.16	347.88	19.	Meghalaya	378.89	0
6.	Delhi	1658.71	1138.10	20.	Nagaland	718.76	7208.89
7.	Goa	5555.07	176.44	21.	Orissa	26387.86	67503.39
8.	Gujrat	1627.06	3662.62	22.	Punjab	3779.31	1600.42
9.	Haryana	3437.52	3403.57	23.	Rajasthan	22104.91	24868.23
10.	Himachal Pradesh	3647.41	9569.24	24.	Sikkim	177.64	1640.44
11.	J & K	22315.92	321.65	25.	Tripura	20.87	0
12.	Jharkhand	5.00	2263.85	26.	Tamilnadu	5423.63	3620.94
13.	Karnataka	4117.17	6016.13	27.	Uttar Pradesh	3033.976	2215.79
14.	Kerala	15474.47	11141.54	28.	Uttaranchal	5915.85	6093.87
29.	West Bengal	6732.43	5029.16				
30.	Other	824.13	64.00		Total	173682.54	413850.25

Source: NCOF, 2008.

Organic farming is catching up fast with the Indian farmers and entrepreneurs, especially in low productive rainfed regions, with the increasing realization that are no diminishing returns to the continued use of production inputs under organic regime. Organic farming could also be easily adopted in agro-eco regions like rainfed zones, hilly areas and North states where fertilizer consumption is less than 25 kg/ha/year. Most of these areas are characterized with low productivity. Green manuring is practiced rarely due to farmer's unwillingness to sacrifice crops. Conversion of farms into organic production system in such areas is a real challenge worth attending.

Nine states of India have promoted policies and programme on Organic Farming. While, Uttarakhand made organic a thrust area for improving mountain agriculture based farm economy and livelihood. Mizoram, and Sikkim declared the intentions to move forwards total organic farming, even without necessary infrastructure, human resources and programmes. Karnataka formulated organic policy and Maharashtra, Tamil Nadu and Kerala supported public private partnership for promoting organic farming.

From the view point of commodities, India today produces range of organic products from fruits and vegetables, spices to food grains, pulses, milk and organic cotton. In addition India produce also includes wild harvest of medicinal, aromatic and dye plants. Some of these are organically cultivated and exported. While data for export commodities is available the data on domestic availability of organic commodities still relies on best guesses of stakeholders. Organic foods are definitely emerging, with organic retail stores, super markets and packaged deliveries visible in big cities and small towns.

6. Accreditation and Certification

Organic agriculture systems and products are required to be certified by accredited agencies if they have been produced, stored, processed, handled and marketed in accordance with technical specification and certified as organic by a certified body. Certification of organic produce is an important issue central to organic production and was therefore, developed firstly, to provide an identifiable label for organic food, and secondly, to assure consumers that food bearing such a label were truly from sowing to consumption. For desired economic gains from organic farming, the upcoming global market on organic produce has to be exploited, for which strict phyto-sanitary measures have to be followed. The organic label is therefore, a production process claim as opposed to product quality claim. India is the first to introduce Group Certification. It is accredited 11 certification bodies. Indian certification with India Organic Logo is recognized by the US (Feb. 2006) and has equivalence with the EU (June 2006) and Switzerland (Dec. 2006).

The major question which arises is whether organic agriculture needs to be necessarily **"certified"** to be called **"organic"?** Certification is in principle a step to make organic products distinguishable from the conventional products to enable these to fetch premium price in domestic and international market for the producers. However, organic certification system is complex and costs associated with it are resulting in marginalizing of small Indian Farmers. In addition regulations and requirements of different countries for organic certification are creating trade barriers, especially for developing nations. The certification conditionality on the organic claim is also hindering mass farm conversion into this sustainable system.

It is felt that certification flexibility will enhance development of short supply chains as well as an expanded participation of small holders in the organic sector incase certain less complex and less costly systems of certification are developed. The USDA/NOP certification' exemption for producers selling organic commodities for less than US$ 5000/year, access of Swedish organic farmers to public support without necessarily being under a certification scheme and the Government of India's promotion of participatory guarantee systems for the development of organic domestic markets are some prominent examples of flexibility provided to organic produce.

Currently, 39 countries have fully implemented regulations, 8 countries have finalized regulations but not yet implemented and 15 countries are in process of drafting regulations. India is among the 39 countries that have fully implemented organic regulation. Agricultural and Processed Food Products Export Development Authority, Department of Commerce, Ministry of Commerce and Industry, Government of India has developed these standards and regulations for India in March, 2000.

It offers several advantages namely, a positive distinction for organic products thus avoiding confusion; helps in market planning and lobbying, requires transparency, keeping the middleman more honest in the process and strengthens the position of the primary producer and gives agriculture a more credible position to the outside world, other than just to consumers. However, on the other hand, it adds another layer of value added service to the cost of organic producers.

7. Potential Crops and Products

Most of the area under organic farms in the world is under fresh fruits, vegetables and plantation crops. Commodities produced organically and traded world wide are avocado, banana, pineapple, walnut, cashew-nut, vegetables, spices and most plantation crops e.g., Tea, coffee and coca. Among other produce, rice, cotton, sugarcane, sesame seeds, horticultural products like dried fruits, herbs, aromatic plants like vanilla and honey are also important.

From a recent study conducted by Agriculture Department Western Australia various commodities identified as having excellent potential for organic production were carrots, wine, beef, wheat, sugar and canola. Other products with strong opportunities included: apples, dairy products, eggs, herbs, oranges, oats, poultry, rice, and safflower. These were followed by asparagus, banana, broccoli, fish, sunflower, grape, honey, nectarines, pear, onion, palm, and potato.

8. Input Management

The major inputs in organic farming required to be addressed are nutrient management, insect pest management and disease management. Soil is the most important organic crp production factor. A healthy soil contains millions of organism the most important being earthworm, bacteria, fungus, algae and protozoa. These main macro and micro organisms help to decompose organic material and build up humus. Earthworms dig tunnels which encourage deep rooting of the plants and good aeration of the soil. Micro-organisms act as bio-fertilizers and bring the insoluble and bound forms of nutrients into soluble forms. Some microbes also control pests and disease organisms affecting the roots of crops. It is important that soils are harnessed properly to maintain proper yield and quality of produce under organic system. Farmers can improve the fertility of soil by various management practices which include soil cultivation and tillage, use of green manure and cover crops, mulching, crop rotation and mixed cropping, appropriate tillage, sound nutrient management strategies and protection of soil micro-organisms etc.

One of the major ingredients of nutrition in organic farming is compost which is the result of bioconversion of heterogeneous organic substrates under controlled conditions. It offers several advantages such as improved soil texture and fertility, biodiversity, higher productivity and safe environment. A wide range of composting technology is available, The choice of composting methods depends on the substrates. Traditional methods are based on passive composting approach, which involve stacking the material in pits or heap to decompose and take longer time (1-2 years) for the final product. The rapid composting involves pulverization, addition of water, mechanical stirring, aeration and use of microbial inoculants. The prepared compost should have no particles of the original residue present. The organic matter content of the compost should be at least 80% (on oven dried basis), moisture content between 30-35%, water holding capacity of (150-200%), and a pH 5.5 to 6.5. The nutrient content would vary depending upon the raw material used for composting, however, 1.0 to 2.5%N, 1.0 to 1.5% P_2O_5 and 1.0 to 1.5% K_2O is ideal composition of the compost.

Phosphate rich organic manure (PROM) is prepared by using indigenously available rock phosphate, cattle dung and farm waste etc. A range of diverse types of substrates viz., paddy/wheat straw, farm wastes from sugar or fruit juice industry, FYM, oil cake, green manure, distillery waste etc. are used for preparing PROM.

Vermi-composting refers to ex-situ conversion of degradable organic residues into nutrient rich composted material for external application by the activity of selected species of earthworm. This technology allows quick transformation of substrate/organic residues/waste into mineral-rich compost (vermin compost) that could readily be used in the field. Vermi-compost is highly mineralized organic manure based on off-soil earthworm action.

Then there are biodynamic preparations. Cow dung forms the basis of various biodynamic preparations. The following biodynamic preparations are used for nutrient management which includes Cow Pat Pit, Biodynamic Composts, Liquid Manures, BD 500, BD 501 and Vedic Preparations. In addition several bio-fertilizer preparations are now available which can effectively reduce the requirement of fertilizer as well as to some extent, the pesticides by inducing plants tolerance to certain diseases and pests.

9. Safety and Quality of Produce

There is a growing demand for organic foods primarily driven by the consumer's perceptions of the quality and safety of these foods and to the positive environmental impact of organic agriculture practices. However, the organic label should not be considered as a health claim as it is only a process claim. It has been demonstrated that organically produced foods have lower levels of pesticides and veterinary drug residues and in many cases lower nitrate contents. No differences have, however, been established in terms of organoleptic quality between organically and conventionally grown foods. Reviews of organic vs conventional product sensory analysis studies have reported results that do not clearly substantiate claims of superior organic product tastiness. There is, therefore, scope to generate reliable information on the quality of produce generated on organic farms in future studies. There are, however, claims that food produced by organic methods contains a better balance of vitamins and minerals than conventionally grown food. However, there is no clear scientific evidence. Even quality after storage has been reported to be better in organic products relative to conventional products.

To reap the desired economic gains from organic farming in the global market strict phytosanitary measures have to be followed. To achieve this, weak links in the certification system have to be identified and removed. These are also need for proper grading, packaging, and storage of organic produce before marketing. There is thus need to set up laboratories to monitor the quality of organic produce so as to prevent the sale of sub-standard material to promote confidence in organic produce.

10. Marketing Opportunities

Since mid 1990s, the market for organic foods has been expanding rapidly and retail sales have been increasing. However, to create the domestic market for organic produce, is still missing decrease of low awareness among consumers about food safety and also because of lack of access to organic products, not because of production per se but because of weak supply chain systems, at present. Organic producer is looking for the market to get good price and quick sales but is partially successful. The organic retailers are looking for constant supplies, specially of fruits, vegetables, and food grains, which they do not know from where to get, and the prospective organic consumer does not have an easy access to organic products so far. In Asia, while most sales take place in Japan, other countries are witnessing a rapid expansion of their organic market. These countries include China, India, the Republic of Korea and Singapore. The main markets of organic produce or products are the United States, followed by Germany, the United Kingdom, Italy, France, Switzerland and Japan. The fruit and vegetables rank first in total organic sales and offers opportunities for developing counties.

11. Organic Farming and Biodiversity

Organic farming increases biodiversity at every level of the food chain – all the way from lowly bacteria to mammals. This is the conclusion of a review of studies from around the world comparing organic and conventional agriculture. The study reviewed data from Europe, Canada, New Zealand and the US. According to the researchers, organic farming aids biodiversity by using fewer pesticides and inorganic fertilizers, and by adopting wildlife-friendly management of habitats where there are no crops, including strategies such as not weeding close to hedge, and by mixing arable and livestock farming. Mixed farming particularly benefits some bird species. One of the reviewed studies from the UK also points to benefits for bats.

12. Organic Agriculture in Context to Indian Farming

A lot of debate has been going on about various aspects of organic farming in India at various forums. The total arable land in India is about 144 million hectares. Out of this only 35% land is irrigated and remaining 65% is mainly rainfed. In the rainfed regions, fertilizer use is very low and in most of the areas negligible amount of fertilizers is used or organic manures are used as source of nutrients. Similarly in north and north-eastern region of country where low quantity of fertilizers and pesticides are commonly used, there is scope and opportunity for continuing organic farming for improved production and quality through suitable interventions. About 18 million hectare of land is estimated in North-East for exploitation under organic production. There is thus ample scope in India to produce organic produce through a cluster approach in the above mentioned areas of the country without loss in productivity.

Since organic farming relies on the inputs produced at farm, it cuts down the cost of inputs. Further, the expenditure on pesticides is reduced as there are mostly preventive and botanical and microbial pesticides are cheaper. Some times higher input costs due to the purchase of compost and other organic manures have also been reported. The production or yield in most of the cases is at par with the conventional system after the few initial years while the price premium is related with organic farming. Organic agriculture thus offers comparative advantages in areas with less rainfall and soil fertility levels. The products are sold at a premium over the conventionally grown products.

13. Researchable Issues

National Commission on Farmers has identified it as a tool for second green revolution in the rainfed and hilly areas of the country. Keeping in mind the need for scaled up public interventions to promote organic farming the commission estimated the need for investing over Rupees 2500 crores during 11th Five Year Plan. The Commission has made a recommendation that 2500 Organic Villages with niche commodities should be developed in the hills during the next five years, and the local farm graduates and youth be involved in it. The Commission has observed the institutional infrastructure for promoting organic farming in the country is as yet for from satisfactory. Commission has recommended that a focused national movement on organic agriculture be launched, for main streaming organic farming in the national agriculture economy.

Institutional Setup: However, this investment and other initiatives in this sector may not yield desired results because they face a key constraint in the absence of proper national institutional set up that can take care of all aspects-promotion, research, development and regulation of standards and trade. Today there are many players and as many stakeholders. Ministry of Agriculture, ICAR, Ministry of

Commerce, National Steering Committee on Organic Farming, organic farmers associations, NGOs and above all new business houses coming up as investors for organic retail and exports. Further, International Developments in organic regulations, trade and development promotion demands that India creates such a body so as to represent and take care of national interests in such for a and formulate comprehensive national strategy. This recommendation had already been made in the Report of the Working Group on Horticulture Plantation Crops and Organic Farming of the Planning Commission for 11th Plan changed by this speaker.

Agricultural Research in India has been emphasizing mainly on improving production and productivity of different crops and animal species through intensive agriculture practices. The technologies so generated have helped in attaining self sufficiency in food production and significant increase in horticultural commodities, milk and fish production. However, emphasis on natural resource management, biodiversity and environmental conservation has been wanting. While area under organic farming, science based recommendations are not available in respect of quantity and quality of various organic inputs. In many cases, even their availability is inadequate and uncertain. There is therefore, an opportunity and need to reorient the entire agricultural research programmes for development of eco-friendly technologies. Some of the aspects requiring attention are:

- Replacing external input with a farm input from easily available local resources.
- Optimizing production and improving soil water quality.
- Reducing cost of production.
- Socio-economic improvement and food security.

This has to be achieved recognizing that small and marginal farmers over centuries have experimented, innovated adopted and standardized agriculture practices suited to their agro-climatic regions and socio-economic conditions. Further, the development of sustainable farming will require inter-disciplinary approach to conduct research on crop improvement, resource conservation, nutrition, pest and disease management, crop rotation, mixed cropping and tree farming. Besides this, National and State institutions engaged in agricultural research, teaching and extension will have to make structural and functional changes to reorient their activities to suit the biotic principles of organic agriculture.

14. Conclusion

Organic Farming is emerging as a dynamic alternate farming system for crop production in India very fast. It is also gaining quick acceptability among farmers and consumers. It offers environmental protection, biodiversity enhancement, reduced energy used and CO_2 emission, sustainable resource use, increased crop yields without over-reliance on costly external inputs, environmental and biodiversity protection, etc. are indirect benefits or organic farming which further adds to the economy of the country.

However, while the growers are quick in adoption the scientific community is divided while accepting all the benefits which organic farming is reported to be yielding. Many consider that organic practices simply can not withstand the regours of growing population because of their limited yield potential without adequate inorganic nutrition, proven plant protection practices and inadequate organic input availability.

Those in favour of organic farming are convinced by the likely changes, which these practices are likely to result in soil fertility, productivity, quality and safety of produce besides positive environmental effects. The increasing area world wide under organic agriculture, its wider acceptability by consumers and the higher prices received by producers all go to strengthen belief in this approach.

The technology available for organic farming though based on individual experience is rather diverse and the many forms of organic farming system are leading farmers to confusion rather than providing clear directions. The effectiveness of several organic products recommended has not been systematically tested. Several input recommended in organic farming are not available in adequate quantities. There is very little scientific research being carried out and on this aspect by the public system.

From a consumer point of view the future organic farming looks to be bright. It is becoming apparent that consumers are ready to pay premium price of organic produce for health or ethical reasons.

There is going to be a bigger spurt in organic products trade and India is expected to be a bigger beneficiary. Presently, India exports tea, fruits, spices, honey, rice, cotton and pulses. There is a tremendous scope of adding more commodities.

COMPOSTING OF AGRICULTURAL AND INDUSTRIAL WASTES

Introduction

The organic matter content of cultivated soils of the tropics and sub-tropics is low due to high temperature and intensive microbial activity. Therefore, soil humus has to be replenished through microbial activity and periodic additions of organic manures for maintaining soil productivity as increased cost of fossil fuels, shortages in energy and production of chemical fertilizers are expected to accentuate in the near future. Availability of chemical fertilizers to the farmer at a reasonable cost will be a real problem. There is a need to utilizing organic manures/wastes for supplementing chemical fertilizers. Preparation of organic manures from rural and urban wastes would provide not only plant nutrients and humic materials but also will result in hygienic disposal of the organic wastes which, otherwise, may cause pollution problems.

Farmers in developing countries need adequate resources for augmenting the crop productivity. Proper soil management ensuring continued maintenance and building up of soil fertility is indispensable for greater productivity from agricultural land. There is no need to emphasize that organic manures influence favourably plant growth and yield. Their indirect effects include augmentation of beneficial microbial population and their activities in the soil of organic matter decomposition, biological nitrogen fixation, solubilisation of insoluble phosphates and availability of plant nutrients. The direct effect is through addition of nitrogen, phosphorus, potassium and small amounts of micronutrients such as manganese, zinc, copper, and iron. In the use of organic materials as manures for crop production, farmers are faced with the problems of organic materials being bulky, with a low nutrient content in relation to their volume and being often messy and smelly. Further more, application of of fresh organic material may reduce the soil fertility and produce substances harmful to plant growth.

Sewage sludge, biogas slurry, waste water, fish pond effluent, pressmud, coir pith, seaweed residue, areca-nut waste, tannery waste and wastes from different industries and mined areas are some of the wastes which can be effectively utilized as organic fertilizers after giving proper treatment. With the mounting disposal problems and widening gap between demand and supply of nutrient sources, the concept of composting of organic (agricultural/industrial) wastes the organic component in biologically decomposed until the material forms a stable, inoffensive organic fertilize.

Definition

Composting is a process by which organic wastes are converted under controlled conditions (Kiyonori Haga, 1990).

Composting is defined as a method of solid waste stream is biologically decomposed under controlled conditions to a state in which it can be handled, stored and/or applied to the land without adversely affecting the environment (Golucke, 1977).

It is an important technique for recycling organic (agricultural/industrial) wastes and for improving the quality and quantity of organic fertilizers. The objectives in composting are to stabilize the putrescible organic matter in raw agricultural/ industrial wastes in reduce offensive odours, to kill weed seeds and pathogenic organisms and finally to produce a uniform, slow release organic fertilizer which stimulates soil life, improve soil structure, helps plants to tolerate/resist pests and diseases.

Principles of Composting

Composting is a self heating, thermophilic and aerobic biological process which occurs naturally in heaps of bio-degradable organic matter such as manure, moist hay and straw. This bio-degradation process is carried out by different kinds of heterophytic micro-organisms, bacteria, fungi, actinomycytes and protozoa, which derive their energy and carbon requirements from the decomposition of carbonaceous materials. For every 10 parts of carbon, one part of N is required for building up their cell protoplasm. Fungi are most efficient in carbon assimilation than bacteria and actinomycytes. Under aerobic conditions, heterotrophic organisms derive energy from the decomposition of organic matter, resulting in the production of CO_2, humic substances and release of available nutrients. Under aerobic conditions, the following major reactions are likely to occur.

Sugars, cellulose, hemi-Cellulose, lignins, proteins

$$(CH_2O)\,X + XO_2 \rightarrow XCO_2 + XH_2O + \text{Energy}$$
$$\text{Organic N} \rightarrow NH_3 \rightarrow NO_2 \rightarrow NO_3$$
$$\text{Organic S} + XO \rightarrow SO_4$$
$$\text{Organic Phosphate} \rightarrow H_3PO_4 \rightarrow Ca\,(HPO_4)_2$$
$$\text{(Phytin, Lecithin)}$$

Under anaerobic conditions, microorganisms break down organic materials by a process of reduction in the absence of O_2. First, special group of acid producing bacteria and facultative heterotrophs degrade organic matter into fatty acids, aldehydes and alcohols. Then a group of bacteria convert the intermediate products to methane, NH_3, CO_2 and H_2.

$$CH_2O + O_2 \rightarrow CH_3COOH \rightarrow CH_4 + CO_2$$
$$\text{Organic N} \rightarrow NH_3$$

Agricultural Wastes

India has a vast potential of agricultural and industrial wastes. The major agricultural sources are livestock and human wastes, crop residues, tree wastes and aquatic weeds, green manures, urban and rural wastes, agro-industrial by-products, marine wastes and tank silt. With the introduction of high yielding varieties and adoption of intensive systems of cropping, large quantities of crop residues such as straw, leaves, twigs, and stubbles along with huge amounts of grasses and weeds are readily available on farms. Potential hazards related to nutrient management, trace elements, trace organic chemicals, pathogens and soil physical problems may be encountered when the above wastes are applied directly in the soil plant system for crop production. Moreover, because of their wide C : N ratio, these agricultural wastes are known to reduce the availability of important mineral nutrients to growing plants through immobilization into organic form and also produce some phytotoxic substances during their decomposition. The traditional method of composting is a slow process and takes more than six months. Also the product thus prepared contains very low quantity of nutrients. In addition, farmers generally prefer burning these waste materials in situ (eg., cane trash). This adversely affects the environment and causes deterioration of the biological properties of the soil. All these drawbacks can be only set right by developing a technique for a rapid treatment of these wastes by composting.

Moreover, in India, a major part of residues is used as animal feed, while some find use as fuel and some may make housing materials in the rural areas. Thus, only a small part of the crop residues would be available for direct incorporation into soil for composting. The average nutrient status of some common organic manures and composts from agricultural wastes is given in Table 11.3.

TABLE 11.3: Average nutrient content of some common composts of animal and plant origin.

Manures/Composts	Nutrient content (%)		
	N	P_2O_5	K_2O
Farm yard manure (FYM)	0.80	0.41	0.74
Poultry manure	2.87	2.93	2.35
Rural Compost	1.22	1.08	1.47
Urban compost	1.24	1.92	1.07
Mixed farm compost	0.87	0.59	2.22
Mixed dry residues	0.90	0.45	1.95
Cotton stalks	1.61	0.48	3.38
Water hyacinth compost	2.00	1.00	2.30
Paddy straw	1.59	1.34	3.37
Dry grass/weeds	1.90	0.55	1.09
Cotton wastes and groundnut husk	1.62	1.04	1.26
Lantana stalks and leaves	1.55	0.52	1.07
Wheat straw	2.90	2.05	0.90
Maize stalks	1.99	1.30	1.01
Sugarcane trash	2.73	1.81	1.31
Vermi-compost	1.60	2.20	0.67

Source: Misra and Hesse, 1982; Hand book of Manures and Fertilizers.

Agricultural Wastes

Methods of Composting of Agricultural Wastes

1. Indore Method

The waste materials such as plant residues, animal wastes, vegetable wastes and weeds can be composted with the Indore method. First the wastes materials are chopped into small pieces of 5-10 cm size and are dried to 40-50 per cent moisture before stacking. Then, they are spread in layers of 10-15 cm thickness either in pits or in heaps of one m width, 4-6 m length and one m depth. The heap is properly moistened with dung, using earth or night soil. Sufficient quantity of water is sprinkled over the heap to wet the composting materials to the level of 50 per cent moisture. Periodical turnings, usually three times at 15, 30 and 60 days after filling, are given to aerate and the material is covered with a thin layer of soil of about 2-3 cm thickness. Under the aerobic process of decomposition, losses of organic matter and nitrogen are heavy (40-50 per cent at initial stage). This process, however, involves considerable labour in the preparation of the heap and periodical turnings and so becomes expensive and impracticable when large quantities of materials are to be processed. The site selected must be at a high level to prevent rain water stagnation. It should be nearer to cattle shed and water source for easy transporting. The compost obtained by this Indore method would have a composition 0.8 per cent N, 0.3 per cent P and 1.5 per cent K_2O.

2. Activated Compost

Fowler developed this technique in which fresh materials are incorporated in an already fermenting heap so that quicker decomposition can be obtained with already established microbial population. This method is useful, particularly when offensive materials like night soil are to be quickly and effectively disposed off.

3. Bangalore Method

Pioneering work in preparation of manure in pits was carried out by Acharya (1939), particularly on the utilization of town residues and night soil. This process is otherwise called as hot fermentation method of manure production. In this method, the compost production depot is located on the city outskirts to transport town refuse and night soil to the pits. The depots normally accommodate about 200 trenches with a spacing of 1-5 m between trenches. First the refuse is to make 15 cm height layer. Then night soil is discharged over this and spread to a layer of 5 cm. After filling the pit with refuse and night soil in alternate layers, the pit is filled to 15 cm above ground level with a final layer of refuse of 15 cm on the top. This may be dome-shaped and covered with the thin layer of soil with red earth or mud to prevent moisture loss and breeding of flies. Sullage water, if collected in carts as in some towns, may be emptied over the layer of refuse. This system provides a method of disposal of any kind of waste, including slaughter house waste, carcasses of animals, sewage, etc.

The materials are allowed to remain as such without any turning and pot watering for about three months. The decomposition of dumped materials in pits takes place largely in the absence of sufficient air except in the surface layer. Though the decomposition is comparatively slow, high temperature is not developed in the lower layers. Since, the material does not receive any turnings, decomposition into a homogenous mass of manure does not take place. Even then, the C?N ratio is reduced to less than 20:1 in about six months and the manure is ready for use. As there is no watering and turning, it is suitable to areas having low water availability and with scarce labour. When pits or trenches are not available for composting, town refuse and night soils can be composted in above ground heaps of 1 m height,

1 m width and of any convenient length by placing refuse and night soil in alternative layers as in trenches and adding the final refuse on the top (Gaur et al., 1984). In this method, the material decomposes more quickly than in pits and can be used after 3-4 months. The compost obtained by this method would contain 1.5 per cent N, 1.0 per cent P and 1.5 per cent K_2O.

4. Nadep Compost

In this method, plant wastes, dung slurry and clay soil are used as raw materials for composting. The process is similar to heap method of composting, but is one in brick lined enclosures provided with air holes on all sides. However, this method has the disadvantage of using large proportion of soil which is not desirable or acceptable under all situations.

5. Coimbatore Method

It is anaerobic degradation followed by aerobic process. First, pits of 4 m length, 2 m width and 1 m depth is formed in which crop residues or farm wastes are filled to a thickness of about 15 cm. Over this layer, cow dung slurry to enhance the rate of biodegradation is applied to a thickness of 5 cm. Above this layer, 1 kg of bone meal, or rock phosphate to minimize the nitrogen loss and to add phosphorus, is applied. Thus, application of crop residue/farm waste, cow dung slurry, bone meal and rock phosphate in alternate layers is repeated till the height reaches 0.5 m above the ground level. Then the above ground portion is covered with red earth or mud to prevent the rain water entry and it becomes an anaerobic process. After 30-35 days, the material is turned and it becomes an aerobic process. The compost will be ready within five months.

6. Synthetic Compost

The principle behind this procedure is that addition of dung to supply organic nitrogen required by microorganisms, can be completely substituted with fertilizer N compounds like urea and ammonium sulphate which are utilized effectively for decomposition of carbonaceous compounds. The basic principle of C/N ratio in manure preparation can be applied to synthetic composting as well by adding nitrogenous fertilizers in sufficient quantity to reduce the ratio to about 30-40 per cent and then allowed to decompose. This method is highly suitable to the areas where supplies of dung are either short of requirements or not available at all and large quantity of organic waste materials is available in the farm.

Either pit or heap method can be used for composting where the organic waste material is spread out in thin layer and sufficiently moistened. Super phosphate may be added to fortify the phosphorus content of the manure. Sprinkling of fertilizer solution followed by lime application may be done. This process is continued layer-wise until the heap or pit is filled and then, the heap or pit is allowed to decompose for a period of 4-6 months by which time, it is ready for field use. It resembles FYM in its effect on soil and plant growth.

7. Windrow Composting

Windrows are prepared as they allow efficient material handling, provide good aeration and allow sufficient absorption of water. First, windrows of 2.5-4 m width at the base, 2.5-3 m height and of any convenient length based on the availability of leafy materials are formed. It is better to use higher windrows for better composition of leaves at the base resulting in aerobic conditions. When the moisture content of incoming leaves is low, it is desirable to add sufficient quantity of water to maintain 40-60 per cent moisture. If the C/N ratio of the leaves is high (30), amending with sewage sludge, urea or grass clippings may be done. Periodical mixing may be given for good aeration. The leaf compost with a neutral pH (6-7) will be ready within 6-9 months under optimum environmental conditions. The use of finished compost (leaf mold) as covering material (10-15 cm) in the subsequent preparations of leaf compost is good to supply a heavy inoculam of microorganisms as in the case of activated compost (Eberhard and Pipes, 1972).

8. Accelerated Composting and Enrichment

For getting better compost, it is necessary to add nutrients, N, P and K which hastens the process of composting. Addition of mesophyllic cellulolytic fungi, or *Coprinus ephemerus,* a cellulose decomposer to compost pit may reduce the composting period by one month with an improvement in compost quality. Inoculation with *Aspergillus niger* or *Pencillium sp.* increases the total N content, available P and humus content of jowar stalk plus wheat straw compost and Jamin leaf compost (Gaur *et al.,* 1982). Further, this may also prove beneficial in the case of in *situ* application of plant residues where thermic rise is unlikely to be more than 40-45°C. Similarly, inoculating the compost previously amended with rock phosphate with cultures of *Azotobector chrococcum* and the phosphate solubilizing strain *Aspergillus arvamori* increases the total nitrogen and humus content.

9. Vermi-Composting

In Asia, vermin-composting is being effectively practiced in Philippines. The earthworms, *Lumbricus rubellus* and/or *Perionyx excavator,* are used for composting organic residues where 1000 t of moist organic matter can be converted by earthworms into 300 t of compost, rich in N, P, K, Ca and Mg contents with increased bacterial and actinomycete population. This method of composting can be suggested for sludge management. Method of vermin-composting, its advantages and applicability are discussed earlier.

10. Animal Waste Composting

Animal manures of different kinds are available in huge quantities in India. If they are not used properly, it will lead to water pollution through rains and runoff water, air pollution in the form of disagreeable odours and human health hazard through multiplication of flies and other insects. Animal waste composting is an effective and environmentally safe method to eliminate odour, to control flies and other insects, to reduce mass and volume, and to handle the material easily with agreeable smell. The type of animal, quantity and quality of feed, housing and waste handling techniques will decide the characteristics of the manure. In general, beef and poultry manures are rich in nitrogen and undergoes decomposition more effectively with lower moisture level content. Addition of carbonaceous material such as straw, wood bark saw dust, or corn cobs can often help the composting characteristic of manure. They can be used to reduce moisture content and to raise the C/N ratio, as it is usually practiced in all villages. However, under Indian conditions, conjoint composting of the manure slurries with plant residues is more viable and profitable.

Mutsuzaki (1977) produced animal waste compost with 30-40 per cent moisture content in about 2 weeks, by regulating the moisture content of raw materials to 55-65 per cent. Using this low moisture compost (seed compost) as a moisture regulator and an inoculam of useful microorganisms, he devised a method of continuous composting of fresh animal wastes in which high quality compost can be obtained every two weeks, regardless of weather conditions and with no use of supplementary energy.

11. Oil Palm Waste Composting

The palm oil industry plays a major role in the economic development in South East Asia, and is slowly gaining momentum in Southern India. Compost can be prepared by using equal proportion of oil palm waste and wheat straw mixed with small quantity of deep litter chicken manure. Before composting, the waste is analyzed for its physical and chemical characteristics. The ingredients mentioned above are thoroughly mixed and wetted to give compost mixed with no draining liquid. The prepared compost mixture (800 g) is then placed on a perforated stainless steel platform within each flask and the flasks immersed in the water bath such that the water level is above the enclosed substances. The thermostat is initially set at 45-50°C, to allow a gradual build up of the naturally occurring micro flora and each flask in independently aerated with humified air at 4 l/hr throughout the composting period. When the compost temperature has attained 50°C the water bath thermostat is raised to 60°C to pasteurize the material within the flasks. After 6 hrs at this temperature, the compost temperature is reduced to 53-55°C where it is held for a further 2-3 days until the concentrations of ammonia fall below 20 ppm which is determined using a Draegar gas detector. The purpose adding wheat straw is to improve the water holding capability and chicken manure to increase the total N content of the compost. Compost mixture of 500 g wheat straw + 500 g oil palm waste + 250 g chicken manure + 30 g gypsum will have total N 1.1 per cent, 36.6 per cent total C, and a C/N ratio of 33.3:1. It can be supplemented with carbohydrate sources such as sugarcane baggase or molasses to activate the resident micro-flora.

12. Phospho-Compost

Phospho-compost can be prepared by mixing farm wastes/crop residues, cattle dung, soil, compost, chopped grasses and tree leaves with mussoorie rock phosphate at the rate of 30 per cent of the compostable material. This mixture is made into slurry so as to provide adequate moisture and after uniform mixing; the slurry is allowed to decompose in a compost pit for about 60 to 90 days (Mishra et al., 1984). Moisture is maintained at 70 per cent throughout the period of composting and the compost will be ready for use in 60-90 days. Normally, this phosphor-compost is prepared by mixing organic wastes and mussourie rock phosphate in the ratio of 7 to 7.5:2.5 to 3.0.

13. Reinforced Compost From Sugarcane Trash and Pressmud

First, sugarcane trash is spread to a thickness of 15 cm over an area of 5-7 m x 3 m depending on the area available. Over this, pressmud is spread to a thickness of 5 cm. The fertilizer mixture containing mussorie rock phosphate, gypsum, and urea in the ratio of 5:4:1 is sprinkled over these layers at the ratio of 10 kg/100 kg of trash. Enough water is sprinkled to keep it moist. This has to be repeated till the height of heap rises to a height of 1.0-1.5 m above the ground instead of water to moisten the layer. This is covered with a layer of soil and pressmud at 1:1 ratio to a thickness of 15 cm or with cow-dung paste. Water is sprinkled once in 15 days. A thorough turning is given after 3 months and again heaped. Another turning is given after a month. The trash is composted and is ready in five months. The method increases the material value of trash compost by increasing N, P and Ca contents. It also brings down the C : N ratio by 10 times as compared to raw cane trash.

14. Enriched FYM (EFYM)

FYM is bulky and low in major plant nutrients such as N, P and K. Hence, there is a need to improve its quality. EFYM is recommended to rainfed crops which require available P for their root proliferations withstand the initial growth stages under dry land conditions. First, 750 kg of well decomposed FYM is taken. After sieving, the recommended dose of P and K_2O for the crops to be grown is mixed with the sieved FYM. The mixture is spread in the form of heap and plastering is done with red-earth paste. This anaerobic process is maintained for 30 days. Then, the nitrogenous fertilizer recommended for basal dose of the crop to be grown is mixed and it should be applied immediately before sowing.

15. Weed Composting

Though weeds certain large quantity of nutrients, they affect the farm operations considerably, compete with crop plants for nutrients, space and water, poses health problem to human beings. This problem can be overcome by bio-converting the weeds using lignolytic and cellulolytic organisms like *Trichoderma viridi, Fusarium sp.* and *Pleurotus sajorcaju* through composting which gives a clean environment in addition to improved moisture conservation, water use efficiency, nutrient use efficiencyand soil health. The weeds can be biodegraded individually or by pooling all the weeds together depending upon the quantity available and necessity. The arable and waste land weeds like Maxican poppy, thorn apple, nut grass, water grass, crowfoot grass, wild grass, carpet weed Parthenium, Cassia etc. and the aquatic weeds like Ipomea, water hyacinth and sea weed can be used for composting. A common bio-conservation technology is suggested for all the weed materials. Composting of parthenium weeds by this bio-conservation technology is given below as an example to other weeds.

Composting of Parthenium

Parthenium, a problematic introduced weed prevailing in cultivable and fallow areas, affects crop yield and poses health hazards to human beings by causing skin and eye irritation, asthma, fever, etc. Parthenium seeds have long storage life and higher germination percentage and spread quickly by wind. For composting purpose, the parthenium plants are to be cut into small pieces of 10 cm size by using chaff cutter. These chopped materials are spread to a thickness of 10 cm. Over this inoculum viz., compost cultures (*Trichoderma viridi, Fusarium sp*) are spread uniformly. Then, weed material is spread over this to thickness of 10 cm. Over this layer, 0.5 per cent urea is spread (5 kg urea/t of chopped materials). This has to be repeated till about 1 m height is obtained. Then plastering with mud is done. Periodical sprinkling of water is done to mainatain 50-60 per cent moisture. After 2 weeks, a through mixing is given. In 40-45 days, the parthenium compost is ready for field application. This compost contains 2,49 per cent N, 0.73 per cent P and 1.37 per cent K with a C : N ratio of 21:1. The same method is followed for other weeds also. The Ipomea weed compost contains 2.49 per cent N, 0.32 per cent P and 0.50 per cent K with a C : N ratio of 20:1 whereas the water hyacinth compost will have 2.11 per cent N, 0.3 per cent P and 0.73 per cent K with a C : N ratio of 22:1.

Hints for Composting Agricultural Wastes

From the several composting methods, the following few practical hints are offered for proper composting of agricultural wastes.

- Collect plant wastes and make the composting mass heterogeneous by mixing two or three types of materials.
- Shred the material is 5-7 cm size.
- Mix the material with cow-dung slurry in the ratio of 70:20:10 (agricultural waste: animal dung: soil).
- Add super phosphate or rock phosphate at 5 per cent (w/w).
- Maintain the medium moisture content at 50-60 per cent by sprinkling water.

- Do plastering with red earth or mud.
- Give periodical turning depending on the method employed.
- Within six months, the compost is ready for field application.
- If needed, retain a part to serve as mother and seed compost for subsequent cycle of composting.

Industrial Wastes

The industrial wastes are those which come out as wastes from industries and factories, commonly known as effluents. These wastes contain considerable quantity of organic matter and nutrients in addition to non-compostable materials. The quantity of wastes available depends on the size of industries/factories. In the case of big industries, the quantity of waste available will be large enough even to maintain their own farm. The sources of industrial and factories wastes are given below.

1. Textile industry
2. Paper factories
3. Cordite and ammunition factories
4. Fertilizer plants
5. Sugar factories and distilleries
6. Food processing units
7. Coir industries
8. Tanneries
9. Lignite and coal mines
10. Petroleum factories
11. Cosmetic factories
12. Mushroom centres
13. Chemical plants
14. Electronic industries
15. Photofilm industries
16. Tinning plants
17. Foundries
18. Electroplating centres
19. Cement factories
20. Dyeing factories
21. Automobile works
22. Sago industry
23. Heavy metal/electrical plants.

The above list of available industrial wastes given a good picture but it is not exhaustive. Among the different industrial and factory wastes, tannery effluent causes serious hazards and environmental problems by affecting streams, fresh water bodies, ground water and agricultural lands. This effluent contains sodium, chlorides, sulphate, bicarbonates and chromium (300-465 ppm). The rice grain produced from tannery polluted soil contains 1.02 to 2.8 ppm chromium while the straw contains 96-200 ppm (Nanda Kumar and Rajendra Babu, 1984). The milk from cow and buffalo and goat which feed on the fodder grown in the tannery effluent affected lands show a chromium concentration of 0.079 to 0.390 ppm.

The paper factory effluent has pH 8.4, EC 1.6 dsm^{-1}, 2.1 CO_3, 10.8 HCO_3, 8.0 SO_4, 21.0 CL_2, 10.0 Ca, 5.5 Mg, 25.0 Na mg/litre. Besides it also contains 774 mg/l of suspended solids and 1044 mg/l of dissolved solids. The colour of the effluent is brown due to the presence of phenolic compounds. The sago factory effluent is rich in coloured starch and organic material which affects soil drainage leading to pasification of yeasts and causes severe environmental hazards. The cement dust from cement factories from a thick coating over plant canopy and affects the respiration and photosynthesis. It also accumulates continuously with soil and induces soil crusting. Most of the urban areas are polluted by the excess presence of sewages and sludge wastes. In India, sewage production is 12,500 ML/D. Sewage sludge from highly industrial areas contains heavy metals in the range of 3.5-8.3 mg/kg of sludge which poses toxicity problems. However, sewage sludge can supply 5.1, 2.9 and 2.8 thousand tones of N, P and K_2O per annum in India.

Pressmud, an organic waste material from sugar industries, contains about 35 per cent organic carbon, 0.8 per cent N, 2.0 per cent P, 1.1 per cent K, 4.45 per cent Ca, 1.12 per cent Mg, 3100 ppm Fe, 80 ppm Cu, 150 ppm Zn, with a C: N ratio of 41:1. However, it can supply 33.3, 79.4 and 55.4 thousand tones of N, P and K_2O respectively every year. It has a manorial value equivalent to 54.4 kg NPK/t besides contributing micronutrients. Slurry from biogas unit provides 44 kg nutrients as composed to 19 kg by FYM and 27 kg by compost. It contains 1.6 per cent N, 1.5 per cent P_2O_5 and 1.2 per cent K_2O in addition to some quantity of micronutrients.

The sea weed residue left in the industries after extracting agar contains nutrients at appreciable level viz. 1.596 per cent N, 0.06 per cent P and 0.076 per cent K. Application of sea weed residue at 12.5 t/ha to groundnut resulted in increased soil moisture conservation, shelling percentage and oil content besides higher uptake of N, P and K. Coir pith production potential of India is 0.66 mt. It has 0.2 per cent N, 0.05 per cent P, 0.8 per cent K, 0.4 per cent Ca, 0.36 per cent Mg, 0.07 per cent Fe, 30 per cent cellulose, 115:1 C/N ratio and 5.4 pH with a water holding capacity of 50 per cent. The nutrient content status of the industrial wastes is given in Table 11.4.

TABLE 11.4: Nutrient status of some industrial wastes

Source	Nutrient content (Per cent)		
	N	P_2O_5	K_2O
Coirpith	0.26	0.05	0.84
Pressmud	1.15	2.40	1.98
Sea weed residue	1.50	0.50	4.20
Cotton meal waste	1.40	0.60	1.20
Saw dust	0.25	0.20	–
Sewage sludge	1.50	2.00	1.50
Bagasse	0.25	0.12	0.40
Biogas slurry	1.60	1.50	1.20
Mushroom spent waste	1.68	0.60	0.63

Source: Thangavel, P. 1995. Composting of industrial wastes,
In: Manual on Bio-conversion of agricultural wastes. Published by TNAU, Coimbatore, pp: 27.

When these wastes are efficiently processed and utilized, they can serve as good source of nutrients and can replace the inorganic fertilizers to a significant extent. Composting, a better way of disposal of waste rather than incineration or land filling, reduces the bulk volume of materials (solid waste pollution abatement) and at the same time, the nutrients are added to the soil (an efficient recycling process). Composting technique is available for coirpith, pressmud and effluent. To manage other effluents/wastes, the following measures should be adopted for better disposal and to ensure clean environment.

- Using treatment plant (with chemical, physical and bio-organisms treatment), the effluent let out from tannery, textile, sago, sugar, paper and chemical industries could be treated to reduce the intensity of pollution.
- Site evaporation tanks and tagoor with cement lining should be constructed to prevent contamination of ground water resources.
- Spraying water through fine nozzle droplets in chimneys prevent the dust pollution by cement and granite factories.
- The diluted sugar factory and proper factory effluent (to an extent of 40-50 times) could be successfully used for growing tolerant sugarcane varieties.
- The soil affected by tannery effluents could be reclaimed by organic manure application and growing resistant crops like maize, ragi, NB2 grass, Desmanthus and Korai.

1. Composting of Coir Pith

India ranks third in the world in the production of coconuts with an annual production of 100.43 m nuts from an area of 1.5 m ha. The area under coconut in Tamil Nadu is 2.40 lakh ha with more than 1.7 lakh t of coir pith production. The coir pith or dust is a waste product in the coir industry after extracting the coir. The coir wastes (dust and bits of fibres of lesser length) accumulate in large quantities near the location of coir industries and its disposal is posing a serious environmental problem. However, this material can be converted into valuable organic manure by composting. The material contains high amount of lignin and carbon compounds with small amount of nutrients like sulphur and phosphorus. First, an area of 5 m length and 3 m width, preferably in a shady place, is selected. Spreading of 100 kg of coir pith is done. One bottle (300 g) of *Pleurotus sajorcaju* spawns is spread uniformly on the surface of the coir pith. After applying fungus, sprinkling of water is to be done to maintain 200 per cent moisture content. Above this, another layer, is formed with 100 kg coir pith on the surface of the second layer one kg of urea is uniformly spread and sprinkling of water is done as that of first layer. This process of sandwiching *Pleuertous* and urea in alternate layers of coir pith is repeated till the heap reaches one metre height. The heap is periodically sprinkled with water to maintain 200 per cent moisture content which should be felt with hands. At the end of 30 days, coir pith turns into dark brown mass of compost. The chemical composition of raw and composed coir pith is given in Table 11.5.

TABLE 11.5: Composition of raw and composted coir pith

Properties	Raw coir pith	Composted coir pith
Lignin (%)	30.00	4.80
Cellulose (%)	26.52	10.10
Organic carbon (%)	29.00	24.90
N (%)	0.26	1.06
P_2O_5 (%)	0.01	0.06
K_2O (%)	.0.78	1.20
Ca (%)	0.40	0.05
Mg (%)	0.36	0.48
Fe (%)	0.07	0.09
Mn (ppm)	12.50	25.00
Zn (ppm)	7.50	15.80
Cu (ppm)	3.10	6.20

Properties	Raw coir pith	Composted coir pith
C : N ratio	112.1	24.10

Source: Nagarajan, R, Ramaswamy, P.P., Savithri, P. P. and Manickam 1988.
Coir waste in crop productions. Tamil Nadu Agricultural University, Coimbatore, 30 pp.

The composted coir pith reduces the volume by 42 per cent besides narrowing the C/N ratio. This composted coir pith can effectively be used as manure for both dry land and irrigated crops to increase the crop yield, as an amendment in sodic soils and as an alternative to K fertilizer.

2. Composting of Pressmud

It can be done by mixing with distillery effluent which is also a rich nutrient source. First, the moisture content of the pressmud is reduced from 75 per cent to 50 per cent using aerotiller. Then, additive containing materials like composted coir pith 5 per cent, sugarcane trash 2.5 per cent, water hyacinth 2.5 per cent and rock phosphate 2.5 per cent are mixed with pressmud. Then, one tonne of pressmud is inoculated with 1.5 litre of composite culture containing thermophilic microbes *Bacillus sp.* and *Pseudomonas sp.* Above this, distillery effluent is added to bring 60 per cent moisture level. Then, heap is reformed with aerotiller until a temperature of 60-70°C is reached. After 4-5 days, adding of next dose of effluent is done when the moisture content goes below 50 per cent. Again, the heap is reformed with aerotiller. Likewise, composting of pressmud and adding effluent are continued. The addition of effluent is about 4000 litre which is added in 1-2 months. For better enrichment of pressmud, super phosphate is added at 2.5 per cent and mixed well. The heap is allowed for curing for about 15 days. Then, $FeSO_4$ at 20 kg and $ZnSO_4$ at 10 kg/t of pressmud is added and allowed for curing for a period of 15 days. To the composted materials bio-fertilizers viz., Azotobacter, Azospirillum and phospho-bacteria each at 2 kg/ha is added at the time of application, which enables N fixation and solubilization of native and added phosphorus.

Using Microbial Inoculum

First pressmud is spread to a height of 10 cm over an area of 1 x 1 m^2. Over this surface, urea and cow dung slurry are sprinkled. Then inoculation of pressmud is done with microbial inoculum (*Aspergillus fumigateus* and *Bacillus sp.*). This process is repeated several times based on the availability of pressmud. Then, plastering is done with sand. Sprinkling of water is done to maintain 50 per cent water holding capacity, as and when required. The compost is ready within 45 days, which is rich in nutrients and micro-flora.

Using Pressmud and Distillery Effluent

Distillery effluent based compost is prepared by using pressmud, a bye-product obtained from sugar factory. First, pressmud is spread in the compost yard to form a heap of 1.5 m with 3.0 m length and 3.5 m breadth. Ten litres of bacterial culture diluted with water in the ratio of 1:10 is sufficient for a tonne of pressmud. After 3 days, distillery effluent is sprayed on the heaps to a moisture level of 60 per cent and the pressmud heaps are allowed for 4-5 hours to absorb the effluent, then the heaps are thoroughly mixed by aerotiller. When the moisture level drops below 30-40 per cent, again the effluent is sprayed mixed with pressmud, and the heaps are formed. Effluent should be sprayed once or twice in a week depending on the moisture content of pressmud heaps. Mixing of effluent and heap formation should be repeated for 8 weeks so that the pressmud and effluent proportion reaches an optimum ratio of 1:3. Then the heaps are allowed for curing for a month. During the process of composting, the temperature of the heaps may raise 60 to 68°C on 10th day and may reach a maximum of 80°C on 50th day. The compost obtained from this process is neutral in pH with a EC with a EC of 3.12 to 6.40 dS/m. It contains 1.53 per cent N, 1.50 per cent p, 3.10 per cent K, 130 ppm Cu, 180 ppm Mn, and 220 ppm Zn. The organic carbon and C/N ratio are reduced from 36 to 18 per cent and 28.06 to 16.3, respectively.

CONCLUSION

Agricultural and industrial wastes are available in large quantities in India. They contain valuable plant nutrients. Composting of these wastes gives a good solution as a waste disposal method, besides increasing the availability of plant nutrients, destruction of pathogens, elimination of unfavourable odours and easy handling. As a result, pollution problems and bio-magnification of toxic chemicals in the various biological systems can be minimized. They have the potential to raise soil productivity by improving physical, chemical and biological properties of the soil, thereby improving the plant growth. These wastes, after proper treatment, can be used as good amendment in reclaiming the problem soils like sodic soil. In the days of increasing fertilizer price and in view of the necessity to improve the yield of rainfed agriculture, it is imperative to look for these agricultural wastes recycling by composting. With these refinements, in the not so distant future, composting may well be as much an artful science as wine making.

Large scale conversion to organic agriculture would result in food shortage with the present state of knowledge, as the yield reduction of organic system relative to conventional agriculture average 10-15%, especially in intensive farming systems. However, in traditional rainfed agriculture, organic farming has the potential to increase the yield since 70% of total cultivable land falls in this category. More 5-10% increase in farm production would definitely help achieve the targeted growth rate of 4.5% in agricultural production during tenth five-year plan.

Organic manure is an alternative renewable source of nutrient supply. A large gap exists between the available potential and utilization of organic wastes. However, it is not possible to meet the nutrient requirements of crops entirely from organic sources, if 100% cultivable land is converted to organic farming.

Organic farming systems can deliver agronomic and environmental benefits both through structural changes and tactical management of farming systems. The benefits of organic farming are relevant both to developed nations (environmental protection, biodiversity enhancement, reduced energy use and CO_2 emission) and to developing countries like India (sustainable resource use, increased crop yield without over reliance on costly external inputs, environment and biodiversity protection, etc.).

Organic foods are proved superior in health and safety, but there is no scientific evidence to prove their superiority in taste and nutrient, as most of the studies are often inconclusive.

Organic farming is a holistic production system that has the advantage of efficient use and recycling of locality available resources. It is the best suited to rainfed areas with scarcity of water and light soils. Some monopoly high value crop of rainfed areas like seed spices have great international demand, if produce organically. Organic production in rainfed areas not only boost the economy of this region but also sustain the productivity of natural resources. The need is to strengthen research on development of processing and marketing infrastructure, and financial as well as technical support for quality organic production.

FUTURE NEEDS

- Methods of combining bio-fertilizes with the composted organics have to be taken up, so that the efficiency of both can be combined
- Research has to be carried out to hasten and stabilize composting.
- Organics have to be composted and produced in forms that will be easy for transportation.
- Collaborative work with engineering professionals is needed to design various bio-reactors for efficient and large scale composting.
- Recommendation is to be given on the dosage of the composted organic materials for various crops.
- Awareness on the use of composted organic manures is to be created among farmers.

Organic Farming is developing rapidly and is now practiced in more than 120 countries of the world. Globally 30.4 million ha of the agricultural land, are currently managed organically. India ranks 33rd position in total area under organic cultivation. India has now a leading supplier of organic herbs, organic species, organic basmati rice, etc. to the developed nations. The export realization during 2007-08 was around 78 million US $registering a 200% growth over the previous year. Though organic farming is an inevitable phenomenon, it alone cannot feed the billion population of India. To ensure the food and nutrition security, rather than promoting organic farming *enmasse* and universally, it would be desirable to carefully delineate areas for organic farming. In India, there is ample scope for pure organic farming in the rainfed areas, where there is little or no use of fertilizers and other agro-chemicals. The Government of India has taken considerable efforts to promote the organic farming without compromising the food grain production to feed the ever growing population in India. Launching of "National Programme for Organic Production" (NPOP) and the dedication of "National Institute of Organic Farming" were the landmark achievements in the arena of Organic Farming in India.

"The battle is not to the strong alone, it is to the vigilant, the active, the brave."

12 Precision Farming

> ***"Increasing yields and input use efficiency will require new farming concepts that focus on fine-tuning of seed, nutrients, water, pesticides, energy and labour inputs for smaller management units".***
>
> ***P. C. Robert, 1995***

Many farmers have achieved tremendous yield grains during the past 30 years in irrigated rice systems of South and South East Asia, often approaching the economically attainable yields. However, average yields of irrigated rice have to increase from 4.9 t/ha in 1991 to about 8 t/ha in 2025 (Cassman and Pingali, 1995). Therefore, irrigated rice farmers will have the greatest need to fine-tune their farm management to produce the bulk of the future increase in rice supply. At the same time, the positive awareness of environmental aspects of farming needs to be ensured for the farmers, future yield increase will require improved crop care, integrated resource management approaches and more knowledge intensive strategies for the efficient use of all inputs. They will have to move from simple general decision and recommendations to much more knowledge of intensive, site-specific crop management. Increasing yields and input use efficiency will require new farming concepts that focus on fine-tuning of seed, nutrients, water, pesticides, energy and labour inputs for smaller management units. Precision farming (PF), prescription farming or site-specific crop management are frequently used new terms to describe such emerging technologies (Robert *et al.*, 1995).

OBJECTIVES

The objectives of the precision farming are given below:

1. To improve crop performance and environmental quality.
2. To manage spatial and temporal variability associated with all aspects of agricultural production.
3. To match resource application and agronomic practices.

CONCEPT

Two Phrases

1. 'Spatial Variable', 'GPS based', 'Prescription', 'Site-specific' or 'Precision'.
2. 'Farming': 'Agriculture' or Crop production'.

   Variations occur in crop or soil properties within a field:

- These variations are noted, and often could be mapped.
- Management actions are taken as a consequence of the spatial variability within the field.

COMPONENTS

1. Geographical Information Systems (GIS)
2. Global Positioning Systems (GPS)
3. Sensors.
4. Variable Rate Technology (VRT).
5. Yield Monitoring (YM).

1. Geographic Information Systems (GIS) and its Uses

- GI is a software that provides data storage, retrieval, and transformation of spatial (field) data.
- It provides data such as soil types, nutrient levels, etc., in layers and assign that information to the particular field location.
- A fully functional GIS can be used to analyze characteristics between layers to develop application maps or other management options.
- Field location is stored by the latitude and longitude of that position.
- Several maps can be created showing the variability of nutrient levels, soil type, topography, pest incidence and yield.
- Most GIS software packages also have some form of data smoothing or contouring.
- Simulation models for different purposes like the flow of water, crop growth, soil erosion, nutrient and pesticide leaching.
- GIS helps in integrating geographical data on various aspects such as soil, weather and field history along with simulation models.
- GIS support in agriculture is the engineering component, in which the research findings are translated into operational systems for use at farm level.
- GIS can support this engineering activity by providing a good platform for storage of base data, simple modeling, presentation of results, development of a user interface, and in combination with a GPS, controlling the navigation of farm.

2. Global Positioning Systems (GPS)

The ability to know where you are and where you are going is a critical to so many; to data Global Positioning Systems (GPS) achieved it. Today, Agricultural demands improved productivity and efficiency, cutting costs, saving time and ensuring the entire agricultural enterprises is more efficient and accountable, and it is essential to compete in domestic and global markets. Positioning wireless and information technologies are changing the farmer relationship with land, bringing them quite naturally down to earth. Farmers can now manage every aspect of an agricultural operation to improve over all productivity and efficiency from planting to harvesting, land leveling and literally work their land by square foot instead of square miles.

i) **GPS** farming systems provide precise guidance for field operations or collect data on tillage applications, planting, weeds, insects and disease infestations, irrigations etc.,
ii) Increase farm and farm productive area,
iii) Precisely plant seeds at optimal rates at different areas of the field,

iv) Saves time by performing multiple operations in one pass over the field,
v) Create and maintain detailed records of crops, weeds, insects, diseases, topography for prescription farming and water management.

3. Sensors

Sensors are being developed to determine crop stress, soil properties, pest incidence, etc. Sensors mounted on a tractor are used to measure soil and crop properties. As the tractor passes over the field, as a scout goes over the field on foot, or as an airplane or satellite photographs the field from the sky. Measurement of plant and soil properties as the tractor or combine travels over the field is a developing area in precision farming. Currently, yield monitors are the primary sensing system that makes measurements "on-the-go". Commercial sensing systems are available that claim to measure soil properties and make application rate changes "on-the go".

There are also sensors that can be carried by a scout to the field and used to spot-check the health of plants and soil properties. They use light reflectance on the leaf to determine chlorophyll production. Several soil chemistry kits are also available to measure soil pH, nitrogen, phosphorus and potassium in the field without waiting for soil sample results.

4. Remote Sensing

PF needs information about mean characteristics of small, relatively homogenous management zones.

i) These may be obtained from soil tests for nutrient availability, yield monitors for crop yield, soil samples for organic matter content, information in soil maps, or ground conductivity meters for soil moisture.
ii) Remote sensing technology is to obtain spatially and temporally variable information for PF.
iii) Remote sensing imaginary for PF can be obtained either through satellite-based sensors or video digital cameras on board small aircraft.
iv) In the first approach, the multi-spectral images used for anomaly detection. These anomalies can be in the forms of disease/pest, weed growth, water stress etc.
v) Using the reflectance in the visible part of the spectrum, it is possible to detect diseases and identify weeds from crops.
vi) The difference between remotely sensed surface temperature and ground-based measurements of air temperature has been established as a method to detect stress in plants.
vii) Correlating variation as in spectral response to specific variables such as soil properties or crop yield. Soil physical properties such as soil water, organic matter and soil texture can be correlated to spectral reflectance.
viii) Vegetation spectral response used to infer other soil conditions.
ix) Crop yields for many crops like rice, wheat etc. have been found to be highly correlated with spectral vegetation index during maximum vegetative cover.
x) The yield map generated from spectral images can be used to farm management units.

5. Variable Rate Technology

Variable Rate Technology (VRT) includes computers and associated hardware to vary the output of fertilizer, lime and pesticides. Controllers are manufactured by several companies and can generally take an application map, along with the GPS to locate field position, and control the hardware that varies the application rate.

6. Yield Monitoring

Yield Monitoring is the most important direct method to assess the field production and how it should be better managed. A yield monitor measures the crop as it is harvested. YM usually measure crop weight, impact forces, or the time an array of light beams is broken. As the yield is measured, data are stored on a computer along with the GPS coordinates at the point where the yield was measured. Mapping software can then create a yield map. The yield map can immediately provide two important pieces of information viz., yield variability, and yield production. Yield variability is illustrated on a map by a change in colour, where each colour represents a range in yield.

FARM AND SITE-SPECIFIC CROP MANAGEMENT

We consider "Precision farming" or SSCM as a scientific concept that is applicable to farm and fields of all sizes, including those found in Asia. It is proposed to characterize three terms for modern farming operations.

1. Total Farm Management (TFM)
2. Technology Application Domain (TAD)
3. Site Specific Crop Management (SSCM)

1. Total Farm Management (TFM)

This TFM is an information-based agricultural management system to provide an optimum of the balance between profitability, food production and sustainability within a **single farm.** This definition emphasizes optimization of the management of the whole farm, including various cropping system-and, if possible, other types of agricultural production. At this level a farmer has to make decision about the allocation of major inputs and any optimization attempts should also consider relationships with other farms in a neighborhood of varying size. The later includes buy and sell operations, rental of machinery and labour, credits, distribution of seeds, manure and post harvest activities.

2. Technology Application Domains (TAD)

These are for a **specific technology,** the **smallest uniform spatial units** that can be treated differently. Within a SSCM approach, the size of TADs can be different for different crop management operations. This definition of TAD emphasizes a **specific technological solution for implementing a certain crop management operation according to site characteristics.**

3. Site-Specific Crop Management (SSCM)

This is the crop-specific use of all local soil, crop and climatic parameters to make pricise application of production inputs to technology application domains (TAD) with different characteristics. The typical SSCM components of a particular crop are varying the depth of soil tillage based on topography, changing varieties or sowing rates according to soil types, adjusting the rate of fertilizer application according to variation in soil test values, varying rates of pesticide spray based on actual crop stress. Therefore, this definition emphasizes management of a **single crop** according to site characteristics.

Therefore, SSCM decisions and operations may include **continuous adjustment,** applications specific to patches within a single field, uniform application to single field, or uniform applications to groups of several fields with similar properties.

1. Practices of SSCM

A typical SSCM application includes the following steps:

i) **Knowledge Capture:** Identify and quantify (map) the variability of the key input parameters at the scale needed to make a decision about the specific SSCM operation.
ii) **Knowledge Interpretation:** translate the information into decision about management (application) by understanding the sources of variability and its impact on yield.
iii) **Applications:** Identify the available application technology and the optimal size of the technological application domain (TAD). Apply differential amounts of inputs to TADs.

2. Potential Benefits of SSCM

The basis of SSCM is that fields are highly variable (both between and within) and new technologies are available that can characterize that variability and delineate meaningful management zones to optimize supply and demand of nutrients, water, energy and other resources according to their variation in time and space. Where there is less supply (e.g. certain fields or certain locations within a field) application has to be greater than in physical land units with higher supply and vice versa. Hence, the degree of variability, the quality of the input information, the approaches used to translate it into application decisions and the costs associated with differentially managing a field determine the benefits obtained from SSCM.

3. Opportunities for SSCM

i. Need

Crop management recommendations over the past four decades in Asia were driven by increasing use of externally provided inputs and the so-called **"Package approach"** based on blanket recommendations over wide areas (Byerlee, 1996). If a question is raised whether we should and can apply the principles of SSCM to manage rice fields in Asia. We believe yes, because recent research in lowland rice areas has demonstrated that

- There is large variability in stable soil properties, soil nutrient supply, nutrient use efficiency, other production factors, grain yield, and economic performance between rice farms or between singe rice field (Cassman et al., 1996; Dobermann *et al.*, 1997; OLK *et al.*, 1996; Dobermann and Oberthuer, 1977) and
- Even within very small rice fields, tremendous variation in yields and yield components exists that is caused by micro variation in soil nutrients, land leveling, crop emergence, weeds and other pests (Baki, 1993; Dobermann *et al.*, 1997, Dobermann *et al.*, 1994).

This will lead to achieve the significant gains in productivity and input use efficiency by soil and crop management technologies that are much tailored to the specific characteristics of individual farms, rice fields and variation within fields (Cassman et al., 1997; Dobermann *et al.*, 1996). Farmers already integrate a lot of knowledge and often already have a sense of variability they have in fields and crops.

ii. Feasibility

While much of the purpose of PF in developed countries is to breakdown variability across large fields to small uniform units, in Asia, much of this division has already taken place due to the typically small farm size. Although India has made much advances in agricultural research, but still the blanket recommendations are very much in vogue for adoption over larger areas. These blanket recommendations are no more useful to enhance productivity gains, which were witnessed between 1960s and 1980s. Now, to enhance growth rate in productivity, Site Specific Crop Management (SSCM) practices have to be developed.

By definition, SSCM practiced in developing countries focuses on managing spatial and temporal variability within a single and usually large field. In a broader sense, however, SSCM should be viewed as farming philosophy aiming at replacing prophylactic application of agricultural inputs by demand base application. Because of this, SSCM is relevant to Indian conditions, which have varied type of soil and climate. Conceptually adjusting tillage, sowing, fertilizer or pesticide rates separately for many small fields or farms (< 1 ha to < 10 ha) in an Asian domain are similar to adjustment according to soil variation within a large field (>10 ha to > 100 ha) found in North America. There is a continuum, however, as there is still variability within these small fields. We feel that first major step is to refine regional recommendations to individual field level recommendations-further advances can be made in the future to within field recommendations. Generally, SSCM in Asia can be built around much less sophisticated technology than implementing SSCM in large fields, where GPS (Global Positioning System), mapping systems, computer technology, and VRT (Variable Rate Technology) are minimum requirements.

The practical options for SSCM in irrigated rice are very much determined by the general shifts in these production systems, including:

a) **Labour Shortages:** Fewer agricultural field workers
b) **Water Shortage for Rice:** Less water for agriculture, urban demands and alternate uses increase.
c) **Less Land for Rice Production:** As cities grow over production areas and alternate land use increases
d) **Shifts in rice quality Preferences:** As economy develop through out the region, quality preferences will become more important in some countries.

Generally land preparation, crop establishment, harvest and post harvest activities are labour-intensive farm operations and farmers will increasingly seek ways to reduce costs associated with them. They will also seek ways to add value to their production by improving the quality of harvest products or improving their use for various by-products.

iii. Technology

Although precision farming (PF) or SSCM stated as a technology led development, it is not just synonymous with yield mapping and variable rate technology (VRT) for managing spatial variability within the field. Instead, PF should be viewed as a system approach to crop production in which the goal is to reduce decision uncertainly through better understanding and management of uncontrolled variations. SSCM aims at improving input-output characteristics of soil and crop system as they vary in space and time.

iv. Sscm Decision on Region Basis

Even in a SSCM approach, recommendations and decisions at regional scales play a role. At scales such as a district, important agronomic advice can be given to a farmer. Information captured may include:

a) Information about most suitable crops and varieties for a given environment.
b) General recommendations for soil fertility management based on delineation of soil types.
c) Weather forecasting and real-time variation in weather, and
d) Inputs application, Pest forecasting and seasonal variation in pest populations

This would help the farmer to make decisions about varieties, seedling age, planting date, fertilizer application, pest management and harvest time. The major tools involved in this are remote sensing (e.g., high resolution radar images), GIS crop models (Mathews et al., 1997; Singh et al., 1991) and post models (Kropff et al., 1995), Mass media (TV, radio, newspapers), HAM Radio communication and the existing extension systems are the main information provides.

1. Selectivity of Variety

Most rice farmers already select their varieties based on knowledge about adoption to site characteristics and agronomic fitness. In general, choice of the variety for a place as well as season is very important. Sowing the crop at appropriate time in the season will give higher yield. The production and productivity are generally higher in dry season (Kuruvai/Kar/Sornavari) due to favourable weather parameters. Bright sunshine hours, favourable temperature regime, optimum moisture, marked response to nutrients application, relatively less pest and disease incidence are some of the factors contributing to higher grain yield in dry season.

During the wet season (Thaladi/Samba/Pishanam), factors like reduced sunshine hours during rainy season, cloudy weather, cool temperature, flat terrain fields, excess moisture, ill drained conditions due to continuous rain, poor response to applied nutrients, incidence of pest and diseases etc., are responsible for lesser yields. Most modern varieties released are resistant to some common pests and they are also screened for adaptation to some soil stresses such as severe P and Zn deficiency or Fe toxicity. Among the various research options to break the yield barrier, exploitation of heterosis and restructuring of rice plant ideotype for better harvest index are important. Hybrid rice is found to be a viable technology and it has been proved successful in China. In Tamil Nadu three hybrids viz., CORH 1 (1994), CORH 2 and ADTRH 1 (1998) were released for cultivation. The ADTRH 1 rice hybrid has got the yield advancement of 10 t/ha over the best high yielding varieties like ADT 36, ADT 43 and ADT 45. Ever since the evolution of dwarf varieties that helped to break the centuries old genetic yield barrier, rice scientists have been looking for still more efficient technology for raising genetic yield potential. There is probably some scope for refining selection of varieties in a SSCM approach with regard to

- Fitness for dry or wet seeding,
- Resistance to pests,
- Nutrient requirements,
- Grain quality and
- Seed health.

Though biotechnological approach and transgenic lines may be developed for enriching with vitamin A and iron, it is felt necessary that more variety-specific information should be jointly established by breeders, agronomists, plant protection specialists and post production specialists (e.g. millers) and released through the national seed distribution and extension systems.

2. Land Preparation

SSCM options for land preparation vary widely. In many small to medium size farms, the lowest feasible TDA for plowing with a four wheel tractor would be a whole field, i.e., ploughing depth should be uniform. Examples of field-specific SSCM decisions include:

- Is there need for ploughing and puddling in each rice crop?
- Is there need for occasional deep ploughing?
- Is precision leveling feasible?

In the last years, many farmers have switched to shallow tillage machinery such as hydro tillers. This may lead to formation of shallow plough pans and a reduction of the rooted layers.

3. Crop Establishment

Two types of site–specific crop establishment decision and operations are important: (1) asses whether a (whole) field is suited for a particular crop establishment method, and (2) vary sowing rates or transplanting density according to variation in soil properties and micro relief within a field. Decisions about suitability for a specific crop establishment technology and varying the sowing rate require expert knowledge and basic soil and climatic information. If this information is available, farmers could do field specific or spot-based variation within a field in either sowing/planting by hand or modified mechanical transplanters or seeders that allow easy adjustment on the go. For large farms, drill seeders with VRT features are available and can be modified for use in dry-seeded rice. Although many farmers will do their own judgment, a more quantitative approach is warranted, which would probably be under the responsibility of the existing extension service or, if they exist, crop consultants. Access to specialized equipment will likely be through contractors.

4. Precision and Sowing

Determination of appropriate seed sowing rate and depth at which the sowing should be done will improve the uniformity in the plant population hence reduction in the spatial variation of yields.

5. Precision Weed Control

The following three components are involved in precision weed management:

a) Preventing weed effects through adapted crop management
b) Improvised decision making on weed control
c) Developing precision control technology.

The improvement of the weed management systems, on the basis of the above three component strategy, requires quantitative insight into both crop-weed interactions and the dynamics of weed population in space and time. Some of the recent advances like 'Variable weed control techniques' minimizes the pollution and increases the efficiency of the chemicals and reduces the cost. The other areas concerned with precision farming are irrigation, farm machinery, control systems and spatial database of all crop and field attributes.

6. Precision Water Management

The water management technology is highly located specific, and as such the various decisions for irrigation scheduling vary with soil characteristics, crops to be grown, climatic conditions, quality of water and social and other factors. In arid and semi-arid regions owing to water shortage and multifarious competing demands for the limited supply of available water, there is increasing need to irrigate as efficiently as possible. This situation emphasizes the need for using scientifically sound technology for deciding when and how much to irrigate. Most of the conflicting reports concerning the effects of irrigation on plant growth resulted from irrigation scheduling not always based on sound knowledge of soil-plant-atmospheric water relations. There is an utter lack of literacy about real value of scarce water resource among farmers. There is also a mismatch between water delivery and crops needs. These deficiencies result in over or under irrigation and sub-optimal production efficiency and degradation of natural resources. A major emphasis is placed to describe interactive effects of soil, plant, atmospheric factor and resources like water, fertilizers on plant water status, crop water use, and possible alterations and adjustments in various management strategies to ensure high input efficiency of various resources. Crop water use is primarily influenced by evaporative demand of atmosphere, but plant and soil factors also play their important role, not so much so in altering water use, but to a great extent in improving water use efficiency. In this aspect, the controlled irrigation depth, optimum dose of fertilizers, suitable crops/ their cultivars and crop sequences under limited supply of water, proper agronomic practices for efficient resources management under problematic soils and water conditions are important factors to increase the proportion of water passing through the crop in transpiration, and minimizing evaporative and other losses of water from soil system. Water and fertilizers improve not only internal plant water status but also each other's efficiency. Recently, the crop coefficients and irrigation management through simulation modeling using the concept of water management response indicators have been developed for Indian conditions. However, there is an urgent need to develop optimum crop plans and irrigation management in canal command areas by adopting suitable combinations of water, fertilizers and other inputs to ensure nt only high use efficiency of these resources but also high economic returns on sustainable basis to the farming community from agriculture. A multidisciplinary and farmers' participatory approach for efficient irrigation management is required to achieve desired results from irrigated agriculture.

Site-specific water management is closely linked with the technologies available for soil tillage (e.g., puddling requirements). In particular, precise leveling is the most important factor in efficient irrigation management (Hill *et al.*, 1991), but it would also reduce variability in weed growth, soil proportion and rice growth caused by heterogeneous water flow patterns within fields (Dobermann *et al.*, 1997). Various kinds of reduced irrigation, including reduced water depth, periodical flush irrigation, or sprinkler irrigation are options, but in most of them a single field would be treated homogeneously. Bypass flow in cracking clays may cause huge unproductive water losses during land soaking, i.e., during the initial irrigation flush to achieve water saturation (Tuong *et al.*, 1996). Dry shallow tillage soon after harvesting reduces soil drying and cracking during the fallow period and the water need for the subsequent rice crop. In SSCM, soils and fields where this maybe beneficial need to be identified. The same relates to decisions about need and type of drainage. The primary focus is on capture of suitable field or farm specific information to make a decision about the most appropriate water management technology. At this stage, we do not know who would be the best choice for this.

In Tamil Nadu, the area under rice is about 2 million ha and 70 per cent of which is irrigated. The dominant system of water management in rice cultivation is irrigation up to 5 cm depth one day after disappearance of previously pounded water. However, many farmers are unable to follow this recommended water management practice due to difficulties in water flow control and uncertain water availability.

Typically water input in rice fields varies between 1000 and 3000 mm depending on environmental conditions and the length of the growing period. It is estimated that the water supply demand gap for irrigated crops will be about 21 billion m^2 in 2025. Water shortage in Tamil Nadu has already resulted in reduction of the irrigated rice area and in a shift towards less water demanding cropping activities. To sustain present food sufficiency and to meet future food requirements, India has to realize an annual growth rate of at least 3 per cent in productivity of rice. Considering the future food requirements, new methods of rice cultivation must be identified to increase water and crop productivity. The practices of system of rice intensification (SRI) developed at Madagascar has been reported to increase grain yields of rice and to save water based on the synergistic effects of adopting various cultivation practices simultaneously, i.e., the use of young and single seedling per hill, wider spacing, square planting, cone weeding, limited irrigation and frequent loosening of the top soil to stimulate aerobic soil conditions.

7. Precision Pest Management

Field-specific decisions may be based on both qualities and quantitative information. There are three major agronomic options for site-specific management of weeds, insects, diseases, and other pests like rodents, snails, nematodes etc.,

- Control via combination with other SSCM measures
- Homogeneous, prophylactic control
- Variable rate control based on observation or sensing (spray only in needy places)

Hand-held sprayers dominate for applying pesticides, but there are improvements in their design allowing more accurate and variable adjustment of rates during field operations. The heavy machinery (or even aircraft) used for variable rate application of pesticides in regions such as North America is only of interest for few large farms of Type III.

Ecologically sound site-specific pest management includes measures such as selection of resistant varieties, pest specific bio control agents, reduction in the amount of pesticides used, substitution of less hazardous chemicals for the more hazardous ones, and use of pesticides or non-chemical control measures based on knowledge of pest pressure. Following an integrated pest management (IPM) concept, application of insecticides could be restricted to spots with high infestation only, i.e., the farmer would use a TAD of much smaller size to manage within–field variation with simple means. With the farm yields in tropical Asia expected to rise to 7-8 t/ha and more, we will see an increasing need for disease control using fungicides, cultural practices, or improved host plant resistance (Heong et al., 1995). Presumably, most of these are prophylactic measures applied to whole fields with loss scope for managing variability within a field. Unlike fertilizer management, much of the information required can be captured by observation so that site-specific pest management would mainly be the responsibility of a farmer or farm manager. On these situations, development of location specific insect forewarning system and pest forecast modules assumes significance. Training to improve farmers' knowledge about capturing the information and translating it into application decisions plays a vital role.

Precision leveling allows precise water management as one important measure for field-specific weed control, particularly in direct-seeded rice (Williams et al., 1990). Weeds, particularly annual species tend to occur in patches (Baki, 1993) so that spot that applications or variable rates of herbicides are feasible even when farmers use hand-held sprayers. Identification of suitable post-emergence herbicides will be required for this approach.

8. Precision Post-Production Harvest

SSCM can be interpreted in various ways after harvest. Firstly, yield maps such as those presently being developed automatically by yield monitors (such as in the USA etc.) can be used as guides to refine management both of fields and within a field. They can be used to identify management effects (e.g., variety, fertilizer, rotation, etc.) on yields, and so management can be refined. While yield monitors are not expected to be common, good record keeping of yields on a field or parcel-specific basis could be used to provide such management guidance in small farms.

The second aspect relates to maximizing returns during the post-production system. While farmers can see their yield, much of the subsequent effects of their harvest and losses are unseen. For example, variety, harvest date and post-production management (e.g., drying, storage) can have large effects on head rice yields and quality (e.g., discolouration). These are factors that are essentially hidden from the farmers, who will likely lose interest once the grain leaves their farm gate. However, as premiums for quality (tast and head rice yields) become important, they will drive for improvement in the post-production systems, and add incentives for farmers (if credit systems allow) to maximize the added value of their crops. A major consideration in this is the method of harvest and handling. The major SSCM aspect of this is therefore, variety selection and timely (i.e., optimum moisture content) harvest. Although options for byproduct use (e.g., hulls, straw) are generally known, they will require system (total farm) management. For example, straw removal (e.g., integrated animal system, mushroom production) will require fertilizer substitution and straw enrichment for animals. Farmers will need greater information on variety effects on quality and the effects of harvest time (grain moisture content) and byproduct use options. This will require training to improve farmer's knowledge and incentives for the multiple players in the post-production chain. Mechanized harvest will likely be done by contractors.

9. Ham-Wireless Communications

Use of internet, internet and wireless communication in villages for knowledge revolution among farmers are made through high order transfer of technology (TOT). The web camera addition saves time travel etc., in getting the quick remedial measures from the research station or scientist for distant places.

10. Precision in Nutrient Management

Precision in nutrient management aims at optimization of nutrient resources and increasing the nutrient use efficiency through various means like soil testing, balanced application of primary, secondary and micro-nutrients, proper placement of fertilizers, water management, pest management, maintenance of optimum plant density, conjunctive use of bio-fertilizers and crop residues etc. The most accurate nutrient management is obtained in PF, also known as site-specific management. The PF technology allows the farmers to fine tune the application of fertilizers according to the needs and conditions within the field. This has become possible with the use of computer for management of large amount of data. Using the GPS, soil sample can be taken from within small areas and be accurately identified in a GIS. Fertilizer applicator or drill can use the GPS in conduction with soil test data stored in a GIS to apply needed nutrients to individual areas using variable-rate application equipment. Planting machinery can be equipped to change seeding rates in response to soil parameters in GIS, and irrigation scheduling can be linked to the GIS to apply water in response to soil infiltration rates or water-holding capacity. These systems are expensive and need further development, but they provide valuable tools that allow farmers to manage land, nutrient inputs and crop productivity more efficiently.

v. Nutrient Management in Asia

In irrigated rice systems of south and southeast Asia, farms and individual fields are small. Only small equipment can be used. In the early stages of the Green Revolution, response of N, and later also P, was nearly universal. Blanket approaches have dominated rice nutrient management in the past. Fertilizer N recovery efficiencies average only 30 to 40 % unbalanced plant nutrition occurs, and yield growth has stagnated since the mid-1980s in some of the most important regions (Dobermann and Cassman 2002) Revitalizing yield growth, increasing profitability, and sustaining soil fertility are key targets for farmers in these environments. On-farm studies demonstrated that major uncertainties were caused by variability occurring among small fields in the supply of nutrients from indigenous sources and, consequently, the response to fertilizers (Angus *et al.*, 1990; Cassman et al., 1996; Dobermann *et al.*, 2003). Soil tests are of less use for predicting this variability because they fail to account for the dynamics of nutrient supply under submerged conditions and do not measure nutrient inputs from other indigenous sources such as irrigation or N_2 fixation (Dobermann *et al.*, 2003). Site-specific management in these environments has focused on managing nutrients at the scale of a single, small field, including in-season N management decisions. One line of research has focused on corrective, in-season N management using tools such as a chlorophyll meter (Peg et al., 1996). Typically, this involves no or a small application of N before planting and assessing leaf N status ("greenness") at interval s of 7 to 10 days beginning from mid-tillering stage of rice. Each time, a decision on whether to apply N is made, which requires local calibration to empirically define thresholds and suitable N amounts. A simple leaf colour chart (LCC) can replace the chlorophyll meter in this approach (Balasubramaniam *et al.*, 1999, Furuya, 1987; Singh *et al.*, 2002, Yang et al., 2003). Evaluation of these methods has generally shown that the same rice yield can be achieved with about 20 to 30 per cent less N application, whereas increase in yield is less common or relatively small. The gains in NUE are usually large.

Some of these uncertainties were addressed in a broader concept for site-specific nutrient management (SSNM), which integrated information collected at different scales fir making field-specific decisions on N, P and Kmanagement (Dobermann *et al.*, 2004). Key components are measurement of grain yield in nutrient omission plots to obtain field-specific estimates of the indigenous supply of N, P and K (Dobermann et al., 2003), a decision support system for predicting total NPK requirements and N splitting patterns before planting, and in-season upward adjustments of pre-determined N doses at critical growth stages, based on chlorophyll meter or LCC readings. Variations in crop nutrient managements due to regional differences in yield potential, location-specific crop management practices, and the overall nutrient input-output balances are taken into account in addition to spatial variation in indigenous nutrient supply. From 1997 to 2000, this strategy was evaluated in permanent on-farm experiments at 179 sites in eight irrigated rice domains of Asia (Dobermann et al., 2002). On average grain yield increased by 11% and the N fertilizer rate decreased by 4% as compared to the baseline farmers' fertilizer practice. Average profit increased by US$ 46/ha per crop cycle (12%) and the recovery efficiency of fertilizer N increased from 30% to 40%. Benefits of SSNM varied widely among sites and were strongly affected by the overall crop management quality. Average yield increases ranged from 0.1 to 0.6 t/ha among the eight rice domains studied. SSNM was the most profitable at cites in China, southern

India, and the Philippines, with average profit increases ranging from US$ 57-82/ha per crop in those domains. The performance of SSNM improved over time due to a gradual learning process and re-capitalization of investments in P and K applications made in the first year. Extra costs and other socio-economic issues may complicate the adoption potential of SSNM, but the gains in profitability appear to be attractive enough for farmers. More recent data show remarkable consistency of increases in yield, profit and NUE.

1. Site Specific Nutrient Management (SSNM) for Irrigated Rice

I. SSNM

Site Specific Nutrient Management (SSNM) is an approach for feeding rice with nutrients as and when needed. The SSNM approach ensures application of fertilizers for location and season-specific conditions. With the SSNM approach, fertilizers are applied.

1. at right time
2. in optimum amount
3. in the optimum ratio and
4. with the appropriate method
5. to meet the deficit between rice demand for nutrients and the supply of nutrients from soil and organic inputs.

II. Seed

The SSNM approach is needed for proper management of fertilizers in rice because,

- Soil fertility varies from field to field, the demand for nutrient inputs therefore, varies among fields.
- Demand for nutrient inputs varies from season to season, year to year and at different crop growth stages within a season.

Such variations in soil fertility and demand for nutrients are not effectively handled with blanket fertilizer recommendations, consisting of fixed rates and timing of fertilizers for large areas. The SSNM approach provides farmers with guidelines for dynamically adjusting fertilizer application to local conditions.

III. Benefits

- Increase profit through optimal fertilizer use.
- Sustain soil fertility through balanced application of nutrients (fertilizers).
- Reduce pest and disease incidence through balanced nutrient use.
- Protect soil, water and environment from contaminations by avoiding excess fertilizer use.

IV. Components

The following tools and techniques are promoted for proper management of fertilizers.

a) Leaf colour chart (LCC) to determine the correct time and amount of N fertilizer.
b) Omission plot technique to calibrate the P and K fertilizer rate.

A. LCC

Using LCC, farmers can determine the right time of application of N fertilizer by matching for leaf colour with strip of LCC. Farmers can decide on N fertilizer application, when leaf colour measurement falls below the critical value.

Guidelines

1. Start leaf colour measurement by LCC is transplanted rice at about 14 days after transplanting (DAT) and 21 days after seeding (DAS) in wet (direct) seeded rice.
2. Measure the colour of the top most youngest fully expanded leaf in ten randomly selected healthy hills as follows:

- Select the top most fully expanded leaf in the selected hill.
- Place the middle of the selected leaf on the LCC colour strip.
- Do not detach the leaf from the plant.
- Place the leaf under the shade of your body to prevent falling of direct sunlight on LCC for accurate reading.
- Match the leaf colour with the LCC colour strip and record the number, matching the colour strip. The recorded number is the LCC reading for the leaf measured.
- If the colour match falls between two-colour strips, take the mean of two colour strip numbers as LCC reading. For example, if the leaf colour matches with LCC shade numbers 2 and 3, then the LCC reading will be taken as 2.5.
- Measure the leaf colour from each of the 10 selected hills in the same way and record the LCC readings.
- Repeat leaf colour measurement at 7-day intervals up to the flowering stage in respect of short duration varieties and 10 days interval up to flowering in respect of medium and long duration varieties. The last LCC measurement is at about flower initiation.
- Whenever more than 5 out of 10 LCC readings fall below the threshold LCC value (Threshold LCC value to 4 for varieties other than Improved Ponni and hybrids, and 3 for Ponni and direct seeded rice), apply 35 kg N/ha in kuruvai and 30 kg N/ha in Thaladi season.

B. Establishing and Managing Omission Plots

The objective of P and K omission plots is to develop on site-specific P and K fertilizer recommendation. The difference in grain yield between a fully fertilized plot and a P or K omission plot illustrates the deficit between the crop demand for P or K and indigenous supply of P and K, which must be met by fertilizers. The layout for omission plot is given below:

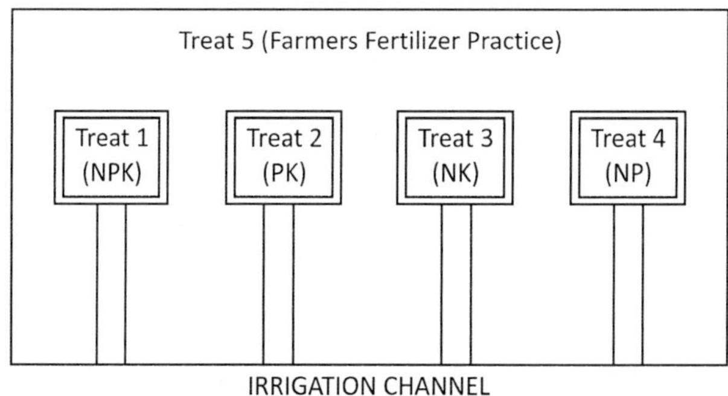

Plot size for each of treatments 1, 2, 3 and 4 = 5 m x 5 m (25 m^2)
Plot size for treatment 5 (remaining area of the field)
Field lay out of Omission plot trial

- Establish omission plots both in wet season as well as dry season.

- Measure four plots of 5m x 5m size each within a farmer's field. Do not place the plots in a corner of the field. Separate each plot by about 3 m distance from the next plot to minimize risk of fertilizer movement between plots.
- Construct bunds of sufficient height surrounding each plot to prevent water flow between the plots and the adjacent field. Maintain the bunds nicely throughout the crop season to avoid fertilizers and water contamination from one plot to another or from outside the plot. Provide guard channel all around the plot. It is essential to prevent movement inward and outward of fertilizer with flood water and irrigation water.
- All the crop management practices including pest management should be under optional conditions and uniform in the treatment plots.
- Apply fertilizers to the plots strictly as per the treatment with the involvement of the farmers.

Treatment Details

T_1: NPK: Recommended N, P. K, application without limitation.
T_2: PK: Require P and K without limitation (ON) – N omission plot.
T_3: NK: Required N and K without limitation (OP)- P omission plot.
T_4: NP: Require N and P without limitation (OK)- K omission plot.
T_5: FFP: Farmers Fertilizer Practice in the remaining area.

Determination of Grain Yield

At crop maturity, harvest the sample from all the five treatment plots in the following way:

- Harvest all the hills of a central 5 m^2 area of each treatment plot separately and properly label each harvested sample.
- Thresh and clean the harvested grain. Record the grain yield at 14% moisture level.
- Convert the sample weight to yield in t/ha.
- Determine the P and K response.

P response = Yield in NPK plot – Yield in P omission plot.
K response = Yield in NPK plot – Yield in K omission plot

- From the P and K response, P and K requirement of the field will be calculated by the researchers based on the ready reckonar table as suggested by Fairthurst and Witt (2002) (Table 12.1 and 12.2).

TABLE 12.1: Recommended P_2O_5 rates according to target yield and P-limited yield.

Target yield (t/ha)	4	5	6	7	8
Yield in O-P plots (t/ha)		Fertilizer P_2O_5 rate (kg/ha)			
3	20	40	60		
4	15	25	40	60	
5	0	20	30	40	60
6.	0	0	25	35	45
7	0	0	0	30	40
8.	0	0	0	30	35

TABLE 12.2: Recommended K_2O Rates at medium straw input*

Target yield (T/ha)	4	5	6	7	8
Yield in 0-K plots (T/ha)		Fertilizer K_2 rate (kg/ha)			
3	30	60	90		
4	0	35	65	95	
5	0	20	50	80	110
6	0	0	35	65	95
7	0	0	0	50	80
8	0	0	0	0	65

* 2-3 t/ha straw input.
Source: Fairthrust and Witt, 2002. Rice: A practical guide to nutrient management.

vi. Advantages of Ssnm

The SSNM strategy aims to achieve sustainable, large and economic yield through proper nutrient and crop management by:

- Following plant need based N management strategies using the Leaf Colour Chart (LSS) and avoiding over dose of N coupled with pest and disease incidence.
- Using nutrient omission plots to determine soil indigenous nutrient supply (Particularly for P and K).
- Providing the crop with a balanced supply of nutrients (N, P, K and micro-nutrients).
- Replacing nutrients Particularly P and K) removed with grain and straw to avoid depleting and nutrient reserves.

vii. Economics

Adoption of SSNM always recorded higher grain yield viz., averaging 500 to 650 kg/ha in Cauvery Delta of Tamil Nadu than in Farmers Fertilizer Practice (FFP). Therefore, the farmers can get Rs. 2250 to Rs. 3000/- per hectare as profit, besides a reduction in N fertilizer dose and avoiding excess cost on pesticide spray in SSNM plots.

Thus, SSCM has to include land preparation, crop establishment, water management, pest management, and nutrient management plus all their interactions as the basic components. Such SSCM will help to increase production and profitability while protecting the environment. The new system will, however, come with new demands in terms of management and extension expertise. In countries where SSCM has evolved, the technology is advancing fast, but the theoretical work has lagged behind. Development of sound procedures for information collection and accurate decision-making is insufficient and the long-term biophysical and economic benefits of SSCM remain to be demonstrated. Further many of the input response functions still have to be verified under a range of conditions. Nevertheless, we believe that the concept of SSCM is right and that many of the initial difficulties can be overcome as the technologies developed. In China, SSCM may provide the basis of sustaining high rice yields in the coastal regions as well as achieving substantial yield increase in the central and western parts of the country where yields are currently lower. We believe SSCM can lead to substantial improvements in input use efficiency (i.e., product/unit input). One important opportunity will be to design integrated farms (e.g., rice-animal) where rice byproduct use (e.g., straw) is examined in terms of the optimal flows of nutrients, water, energy, and labour at the whole farm level. SSCM aims at integrating the knowledge generated by various scientific disciplines into more complex but site-specific guidelines for action at the farm level. The principles of **"Precision Farming" or** SSCM are applicable to farms and fields of any size, including those found in Asia, but the specific technological solutions differ from case to case.

"What we observe is not nature itself, but nature exposed to our method of questioning."

13 Dry Land Farming

"Rain is essential for cultivation and the latter is essential for life, so one should first acquire carefully the knowledge about rainfall".
Atharveda

"An increase in the supply of milk and butter depends upon the nature of the soil and the quality of fodder and water".
Arthashashtra

"A large part of the country depends on rains for agricultural production. It has, therefore, been felt that dry farming practice should be undertaken in such areas".
Dr. M. S. Swaminathan

India has about 108 million hectares as rain fed area out of a total of 143 million hectares of cultivated land which constitutes about 75 per cent of the agricultural land. In such areas crop production becomes relatively difficult as it mainly depends upon intensity and frequency of rainfall. The crop production, therefore, in such areas is called as rain fed farming as there is no facility to give any irrigation and even protective or life saving irrigation is not possible. These areas get an annual rainfall between 400 mm to 1000 mm which is unevenly distributed, highly uncertain and erratic. In certain areas the total annual rainfall does not exceed 500 mm. The crop production, depending upon this rain, is technically called as dry land farming and areas are known as dry lands. India contains about 47 million hectares as dry lands out of 108 million hectares of total rain fed area which amounts to nearly 43 per cent. Besides being water deficient, such areas are characterised by high evaporation rates, exceptionally high day temperature during summer, low humidity, high run-off and soil erosion. The soil of such areas often turns to be saline and poor in fertility. As water is the most important single factor of crop production, inadequacy and uncertainty of rainfall often cause partial or complete failure of the crops which leads to periodic scarcities and famines. Thus the life of both human being and cattle in such areas becomes difficult and unsecured.

While defining the dry farming it may be expressed as: " Dry farming or dry land farming is a practice of growing profitable crops without irrigation in areas which receive an annual rainfall of 500 mm or even less." Presently the dry farming areas contribute about 42 per cent of the total food grains, especially all coarse grains, and 75 per cent of oil seeds and pulses. Thus about more than 90 per cent of sorghum, pearl millet, peanut and pulses are grown in semi-arid and arid areas of the country. Besides these, about 60 to 70 per cent of the rice, rapeseed and mustard and 30 to 40 per cent of the wheat are grown in rain fed areas. Although a low intensity cropping, ranging between 100-200 per cent or lower, is hardly being adopted, efforts are being made to bring more area under irrigated agriculture and thereby to increase cropping intensity, but even when we achieve our target of 133 million hectares by the year 2000 A. D. we would still have about 45 per cent area under rain fed culture. We continue to stress on intensive agriculture on irrigated lands but we cannot afford to be complacent without dry lands and agriculture therein as its contribution towards total food production is very significant. Thus intensive irrigated farming is an imperative for survival but improved dry farming is necessary for equity and prosperity. As such we cannot achieve stability in food production with unstabilized dry land agriculture. Therefore, we are required to adopt improved agricultural technology especially developed for dry land agriculture.

DRY LAND AGRICULTURE

Dry land Agriculture means cultivation of crops entirely under rainfed condition. Dry land Agriculture may be classified into three groups on the basis of annual rainfall.

Dry Farming

Cultivation of crops in areas where annual rainfall is less than 750 mm and crop failures due to prolonged dry spells during crop period are most common. Dry farming is practiced in arid regions with the help of moisture conservation practices. Alternate land use system is suggested in this region.

Dry Land Farming

Cultivation of crops in areas where annual rainfall is more than 750 mm but less than 1150 mm is called Dry land Farming. Dry spells may occur, but crop failures are less frequent. Higher Evapotranspiration (ET) than the total precipitation is the main reason for moisture deficit in these areas. The soil and moisture conservation measures is the key for dry land farming practice in semi-arid regions. Drainage facility may be required especially in black soils.

Rainfed Farming

Cultivation of crops in regions where annual rainfall is more than 1150 mm. There is less chances of crop failures due to dry spells. There is adequate rainfall and drainage becomes the important problem in rain fed farming. This farming is practiced in humid regions.

United Nations Economic and Social Commission for Asia and the Pacific differentiated between dry land farming and rainfed farming with the following constituents in the given Table 13.1.

TABLE 13.1: Dry land vs Rainfed Farming

S. No.	Constituent	Dry land Farming	Rainfed Farming
1.	Rainfall (mm)	< 800 (less than)	> 800
2.	Moisture availability	Shortage	Enough
3.	Growing season	< 200 days	> 200 days
4.	Growing regions	Arid and semiarid as well as uplands of sub-humid and humid regions	Humid & Sub humid region
5.	Cropping system	Single Crop or Intercropping	Intercropping or double cropping

S. No.	Constituent	Dry land Farming	Rainfed Farming
6.	Constraints	Wind & Water erosion	Water erosion

CHARACTERISTICS OF DRY LAND AGRICULTURE

Dry land areas may be identified by looking at the following features:

1. **Uncertain and limited annual rainfall.** Dry farming areas are characterised by low annual rainfall and according to the definition the dry tracts are those which receive an annual rainfall of 500 mm or less. Even this much rain is not well distributed and due to that the crop production becomes very poor. The crops very often suffer from aberrant weather like delayed onset of monsoon, long gap in rainfall and early stoppage of rains.
2. **Extensive climatic hazards.** As per distribution of rains the dry farming areas used to get uneven distribution of rains i. e. sometimes intensity of rains is too much and rest of the year there is no rain at all. This leads to an accelerated run-off water and drought respectively. Thus the area has to face flood, water-logging or drought. Since there is no facility for irrigation, therefore, the crops suffer badly from the frost also during winter. This way there are extensive climatic hazards leading to the partial or total crop failure.
3. **Undulating soil surface.** Because of improper distribution of rains there is soil erosion caused by both water and wind. During rainy season water erosion and during summer season wind erosion are noticed as in the former case there is too much of water but in the latter case the surface is completely dry. Therefore, every year soil is eroded which leaves behind an eroded gulley or rill and it requires levelling but due to poor economy of the farmers the soil is left as it is and the crops are grown there in. Thus the surface gradually becomes highly undulating.
4. **Extensive holding.** Dry farming areas use to face uncertainty of rains which are erratic and unevenly distributed resulting in crop failure either partially or completely. Under such conditions the cattle and human lives become very difficult, thus the population density is found to be very poor and villages are located at distant places. Therefore, the farmers of dry farming tracts use to have a large holding.
5. **Extensive agriculture.** The type of agriculture adopted in dry land areas always remains extensive due to lack of irrigation facilities. In such areas most of the farmers use to grow hardly one or two crops annually while in some cases the farmers use to keep fallow during rainy season with an object to conserve more moisture and to control weeds; then they grow one single rabi season crop or some farmers use to grow pegionpea based mixed crop that takes whole year. Thus the cropping intensity remains only one hundred per cent.
6. **Relatively larger plot size.** Since there is no irrigation facility, therefore, smaller plots are not required and farmers use to keep much bigger plot sizes which make it easy to identify as a rain fed area. However, bigger plots are also made in case of fully mechanized farms but the area remains under intensive cropping and the plots are always engaged under some or the other crop.
7. **Similar types of crops.** In case of dry land agriculture only drought resistant crops can be grown and almost all the farmers of the region use to grow similar crops like arhar, castor, millets, oilseeds, etc. Thus the entire area can be seen under the same crop having poor growth and development and it can be identified as a rain fed area.

8. **Lower crop yields.** Yields of various crops grown in dry lands are found to be very low as the farmers being very poor can-neither apply full doses of inputs nor the improved seeds of high yielding varieties. Sometimes the yield becomes so low that it does not meet out the cost of its cultivation but the farmers have to do the farming because of their livelihood and they do not take the family labour into account while calculating the cost of cultivation.
9. **Poor market of the produce.** Since everyone uses to grow the same crops and they become ready for sale at the same time; hence the farmers have to face a hard time and they get less price of their produce because of poor demand and abundant supply rather a glut in the market. Very often the people of the locality do not want to buy the commodity and the farmers fall in the grip of small grain traders who use to purchase at cheaper price and finally sell to herders. Thus the farmers do make any profit.
10. **Poor farmers' economy and cattle health.** The farmers do not get proper return of their efforts hence their economy remains poor. Due to scarcity of fodder and feed the cattle are not stall – fed, they are loose and depend on grazing. The poor economy of the farmers affects their education, brought up of children and productivity of the cattle.
11. **Widespread deficiency diseases in human beings.** Surplus of low calorie food and short supply of vegetables, fruits, milk, etc. cause deficiency of vitamins and minerals which result into malnutrition and deficiency diseases in human being. The most important example is: consumption of Lathyrus pulse in place of Arhar pulse in rain fed areas cause lathirism in people. Lathyrus grows very vigorously and farmers get plenty of it to feed their cattle and people, though its production is banned by the Government. Likewise many other diseases are found common in such areas.

PROBLEMS AND PROSPECTS OF DRY LAND AGRICULTURE

In our country, irrigated area constitutes 33% and 67% is dry land and rain fed, out of 143.8 million hectares of cultivated land. Total cropped area is 180.36 mha. Such dry land agriculture contributes about 45% of national food grain production. The maximum irrigated area will be achieved 50% after the full utilization of all sources of irrigation. In this way 50% cultivated area will remain unirrigated. After full exploitation of dry land, it may contribute up to 75% of total food grain production. Pulses and oilseeds are mainly grown in such areas. Important commercial crops viz Cotton, Castor, Groundnut and all coarse grains viz Jowar, Bajra, Maize crops are rain fed. The major part of milk, meat, wool hides, bone meal etc are also from this area.

Dry farming areas are characterised by very low and highly variable and uncertain yields. Problems of dry land agriculture are:

Inadequacy and uncertainty of rainfall and its erratic distribution. Expected rainfall in a year is 650 mm but 80% of it is received in 75-90 days of the monsoon season.

a) Late onset and early cessation of rains.
b) Prolonged dry spells during the crop period.
c) Low moisture retention capacity.
d) Poor soil fertility condition.
e) Socio – economic constraints particularly because of the pre–dominance of small and marginal farmers. 54% of the holdings is less than one hectare.
f) Technological and developmental constraints.

g) Limited infrastructure development and unproper and untimely availability of credits and agricultural inputs.

There are three types of agriculture possible in Dry land areas viz.

a) Crop Production
b) Animal husbandry with pastures management and
c) Agro forestry.

PROBLEMS OF DRY FARMING

The vital problem which the farmers have to face very often is to keep the crop plants alive and to get some economic returns from the crop production but the aforesaid problem is governed by a number of associated problems that are described below:

1. **Moisture stress due to scarce and uncertain rainfall.** As per definition the dry farming areas receive an annual rainfall of 500 mm or even less. The rains are very erratic, uncertain and unevenly distributed. Therefore, the agriculture becomes a sort of gamble with the nature and very often the crops have to face quite a number of climatic hazards because of which they fail all–together. The farmers also take up the farming half heartedly and used to get nominal yield. Thus water scarcity becomes a serious bottleneck in dry land agriculture.
2. **Effective storage of rain water:** According to the characteristics of dry farming either there will be rain at all or there will be torrential rain with very high intensity. Thus in the former case the crops will have to face severe drought and in the latter case they will face either flood or water logging and they get spoiled. In case of very heavy downpour the excess water gets lost as run-off which goes to the ponds, ditches, etc. this water could be stored for providing life saving or protective irrigation to the crops grown in dry land areas. The loss of water takes place in several ways viz. run – off, evaporation, uptake though weeds, etc. The water could be stored for short period or long period and it can be preserved either in soil, pond or ditches based on situation. Following types of storages are required in dry tracts:

   a) Year round storage of water in soil: From a water reservoir it is lost through percolation, seepage, evaporation / transpiration or evapo – transpiration. For a profitable cropping the moisture in soil should be regulated in such a way that the plants may be able to absorb it as and when they need. Therefore the water storage should be done to meet year round need of the crop. Generally dry farming areas face a serious problem of water loss because of undulating soil surface, barren lands, excessive growth of weeds in the field.

   b) Retention of moisture till need for the crops: To meet the water requirement of the crop plants it is essential to adopt such management practices which may help in retention of moisture in the soil at least till the crop needs it. The technique of moisture retention requires special skill, care and too much expense which majority of the farmers cannot fulfil. Therefore, this problem remains untackled and becomes the main reason for crop failure.

   c) Prevention of direct evaporational loss of moisture: Evaporation is the major type of water loss from the soil or water surface. Mostly the agriculturist does not bother for this loss and after sowing the crop he does not visit the plot or the water collected in some ponds or ditches is left uncared and slowly it gets lost through evaporation. Prevention of this loss needs lot of expenses which the farmer can hardly meet and crops fail due to lack of moisture. Sometimes farmers do not do it because of their ignorance and carelessness, otherwise the collected water, if well preserved, may be used for fish raising, chest nut production, irrigation purposes, etc. and they can get best return from it.

d) ***Regulation of water withdrawn by crop from soil:*** The soil moisture is transpired by the crop plants at a very high rate from the soil which hardly gives any positive result. This rate of transpiration may be reduced to half or even low in certain cases and stored water may be absorbed by the plants till they need it, otherwise in surplus months they get too much moisture but in deficit months they do not get at all. This type of supply helps in a profuse growth and succulence in early growth period and wilting during reproductive growth which ultimately result into a poor yield and quality of the produce. Thus it becomes essential to adopt some practice which may help in a proper regulation of water throughout the crop growth period.

3. **Disposal of dry farming products:** In dry farming all the farmers grow similar type of crops which mature at the same time and the growers like to dispose their products soon after the harvest because due to poverty they cannot hoard or store for a longer period in anticipation for an appreciation in the market price. This results into a glut in the market and the grain traders or the middle man use to have upper hand in the deal. Thus the product is sold with a great difficulty and at poor price. Therefore, marketing becomes a serious problem in dry farming tracts.
4. **Selection of limited crops:** Only drought resistant crops viz. oil seeds, pulses and coarse grains like millets can be sown. Thus the farmers have to be satisfied with above mentioned crops and they have to purchase rest of the commodities to fulfil their household demands.
5. **Careful and judicious manurial scheduling:** In case of irrigated farming the farmers are at liberty to apply manures and fertilizers according to their availability and facility but in case of dry farming they have to be very careful in fertilizer application because due to lack of moisture broadcasting either for basal or for top – dressing becomes a wasteful and meaningless. There can be deep placement and foliar spraying of the nutrients for an improved crop production.
6. **Utilization of preserved moisture:** Purposeful utilization of preserved water depends upon soil type, plant type and other factors. The amount of available water to the plants depends upon depth of their roots, their proliferation and density. In case of limited moisture condition the yield of the crop is directly related to their rooting depth. The rooting depth can be substantially increased by mechanical manipulation of the soil. If the planting is very dense and all the plants have same type of rooting then there will be a tough competition of roots for moisture and scarce moisture condition will result in wilting of plants. Therefore, utilization of preserved moisture is an art with which best use of moisture can be made. The water collected in ponds or brooks may be used to give protective or life saving irrigation to the crops or for raising fish or for growing chestnut along with fish culture. The widely spaced crops can be intercropped with oilseeds or pulses for increasing the productivity of land per unit area per unit time. Thus the water collected during the rainy season and stored either in soil or in some ponds need special technique and skill for its purposeful utilization.
7. **Quality of the produce:** Invariably the quality of the produce from dry farming areas is found to be inferior as the grains are not fully developed or they are not filled properly, often mixed with other crop seed due to mixed cropping system prevalent in the area and the fodder becomes more fibrous. These all reduce the market price and the farmers do not get profit of their labour and investment.

History of Dryland Agriculture

First systematic scientific approach to tackle the problems of dry farming areas was initiated by Tamhane in 1923 on a small plot at Manjri farm near Pune and the work passed on to Kanitkar in 1926. A comprehensive scheme of research was drawn up by Kanitkar with financial support from the ICAR. Realising the importance, the ICAR launched a comprehensive project on dry land farming at five centres: Sholapur and Bijapur in 1933, Hagari and Raichur in 1934 and Rohtak in 1935. A decade of work upto 1933 – 44 mainly on rainfall analysis, physico–chemical properties of soils, physiological studies on millets and on agronomic aspects resulted in a series of dry farming practices commonly known as the Bombay dry farming practices, Hyderabad dry farming practices and Madras dry farming practices. These practices stressed the need for contour bunding, deep ploughing, and application of FYM, low seed rate with wide spacing, mixed cropping and crop rotation. These recommendations could not motivate the farmers to adopt them as the yield advantage was about 15 -20 per cent over a base yield of 200 – 400 kg ha $^{-1}$.

By the mid 1950s, importance of soil management (soil and moisture conservation) was realised for improving the productivity of dry lands and the ICAR established eight Soil Conservation Research Centres in 1954. However, yield improvement was not more than 15 – 20 per cent over the base yield of 200 – 400 kg ha $^{-1}$. Importance of short duration cultivars maturing within adequate soil moisture available period (crop growing period) was recognised during 1960s. The place of high yielding varieties and hybrids for yield advantage in dry land agriculture was realised in mid 1960s. With the establishment of All India Coordinated Research Project for Dry- land Agriculture (AICRPDA) in 1970, emphasis was shifted to multidisciplinary approach to tackle the problem from several angles. Similar efforts were initiated at ICRISAT, Hyderabad in 1972. Major events in dry land agricultural research are:

- 1920: Scarcity tract development given importance by Royal Commission Agriculture
- 1923: Establishment of Dry Farming Research Station at Manjri
- 1933: Research Stations established at Bijapur and Sholapur
- 1934: Research Stations established at Hagari and Raichur
- 1935: Research Station established at Rohtak
- 1942: Bombay Land Development act passed
- 1944: Monograph on dry farming in India by Kanitkar
- 1953: Establishment of Central Soil Conservation Board
- 1955: Dry Farming Demonstration Centres started
- 1970: Twenty three Research Centres established under AICRPDA
- 1972: Establishment of ICRISAT
- 1976: Establishment of dry land Operational Research Projects (ORPs)
- 1977: Krishi Vigyan Kendra (KVK), Hyatnagar
- 1983: Starting 47 model watersheds under ICAR
- 1984: Establishment of Dry land Devlopment Board in Karnataka and Word Bank assisted Watershed Development Programmes in four states
- 1985: Central Research Institute for Dry land Agriculture (CRIDA), Hyderabad
- 1986: The NWDPRA programmes in 15 states by Government of India.

The ICAR collaborated with Canadian dry land research (Indo-Canadian Dry land Research) form 1970 to 1987. The ICAR – ICRISAT collaborative research programmes are working in areas of biological nitrogen fixation, agro forestry, soil and moisture conservation, economics of watershed technology and vegetative barriers in soil and water conservation. The Indo – US collaboration started in 1980. Currently, CRIDA – US collaborative research projects are working on fertiliser use efficiency, germplasm enhancement, evaluation of legumes for high nitrogen fixation, crop simulation modelling and implements for dry land agriculture.

Crop improvement brought out very promising strains during sixties which received our interest in finding out ways and means through which the crop production can be maximised in the semiarid and periodically dry areas of the country. Keeping in view the ICAR launched the All India Co-ordinated Research Project for Dry land Agriculture in June 1970 which was in active collaboration with Government of Canada (Research Branch, Canada Agriculture). The project started with multidisciplinary research units at 23 co-operating research centre located in various typical agro-climatic region of India viz. Hyderabad (headquarter), Anantapur, Bellary, Bangalore, Kovilpatti, Bhubaneswer, Ranchi, Varanasi, Rewa, Jhansi, Dehra Dun, Ludhiana (formerly at Samba, Jammu & Kashmir), Hissar, Agra, Jodhpur, Udaipur, Rajkot, Anand, Indore, Akola, Sholapur and Bijapur. The project started with identification of the constraints responsible for lower yields in different regions and then to develop a relevant location specific research programme to solve production constraints. Presently, a joint team of senior scientists of different agricultural Disciplines of India and Canada are working together at Hyderabad. To make the programmes more effective it is expanded to 16 Agro-Economic Research Centres, 5 centres for Pasture Research and Development Studies, 4 centres for Operational Research Projects, 1 centre for K.V.K. for Dry land Agriculture and 1 centre for Trainers Training.

These centres have been established with an obvious objective leading to the conservation, development and efficient long-term use of basic resources of soil and water for a self-sustaining system yielding enhanced and stable production. The main areas of investigations centre around the following:

1. Identification of different crops and selection of high yielding varieties for different agro-climatic zones of the country.
2. Developing cropping sequences and cropping systems viz. mixed, inter-crops, etc. for increasing intensity of stabilized crop yields.
3. Determining the optimal crop geometry including population and planting pattern.
4. Evaluating tillage implements and practices for improving water intake and storage in the soil profile, establishment of better crop stand and control of weeds.
5. Designing and developing animal drawn implements for speedy and efficient cultural and seeding operations.
6. Evaluating use of surface mulches, both organic and inorganic, for short-term moisture conservation.
7. Determining ideal fertilizer doses and improving fertilizer use efficiency.
8. Testing new planting materials for introduction/substitution after they have proved their superiority over existing ones.
9. Harvesting and storing inevitable run-off and recycling it as life saving or protective irrigation, and
10. Developing strategies for meeting the challenges of aberrant weather like skip or catch cropping etc.

DROUGHT PERIOD

Droughts do not occur in Assam, South Kerala and eastern part of West Bengal. Severe drought begins on 1 October in the northwest arid zone and even much earlier in the western part. In the southern arid zone and adjoining interior portion of Maharashtra State, the severe drought begins by the end of November. However, in most of the central portion of the country to the east of the line joining Delhi, Udaipur and Baroda, the commencement is only in the month of February or later. This is due to high water holding capacity of the black soil region. In the western coastal region of Maharashtra and Karnataka states, the rainfall is very high. In spite of this, severe drought begins by December-January, probably because of the lower water holding capacity of the soil. Severe drought commences only after April in Gwalior, Guna, Jabalpur, Pendra, and Satna region of Madhya Pradesh.

On the average, severe drought ends outside the regions of east Bihar, Tamil Nadu, Karnataka, and southern Andhra Pradesh only by 1 May. In most of these regions it ends mainly after 15 May. In the arid zone of northwest India, severe drought ends normally during the second fortnight of June, except in the Jaisalmer and Bikaner regions where normal cessation of severe drought is only by the first week of July.

DROUGHT

Low rainfall or failure of monsoon rains is a recurring feature in India. This has been responsible for droughts and famines. The word drought generally denotes scarcity of water in a region. Though, aridity and drought are due to insufficient water, aridity is a permanent climatic feature and is the culmination of a number of long term processes. However, drought is a temporary condition that occurs for a short period due to deficient precipitation for vegetation, river flow, water supply and human consumption. Drought is due to anomaly in atmospheric circulation.

DEFINITION OF DROUGHT

There is no universally accepted definition of **drought.** Early workers defined drought as prolonged period without rainfall. According to Ramdas (1960) drought is a situation when the actual seasonal rainfall is deficient by more than twice the mean deviation. American Meteorological Society defined drought as a period of abnormally dry weather sufficiently prolonged for lack of water to cause a severe hydrological imbalance in the area affected.

Prolonged deficiency of soil moisture adversely affects crop growth indicating incidence of agriculture drought. It is the result of imbalance between soil moisture and evapo-transpiration needs of an area over a fairly long period as to cause damage to standing crops and to reduce the yields.

CLASSIFICATION OF DROUGHT

Drought can be classified based on duration and nature of users. In both the classifications, demarcation between the two is not well defined and many a time overlapping of the cause and effect of one on the other is seen. Droughts are classified into four kinds.

Permanent Drought

This is characteristic of the desert climate where sparse vegetation growing is adapted to drought and agriculture is possible only by irrigation during entire crop season.

Seasonal Drought

This is found in climates with well defined rainy and dry seasons. Most of the arid and semiarid zones fall in this category. Duration of the crop varieties and planting dates should be such be such that the growing season should fall within rainy season.

Contingent Drought

This involves an abnormal failure of rainfall. It may accur almost any where especially in most parts of humid or sub humid climates. It is usually brief, irregular and generally affects only a small area.

Invisible Drought

This can occur even when there is frequent rain in an area. When rainfall is inadequate to meet the evapotranspiration losses, the result is borderline water deficiency in soil resulting in less than optimum yield. This occurs usually in humid regions.

Droughts are also classified based on their relevance to the users.

Meteorological Drought

It is defined as a condition, where the annual precipitation is less than the normal over an area for prolonged period (month, season or year).

Atmospheric Drought

It is due to low air humidity, frequently accompanied by hot dry winds. It may occur even under conditions of adequate available soil moisture. Plants growing under favourable soil moisture regime are usually susceptible to atmospheric drought.

Hydrological Drought

Meteorological drought, when prolonged results in hydrological drought with depletion of surface water and consequent drying of reservoirs, tanks etc. This is based on water balance and how it affects irrigation as a whole for bringing crops to maturity.

Agricultural Drought

It is the result of soil moisture stress due to imbalance between available soil moisture and evapo-transpiration of a crop. It is usually gradual and progressive. Plants can therefore, adjust at least partly, to the increased soil moisture stress. This situation arises as a consequence of scant precipitation or its uneven distribution both in space and time. It is also usually referred as **soil drought.**

Relevant definition of agricultural drought appears to be a period of dryness during the crop season, sufficiently prolonged to adversely effect the yield. The extent of yield loss depends on the crop growth stage and the degree of stress. It does not begin when the rain ceases, but actually commences only when the plant roots are not able to obtain the soil moisture rapidly enough to replace evapotranspiration losses. Important causes for agricultural drought are:

- Inadequate precipitation,
- Erratic distribution,
- Long dry spells in the monsoon,
- Late onset of monsoon, and
- Early withdrawal of monsoon.

PERIODICITY OF DROUGHT

The India Meteorological Department examined the incidence of drought for the period from 1871 to 1967, utilising the monthly rainfall of 306 stations in the country. It was seen that during 1877, 1899, 1918 and 1972 more than 40 per cent of the total area experienced drought. General observation on the periodicity of drought is given in Table 13.2.

TABLE 13.2: Periodicity of Droughts in India

Meteorological subdivisions	Period of recurrence of drought
Assam Very rare,	once in 15 years
West Bengal, MP, Konkan, Coastal AP, Kerala, Bihar, Orissa	Once in 5 years
South interior Karnataka, Eastern UP, Gujarat, Vidharbha, Rajasthan, Western UP, TN, Kashmir, Rayalaseema and Telangana	Once in 3 years
Western Rajasthan	Once in 2½ years.

PLANT ADAPTATIONS TO DROUGHT

Plants can grow and survive in dry habitat by escaping drought and drought resistance (Table 13.3).

Escaping Drought

Many short duration desert plants (ephemerals) germinate with rains and mature in five to six weeks. They have no mechanism to overcome soil moisture stress and are not drought resistant. In cultivated crops, early maturity before soil moisture stress is the main adaptation to grow in dry regions.

Drought Resistance

Plants can adapt to drought conditions in two ways: avoiding stress and tolerating stress. Stress avoidance is the ability to maintain favourable water balance and turgidity even when subjected to drought thereby avoiding stress and its consequences. Favourable water balance can be achieved either through conserving water by **restricting transportation** (water savers) or by **accelerating water uptake** (water spenders). The mechanisms for conserving water are regulating stomatal opening, increased photosynthetic efficiency, low rates of cuticular respiration, decreasing transpiration by lipid deposition on leaves, reducing leaf area, stomatal frequency and location and presence of awns. Water uptake can be accelerated by efficient root system, high root to top ratio, differential osmotic potential of plants and change of water spenders to water savers.

Drought Tolerance

Plant can tolerate drought either by mitigating the actual stress or by showing high degree of tolerance to stress. **Mitigating the stress** by resistance to dehydration and by preventing leaf collapse permit the plants to maintain a high internal water potential in spite of drought conditions. **Tolerating the stress** by resistance to metabolic strain (starvation acid protein loss) and plastic strain (increased resistance to stress due to exposure to sub lethal stress for long period) can increase the plant ability to resist and survive under conditions of soil moisture stress.

TABLE 13.3: Crop Adaptations for Dry Growing Conditions

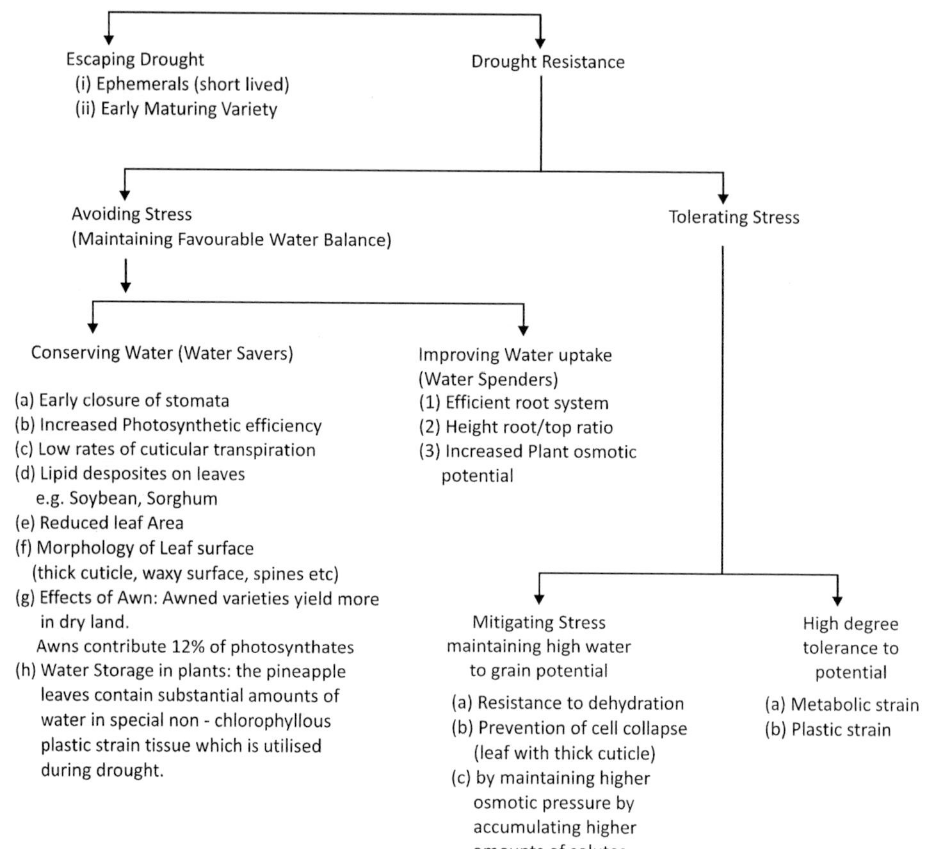

Drought Prone Areas

In the past, one or more of the following four criteria were used to identify drought prone areas:

1. Meteorological data,
2. Revenue Commission,
3. Frequency of famine or scarcity, and
4. Availability of irrigation facilities.

Some states used other criteria also. Tamil Nadu identified 311 taluks in which rainfall was less than 900 mm or less than 35 per cent of the cultivable area irrigated as drought area. Rajasthan considered an area to be drought prone when the ratio of good crop year to scarcity year was 2: 1 or when not less than 50 per cent of the village of the area affected by drought. Karnataka considered those areas as drought areas which received less than 400 mm rainfall during *kharif* and less than 150 mm during *rabi* with a variability of more than 30 per cent during each season and rainfall deficiency of more than 20 per cent at the critical stages of crop growth. Thornthwaite used water balance approach for evaluating drought and proposed the aridity index.

Departure of aridity index from normal	Drought classification
< 0.5	Moderate
0.5 – 1.0	Large
1.0 – 2.0	Severe
> 2.0	Disastrous

The Irrigation Commission relied on only two criteria: meteorological data and available irrigation facilities and demarked the areas as drought or chronic drought zones. Drought zones are areas with 25 per cent probability of rainfall departure from the normal. Chronic drought zones are areas with 40 per cent probability of rainfall departure or more than – 40 per cent from the normal.

STRATEGY FOR DRYLAND AGRICULTURE

Any technology for dry lands should be able to maximise the benefits of a good season and minimise the adverse effects of an unfavourable season. Attempts were made to develop a **good weather code** and a **drought code.** When the monsoon is normal, it should be used most effectively, using a best variety and the recommended package of practices.

Drought code come into operation with aberrant weather and the drought programmes indicate:

- Maximising production through alternate cropping pattern, if necessary,
- Mid season correction to standing crops,
- Crop life saving irrigation, and
- Build up of seed and other inputs to implement the drought complex strategies.
- Measures necessary for counteracting aberrant weather are:
- Thinning the plants or rows in a sole crop,
- Removal of the more sensitive crop in intercropping system (sorghum in sorghum-pigeon pea system),
- Ratooning on receipt of rain if the damage is not beyond recovery,
- Sowing a new crop suitable to the remaining part of the season,
- Urea spray, and
- Life saving irrigation.

DROUGHT AND AGRICULTURAL DROUGHT

Indian Meterological Department, Poona defined: Drought as situation occurring in any areas in a year when annual rainfall is less than 75% of the normal. When deficiency of rainfall is above 50% of the normal, is called severe drought. However, agricultural drought is the moisture deficit which results when amount of water available in the soil is not sufficient to meet the demands of potential evapo-transpiration (PET)

Areas where drought has occurred in 20% of the years during the period are considered Drought areas and where it has occurred in more than 40% of the years, called chronic drought areas.

ICRISAT: International Crop Research Institute for the Semi-Arid Tropics: established on Oct 11, 1972 near Hyderabad (Patancheru)

ICARDA: International centre for Agl. Research in Dry Areas in 1977 at Aleppo (Syria)

Moisture Availability Index (MAI)

Moisture Availability Index (MAI) is the ratio between rainfall (weekly/monthly) at 50% probability level to Potential evapotranspiration of the corresponding period.

$$\text{MAI} = \frac{\text{Rainfall at 50\% probability}}{\text{PET}}$$

Aridity Index (A.I.)

$$\text{Ia} = \frac{\text{WD}}{\text{PET}} \times 100$$

Where WD : Water Deficit
PET : Potential evapo-transpiration.

Moisture Deficit Index (MDI)

$$\text{Im} = \frac{\text{P - PET}}{\text{PET}} \times 100$$

Where p = Precipition

Thornthwaite Moisture Index (TMI)

$$\text{Im} = \frac{\text{AE}}{\text{PET}}$$

Where AE: Percent probability of Actual evaporation Ima is much superior to Im in explaining the variability in crop yields from year to year.

WATERSHED MANAGEMENT

Watershed means an area of land whose runoff drains to a common point i.e. in the river or waterpool. Watershed may be defined as a natural unit of land whose runoff collects & flows out of the area through a single common outlet into another water body. It is a drainage basin which is demarcated by ridges or gullies. Watershed, drainage area or catchment areas are generally interchangeably used meaning for the same. Ordinarily the cachment area (water collecting area) is larger but the area of watershed is comparatively smaller. Watershed area is the unit from the both angles i.e. land unit and water body unit; but there are separate units for both land and water in the catchment area. Therefore catchment area comprises of all the areas from which water flows out into a common river or water pool. The different river's tributaries having different water cachment areas flow out into a common lager river. The larger drainage area is called river basin. The water catchment area of a tributary is called sub-catchment area and the different cachment areas may be identified in it for it's certain parts. The catchment area of this certain part is called watershed area. There may be variations in the watershed area. When these variations are more in number it means the area is hilly, undulating or rolling and we get smaller watershed area where as in the plain we get larger.

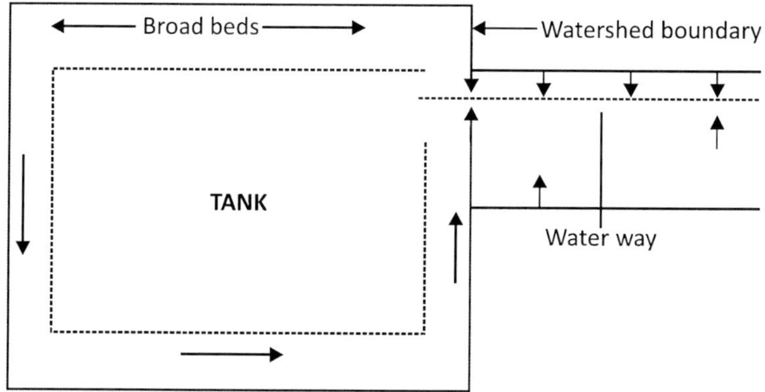

FIG. 13.1: Water shed.

The watershed for the agricultural purpose may be several fields of farmers that contribute runoff water to the flow at the lower point. The runoff movement depends on the land topography. Runoff from the upper fields enters into the lower fields in watershed causes soil and water erosion (Fig. 13.1).

Soil conservation measures in an isolated manner are not sufficient to check the erosion. Therefore, it is necessary that soil conservation measures should be adopted on the watershed basis instead of personal individual field. In the dry land areas, soil and water erosion, flood etc. are the major problems. Considering these point, the Govt. of India launched a Watershed Management Programme in the seventh five year plan to develop dry land areas on the basis of watershed. For soil and water conservation measures, watershed is demarcated into sub-watersheds & micro-watersheds. Micro-watersheds are the basis for planning and execution. In the beginning, there was a plan to develop 4000 watersheds in different agro-climatic regions. The Department of Agriculture provided the funds for this programme and the responsibility of evaluating 47 watersheds management programmes initiated in different states was given to ICAR. This programme was accelerated in 1983 when world bank gave the assistance in the development of four watersheds. Watershed management programme has following features:

a) **Soil & Water Conservation Measures:** Soil and water conservation measures on the watershed basis include all those measures which are effective in preventing or delaying the movement of soil and rock particles. In it the emphasis is given on soil surveying, contour and graded bunding, repairing of old bunds and dams, land reclamation, digging of farm ponds etc.
b) **Scientific Dry Farming:** The agricultural activities like contour ploughing, pre-monsoon ploughing, deep ploughing, mulching, development of new varieties according to agro-climatic zones, weed control, integrated nutrient management and integrated pest management improve the production in dry land areas.
c) **Forestry and development of pasture; and**
d) **Animal husbandry and development of dairy:** Proper animal husbandry, artificial insemination, economic assistance to the milk producing cooperative societies etc. are to be adopted.

PRINCIPLES OF WATERSHED MANAGEMENT

Main principles of watershed management based on resource conservation, resource generation and resource utilisation are:

- Utilising the land based on its capability,
- Protecting fertile top soil,
- Minimising silting up of tanks, reservoirs and lower fertile lands,
- Protecting vegetative cover throughout the year,
- *In situ* conservation of rain water,
- Safe diversion of surface runoff to storage structures through grassed water ways,
- Stabilisation of gullies and construction of check dams for increasing ground water recharge,
- Alternate land use systems for efficient use of marginal lands,
- Water harvesting for supplemental irrigation,
- Ensuring sustainability of the ecosystem,
- Maximising farm income through agricultural related activities such as dairy, poultry, sheep and goat farming,
- Improving infrastructural facilities for storage, transport and agricultural marketing,
- Setting up to small scale agro industries, and
- Improving socio-economic status of farmers.

AIMS AND OBJECTIVES OF WATERSHED MANAGEMENT

1. To protect, conserve & improve the land resources for efficient and sustained production;
2. To protect and enhance water resource, moderate floods & reduce silting up of tanks, increase irrigation and conserving rainwater for crops and thus mitigate droughts.
3. To utilise the natural local resources for improving agriculture and allied occupation or industries so as to improve socio-economic conditions of the local residents.

Thus, watershed management is the planned use of watershed lands in accordance with pre-determined objectives such as the control of erosion, stream flow, sedimentation, and the improvement of vegetative cover and other related resources. Such watershed management integrate plans for soil conservation, soil improvement, irrigation, drainage, flood prevention and development of water supplies for all purposes. In case of agricultural micro-watersheds, important aspects are soil conservation measures, soil improvement, storage and utilisation of runoff water. In dry farming areas, watershed approach provides an ideal means for integrated development. There is a need for multipronged approach to maximise crop production & also to ensure stability in rural income for integrated watershed management having following steps:

i) Water harvesting
ii) Location specific technology for crop production.
iii) Adoption of intercropping and crop substitution according to soil suitability.
iv) Afforestation on cultivable wastes and marginal lands.
v) Practice of dry land horticulture to minimise risk. Mango, guava, ber, pomegranate, sapota, and tamarind have good potential on marginal land.

vi) More efficient exploitation and use of ground water and irrigation.
vii) Development of dairying and pasture management.
viii) Stall-fed goat and sheep farming needs to be popularised.
ix) Popularisation of sericulture and mulberry cultivation.
x) Processing of farm and horticultural produce.
xi) Sound extension and training programmes for the farmers.

WATER HARVESTING / RUNOFF CONCENTRATION / RAINFALL PRECIPITATION

In humid climates, runoff usually occurs only when rain falls on a saturated soil. Under semi-arid conditions, the total annual precipitation may or may not be sufficient, but most of the rain falls in a short period causing humid climate conditions and resulting into runoff. In both conditions, the excess rain water should be safely guided and collected in nearby ponds for recycling in drier period for life saving irrigation.

Water harvesting is defined as the collecting and storing water on the surface of the soil for subsequent use. It is a method to induce, collect, store and conserve local surface runoff for agriculture in arid and semiarid regions. Water harvesting was developed to a fine art about 2000 years ago by the Nabateans. The term 'water harvesting ' was first time used by Myors.

Normally water harvesting is more effective in areas situated near hill side or in areas where large portion is bare soil and cultivation is done on small portions. The technology of water harvesting differs according to soil types, topography, annual rainfall and the crops to be grown. In India, three types of water-harvesting techniques are generally practised:

(i) Inter-row water harvesting: Practised in arid areas with light soil where annual rainfall does not exceed 400-450 mm. The crop is sown in narrow strips between wide intervals that are ridged as artificial miniature watersheds. Later on these are compacted to increase runoff to the crop rows. It is more practical since no land is sacrificed for harvesting water (Fig. 13.2).

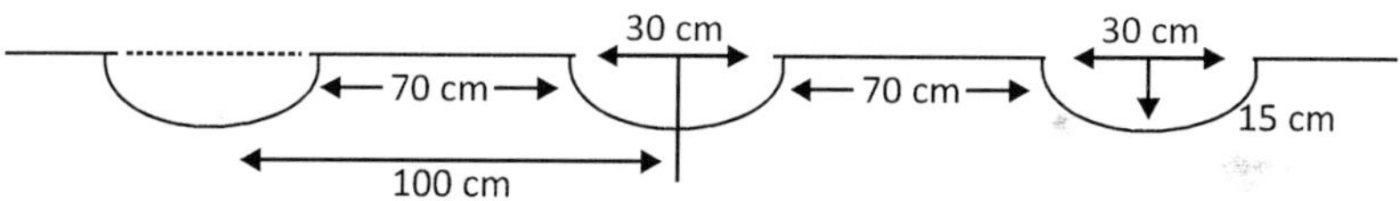

FIG. 13.2.: Inter-row water harvesting system.

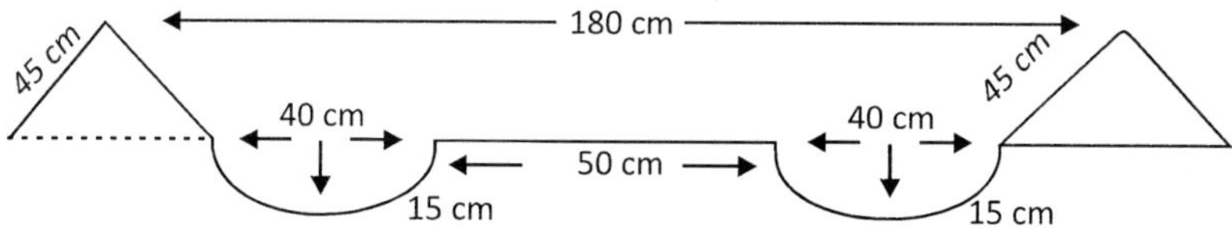

FIG. 13.3: Modified inter-row system.

(II) Inter-plot or Micro plot water harvesting: In this case water is harvested in the passages or furrow between the plots when rainfall is comparatively more. Runoff from the sloping area supplements rainfall for raising crop on level land (Fig. 13.3),

(III) Water harvesting in farm ponds and reservoirs: Surface runoff fromsmall watersheds is stored in farm ponds & reservoirs for utilisation as supplemental or life saving irrigation. This harvested water can either be used as life saving irrigation to kharif crops at the time of drought due to irregular monsoon or as supplemental irrigation to winter crops. Since the stored water is limited, steps should be taken to reduce losses from seepage and evaporation. Suitable lining materials for pond beds and anti-evaporantes should be found out and used.

Runoff can be increased either by land alteration or by chemical treatment. Impermeable catchments are prepared by spraying asphalt compounds. Thin plastic films or metal sheets can also be used to cover ground surface for water collection. Waterborne asphalt emulsions are much effective in controlling seepage loss. Bentonite has excellent sealing properties if kept continuously wet.

In the case of Broad Bed & Furrow system, excess water during rainy season is collected through the furrows and passed through grassed waterways into farm pond or reservoirs.

The stored water should be recycled most judiciously so as to get maximum return out of it. Only one or two irrigations are possible with the limited water stored. Therefore supplemental irrigation can be scheduled by two approaches viz. soil moisture depletion approach and critical stage approach. Irrespective of the crop stage, irrigation is scheduled when soil moisture approaches permanent wilting point to save the crop. This supplemental irrigation is known as life-saving irrigation.

Jalshakti: A chemical which when applied (mixed) in soil, improves the aeration, infiltration and water holding capacity of the soil.

Evapo-traspiration: Soil moisture is the most limiting factor in dry land agriculture. About 60-75% of the rainfall is lost though evaporation. The evapo-transpiration is the evaporation from the soil surface and transpiration from the plant surface. Such losses can be reduced by (i) mulches (ii) antitranspirants (iii) wind breaks and (iv) weed control.

Antitranspirants: Such materials are applied to transpiring plant surfaces to reduce water loss from the plant. Antitranspirants are of four types:

1. **Stomata Closing Type:** Such antitranspirants reduce water loss through stomatal closing. e.g. phenyl mercuric acetate (PMA), herbicides like **Atrazine** in low concentrations, **ABA** and CO_2.
2. **Film Forming Type:** Retard moisture loss due to formation of thin film as physical barrier e.g. Mobileaf, hexadecanol, silicone, oils, waxes.
3. **Reflecting Type:** Such materials reflect the radiation and thus reduce leaf temp and vapour pressure gradient from leaf to atmosphere and ultimately transpiration is reduced e.g. celite (a diatomaceous earth product) and Kaolin. Application of 5% Kaolin spray reduces transpiration loss.
4. **Growth retardants:** Such chemicals reduce shoot growth and increase root growth and thus enable the plants to resist drought. It also inducesstomatal closure e.g. cycocel (ccc).

Antitranspirants generally reduce photosynthesis. Therefore, their use is limited to save the crop from death under severe moisture stress. They have some practical use in nurseries and horticultural crops.

WINDBREAKS AND SHELTER BELTS

Windbreaks are such structures which break the wind-flow and reduce wind speed while shelterbelts are rows of trees or shrubs planted for protection of crop against wind. They provide a protective shelter against wind and suitable for birds and honeybees as well as produce cattle feed and fuel wood. In the wind erosion areas, windbreaks & shelterbelts are to be planted. They prevent soil erosion, arrest the march of desert and protect the agricultural and residential lands from dust storms. A rows of trees and shrubs planted across the wind-direction is the most effective.

It reduces the wind speed up to 60-80% on leeward side. The height of tall tree and length of windbreak determine the extent of protection provided to soil. Windbreak & shelterbelt provide the protective shelter against desiccating winds to the extent of 5-10 times the heght of tall tree on windward side and upto 30 times on leeward side. For example, a 10-11 m tall windbreak when encountered by 45- 50 km/hr. wind, it reduces on windward side to 20-30 km/hr and to 10 km per hour on just leeward side which is illustrated by the Figure 13.4

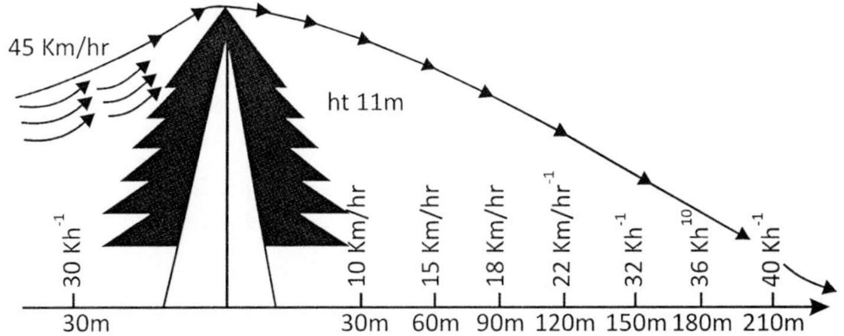

FIG. 13.4: The effect of windbreak on wind-speed.

The wind-break reduces the wind speed on the leeward side at 200 m away up to 20% only and the areas 300 m away, there is no effect at all. Therefore at a interval of 300 m, such windbreaks should be repeated.

To control weed erosion, the capacity of shelterbelt depends upon the speed and direction of wind. In case of high wind speed, the protective area is reduced and in such areas, the interval between two shelterbelts is to be reduced. To counter wind's direction, it is necessary to be long length of windbreak so that wind blows across the windbreak.

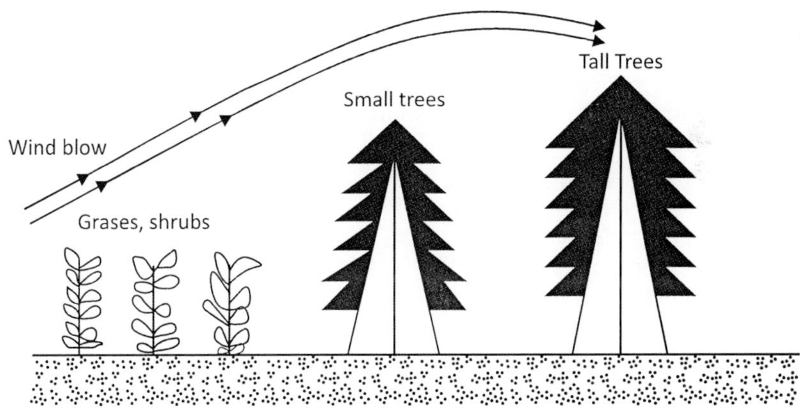

FIG. 13.5: Planting of trees & shrubs as sheltrbelt.

Depending upon the porosity of shelterbelt, certain amount of wind passes through it and some deflects & crosses over it. Thus it does not produce turbulence of air. Reduction in wind speed, reduces the evaporation losses and makes available more water to crops. According to planting pattern of trees & shrubs as shelterbelt, the grasses and shrubs are planted on the outer rows which train the wind to rise much above the ground surface. The inner rows are of small trees and shrubs and the outer rows are planted by tall trees which further raise the wind level (Fig. 13.5).

The beneficial effects of shelterbelts are more clearly seen in drought areas. In such areas windbreaks of 3-7 rows and 15-30 m wide are more effective. It modifies the micro-climate favourable for crop production and shelters for birds, honeybees & pet animals.

AGRONOMIC APPROACHES FOR DRY LAND AGRICULTURE

The main aim of dry farming programme is to conserve the soil and moisture and to achieve maximum production from the dry farming areas. In the past two decades we have been able to solve many hurdles in the aforesaid areas but there had been no break through as in case of irrigated crop production. Now we have the promising crop varieties and technology available with us about the maximum soil and water conservation. There should be an appropriate agro-technique to utilize these improved strains and other devices for maximising our crop yields. If somehow we can increase the productivity of dry farming tracts by even 25 per cent, we can be a surplus nation.

Several agronomic reseaches have been under study with remarkable achievements. Some of the important ones are described below:

For an easy approach to the problems it is thought that the agronomic researches must be problem based. In accordance with these the problem areas have been divided into following groups:

1. Highly undulating lands,
2. Marginal lands,
3. Diara land, and
4. Levelled lands.

Agronomic approaches for highly undulating lands. These lands are confined to hills or the locations which have suffered serious soil erosion problems and have been divided into various gullies. The soils of these areas are more prone to further erosion set backs if they are not managed tactfully. Therefore, the crop management practices are entirely different from those of other groups mentioned above. The object of soil and management under such situations would be:

a) to stabilize the soil by forestry, pasture management with a regulated grazing/ rotational grazing or no grazing at all.
b) to level the land gradually by the help of contour bunding, terracing, etc.
c) to practice strip cropping and pitcher farming.

The lands subjected to heavy erosion and slides are often unfit for crop production as tilling or smoothening the surface will accelerate the process of soil erosion resulting into a still bad topography. The best measure for such areas is to adopt such practices which will stabilize the river banks or the soil of the hill top or the soils of the gully catchment areas. Such approaches include forestry by planting saplings of different categories which are characterized with – faster growth habits, extensive and deeper root systems, extensive branching habits, profuse canopy for maximum ground coverage, habits of producing more rooted side shoots that may grow up into a tree after the main trunk is cut off, timber should be of higher economical values, etc.

These trees are interplanted with fast growing grasses like Anjan (*Cenchrus ciliaris*, *C. setigerus*), Dhaulu (*Chrisopogon fulvus*), Sain (*Sehima nervosum*), Dinanath (*Pennisetum pedicellatum*), *Saccharum spp*., Johnson grass, barmuda grass, etc. for having more soil binding effects and providing greater soil-surface coverage for least beating of rain drops. These grasses may be cut for the use as fodder, if the land does not permit grazing, other-wise cattle grazing is allowed which can be under either constant grazing or rotational grazing according to the situations.

The land surface may be given a shape of contour or terrace, if the surface dose not have steep slopes and it can tolerate mechanical manipulations. After surface has been divided into various terraces the land is used for crop production. The type of crop production adopted in such area is termed as strip cropping.

Strip cropping. Strip cropping is a practice of raising erosion permitting and erosion resistiong crops in a sequence with an object of levelling the land slowly, regulated flow of run-off water, better and purposeful use of lands and increasing the soil productivity. The uppermost terrace is sown with erosion permitting crops like millets and cereals, etc. which grow erect without providing cover to the ground. These crops allow the soil surface to get beaten by the rain drops and the soil particles are loosened and finally transported alongwith water to the second strip. The second strip is planted with erosion resisting crops like urd, mung, cowpea, soybean, moth bean, etc. which cover the soil surface and the soil particles do not get detached by rain beating. These crops also trap and hold the soil particles being carried by water from the first strip to the second strip. Thus after some time both the strips are nearly levelled and gradually both the strips form a single terrace.

Sometimes skip strip cropping is also practised in which the strips of 1-2 metre width are made and alternate strips are planted with an object of gradual levelling and increaserd crop productivity.

Agronomical approaches for marginal lands. Usually marginal lands are very poor in their fertility status. The crop management is done exactly in a similar pattern as that of plain lands.

Agronomical approaches for diara lands. Diara lands are the areas located on either sides of the rivers or between river and riverine which are flooded every year due to overflow and swelling of the rivers. Thus diara lands are formed due to flood as under usual course of water flow, the area may have deposition of fine to coarse sand but the deposits will tend to be finer clay to silt, particularly when back waters spread out thinly and such deposits make the patch exceptionally fertile.

These areas often lack irrigation and need a careful crop management. Since the land of this area is prone to get flooded and hence no crop is possible during kharif season except in some cases where fodder crops are grown soon after onset of monsoon and it is harvested depending upon position of flood. The harvesting or cutting of the crop is started from close to the river bed and as the water spreads the harvesting is also advanced. The life saving irrigations may be given by lifting water from the river or by drilling cavity wells or bamboo borings; in most of the cases these wells go out of order after a flood occurs. Therefore, the cavity wells or bamboo borings are made at a sizable distance from the river.

Agronomical approaches for levelled lands: This area forms the main dry land tract of the country. There has been major emphasis to find out ways and means through which the total soil productivity may be increased. Following aspects of research have been under study for an efficient soil and water conservation and improved crop productivity.

1. Tillage Requirement of the Crops

Tillage starts with the seed bed preparation ends with mulching and control of weeds. The seed-bed preparation depends upon the nature of crops to be grown and soil type, though it is slightly influenced by the previous history of the tillage given in to the field. The land ploughed during summer needs less labour and lesser number of tilling due to soft surface and weed free situation. Researches conducted so far have concluded that deep summer ploughing results in destruction of weeds, insect/pests, diseases or its absorption is also very much improved. The seedbed preparation under such conditions becomes very easy and quick and the farmers can easily sow their crops soon after onset of monsoon. Deep ploughing during summer also helps in an efficient root penetration very deep into the soil. It is further reported that the best results of water conservation and crop production may be achieved by giving a deep summer ploughing followed by shallow surface tilling during rest of the year.

Simultaneous placement of seed at 5 cm and fertilizers at 7.5 cm in the same furrow followed by soil compaction have resulted in better germination, plant vigour, extensive root development and higher crop yields. The soil compaction results in a good moisture conservation also. Removal of weeds through intercultural operation like running a cultivator between the crop rows or removal of weeds though khurpi has been found a best way for control of weeds.

2. Selection of Most Adaptive Crops and their Varieties

Under our earlier concept we had only a few crops which had better drought resistance but now we had numerous improved varieties of most of the crops that have either drought tolerance or resistance. The evaluation trials with various and their cultivars have been conducted at the co-operating dry land research centres of various agro-climatic zones of the country with an object to select the best crops and their genotypes of higher yield potentials which may replace the existing ones in the region. The results conducted in this field conclude that the most commonly grown compatible crops under dry land are rice, maize, sorghum, pearl millet, finger millet, minor millet, wheat, barley, triticale, pulses and oil seeds, etc. The improved genotype of these crops has been described under the heading of genetical approaches for dry land. The crops to be grown on rain fed lands should possess morphological plant attribute combinations and development rhythm conducive to a consistant level of production under water scarcity conditions.

3. Sowing of Crops

It is one of the most important operations in the crop production because it affects the plant emergence pattern, planting density etc. Thus the proverb of 'well begun is half done' holds true because once the ideal plant population is achieved the crop is bound to give good yield. Sowing of the crop, deals with several associated factors, viz. time of sowing, seed rate and spacing etc. which are described below.

a) **Time of Sowing of the Crops:** Sowing time alone can make lot of difference in production and productivity of dry land crops. Sowing time is not only crucial from moisture harvesting, crop stand establishment and land productivity point of views but it has long range effect on total productivity also. An increase in yield has been found by advancing the date of sowing of *kharif* and *rabi* crops. Besides better yield, an early sowing of *kharif* crops helps in proper establishment of plants to safeguard against torrential rains or temporary water logging. Early sowing of *kharif* crops results in their early maturity and thereby it facilitates an early planting of *rabi* crops. Apart from this, early planting prevents the attack of insect-pests and disease like shoot-fly in hybrid jowar, downy mildew in hybrid maize, aphids in mustard, etc. Early planting of *rabi* crops, after an early harvest of *kharif* crop, helps in overcoming the moisture stress at later stages of plant growth, particularly at grain filling stage. Later sowing of jowar var. CSH-5 on 3 & 5 th August at Jhansi and Udaipur resulted in a reduction of 76.8 per cent and 72.6 per cent respectively which is in addition to late planting of *rabi* crops that gave about 90 per cent lower yield than the normal planting dates (Singh and Das, 1984).

b) **Sowing Method and Seeding Depth:** Both of these factors are more or less interrelated because seeding depth is governed by method of sowing. Broadcasting, a very common method adopted in dry land areas, spreads seeds on the soil surface which are subjected to various losses like bird and ant damages, the seed does not reach to the moist zones; hence a poor seed-moisture contact is achieved. This results in a poor sprouting, lower plant population, poor root growth, finally lower drought resistance and lower yields. Placing the seed at about 5 cm depth and fertilizers at about 7.5 cm depth through malabasa/pora or through seed-drill has given best drought tolerance/resistance and higher yields. This has also resulted into a better initial growth, vigor of plants and ultimately into a week's early maturity.

c) **Seed Rate and Plant Population:** Various seed rates of crops have been tried in order to arrive at the optimum plant population of the crops sown/single or mixed/intercropped. The objects of such experiments have been to increase and stabilise the crop yields with least competition between the plants.

The experiments with varied plant population and row spacings have given different results in different crops viz. increasing inter row spacing from 45 to 60 cm in groundnut and from 35 to 45 cm in finger millet significantly decreased yields at Bijapur and Ranchi respectively but at Solapur an increased inter row spacing of bajra from 30 to 45 cm resulted in an increased yield of 1.8 q/ha. At Bijapur the cotton yield was not influenced either by increased or decreased spacing. The spacing of jowar between 45 and 90 cm had no effect on grain yield.

It is concluded from most of the co-operating centres that an ideal plant population is a must for an ideal crop yield. To achieve the highest yield it is, therefore, essential to use about 25 per cent higher than the required seed rate and the excess plants should be thinned out about 2-3 weeks after the sowing depending upon the crops.

4. Fertilizer Management

Use of fertilizers in dry lands is very limited compared to irrigated areas. Today India is using 10 kg/ha fertilizers in dry lands as against 36.6 kg/ha as National average of fertilizer consumption. There are many reasons for this lower consumption like – (i) poor response because of faulty method of application, (ii) poor financial condition of farmers for purchase of fertilizers, (iii) unavailability of fertilizers at required time, (iv) ever increasing cost of fertilizers, (v) wrong concept of farmers that fertilizers will burn the seedlings and destroy the soil structure, and (vi) application of organic manure only which dose not meet the requirement of ideo-types without supplementary dose of fertilizers.

There are two schools of thought about the response of crops to fertilizer application in dry lands. According to some workers the fertilizer application in dry lands results in a better crop yield as compared with irrigated areas because in case of irrigated lands the soils have got very poor residual fertility due to high intensity of cropping and the soil is robbed off minor and major plant nutrients, whereas the dry lands still preserve the nutrients as they are not depleted badly. Thus the crop productivity/unit fertilizer use is manyfold greater in dry lands than corresponding dose in irrigated areas. While others have mentioned that the dry lands have lost their inherent fertility because of serious problems and conversion of soils to saline and alkaline conditions and they need higher doses of fertilizers for better response. Some people think that fertilizer application in dry land areas aggravates moisture problem (which is true in case of false method adopted for application) but truly speaking deep placement followed by foliar application results in extensive rooting up to deeper layer from where the plants can draw moisture to meet their demands.

Spreading of fertilizers at the surface or applying them at the seed layer is less beneficial and at times it is hazardous too. Therefore, a careful and judicious fertilizer scheduling is a must for achieving higher and better crop productivity in dry lands. Numerous nutritional trials have been conducted with varied results which are described below:

a) Use of organic manures is always better but they do not provide sufficient available nutrients to meet the requirement of new plant types; therefore, the nutrient requirement must be met by providing 1/2 quantity though fertilizers. This improves water use efficiency and also gives higher crop yields.
b) The response pattern is more spectacular in case of new plant types rather than local varieties. Singh (1977) reported from All India Co-ordinated Sorghum Improvement Project that the local sorghum variety gave only 7 kg grains /kg N application but CSH-1 gave a response of 21.5 kg grains/kg N used. Apparently the water use efficiency was very much enhanced in the case of improved varieties.
c) Green manuring in situ or green leaf manuring followed by rabi planting or a catch crop and cropping economies fertilizer application as it supplies nitrogen ranging between 40-60 kg N/ha.
d) Inclusion of one legume crop in rotation rather than cereal rotation adds about 20-25 kg N/ha, increases water use efficiency, reduces weed intensity and thereby increases crop.
e) Inoculation of legumes though Rhizobium culture and non-legumes through *Azotobacter* adds about 30-35 kg N/ha, reduces cost of fertilization and increases crop productivity.
f) Deep placement of fertilizers 5 to 7.5 cm below seed layer, particularly during rabi season, increases water use efficiency by producing longer or deeper roots, discourages weed growth (in case of surface application of fertilizers through broadcasting, the weed growth is very prolific and crop suffers from a sever weed competition) and increases crop yields.
g) During rainy season deep placement of first split of N (75 per cent in case of heavy soils and 50 per cent in case of loams and 25 per cent in light sandy loams of the total nitrogen) along with full doses of P and K resulted into best fertilizer use efficiency by plants. Rest of nitrogen should be applied at critical growth stages depending upon rains but the top dressing should also be done by side or band placement. During winter season in all the cases 75 per cent of N + full P and K should be basal placed and rest of N should be sprayed over foliage.
h) Foliar application of urea having lowest biurate content, zinc as zinc sulphate and potash has resulted in increased water use efficiency and grain yields.
i) Ley farming i.e. growing forage legume in pearl millet rotation added 100 kg nitrogen/ha under semi-arid conditions of Hyderabad.

Above results indicate that with proper fertilizer scheduling but the drought resistance/tolerance and higher stabilized yields may be obtained. This way a higher net return per rupee investment on fertilizers in particular and on total cost of cultivation in general can be achieved by following a careful and judicious application of fertilizers under arid and semi-arid tracts of India.

5. Cropping Systems

Intercropping

Intercropping is a way in which various crops are grown together in the same field whereas cropping pattern is the sequence of crops i.e. rotation of crops prevalent in a region. The cropping pattern of most of the rainfed areas of the country are based on the traditional systems of subsistence farming where farmer tries to grow every thing he needs, though big farmers do produce for the market. With increase in population, availability of new plant types and knowledge of modern crop production technology subsistence farming is no longer practiced. The subsistence farming included mixed cropping which has become increasingly difficult to manage, particularly in the age of mechanization. The mixed cropping used to ensure some produce to the farmers against various risks but now better control over various production factors is achieved and chance for total crop failure is nearly eliminated. Therefore, the farmers are giving more emphasis on monoculture rather than raising several crops mixed together

Mixed Cropping

Some of the distinct advantages of mixed cropping have been identified but they are negligible if problems are taken into account. Inter-cropping, slightly different in its approach than the mixed cropping, provides all the advantages of mixed cropping and is convenient to manage efficiently. Thus intercropping is more refined than the mixed cropping. Considering the definition of these two terms – the mixed cropping may be defined as sowing of mixed seeds of two or more crops within the same row whereas intercropping refers to growing of two or more crops simultaneously on the same piece of land in separate rows. Earlier the concept of growing intercrops was to raise any less important crop (subsidiary or secondary) between two rows of widely spaced long duration main crop. With recent concept the idea of main and secondary crops is no more in existence.

Present need is to conduct intensive research on various agronomic aspects of scientific inter-cropping on dry lands aimed at maximum production of food, fodder, oilseeds and pulses/unit area/unit time. The research strategy for different agro-climatic regions consists of the following steps:

1. Screening of different crops for compatibility in intercropping systems. The emphasis is given to choose such crops which differ in their growth rhythms so that periods of peak demand for moisture, light and nutrients do not coincide.
2. Modifying existing planting pattern and change it to double rows.
3. Identifying the optimum row ratio so that the component crops may utilize moisture and nutrients in an efficient way
4. Identifying the best genotypes of the component crops from intercropping point of view. A high yielding variety of a crop may not tolerate the competition of a companion crop.
5. Determining the optimum doses of fertilizers for most profitable production of crops in an intercropping system.

Relay Cropping

It is a practice of growing succeeding crop in the standing or maturing preceding crop so that there is no gap between two successive crops. This type of cropping is limited to some diara lands or low lying dry lands where water gets accumulated during rainy reason and it is expected that sowing will be delayed e.g. lathyrus in standing rice crop or gram in maturing rice crop.

Farm Forestry System

Tree plantations with *Casuarina* and *Eucalyptus* were studied at Bangalore where it was found that plantation of *Eucalyptus* was more beneficial than *Casuarina.* Tree plantations were compared with profitability of cultivation of finger millet in dry land areas and it was found that tree plantations were highly profitable.

Agro-Forestry System: Agro-forestry is one of the several types of integrated systems for combined use of natural resources of arid and semi-arid regions and capable of giving higher productivity on sustained basis. Agro-forestry is to be termed as an umbrella term which embraces the following systems:

1. Agro-silvicultural system,
2. Silvi-pastoral systems,
3. Agro-silvi pastoral system, and
4. Multi-purpose forest tree production system.

The aims of agro-forestry systems are: (a) judicious land use for higher productivity, (b) better soil and water conservation, (c) improving soil fertility, (d) providing protection to crops, (e) creating greater employment opportunities, (f) rehabilitating environment etc.

The objectives of such studies are:

1. To study about the tree-crop and grass/legumes compatibility.
2. To study the relationship between tree density and forage yield.
3. To study about the tree canopy characteristics and lopping practices.
4. To study the economic feasibility of agro-forestry systems.

The trees to be included in the agro-forestry must have the following characteristics:

1. The trees must be of multi-purpose use which can give loppable fodder, fuel and timber
2. They must be wider adaptability for growing in a wide range of climatic conditions.
3. They must be exotic in nature. They should establish easily with least care.
4. They must be of rapid growth, ability to coppice, vertical canopy and preferably should have nitrogen fixing ability.
5. They must suit the problem environments such as arid zones, as saline and alkaline soil conditions, steep hill slopes, ravines etc.

The experimental findings of various agro-forestry systems conducted at different research stations in India concluded that:

1. At Hyderabad the yields of crops like pearl millet, sorghum and castor increased when grown in association with multipurpose leguminous trees (*Leucaena leucocephala)* as compared to these crops grown pure.
2. At Bijapur *Leucaena and Sesbania grandiflora* were planted and trenches were made on both the sides of the tree-rows just to eliminate root expansion after which safflower, bengal gram and other pulses were grown. The grain yields were found to be encouraging as compared to their single or pure cropping.
3. At Hyderabad plantation of *Eucalyptus* at a spacing of 27.2 metre and grass below resulted into a superior fodder production.
4. Tree plantation changes micro-climate in relation to relative humidity and temperature. Their leaves fallen on ground used to add organic matter, nitrogen, phosphorus, calcium contents to the soil and thereby the physical condition of

the soil upto 120 cm depth was improved at Hyderabad under Prosopis cineraria plantation.

5. Agri-horticultural systems are similar to agro-forestry. The trees also yield fruits in addition to fuel wood. Some small fruit trees like *Zizyphus spp. Psidium guajava, Mangifera indica, Tamarindus indica, Emblica officinalis* etc. can be intercropped with crops, legumes and grasses.

6. Crops and Cropping Patterns

The cropping pattern is an area based cropping sequence which is more technically called as crop rotation. The cropping pattern depends upon agro-climatic variations, food habit of the local people, social traditions, socio-economic structure of the region and many alike factors.

Traditional crops and cropping systems in arid and semiarid regions are need based. They may not be the most efficient in terms of productivity and profitability. Single *kharif* crop or intercropping is common in Alfisols of south India. Majority of Vertisols are left fallow during rainy season. A post rainy season crop is taken on stored soil moisture. Sorghum, chickpea and safflower are grown in central India and wheat in north central plains. Cotton as sole or intercrop with few rows of pigeon pea, is more common in Vertisols of relatively low rainfall.

(1) Crops and Cultivars

Crops and varieties ideal for dry lands should have certain characteristics which enable them to withstand harsh environment during the crop period. The following characters, among others, assume significance.

- Shorter duration to escape terminal drought,
- Faster growth during initial period to withstand environment,
- Strong and penetrating root system to withstand soil moisture stress,
- Response to fertilizers, and
- Potential for high yield.

Experiments all over the country for over three decades indicate superiority of hybrids over traditional varieties. Safflower is more efficient crop than rain fed wheat in Bellary region of Karnataka. Sunflower is more remunerative than setaria in relatively deeper red soils of Rayalaseema (AP). Efficient crops and varieties have been identified for all the dry lands in different agro climatic conditions.

In areas like western Maharashtra, North Karnataka and Telangana and Rayalaseema of AP, sorghum as rabi crop is still popular inspite of more remunerative crops like sunflower and soybean. Hybrid rabi sorghum is more productive than traditional and improved varieties. However, hybrids could not replace the varieties, since varieties satisfy both grain and fodder requirements.

(2) Crop Planning

Crops and varieties selected should match the length of growing season during which they are not subjected to soil moisture stress. Climatological analysis helps to identify cultivars suitable for different regions. Feasibility for intercropping, sequence cropping and double cropping can also be known from such ananlysis. For regiouns with cropping season less than 20 weeks, single crop during *kharif* or *rabi* is recommended (Table 13.4). Regions with more than 30 weeks and above have no problems for double cropping. In regions with 20-30 weeks cropping season, double cropping may be risky. Such areas are ideal for intercropping.

TABLE 13.4: Length of Effective Cropping Season in Different Areas (Randhawa and Venkateswarulu, 1979.)

Category (weeks) in different areas				Effective cropping season system
Less than 20 weeks	– Bellary (8) – Hissar (17)	– Jodhpur (11) – Rajkot (17)	– Anantapur (13) – Bijapur (17)	Sole cropping
Between 20 and 30 weeks	– Jhansi (21) – Udaipur (22) – Anand (25)	– Kovilpatti (21) – Solapur (23) – Akola (27)	– Hyderabad (22) – Agra (24)	Intercropping
More than 30 weeks	– Bhubaneswar (32) – Bangalore (36) – Rewa (36) – Samba (44)	– Varanasi (32) – Hoshiarpur (35) – Ranchi (35)	– Indore (36) – Dehradun (44)	Sequence cropping

Water balance for different agro climatic regions has been calculated and water availability periods worked out (Table 13.5). Regions with 350-600 mm rainfall having 20 weeks effective growing season are suitable for single cropping in *kharif* (red and shallow back soils) or *rabi* (deep black soils). Intercropping is possible in regions receiving 600-750 mm rainfall and having 20-30 weeks of effective growing season. Areas with more than 750 mm rainfall or with more than 30 weeks are suitable for double cropping.

TABLE 13.5: Suitable Cropping Systems based on Rainfall and Water Availability Period

Rainfall (mm)	Water availability Soils	Period (weeks)	Potential cropping system
350-600	Alfisols & Shallow Vertisols	20	Single *kharif* cropping
350-600	Aridisols & Entisols	20	Single cropping either in *kharif* or rabi
350-600	Deep Vertisols	20	Single *rabi* cropping
600-750	Alfisols, Vertisols & Entisols	20-30	Intercropping
750-900	Entisols, Deep Vertisols, Alfisols & Inceptisols	> 30	Double cropping with monitoriing
> 900	Entisols, Deep Vertisols, Alfisols & Inceptisols	> 30	Assured double cropping

7. Efficient Cropping Systems

Vertisols and Alfisols are major soil groups in Indian SAT. Results of experiments have proved scope for sequential cropping in deep Vertisol. Maize followed by postrainy season sorghum or chickpea is more remunerative than a single crop in either of the seasons. Similarly, a short duration mung (75 days) followed by *rabi* sorghum is remunerative at several places. However, two crops may not be possible if the onset of monsoon is delayed. Intercropping systems are more assured than two crops. A cereal and pigeonpea combination has been more successful on Vertisols. A very promising system in Madhya Pradesh is soybean and pigeon pea intercropping system. In drier Vertisols, efforts should be made to improve the productivity of *kharif* crop since the scope for a post-rainy season crop is limited.

Alfisols are mostly suitable for one crop during rainy season. Double crop systems are rarely successful. Intercropping low canopy cereals with legumes are common system. Intercropping systems involving short and long duration crops are successful. Groundnut with pigeonpea in Rayalaseema and sorghum with pigeonpea in Telangana regions of AP are the promising intercropping systems in Alfisols. Efficient cropping systems for different dryland areas are given in Table 13.6.

TABLE 13.6: Efficient Cropping Systems for Different Situations (Singh 1995)

Soil and region	period (days)	Water availability	
		Double cropping system	Intercropping system
Vertisols and related soil zones			
Malvaplateau (Madhya Pradesh)	210-230	Maize-Safflower-Chickpea	– Maize + Soybean
	191-210	Sorghum-Safflower/Chickpea	– Soybean + Pigeon pea
Bundelkhand (Uttar Pradesh)	190-220	Sorghum-Chickpea, Black gram- Pearl millet Fodder Mustard / Safflower / legume	– Sorghum + Pigeon pea
Vidharbha (Maharashtra)	190-210	Groundnut-Safflower Sorghum-Safflower	– Sorghum + Pigeon pea – Cotton + Pigeon pea
	170-190	Green gram-Sorghum/ Safflower	– Pearl millet + Green gram – Sorghum + Green gram
Southern Maharashtra	160-180	Green gram-Sorghum/ Safflower	– Pearl millet + Pigeon pea – Sunflower + Pigeon pea – Groundnut + Sunflower
Southern Rajasthan	160-180	Greengram-safflower	– Maize + Pigeon pea – Groundnut + Pigeon pea – Chickpea + Mustard
Northern and Central Karnataka	130-150	Cowpea-Sorghum	– Pearl millet + Pigeon pea – Green gram-Safflower – Groundnut + Pigeon pea – Sunflower + Pigeon pea
	100-120	–	– Sorghum + Coriander – Safflower + Coriander
Saurastra (Gujarat)	130-140	–	– Groundnut + Pigeon pea – Pearl Millet + Pigeon pea
Southern Tamil Nadu	120- 130	–	– Sorghum + Black gram – Cotton + Back gram
Alfisols and related soil zones			
Southern Karnataka	190-220	Cowpea-Finger millet Soybean-Finger millet	– Finger millet + Pigeon pea – Groundnut + Pigeon pea – Finger millet + Soybean
Telangana (Andhra Pradesh)	140-160	–	– Sorghum + Pigeon pea – Castor + Clustrbean
Rayalaseema (Andhra Pradesh)	110-130	–	– Groundnut + Pigeon pea – Groundnut + Pearl millet
Arid soil zone			
North western Gujarat	100-120	–	– Pearl millet + Green gram – Pearl millet + Cluster bean – Cluster bean + Green gram
	75-90	–	– Pearl millet + Black gram – Pearl millet + Cluster bean – Cluster bean + Black gram – Pigeon pea + Black gram

8. PASTURE MANAGEMENT

Lands belonging to the capability class of 'V' and above are characterized with very low productivity and high undulation. These lands are fit for forestry, and grassland or pasture. Such lands are dominated by less productive and less palatable grass species like *Eragrastis*,. *Aristida* and *Heteropogan* etc. Because of low profitability the farmers do not take much care and they do not apply any fertilizer. It is, therefore, very essential to evaluate grasses having productivity and palatability and their response to various doses of nitrogen for an increased tonnage and socculency.

Experiments conducted at Hyderabad concluded that *Cenchrus setigerus* gave the highest yield followed by *Urochlo mosembensis* and *Cenchrus ciliaris*. *Panicum antidotale* yielded the least. All the species responseded to various doses of nitrogen application and yield increased up to 80 kg N/ha dose.

Studies on the ley farming were also conducted in which the attempt has been made to cultivate the pasture lands (by raising most suitable pasturre legumes), particularly improved pastures and to grow crops thereafter. The experimental findings indicate that *Stylosanthes hamata* influenced the yield of castor and highest yield of castor was obtained from the plots which were under *Stylosanthes hamata* regularly for three years followed by two years and least was from the plots where it was grown only for one year. The yield of *Stylosanthes spp*. was also in similar pattern which was probably due to establishment effect.

Companion cropping of *Stylosanthes* spp. *with* pearl millet and sorghum (as inter-cropping of pearl millet in pasture legumes spp.) was studied but the grain yield of both sorghum and pearl millet were less than their respective sole crops. There was severe mortality of seedlings of pearl millet and sorghum due to rabbit grazing.

9. Cropping Plans to Meet Aberrant Weather Conditions

The unpredictable weather in dry farming tracts requires certain adjustments to meet an aberrant weather. Normally aberrant weather comprises delayed onset of monsoon, long gaps in rainfall, early stoppage of rains and early onset of monsoon.

An appropriate strategy for conditions of late or delayed onset of monsoon comprises alternate crops and varieties, dry sowing, pre-sowing, seed soaking or treatment, transplanting of month old bajra seedlings early and complete weed control. Results of Central Arid Zone Research Institute, Jodhpur indicate that cereals/ millets under late sown conditions gave very poor yields while grain legumes and oils seeds, on the other hand, gave reasonably good yields. Crop seeds with harder seed coat like castor, cotton, sunflower should be sown after an overnight soaking in water.

In case of long gaps in rainfall a high rate of seedling mortality has been observed when gap occurs early in season but when gaps take place at later growth stages it becomes more devastating for crop yields. Thus it has been found beneficial to resort to gap filling for recouping the loss in plant population due to seedling mortality. Other beneficial operations are spraying of urea solution for boosting crop growth rate, providing life saving irrigations at critical growth stages from the farm ponds, weeding and inter-culturing to keep the plots clean etc.

For situations like early stoppage of rains, lower plant population, use of short duration varieties of crops, creation of soil mulch by repeated interculture and use of life saving irrigations have given positive results.

In situations like early onset of monsoon, crops like bajra and sesame could be sown with advantage. Grain legumes, in particular green gram, cluster bean and moth bean etc. when sown early grow very vigorously and give poor yield.

10. Weed Control

Presence of weeds in the crop field especially in case of dry lands, cause a severe crop- weed competition for moisture and nutrients. The reduction in yield due to presence of weeds varies from 30-75 per cent depending upon the crop and nature and extent of weed infestation. Weeds may be controlled by hand weeding, intercultural operations and herbicidal application or by adopting an integrated approach by combining all of them together. The results of weed control experiments conducted at various centres indicate that at Dantiwada the yield of castor was higher by seven times than unweeded check when field was kept weed free up to 60 days after sowing. Weed control up to 35 days after sowing gave 218 per cent higher rice yield (12 q/ha) under upland condition of sub-humid red soils of Ranchi. A weed free condition in maize up to 45 days of sowing in black soils of semi-arid condition of Udaipur resulted in 51 per cent higher yield (10.4 q/ha) over unweeded check. At Varanasi application of Benthiocarb @ 1.5 kg/ha increased the yield of upland rice upto 19.2 q/ha whereas Benthiocarb @ 1.25 kg/ha+weedding 35 days of sowing gave yield of 22.9 q/ha over 14.1 q/ha in unweeded check. Groundnut pod yield at Solapur was found to be increased over control by 4.1 q/ha and 4.9 q/ha by interculture+ hoeing and Nitrofen @ 0.75 kg/ha+one weeding respectively. Thus it is essential to control weeds for good yield.

11. Plant Protection Measures

In light textured soils of arid and semi-arid regions, termites and white grubs cause extensive damage to emerging seedlings and also to grown up plants. Use of 25-30 kg BHC 10% dust or 15-20 kg Aldrin 5% dust in the soil and incorporating it well into the soil at the time of last ploughing or at sowing controls termites. The white grubs may be controlled by drilling of Thimet 20 G granules @ 15 kg/ha along with seeds. Aphids in mustard do not leave any productive part to yield any thing and it may be controlled effectively by spraying of 0.02% metasystox or dimecron. Similarly the pod borers in pulses and semiloopers in castor bean take a heavy toll of yield. These should be controlled by spraying 0.05% endosulphan. The viral diseases of pulses should be controlled effectively by seed treatment and foliar spraying with suitable fungicides.

12. Mitigating Adverse Effects of Aberrant Weather

Long term historical data on rainfall indicate that the rainfall is seasonal, erratic and highly variable in time and space. Three important aberrations in the rainfall are more common.

1. Commencement of rains may be early or considerably delayed,
2. There may be prolonged dry spells during southwest monsoon, and
3. Rains may terminate earlier than normal cessation date or may continue beyond normal rainy season.

Management options to mitigate the adverse effect of aberrant weather can minimize the risk of total crop failure.

13. Late Onset of Monsoon

Crops and varieties recommended for normal onset of monsoon may not be remunerative for delayed sowing due to late onset of monsoon. Two management options under these circumstances are transplanting and alternate crops/varieties.

Transplanting

Community nurseries under irrigation are the sure way of compensating delay in monsoon. Transplanting can be taken up with the receipt of rains. Finger millet, pearl millet and setaria (foxtail millet) are well adapted to transplantation.

Alternate Crops / Varieties

Certain crops and varieties can perform better even if sown late in the season. Depending on the receipt of rains, such crops and varieties can be selected. Castor (Aruna) is more remunerative in red soils of Telangana (AP) than pigeon pea under late sown conditions. Sunflower, because of its shorter duration, performs better than groundnut when sowings are delayed beyond September first fortnight in relatively deeper red soils. In western Rajasthan, short duration green gram and cowpea are more remunerative than pearl millet when sown late. Crops and varieties for different dry land regions have been identified for late onset of monsoon.

14. Dry Spells During Crop Period

Dry spells during crop period are common in dry land agriculture. The adverse effect of moisture stress should be mitigated to avoid total crop failure.

Ratooning / Thinning

The rate of soil moisture depletion increases with increasing leaf area. If drought occur at 40-50 day after sowing, reduction in leaf area either by ratooning or thinning the plant population can mitigate the adverse effects of drought. Sorghum and pearl millet responds to ratooning. Two per cent urea sprays after drought period is useful for indeterminate crops like castor, pigeon pea and groundnut.

Mulching

Surface mulching with organic materials can extend the period of water storage in soil profile by minimizing the heat load on soil leading to reduction of evaporation loss of stored water. Repeated intercultivations in black soils result in formation of soil mulch on the surface. It act as a barrier to evaporation loss from soil profile. Deep cracks formed due to drought increase the loss of stored water by evaporation. Soil mulch minimizes deep cracking leading to reduced evaporation losses.

Weed Control

Weeds compete with crop for nutrients, water and other natural resources. Such competition for soil water from weeds can be reduced by timely weed control.

Water Harvesting and Protective Irrigation

In situ water harvesting and protective irrigation are the potent measures of crop life saving during periods of soil moisture stress. These aspects have been discussed.

15. Early Withdrawal of Monsoon

Early withdrawal of monsoon before *rabi* seeding leads to problems of crop stand establishment and terminal drought. Therefore, *rabi* seeding should be as for as possible in early September to avoid the above two problems. All the suggested measures to mitigate the adverse effect of prolonged dry spells during crop period (5.8.2) are necessary to overcome the soil moisture stress due to early withdrawal of monsoon.

16. Extended Monsoon

Extended monsoon is seldom a problem in dry land agriculture. It benefits long duration *kharif* crops taken as a component crops in intercropping system and also advantageous for *rabi* crops in Vertisols.

17. Alternate Land Use System

Crop production in dry lands is a gamble with monsoon. The per capita land, ideal for crop production is decreasing progressively. Therefore, cropping has been extended to marginal lands leading to their fast degradation. According to National Remote Sensing Agency, culturable waste lands in India occupy 38.8 M ha. Income from dry lands could be increased through alternate land use systems adopting improved technology. The term **alternate land use system** is applicable to all classes of land to generate assured income with minimum risk through efficient use of available resources. It aims at a suitable farming system, matching the land capability classes. Eight well recognized land capacity classes have been discussed in Chapter 3. Need based alternate land use systems are:

Food (Class II & III)	Alley cropping, agrihorticulture, intercropping with nitrogen fixing trees
Fodder (Class IV& V)	Hortipastural, silvipastural, ley farming, pasture management
Fuel/timber/fibre (Class VI & VII)	Tree farming, timber cum fibre farming

18 Alley Cropping for Arable Lands

Food crops are grown in alleys formed by hedge rows of trees or shrubs in arable lands. It is also known as **hedgerow intercropping.** Hedgerows are cut at about one meter height at planting and kept pruned during cropping to prevent shading and to reduce competition with food crops. It is recommended for humid tropics, primarily, as an alternative to shifting cultivation. In semiarid regions of India, alley cropping provide fodder during dry period since mulching the crop with hedgerow prunings usually does not contribute to increased crop production. Advantages of this system of this system are:

- Provision of green fodder during lean period of the year,
- Higher total biomass production per unit area than arable crops alone,
- Efficient use of off season precipitation in the absence of a crop,
- Additional employment during off season, and
- It serves as a barrier to surface runoff leading to soil and water conservation.
- Based on the objectives, three types of alley systems are recognized.
- Forage-alley cropping,
- Forage-cum-mulch system, and
- Forage-cum-pole system.

In all the three systems, crops are grown in alleys and forage obtained from loopings of hedgerows.

Forage-Alley Cropping System

In this system, both yield of crop and forage assume importance. Only a few tree species are suitable for hedge rows: *Leucaena leucocephala,* Colliendra and Sesbania. Pigeon pea or castor crops are suitable for growing in the allies of *leucaena.* Crop yield decreases with decrease in row width. Increase in cutting height of hedgerows decrease the crop yield due to shading effect. Longer harvest intervals narrow the yield differences between the cutting heights.

Forage-Cum-Mulch System

In this system, hedgerows are used for both forage and mulch. Loopings are used for mulching during the crop season and used as fodder during off season. Substantial reduction in crop yields of sorghum, groundnut, green gram and black gram have been observed at several places.

Forage-Cum-Pole System

Leucaena alleys are established at 5 m intervals along the contours. Hedge rows are established by direct seeding and topped every two months at 1.0 m height during crop season and every four months during the off season. A *Leucaena* plant for every 2 m along hedgerows is allowed to grow into a pole. Crop yield is usually reduced due to competition from hedgerows. However, gross returns are higher in all alley cropping system than with sole crop.

19. Agri-Horticulture for Arable Lands

It is one form of agro forestry in which the tree component is fruit tree. It is also called as **food-cum-fruit system** in which short duration arable crops are raised in the interspaces of fruit trees. Some of the fruit trees that can be considered are guava, pomegranate, custard apple, sapota and mango. Pulses are the important arable crops for this system. However, depending on the requirements, others like sorghum and pearl millet can be grown in the interspaces of fruit trees. Reasons for this system not being widely adopted are:

- Economic position of farmers may not permit awaiting income after 5 or 6 years,
- Watering of fruit trees, till their establishment is a problem in summer period, and
- Marketing problems for perishable horticultural produce.

20. Horti / Silvi-Pastural System for Non-Arable Lands

Class IV and above soils, uneconomical for arable crop production are termed as non-arable lands. **Horti-pastural system** is an agro forestry system involving integration of fruit trees with pasture. When fruit tree is replaced by a top-feed tree, it is called as **silvi-pastural system.**

Guava, custard apple and ber suits well in a horti-pastural system with grasses like *Cenchrus ciliaris* (anjan), *C. setigerus* (bird wood), *Panicum antidotale* (blue panic), *Dicanthicum annulatum* (marvel) and *Chloris gayana* (rhodes) and legumes like *Stylosanthes hamata* (hamata) *S. scarab* (stylo) and *Macroptilum atropurpuream* (siratro).

Top-feed trees ideal for silvi-pastural system are:

- Acacia nilotica (babul)
- Acacia senagal (gum arabica)
- Dalbergia sassoo (shisham)
- Harwickia binata (anjan)
- Leucaena leucocephala (subabul)
- Sesbania grandiflora (agathi).

Grasses and legumes indicated under horti-pastural system are also suitable for silvi-pastural system.

21. Ley Farming for Non-Arable Lands

A rotation system which includes a pasture (ley) for grazing and conservation is called **alternate husbandry** or **mixed farming.** A farm or part of it, entirely cropped with leys which are reseeded at regular intervals is termed **ley farming.** It helps to conserve soil and improve its structure and fertility. For dry lands, it is a low risk system. A 4 year rotation system involving *Stylosanthes hamata* and sorghum has yield advantage of sorghum besides improvement in soil physical properties.

22. Tree Farming for Non-Arable Lands

Trees can flourish and yield abundantly where arable crops are not profitable. Farmers of dry lands are inclined to tree farming because of labour cost, scarcity at peak periods of farm operations and frequent crop failure due to drought. A number of multipurpose tree systems (MPTS) have been tested for their suitability and profitability under different situations.

Annual rainfall less than 500 mm	Annual rainfall 500 to 750 mm
– *Acacia nilotica*	– *Acacia nilotica*
– *Acacia aneura*	– *Acacia ferruginea*
– *Acacia tortilis*	– *Albizia lebbek*
– *Acacia albida*	– *Azadirachta indica*
– *Prosopis cineraria*	– *Casuarina equisetifolia*
– *Prosopis juliflora*	– *Cassia sturti*
– *Pithecallobium dulce*	– *Dalbergia sissoo*
	– *Leucaena leucocephala*
	– *Tamarindus indica*

23. Timber-Cum-Fibre System (TimFib)

Subabul intercropping with agave appears to be more remunerative at Bijapur area of Karnataka. High density plantations (10,000 plants ha^{-1}) and short rotation intensive culture (4-5 years) have gained popularity in Nellore and Prakasam districts of Andhra Pradesh. About 50 per cent of rainfed area in these districts is under *Casuarina* and *Leucaena*. Such system will not be viable in marginal land degraded lands due to problems associated with survival of seedlings.

24. Agroforestry

Agro forestry may be defined as an integrated self-sustained land management system, which involves deliberate introduction/ retention of woody components with agricultural crops including pasture/livestock, simultaneously or sequentially on the same unit of land, meeting the ecological and socio-economic needs of people. It is also defined as a collective name of land use systems and technologies where woody perennials are deliberately used from the same land management units as agricultural crops and/or animals in some form of temporal sequence. In agro forestry systems, there is both ecological and economic interaction between different components.

Based on the kind of associated agricultural products, major function of the tree component, spatial arrangement of trees and duration of combination, several systems have been identified. In India, agroforestry systems are classified as:

- Agri-silviculture: Crops + trees

- Silvi-pasture: Trees + pasture
- Agri-silivi-pasture: Crops + trees + pasture/animals
- Horti-pasture: Fruit trees + MPTS + pasture/animals
- Agri-horti-silviculture: Crops + fruit trees + MPTS

RECOMMENDATIONS FOR DRY FARMING AREAS

The research programmes of the All India Co-ordinated **Research Project** for Dry Land Agriculture have concluded into certain justified claims of research achievements which have made the following distinct recommendations:

1. Bunding across the slope and levelling the land should be done before onset of monsoon.
2. Deep summer ploughing should be followed by surface tillage during monsoon months and also rest of the year.
3. Application of organic manures like F.Y.M., compost @ 15-20 tonnes/ha or green manuring should be done. These manures should be applied about 20-25 days before sowing and should be well mixed in soil.
4. Fertilizers should be basal placed at a depth of 7.5 to 10 cm in the soil and the seeds should be sown in the same furrows about 3 cm above the fertilizers. This is important especially during winter season. The nitrogen (25-50% of the total) should be top-dressed by side or band placement method at about 10-15 cm apart the crop rows which should be done soon after the rains but if there is no moisture in the soil the nitrogen should be sprayed over the foliage with urea solution containing 5-15% nitrogen. Zn and sulphur should be applied as basal, if needed.
5. Soil applications of BHC (10%) dust @ 25-30 kg/ha for termites and Thimet 20 G @ 15 kg/ha for white grub should be done and these chemicals must be mixed well in the soil at the time of final field preparation.
6. Selection of suitable crops and their varieties, as mentioned in genetic approaches of the text, must be done.
7. Seeds must be treated with a suitable fungicide and that of legumes with a bacterial culture before sowing. Soaking seeds in plain water for rabi sowing helps in getting higher germination, better seedling vigor and an early maturity by a week's time.
8. Proper crop rotation should be followed which should preferably have at least one legume crop every year.
9. For better seed-soil-moisture contact thorough soil compaction should be done by running a planker or a roller, especially after rabi planting.
10. At the event of total crop failure during kharif season a suitable catch crop like urd T.9 or toria etc. should be sown.
11. Inter-cropping of oil seeds and pulses should be done with jowar, bajra and maize crop for the purpose of making best use of soil and inter row moisture harvesting. This also increases crop productivity/unit area/unit time.
12. Line sowing by drilling the seed at a depth of 7.5 to 10 cm or even more depending upon the situation should be practised because it helps in better seed germination, high drought resistance, etc. line sowing also facilitates the agricultural operations. On terraces strip cropping should be practiced.
13. Twenty five per cent more than the recommended seed rate should be used and thinning of excess plants should be done about 2-3 weeks after germination. This helps in stabilizing the required plant population and thereby in getting better yield.

14. Proper weed management practices should be followed by adopting integrated weed control measures like one hand weeding about 20-25 days after sowing, then one hoeing about 5-10 days later followed by use of appropriate herbicide for an effective weed control.
15. Mulching should be done by providing frequent inter culture and pulverizing the soil. If intercultural operations are not possible then use of artificial mulches like covering the surface by organic refuse (tree leaves, uprooted weeds, sugarcane leaves, saw dust or polythene sheets) are used to check the evaporation loss of water from the soil.
16. Water harvesting between the rows should be done by growing pulse crops and run-off water should be collected in some nearby located pond which may be used for recycling in the form of protective or life saving irrigation to the crop.
17. An efficient plant protection measure should be adopted to protect the plants from various insect/pest and disease damage.
18. Crops should be harvested at physiological maturity so that the following or succeeding crop may be sown slightly earlier than the scheduled time and best use of rain water or residual moisture may be made for crop production.
19. Crops like cotton, chillies should be sprayed with CCC or Cycocel and groundnut should be sprayed with Plano fix for modified growth, higher drought resistance and better yield.

RAINWATER HARVESTING AND WATER SAVING TECHNOLOGIES

Annual precipitation including snow fall in our country has been 4,000 km^3 for centuries and will remain nearly the same for centuries to come. Out of 4,000 km^3, India's average annual runoff generated by rainfall and snowmelt is estimated to be about 1,869 km^3, while the lost in evaporation etc. or escapes into sea during the floods. However, the present availability of utilizable annual surface water (690 km^3) and ground water (432 km^3) of 1,122 km^3 is just sufficient to meet the estimated demand of 1,093 km^3 by 2025 AD. Present utilizable water resources may just sufficient to meet the total water needs for the country by the year 2025 AD. Moreover, the entire replenishable groundwater is assumed to be utilized by 2025. The present population of India is about 101 crores which is projected at 170 crores by 2025 AD. The country is producing today around 200 mt of food which is estimated of the order of 450 mt by 2050 AD. This has to be produced from available net cultivable area which is 143 m ha by 2050. So, there will be a gap between the water availability and requirement by 2050 AD. In India, the water situation is steadily worsening. The average annual per capita fresh water availability has come down to 1,869 m^3 in 2000, from 5,000 m^3 in 1950. The average water availability though looks comfortable up to 2025 AD, the water scarcity condition (availability less than 1000 m^3/capita/annual) already exists in 8 of the 20 major river basins in the country. Some of these regions suffer absolute water scarcity, with availability below 500 m^3. Many basins are either physically water scarce or have pockets of unsustainable ground water use or have high food dependency ratios (Rao, 2007).

Water harvesting is the activity of the direct use or can be recharged into the ground water. Water harvesting is the collection of runoff for productive purposes. Rain is the first form of water, which enters into hydrological cycle and it is a primary source of water for us. Presently rivers, lakes and ground water are the secondary sources of water in which we depend entirely on them.

Challenges in Sustaining the Water Resources

There are some challenges in sustaining the water resources. The foremost challenge related to water scarcity in developing countries like India is the need to increase the deficiency of water use in agriculture, urban areas, and industry. Irrigated area accounts for over two-thirds of world rice and wheat production, so growth in irrigated output/unit land and water is essential. Improved efficiency in agricultural water use is required both to maintain productivity growth and to allow reallocation of water from agriculture to urban and industrial uses.

New sources of water are increasingly expensive to exploit. Water to meet growing household and industrial demand may come through water saving from irrigated agriculture, which generally, accounts for 80 % of water diverted for use in developing countries. To truly contribute to reduce the water scarcity, improved efficiency in agricultural use should be accompanied by improved efficiency in urban and industrial use. The quality of land and water must be sustained in the face of mounting pressure to degrade these resources through water logging, salinization, groundwater mining, and water pollution.

Experts feel that with the present rate of water consumption, India is poised to have more Cauvery like disputes among states. Worldwide, the crisis can be so grave as to trigger a Third World War. According to the data presented with the standing sub-committee for assessment of availability and requirement of water for diverse uses, the total demand for water at the start of the millennium stood at 634 BCM, which is likely to soar to 1,447 BCM at the present rate, whereas the total utilizable water is 1,123 BCM. Central Water Commission has stressed on the need of water conservation and judicious use of water. **"If India has to sustain its growth rate, it is very imperative that we use water efficiently. Even 10 % of the water saved during the course of irrigation can meet 40 % of the other demand".**

Water use efficiency is presently estimated to be only 38 to 40 % for canal irrigation and about 60 % for groundwater irrigation schemes. On the basis of 1991 census, our country's per caput water availability/year was at 2,214 m^3 against the global average of 9,231 m^3. In 1990, India was ranked at the 42th position among 100 countries by per caput water availability. In the total water use in 1990, the share of agriculture was 83 %, followed by domestic (4.5 %), industrial (2.7 %) and energy (1.8 %). The remaining 8 % was for other uses including environmental requirements. The projected total water demand by 2025 AD is around 1,050 km^3 against the country's utilizable water resources of 1140 km^3. The share of agriculture in total water demand by 2025 AD would be about 74 %. Thus, almost the entire utilizable water resources of the country would be required to be put to use by 2050. Irrigation being the major water user, its share in the total demand is bound to decrease from the present 83 % to 74 % due to more pressing and competing demands from other sectors by 2025 AD. It is estimated that with 10 % increase in the present level of the water use efficiency in irrigated projects, an additional 14 m ha area can be brought under irrigation from the existing irrigation capacities which would involve a very moderate investment as compared to investment that would be required for creating equivalent potential through new schemes (planning commission Gov. India/plans/annual plan/1999-00).

Rain Water Harvesting

The collection and storage of rain water for later productive use is defined as **rain water harvesting.** Any rain water harvesting system has **three components** viz., **Catchment, conveyance** and **storage.** The benefits of rain water harvesting include:

1. Rain water harvesting (RWH) is the ideal solution for all water requirements,

2. Increase in ground water level,
3. Mitigates the effects of drought,
4. Reduces of runoff, which otherwise flood storm water drains
5. Reduces flooding of roads and low lying areas,
6. Reduced soil erosion,
7. Improves the ground water quality,
8. Low cost and easy to maintain and
9. Reduces water and electricity bills.

The approaches pertaining to rain water harvesting are:

1. Planning houses with rain water harvesting structures, modification of house, existing house, etc., from rooftops,
2. Government buildings, institutions, hospitals, hotels, shopping malls etc. from rooftops and open areas,
3. Farm lands, public parks, play ground, etc. and
4. Paved and unpaved areas of a layout/city/town/village.

Rain Water Harvesting Techniques

There are 2 techniques of rain water harvesting viz., rooftop rain water harvesting and storm water recharging through open drains.

Rooftop Rainwater Harvesting

The rooftop rainwater harvesting system requires **four elements** i.e., (i) Catchment area, (2) conduits, (3) Settlement tank and Recharge or storage facility. For rain water harvesting system, a dried bore well, an underground tank or soak pit are required along with conduits or the pipes carrying rainwater from the catchment or roof tops to harvesting/storage system.

Storm Water Recharging through Open Drains

In urban areas, natural way rain water harvesting occurs only in areas left out for open spaces and parks. In the new areas which are being developed by the urban development authorities/civic bodies, the storm water is disposed off through underground channels of sewer lines, thus leaving no scope for recharging of this rainwater into the earth. The old technique of disposal of storm water through open drains is considered more environment friendly as through these open drains storm water charging take place in natural way.

Water Harvesting and Recycling

Collection of rain water through check drains, farm ponds and recycling of saved water through application devices (rain gun) increase the productivity of water. In on-farm, overhead sprinkler and rain gun systems are used to give supplemental irrigation under rainfed cultivation through farm ponds to mitigate the intermittent drought.

On-Farm RWH and Conservation Technologies

Community-Based Technologies

Community-based technologies include underground pipe systems, lined –canals and sprinkler systems. In most applications, they have large fixed costs and require collective action or ongoing co-ordination of many house–holds. Sprinkler systems, for examples, require substantial water pressure to operate. To attain sufficient pressure, some villages need construct water towers and elaborate piping networks. Besides, the small size of plots and fragmented nature of most farm holdings in most of the Indian states mean that operating a sprinkler system requires coordination for use. It is difficult to use a sprinkler that irrigates in a large circular pattern on 1 plot without irrigating the plots of other households around it.

Water Conservation Through Watershed

The watershed approach has been applied for the purpose of arresting rainwater runoff, rainwater harvesting, and ***in-situ*** soil and moisture conservation. Community based watershed development has become the guiding principal for rejuvenation of natural resources, especially land and water.

The engineering measures for rain water harvesting adopted differ with location, slope of land, soil type, amount and intensity of rainfall. Depending on these parameters, the methods commonly used are contour trenching, contour stone walls, construction of temporary and permanent check dams and gully plugging structures. Additionally, percolation ponds, silt detention tanks and irrigation tanks are constructed to harvest water and recharge it to the ground water for use in agriculture (irrigation). Farm ponds can also be constructed for every 4-5 ha in the watershed to provide protective/supplemental irrigation.

Water Harvesting Structures

Contour increase bunds, contour barriers (vegetative and stone), contour trenches and contour stonewall will not only prevent soil erosion but also obstruct the flow of runoff water. Consequently, the obstructed water will increase the soil moisture and recharge the ground water in the area.

Check Dams

Check dam may be a temporary structure constructed with locally available materials. The main function of the check dam is to impede the soil and water removed from the waterbed. This structure is cheap, but lasts for about 2-5 years only. The cost of the structure is about US$ 200-400 depending on the materials used, the size of the gully and the height of the dam. A permanent check dam can be constructed using stones, bricks and cement Small earth work is also needed on both sides. Costs vary from US$ 1000-3000 depending upon the length and height of the dam. A little water is also stored above the dam. This water recharges the ground water.

Percolation Pond

The percolation pond is a multipurpose conservation structure. It stores water for livestock and recharges the ground water. It is constructed by excavating a depression, forming a small reservoir or by constructing an embankment in a natural ravine or gully to form an impounded type of reservoir. The cost of this type of structure is estimated at around US$ 5000-10,000. The capacity of these ponds or tanks varies from 0.3 to 0.5 mcft (10,000-15,000 m^3). Normally 2 or 3 fillings are expected in a year (season) and hence the amount of water available in a year in such a tank is about 1 mcft (30,000-45000 m^3) The cost for one mcft (30,000 m^3) varies from US$ 5000 to 15,000. This quantity of water if it is used for irrigation, is sufficient to 4-6 ha of irrigated dry crops (maize, cotton, pulses, etc.) and 2-3 ha of rice crop.

Irrigation Tank

The main function of this storage structure is irrigating crops. It is constructed below the percolation pond in a watershed. In Tamil Nadu, each tank irrigates 10 to 5000 ha. In south India, there are about 2,00,000 tanks, irrigating about 3.5 million ha. Earthen bunds are reinforced with masonry to collect and store rainwater for irrigation. The cost of this tank varies from US$ 1,500-2,500 to irrigate 1 ha of land. Water from the tanks is normally used to grow rice.

On-Farm Rain Water Harvesting

The farmers of rainfed and dry farming regions, are poor because of lower and unstable yield. In addition, it has uneven and erratic distribution which results in moisture stress to the crops. To overcome this acute problem of water scarcity, the natural rain water is stored in augmented storage structures and being used for domestic and irrigation purposes.

The most efficient and cheapest way of conserving rain water at the agricultural farm was found to be *in-situ* runoff management, which also reduces soil losses and increases the opportunity time for ground water recharging. The average depth of runoff at 80 % probability level was found to be 14 cm. It is observed that earthen embankment for rain water harvesting has cost benefit ratio of 1.38:1. Hence, it can economically be replicated. It is also observed that earthen check dam/reservoirs which are recommended across the streams to regulate runoff for flood control, trapping silt, collecting water for irrigation and fish production and to recharge ground water has good potential for replication. Excavated dug-out ponds tanks are found most suitable for storing runoff in cultivated lands with inverted truncated pyramid shape having 1:1 side slopes with lining of polyethylene sheet of 200 micron buried under 20 cm thick soil at bottom and pitched with bricks. The rain water harvesting during monsoon and its use for irrigation during scarcity period was found to increase the crop yield by 20-30 % during *rabi* and additional water for population use by 60 % in the area (Sharma *et al.*, 2004).

In-Situ Soil Moisture Conservation

A part from the above, to increase moisture availability to agricultural and tree crops, *in-situ* moisture conservation techniques must be adopted in addition to the large scale soil and moisture conservation and water harvesting structures in the watershed. The following are some of the *in-situ* moisture conservation measures, which can be practiced in the watershed to increase the production.

For agricultural crops, the measures adopted are forming ridges and furrows, broad-bed and furrows, basin tie ridging (random tie ridges) and water spreading. For tree crops micro-catchment, saucer basin, semi-circular bund, crescent-shaped bunds, V ditch technology, catch pits and deep pitting can be practiced.

The small storage structures with a water storage capacity for an area of about 0.4 to 0.5 ha can be constructed in large numbers for every 10 to 20 ha catchment or watershed at the foot hills slopes and hilly areas. These storage facilities would attenuate the floods during storms. These measures will also ensure soil moisture for good growth of trees grown down stream recharging the ground water in the rewgion and making available more water for drinking and irrigation water.

Contour Farming

During intense rain storms, the soil cannot absorb all the rain as it falls. The excess water flows down the slope under the influence of gravity. If farming is done up and down the slope, the flow of water is accelerated, because each furrow serves as a rill. The major part of the rain is drained away without infiltrating into the soil. The top fertile soils are washed off. All this results are in a scanty and uneven growth of crop.

A simple practice in farming across the slope keeping the same level as far as possible (which is technically called contour farming) has many beneficial effects.

The ridges and the rows of the plants placed across the slope form a continual series of miniature barriers to the water moving over the soil surface. The barriers are small individually, but as they are large in number, their total effect is great in reducing runoff, soil erosion and loss of plant nutrients.

It has been experimentally proved that contour farming reduces run-off and prevents soil erosion as compared with the up-and-down cultivation in the major groups of soils in India, viz., alluvial soils, black soils and deep lateritic soils. Apart from conserving the water and soil, contour-farming conserves soil fertility and increases crop yields. Contour farming has also given 490 kg of sorghum grain and 273 kg stalks more than the up-and-down cultivation at Kanpur. Thus, 1 mm of rain water conserved by contour farming gave 22.5 kg of sorghum grain and 12.5 kg of straw than the up-and-down cultivation. It has been proved that much less power is required to be exerted by men, animals and machines, if the cultivation is done in the contour instead of up-and-down the slope. There is less wear and tear of the implements and the same job is done in less time in contour farming.

Broad Bed Furrow (BBF)

The BBF system consists of a relatively raised flat bed or ridge approximately 95 cm wide and shallow furrow about 55 cm wide and 15 cm deep. The BBF system is laid out on a grade of 0.4-0.8 % for optimum performance. It is important to attain a uniform slope without sudden and sharp edges because of the need in many crops and cropping systems to plant rows also on the shoulder of the broad bed. This BBF system is most effectively implemented in several operations.

Benefits

1. The raised bed portion acts as an *in-situ* 'und' to conserve more moisture and ensures soil stability, the shallow furrows provides good surface drainage to promote aeration in the seedbed and root zone, prevents water logging of crops on the bed.
2. BBF design is quite flexible for accommodating crops and cropping systems with widely differing row spacing requirements.
3. Precision operations viz., seed, fertilizer placement and mechanical weeding are facilitates by the defined traffic zone (furrows), which saves energy, time, cost of operation and inputs.
4. Facilitates double cropping and increases crop yields.
5. It can be adopted for groundnut crop in red soils with a reduced gradient along the bed (0.2-0.4 %).

The BBF formed during the first year can be maintained for the long term (25-30 years). This will save considerable cost as well as improve the soil health. The increase in yield in improved system is give in Table 13.7.

TABLE 13.7: Crop yields in improved and traditional system in Vertisols Waterbeds, ICRISAT, Patancheru (1976-2000)

Crop	Mean crop yield (t/ha)
Broad Bed furrow system with double cropping (Maize or Sorghum – Chickpea or Pigeon pea)	3.9 – 1.3
Traditional flat system (*rabi* crop) Sorghum or Chickpea	1.1

Mulching

Surface mulches are used to prevent soil from blowing and being washed away, to reduce evaporation, to increase infiltration, to keep down weeds, to improve soil structure and eventually to increase crop yields. Inter-culture kills weeds and produces 5-7 cm thick soil mulch which helps to reduce evaporation from the top soil. It also breaks the surface crust which forms after each downpour. Studies on mulching carried out in India under rainfed agriculture have concentrated on the measurement of crop responses rather than on the manner in which crop responses are influenced.

Mulching with maize residues did not significantly influence the yield of the succeeding crop. Mulching, in general, gave a higher yield of wheat grain and stalk than no mulching. At Bellary, mulching with different materials (e.g., rice husk, grass mulch, jower stubble) did not give encouraging results in increasing the yields of cotton and jowar. Mulching with straw on Encap Esso mulch increased the infiltration rate, improved soil moisture and increased the yields of wheat, barley, gram and linseed, succeeding the maize crop on eroded soils at Ranchi.

Growing of Crops Which Provide the Maximum Cover, Reduce Run-Off

Cultivated legumes in general furnish a better cover and hence better percolation to cultivated land against erosion than ordinary cultivated crops. The crops and the cropping systems will naturally vary from region to region, depending on the soil and climatic conditions. At Dehra Dum, cowpeas provided the maximum canopy with or without the application of phosphate fertilizer. This was followed by ming, urd and dhaincha. At Vasad, cowpeas provided the best vegetative cover for the soils. Sunhemp was the next best; mung and groundnut were effective to some extent.

At Kota, velvet bean formed the canopy effective to the maximum extent. Cowpea was the next best. At Kanpur, mung served as the most effective canopy, followed by urd and cluster bean. The application of phosphorus increased the canopy. At Rehmankhere, cowpea formed the best canopy, followed by *Styzolobium mung,* groundnut, urd, moth and soybean. Sunhemp, mung and *Styzolobium,* helped to conserve soil moisture well, leading to a higher yield in the succeeding barley crop. All the legumes were equally effective in reducing the loss of soil and nutrients. Thus among the legumes, cowpea and mung proved to be important crops for providing a good cover for the land during the rainy season.

Vetiver

Vetiver grass, or khus khus (*Vetiveria zizanioides* L.), roots and tops have also been harvested. Traditionally, these roots were woven into mate, fans and fragrant screens, while the tops of the grass were used for handicraft, thatch, mulch, fodder and animal bedding. In the last 50 years, attention has been focused on vetiver's unique soil conservation properties. When planted in single lines along the contour, hedges of vetiver grass are found to be very effective in soil and moisture conservation. The stiff stems of the thick hedge slow the movement of run-off water and spread it out, trapping silt behind the hedge. This allows more water to be absorbed into the soil, thus reducing run-off and erosion.

Vetiver hedgerows will bring soil losses down in responsible and acceptable levels, will prevent organic matter loss (i.e., FYM and soil organic matter), and will reduce rainfall runoff resulting in improved soil moisture, increased ground water, and better yields. Because of its deep root system vetiver can recycle nutrients, not normally available to crops, by treating the vetiver leaves as a crop residue and incorporating them in the soil.

The water used by crops accounted for 33-67 %, soil water storage accounted for 0-14 % of rainfall drained to ground water recharge was quite high (31-65 %) due to the recharge season (rainy season that rainfall amount was input to the study sites and infiltrated into the soils while the excess water runoff. However, rainfall is found to be for exceeding evapo-transpiration during this period which may cause more excessive soil water content than the soil can store. Soil water would subsequently be drained freely from the soil profile through gravity. In particular, sandy soils may be capable of draining soil water within a few hours compared to few days in fine texture soils.

Comparing to the control plots, the maize plots where vetiver hedgerows were applied showed higher recharge values with 20 % increased ground water recharge (Table 13.8). Meanwhile there were no difference in recharge values found between control and vetiver hedgerows applied plots at Longan and soybean cultivation sites. Vetiver hedgerows planted across cultivated slopes at Yam Watershed, Northern Thailand were found to increase water stored in soil profile while reducing 9-56 % of water loss by runoff by diverting it perpendicularly along the contour hedgerows and allowing smaller amounts to pass through, while other amounts are seeped through the soil and retained by soil profile. In the cropping season, the major proportion of rainfall amount was used by crops and evapo-transpiration has greatly affected the water balance. Vetiver hedgerows acted as intercrops that withdraw more water compared to the sole crop. Thus, the water output as deep drainage was not obviously enhanced at high evapo-transpiration field. The finding is brought to the conclusion that with fine soils, low soil hydraulic conductivity, and high runoff, vetiver hedgerows could increase ground water recharge up to 20 per cent.

TABLE 13.8: Soil Water Recharge Calculated through Water Balance Equation

Site	Plots	Rainfall	Evapotranspiration		Runoff		Change in soil moisture		Recharge	
		(mm)	(mm)	(%)	(mm)	(%)	(mm)	(%)	(mm)	(%)
Longan	Control	1043.2	666.9	63.9	57.7	5.5	30.0	-	348.6	33.4
	Vetiver		697.7	66.9	34.2	3.3	136.3	-	347.7	33.3
Maize	Control	967.4	404.7	41.6	125.3	12.9	139.4	14.4	298.0	30.8
	Vetiver		437.3	45.2	55.2	5.7	116.9	12.1	357.9	37.0
Soybean	Control	1242.2	408.1	32.9	38.9	3.1	- 9.0	-	804.2	64.7
	Vetiver		442.3	35.6	31.4	2.5	- 29.4	-	798.0	64.2

Source: http/www.vetiver.com/ICV4pdfs/DAS01.pdf

Recommendation deriving from this study is that, within the head water areas which are normally high slope, vetiver hedgerows can be used as an effective conservation treatment in agricultural practices. The potentials of vetiver hedgerows in holding water, reducing runoff and increasing ground water recharge are, therefore, the ability to help watershed to store and provide water properly in quantity and timing. This may be an alternative way in watershed conservation that is similar to the function as provided by natural forest.

Water Saving in Farms

Field level water savings and real, basin-wide water savings may differ due to several agronomic and hydrologic factors. Water saving technology adoption will increase in response to water shortage only if (farmers) perceive that that adoption will lead to water saving or generate other benefits. Micro-irrigation is suitable for all now crops and especially for wide spaced high value crops. The required quantity of water is provided to each plant at the root zone through a pipe net work. Hence, there is little loss of water from the soil surface. Micro-irrigation is well adopted for undulating terrain, shallow soils, porus soils and water scarce areas. Saline/brackish water can also be used since water is applied daily, which keeps the moisture and salt stress at a minimum. Water use efficiency in crops can be increased by using water saving irrigation methods. The modern improved irrigation methods give overall efficiency as high as 90 % as against only 30-40 % under the conventional methods. Pressurized irrigation systems (sprinkler, surface and sub-surface drip) have the potential to increase irrigation water use efficiency by providing water to match crop requirements, reducing runoff and deep drainage losses, and generally keeping the soil drier reducing soil evaporation and increasing the capacity to capture rainfall. Agronomic practices for water conservation help to intercept rain drops and reduce the splash effect, help to obtain a better intake of water rate by the soil by improving the content of organic matter and soil structure, help to retard and reduce the overland runoff the use of contour cultivation, mulches, dense growing crops, strip cropping and mixed cropping.

Micro Irrigation

The decade of 1990-2000 witnessed a quantum leap in expansion of micro-irrigation technology (Table 13.9), both in developed and developing countries. In micro-irrigation, the highest coverage is in America (1.9 m ha) followed by Europe and Asia (1.8 m ha each), Africa (0.4 m ha), and Oceania (0.2 m ha). The area under micro irrigation increased almost 6 fold during last 20 years i.e., from 1.1 m ha in 1986 to 6.1 m ha at present.

TABLE 13.9: Utilization of Micro-Irrigation in World

Year	1981	1986	1991	2000	2006
Area (ha)	4,36,590	10,30,578	18,26,287	32,01,300	60,89,534
% increase		136.1	77.2	75.3	90.2

Drip Irrigation

A new technology of drip irrigation was then introduced in Israel by Simcha Blass and his son Yeshayahu. Instead of releasing water through tiny holes, blocked easily by tiny particles, water was released through larger and longer passage ways by using friction to slow the water flow rate inside a plastic emitter. The first experimental system of this type was established in 1959 in Isael by Blass, where the developed and patented the first practical surface drip irrigation emitter.

Drip irrigation is an effective tool for conserving water resources and studies have revealed significant water saving ranging between 40 and 70 % by drip irrigation compared with surface irrigation, with yield increases as high as 100 % in some crops in specific locations (Table 13.10). Drip irrigation is very popular in wide spaced horticultural crops viz., coconut, mango, guava, pomegranate, lime, oranges, grapes, banana, tapioca, turmeric, close spaced crops like vegetables, potato and flowers such as jasmine, rose, and commercial crops like sugarcane and cotton. In view of its higher water use efficiency, wider adaptability to diversity of soils, crops, climate etc. and other benefits, the Government of India, is providing high subsidy to its large scale adoption. Increased efficiency in drip irrigation is mainly due to maintaining optimum soil moisture for plant growth throughout the crop period. Relative soil moisture varies most in furrow irrigation and least in drip irrigation system.

TABLE 13.10: Water Requirement (WR), Yield and Water Use Efficiency (WUC) Of Crops in Micro-and Conventional Irrigation Methods

Crops	Irrigation Methods	WR (cm)		Yield (t/ha)		WUE (kg/ha/mm)
Banana	Surface	176.0		57.5		32.7
	Drip	97.0	(45.0)	87.5	(52.0)	90.2
Sugarcane	Furrow	215.0		128.0		59.5
	Drip	94.0	(56.0)	170.0	(33.0	180.9
Grapes	Surface	53.2		26.4		49.6
	Drip	27.8	(48.0)	32.5	(23.0)	116.9
Cotton	Furrow	83.0		26.4		31.3
	Drip	28.0	(66.3)	32.5	(25.0)	116.1
Sugarbeet	Furrow	49.5		41.8		850.0
	Drip	37.1	(25.0	49.0	(17.1)	1,320.0
Sweetpepper	Furrow	49.0		11.9		242.0
	Drip	48.0	(2.0)	12.0	(0.8)	249.0
SweetPotato	Surface	61.1		4.2		233.7
	Drip	25.2	(60.1)	5.9	(38.7	233.7
Beetroot	Surface	85.7		0.6		6.7
	Drip	17.7	(79.3)	0.9	(55.3)	50.1
Radish	Surface	46.4		1.1		22.5
	Drip	10.8	(75.7)	1.2	(13.5	109.8
Papaya	Surface	228.0		13.9		0.06
	Drip	73.8	(67.8)	23.5	(69.5)	0.32
Mulberry	Surface	50.0		69.3		1,386.0
	Drip	20.0	(60.0)	71.4	(3.0)	3,570.0
Tomato	Surface	30.0		32.0		106.7
	Drip	18.4	(39.0)	48.0	(50.0)	260.9

Source: WTC Annual Reports 1985-2003
Figures in parenthesis are increase over conventional irrigation.

Sprinkler Irrigation

Sprinkler irrigation is a method of applying irrigation water, which is similar to rainfall. Water is distributed through a system of pipes usually by pumping. It is then sprayed into the air and irrigated entire soil and surface through spray heads so that it breaks up into small water drops, which fall to the ground. Sprinklers provide efficient coverage for small to large areas and are suitable for use on all types of properties. It is also adaptable to nearly all-irrigable soils since sprinklers are available in a wide range of discharge capacity. Sprinkler system is an unique irrigation system. It is designed ensure maximum water saving, combining high quality, affordability and ease of installation. All the products are made out high strength and chemical resistance engineering plastics to achieve functional satisfaction and to maintain cost economics. All sprinklers undergo extensive quality testing in our well equipped state of the art lab. Moreover, performance of the products is also tested in the field to ensure uniform water distribution and higher efficiency.

It is suitable for almost all field crops like wheat, gram, groundnut, pulses, as well as vegetables, soybean, tea, coffee, and other fodder crops.

Some Unique Benefits of Sprinkler Irrigation Include:

1. Good control of water application, which can reduce vegetative growth and improve the product quality (eg., onion bulbs and taproot quality of carrots),
2. Can be used to cool the crop during heat waves, to protect against frost and to manage plant health (eg., reduces the impact of *Aspergillus* black mould near harvest in onion),
3. Has the ability to keep soil soft for emerging seedlings with frequent short irrigations,
4. With proper drainage, sprinklers can be used efficiently to flush accumulated salts down the profile.

Collection of rain water through check dams, farm ponds and recycling of saved water through application devices (rain gun) increase the productivity of water. In on-farm, overhead sprinkler irrigation and rain gun systems are used to give supplemental irrigation under rainfed cultivation through farm ponds to mitigate the intermittent drought.

It is concluded that water is likely to become an increasingly scarce resource in future. Therefore, its conservation and utilization has necessarily to achieved through water harvesting and recycling technologies on one side and reducing losses and wastages on the other. People at large who are the users of water realize the scarcity of water and treat it as a precious resource and participate in the water conservation process by avoiding wastage of water. The Government has to play a major role in water conservation along with the stakeholders to achieve the best results.

"Perfection comes from practice"

14 Farming Systems

> *"Agricultural production and efficiency largely depends upon the inputs applied and the method adopted. In India, while population grows, the land surface is fixed and of this only a certain portion is available for cultivation. Further scope for bringing extra land under the plough is limited. If more production is to be got out of the existing area, the problem has to be tackled on a wide front. This can be done by applying inputs in a more intensive way and by adopting modern methods of production through use of improved technology, besides making an adequate provision for institutional financing, better methods of marketing, etc."*
>
> *Planning Commission*

INTRODUCTION

World population has been increasing by leaps and bounds. India's population is expected to reach 1370 and 1660 m in 2030 and 2050 AD, respectively. The country's food production has reached an all time high of 204 m. t during 2000. A food production level of 289 and 349 m. t is needed to satisfy the projected population in 2030 and 2050 AD, respectively. The net cultivable area is 142.8 m ha. Unlike the population spurt and corresponding food need for 2050, there is every chance that the land area under cultivation will decrease due to diversion of some of the cultivable area to buildings and industrial purposes. It is anticipated that the land area available for cultivation in 2050 would be 137 m ha. So, it becomes necessary to increase the productivity almost to double of what we are producing today. This could be made possible by putting the land, both irrigated and rain fed, under intensive cultivation. Fortunately most of our country lies in tropics and so is blessed with abundant solar energy, thus making cropping round the year possible.

Time concept relates to increasing the intensity of cropping under assured irrigated conditions, whereas, space utilization pertains to building up of vertical dimension though multi-tier cropping and farming system approach. By adopting these two concepts, the productivity per unit area per unit time can be substantially enhanced. Farming system approach is one of the approaches wherein the risk in dealing with single component can be minimized, and at the same time increase the productivity through effective recycling.

The Indian economy is predominantly rural and agriculture oriented. In agriculture, 85 per cent of the holdings are less than two hectares and the declining trend in the average size of the farm holdings, poses a serious problem. Majority of them are dry lands which depend on erratic monsoon rains. The rest of the area is cultivated with supplemental irrigation. The farmers concentrate mainly on crop production which is invariably subjected to a high degree of uncertainty in income and employment. To sustain the income and productivity, the farmer has to integrate ancillary propositions with crop production.

India with 2.2 per cent of global geographical area supports more than 15 per cent of the total world population, 70 per cent of whom depend on agriculture. It also supports nearly 15 per cent of the total livestock population of the world. One-third of the gross national product comes from agricultural sector. During 2050 AD, 349 million tones of food, 25 million tones of vegetable oil and 92 $m.m^3$ of industrial wood shall be needed for approximately 1667 million people.

As of now, out of 328.73 million ha geographical area approximately 18 per cent is under forest; only 13.5 per cent is not available for cultivation. Total problem areas constitute 173.65 million ha which include areas subject to wind and water erosion (145 million ha), water-logged areas (8.53 million ha), alkali soil (3.58 million ha), saline and coastal sandy areas (5.50 million ha), ravines and gullies (3.97 million ha), shifting cultivation (4.91 million ha) and reverie torrents (2.73 million ha). Besides 40 million ha are prone to flood and 260 million ha are drought prone. Thus the net sown area is 136.18 million ha (41.42 per cent of the total geographical area) (Subbian et al., 2000).

Unlike industries, agriculture is practiced by 150 million farm families who live in 0.6 million villages. More than 40 per cent of them are below the poverty line. Nearly 85 million farm families belong to small and marginal categories. In spite of increase in food production, after the independence of the country, only in northwestern India, per capita food production has increased and it has declined in other parts of the country. The per capita availability of land during the same period has declined from 0.48 ha to 0.15 ha by 2000 A.D. Per capita investment in agricultural infrastructure is the lowest in eastern India, where the density of population is the highest. Only 25 to 30 per cent of the modern agricultural technology has reached the farmers. This is often because the technology has not been consistent with conditions of the farm situations.

The benefits of modern technology have, however, been restricted to favorable farming situations. Only 44 out of 453 districts are contributing half of the total food grain basket of the country. This clearly suggests that the technology supposed to be scale and resource neutral has been confined to the districts with favorable farming conditions. Since there is no further scope for horizontal expansion of land for cultivation, the only alternative left is for vertical expansion and diversification of farm enterprises with less demand on space and time particularly for small and marginal farmers (constituting 76 per cent of the farming community) who do not have much of resources, specially in rain fed areas (70 per cent of the total cultivated land). The new farming system research strategy should, therefore, necessarily concentrate on developing technology with participatory approach within the biophysical and socio-economic environments in which the farmers operate.

The period of late sixties is commonly referred as the golden period of Green Revolution in India. The attainment on food front during green revolution period resulted in spectacular increase in food grain production synchronizing with population growth. It reached as high level as 202 million tones in 1999-2000. The green revolution, however, remained confined to irrigated tracts and that largely to wheat alone. Wheat received maximum stimulus of varietal improvement, shared major portion of irrigation resources, and benefited most from the national research, extension and development infrastructure and government policy of support price and procurement. Conversely, rainfed crops occupying 70 per cent of the cultivated area remained deprived of these benefits and they are being pushed to marginal areas further (Singh *et al.,* 1990). Singh (1990) reported that growth in food grains was ahead of population growth during 1967-68 to 1985-86, in only 5 states (Punjab, Haryana, Uttar Pradesh, Maharashtra and Andhra Pradesh) while in 11 states, it lagged behind the population growth.

In addition to creating regional disparity in agricultural growth, green revolution resulted in overexploitation and degradation of land and water resources, deforestation, near extinction of valuable flora and fauna, advent of newer pests and diseases, and ecological problems. All these have resulted in platening or even decline in production and productivity even in the green revolution areas. Experiences have shown that every farming situation has specific mix of farm enterprises and their problems, and they require need-based specific solutions. So far major emphasis has been laid on crop production, which is one out of many farm-based activities. Even national thrust area has been simply changing from the group of food crop to the other. Agricultural development is no more a commodity or crop oriented programme. Today, there is a need to improve the overall agricultural development with multiple goals of growth, equity, employment and efficiency. It should aim at the twin objectives of developing packages of technologies that will result in sustained productivity and economic stability on the one hand and ecological stability on the other. This gigantic task is to be undertaken when national resources are shrinking, fragmentation of holding is continuing, incidence of rural poverty and unemployment are increasing, and uncertainty of weather is more certain. These have greater impact on small and marginal farmers. Will the crop production be able to sustain the pressure of decreasing farm income any more? It is doubtful. The solution lies in farming system management: a holistic approach.

Every farmer tries to choose the farm activities / enterprises depending upon physical and economic conditions prevailing in his ecosystem. Integration of various farm enterprises in a farm ensures growth and stability in overall productivity and profitability. It also ensures recycling of residues, optimization of resource use, and minimization of risk and generation of employment. Various enterprises that could be included in the farming systems are crop production, dairy, poultry, fish farming, vegetable and fruit growing, goat rearing, pig farming, sericulture, mushroom cultivation, agro forestry, bee keeping, flower cultivation, silvi-culture, agro-based industries and food processing etc. A judicious mix of enterprises, complimentary to cropping and suited to the given farming situation and the farmers' preferences would bring prosperity to rural areas. Sporadic exercises have to be taken on a large scale in systematic order in various agro-climatic zones and dry farming situation to identify suitable farming systems.

DEFINITION OF FARMING SYSTEM

Farming system is a complex inter-related matrix of soil, plants, animals, implements, power, labour, capital and other inputs controlled in part by families and influenced to varying degrees by political, economic institutional and social forces that operate at many levels.

Farming system is the scientific integration of different interdependent and interacting farm enterprises for the efficient use of land, lab our and other resources of a farm family which provides year round income to the farmers specially located in the handicapped zone.

The farming system, therefore, refers to the *farm as an entity of interdependent farming enterprises carried out on the farm.* The farm is viewed in a holistic manner. The farmers are subjected to many socio-economic, biophysical, institutional, administrative and technological constraints. The farming system conceptually is a set of elements or components that are interrelated which interact among themselves. At the centre of the interaction is the farmer exercising control and choice regarding the type and results of interaction.

The income from cropping alone in small and marginal farms is hardly sufficient to sustain the farmers' family. With the decline in farm size (0.15 ha/person) due to population explosion, it would be increasingly difficult to produce enough food for the family by the end of the century. Therefore, the farmer, to be assured of a regular income for a decent living (above the poverty line), a judicious mix of any one or more of these enterprises with agronomic crops should complement the farm income and help in recycling the farm residue / waste. The selection of enterprises must be based on the cardinal principle of minimizing the competition and maximizing the complementarities between the enterprises (Annadurai et. al., 1994).

Farming systems represent integration of farm enterprises such as cropping systems, animal husbandry, fisheries, forestry, etc. for optimal utilization of resources bringing prosperity to the farmer.

Farming systems approach introduces a change in farming techniques for higher production from the farm as a whole with the integration of all the enterprises. The farm products other than the economic products, for which the crops are grown can be better utilized for productive purposes in the farming systems approach. A judicious mix of cropping system with associated enterprises like dairy, poultry, piggery, fishery, sericulture, etc. suited to the given agro climatic conditions and socio-economic status of farmers would bring prosperity to the farmer.

LOWLAND FARMING SYSTEMS

Common cropping system in rice based systems, especially under agro climatic conditions of south India is rice-rice-pulse. Modified cropping system includes crops like maize, groundnut, and sesame. Rice-poultry-fish culture system appears to be more remunerative. Poultry droppings from the poultry shed placed well above the farm pond meets the needs of fish in the ponded water. Water in the pond can be used for irrigating the crops. About 500 layer chicks are sufficient to void required quantum of excreta to meet the feed requirement of 7,500 polycuture fingerlings in one ha of ponded water. Rice-fish system is also remunerative.

UPLAND IRRIGATED FARMING SYSTEMS

Additional income can be generated by adding enterprises like dairy, biogas and silviculture to the usual cropping systems. Two to three milch cows can be maintained from one ha straw. Recycling of farm and animal wastes through biogas unit can produce cooking gas for family use. Several such integrations can greatly increase farm income, besides providing work to family members throughout the year

RAINFED FARMING SYSTEMS

Principles of relevant components of environmentally sustainable farming systems should include:

Environmentally sustainable dryland farming systems emphasize conservation and utilization of natural resources. Agronomic practices of conservation tillage and mulch farming, rotational cropping, use of legumes and cover crops for improving soil fertility and suppressing weeds and efficient use of cattle manure are some of the components of sustainable farming systems. Following are the scientifically proven components that go into sustainable farming systems (Lal and Miller, 1990).

Objective	Components/Systems
Controlling erosion	Conservation tillage, strip-cropping, cover crops, inter-cropping, agro forestry, mulch farming, terracing, tied ridging, shelterbelts and windbreaks.
Conserving water	Mulch farming, ecofallow, weed control, tied ridges and supplemental irrigation.
Alleviating compaction	Growing deep rooted legumes, deep ploughing, application of organic manures, guided traffic.
Enhancing soil fertility	Legumes in cropping systems, cover crops, organic amendments and chemical fertilizers.
Improving soil fauna	Mulch farming, crop diversification, maintaining soil tilth, reduced tillage, organic amendments, planted fallows.
Plant protection	Crop rotations, crop diversification, residue manage-ment, need based pesticide use.

Sustainable farming systems for different ecological regions are:

Soils		Sustainable systems
Structurally active soils	A.	1. Grain-cover crop rotation using improved cultivars
		2. No tillage or minimum tillage
		3. Residue mulch
		4. Chemical amendments
		5. Planted fallow
	B.	A with agro forestry using woody perennial or shrubs
	C.	B in association with cattle at low stocking rate
	D.	Strip-cropping with alternate strips of shrubs.
	E.	Cover crops with grain crops.
Structurally inert soils	A.	1. Intercropping
		2. Tied ridges, fallowing, ploughing at the end of rains
		3. Compost or FYM
		4. Chemical amendments
		5. Planted fallow
		6. Supplemental irrigation
	B.	A with contour hedge rows of woody perennial at oneto two m vertical interval
	C.	B integrated with livestock based on light grazing.

A sound land use policy is necessary to tackle the problem of deteriorating natural resources, especially soil and water. Land use according to land capability and minimizing human and livestock pressure on land in accordance with carrying capacity should form central theme of land policy. Agriculture on steep slopes be stopped. Alternate land use systems, in place of arable cropping, should be popularized among small and marginal farmers. A strong political will and commitment is needed to combat forces leading to land degradation.

By and large, many medium and marginal farmers are used to farming systems since they started cultivating crops. Majority of the farmers in rural areas are maintaining work animals, milch animals, chicks, sheep, goats, etc. in addition to crop production. More than 75 per cent of farming community is used to farming system approach. The usual farming systems are agriculture along with milch animals or chicks or both milch animals and chicks. What is required for the farmers at present is not educating them on the benefits of farming systems for which they are used since they started crop production, but improving the productivity of the existing farming systems through latest technological innovations.

FARMING SYSTEM CONCEPT

'Farming' is a process of harnessing solar energy in the form of economic plant and animal products. 'System' implies a set of interrelated practices and processes organized into functional entity, *i.e.* an arrangement of components or parts that interact according to some process and transforms inputs into outputs (Fresco and Westphal, 1988). Farming system is therefore, designated as a set of agricultural activities organized into functional unit(s) to profitably harness solar energy while preserving land productivity and environmental quality and maintaining desirable level of biological diversity and ecological stability. The emphasis is more on a system rather than gross output. In other words 'farming system' is a resource management strategy to achieve economic and sustained production to meet diverse requirement of farm household while preserving resource base and maintaining a high level environment quality (Lal and Miller, 1990).

Thus, in farming system all the activities, decision, management, input/output, purchase/sale and resource(s) utilized make the matrix of farming system which interact with socio-economic and bio-physical environment for purchasing the necessary inputs and disposing the outputs by utilizing the natural resources (land, water, air, sunshine etc.) effectively. Sustainability is the objective utilization of inputs without impairing the quality of environment with which it interacts. Therefore, it is clear that farming system is a process in which sustainability of production is the objective.

This overall objective is to evolve technically feasible and economically viable farming system models by integrating cropping with allied complementary enterprises for irrigated, rainfed, hilly and coastal areas with a view to generate income and employment from the farm. The specific objectives are:

1. to identify existing farming systems in specific area and assess their relative viability;
2. to formulate farming system models involving main and allied enterprises for different farming situations;
3. to ensure optional utilization and conservation of available resources and effective recycling of farm residues within system;
4. to maintain sustainable production system without damaging resources base/ environment;
5. to raise overall profitability of farm household by complementing main/allied enterprises with each other.

SCOPE OF FARMING SYSTEM

Farming enterprises include crop, livestock, poultry, fish tree crops, plantation crops, sericulture, etc. A combination of one or more enterprises with cropping, when carefully chosen, planned and executed, gives greater dividends than a single enterprise, especially for small and marginal farmers. Farm as unit is to be considered and planned for effective integration of the enterprises to be combined with crop production activity. Integration of farm enterprises depends on many factors such as,

FARMING SYSTEM RESEARCH (FSR)

FSR programme implies the concept of total environment consisting of both natural and human elements, the methodological complexities may be great. The whole research process is recognized as dynamic and interactive with both 'backward' and 'forward' linkage between farmers and researchers. The "feed back" from the farmers help the scientists to eradicate the problems in achieving the useful results.

The term FSR in its broadest sense can be defined as any research that views the farming in a holistic manner and considers interaction between the components i.e., soil, plant, animal, labour etc. and also of the components with the environment (the bio-physical factor) in the system (CGIAR, 1978).

This type of research is most appropriately carried out by interdisciplinary team of scientists who, in association with extension specialists, continually interact with farmers in the identification of problems and in devising ways of solving them.

The farming system research (FSR) originates from recognizing the inter-dependence and inter-relationships of natural environment within the farming system. In FSR the farmers by participating in the research process help in the identification of the research problem as well as take part in testing the possible solution. Farmers' participation in FSR means that the system itself can be incorporated in the experiment and realistic results can be obtained through farmer-managed trials.

Improved agricultural technologies, even when considered as technically sound, are of limited value if they are not adopted by the farming community. In absence of adequate attention to the understanding of the agro climatic and socioeconomic milieu in which the farmers operate, generated and transferred technologies are found inappropriate to the needs and circumstances of the practicing farmers. The potential beneficiaries, particularly those farmers with limited resources operating in less favourable natural environment, often do not adopt the new technologies due to various reasons:

A less frequently heard explanation is that the recommended technologies themselves are simply not appropriate to farmers and their environment. Farmers, generally, seek those technologies which increase their income while keeping risks within reasonable bounds under their own circumstances and management practices.

The green revolution strategy mainly concentrated on better-off farmers and better-endowed areas with greatest potential for increasing agricultural production. The traditional technology generation and transfer models adopted have been found inadequate to meet the specific needs of a vast majority of farmers in developing countries. Research is mostly conducted at research stations under the conditions not representative of farmer's fields, with little or no farmer involvement. Inadequacy of the discipline oriented research and extension strategies to solve the problems of small, resource poor farmers operating in the complex, diverse and risk prone environments have led to the development of a more systemic, farmer oriented and interdisciplinary approach, known as **farming systems research (FSR).** The crux of the FSR approach is to develop and transfer appropriate technologies based on a clear understanding of the existing farming systems. Many programmes have been initiated in recent years with an orientation on farming systems, with limited success.

Principles Involved

The FSR approach includes: (a) viewing the farm as a whole, (b) identifying the farming systems, the interacting components and delineating boundaries, (c) systematically investigating the nature and extent of interdependence among the enterprises and identifying the constraint, (d) applying the modern technical know-how to the system so as to make it yield the optimum results, (e) studying the equity gender income, employment and resource use efficiency, and (f) dealing with the issues at integration level through analysis and solution of problems towards sustainable farming systems development.

Scope and Objectives of Farming Systems Research (FSR)

FSR aims to meet the following interrelated objectives (Pluknett et al. 1986):

1. To understand the physical and socio-economic environment within which agricultural production takes place.
2. To gain an understanding of the farmer in terms of his or her skills, constraints, preferences and aspirations.
3. To comprehend and evaluate existing important farming system, in particular the practice and performance of these systems.
4. To improve the identification of problem and opportunities for change in existing farming system and thereby better focus research on specific key aspects.
5. To enhance the capacity of research organizations to conduct research on priority problems.
6. To conduct research on new or improved practices or principles and to evaluate these for possible testing on-farms.
7. To evaluate new or improved systems or system components on-farms in major production areas under normal farm conditions.
8. To assist the extension, monitor the adoption and assess the benefits of improved farming systems.

Evolution of the Concept FSR

Farming system as such, is not a new concept to our country. The farmers are having different enterprises since long without any scientific base. The modern FSR gives scientific thought to the existing practice and finals out ways and means to make sustainable.

FSR evolved in the post "green revolution" era, during 1970s when it was observed over the years that small and marginal holders/farmers (constituting 76% of the farming community) toiling hard in adverse and risk prove areas (70% of the total cultivated land) were not benefiting from conventional agricultural research and extension programme and to generate and disseminate technologies widely adopted by small scale resource poor farmers as it is crystal clear that technology albeit sound by scientific standards of limited value if it is not adopted by farming community. The benefits of modern technology have however been restricted to favourable farming situations, only 44 out of 453 districts contributing to the half of the total food grain basket of country, which is a matter of great concern.

Since 1978, major reviews of FSR have voiced concern about the imprecision of terminology. They have called for a systematic clarification, ordering and labeling of concepts, activities and methods in FSR (Dillon et al., 1978) and conceptualization of FSR persists. In fact, it has become more acute as the range of activities encompassed by the term FSR has broadened. If the lack of clarity continues, confusion and misunderstanding about the objectives, products and role of FSR are likely to discredit research executed under the name of FSR and jeopardize donor support (Byerlee et al., 1982; CIMMYT and ISNAR, 1984; Dillon et al., 1978; Gilbert et al., 1980; Stoop, 1984 and Zandstra, 1982).

In response to this situation, FSR argued that:

1. Development of relevant and viable technology for small-farmers must be grounded in a full knowledge of the existing farming system, and
2. Technology should be evaluated not solely in terms of its technical performance, but in terms of its conformity to the goals, needs and socio-economic circumstances of the targeted small-farm system as well. The diagnosis of the problem and proposed solution reflected several critical changes in the scientific research community's assumptions and perceptions of small-farm agriculture. These became fully expressed in the development of FSR.

The most fundamental change was abandonment of the idea that smallholder-farming systems in the tropics and sub-tropics were static and primitive. Rather, it was realized that they are complex, dynamic systems, which had evolved in response to particular agro-climatic, ecological and socio-economic conditions (Collinson, 1983; Harwood, 1979; Hildebrand, 1981 and Norman, 1974). Similarly, researchers gradually realized that small-farmers were not rejecting technologies out of sheer ignorance. They recognized that small-farmers are rational decision-makers, but often pursue goals and employ criteria for evaluating technologies distinct from those used by agricultural scientists (Collinson, 1983; Norman et al., 1976 and White, 1981). Small-farm households, operating diversified production systems, are striving to meet both consumption and production goals under what are often marginal conditions for agriculture. They have to make complex decisions about the allocation of scarce resources and factors of production among numerous enterprises. Furthermore, small-farm households formulate management strategies and make decisions within the context of the whole economic system exploited by the household, including off-farm enterprises (Byerlee and Collinson, 1983 and Shaner et al., 1982). Consequently, neither yield maximization nor even profit maximization can be assumed to be the sole appropriate measure for the potential acceptability of a new technology. Under these conditions, optimum management of one of the several enterprises exploited within the system is rarely feasible in the small-farm situation (Cornick, 1983). Yet optimum management had been, and often still is, assumed in research station trials and evaluation of new component technologies.

As agricultural scientists strive to reach small-farmers with improved technologies, it also became clear that many broad-based technologies were endured inappropriate by the significant diversity in socio-economic circumstances and physical environments in which small-farmers operate. CIMMYT studies of adoption of high-yielding varieties of wheat and maize revealed that agro-climatic zone and socio-economic circumstances were the primary factors affecting adoption patterns. CIMMYT concluded that if technology is to be successfully transferred to small-farmers, more adaptive research is necessary (Byerlee and Collin son, 1980; Byerlee et al., 1982; Perrin et al., 1976 and Winkelmann, 1984). Finally, the development of ecological concepts and system theory and modeling techniques provided a framework and a tool for analyzing the multiple variables and complex interactions characteristic of small-farm systems (Simmonds, 1985). Drawing on these new insights and concepts, the FSR is evolved as an alternative to the stop-down approach of mainstream agricultural research and development.

FSR Concept

The FSR concept was developed in 1970s in response to the observation that groups of small-scale farm families operating in harsh environment were not benefiting from the conventional agricultural research and extension strategies.

The farming system, as a concept, takes into account the components of soil, water, crops, livestock, labour, capital, energy and other resources with the farm family at the centre managing agricultural and related activities. The farm family functions within the limitations of its capability and resources, sociocultural setting and interaction of these components with physical, biological and economic factors. The term FSR in its broadest sense is any research that views the farm in a holistic manner and considers interactions (between components and of components with environment) in the system. This type of research is most appropriately carried out by interdisciplinary teams of scientists, who, in association with extensionists, continuously interact with farmers in the identification of problems and in advising ways of solving them. It aims at generating and transferring technologies to increase the resource productivity for an identified group of farmers.

FSR Activities

The FSR approach requires that scientists should take up the technology generation by learning about the farming system and problems of representative farm families in a wide variety of agro climatic and socioeconomic contexts. On the basis of this learning, on-farm research activities are to be designed, implemented and evaluated by the scientists in association with the farmers. Feedback from the farmers will provide the agenda for further on-station research, as well as for the agricultural extension services.

The FSR process involves a set of integrated activities.

Farming System as Foot Stone of Sustainability

In order to combat the complex situations arising now-a-days, attempts are to be made to reorient our research programmes in a holistic approach for achieving sustainability in production and maintaining the ecological balance by a harmonic combination of interdependent and inter linked analysis of farm enterprises.

Sustainable Agriculture

The word "sustainable" is very easy to speculate than to define and is easy to conceptualize than operationalise.

Economic and ecological sustainability is enhanced when the available land, water, energy and other resources are utilized in a mutually reinforcing manner. A systems approach involving integrated attention to crop and livestock farming and to agro-forestry and aqua-culture will be helpful in generating more jobs and income and at the time protect soil health.

The definition of "sustainable agriculture" as a concept or as a process it self is not simple. It is possible to define it generally more in qualitative terms and less in quantitative terms. As a development process "sustainable agriculture" can be defined broadly as a process that ensures or result in the following:

1. Increase agricultural production in terms of less reliance on application of chemical fertilizers for yield, multiple cropping should be practiced that results in improving land quality.
2. Generation of additional or alternate employment to gainfully utilize the available human resources.
3. General improvement in quality of life without causing permanent or irreversible damage to the environment.
4. Should not allow natural resources (soil and water) degradation, but promote ecologically sound soil and crop management practice.

5. It should be economically sound.
6. Efficiently utilize all available natural resources and energy use efficiently should be high.

So far sustainable agriculture production farming system research must consider the following parameters.

1. Production efficiency
2. Economic incentive
3. Energy use
4. Employment generation
5. Environment friendly technology
6. Equity (The production and productivity we are getting at present should not jeopardize the likely potential for similar benefits in the future).

Harwood (1988) defined "sustainable agriculture" as an agriculture that can evolve indefinitely towards greater humanity, greater efficiency of resource use and a balance with the environment that is favourable to human and to most other species.

Comparison Between Existing Research and FSR

Existing Research	Farming Systems Research
– Existing one is completely component based research and there is no system	– Unlike component research it is system based research.
– Inappropriate technology of needs and circumstances of practising farmer	– It focuses on generation of low cost, location specific relevant technology.
– Faulty top-bottom conventional research extension method.	– It lays emphasis on farmers participatory research and ensures smooth diffusion and adoption.
– Do not consider biophysical, socio-economic and other production constraints of farmers.	– It enhances productivity provides full employment to farm family member and focuses mainly on benefit of resource poor farmers.
– Not eco-friendly and sustainable	– It is based on environment friendly technology and sustainable.

Goals

The growing concern on sustainable development has led FSR to emphasize on sound management of farm resources to enhance farm productivity and reduce the degradation of environmental quality. It advocates a process of change to meet the needs of the growing population and to promote their quality of life.

Farming enterprises include crop, livestock, poultry, fish, tree crops, plantation crops, sericulture, etc. a combination of one or more enterprises with cropping, when carefully chosen, planned and executed, gives greater dividends than single enterprises especially for small and marginal farmers. Farm as a unit is to be considered and planned for effective integration of the enterprises to be combined with crop production activity. This integration of farm enterprises depends on many factors such as:

Objectives of FSR

For any farming system research, decision matters have to understand the physical and socio economic condition within which farm business is to be conducted. It should help to allocate the existing resources to meet the objectives of farming. In adopting a farming system, the following important objectives must be kept in mind.

1. Identifying limitations to increase farm productivity.
2. Provide technological intervention options for improving farming systems with a resource base technique through "On-Farm Adaptive Research (OFAR)".
3. Define and develop appropriate farm technologies so as to provide stability in production, employment throughout the years, income at periodic intervals, improved management skills of the farmers and descent living.
4. Develop methods for measurement and monitoring of sustainability of farming system.
5. Efficient utilization of available resources.

TYPES OF FARMING SYSTEM RESEARCH (FSR)

There are many different methods adopted for FSR by different institutions and individuals. Simmonds (1984) with a detailed study on these variations grouped them into 3 basic categories.

FSR in Strict Sense

It refers to the study of farming system as they exist. It is strictly an academic activity oriented towards description and analysis of farming systems including in depth understanding of how they function.

New Farming System Development

This type of research often undertaken at experiment stations, using state-of-the-art technologies and integrating crop, livestock and trees in a synthetic farm.

On-Farm Research with a Farming System Perspective

Randhawa and Venkateswaralu (1986) suggested that in the early stages, farming systems research consisted mainly of testing research results through on-farm trials. The on-farm trials are aimed to identify and quantify the effects of improved practices. Then it follows the sequences of national demonstrations and operational research to confirm the production potential of new crops and cropping systems. Farming systems research as a major toll aims to close the gap between agricultural research and the performance in actual farms.

Farming by tradition, characterized by the mix of enterprises and the farmer as a manager discovers the best way of organizing the individual enterprises in an over all farming system. He does so by seeing how they compete against or complement one another in the use of scarce resources and chooses the mix which will maximize his net return. Farming systems approach particularly in this context of constantly changing technologies can help the farmer in optimizing the use of farm resources and maximizing the net income. Production potential and economics of intensive farming system have been well documented by several workers (Malviya *et al.* 1986 and Sing *et al.* 1988).

On–farm research provides the context for collaboration between farmers and researchers. Researchers gain a deeper understanding of the farming systems and decision making context of the farm family. On–farm research also permits technology to be evaluated under the environmental and management conditions in which it will be used. Farming systems research channels feed back on farmers' goals, needs, priorities and criteria for evaluating technologies to station based agricultural scientists and national policy makers.

It is a problem oriented research which recognizes that changes to farming systems must be adapted to circumstances of the intended user.

The majority of FSR to date can best be described under this, which has increasingly found favour with the FSR practitioners' world wide.

FSR Provides Feedback from Farmers

This is the FSR channels feedback on farmers' goals, needs, priorities and criteria for evaluating technologies to station-based agricultural scientists and national policy makers.

The types of research activities to which the term is applied: (1) the precision in the use of the concept, (2) the definition of the boundaries of the system under analysis, (3) the level of specificity employed to define a farming system, and (4) the degree of socioeconomic analysis incorporated.

Precision in the Use of the Concept

There is significant variation in the precision with which the concept of FSR is used in the literature. FSR has been interpreted to mean a framework for agricultural research, which allows problem definition to transcend conventional commodity or disciplinary boundaries. It has also been used to refer to strategies for rural development, which target a specific type of farming system. For others FSR has a more precise meaning, indicating a specific methodology for adaptive research. In its strictest sense, FSR is conceived as true systems analysis resulting in a detailed quantitative model of an existing farming system.

Definition of the Boundaries of the System

Defining the boundaries of a system is a fundamental step in system analysis because it determines the factors are to be treated as endogenous variables and as exogenous parameters. Yet this has been one of the principal areas of ambiguity included under the broad label of FSR.

All practitioners of FSR assert that the whole-farm system is the framework for their research. In practice, however, the concept of the farming system and the definition of its boundaries often recede into the background. The actual system under analysis may be defined as broadly as an agro-climatic zone, or as a specific set of biophysical interaction between elements of a generalized farming system. While all are related to FSR as it was originally conceived, these are clearly distinct areas of research involving equally distinct objectives, theoretical frameworks and methods, which result in very different products. For example, running on-farm experiments on the competition for nutrients within an intercropping system is radically different from collecting base-line climatic zone. Yet both have been thought of as FSR.

Level of Specificity Employed to Define Affarming System

FSR researchers use the concept of the recommendation domain to resolve the tension between the diversity among individual farms and the need to design technology applicable to as large a number of farms as possible. The recommendation domain is a grouping of relatively homogeneous farming systems for which the technology developed in FSR should be relevant and appropriate.

The degree of specificity in criteria used to classify farming systems into recommendation domains will reflect the scale on which the research institute is operating, its mandate and its specific research objectives. An International Agricultural Research Center (IARC), for example, may develop very generalized categories of farms divided largely according to agro-climatic criteria. In contrast, a

National Agricultural Research System (NARS), working in a specific sub-region, may categorize farms according to a set of much more specific criteria such as product mix, presence of draught power and socio-economic status of the farming hot behold. Clearly, the research processes employed for constraint analysis, technology development and evaluation will be markedly different in the cases. The question then arises, should they both be considered FSR?

Level of Socio-Economic Analysis Incorporated

The integration of socio-economic analysis into agricultural research and technology development was a central concept of FSR as originally conceived. Today, however, it is not uncommon that socio-economic factors are completely ignored in research conducted under the name of FSR.

This is due to several conceptual and organizational difficulties in uniting socio-economic and technical forms of inquiry and analysis into a coherent research process. First, the 'body of knowledge' on the socio-economic aspects of small-farms upon which FSR can draw is limited. Second, a new conceptual and methodology had to be developed to incorporate the socio-economic sphere of agriculture into the research and technology development process. Third, the time required to collect and analyze socio-economic data often does not conform to the pressured pace of funding cycles in agricultural research and development. Fourth, developing effective inter-disciplinary collaboration between social scientists and technical agricultural scientists can be difficult and consuming because it requires expanded training and a strong commitment on the part of scientists (Hildebrand, 1981; Horton, 1984 and Rhoades and Booth, 1982). Fifth, many socio-economic factors influencing agricultural production is quite location-specific. Consequently, researchers often do not grant them importance until the final testing stage of a new technology. Lastly, a strong technological bias prevails in research institutions, the assumption is that extension and development agencies, not research institutions, would bear the responsibility for demonstrating the utility of an 'improved' technology to farmers or for adjusting farmers' circumstances to accommodate the technology.

The terminology proposed in the following section attempts to eliminate these areas of ambiguity in the conceptualization of FSR by systematically categorizing the diverse research activities and approaches which have been labeled FSR into six more precisely defined types of research endeavors.

Classification of Farming Systems Research

1. Farming System Analysis (FSA).
2. Farming System Adaptive Research (FSAR).
3. Farming System Component Research (FSCR).
4. Farming System Base-line Data Analysis (FSBDA).
5. New Farming System Development (NFSD).
6. Farming System Research and Agricultural Development (FSRAD).

Farming System Analysis (FSA)

FSA is in-depth, quantitative analysis of an existing farming system. The objective of FSA is to understand the structure of interactions within the system and to quantity stocks and flows. The typical product of FSA is a model of the system. FSA is basic research on the structure and functioning of a farming system, which contributes to the general 'body of knowledge' upon which FSAR depends. FSA is what Simmonds (1986) has called Farming System Research in strict sense.

Farming Systems Adaptive Research (FDSAR)

FSAR is restricted to research, which integrates all of the key concepts, which originally defined FSR as a distinct approach to agricultural research. FSAR is interdisciplinary, on-farm, adaptive research targeting a defined group of relatively homogeneous farming system characterized by similar constraints. It draws on the existing 'body of knowledge' generated by strategic and applied research and on farmers 'knowledge of their environment to develop and fine-tune short-term technologies for the designated farming system. FSAR is location-specific and is most appropriate for the national and regional level of agricultural research. The objective of FSAR is to increase productivity of the farming system through the development of relevant technology which conforms to farmers' circumstances. This means that the technology must be (1) possible, and (2) attractive for the farmers to adopt in that it conforms to farmers' determination of their needs. (Collinson, 1982).

Farming System Component Research (FSCR)

As noted above, this term designates the more focused, station-based, applied and adaptive research on farm sub-systems or components designed to support FSAR. It is distinct form mainstream agricultural research because it responds specifically to problems identified in small-farm systems and endeavors to produce prototype technologies relevant to the circumstances and needs of small-farmers. FSCR is distinguished, however, by its tight definition of the boundaries of the system under analysis. FSCR focuses on a specific sub-system or the management of specific resource. The unit of analysis is the field or plot, not the farming system, and the number of variables considered is limited. The emphasis is on biophysical processes and interaction within the designated sub-system.

Farming Systems Base-Line Data Analysis (FSBDA)

PSBDA takes a key idea of FSR as originally conceived— that it is necessary to understand the farming system and environment before trying to induce change-and applied it to a much larger unit of analysis, the agro-climatic zone. In FSBDA, basic data on an agro-climatic zone is collected and analyzed with the objective of diagnosing major constraints and developing a classification of major types of farming systems. Collection and analysis of secondary data coupled with the execution of large-scale surveys are the principal methods employed in FSBDA. FSBDA is comparable in its objective to the Description and Diagnostic Stage of FSAR. However, it is executed on a much lager scale and greater emphasis is placed on biological and physical, rather than socio-economic, variables.

New Farming Systems Development (NFSD)

NFSD is fundamentally a 'top-down' approach to agricultural research and technology development, which seeks revolutionary, rather than evolutionary, change. It requires multiple and complex changes instituted simultaneously with the ultimate product being a radically new farming system (Simmonds, 1986). It is neither location-specific nor farmer-oriented. Rather, it attempts to generate a broad-based technology designed to overcome major constraints in large agro-climatic zones. Although NFSD is distinct from mainstream agricultural research in that it cuts across commodity and disciplinary boundaries in seeking solutions to major environmental constraints, it embodies only a few of the key concepts of FSR as originally conceived. NFSD should, therefore, be viewed as an entirely different research activity, despite its use of the farming system as the unit of analysis.

Farming Systems Research and Agricultural Development (FSRAD)

FSRAD refers to farming system programme, which integrates agricultural research and development strategies. The farming system is placed in the broader context of the economic and policy environment of a targeted sub-region. The objective of FSRAD is the long-term transformation and development of agriculture in a designated sub-region through both technological development for major farming systems and institutional and economic reform in the environment. It, thus, unites, mainstream agricultural research, FSCR, FSAR and sometimes NFSD with development of agrarian infrastructure and services as a means of increasing overall agricultural productivity of the sub-region. FSRAD embodies all of the key concepts of FSR as originally conceived but expands on them further by incorporating an additional level of system analysis. Furthermore, it not only facilitates communication between farmers and agricultural researchers, but also between farmers and development experts, policy-makers, and extension personnel

Farming system perspective (FSP) research is broken down into six more precisely defined types of research. They are all incorporated under FSP because they share some of the key concepts of FSR as originally conceived. In essence, they are all products of the post-Green Revolution conceptual shift in agricultural research and development. The new conceptual framework recognizes that small-farm agriculture in developing countries is fundamentally different from western capitalist agriculture. It thus requires a distinct institutional structure and experimental framework if agricultural research and technology development is to be effective. The research activities incorporated under FSP are manifestations of this realignment.

The primary objective of all the research activities described above is to increase the productivity of small farms. All, except NFSD, attempt to do this through the generation and transfer of technology appropriate for the goals, needs, priorities and circumstances of small-farmers. They all employ a systems approach in their analysis emphasizing interactions between elements and components of the system. They all require interdisciplinary collaboration since problem definition cuts across conventional commodity and disciplinary lines. And, finally, they all complement and build upon strategic and applied commodity and disciplinary research.

TABLE 14.1: Comparison of Farming System Perspective (FSP) research activities in terms of inclusion of key concepts from Farming System Research as originally conceived

Key concepts	FSP Research Activity					
	FSA	FSAR	FSCR	FSBDA	NFSD	FSRAD
Farmer-oriented	X	X	(X)		X	
Systems-oriented	X	X	(X)	X	X	
Problem-solving		X	X	X	X	
Interdisciplinary (including social sciences)	X	X	(X)	(X)		X
Complementary to commodity and disciplinary research		X	X	X	X	X
On-farm experimentation		X	(X)			
Farmer participation and Feed-back	(X)	X	(X)			

X = Full Integration of key concept,
(X) = Moderate integration,
FSA = Farming System Analysis,

FSAR = Farming Systems Adaptive Research,
FSCR = Farming System Component Research,
FSBDA = Farming Systems Base-line Data Analysis,
NFSD = New Farming Systems Development,
FSRAD = Farming Systems Development,
FSRAD = Farming Systems Research and Agricultural Development.

The six research activities do, however, differ significantly, even within these common dimensions. Their specific objectives, functions and products are quite distinct. It is these differences, which require that a more precise terminology be employed to distinguish among the six types of research. The differences, as well as the similarities, among the six types of research are clearly seen when they are compared in terms of the degree to which they incorporate the key concepts of FSR as it was originally conceived.

Basic research: that designed to generate new understanding (*e.g.* how the partitioning of assimilates is influenced by plant height);

Strategic research: that designed for the solution of specific research problems (*e.g.* technique for detecting dwarfing genes in wheat seedlings);

Applied research: that designed to create new technology (*e.g.* breeding new varieties of dwarf wheat that can respond to high levels of nitrogen without lodging);

Adaptive research: that designed to adjust technology to the specific needs of a particular set of environmental conditions (*e.g.* incorporating dwarf wheat into farming systems of the rain fed areas of the Pampean region of Argentina).

Priorities of Fsr and Development (D)

The intervention may involve in one or more of the following enterprises:

Improved technology and knowledge must be transferred within recommendation domain. The participation of extensionists in farming system is important and their contribution can be from stage of technology evaluation on farmers' field to transfer in the recommendation domain. Extension specialists must be involved in technology dissemination and technology verification on farmers' fields. Technology recommendation domain must consider physical, soil-water regimes, biological (pests/diseases) constraints and socio-economic level of farming community. The availability of critical inputs of farming and the infrastructure facilities and associating the allied services agencies, are essential for successful adoption of farming system technology.

It must be judged by six parameters of performance. These include:

Improved farming system must enhance one or all of these parameters that are foundation for sustainable development. The technology intervention must not degrade the environment. All farmers are not equal in technology adoption and their resource base and management skill differs and thereby different levels of FSD may be needed. Technology adoption by a farmer depends on his resource base, reliable of technology, input and credit availability, support price and infrastructure. The risk bearing capacity, awareness, management skill and opportunity make the farmers to adopt technology to various degrees resulting in different levels of productivity.

FSR METHODOLOGY

A lot of emphasis has to be put on the methodology of FSR for its successful adoption and sustainability. Methodology of FSR is a 5-step process:

1. Target and Research area selection (Descriptive stage).

2. Problem identification and development of a research base (Diagnostic stage).
3. Planning on Farm Research (Design stage).
4. On Farm Research analysis (Testing and Evaluation stage).
5. Extension of results.

Crop, livestock, fish, tree crops and other farm enterprises practiced in various parts of the country differ widely due to climate, soil and socio-economic environment. The focal point in FSR being the farmer, research in farming systems perspective must be done in each micro-farming situation. It is, therefore, suggested that to define FSR domain, the country should be broadly divided in to 7 zones, *viz.*

1. Areas falling under green revolution,
2. Areas potentially rich but have not undergone green revolution,
3. Dry lands,
4. Arid zones,
5. Coastal areas
6. Hilly areas, and
7. Tribal belts.

The methodology of FSR consists of 5 stages, *viz.* (a) descriptive and diagnostic, (b) design, (c) testing, (d) evaluation, and (e) extension.

Multidisciplinary efforts of a team of scientists combining agronomists, social scientists, economists, and specialists in livestock, fishery, forestry, tree and plantation crops and farm management are essential to carry out the Farming System Research (FSR) at all the above mentioned stages.

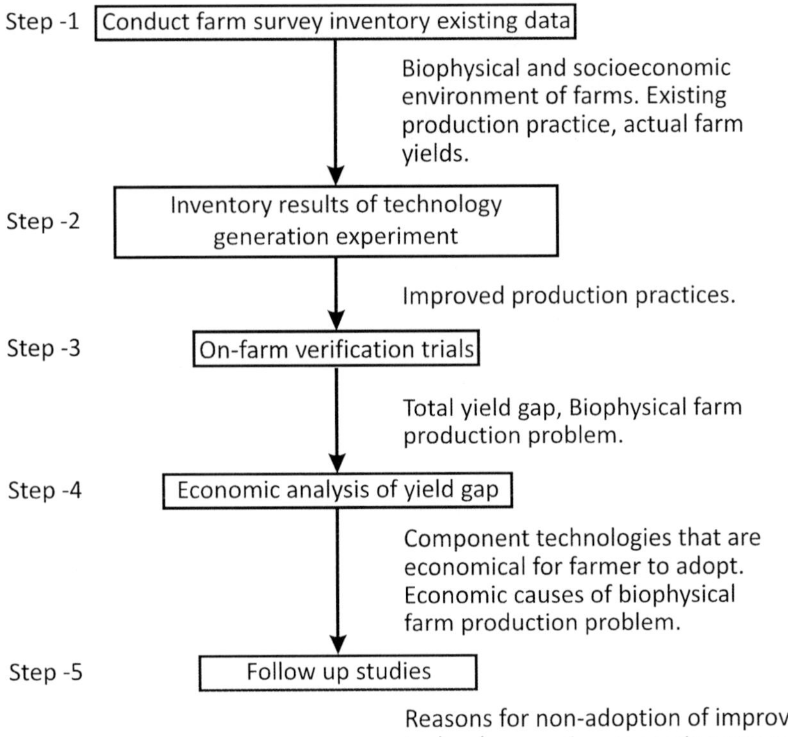

Steps in Problem Identification

Multidisciplinary efforts of a team of scientists comprising agronomists, social scientists, economists and specialists in livestock, fishery, forestry, tree and plantation crops and farm management are essential to carry out the FSR programme at all the above-mentioned stages.

Descriptive and Diagnostic Stage

The aim of the diagnostic survey is to understand the farming systems that are practised in the selected target areas. This enables the scientists and the farmers to determine the constraints that the decision-making households face and the flexibility in the current farm enterprises could adopt. To carry out the diagnostic survey, the following sampling procedure is suggested.

Step I

All the districts falling in a particular farming system zone should be grouped based on 4 criteria, *viz.* (i) high productivity and low spread, (ii) high productivity and high spread, (iii) low productivity and low spread, and (iv) low productivity and high spread. Further, each of the farm enterprises such as production, livestock, poultry and fish, having been subjected individually to above 4 criteria would be factorially combined to arrive at total number of groups in a zone. For instance with 2 farm enterprises (say crop and livestock) there will be 16 groups, with 3 enterprises the number of groups will be 64, and so on. After having classified all the districts as above, select randomly adequate number of districts form each group.

Step II

In the district chosen for the study, prepare a list of all the blocks/taluks and select a few blocks randomly for the study.

Step III

From each of the selected blocks/taluk, select adequate number of villages say, ten, randomly from the totality of the villages in a block/taluk.

Step IV

Carry out the diagnostic survey by rapid rural appraisal for each of the ten selected villages and prepare an inventory of farm resources, production constraints and support services. Using the information so obtained the entire population of farm holdings may be grouped, based on the 4 criteria: (a) large, (b) medium, (c) small and (d) marginal holdings. Further, each of the enterprises such as crop, livestock, poultry and fish, having subjected individually to the above criteria, would be factorically combined to arrive at the number strata as has been done above, for two enterprises the number of strata will be 16, for three, it will be 64, and so on.

Step V

From each of the stratum so obtained, select randomly adequate number of farm holding, say 6. Four out this, will be the sites on which FSR will be conducted while the remaining 2 would serve as control.

Design Stage

In the stage suitable action plan for the selected farmer is formulated. The improved technologies relevant to overcoming or avoiding constraints identified in the first stage are specified. The cumulative store of knowledge at the research station is the best source of ideas for potential improved technologies that might be appropriate. Equipped with this knowledge and that obtained from the diagnostic survey, a farm management design for the selected farmer, taking into consideration his preference, resources and overall production benefits, is to be prepared. All intervention plan taking into consideration farmers' preference, risk involved, labour resources, gender distribution of labour and family age structure etc. ensures that the intervention fits in the farmers' existing skill as far as possible.

Testing Stage

The objective of the testing stage is to evaluate a few of the more promising technologies arising from the design stage on the farmers' farms. First part consists of trials at the farmers' level that use farmers' land, labour and other resources but the research team provides the managerial input skill. The second part involves farmers' testing, providing their own land, labour and other resources. The second part on-farm experimentation under farmer-managed trial technique will provide the potential for substantial research worker/farmer interaction and result in assessing the technologies under conditions as close to reality as possible. In this stage, the involvement of extension personnel can particularly strengthen the programme

Evaluation Stage

The research team by visiting the farmers periodically will collect the necessary data for analysis and drawing conclusions. The data needed would consist of (a) input use (both available on the farm and purchased), (b) labour utilization pattern, (c) output of crops, animal, birds etc., (d) production of by-products like crop residue and manures, (e) soil fertility in the case of crops, and (f) income farm enterprises and the build up of assets. The data would then be subjected to economic analysis with a view to work out the profitability, technical feasibility, social and cultural acceptability and sustainability of intervention with existing resource base and infrastructure. An index of acceptability is calculated and if acceptability is low, the research team reconsiders the benefits of the improved technology. The farmer is to finally judge the appropriateness of the proposed technology and if the farmer rejects the technology, the research process can be repeated to determine the reasons and ways to overcome it.

Extension Stage

The acceptable new technology is promoted in collaboration with Departments of the State Governments and other non-government voluntary agencies. To achieve this, it is necessary to bridge the gap between the knowledge of the farmer and the know-how of the technology he is going to adopt. Since all identified problems are not research issues but are on a whole farm basis, in terms of cash, labour and land allocation, imparting training to the farmer is essential before technology is promoted. So also is the availability of critical inputs of farming by increasing the infrastructure facilities and by associating the different services and agencies involved.

Strategy

In view of serious limitation on horizontal expansion of land for agriculture, only alternative left is for vertical expansion through various farm enterprises requiring less space and time but giving high productivity and ensuring periodic income, especially for the small and marginal farmers belonging to handicapped zones (rainfed area, dry lands, arid zone, hilly areas, tribal belts and problem soils). For this purpose, several farm enterprises could be combined, *viz.*, crop + livestock, crop + poultry, crop + dairy, agri-horticulture system, crop + horticulture +sericulture system, agro-forestry, silvi-pastoral, rice + fish, rice + fish + mushroom, floriculture, apiary, fish + duck +poultry, etc.

FSR Programme Implementation

For meaningful execution of integrated farm enterprises, the following activities should be undertaken by a multidisciplinary team of scientists with farmers' participation and involvement at all stages:

1. Thorough understanding of existing farming system and their components.
2. Assessment of resource availability in the farm environment and identification of biophysical, socio-economic, institutional, administrative and technological constraints.
3. Development of economically viable and efficient integrated farming systems suitable for various domains.
4. Diffusion of improved technologies and receiving 'feedback' for further improvement of the system as a whole.
5. Continuous improvement in component technology to fit into a given farming system.
6. Improvement in quality of farming systems research.
7. Research-extension linkage through 'On-farm Adaptive Research'
8. National and internal linkages.

Policy Orientation

Farming System Research and Development (FSR and D) would need a major shift in programme, planning since it will be a multi-enterprise management system. A multidisciplinary team would be needed to implement the programme. Since major emphasis is on on-farm trials, frequent mobility in the operational area will be required for closer interaction with the farming community. Pruning of routine and unproductive research and development projects will have to be taken up to stengthen the FSR and D programme.

To protect the natural resources (land, water and forests) it will be essential to develop ecologically safe energy policies. Efficient farming system needs efficient energy management. The benefit likely to be harvested from farming system would be much higher than from expenditure incurred on it, as the quality of life would improve in rural sector and would have long-term bearing on the survival of mankind.

Since Farming System is a developing concept, courses have to be designed and offered at the undergraduate level in the Agricultural Universities and Farming System Research methodology has to be taught at the post-graduate level. Pending finalization of such academic curriculum, the scientists belonging to various disciplines but involved in system research should be given crash training on FSR without further delay for immediate manpower requirement.

Farmers' training on various aspects of farming systems is also equality important for successful.

Farming System Research and Development

FSR and Development programme should be such that it

1. Integrates the consideration of sustainable productivity, profitability, efficiency, equity and ecology in its research strategy;
2. Consider are the enterprises being handled by the farmers' family at one point of time;
3. Gives more emphasis on on-farm trials than on on-station research;
4. Places as premium on team work and community approach in the system
5. Involves researcher, development/extension agencies and the local farmers at all stages of operation of FSR and D;
6. Promotes creativity and commitment amongst the people involved;
7. Involves social scientist and environmentalist in the programme; and
8. Enables women to play a pivotal role in system research and development.

Orientation of Present-Day Research and Development Towards FSR and (Implementation)

The status of system research being conducted was reviewed by Sankaran and Rangaswamy, 1990 and Singh *et al.*, 1990. The major problem is how to tune the single crop/commodity/enterprise based present day research and development approach to multi-enterprise research and development approach. Since, the characteristics of the FSR and D mentioned above, and the concept, objective and working modalities of the National Agricultural Research Project (NARP) already in operation in the country are similar to a great extent, hence, it seems to be feasible to integrate the FSR and D approach in the working System of the NARP. It may function as a major complement of the NARP rather than an independent new project. It will simply require little strengthening of the centres of NARP and turning its research and extension mandate towards FSR and D while it will continue to be a part of the State Agricultural Universities (SAU's) research system.

1. NARP has already established/strengthened 120 zonal research stations in 17 states covering all the 26 Agricultural Universities in the country. There are altogether 343 zonal research stations and sub-stations engaged in conducting location specific, problem – oriented, and need-based agricultural research to develop appropriate technology for a particular agro-ecological situation. Thus, NARP provides a strong national network of regional research stations to initiate Farming System Research on moderate scale.
2. It possesses site-specific approach.
3. It is provided with multidisciplinary team at each centre (crop production, soil science, agro forestry, horticulture, fishery, animal science, economics, and extension) to address a variety of problems confronted by local farmers.
4. Research programme is framed around the location-specific production constraints.
5. Every zonal station is guided by, and programme is periodically discussed in, the Zonal Research and Extension Advisory Committee (ZREAC), which includes Zonal Research Scientists, local development and extension personnel.
6. Each Zonal Station has 25-50 ha of research farm for on-station research to develop location-specific and appropriate component technology to feed the local farming systems.
7. On-farm trials are being taken up in a large scale to keep close rapport with the local farming community.
8. Every zonal station is equipped with office-cum-laboratory building, residential quarters, lab and farm equipments, library facilities and training hostel.
9. NARP has created training facilities for working farm scientists in their relevant subject to update them with the latest advances made in their fields of specialization at national and regional level.
10. NARP has a mandate to strengthen research and extension linkage at the regional level.

While integrating FSR and D in NARP functioning, the following facts may be kept in mind for its effective implementation:

1. FSR and D may be initiated on a moderate scale at selected NARP zonal stations situated in agriculturally difficult and backward region of the country, *viz.*, dry farming/rainfed, and desert, hilly, flood prone and coastal ecosystems.
2. Multidisciplinary team at zonal station may initially form the core group for FSR and D as per existing enterprises in the tract. Scientists of the relevant discipline may be deployed from the SAU headquarters, if need be.
3. Inclusion of social scientist, an environmentalist and an economist in the core group should be the pre-requisite for FSR and D.
4. 70 per cent of the FSR and D activities are to be undertaken in farmers' fields.
5. Facilities for mobility of the scientists and development staff are to be strengthened.
6. Provision of devoted field staff for conduct and continued supervision of the on-farm farming system trial must be ensured. The field staff will have to stay nearer to the site of trial.
7. Emphasis must be given on the full involvement of farmers belonging to poor and marginal category as well as farm women in decision making and implementation of the programme.

8. Researcher, extension and development personnel and farmers of the area will have to work as a unit, each participating in activities of the other.
9. Feedback mechanism of Extension Education, which is completely missing in present-day development strategy, needs to be made functional and effective.
10. Development of enterprise-based component technology will be a continued activity of the research station and SAU headquarter to feed appropriate technology to the farming systems.
11. SAUs have developed status report of each agro-ecological zone wherein each zone has been sub-divided into different farming situations. This will enable to plan farming system programme down to a micro-level in the zone.
12. At NARP station, several research projects funded by different organizations other than NARP may continue their mandated research programme provided it is of relevance to the local situation. But, they will orient their plan of research towards the objectives of FSR and D programme of the area.
13. As far as on-farm trial on Farming System is concerned the existing farming system of the area will form the check treatment. The experimenter may superimpose one or two enterprise combinations/well proven component technology over and above the treatments. The major objective of such trial will be to study the relative productivity and ecological sustainability of different enterprise combination put under test, in a particular farming situation.
14. There will be an immediate need to develop appropriate methodology for conduct of farming system research, particularly in farmers' fields.
15. All development agencies including agriculture should work in collaboration with each other, preferably under one umbrella.

DIFFERENT COMPONENTS OF FARMING SYSTEMS

Field Crops

So far as field crops are concerned cereals like rice, wheat, maize, sorghum, ragi; pulses like black gram, green gram, red gram, horse gram, soybean; oilseeds like groundnut, mustard, niger and cash crops like sugarcane, tobacco, jute, etc. are grown in Indian fields. Crop production constitutes a sizeable portion of farm income followed by livestock and Agro-horticulture system or Aqua-culture.

Development of efficient need based cropping system to match the growing needs of farmers' family and community will continue to find high priority in agricultural research. To support cropping system research the component technology, improved cultivar, right kind of farm machinery, pesticides and better soil and water management practices must be developed to match the systems in Northern India. Where rice cultivars must be water regime specific-upland, lowland (irrigated, shallow, rainfed and deep water) and wheat cultivar must fit in different dates of planting when soil moisture becomes favourable for wheat planting after harvest of rice.

A short duration legume such as mung bean and bean or cowpea can be planted with minimum tillage and irrigation to maximize the economic return. Similar example exists in rice-legume system in South-eastern India. A suitable mix of grain and fodder crop is essential in mixed farming. The income from this unit may be upto Rs. 1.80/Rupee invested.

Vegetables

Vegetables form a very important component specially for marginal farmers to provide income at frequent short intervals round the year. Cash crops like tomato, brinjal, chillies, greens, cole crops, gourds, mellons, pumpkin, colocassia etc. could be grown. The income from this unit may be Rs. 1.60/Rupee invested.

Fruit Cultivation

Fruit plants like coconut, banana, papaya, mango, pineapple, etc. may be planted by small farmers in a systematic manner. Multistoried cropping by taking coconut, pineapple, banana, papaya, and ginger/turmeric can be practised. The returns per rupee invested may go beyond Rs. 2.50/Rupee invested.

Livestock Integration

Crop-livestock integration is an old age practice of Indian farmers. In present context, crop-livestock integration must consider continuing need of growing family size of small farmers. Small farm of 0.5 to 1.0 ha must adopt a cropping system that support a reasonable number of animals that generate enough draught power, provide organic nutrient for growth and add to the income of small farmers.

Integrated farming system involving crops and animals must consider an appropriate combination of animals (cow, buffalo, pig, goat, sheep, etc.) and crops that maximize the return per unit of investment to small farmers. This should also optimize the farm family labour and skill.

Integrated farming system involving livestock must also harmonize the land use for other economic activities of small farmers for sustainable development.

About 12 litres of milk can be obtained from the 4 cows in Orissa condition.

Poultry Farming

Poultry farming can easily be combined with integrated farming. The battery system and deep-litter systems are generally followed by constructing poultry farming. The battery system may be followed by constructing a perforated platform over the deeper portion of the pond. The structure can be raised by split bamboo sticks or wooden planks or wine mesh. Small pillars support this platform. The cages have to be installed over the platform with some suitable shade structure. Initially about 100 birds in 4 rows may be accommodated. Poultry droppings and waste food would be directly available to the heavily stocked fishes in the pond through perforations created on the platform.

Rhode Island and White Leghorn breeds are suitable for this purpose. Egg production starts at the age of five and a half months and attains the peak by 18 months. Thereafter production gradually declines. At this stage layers may be replaced by fresh stock and old ones marketed for meat. Poultry feed can be prepared from locally available food mixed with fish meal. The profit per layer bird/yr is Rs. 26.56.

Duckery

Ducks and geese are preferred to continue with fish farming. Their meat is in much demand and their excreta serve as an effective pond fertilizer. A duck house is constructed adjoining the shallow fish breeding ponds when the fingerlings become more than 12 g in weight otherwise the ducks will prey upon them. Duck excreta indirectly utilized by fish. The Indian runner duck is reported to produce 45-55 kg of excreta in a year. If required wastes from kitchen and green vegetables may be used for supplementary feeding. The profit per duck/yr is Rs. 39.52.

Aquaculture

Pond serves various useful purposes like domestic requirement of water, supplementary irrigation to adjacent crop field, and pisciculture. With traditional management farmers obtain hardly 333-444 kg of wild and culture fishes per ha annually. The bright prospect of boosting the production of culture fishes by different methods was impressed upon the client through training meets and by demonstration of compost fish culture. The method includes elimination of wild and predatory fishes by using Mahua (*Madhuka* spp.) cake at the rate of 250 kg/ha one month before stocking. Analysis of pond soil and water; having at the rate of 600 kg/ha; application of organic manure and chemical fertilizers; stocking density of 7500 fingerlings such as Roho (*Labeo rohita*), Catla (*Catla catla*), Mrigal (Cirrhiness mrigala) Silver carp (*Hypophthalmiclithys molitrix*), common carp (*Cyprinas carpio*) and grass carp (*Ctenopharyndodesh idealla*) in the ratio of 2.1:2.5:1.5:1.5 and supplementary feeding of growing fingerlings at the rate of 2 per cent of the composition of fish feed was mustard oil 50 per cent, rice bran 40 per cent and fish meat 10 per cent.

Besides aquaculture in whole some ponds, it is also possible to undertake rice-cum-fish culture in low and deep water situation or in dish shaped depression (Bil) using latest technology. It has been observed that rice and fish grown together are mutually benefited and if supplementary fish feed, fertilizers and cultural management are provided, it could be a profitable avocation.

Large tracts of land inundated by monsoon or tidal wave to variable depths with submergence tolerant rice growing there form the natural habitat or a variety of fish species and prawn juveniles. They enter the rice field from nearby perennial water sources or rivers. In lowland rice area every rice farmer is also a potential fish farmer. With motivation, training, finance and co-operative effort fish farming may become a major income generating enterprise.

Agroforestry

Agro forestry is a system where woody and non-woody interacting plant combinations are carried out for achieving certain objectives like (1) conservation of soil and water (2) recycling of nutrients (3) maintenance of organic matter and some important physical properties of soil (4) sustained crop productivity, In addition to the production of timber and green leaf forage for cattle.

Fuel trees like Acacia, Subabool and Eucalyptus etc. can be planted to meet the fuel and forage requirement of farm family and cattle respectively. Intercrops with pulses and oil seeds can be taken up initially, till the canopy of the trees obstructs the penetration of sunlight.

Beekeeping

This is a very good supplementary enterprise in a farming system. Investment required is very low. Indian species *Apis cerana indica* yield 1.5 kg honey/box/ collection with 4 collections per year, whereas Italian species *Apis mellifera* yield 10 times more *i.e.* 15 kg/collection. The boxes can be maintained for 5 years during which profit from country breed will be Rs. 1238/box and for Italian breed Rs. 5831/ box. Six boxes can be maintained in a ha of land.

Mushroom Cultivation

It is also a good supplementary enterprise in a farming system which provides engagement of farm women and children. There is a good demand of this product in the market and it can also be consumed by the farm family as a delicious dish. Dhingri mushroom in winter season and paddy straw mushroom in rainy season in addition to white milk mushroom could be grown successfully. The investment per bed is Rs. 30.00 in case of paddy straw mushroom and Rs. 20.00 in case of Oyster mushroom with a net income generating Rs. 37.00 and Rs. 28.00 respectively.

Biogas Plant

The excreta of cattle is used directly or indirectly through biogas plant as slurry for fish culture, for horticulture and animal provides sufficient manure and feed for a pond of 0.13 ha.

The cattle shed measuring 8m x 4m are sufficient for 4 cattle. Washings from the cattle shed goes directly to the biogas plant. The digested slurry can be utilized for fish culture pond. Excess of slurry may be stocked in a compost pit. The biogas generated may be used for domestic purposes and also for lighting the fish pond to attract insects which form a choice food for fishes.

The net return from a 2 cum. plant could be Rs. 12/day with daily investment of Rs. 9.00 and total return of Rs. 21/day.

Miscellaneous Enterprises

Other enterprises like sericulture, lac culture, homestead vocations like coir work, bamboo work, rope making etc. can also be included in a farming system by taking into account the prevailing demand of these finished products in the village or in local markets.

Economics of Different Components of Farming System

Economics of different components of farming system differ with time, (season), region (agro-climatic zones), and with the combination of different enterprises. A comparison of return/rupee invested or economics of different components of farming system are given in this chapter at appropriate place.

Inter Relationship among Different Enterprises

The components of a farming system depend on one another and interact with each other, as a result of which there is recycling of available resources. This character contributes towards sustainability, environment friendly technology and economic feasibility of farming system by the following ways:

1. **In aquaculture crop integration**: A large quantity of silt and manure is added to the fish ponds every year. This along with animal excrement and bodies settle down at the bottom of the pond forming 30-40 cm depth each year. On an average 25 kg of pond silt (rich in NPK) produce 1 kg of green fodder.
   Due to synergistic effect of fish on rice, the rice yield increased by 10% and weeds and insects are controlled by fish. In coastal Orissa, a production of 50 kg/ha of tiger prawn (*P. monodon*), 250 kg/ha of mullets, 3000 kg/ha tilapia besides 2.4 tonnes/ha of rice have been achieved.
   Aquatic plants like Makhana (*Buryale ferox*) and singhara (Trapa spp.) have been demonstrated specially in swampy areas.

2. **Aquaculture biogas livestock integration**: Cattle in aquaculture system derive benefits either directly by bathing in the pond and drinking the pond water or by utilizing the fodder generated through application of silt to the crop.
   In contrast to this fish in pond feed on the zoo-planktons and phytoplankton, bacteria etc. produced by application of cow-dung and poultry manure.
   In addition to this ducks act as bio-aerators who aerate the pond water by their legs. Biogas plants set up in the farm generate energy which is used for cooking and lighting, light traps over the pond attract the harmful insects of the crop plant and kill them and make available for consumption of fish in pond.
   In general in integrated farming system, the chances of diseases are less, control of insects and pest is better but proper vigilance and timely vaccination are necessary toward off diseases.

FARMING SYSTEM RESEARCH IN INDIA AND ABROAD

Income and Employment Generation

Crop and Employment

Increased irrigation facilities and consequent increase in cropping intensities have positive impact on demand for farm labour. There is need for the farm sector to grow @ 3.5 to 4 % per annum to bring an end to the rural unemployment and under employment problem where a higher proportion of the rural population are dependent upon farm sector in one farm or the other. The number of rural unemployment is growing at a rate exceeding 2 per cent per annum. So a high and sustained rate of agricultural growth is necessary and sufficient condition for generating high rates of growth of non-farm employment opportunities in agro based industries. These problems will never be solved when there is stagnation in farm sector. A stagnant agriculture leads to increase in rural unemployment and poverty **(Singh, 1979).**

Goud (1988) reported that income to an individual flows from ownership of assets and farm employment. The basic reason why the rural people have low income is that they posses little or low cultivable land due to the process of continuous fragmentation of holdings which is the single most productive assets in rural areas.

Nagaraja (1989) reported that effect of crop mixture not only give relatively higher monetary return but also provide more opportunities of employment. He therefore emphasized on reorientation of cropping pattern in favour of crop mixtures. For a favourable output- input ratio, the author suggested raising of groundnut + red gram in Anantpur district of Andhra Pradesh to realise more income and create more employment for the work-force.

Poultry and Employment

Rao (1991) reported that poultry farming was not only labour intensive but also labour distributive and the family labour (both female and child) can be productively utilised throughout the year. Among livestock activities, poultry was identified as second major activity. The landless and the tenant house holds that depend upon wage earning, found poultry as an effective tool for improving their economic condition. It was further reported that the average human labour utilisation on poultry farms was 150 man-days for 100 layers. Thus poultry units have the potential for creating employment opportunities in the farm sector and should, therefore, be given emphasis for expansion.

The findings of **Pati (1992)** on labour utilisation pattern revealed that rice based farming system with dairy and pisciculture had relatively larger employment generating potentialities than the farming system with dairy enterprise alone. The author further observed that of the total employment, female labour employment in case of the latter outweighed (per cent) than the former. Labour employment pattern under two different farming systems made it abundantly clear that dairy and pisciculture units were the two main enterprises where it was relatively higher than other enterprise. In case of rice based farming system with dairy, the female labour employment per annum was 275 man-days while in case of the second system with dairy (200 man-days), and pisciculture (143 man-days), it was 343 man-days. Among crop enterprises, HYV paddy in both *kharif* and *rabi* seasons offered more employment to female work force than the local paddy, pulses and oil seed crops under two different farming systems and the intensity of such use appeared to maintain an inverse relationship with the farm size classes.

Integration of Farm Enterprises

The basic necessity of integration of enterprises is not only to make the farm an independent unit but also fulfil the input demands of other structural units. The major benefit in integrated aquaculture is the comprehensive utilisation of various farm products and recycling of wastes within the farm itself leading there by to effective, utilisation of resources, sharing of energy and finally maintaining the ecological balance **(Rath, 1990, Sharma and Das, 1985 and Sinha, 1989).**

The system of integrated fish culture and livestock raising is very much well developed in China, Malaysia and in certain European countries (**Hickling, 1960 and Rath, 1989**). However, little attention is paid on this line in India (**Jhingran and Sharma, 1978**). It is apparent that integrated fish farming can provide an excellent opportunity for increasing farm produce; farm income and can give rich dividends to farmers for their efforts and investment.

A case study conducted under farmers' condition through adoption of pond-based technology in an area of 1.8 ha in the village *Sanghalei* (Pipili block Puri district of Orissa) showed that by integrating fish, field crops and vegetables the gross income and net income could be increased to the tune of Rs. 45,500 and Rs. 36,000 respectively for an operational expenditure of Rs. 9,500. The return per rupee invested was as high as Rs. 4.79. (Project Report of Farming System Research, 1989, OUAT, Bhubaneswar).

Similarly a case study undertaken in Keonjhar district of Orissa with enterprise combination of Field crop- vegetables – Fruit- Plantation Crops – fishery – revealed that the gross income and net income of Rs. 25,949 and Rs. 14,779 respectively would be obtained with an investment of Rs. 11,170. The return per rupee invested was Rs 3.32.

Singh (1992) reported a net profit of Rs. 41,000 through adoption of farming system involving *Magur* fish, duckery, poultry and dairy. From a holding size of 0.12 ha, the percentage return of capital was 29. However, the expenditure on construction of integrated farming complex was Rs. 141,000 and the recurring cost, Rs. 43,000. the annual investment was Rs. 85,000.

Rangaswamy *et al.* (1996) obtained a net profit of Rs. 11,755 per year from rice-poultry- fish- mushroom integrated farming system (IFS) in 0.40 ha area. Conventional cropping system (CCS) with rice – rice – green manure/pulses gave a net income of Rs. 6334 per year from the same area. By inclusion of poultry- fish-mushroom components, the net income per day increased to Rs. 32.20 as compared to Rs. 17.35 in the CCS. An additional employment generation to the tune of 174 man-days per 0.40 ha per year was also possible in IFS.

Behera and Mahapatra (1999) from pond based farming system research studies comprising enterprise like field and horticultural crops, fishery, poultry, duckery, apiary, mushroom, dairy and agro forestry at Bhubaneswar observed net income of Rs. 58,360 and employment for 573 man-days, on a small piece of land (1.25/ha), ensuring high standard of living for small and marginal farmers

Panda and Patro (1998) from the integrated farming system (IFS) research studies for the marginal farmers having 0.6 ha of land comprising enterprise like field crops and dairy (IFS) and conventional cropping system (CCS) at Bhubaneswar observed that the total net income under rice-dairy IFS was Rs. 12, 359.00 and the net income in CCS was only Rs. 9895/year from 0.6 ha/year over CCS. Moreover, it generated an additional employment of 398 man days per year over CCS from the specified area and thus helped in the recycling of farm wastes and by-products round the year. It also minimized the risks associated with mere crop production.

The future of Indian agriculture depends heavily on the development of appropriate farming system as it fits to resource poor farm families of different agro-ecological zones. The endowment of abundant sunshine, long growing season, responsive soil type, and combination of surface water, ground water and seasonal rains and above all a progressive peasantry offer vast scope for an intensive farming system through multiple cropping and diversified farming including animal husbandry, forestry, sericulture, fisheries etc. The challenge is to upgrade technological and social disciplines on a continuous basis and integrate these disciplines to suit the region and the farm families in a manner that may ensure increased production with stability, ecological sustainability and equitability.

Research studies carried out in lowlands, uplands with supplemental irrigation and dry lands have demonstrated the technical feasibility and economic viability of the integrated farming systems attempted. Besides facilitating cash income, these models generate additional employment for family labour and minimize the risks associated with mere crop production. In order to make the Integrated Farming Systems Management more effective, the lead farmers are to be trained in the Agricultural University farms. If the farming system is linked with agro-processing and other post harvest technology, the farm economy could be further stabilized and strengthened. This requires a massive extension approach and institutional support.

FSR for Different Agroclimatic Zones in India

1. High altitude cold deserts: Pastures with forestry, goats, rabbits and settled agricultural crops like millets, wheat, barley, and fodders.
2. Arid and Desert region: Centering mainly in animal husbandry with the camels, sheep and goats, and growing with fodder and field crops.
3. Western and Central Himalayas: Horticultural crops as a major component and have a less intensive agriculture mainly on the hill terraces and slopes with maize, rice, wheat, pulses and fodder crops.
4. Eastern Himalayas: Primitive crop husbandry with rice, millets, pulses etc. Agro forestry systems are also common. Piggery and poultry are the chief livestock activity.
5. Indo-gangetic plains: Intensive crop husbandry like Rice-wheat-maize/mustard/ pulses and livestock, dairy, cattle and buffaloes.
6. Central and Southern Highlands: Cotton-sorghum-millets/pulses with dairy cattle, sheep and goats and poultry are the secondary livestock and animal husbandry enterprises.
7. Western ghats: Major activity on plantation crops, cultivating rice and pulses are the secondary agricultural activity. Cattle sheep and goats are the livestock

components, which in most parts, are maintained as large herds and allowed to range.

8. Delta and Coastal Plains: Rice cultivation with other enterprises like fishery, poultry and piggery, etc. Capture fisheries of the marine ecosystem is a specialized enterprise and does not mix with cropping activity.

Farming System Research in the Sub-Humid and Humid Tropics of Northeast India

Shifting cultivation, locally known *'Jhumming'* is widely practiced in the hills of northeast India and in tribal areas of Andhra Pradesh, Madhya Pradesh and Orissa. The traditional system followed a 20-30 year rotation cycle, but population pressure and the increasing demand for food have shortened the rotation cycle to between three and six years (ICAR, 1983). Research on farming systems with improved technology to replace shifting cultivation by permanent agriculture was initiated by ICAR in 1975 with the establishment of an Agricultural Research Complex for the North Eastern Hill Region at Shillong. In this region, upland rice is the main crop and is grown in mixture with maize, finger millet (*Elusine coracana*), Setaria, beans, tapioca, yam, banana, sweet potato, ginger, cotton, tobacco, chillies, sesame and vegetables. After two or three years of cultivation the land is left fallow to allow the regeneration of forest vegetation and improvement of soil fertility, but the productivity of the land is declining and food shortages are increasing.

A comparison of shifting cultivation with alternative improved farming systems conducted by ICAR from 1975 to 1980 indicated that bench terraces and contour bunds for conservation, combined with a permanent agricultural system, provide a more stable and renumerative system than shifting cultivation, provided the conservation measures are properly maintained and improved crop management practices and fertilizer are used. Research also showed that the best land use pattern for hill slopes is agro forestry for the upper third of the slope, a horticultural-pastoral system for the middle slope, and agriculture or crop-based systems for the lower part of the slope. The best land treatment is contour for the upper slope, half-moon shaped terracing for the middle slope, and bench terracing for the lower slope.

Farming System Research in the Humid Tropics

Plantation crops such as coconut, arecanut, cashew nut, tuber crops, banana and pepper are the main crops of the humid tropics in India, especially in Kerala, parts of Tamil Nadu, Karnataka, and Andhra Pradesh. The Central Plantation Crops Research Institute and the Coordinated Project on Plantation and Tuber Crops have been leaders in research on mixed cropping and farming systems based on the dominant commodity of interest to small farmers. According to the ICAR Review Team's report for the years 1982-87 intercropping with tuber crop, pineapple and banana is the most promising system in coconut-based farming, while in the mixed cropping system, a combination of coconut with cocoa, pepper and pineapple is the most efficient mixture. A net return of Rs. 23,000/ha was obtained from an irrigated intercropped plantation, compared with Rs. 11,400 from a sole crop of irrigated coconut, and Rs. 6,900 from a sole crop of rain fed coconut. Growing grasses and forage legumes between the coconut palms and raising milk cows is another promising farming system. Besides, the normal crop of coconut, five cows can be raised successfully on one hectare of land, giving Rs. 14,500 /ha as net profit. The sustainability of the various systems is still being evaluated, but the farmers need better genotypes and improved components of the technology to increase productivity, sustainability and profitability.

Farming System Research in the Arid Regions

The main limitations to crop production in arid regions are low precipitation and high evaporative demand. The soils are sandy, and wind erosion and the frequent drought are serious problems. To study these problems and to develop a sustainable agriculture, the Government of India established a desert afforestation station at Jodhpur in 1952, which was later converted into the Central Arid Zone Research Institute (CAZRI), in 1959 with support form UNESCO and the Australian government. It is responsible for developing agro forestry, silvipasture, agro-horticulture and crop-animal based farming systems for the arid zones of India, with an emphasis on socio-economic constraints and the development of farming systems that ensure the efficient management of resources.

Before the Green Revolution, the research at CAZRI concentrated on improving subsistence systems of farming, the stabilization of sand dunes, shelter belt establishment, the selection of shrubs, trees and grasses suitable for different ecological zones, and improving the productivity of crops and animals suited to these environments. Progress was limited by the lack of biologically superior plant material. However, after the introduction HYVs, especially hybrid pearl millet in 1964, the institute started introducing more productive genotypes of crops and packages of practices to improve the traditional arid zone farming systems. Emphasis also shifted to the use of water harvesting and recycling to stabilize and increase productivity. Crossbred sheep and scientific range management systems were effectively used for developing improved crop and animal-based farming systems. In the 1980's silvipastoral, agro forestry and agro-horticultural systems of farming started to receive greater research attention.

Farming System Research in the Semi-Arid Tropics

The reorganized ICAR formulated plans for dry land agricultural research in India following the Green Revolution of the mid 1960s and developed an All India Coordinated Research Project for Dry land Agriculture (AICRPDA) in 1967, comprising a network of 24 research centres. The project was approved by the Government of India, which initially allotted a budget of Rs. 5 million to introduce dry land farming research at the Soil Conservation Research and Training centre directly under the ICAR management. However, to strengthen and extend the support for research into dry land agriculture to all 24 research centres, and to give it a national priority, the budget was enhanced to Rs. 18.7 million from 1970, and headquarters for the coordinating centre of AICRPDA was established at Hyderabad. The centres included in the network of this important research project catered to the need for technology development for different agro- ecological regions. The work in the semi-arid tropics (SAT) was relevant to nearly 80 per cent of the geographical area of the country and so this formed the major thrust of the project. The SATs are defined as areas where the mean monthly temperature exceeds 18 °C and where precipitation exceeds potential evaporation for two to seven months a year (Troll, 1965). Areas with two to four-and-a-half wet months are called dry semi-arid. The SAT environment is characteterised by strongly erosive and variable rainfall, by soils that are generally of poor fertility, by numerous biological constraints and by complex risk-reducing farming systems which combine crops and livestock.

The Government of India also allotted further Rs. 200 million for pilot projects to verify, validate and adapt dry land technology on farmers' fields. The research scientists were encouraged to use the pilot project areas as 'on-farm research laboratories'. Financial and technical assistance from Canada enabled AICRPDA to operate an interdisciplinary research team at every centre, consisting of an agronomist, a plant scientist, a soil scientist and an engineer, with supporting scientific staff and with other specialists at the central Coordinating Centre at Hyderabad. In 1985, the project headquarters was upgraded to form the Central Research Institute for Dry land Agriculture (CRIDA). Most of the ICAR-funded FSR relevant to dry farming is carried out at ICAR central agricultural research institutes at Hyderabad, Jodhpur, Dehra Dun, Shillong, the Andaman and Nicobar Islands, and at the SAUs. A major contribution to FSR in the SAT has also been made by ICRISAT. The international centre set up in 1972, which combines its efforts with the NARS in producing dry land farming technology for use in the semi-arid tropics. FSR at ICRISAT had strong implications for research organization and management but was highly relevant to the SATs of India. Because the national programme is more suited to the development of location-specific applied and adaptive research, FSR at ICRISAT gave priority to strategic and basic research, with the aim of understanding principles and processes, and developing conceptual models through path-breaking research for improving farming systems. It was directed at the identification of constraints, diagnostic surveys, the development of components and packages of component technology for enhancing productivity, sustainability, and the economic viability of a soil-and climate-specific farming systems using base data analysis, on-station and on-farm research, but excluding diagnostic surveys and benchmark village level studies.

FSR at AICRPDA / CRIDA

Reviews of the achievements to ten years (AICRPDA, 1982) and fifteen years of dry land agricultural research in India (CRIDA, 1986) provide an insight into the development of farming systems research in the national programme. The change from simple experiments to systems research has been an evolutionary process, with a series of phases of research development (Singh, 1983). At first, AICRPDA attempted to change the traditional farming system by replacing the long-duration photosensitive varieties of crops with the higher-yielding short-duration and non-photosensitive varieties (HYVs). It also mounted a programme for screening and modifying landraces and locally improved varieties for their adaptability to a range of environments typical of the rain fed areas in India.

In the first phase, which coincided with the world oil crisis, which made fertilizers and pesticides, the two important inputs in agriculture, expensive and unattractive to risk-prone agriculture, research was aimed at developing technologies involving low monetary inputs. Extensive investigations showed that plant nutrient deficiency was as serious a yield-limiting factor as moisture stress; that matching of genotypes and nutrients with moisture availability in the soil would make the farming system more efficient; and that the short duration HYV and hybrids, though developed for more favorable environments, showed equally good potential in the moisture stressed environments of SAT. Thus development of dry land technology all over India was aimed at improving the productivity of existing crops and cropping systems, and adapting the new genotypes to those cropping systems.

In the second phase (1975-80), much effort was directed at studying long term climatic data (Singh, 1987) in order to minimize risk during crop growth. During the same period ICRISAT worked on concepts for intercropping and on environmentally adapted cropping systems. ICRISAT also started research on strategies for changing the physical, chemical and biological environments and their management, resulting in the concepts of watershed management technology.

The third phase (1980-85) considered integrated farming system development. This period was marked by an appreciation of the importance of soil and rainwater management technology and of integrated nutrient management techniques and their integration with cropping systems. Growing awareness of agro-climatologically and socio-economic factors, agro-forestry, and tillage and land-use system research influenced research on dry land agriculture during this period.

In the final phase (1985-90) the concept of watershed management was appreciated, with acceptance of responsibility for research on watersheds in different agro-climatic regions. In this, CRIDA/ICRISAT collaboration has stimulated greater research effort on farming systems and sub-systems and on the components of integrated dry land technology. The development of farming systems research in India has thus been an evolutionary process in which the national and international agricultural research organizations have played complementary roles.

INTEGRATED FARMING SYSTEMS (IFS) UNDER VARYING SITUATION

Integrated Farming

Integrated Farming is defined as biologically integrated farming system which integrates natural resources and regulation mechanisms into farming activities to achieve maximum replacement of off-farm inputs; secures sustainable production of high quality food and other products through ecologically preferred technologies; sustain farm income; eliminates or reduces sources of present environment pollutions generated by agriculture, and sustains the multiple function of agriculture (IOBC, 1993). Thus integrated farming systems must be built on the sound foundation of ecological science. It emphasizes a holistic approach. Such an approach is essential because agriculture has a vital role to play that is much wider than the production of crops, including providing diverse, attractive landscapes and encouraging biodiversity, and conserving wild life. Sustainable development in agriculture must include integrated farming systems with efficient soil, water, crop and pest management practices, which are environmentally friendly and cost effective. The future agricultural system should reorient from the single commodity system to food diversification approach for sustaining food production and income. Integrated farming systems, therefore, assume greater importance for sound management of farm resources to enhance farm productivity which will reduce environmental degradation and improve the quality of life of resource poor farmers and to maintain agricultural sustainability. The aim of the integrated farming system can be achieved by:

The Role of IFS in Crop-Livestock Interaction

A review of Mc In tire and Gryseels (1987) reveals that crop-livestock interactions take place in two main systems. First, there is interaction without full integration, defined here as a segregated system. Farmers own animals, and herders might grow crops, but the integration is not complete because power and manure are not fully used by either group. This type of interaction is dominant in the West African Sahel and Savannah.

The second is complete crop-livestock interaction, defined here as an integrated system. The cultivator owns and manages animals, uses draft power, feeds crop residues or fodder crops and recycles manure for soil fertility or for fuel. This type of interaction is dominant in the Ethiopian and Malgache high lands and in parts of Zimbabwe, Malawi, and Senegal.

It also reveals that the goals of Farming systems research in the study of crop-livestock interactions are to increase the direct physical contribution of animals to crop production via inputs of power and manure; to increase the indirect contribution of one sector to the other via cash income; and to increase the direct physical contribution of crop production to animal production, especially from crop residues. For each goal, there are different strategies according to the current degree of integration and according to the constraints to crop and animal production.

Integrated FSR Activity and Models

Integration of various agricultural enterprises *viz.*, cropping, animal husbandry, fishery, forestry etc. have great potentialities in the agricultural economy. These enterprises not only supplement the income of the farmer but also help in increasing the family labour employment.

The integrated farming system approach introduces a change in the farming techniques for maximum production in the cropping pattern and takes care of optimal utilization of resources. The farm wastes are better recycled for productive purposes in the integrated system. A judicious mix of agricultural enterprises like dairy, poultry, piggery, fishery, sericulture etc. suited to the given agro-climatic conditions and socio-economic status of the farmers would bring prosperity. Farm as a unit is to be considered and planned for effective integration of all enterprises.

A review of an article by Obias (1986) reveals that the research and development team of Maya Farms, an integrated livestock farm and meat processing enterprise on the outskirts of Metro Manila, Philippines, developed a number of integrated farming systems, in which the biogas plant provided a vital link between crop and livestock farm, between coconut plantation and cattle raising, between pig and poultry production and between livestock and fish farming.

In an integrated crop and livestock farm, a demonstration model for 1.2 ha developed at Maya Farms, had one hectare under crops and the remaining area for piggery unit. Manure from pigs and water buffalo (a draft animal for land preparation), processed in a bio-gas plant, which supplied the farm house with enough fuel for cooking, ironing clothes, lighting and running a small gas refrigerator. The sludge was sufficient to fertilize the one hectare field, producing one crop of rice during the wet season and two crops of corn during the dry season. The corn and the rice bran were used to supplement the feed for the pigs and the rice straw and corn stover were fed to the water buffalo. The farmer was able to sell five or six porkers every month.

In the integrated coconut and cattle production program, the farm was able to increase the nut production to triple times by utilizing the sludge from the biogas plant and through the irrigation and fertilization of the forage crops intercropped in the coconut plantation while the cattle bring additional profit.

Obias (1986) also indicated the possibility of integrating piggery and poultry so that the biogas produced over and above the requirement of pig farm is sufficient to cover the biogas deficiency in the poultry operation. Moreover he attributes that by combining pig production with poultry raising, chicken manure helps in increasing the consistency of pig manure slurry for the biogas unit. It has been identified that a 500-sow piggery would produce enough excess biogas to support a 12,000 layer operation reared on the farm itself.

In another attempt it was reported that by combining pig and fish, the pig manure supplements the fertilizer need of the fish pond. To avoid the direct effect of pig manure on fish through infection of pathogens and the tainted taste of fish, the processing of manure through a biogas plant is being effectively attempted. An additional advantage in the latter is that the system produces a fuel gas before the waste is used for the fish pond. Thus biogas can serve as fuel for the farm house and the piggery.

Obias (1986) concludes that recycling of farm wastes though biogas plant relieves the farmer of the problem of waste disposal and environmental pollution. According to him it also provides some relief for problems caused by the high cost of three major farm inputs – fertilizer, feed and fuel. By using integrated farming systems that optimize the utilization of farm wastes, the farmer reduces his production costs and increases his productivity.

Mehla *et al.* (1986) reported that in Northern India, with the triple cropping intensity of fodder crops and recycling of feed nutrients in a milch animals and biogas system, from one hectare of land, it is possible to produce 82.63 GJ of energy and 906.11 kg of protein from milk; 263.93 kg of N; 148.93 kg of P_2O_5 and 90.94 kg of K_2O fertilizer from biogas slurry and 2897 kilowatts of power per annum for farm operations from biogas using the Kirloskar biogas engine. Traditional systems of wheat and rice cultivation can produce 93.87 GJ of energy and 654.25 kg of protein from food grain per hectare per annum. The recycling of feed nutrients in the available straws of rice and wheat crop can yield an additional 16.31 GJ of energy and 166.67 kg of protein from milk; 81.23 kg of N; 56.00 kg of P_2O_5 and 25.14 kg K_2O fertilizer from biogas slurry and 1097 kilowatts of power per hectare annually from biogas.

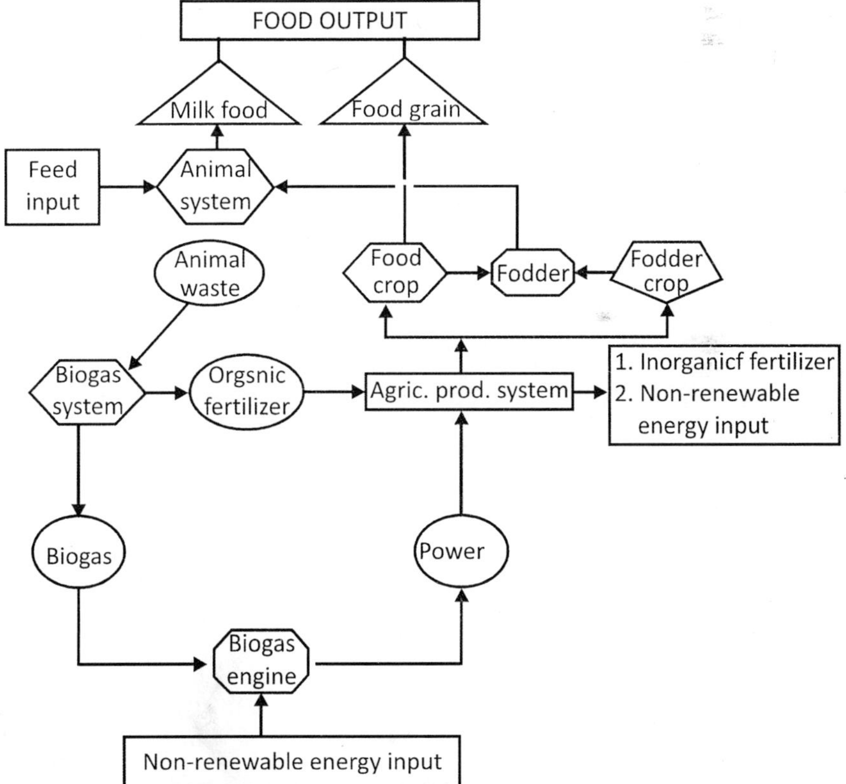

FIG. 14.1: Model for recycling feed nutrients in crop, animal and biogas systems. (Mehla et. al 1986).

A model as constructed by Mehla *et al.* (1986) to integrate the process of recycling of feed nutrients through a crop, animal, biogas system is given in Fig. 14.1. The model was constructed on the basis of the following observations.

1. There is a linear relationship between availability of feed and fodder inputs and milk production system with a provision of horizontal growth of herd size in relation to feed and fodder inputs (Mishra, 1975). This observation holds true under Indian conditions.
2. During anaerobic digestion of animal waste (dung) in a biogas system, the carbon content of organic matter is degraded into carbon dioxide and methane (biogas). There is a linear relationship between degradation of the carbon content of organic matter and biogas production (Prakash, 1976).
3. The biogas energy can be burnt to produce energy in the form of heat and can be used as fuel to generate power through a biogas engine in which 80% of diesel oil is replaced by biogas.
4. The residues available from the biogas system contain organic manure and can be used for replenishing nutrients in the soil.
5. In the model, the flow of uncontrolled inputs and losses are not accounted.

Integrated Approach in Wetland Situation

A model was developed at TNAU, Coimbatore for 0.4 ha (one acre) farm considering small and marginal farmers of Tamil Nadu. The components integrated in the rice based farming system are (a) cropping (b) fish culture and (c) poultry. While programming for cropping system, due consideration was given to crops under cultivation in the locality, the crop rotation that is being followed, and water requirement for fish culture included. An area of 0.36 ha was allotted for cropping and remaining 0.04 ha for fish rearing.

1. Rice Based Cropping With Poultry-Cum-Fish Culture

In wetland, the normal cropping programme followed is rice-rice-green manure/pulse. But in the integrated approach, a modification was made in the existing cropping pattern by including crops like maize, groundnut and gingelly. The cropping intensity increased to 300 per cent as against 240 per cent recorded earlier.

The polyculture fingerlings are preferred for this system as the growth rate within a year is found more economical than monoculture. The fingerlings were released in the pond as per the recommended stocking density of 7500 numbers per hectare of ponded water; accordingly 300 fishes in the following ratio were released in the pond.

Grass-carp (grass and algal feeder)	30 (1)
Rohu (column feeder)	60 (2)
Mirgal / common-carp (bottom feeder)	90 (3)
Silvercarp / catla (surface feeder)	120 (4)
Total	300

No separate feed was supplied to the pond, instead, then poultry droppings from the poultry shed placed well above the farm-pond formed the feeding material.

It has been quantified that 500 layer chicks are sufficient to provide required quantum of excreta to meet the feed requirement of 7500 polyculture fingerlings in a hectare of pounded water. As such 20 Bapkok chicks were used in the proposed program for egg yield and to meet the feed requirement of fingerlings released in the

0.04 ha (10 cent) pond. The fingerlings which solely depend on the poultry excreta grew at the rate of one kg per year.

In this integrated system, the waste products of rice formed 33 per cent of the poultry feed and the feed cost could be reduced by 57.5 per cent by substituting rice grain, maize, oil cakes of crop component and the Lucerne grown around the fish pond. In commercial poultry farm, cost of production per egg will be around 42 paise. By way of reducing the feed cost, the cost of production per egg under integrated farming system was reduced to 19 paise.

Thus, the net income under the integrated farming system was Rs. 15,145 / acre (0.4 ha) while Rs. 9,010 acre (0.4 ha) was obtained form the cropping system, rice-rice-green manure. The employment opportunity generated due to integration was 73 man-days and 246 woman-days while in pure cropping it was 68 man-days and 180 women-days (Rangasamy *et al.*, 1988).

The planning strategy for lab our surplus economy in the rural sector have the objectives of maximization of employment opportunities (Thiagarajan, 1977) However, various programs launched and efforts made in the country in the past have not been adequate enough to absorb the growing and existing manpower. It has been stated that about 300 million lived in the rural areas. If the rural poor consisting of the landless laborers, marginal and small farmers to be helped to get better income, there has to be diversification of lab our use through the creation of employment opportunities so that the growing surplus lab our force could be absorbed in the villages (Swaminathan, 1981). This is possible by integrated faming system approach.

2. Crop-Poultry-Fishery in Cauvery Delta

A model was developed at Tamil Nadu Rice Research Institute, Aduthural with poultry-cum-fish culture as a component in mixed farming. The study was conducted in a holding size of 2 hectares. In one hectare conventional cropping (rice-rice-rice fallow pulse) as practiced by farmers was followed. In another holding, modified cropping (rice-rice-rice fallow cotton/maize) was followed in 0.96 ha and an area of 0.04 ha was allocated for fish pond over which the poultry shed was constructed. The poultry droppings collected in the pond were used by fish as feed material.

In the third season, rice fallow pulses fetched a net profit of Rs. 373 in the existing farmers' practice. But in the modified cropping system, a net profit of Rs. 6,660 and Rs. 729 were obtained by introducing cotton and maize respectively instead of rice fallow pulses. The income from the poultry sub system was Rs. 581. The income per day is worked out to Rs. 3.81 for five months rearing period. On an average, one bird voided about 110 g of droppings per day (on net weight basis) and about 794 kg fresh droppings were recycled in the fish ponds. On moisture free basis about 260 kg of droppings were recycled in the fishponds. The results of the experiments showed that higher fish production of 4503 kg/ha in 6 and half months could be achieved in Cauvery delta by recycling poultry droppings in fishponds.

A net income of Rs. 8311 was obtained from conventional method of cropping and a net income of Rs. 15,009 was obtained by the modified cropping system. A net profit of Rs. 2201 was obtained from poultry-cum-fish culture unit. Totally an additional income of Rs. 8,899 was obtained from the modified cropping linked with poultry-cum-fish culture over the farmer's method of cropping. The additional employment generated from the system was 123.9 man-days.

Application of poultry manure in combination with inorganic fertilizer to rice-fallow-cotton recorded the highest seed cotton yield (19.97 q/ha) with a net income of Rs. 10,903/ha (Govindan, 1988).

3. Cropping with Dairy

Two holdings of one hectare each was taken for the study purpose, rice-rice-rice fallow pulse was the normal cropping system followed in Tanjore delta. In other holding, modified cropping system was follow ed with rice-rice-rice fallow black gram, rice fallow cotton and green grass for 3 number of milch animals.

A net return of Rs. 8,422 was obtained from farmers method of cropping and Rs. 10,912 from modified cropping system. Net return from dairy enterprise was Rs. 8,986. The additional net income of Rs. 11,477 was obtained from the dairy based mixed farming over the existing farmer's method of cropping.

Inclusion of dairy unit generated an additional employment of 190 man-days. The total additional employment generated by dairy based mixed farming was 309 man-days over the practiced cropping system. Similar findings have been reported by Pracer et al. (1983) and Rai and Singh (1982). The farm yard manure obtained from dairy unit was applied to rice-fallow-cotton in combination with inorganic fertilizers. The treatment recorded the highest seed cotton yield of 17.6 q/ha and a net return of Rs. 8,573/ha (Rajkumar, 1988).

4. Cropping with Goat Rearing

Field experiments were conducted at farmer's holding in Kabisthalam village at Tanjore district of Tamil Nadu to study the feasibility of rearing Tallichery goats under deep litter system (5 does and 1 buck). The conventional method of cropping rice-rice-rice fallow black gram was compared with modified cropping rice-rice-rice fallow cotton, para grass. The goat manure collected under deep litter system was recycled as a source of nutrient to rice fallow cotton. The annual net income obtained from one hectare of conventional cropping and modified cropping were Rs. 7,654 and Rs. 15,409, respectively. The goat unit resulted in a net return of Rs. 4,330. Totally the additional income generated by introduction of mixed farming with goat rearing was Rs. 12,084.

The farmer's method of cropping generated 305 man-days in a year while 413.5 man-days were generated with mixed farming. The additional employment generated was 108.5 man-days. For landless laborers, rearing of goats and sheep may provide continuous employment throughout the year. This result adds to the findings of Rastogi (1976), Faroda *et al.* (1978) and Singh *et al.* (1986)

Application of goat manure in combination with inorganic fertilizer (50:50) gave the highest seed cotton yield of 18.89 q/ha (Santhanagopalakrishnan, 1988)

5. Cropping with Aquaculture

The farmers depend largely on cultivable crops for their income. They could enhance their farm receipts by adopting integrated fish farming methods such as cropping-cum-fish culture and livestock-cum-fish culture. These practices are already in vogue in several countries of Southeast Asia from the beginning of 19th century although it is yet to be popularized in countries like India.

The integration of cropping with fishery includes rice-fish farming; banana-fish farming and coconut fish farming. Rice-fish farming especially plays an important role in the rural economy as fish culture levels itself well to labour intensive farming operations.

1. Rice-Fish Culture

The different types of exploitation of rice field fisheries are as follows:

Captural Systems

In this system, there is no stocking of fish fingerlings in rice fields. On the other hand, wild fish population reproduced in the flooded rice fields are harvested at the end of rice growing season. In captural systems, sumps of about 40-50 m^2 size and 2 m deep are dug in the lower region of a group of fields. Species of Channa, Clarias, and Anabas are cultivated in such systems.

Cultural Systems

As in fishponds, fish fingerlings are stocked in these systems at a known density depending on the fertility of the systems. The practices involved in cultural systems may be "concurrent culture" in which fish and rice are simultaneously grown and harvested or 'rotational culture' in which rice and fish are grown alternatively.

In concurrent culture system, fish controls weeds, mollusks and harmful insects of rice and helps in securing higher productivity of rice. However, this practice limits the use of modern agricultural techniques like chemical fertilizers, insecticides and herbicides.

In rotational culture system, insect icides and herbicides could be used for better rice production. Normally after one or two weeks of rice harvest, the field is prepared for fish rearing. Bunds of 50 cm height are formed around each field and ditches of 50 cm width and 30 cm depth are dug around the field. Simple bamboo pipe or screens are kept at the inlet and outlet to prevent passage of fish and detritus during the culture period. Common carp is known to be very ideal for this system.

Integrated fish farming with agriculture and livestock is certainly profitable venture as the risks are less in these practices than in either rice culture or livestock. In many farm complexes, excess livestock wastes could be diverted to either biogas plants or to fishponds, thereby fuller and profitable utilization of these could result especially through fish culture ponds.

Integrated Farming Systems for Irrigated Areas

A model on integrated farming system to suit the farmers in irrigated area of Coimbatore was developed. An area of 1.0 ha was taken for the study with the following enterprise mix.

Cropping with Dairy, Biogas and Siliviculture

The normal cropping pattern followed in Coimbatore district of Tamil Nadu in garden lands is finger millet-cotton-sorghum. In the integrated approach, the same cropping pattern was slightly modified by inclusion of intercropping such as green gram in cotton, cowpea in sorghum and sunflower as border crop in finger millet (0.75 ha). The perennial grass fodder (0.15 ha) and perennial legume fodder, lucerne (0.05 ha) were also raised in the holding.

Three graded jersey cross bred milch cows with 2 calves were included under dairy component. For effective recycling of farm and animal waste, a biogas unit 2 m^2 capacity was installed for the production of fuel, light and enriched manure. Sixty kg of cow dung expected out of 5 animals is sufficient enough to produce m^3 of gas everyday which is equivalent to 1.5 litres kerosene.

Two hundred subabool trees were planted all along the boundary of the field for fodder and timber production.

The total grain production in the normal cropping component was 8230 kg (grain + fodder) valued at Rs. 18,184/ha while 9830 kg grain valued at Rs. 26.300/ha, was obtained due to integration of dairy and biogas. This system facilitated continuous flow of income to the farmer throughout the year besides generating additional employment.

Integrated Farming System for Rainfed Areas

One hectare of dry land was selected for the study involving crops, goat rearing and silvipasture (Rangasamy, 1988).

Cropping with Goat and Silvipasture

Generally in dry land, farmers raise only fodder sorghum yielding 3 to 10 tonnes of fodder/ha depending on the rains received. In the integrated approach, the cropping pattern was modified by including both sorghum grain crop (0.2 ha) and sorghum fodder crop (0.2 ha). To meet the fodder requirement of goats, 0.2 ha of land was raised with subabool and *Cencrus ciliaris*, a pasture grass as an intercrop. Trees like *Acacia senegal* and *Prosopis cineraria* were raised in 0.20 ha.

Tellichery goats (1 buck + 5 does) were advantageously integrated into this dry farming situation. The sorghum grain crop with 1830 kg grain and fodder valued at Rs. 1660 / -, 240 kg of goat meat valued at Rs. 6000.00 and manure 3000 kg valued at Rs. 750.00 were obtained in the integrated approach with total value of Rs. 8410.00 while 12000 kg of fodder valued at Rs. 6000.00 only was obtained in the conventional system of cropping alone. Thus, an additional income of Rs. 2410 /ha was obtained by integration. This system had generated additional 54 man-days employment over the conventional system (Rangasamy, 1988).

During last few years, some studies conducted on profitability of various farming systems indicated mixed farming's as assured high income enterprises. (Faroda et *al.*,1984).

Integrated Aquaculture – Agriculture Systems

Small ponds on a mixed enterprise can improve the whole farm productivity and profitability as well as produce fish. Farm ponds could at best be an alternative to the large scale, intensive, specialized fish farms. The scientists at the International Center for Living Aquatic Resources Management (ICLARM) in the Philippines conducted a series of experiments in farmer's holdings in Africa. They have found that wide adoption of small ponds could result in sizable increase in farm fish production in Africa, in addition to facilitating increases in other produces such as vegetables, fruits and condiments from pond dikes and margins and adjacent garden and from pond bed itself in dry periods. This type of integrated farming system practices are spread to a number of other Asian countries. e.g., the Mayan Chinampa system of water channels and raised beds for plants. The small farm ponds make as bio-digesters of waste, providers of fertile soil and water and help in site improvement that permits new operations for rotating and linking enterprises for risk management.

A set of six 'traditional' ponds were integrated with vegetable plots, maize gardens, banana groves, leucaena species, hedge rows and rice fields. The small ponds (50 sq.m) were undrainable and filled with runoff and underground infiltration. Waste materials from the other enterprises were used as pond inputs. Fish were periodically harvested and no special efforts made to control predation from birds and others. Pond muds and water were used to fertilize the vegetable plots. In studies using these traditional ponds, sizable yields of mustard, cabbage, hybrid maize and fish were obtained. The total productive area of 639 sq. m (250 sq. m of ponds and 139 sq. m of vegetables and maize) yielded an equivalent of 43 tonnes of mustard- cabbage, 42,000 cods and 454 kg fish/ha. Various combinations of farming enterprises with aquaculture have proved to be rewarding.

Integrated Farming System for Hill Regions

The North Eastern Hill Regions of India are unique in their physiographic characteristics of land, varied agro climatic conditions, land tenure system and cultivation practices from the rest of the country. The region falls under high rainfall zone and is characterized by hilly terrain with wide variation in elevation from 100 m to 5,500 m. Majority of the farmers in the region are maintaining dairy cattle and the major source of green fodder comes from lopping of the fodder trees and locally available grasses. Crops and cropping patterns of such land uses will differ depending on the type of enterprise such as milk, beef, mutton, wool, pork and poultry production. Combination of cultivated varieties of perennial legumes, grasses, shrubs and trees can extend availability of green fodder upto February at low altitude, thereby shortening the requirement of conserved fodder for lean season feeding. Land upto 100% slope having soil depth greater than 1 m can be used for agri-horti-silvipastoral system. This system has the potentiality of maintaining 1.18 livestock unit/ha of land use.

Integrated Farming System for Island Ecosystems

Integrated farming system models have been developed for the Andaman and Nicobar Islands. Major part of Andaman and Nicobar Islands comes under medium slope category. The topography is mainly characterized by undulating sloppy lands. Present cropping system in this region is raising of fruits trees such as banana, papaya, mango, guava, pineapple, sapota, pomegranate, lemon, lime, sweet orange, mandarin, jack fruit, custard apple etc. Among the plantation and spice crops, coconut, arecanut, cashewnut, black pepper, nutmeg and clove are mainly grown. Rice is predominantly cultivated during the rainy season (May to December) and field crops such as green gram, black gram, cowpea, sesame, groundnut, vegetables, etc. are grown in the dry season on a very limited scale in rice fallows. Animal husbandry is in its infant stage. Fishery resources are very rich but remain unexploited.

Some of the possible integrated farming system (Shyam Singh and Gangwar, 1989) considered to be useful for the socio-agro climatic conditions of Andaman and Nicobar Islands are discussed below:

1. **Coconut-cum-fodder-cum-milch cattle:** Mixed farming by raising fodder grass in coconut has been found profitable. Raising grasses such as hybrid Napier alone or with leguminous fodder crops like Stylosanthes in the coconut garden can support 4-5 dairy animals. The animals supply large quantity of cattle manure which, when applied to coconut garden, improves the soil fertility considerably.
2. **Coconut-cum-fodder-cum-fish or prawn culture:** This system is suitable for tidal marshy area. The coconut palms are grown on platforms raised upto the highest tidal level and fish and prawn in channels between the coconut plantations. On platform with coconut trees, fodder crops like Guatemala grass, thin Napier and guinea grass could be raised successfully with an average forage yield of 290, 190 and 184 q/ha respectively under high salinity conditions.
3. **Coconut-cum-fish culture in salt affected lands:** Coconut grows successfully even in salt-affected paddy fields if the fields are within the approach of brackish water. The bunds of fields at water entrance gate should be raised to maintain the required level of water. Field bunds should also be raised as per requirement. In high tides, water will be exchanged and thus fish rearing can be taken up together with coconut in dry period. In rainy season, salinity-tolerant rice varieties could be cultivated in the same field with coconut and fish culture.

4. **Fruits-cum-fodder-cum-milch cattle:** The space available between the fruit trees can be utilized for growing fodder crops such as cowpea, rice bean, green gram and black gram. The fodder could be used for feeding the cattle and cattle manure could be applied to the fruit trees and fodder crops.

FARMING SYSTEM RESEARCH (FSR) FOR DIFFERENT SITUATION

FSR for Degraded Lands

A chain of soil conservation demonstration and training centres was set up during the 1950s to develop soil and water conservation systems to increase productivity and reduce erosion from the degraded lands of rainfed areas. These centres evolved methods for the mechanical measurement of erosion control, and developed effective land use systems based on the best engineering, agronomic, and soil and water management practices. They also trained staff in soil conservation, although their research was limited in scope and concept. In 1967 these centres were transferred to ICAR and one of the leading centres, at Dehra Dun, was converted into a Central Soil and Water Conservation Research and Training Institute with seven regional stations. The research achievements of these centres over the 15 years from 1956 to 1971 have been reviewed by Tejwani *et al.* (1975). They showed that although the research undertaken was on multidisciplinary lines, research on alternative land use systems and integrated farming systems based on the watershed concept became more prominent after 1971, but has yet to make a significant contribution to real farm situations. The data on national average yields for the period from 1971-84, compared with the average and highest yields from National Demonstration plots is given in Table 14.2.

The watershed management approach has been effective in improving the productivity of land by reducing erosion though the introduction of appropriate land use systems, but it has proved difficult to obtain the cooperation of small farmers who own small parcels of land in the watershed. However, when the social aspects have been effectively managed, as in Sukhomajri and Nada villages in Haryana, the equitable distribution of the benefits of watershed management has ensured the acceptance of the system even by small farmers (Mittla Gurmel Singh, 1990).

TABLE 14.2: National average yields 1971-84 (t/ha), compared with the average and highest observed yields from National Demonstration plots

Crop	National average yield	National demonstration average yields from national demonstrations	Highest average
Paddy	1.8	5.1	11.5
Wheat	1.5	3.7	7.6
Maize	1.1	3.3	7.5
Sorghum	0.6	3.7	7.1
Pearl millet	0.5	2.6	4.8

FSR for Wastelands

Cultivable Wastelands

The land, which is capable or has potential for the development of vegetative cover and not being used due to different constraints of varying degrees, such as erosion, water logging, salinity etc. The lands can further classified into:

Uncultivable Wasteland

These lands cannot be developed for vegetative cover, for instance barren rocky areas and snow-covered glacial areas. These are classified into:

1. Rocky outcrop or stony waste.
2. Snow or ice covered areas.
3. Salt pans and salt marshes.

Pasture Lands

Reseeding Methods

Natural succession of desired grass species is a very slow process in arid regions. Reseeding of pasturelands with perennial grass species suiting to the region is the only solution for upgrading the degraded lands. The best sowing time is end of June or after the first effective showers. Sowing of seed of *Cenchrus cliaris* at the Central Research Farm, Jodhpur in dry soil in the last week of June gave 36 per cent higher forage yield over post monsoon sowing in the month of July.

Grass-Legume Mixture

Performances of *Stylosanthus hamata* appeared to be promising in association with *Dichanthium annulatum* in rainfall zone of more than 400 mm. The total productivity of grass-legume mixture of D. *annulatum* + *S. hamata* was highest (42.0 q/ha) which was 26 per cent higher over the control. *Prosophis cineraia* is an important drought hardy-to-feed free, which grows in pasturelands without any detrimental, effect on grass grown in association with it. Silvipastural studies conducted with *Acacia tortilis, Azadirachta indica, Albizzia lebbek* and *Holoptelia integrifolia* and four grasses namely *Cenchrus ciliaris, C. settigerus, Dichanthium annulatum* and *Panicum antidotale* proved to give better economic return.

Sand Dune Stabilization

A check-board or parallel hedge system can be adopted depending upon the direction of the wind. If the prevailing wind blows from one direction or in the opposite direction then mechanical barriers can be established in parallel lines. In western Rajasthan locally available brushwood, like *Leptadenia pyrotechnoia, Zizyphus mauritiana, Lelligonum, Polygonoides, Lasiurus sindicus, Panicum turgid and Helianthus mina* have been used by cutting and burying them vertically downwards in lines 2-3 m apart. In areas where wind blows from different directions a check board system (2-3 m^2) has been found to be very effective in arresting the movement of sand.

Species Suitable

1. *Prosophis chinensis.*
2. *Prosophis cineraria.*
3. *Acacia tortilis*
4. *A. raqdiana.*
5. *Zizyphus mauritiana.*
6. *Calliconum polygonoides.*
7. *Crotalaria burbia.*
8. *Aerva javenica.*
9. *Laiurus sindicus.*
10. *Panicum turgidum.*

It was found that sand dune afforestation with *Prosophis chinensis* yielded fuel-wood of about 38 tonnes/ha after 10 years.

Farming System for Arid Lands

Studies conducted at CAZRI Jodhpur, revealed that growing of pearl millet as single crop after 7-9 months fallow period proved to be most productive (17.2 q/ha). Years receiving more than 800 mm rainfall with an extended rainy season favour double cropping. An analysis of 80 years rainfall data revealed that 20 years *i.e.* one out of four were such which experienced more than 500 mm of rainfall with an extended monsoon and showed promise of double cropping. It was found that pearl millet followed by mustard gave the highest total productivity and gross monetary returns. Among the intercropping systems tried, planting of one row of pearl millet in between paired rows of mung bean or cluster bean resulted in highest productivity.

Horticultural Crops in Arid Land IFS

Criteria for selecting fruit crops are:

Fruit crops like ber *Zizipus mauritiana*, Pomegranate (*Punica granatum*), custard apple (*Annona squamosa*) gonad (*Cordia dichotoma*), karonda (*Carissa carandas*), amla (*Emblica officinalis*), and sapota (*Arches sapota*) and guava (*Psidium sativa*) are suitable.

Tree Farming or Wood Lots

In the semi-arid areas (500-1150 mm of rainfall), tree farming as a component of the farming system is a sure way of reasonable productivity and income. The tree species found suitable are:

1. *Acacia abide*.
2. *A. militia*
3. *A. nilotica* var. *cupressiformis*
4. *Annona squamosa*.
5. *Albizia labbek*.
6. *A. indica*
7. *Casuarina equisetifolia*.
8. *Dalbergia sisso*.
9. *Gliricidia sepium*.
10. *Grevillea robusta*.
11. *Leucaena labusta*.

Range Management

Improved grass cultivars with high yield potential and good quality forage are now available. Of the 5 grasses *Cenchrus ciliaris, C. setigerus, Chloris gayana, Panicum antidotale* and *Urochloa mosambicansis* tried for three years under semi-arid conditions at Hyderabad. *Cenchrus setigerus* yielded the highest amount of fodder. *Stylosanthes hamata, S. scabra* and *S. viscose* also performed well.

Contingency Planning

The following aberrations can be observed in many situations on analysis of rainfall:

1. Early onset of monsoon.
2. Late onset of monsoon.
3. Early recession of monsoon.
4. Breaks in monsoon.

Crop selection should be done in such a manner to suit the above situations.

Ley Farming

This system fits well where moderate rainfall (500-700 mm) is received but the soil is deep (770 mm). Studies revealed that inclusion of *stylosanthes hamata* as a pasture legume in the rotation with castor or sorghum improved the yield of sorghum in succeeding style. Grain yield of sorghum can be increased and sustained at a higher level without any application of nitrogen under dry lands in ley-farming system.

Wasteland Development

Soil is basic resource for all agricultural activities. However, over time as biotic pressure is built up beyond the carrying capacity of this resource, there is misuse or over use of crop land. On account of disturbed hormony of natural environment, there are increasing instances of soil degradation. The reasons for land degradation may be due to inherent or imposed disabilities or both such as location, environment, chemical and biological properties and even suffer from management conditions. The explosive increase of population, increasing desire of man to exploit marginal and sub-marginal lands, pressure from industry for its raw material needs, defective land use and improper irrigation management emerged as the main factors causing land degradation. At some places, degradation has reached disastrous proportions and if not corrected in time, will have a devastating effect on our future population.

Failure in the restoration of degraded areas and increase in the population may reduce the available land area for productive use. Due to environmental destruction, the weaker section, particularly the rural poor and tribes, are affected mostly because their life revolves around constant search for food, fodder, fuel and water. As ecological conditions worsen, their survival becomes increasingly difficult.

Late Prime Minister of India, Sri Rajiv Gandhi, in his broadcast address to the nation on 5 th January, 1985 said 'Continuing deforestation ha brought us face to face with a major ecological and socio-economic crisis. The trend must be halted. I propose immediately to set up a National Wasteland Development Board with the objective of bringing 5.0 million ha of land every year under fuel wood and fodder plantations'. A Technical Task Group was constituted by the Planning Commission and National Wasteland Development Board. The primary task was comprehensive estimation of wastelands in the country and to propose standardized definitions for various categories of wastelands.

According to dictionary meaning, wasteland is the area which is unfit for use. The lands which are unproductive and least utilized are termed as **wastelands.** This does not mean that they are unfit for use. Due to neglect or degradation, these lands are not utilized to their full potential. The Wasteland Survey and Reclamation Committee (1961) defined wastelands as those lands which are either not available for cultivation or left out as fallows and cultivable wastes.

According to the National Wasteland Development Board, wastelands are defined as the lands which are currently lying unutilized due to different constraints. These include culturable and unculturable watelands. The wastelands are also termed as **degraded Lands**

Classification of Wastelands

Out of 329 million ha of total geographical area in India, about 158 million ha comes under different kinds of land degradation. Wastelands are broadly grouped into culturable and unculturable wastelands.

1. **Culturable wastelands:** The lands are capable or have the potential for the development of vegetative cover. However, they are not being used due to different

constraints of varying degrees such as erosion, water-logging, salinization etc. Most of these lands can be put to some productive use after proper treatment and reclamation measures. Actually these lands are problematic lands instead of wastelands.

2. **Unculturable wastelands:** The lands that can not be developed for vegetative cover are called unculturable wastelands e.g., barren rocky areas, snow covered glcial areas.

Classification of wastelands based on National Commission on Agriculture (1976), Ministry of Agriculture and Irrigation, Government of India is given in Table 14.3.

TABLE 14.3: Classification of wastelands.

Category	Spread
– Raverine land, Coastal sandy areas, high altitude and steep sloping areas.	– Coastal region, interior mountanous regions, steep slopes in Himalayas steep slopes.
– Salt-affected areas	– Semi-arid and Indo-Gangetic plains and other semi-arid areas in Peninsular region.
– Waterlogged areas	– Low-lying areas in high rainfall regions and irrigated areas.
– Ravine land	– Gujrat, Rajasthan, Madhya Pradesh and Uttar Prodesh
– Shifting cultivation	– North-Eastern region
– Barren hill ridges and rock outcrops	– Distributed in various states of the country.

National Wasteland Development Board (1987) classified as follows:

1. Utilizable/Culturable Wastelands
2. Unculturable Wastelands

Management of Wastelands

The characteristics of the wastelands and their development vary with the categories. Agro-forestry system plays greater role in the development of culturable wastelands.

Raverine Land

Rivers originating from Himalayas have twin actions of erosion and deposition. Depending on the course of rivers and speed of flow, heavy load of sand and silt is carried by the gushing streams. They deposit large quantities of sediments on way to the sea. Gradually, land masses are formed which remain for some part of the season atleast above the flood level. Raverine lands are called as '*khadar* 'in North India, '*diara*' in Eastern Uttar Pradesh and '*char*'in West Bengal. In Punjab and Haryana, they are called '*Rao*' and 'cho' which are most common in branched or multiple channel rivers. Development of such land is mainly based on adoption of suitable soil conservation measures. Planting live hedge is most beneficial in controlling erosion and large deposition of sediments.

Coastal Sandy Areas

India has a long coastal line about 5600 km. Wind erosion is more serious and extensive, especially during monsoon seasons. There is blowing and shifting of sands during the monsoon and cyclonic periods. This causes heavy damage to the standing crops in the neighbouring areas. The problem is acute in Eastern coast of Orissa and Western coast of Keral. In such areas, afforestation is necessary as it acts as a moderator of the effect of cyclone. All along the coastal areas, wide forest belt is to be created for moderating the effects of cyclones. Casuarina is an important tree suitable for coastal belts as it can stand high degree of salinity and moisture stress. It also acts with least attention. Casuarina plantation on the sandy foreshore should be backed by plantation of other adaptable tree species further inland. These plantations can be maintained upto 10 metres width along the coast. The inner areas can be planted with cashew trees interspaced with coconut.

The species such as *Salvadora persica, S. oleoides, Atriplex nummularia* and *Juncus regid* are suitable for saline coastal regions. *Simmondsia chinensis* (Jojoba), *Simaroba spp* and *Prosopis spp* are highly adopted for sand dune areas. Grasses viz., *Panicum maximum, Cenchrus spp, chloris spp, Spinifex squamosus, Sporobolus spp* and *Cynodon dactylon* are most suited.

High Altitude and Steep Sloppy Areas

In high altitude region of Himalayas, the soils suffer from excessive run off during monsoon, causing floods and there is moisture deficiency during the dry period. Indiscriminate grazing by cattle, especially sheep and goat including the nomadic herds, are the major causes for soil erosion in hilly areas. Hedge row planting is required under hilly terrain. Nitrogen fixing tree species and shrubs suitable for agro-forestry may be planted at close spacing. The space between trees and within the rows may be utilized for growing hedges with perennial grasses. Tree species such as *Acaia modest, A. catechu, Prosopis juliflora, Populus alba, P. nigra, P. ciliate* and *P. candicuas* are best suited. In Nilgiris with land slope of above 33% perennial vegetation preferably with silvipasture and tea plantation is recommended.

Salt-Affected Soils

Salt-affected soils may be classified as saline soils and alkali soils. Saline soils are predominant in arid, semi-arid and coastal areas. The alkali (sodic) soils occur in the Indo-Gangetic plain and parts of Madhya Pradesh. In peninsular semi-arid tracts, sodic soils occur where the soils are irrigated injudiciously. Most of these soils have high productivity potential due to abundant nutrient reserve and ideal moisture retaining and release characteristics. Classification based on the United States Salinity Laboratory (USDA) and Soil Science Society of America (SSSA) is given in Table 14.4 and 14.5)

TABLE 14.4: Classification of salt affected soils (USDA)

Class of soil	EC (mmhos/cm)	ESP	pH
Saline	> 4	<15	< 8.5
Saline-alkali	> 4	> 15	> 8.5
Non saline-alkali	< 4	> 15	8.5– 10.0

TABLE 14.5: Classification of salt affected soils (SSSA)

Class of soil	EC (mmhos/cm)	SAR	pH
Saline	> 2	-	-
Saline – sodic	> 2	> 15	> 6.5
Sodic	-	> 15	-

Salt Affected Soils of India

In India, about 7.0 m ha of otherwise productive land is salt affected and a large part of it has got out of cultivation. The problem is acute in semi-arid and sub-humid IndoGangetic plains. The extent salt affected areas in India is given in Table 14.6.

Characteristics of Salt Affected Soils

Saline Soils

Soils containing excess amount of soluble salts are called saline soils. The predominant soluble salts are chlorides and sulphates of sodium, magnesium and calcium. These soils sometimes contain appreciable amount of nitrate. Presence of excess amount of soluble salts induces flocculation of the clay fraction and hence the physical properties of the saline soils are generally good. Plant growth is affected by the osmotic effect of excess soluble salts. Accumulation of specific ions such as B, Cl and Na in plants to toxic levels may often affect plant growth adversely.

TABLE 14.6: Salt affected soils in India

Group	States	Approximate area (m ha)
1. Coastal salt affected soils		
a) Coastal salt affected soils of arid region	Gujarat	0.714
b) Deltaic coastal salt affected soils of the humid region	Orissa, Andhra Pradesh, Tamil Nadu, West Bengal	1.394
c) Acid salt affected soils	Kerala	0.016
2. Salt affected soils of the medium and deep black soil region.	Karnatak, Madhya Pradesh, Andhra Pradesh, Maharashtra	1.420
3. Salt affected soils and semi-arid regions	Gujarat, Rajasthan, Punjab, Haryana, uttar Pradesh	1.000
4. Alkali soils of Indo-Gangetic plains.	Haryana, Punjab, Uttar Pradesh, Bihar, Madhya Pradesh	2.500

Alkali Soils

Alkali soils have high Exchangeable Sodium Percentage (ESP) of above 15 which is generally considered as the limit between the normal and alkali soils. The predominant salts are carbonates and bicarbonates of sodium. The soils are also termed sodic soils. In the moist situation, there are black patches due to the dissolution of organic matter and subsequent evaporation. Surface of the soil becomes very hard during drying and turns muddy with the receipt of rain. The soils are completely disbursed due to high ESP and hence water stagnation occurs in the field. The bulk density of the sodic soils is normally (1.5-1.85 g/cm^3) and hence root penetration is hampered. The tilth of the sodic soils are poor because of high pH and ESP. Seed germination and root growth are affected due to the presence of hard surface layer. Plant growth is adversely affected due to the toxic quantities of sodium or deficiency of other essential plant nutrients like Ca and Zn.

Sources of Excess Salt in the Soil

- Reducing soil erosion and improving soil conservation,
- Inclusion of legumes and cover crops in crop rotations,
- Agro forestry as an alternate land use system, and
- Judicious use of organic wastes.
- Soil and climatic features of the selected area.

- Availability of the resources, land, labour and capital.
- Present level of utilization of resources
- Economics of proposed integrated farming system.
- Managerial skill of the farmer.
- Lack of awareness about the new technology,
- Ineffective extension services,
- Non-representation of their conditions by the research stations where the technologies are developed,
- Lack of resources to invest on the required inputs, and
- Non-availability of inputs in time.
- Target and research area selection,
- Problem identification and development of a research base,
- Planning on-farm research,
- On-farm research and analysis for further testing,
- Scientists designed and scientist implemented trials,
- Scientists designed and farmer implemented trials,
- Farmer designed and farmer implemented trials, and
- Farmer designed and scientist implemented trials.
- Soil and climatic features of the selected area.
- Availability of the resources, land, labour and capital.
- Present level of utilization of resources.
- Returns from existing farming system.
- Economics of proposed integrated farming system.
- Managerial skill of the farmer.

1. Priorities of research problems/production constraints taking into consideration their effort on overall sustainability of production and farmers' preference.
2. Prioritize the interaction, which can be achieved by farmer --- farm scientist dialogue, desk analysis and scientist-scientist dialogue. Farmers --- farm scientist and scientist --- scientist intervention must consider the farmers' preference and identify the risk/problem with each intervention. With farmer priorities initially identify the risk/problem with each intervention. With farmer priorities initially identified and intervention plans prepared, scientists have to take a new approach to the research to solve the farm problems. The technical especially of the participating scientist should not be pushed but rather had to fit within the research priorities of farm problem. A crop scientist has to seek the help of other scientist to solve the problem.
3. Research design and technology generation and adaptation: While developing technology, one must consider the labour resource, distribution of labour and the age structure of the farmer's family. One should also ensure that the technology intervention fits with the farmer's existing skills as far as possible. The technological adaptation and intervention should also offer flexibility in options. The farm's participation in technology generation and adaptation makes technology more relevant to resource poor farmers.
   a) Crop
   b) Animals (cattle, buffalo, pigs, goat, sheep, birds, etc.)
   c) Fish
   d) Sericulture/apiculture etc.

   e) Agro-forestry (with tree/fruit crops).
4. Testing and monitoring of designed intervention:
   a) Involve farmers' participation in intervention.
   b) Ensure that desired intervention is correctly implemented.
   c) Constantly oversee the progress of intervention.
   d) Technology/intervention to be tested must satisfy the following criteria:
   e) Productivity and profitability.
   f) Technical feasibility.
   g) Socio-economic/cultural acceptability.
   h) Sustainability or intervention with the existing resource base and infrastructure.
5. Technology transfer and diffusion of improved farming system within recommendation domain.
6. Impact of technology of improved farming system:
   a) Production efficiency,
   b) Economic return,
   c) Energy-input/output,
   d) Employment,
   e) Equity,
7. Environment.
8. Integration of cropping system research and extension into FSR and extension at ICAR level.
9. Developing multidisplinary teams for OFR at agro-ecological zone level in each University. This can be done by integrating the FCF programme and other subject matter specialists and extension workers.

- Efficient recycling of farm and animal wastes;
- Minimizing the nutrient losses;
- Maximizing the nutrient use efficiency;
- Following efficient cropping systems and crop rotations; and
- Complementary combination of farm enterprises

1. Lands, which could be reclaimed with reasonable investment on inputs and labour:
   a) Gullied and degraded land.
   b) Scrublands with rough pastures: slope and elevation categories (undulating stony graveled wastelands).
   c) Surface waterlogged land (high water table, temporary flooding).
   d) Salt affected land (alkaline).
   e) Shifting cultivation area.
   f) Degraded forest land.
   g) Sands-Coastal lands, semi-desert lands.
   h) Mining and industrial wastelands.
2. Land that could be improved with higher investments of inputs and labour;
   a) Marshes of Permanent nature.
   b) Desert sands.
   c) Ravenous land.
3. Fruit crops chosen must have flowering and fruit development synchronous with moisture availability.
4. They must have deep and strong root system to pierce the hard pan and to extract moisture from deeper layers of soils.
5. They should be tolerant to salinity in soil and saline irrigation water.

a) Eroded lands include gullied and/ ravenous land
b) Undulating upland
c) Waterlogged land
d) Salt-affected areas include saline and alkali soils
e) Shifting cultivation areas
f) Degraded forest lands
g) Degraded non-forest plantation
h) Sand dunes, sand bar and coastal sand
i) Minespoils and industrial wastelands
j) Strip lands
k) Pasture and grazing lands
l) Barren rock/stony wastes/sheet rock areas
m) Steep sloppy areas
n) Snow covered and/ glacial area

6. The salts get deposit due to impeded drainage, as the soil may have developed from the parent material having excess salts originating from weathering.
7. Irrigation water may contain salt and by continuous use of such water, salt accumulates and is deposited in large areas.
8. Salt accumulates on the surface due to capillary movement from the ground close to the soil surface.
9. Excess salt accumulation occurs due to high sub-soil water table, aridity and poor drainage.
10. Back water flow or intrusion of sea water also causes sodicity in the coastal areas

Management of Salt Affected Soils

Management of salt affected soils depends on the climate, soil characteristics, including nature of the soils, available water resources and associated features in different regions.

Reclamation

Reclamation of sodic soils basically requires replacing excess sodium by calcium on the soil exchangeable complex. This is done through application of amendment and leaching out of the exchanged Na from the root zone. Gypsum ($CaSO_4$-$2H_2O$) is the cheapest and the most commonly used amendment for reclamation. Gypsum dissolves to a greater extent in sodic soils due to high ESP. The solubility is also found to increase with the fineness of the particle. The quantity of gypsum to be applied will depend upon the pH and texture. Gypsum @ 15 t/ha may be required to reclaim the affected soil. Combined application of gypsum and organic like green manuring (daincha), is found to be better than inorganic amendments alone.

Leaching with good quality water is necessary to leach down or remove the excess salt at the surface and the sodium released from the exchange complex. Proper drainage is important to keep the salt below the root zone.

Agro-Forestry for Salt-Affected Areas

Utilization of problematic soils for cultivation of forest trees is a desirable alternative since the crop production on the salt affected soil is uneconomical. In places where the ground water is saline and scarce in quantity as in arid and semi-arid regions, agro-forestry could be practiced. Afforestation and pasture development are the two ways of managing the salt affected lands when other measures are considered uneconomical. Putting these lands under multipurpose tree species will fulfil the twin objectives of biological amelioration and harvesting biomass to meet the fuel, fodder and timber requirements. In general plantations for fuel wood are rated better for saline soils than timber wood species.

Choice of tree species The following tree species are recommended for sodic soils:

Prosophis chinensis, P. juliflora, Acacia nilotica, A. catechu, A. auriculiformis, A. tortilis, A. arabica, A. repens, A. aneura, Tamarix articulate, Casuarina equisetifolia, Eucalyptus hybrid, Albezia lebbeck, Azadirachta indica, Terminalia arjuna, Dalbergia sisso, Salbadora persica, Leucaena leucocephala.

Grass species: Grasses can also be grown in saline/sodic soils for reclamation and economic returns.

Common name	Botanical name
Karnal grass	*Diplachne fusca.*
Rhodes grass-	*Chloris gayana*
Bermuda grass	*Cynodon dactylon*
Para Grass	*Brachiaria mutica*
Blue panic	*Panicum species*
Hybrid Napier	*Pennisetum purpureum*
Guinea grass	*Penicum maximum*
Kallar grass	*Leptochloa fusca*
Alkali sacaton	*Sporobolus spp*

Fruit trees: Among the fruit trees, sapota is found to be the most promising. Date palm is highly salt tolerant. Guava, ber, pomegranate and custard apple are also suitable fruit trees for salt affected areas

Management of Trees in Problem Soils

Tree species can be grown successfully on the sodic soils by treating the soil of the planting pit with gypsum (3-4 kg/pit), FYM (25 kg/pit) and fertilizer application. This practice gives better response than replacing the sodic soil of the pit with good soil. The pH value considerably decreases throughout the profile soon after the application of amendment. Incorporation of gypsum and farm yard manure in the soil of the planting pit also helps in checking subsequent deterioration of the soil as aresult of salt movement from the surrounding salt affected areas.

Auger Hole Planting

This technique has been developed for planting trees in alkali soils. In this method, 100-150 cm deep and 20-30 cm diameter auger holes are made with tractor operated auger and trees are planted. It has been found that height, girth, growth and fuel wood production of *Prosopis* were more with auger hole planting.

Sub-Surface Planting

In this method, saplings are planted in the trenches or furrows of 30 cm deep and these trenches are also used for irrigation. A maximum salt builds up in the surface soil, planting saplings in the furrows is found highly suitable fpr saline soils.

Waterlogged and Marshy Lands

In the waterlogged areas, stagnation of water occurs for most part of the year because of imperfect drainage. In the coastal regions, water-logging may be caused by intrusion of sea water in the lands below mean sea level as found in the parts of West Coast. Large extent of fields in the canal irrigated areas become unproductive due to water-logging. The Central Board of Irrigation defined water-logging as an area where the water table rises to an extent that the soil pores in the root zone of the crop becomes saturated resulting in restriction of the normal circulation of air, decline in the level of oxygen and increase in the level of carbon dioxide.

Marshy lands are permanently or periodically inundated by water and characterized by vegetation which include grasses and weeds.

Proper drainage is important to reclaim these areas. Tree species suitable for the waterlogged areas are: *Eucalyptus robusta, Syzygium cumini, Salix spp., Populus nigra, Terminalia arjuna, and Acacia nilotica.* In marshy places, trees species best adopted are: *Baringtonia spp. Bischofia javania, Eucalyptus robusta, E. rudis, Lagerstroemia flosreginea* and *Casuarina equisetifolia.*

Gullied and Ravenous Land

Gully erosion is described as 'Cancer of the land" and is considered as the severest form of erosion. A narrow deep gash formed as a result of localized and concentrated run off by running water is called gully. Very deep (>9 m) and wide gullies with steep sides are known as 'Ravine'. All gullies and ravines contribute considerably to soil erosion. The ravines are characterized by the absence of vegetation of any type. Run off is maximum, slopes are steep and gullies are of varying depth, few metres to 100 metres or more. The ravines are classifieds as shallow, medium and deep based on the gully depth, bed width and area. Ravenous lands are mainly confined to the states of Uttar Pradesh, Madhya Pradesh, Rajasthan and Gujarat along the banks of Yamuna, the Chambal, the Mahi and Sabarmati and tributaries. In addition to the existing areas, about 8 thousand ha are added every year due to faulty management practices.

Afforestation

Small and narrow gullies are put under agro-forestry. Silvipasture is suitable for medium and shallow gullies.

The deep and narrow gullies are best controlled by putting them to permanent vegetation. Deep gullies are utilized as woodlands by planting suitable tree species. Gully heads and sides are stabilized by planting grass species. Reducing the slopes of gullies and diverting the surface flow are essential pre-requisites for any successful afforestation programme. Constructing gully plugs in the bed of main and branch gully is important to conserve soil and moisture. Suitable species for afforestation of gully and ravenous land are given below:

- **Trees and Shrubs:** Acacia nilotica, Acacia catechu, Azadirachta indica, Albizia lebbeck, Dendrocalamus stratus, Dalbergia sissoo, Prosopis chilensis syn, P. juliflora, Eucalyptus spp.
- **Life hedges:** Impomea carnea, Agave Americana, Arundo donax, Vitex negundo.
- **Grasses and legumes:** Cenchrus ciliaris, Pennisetum peiecellatum, Saccharum spontaneum, Dicanthium annulatum, Cynodon dactylon, Heteropogon contortus, Stylosanthes species.

Shifting Cultivation

Shifting cultivation is a primitive cultivation technique and is a transition between food gathering and food production. Shifting cultivation is dominant in the tropical and sub-tropical high land regions of the world, having high rainfall, steep slopes and shallow soils. It involves 10% of population and 30% of exploitable land of our planet. In India, this system is prevalent in several parts of the North-Eastern Hill regions of Arunachal Pradesh, Meghalaya, Mizoram, Manipur, Nagaland and Tripura. It is also prevalent in some parts of Assam, Orissa, Madhya Pradesh, Andhra Pradesh and Bihar. Shifting cultivation is popularly known as *'Jhuming'* in the North Eastern zone of India. It is called as *'Khallu'* or 'Kurwa'in Bihar, *'Dahiya'* or *'Podo'* in Orissa, Madhya Pradesh and Andhra Pradesh. Shifting cultivation is also known as *'Slash and burn'* or *'Swidden'* cultivation. It is estimated to cover an area of 4.36 million ha. Shifting cultivation is defined as the practice of cultivating clearings scattered in the reservoir of natural vegetation (forest, grass, woodland) and abandoning them as soon as the soil is exhausted and this includes the practice of shifting homesteads in order to follow the cultivator's search for new fertile area.

The practice of shifting cultivation consists of three phases: Land clearing, Cultivation phase and Restoration phase.

Land Clearing

Forest vegetation is felled down at various successional stages on the hill slope during winter months of December and January. The slash are then allowed to dry and burnt during February to April for nutrient enrichment, moderation of pH and charcoal. Fields are cleared of boulders and stones from surface in the process on clearing and piled up forming bunds on slope edges and field boundary.

Cultivation Phase

Minimum tillage is generally adopted and practiced on contour, using hand tools and bullock drawn implements. Hill slopes under shifting cultivation are mostly cultivated during *kharif* season, with pure or mixed crops. The crops include Indian millet, finger millet, upland rice, pigeon pea, black gram, rice bean, niger, horse gram, etc. This cycle of cultivation varies from 3 to 5 years. The crop yield sharply declines in the successive years.

Restoration Phase

A short period of cropping generally alternated with **a** long period of fallow ranging **5** to 50 years. Fields are allowed for natural restoration.

Impact of Shifting Cultivation

Burning of vegetation in shifting cultivation alters the plant nutrient supply from organic to inorganic form. Some of the nutrients, C, N and S are lost by volatilization while burning the slash. A sizable quantity of nutrients, particularly K, Ca and Mg are lost before the onset of the first rain through blown off ash due to string winds in the drier period. During the short *'Jhum'* cycle, the soil fertility deteriorates to a very low level.

During the fallow period, soil fertility is lost due to erosion and other causes. Wrong tillage practices in shifting cultivation promote run off and soil erosion. Average soil loss due to shifting cultivation has been recorded to be 40.9 t/ha in Meghalaya. It is reported that erosion rates on 50-60% slope are 147 t of soil loss/ha under first year of shifting cultivation and 170 t/ha in the second year.

Under *'Jhum'* cultivation, some of the weed species, *Eupatorium odoratum* at lower elevation and F. *adenophorum* at higher elevation, become dominant. Shifting cultivation causes large scale damage to the forests and results in deforestation and denudation of hilly slopes. During *'Jhum'* cycle, large number of edible vegetations are cut and burnt. Shifting cultivation and deforestation have been identified as major constraints in developing sustainable crop production system in the North-Eastern region. About 19.8% of the total geographical areaof the region has become degraded or wastelands.

Management of Shifting Cultivation

It is necessary to integrate conservation measures and improved farming to stabilize production at higher level. These measures may help in slowing down resource degradation and accelerating aggravation in the system.

Clearing and Cultivation Management

The land use and farm planning should depend upon the hill land classification. Area should be treated at terrace and inter-terrace level with mechanical and vegetative conservation measures. Improved vegetative cover, land configuration management and alley farming will form the base for sustained production achieving better soil and moisture conservation and improved nutrient recycling. The terrace should be protected against erosion. Contour bunds, bench terraces and half moon terraces are the conservation measures that are required for the treatment of land. Such land uses are expected to retain about 90% of the annual rainfall with negligible soil loss. Plantation of vegetative barrier could be ideal for both erosion control and improved nutrient cycling. It is possible to get higher production of finger millet by 80% under natural barrier.

Restoration Management

Lack of restoration management results in slow build up of soil fertility Normally it takes 5 years to build up vegetation and 8-10 years to restore fertility status. Retained fallow vegetation, planted vegetative barriers and integration of cultivation to restoration phase by perennial pigeon pea are successful in Orissa.

Agro-Forestry Systems for Shifting Cultivation

Suitable farming systems in shifting cultivation include forestry, silvipasture/ horticulture and agriculture. Three tier system has been suggested by the ICAR centre, Shillong. In this system, trees are recommended at the hill top, horticultural crops in the middle portion of the hill slopes and opening of terraces at the lower portion for cultivation of field crops. The forestry systems are primarily based on the production of fruits (pineapple, jack fruit, tamarind, etc.), nuts (arecanut) and betel vine. These should be integrated with woody perennials which supply fuel and timber for domestic use.

Broom grass+schema systems are commonly practiced on most of the degraded boulder lands with high slopes and shallow soil depth. The silvipastural system comprising *Alnus nepalensis,* pine apple and forage crops like Guinea grass (*Panicum maximum)* or Nandi grass (*Setaria sphacelata)* coupled with Stylo (*Stylosanthes gyanensis)* could be the sustainable agro-forestry system for hills with 30-60% slope in Meghalaya.

Lower areas of hilly slope are terraced and utilized for raising field crops according to the requirement of the farmers. Crops such as rice, maize, finger millet, mustard sesame, black gram and root crops like ginger and turmeric may be grown. The terraces may be stabilized with mechanical measures and planted with fodder grasses or legumes on risers. Fodder grasses/legumes such as *Stylosanthes hamata, S. quyanensis, Clitoria ternatia, Setaria sphacelata* and thin napier are suitable for terrace rises.

Taungya System

'Taugya' is a Burmese word meaning hill cultivation. In this system, government lands are temporarily allotted to the cultivators. They are allowed to plant forest trees species and to raise agricultural crops in between the rows of forest tree species simultaneously for about 3-5 years. Thereafter, cultivators are shifted to another patch of land.

This system was practiced as early as 1856 in Burma (Myanmar) in establishing teak plantation. Regular 'Taungya' cultivation as a system of raising forest plantations was started in North Bengal in 1911. In India, it is now followed in Kerala, West Bengal, Uttar Pradesh, Tamil Nadu, Andhra Pradesh, Orissa, Karnataka and North Eastern Hill regions. This system provides proper use of land and maximizes production. It also facilitates employment opportunity to unemploy rural people. Important tree species and field crops suitable for 'Taungya' system are given in Table 14.7.

TABLE 14.7: Important trees and agricultural crops for 'Taungya' system

Trees	Crops
West Bengal	
Tectona grandis	Paddy, maize, millets
Shoerea	Turmeric, ginger
Schima wallichii	Pine apple
Cryptomeria japonica	Hemp
Michelia spp.	
Uttar Pradesh	
Shoerea robusta	Paddy, maize, sorghum
Tectona grandis	Wheat, barley
Acacia catechu	Pigeon pea, soybean
Dalbergia sissoo	Rape seed
Eucalyptus spp	
Populus spp.	
Kerala	
Eucalyptus spp.	Paddy
Tectona grandis	Tapioca
Bombax ceiba	Ginger, turmeric

Sand Dune Management

The soils of the arid zone are mostly sandy to sandy loam with excessive permeability. In India, about 22.7% of the total geographical areas is occupied by sandy soils mainly in Rajasthan, Gujarat, Punjab and Haryana. The major characteristics of these regions are severe wind erosion, high temperature, devoid of perennial river and sparse vegetation. Most of these areas is covered by drifting or semi-established sand dunes.

The movement of sand dunes occurs particularly during dry rainless period with high wind velocity. This seriously affects the houses, buildings, farmlands, roads, railways, canals and water courses. Sand dunes shift from one place to another due to blowing wind at high speed in regions devoid of natural vegetation. The problem is more serious in the western part of the desert in India where the vegetation is negligible and sand dune concentration is maximum. They need immediate attention for their stabilization. Some of the dunes are semi-stabilized but their crest and flanks remain acute because of cultivation activity on them or due to improper management.

Afforestation

The afforestation option varies with respect to degree of aridity, wind velocity and availability of different materials. Shifting sand dunes are managed with the following methods:

1. Protection against biotic interferences
2. Mulch barrier in parallel strips
3. Planting suitable trees, grasses and shrubs to act as wind breaks

TABLE 14.8: Species suitable for sand dunes

Trees	Bushes	Grasses
– *Prosopis cineraria*	– *Clotalaria burhia*	– *Cenchurs ciliaris*
– *P. chilensis*	– *Calligonum polygonoides*	– *Lasiurus sindicus*
– *Acacia tortilis*	– *Aerva javanica*	– *Panicum turgidum*
– *Acacia radiana*	– *Ziziphus spp*	
– *Ziziphus mauritiana*		
– *Erianthus munja*		
– *Dalbergia sissoo*		

The plant species chosen should have deeply buried rhizomes or prostrate growth with stem anchored by adventitious fibrous roots for rehabilitation of sand dunes. Suitable species for afforestation of sand dunes are listed in Table 14.8. Among the tree species *Prosopis cineraria* is considered as the king of trees in desert areas because of its natural speed in this area. However, under plantation condition, some initial care of the tree is required during the first year planting. *Dalbergia sissoo* is found more capable of establishing itself in desert ecosystem. Among the grass species, *Cenchrus ciliaris* performs better.

Wind Breaks

Tall plants of the following species may be used as wind breaks for sand dune management.

Dalbergia sissoo, *Albizia lebbeck*, *Acacia nilotica*, *Tamarix* spp., *Parkinsonia aculeate*.

Barren Rock Hill Ridges and Rock Out Crops

The rock exposure for varying lithology is often barren and devoid of soil cover and vegetation. There will be low canopy coverage at the top of the hill primarily due to poor/restricted growth as there is no soil in the hills. Extensive rock outcrops exist in the Deccan Plateau, Eastern and Western ghats, the Aravallis and Vindhyan ranges. They are generally termed as skeletal areas were the soils are very shallow, coarse, poor, eroded and degraded. The selection of species depends on the rainfall and other climatic factors of the areas. The species suitable for barren rock hill ridges and rock out crops are given in Table 14.9.

TABLE 14.9: Species suitable for barren rock hill ridges and rock outcrops

Trees species	Shrubs	Grasses
Areas of low to medium rainfall		
– *Eucalyptus tereticornis*	– *Carrisa spinarium*	– *Cenchrus ciliaris*
– *Dendrocalamus strctus*	– *Dicrostachys cinerea*	– *C. setigerus*
– *Melia azadarach*	– *Indigofera articulate*	– *Dicanthium annulatum*
– *Albizia lebbeck*	– *Inga dulsis*	– *Panicum antidotale*
– *Acacia catechu*		– *Pennisetum pedicellatum*
– *Hardwickia binata*		
– *Ailanthus excelsa*		
– *Cassia siamea*		
– *Prosopis chilensis*		
– *Dalbergia sissoo*		

Trees species	Shrubs	Grasses
Area of high elevation upto 2000 m		
– *Robinia pseudoacacia*		
– *Populus ciliate*		
– *Pinus roxburghii*		

INTEGRATED FARMING SYSTEM MODELS UNDER AGRO-ECOSYSTEM

Importance of Integrated Farming Systems in the Tropics

Population in the most tropical countries is increasing (Subbian and Palaniappan, 1992 and Jayanthi, 1996) warranting producing more food from the limited area available for cultivation. Compounding with this, the agricultural systems have encountered numerous ecological and environment constraints. The goal of sustainable agriculture can be achieved by seeking the optimal use of internal production inputs in a way that provide acceptable levels of sustainable crop-productivity and live stock production resulting in economically profitable return. It would be possible to reach this level with proportionately less use of inputs through integrated farming systems. One choice of components in the farming systems and their management must ensure that the interaction between components should be the most complementary with least competitiveness. Integrated farming system approach is not only a reliable way of obtaining fairly high productivity, but also a concept of ecological soundness leading to sustainable agriculture (Swaminathan, 1987).

Integration of Sub-Systems in Farming System

The various enterprises that could be included in the farming system are crops, dairy, poultry, goat rearing, fishery, sericulture, agroforestry, horticulture, mushroom cultivation etc. Thus it deals with the whole farm approach to minimize risk and increase the production and profit with better utilization of wastes and residues. It may be possible to reach the same level of yield with proportionately less input in the integrated farming and the yield would be more sustainable because the waste of one enterprise becomes the input of another, leaving almost no waste to pollute the environment or to degrade the resource base. To put this concept into practice efficiently, it is necessary to study linkages and complementarities of different enterprises in various farming system. The knowledge of linkages and complementarities will help to develop farming system (integrated farming) in which the waste of one enterprise is more efficiently used as an input in another within the system.

Generally, there are 3 different farming situations, *viz.*, dry land, garden land and wetland farming systems. The improved cropping systems, which form the major component of farming system were best verified and evaluated in farmers holdings at various locations in different agroclimatic conditions. Depending on the ecosystem, the enterprise mix will vary.

- Wetland system: *e.g.* Rice-based system + fish culture + duck/ poultry.
- Garden land system: *e.g.* Garden land crops with forage + livestock/poultry (crops with fodder, mulberry + dairy+ silkworm rearing).
- Dry land system: *e.g.* Grain/fodder crops + sheep / goat rearing + agroforestry (tree crops/agri-silvi/ agrihorti).

Arable farming results in narrow profit or less margin to the farmers, warranting development and adoption of integrated farming systems with various combinations of components. Integration of milch cow with cropping resulted in higher net returns through maintaining high yield of crop and animal than cropping alone (Arora, 1980; Siddeswaran and Palaniappan, 1990, and Chinnusamy and Purushothaman,1992). Increased net returns due to the introduction of poultry-cum-diary based farming systems over cropping alone were reported by Sirohi *et ai.*, (1980) Mohamed Ali *et al.* (1984) and Govindan (1988). Higher net return could be obtained with duck rearing in the integrated farming systems (Sharma *et al.*, 1985). Higher net income with rice + poultry + fish + mushroom culture under low land rice conditions (Rangaswamy *et al.*, 1990 and Govindan *et al.*, 1990), dairy + biogas + silviculture under garden land conditions (Rangaswamy *et al.*, 1990 and Siddeswaran and Palaniappan, 1999) and cropping + goat rearing under rainfed situations (Rangaswamy *et al.*,1990) were reported. Rice + fish + azolla integrated farming system was found to be the best for maximum net and per day returns over conventional cropping under lowland rice fields in Tamil Nadu, India (Balusamy and Shanmugasundaram, 1994 and Balasubramanian, 1994). Integrated farming systems involving dairy cattle became more relevant under existing ecological and socioeconomic conditions in the tropics and would provide an increased quantity of manure for cropping (Swaminathan, 1975). Vikraman Nair *et al.* (1979) and Sundaresan (1979) found that there was little loss of nutrients from the cattle manure in integrated farming systems. Also by recycling the wastes of dairy cow and poultry birds, the value of nutrients added to the soil could be increased (Mohamed Ali *et al.*, 1984). Rice - poultry - fish farming system was found to increase the nutrient uptake of rice and soil organic matter content (Tiwari, 1993 and Lightfoot *et al.* 1992). Intensive integrated farming systems involving duckery-cum-fish culture provided excellent opportunity to organic recycling and farmers' dependency on market purchased inputs was also reduced (Mishra and Pasupalan, 1994). Integration of enterprises like cattle rearing, fishery, poultry, goat rearing/sericulture and mushroom cultivation with cropping could provide proper recycling of residues for deriving maximum compatibility and replenishment of organic matter.(Rangasamy *et. al.* 1995).

Farming System Models for the Tropics

Development of ecologically stable and environmentally sound integrated farming system model for different eco-region to replace or improve the traditional farming systems would maximize the productivity and profitability of the farming systems. Different farming system models developed for varied agro-ecological region are given in Table 14.10.

Farming systems research came into prominence as a part of agricultural research in the past decades with a main stimulus of growing realization that most farmers are economically rational (Simmonds, 1986). The meaning of FSR varies differently to different authors and to varied agro-ecological situations, but the objectives of the farming systems, in general, are conversing to the development of suitable location specific farm technologies to raise and sustain the total farm productivity in terms of food, fodder, fuel and to meet the felt needs of the farmers within the circle of this agro-eco-socio-political favorities and constraints (Gomez, 1988 and Norman and Colinson, 1985). FSR has been applied to a variety of activities and viewed the farm in a holistic manner and considered interactions in the system (Byerlee et al., 1982). The FSR approach is directed at small farmers in marginal areas who operate complex systems characterized by important interactions between enterprises (Gilbert et al., 1980; Maxwell, 1984; Simmonds, 1985 and Collinson, 1988).

TABLE 14.10: Integrated farming system models for different agro-ecological regions

S. No.	Farming system model	Region	Suggested by
1.	Composite fish culture	India	Jayaseelan et al., (1988)
2.	Duck cum fish farming	Hongkong	Sin (1980), Woyanarovich (1980)
		Taiwan	Chen and Li (1980)
		India	Sharma et al., (1985)
3.	Pig cum fish farming	India	Jhingran and Sharma, (1980)
		India	Paulraj (1984)
4.	Poultry cum fish farming	India	Sharma et al., (1985)
		India	Kumar and Singh (1984)
		India	Govindan (1988)
		Indonesia	Djajadiredja et al., (1980)
		Malaysia	Tan and Khoo (1980)
		Malaysia	Delmando (1984)
5.	Cropping cum Mushroom	India	Rao (1984)
		India	Loganathan (1989)
		India	Rangasamy and Jayanthi (1984)
		India	Vangriensven (1987)
6.	Rice + Pig + Fish + Duck + Azolla farming	Philippines	Govina (1990)
7.	Rice + Fish + Poultry farming	Tamil Nadu (i)	Govindan et al., (1990)
8.	Rice + Fish + Poultry + Mushroom Farming	India	Rangasamy (1992)
9.	Rice + Fish + Azolla Farming	India	Shanmugasundaram et al., (1992)
10.	Rice + Fish + Dairy Farming	India	Veerabudran (1994)
11.	Crop +Diary + Biogas + Silviculture	India	Siddeswaran and Palaniappan (1990)
12.	Crop + Livestock Farming	Himachal Pradesh (I)	Sharma et al., (1991)
13.	Rice + Fish + Shrimp + Vegetable Crops	Vietnam	Lighfoot and Minnick (1990)

FSR was considered to be very important in providing scientific approach for problem identification and technology development aimed at improving agricultural production systems and human welfare through increased but sustainable agricultural productivity (Plucknett et al., 1987 and Spencer, 1991). John (1991) viewed FSR as an concept open to multiple interpretations existing for the numerous local varieties of farming systems activities. FSR is a research method designed to understand the farmers' priorities, strategies and resource allocation decisions, with a view to identify and develop technologies useful for each farm environment of varied agro-climatic conditions (Kanwar *et al.*, 1992 and Pandey *et al.*, 1992).

Composite Fish Culture

Fish culture in inland waters as an economic and paying proposition or trade has been carried on to a great extent in Tamil Nadu, Andhra Pradesh and Kamataka. A pond fish culture was given with the innovation of polyculture or composite fish culture technique in 1970's. The technology developed under the composite fish culture programme has put India on the threshold of aquapulsion. Though this system of farming with different species of carp together was practiced in Bengal even earlier, the availability of the Chinese carps (silver carp and grass carp) and common carp for farming from midseventies changed the potentiality of pond fish production. The principle involved in this culture is fuller utilization of fishpond in a body of water. Three exotic fishes viz. grass carp, silver carp and common carp along with the three indigenous varieties viz. catla, rohu and mrigal are grown in the same pond.

The feeding habits of these fishes are follows:

Catla..............................Surface (microscopic animal or zooplankton) feeder.
Silvercarp ....Surface (microscopic plants or phytoplankton) feeder.
Rohu ....Column feeder.
Grass crap ....Aquatic vegetation feeder.
Common crap ....Omnivorous bottom feeder.

Integrated Fish Farming

Integrated farming approach can be adopted to boost food production by clubbing aquaculture with agriculture, dairy, ducks, poultry, pigs and horticulture. This system combines aquaculture, animal husbandry and agriculture and forms-self-reliant integrated farming. This system recycles the organic wastes such as weeds and crops, by-products from fields, wastes from the farm itself. In this system, substances that may otherwise be considered as waste are utilized to increase fish production. Pond bottom humus is used as fertilizer for crops grown adjacent to ponds or directly on their dykes. Live stock manures (Pig, duck, goose, chicken) are used as pond fertilizers for plankton production. Agricultural by-products from wire processing *i.e.* wheat and rice, coconut or groundnut cake are also put into the ponds as fish feed. Recently anaerobic digesters are being used in integrated fish farms to produce methane gas as an energy source from agricultural wastes and manures. The nitrogen rich slurry residues are then used to fertilizer ponds. Even with the great variety of pond inputs, the limited quantity of locally available nutrients is the central factor restricting pond area expansion where integrated fish farming is, presently practiced. There are various systems of integrated fish farming and the most common are given below:

1. Paddy cum fish culture.
2. Fish cum duck farming.
3. Paddy-fish-azolla farming.
4. Paddy-fish-vegetable farming.
5. Crop husbandary and fish culture (pond system).
6. Fish cum poultry farming.
7. Fish cum pig farming.
8. Fish cum sericulture.
9. Fish cum horticulture.
10. Fish cum biogas plants.
11. Crop-Livestock-fish farming

Integrated fish farming is a diversified and Co-ordinated way to farming or producing agricultural items and livestock in the fish farm with fish as the main product. The item produced are to be used either as source of feeds or fertilizer source of additional income or both. Judicious and wise integration of these items in a fish farm promotes the full utilization of its land area and recycling of wastes and by-products; minimize the operational expenses in feeds and fertilizer; improves the living condition of workers due to the increase of income and maintains a balanced ecosystem.

Paddy-Cum-Fish Culture

This farming system of growing fish in the paddy fields is commonly practiced in the coastal areas in the States of Kerala and West Bengal. However, the paddy-curn-fish culture in the freshwater paddy fields is rare. This production system has great scope in the State of West Bengal, Assam, Bihar, Uttar Pradesh, Orissa and Andhra Pradesh where only a single crop using the deep water paddy varieties is cultivated. Investigations carried out by CIFRI have indicated that the freshwater fishes can be cultivated in the specifically designed paddy fields with trenches or pools constructed within the paddy plots so that two crops of paddy (tall variety of rice in *kharif* and high yielding dwarf variety of rice in *Rabi*) and a single crop of fish, covering both the seasons can be raised. The trenches can be classified as under:

1. Perimeter type.
2. Central pond type.
3. Lateral trench type.

The depth in the trench may vary between 1 m and 1.5 m. Composite fish culture can be undertaken in these trenches with fingerlings of catla, roghu, mrigal and common carp. Deep-water paddy varieties are usually selected for_cultivation during *kharif* season. Enough care must be taken to avoid the spraying of pesticides during the cultivation. In order to overcome this problem of pesticides affecting the fish, integrated pest management may be adopted. Pesticides like neem-based pesticides, carbonates and selective organophosphates may be used in the low doses so that the fishes are not affected. When rice is harvested during November-December, the water recedes from the rice field and the fishes take shelter in the trenches. With the onset of the *rabi* season, the rice field is once again prepared and the seedlings of the high yielding variety of paddy such as Ratna, Jaya, etc., are transplanted. Before the application of the pesticides, an earthen ridge is erected along the periphery of the paddy plot to prevent the drainage of the pesticides into the trenches with the fishes. In this system of integrated farming, a fish yield of 700-1000 kg/ha/year is expected, giving a financial return of 20 per cent over mono-cropping of rice.

Duck-Cum-Fish Farming

The integration of ducks in fish culture is economically advantageous to fish farmers. In this type of farming, the increase in production of both the animals may be met and at the same time the input cost on fish culture operations can be reduced considerably. In duck fish integration the ducks are sources of animal protein in addition to fish. In this system of farming, an ecological balance is maintained and sufficient variety of products is obtained to meet farmer's family needs in terms of food and income. Duck-fish culture is highly renunerative since ducks could be a good addition where the turn over is quick and both fish and duck are really acceptable by a majority of the population. A series of experimental trials have helped to evolve a scientifically sound and economically viable system for integrating freshwater aquaculture with duck farming, with actual benefits to both. The droppings of the ducks can be recycled for manuring the fish ponds @ 10 to 15 t/ha/year. The dropping can either be collected from the duck house or the duck houses can be constructed above the water surface of the pond itself. A total of 200320 ducks (about 1 kg in size) are sufficient to provide enough quantity of manure for fertilizing a hectare of water area of the pond. In this system, a production of 3000-4500 kg/ha/year has been achieved, besides the income derived through the sale of duck meat and eggs.

Rich-Fish-Azolla Farming

A field study was conducted at low lying wetlands of Lower Bhavani Project (LBP) Area, Agricultural Research Station, Bhavanisagar, Tamil Nadu with a view of evaluating the feasibility and economics of rice-fish-azolla farming. Experiment was conducted with two rice crop seasons (*kharif* and *rabi*) planted with rice varieties ADT 36 and ADT 38 with a total duration of eight months with one fish crop harvest. *Azolla microphylla* was applied at the rate of 2.0 t/ha as fish feed cum N fixer five days after transplanting. Calotropis, a wasteland weed, was incorporated (12.5 t/ha) as green leaf manure, 10 days before transplanting. Fertilizers were applied at the recommended level (100: 50: 50 kg NPK/ha). Irrigation was scheduled to maintain 5 cm depth of water throughout the crop growth period. Need based plant protection measures were taken with 5 per cent neem seed kernel extract. Fingerlings of catla, rohu and mrigal were stocked in equal ratio with a total of 3000/ha, 15 days after transplanting. Banana pesudostem with cowdung (1:1) mixture and rice bran were fed at the rate of 5 per cent of the fish body weight. Field trenches were provided with 1.0 m depth and 1.0 m width occupying 10 per cent of rice area to shelter the fish (Shanmugasundarm and Balusamy, 1993).

The results revealed that in low-lying wetlands the most profitable farming system was rice-rice-azolla-calotropis + fish farming. There was a slight decline in rice yield under rice-rice-fish farming due to the presence of fish trenches occupying 10 per cent of the area. However, the yield reduction was made to good when azolla and calotropis were introduced. The soil fertility status indicated that in general there was an increase in the available nutrient contents after completion of cycle. Even after heavy depletion, through crop removal, the nutrient status could be replenished by the addition of organic residues. This might be due to decomposition of azolla, fish manure and fish feed in the system.

Rice-Fish-Vegetable Farming

For effective restoration and recycling of organic wastes, an experiment was started to integrate rice-vegetable and fish culture in wetlands at Agricultural Research Station, Bhavanisagar, Tamil Nadu. The objective is to maximize the return per unit area and also efficient recycling of crop residues along with fish culture. The experiments consisted of rice alone, rice +rice + fish, rice-rice-vegetable + fish farming. The experiment was continued for two rice seasons with a total duration of eight months. The study area was 0.25 ha/ system. Field trenches (1m deep x 1 m wide) were provided for fish shelter. Catla, rohu and mrigal were stocked 15 days after transplanting at an equal ratio of 3000 fingerlings/ha. Vegetable crops viz., bitter gourd, ribbed gourd, beans etc. were raised along the raised bund of fish trenches. The results indicated that rice-rice-vegetable + fish culture was more remunerative and it was possible to get 2 rice crops, 1 vegetable crop in eight months in rice fields under two season rice.

Crop Husbandry and Fish Culture (Pond System)

By integrating the fish culture with rice, increased returns could be obtained. Further, the nutrients leached out from rice fields could be effectively utilized for fish culture. With a view to study the feasibility of rice-cum-fish culture under low-lying wetlands, an experiment was conducted at Agricultural Research Station, Bhavanisagar, Tamil Nadu. In one acre holding, conventional cropping system rice-rice-black gram practiced by the farmers was followed. In another one acre plot, integrated farming system (rice-rice-maize + fish culture) was practiced. An area of 0.85 acre for crop husbandry and 0.15 acre was allotted for fish culture. Fish pond size of 600 m^2 area was dug to a depth of 2.0 m. The fish fingerlings of catla, mrigal and common carp were stocked in the fishpond to a density of 3000 fingerlings/ha. The excess water in the paddy fields was drained once in four days from 10 days after transplanting of rice. After draining, irrigation was given up to 5 cm depth each time. Whenever fertilizer was applied as top dressing, the subsequent drainage was done six days after fertilizer application. The same practice was followed for plant protection chemical application. Supplementary feeding were given @ 5 per cent of body weight of fishes.

The comparative economics showed that rice-rice maize + fish culture recorded maximum net return and it increased by Rs. 6,383 and Rs. 5,795/ha over conventional cropping of rice-rice-black gram during 1991-1992 and 1992-93, respectively. Drained water increased the yield of fishes and this might be due to increased nutrients available in the drained water from the pond as revealed by increased plankton population.

Fish-Cum-Poultry Farming

As in the case of duck farming, the fish ponds can be manured with the litter from the poultry pens @ 50 kg/ha/ day every morning. About 500-600 birds can provide adequate quantity of litter for fertilizing a hectare of water area of the fish pond. A net financial return on variable cost @ 54 per cent is achieved through the sale of fish, chicken meat and eggs.

Fish-Cum-Pig-Farming

The pig sites can be constructed on the embankments of the fish pond. The pig manure (faeces and urine) can be drained directly into the pond. Besides serving as rich manure, the pig excreta is directly utilized by fish as feed, as it contain about 70 per cent digestible food for the fish. Hence, in this system of integrated farming, no supplementary feed or pond fertilization is required, thus reducing the input cost to about 60 per cent for the fish farming. About 30-40 pigs are sufficient to provide adequate quantity of excreta required for manuring a hectare of water area of the fish pond. A high fish yield of 6000-7000 kg/ha/year is generally obtained. A financial return of 52 per cent on the variable cost is achieved through the sale of fish and pig meat.

Fish-Cum-Sericulture

In this system of integrating the traditional method of raising fish and silkworm together, high yield of both is ensured. The mulberry trees are grown on the bonds of the fish ponds. The droppings of the silkworms are fed to the fish as grass carp feeds directly on the chrysalides and the droppings. The excreta of the fish decompose in the bottom of the pond into ammonia, phosphorus and potassium and other nutrients, which are essential for the growth. of plants. Hence, this organically rich bottom soil of the fish pond is utilized for fertilizing the mulberry trees. Thus, the fish and the plants sustain their yield through mutual benefit. Besides, the silkworm pupae and their cake after extraction of the oil can be utilized as the supplementary feed for the fish.

Fish-Cum-Horticulture

It is common to come across raising of some horticultural crops on the embankments of the fish ponds. Fruit trees like papaya, banana, etc. can be cultivated on the bunds of the pond for making the integrated farming system economically more viable. The leaves of the banana can be utilized as supplementary feed for the grass carp.

Fish-Cum-Biogas Plants

Biogas plants are usually located as a component in the integrated farming system. The slurry from the biogas plant is rich in humus and nutrients in available form helps in the production of adequate quantity of Zooplankton, when manured in the fishpond. The water manured with the biogas slurry is less acidic when compared to the ponds manured with raw cow dung.

Integrated Crop Livestock Fish Farming

Integrating crops, livestock and fish in smallholder farming systems has many ecological and economic advantages. Such systems are conducive to nature conservation as they promote habitat stability and diversity for wildlife living on the farms and in adjacent areas. As these integrated systems optimize the use of on-farm and adjacent resources, they encourage habitat conservation rather than destruction. Such systems are productive and profitable because they utilize waste as inputs in other enterprises within the farm, because the fishes are highly nutritious and valuable traditional food. They use microenvironments within a farm system, which add to farm productivity as security (Chambers, 1990). The following bio-resource flow models, prepared by farmers in Vietnam an Malawi, illustrate how farm enterprises are integrated to exploit different land and water resources and to substitute for external inputs. Making these models helps farmers transform their farming systems by clarifying how existing enterprises can be better integrated and how new enterprises can be better fitted into the farming system.

In Vietnam, the bio-resource flow models show how water from the trenches is used to irrigate vegetables, grown in a mulch of rice straw, on dikes made from trench mud. Immediately after trenching, chicken and cattle manure is put into the rice-field trenches to promote phytoplankton blooms for the fish and shrimp to feed on. Although the shrimps' diet is primarily natural, they are also fed during their first two months with farm-grown by-products, such as germinated rice grain, cassava flour, rice bran, coconut and groundnut oilcake, and trash from the irrigation canals. Mango and eucalyptus branches are also put in the trenches to keep out cattle and poachers, and to provide an undisturbed habitat, which shrimps need. Such integration reduces the need for external inputs. As the shrimps and fish eat rice weeds, weeding expenses can be reduced by 30 per cent with no detrimental effect on rice production, because animal manure and fish faeces fertilize the paddy.

In Malawi, bio-resource flow models show how fish ponds are integrated with other farm enterprises. The pond water is fertilized with rotten fruits of guava, papaya and avocado, and manure from cattle, sheep and goats. Leaves of leucaena, pumpkin, wild vegetables and cowpea are also put into the ponds to feed the fish, and coarse maize bran is used as a fish food when available and affordable. Apart from fish, other outputs from the ponds are fertile sediments for the vegetable gardens and water for irrigation. The main purchased input on these farms is maize bran which can be supplemented and partly substituted with yam leaves and napier grass.

IFS Models

In an agricultural country like India, the average landholding is very small. There is no scope for increase in farm size. The population is steadily increasing without any possibility of increase in land area. The income from cropping of an average farmer is hardly sufficient to be assured of regular income for a decent living. He has to think of other enterprises, which will complement his cropping activity. This is made possible through optimal crop and livestock mix consistent with the farm resources immediately available. Farming includes crop, livestock, poultry, fishery and tree culture. A judicious mix of one or more with cropping complements the cropping enterprise through effective recycling of residue/waste. A carefully chosen and planned farming system, keeping in view the soil and environment conditions, will be helpful to the users in greater dividend. Farm as a unit is to be considered and planned for effective integration.

Integrated Farming System

Experience through case studies and experiment conducted at Tamil Nadu Agricultural Uuniversity, Coimbatore, revealed that by integration of complementary enterprises like livestock, fishery, agroforestry etc., additional income and generation of additional employment potential could be achieved. Hence models of integrating different components were formulated for different land types viz., dry land, garden land and wetland. The models of the different farming systems are as follows.

Model 1 (Wetland Farming System)

A model was developed for 0.40 ha (1.00 acre) considering small and marginal farmers of Tamil Nadu State (Rangasamy et. al., 1988 and Ranagasamy, 1992). While programme for the cropping system, due consideration has been given to the crops under cultivation in the locality, the crop rotation that is followed, and water requirement for the fishpond. The details of each system integrated are furnished below:

	Components	Allocation of land
I.	Cropping	0.36 ha
II.	Fisheries	0.04 ha fish pond
III.	Poultry (Poultry shed is placed over the pond)	
	Total	**0.40 ha**

I. Cropping Programme

Rice (IR-50)	-	Rice (IR-20)	-	Maize (CO 1)	0.16 ha
(May-August)		(Aug. – Dec.)		(Jan. – April)	
Rice (IR-50)	-	Rice (IR-20)	-	Groundnut (CO-2)	0.10 ha.
Maize (CO 1)	-	Rice (IR-20)	-	Gingelly (TMV-3)	0.10 ha

II. Fisheries

The number of fingerlings used for the proposed programme has been arrived at based on the recommended stocking density of 7500 numbers per ha of ponded water. Accordingly 300 numbers at the ratio noted against each of the following varieties were released in the pond.

Silver carp/Catla 120 (4)
Rohu (Column feeder) 60 (2)
Mrigal/Common carp (Bottom feeder) 90 (3)
Grass carp (Grass feeder) 30 (1)

III. Poultry

Five hundred layer chicks are sufficient enough to void required quantum to excreta to meet the feed requirement of 7500 polyculture fingerlings in a hectare of ponded water. Accordingly twenty papkok chicks were used in the study.

Control: The normal crop rotation that is being followed in this tract under low land condition is also included in the programme to compare with income under integrated farming system. The details of which is furnished below:

Rice (IR-50)	-	Rice (IR-20)	-	Green gram (CO-3)	0.20 ha
(May-AUG.)		(Aug. – Dec.)		(Jan. - March)	
Rice (IR-50)	-	Rice (IR-20)	-	Sunnhemp	0.40 ha

In rice cultivation, even with best management practice we are getting about 5 to 7 per cent ill filled or partly filled grains. This waste product of rice when it is hulled becomes a substitute of 35 per cent of poultry feed. The poultry feed cost can be reduced by about 57.5 per cent by substituting the ill filled/partly filled rice grain, maize grain, oil cakes of crop component, the lucerne grown around the fish pond and fish waste anticipated through disposal of fish. The reduction in feed cost by way of effective recycling the products of crop component, leads to reduction in the cost of production per egg. In commercial poultry farm, cost of production per egg is around 42 to 44 paise. By way of reducing the feed cost, the cost of production per egg is reduced to 19 paise, thus leading to fingerlings which solely depend on the poultry excreta grows at the rate of one kilogram per year yielding about 250 to 300 kg/year, which on marketing will be a bonus.

The growth of fingerlings was solely due to the dropping of the birds kept over the fishpond. The poultry droppings contain protein 30 per cent and fat 2.5 per cent (Table 14.11). The poultry droppings were found to enhance the nutritive value when compared to periodical fertilization and supplemental feedings of fish with groundnut cake and rice bran. Mortality of birds was almost nil till the end of the program. This may be due to the cool environment created by the fishpond. In rice cultivation even with best management practice about 5-7 per cent ill filled or partly filled grains are obtained. This waste product of rice when it is pooled becomes 35 per cent of poultry feed can be' reduced to certain extent. By converting the groundnut and sesame seed into its by product viz., oil and cake, the cake can be made free of cost to the poultry feed component and thus can reduce the cost of the poultry feed. The poultry feed cost can be reduced by about 57.5 per cent by substituting the ill filled or partly filled grain, maize grain of the farm; oil cakes of the crop component and the alfalfa grown around the fish pond. The reduction in feed cost by way of effective recycling the products of crop components, leads to reduction in cost of production of egg. It works to Rs. 0.19 as against Rs. 0.40 in the conventional poultry farm Gayanthi *et ai.*, 1994).

TABLE 14.11: Nutritive value of poultry manure (Dry matter basis)

Name of the constituent	Percentage
Protein (Crude)	30.00
Fat	2.50
Crude Fibre	13.00
Nitrogen free extract	30.50
Total ash Minerals	24.00
Calcium	4.60
Phosphorus	3.00
Sodium	0.40
Potassium	1.37
Copper (ppm)	61.00

Name of the constituent	Percentage
Zinc (ppm)	325.00
Manganese (ppm)	291.00

The economics worked out for the system as a whole for all the five years reveal that the mean net income under integrated farming system (IFS) and conventional system were Rs. 34067/ha/yr and Rs. 17586/ha/yr respectively. The net profit was Rs. 93.33/ha/day and it was 93.7 per cent higher than conventional system. The employment opportunity in IFS was enhanced to the tune of 102 per cent in man-days over the conventional system. The result of the integrated farming system program was compared with conventional crop activity. The income obtained from the integrated farming system program was compared with conventional farming system 59.20 per cent was from cropping, 8.30 per cent from poultry, 7.50 per cent from fisheries and 25.00 per cent from mushroom. The additional net income realized from the IFS was Rs. 14126 over conventional system. The additional employment generated through farming system over the conventional system was 627 days/ha/yr.

2. Model II (Garden Land Farming System)

A model on integrated farming system to suit the small farmers under upland condition was developed. The following are the components of the model.

Cropping: 1.00 Ha

.... Ragi + Sunflower – Cotton + Maize + Green gram/ Cowpea... 0.56 ha
..............................Summer cotton + Maize + Wheat + Green gram/Cowpea/ .... Sunflower/Radish/Beetroot/Bengal gram/Bellary onion... 0.19 ha
.... Perennial grass fodder (CO-1)... 0.15 ha
....Perennial legume fodder (Lucerne)... 0.05 ha
....Farm shed .... 0.05 ha

Dairy

Three number of graded jersey cross breed milch animals with two calves.

Biogas Plant

For effective recycling of farm and animal waste, and for fuel, light and enriched manure.

Farm Silviculture

Two hundred Subabul trees along the boundary of field for fodder and timber.

A model on integrated input management to suit the small farmers under garden land conditions was studied during 1988 to 1993 at the garden lands of TNAU, Coimbatore, Tamil Nadu (Premsekar and Rangasamy, 1994). The model integrating the components of cropping, dairy, spawn production, biogas and silviculture were contemplated in. this study (Table 14.12). The receipt, expenditures and net income of 1988 to 1993 in Integrated Farming System (IFS) and conventional cropping system (CCS) was Rs. 34584 and Rs. 13496/ha respectively. IFS thus provides an additional mean net income of Rs. 20638/ha over a period of five years by way of effective recycling. The employment opportunity in the new integrated model with proposed components was enhanced to the tune 770 man days per year over CCS. The results of five years study revealed that by way of integrating agricultural allied enterprises with crop activity, the standard of living of the farmer, income per unit area, employment opportunity for the family etc. enhanced over the CCS (Table 14.13 and 14.14).

TABLE 14.12: Treatment details of integrated farming system and conventional cropping system

Treatment details					Area (ha)
Integrated farming system (IFS)					
Cropping					
a. Cotton + Green gram	–	Maize + fodder cowpea	–	Bellary onion	0.56
(Aug,-Feb.)		(Feb. – May)		(May –Aug.)	
b. Wheat + Sunflower/Beet root/Bengal gram/ Radish	-	Maize + fodder cowpea	-	Summer cotton	0.19
c. Grass bajra Napier (CO 1)					0.15
d. Lucerne					0.05
e. 150 trees of Lucerne and Leucanea lecucocephala planted on bunds					
f. Farm shed					0.05
ii. Dairy unit					0.05
iii. Jersey cows + 2 calves					
iv. Spawn and mushroom production capacity 1.5 to 2 kg/day					
Conventional cropping system (CCS)					
Cotton	-	Sorghum	-	Ragi	0.20
(Aug.-Feb)		(Feb.-May)		(May-Aug.)	

TABLE 14.13: Receipt, expenditure and net income for five years (1988-89 to 1992-93) in Garden land IFS.

Years	Receipt (Rs./ha/yr)		Expenditure (Rs./ha/yr)		Net income (Rs./ha/yr)	
	CCS	IFS	CCS	IFS	CCS	IFS
1988-89	42701	79862	23887	34547	27161	10660
1989-90	33977	70252	13910	26400	36276	12490
1990-91	41987	77832	14620	26740	35655	12420
1991-92	28589	45707	10338	37845	17117	27507
1992-93	25665	101105	6975	35615	75440	28640
Mean	**34584**	**74952**	**13946**	**32220**	**40369**	**18343**

Net income in IFS (Rs,/ha/yr) = 20638
Additional net income in IFS (Rs./ha/yr) = 5054

TABLE 14.14: Contribution from different components in IFS (Garden land)

Details	Crop	Dairy	Biogas	Silviculture	Spawn
Receipt in Rs./ha/yr	19488	33728	5247	5996	104493
Expenditure (Rs./ha/yr)	8380	14503	2256	2578	4512
Net income (Rs./ha/yr)	11108	19255	2991	3416	5980

Details	Crop	Dairy	Biogas	Silviculture	Spawn
Percentage to the tolal	26%	45%	7%	8%	14%
Labour consumed					
Man days/ha/yr	668	274	91	15	162

3. Model III (Dry Land Farming System)

A. Problems and Potential in Dry Farming

Dry land farmers are poverty stricken and their resource base is poor. They do not have dependable infrastructure to support their family and livestock. 10 to 40 per cent of rainfall is unutilized and along with it, an estimated 8.4 million tones of NPK are lost. Soil erosion has been and will continue to be a problem. Fodder shortage is another problem and because of this, there is death of draft power, which, in turn, leads to difficulty in timely completion of field operations. Dry lands are impoverished which is further compounded by poor rainfall with respect to quantity and distribution (Sivasankaran, 1994).

Despite this fragile ecosystem, scope to improve the productivity is vast. That means, there is lot of hidden yield potential. Experiments go to show that the yield of rainfed cereals could be increased 6-8 times over the farmers' practice. In marginal lands of SAT, 6 t/ha of *Stylosanthus hamata,* 3 t/ha of *Cenchrus ciliaris* (all in dry state) besides 5 t/ha of fuel wood could be obtained with the incorporation of scientific management practices in silvipasture. The yield of ber could be 5O kg/tree/annum. So much is the potential in dry lands. But unfortunately in dry farming, technologies cannot be uniformly recommended because of wider variability of rainfall from place to place.

B. Crop Intensification

Crop intensification can be by and large, in space dimension (inter cropping) only barring areas where there is bimodal distribution of rainfall. Intercropping is different from the mixed cropping. Depending upon the rainfall, the inclusion of intercrop aids to achieve yield stability and under favourable rainfall, to yield/income per unit area.

C. Alternative Land Use Systems

Dry farming is at best risky. If marginal and sub-marginal lands as per land capability classification is earmarked for appropriate alternative land use systems such as silvipasture, dry land horticulture etc dry farming may be less risky. Such alternatives could rescue the farmers from reaching disaster level of income in the event of crop failures due to poor monsoon. A hint at the economics of dry land use system. While the B: C ratio with annual field crops never goes beyond 2.0, it is always more than that in dry land horticulture (Korwar et al., 1991). Similarly inclusion of subsidiary enterprises like goat or sheep rearing not only rescues the farmers in terms of crop failures but also maximizes the income for poverty stricken dry land farmers, That is why, goats and sheep are considered 'walking banks' of the farmers. Findings of an ongoing IFS experiment at Regional Research Station, Aruppukottai, Tamil Nadu, are given below (Table 14.15).

D. Dairy Project

Studies show that dairy based mixed farming practiced in marginal farmers holdings with 3 milch animals was more renumerative than other mixed farming practices which gave an annual mean net income of Rs. 48000/ha at Coimbatore, Tamil Nadu. In Paiyur, Tamil Nadu, in a 0.8 ha farm, dairy based farming gave the maximum return (Rs. 4,872/ha) and employment (518 man days) per year. Four dairy animals can be maintained in one hectare of land in such a way that two give milk throughout the year. By keeping 4 milch animals, it was clear that the benefit cost ratio of 1.61 and employment generation of 274 man days could be obtained (Chelliah and Chandrasekaran, 1989).

TABLE 14.15: Economics of IFS (Rs./four acres)

Particulars	Expenses	Income	Net income	Additional returns from four acres
1989-90				
Control (Crop alone)	4050	9576	5526	
IFS*	9277	16966	7889	2163
1990-91				
Control	5500	9340	2840	
IFS*	5664	12060	6390	2556
1991-92				
Control	13752	16098	2346	
IFS*	22428	29981	7552	5206

IFS*: Annual crops + Tellicherry goat * fruits trees (all in 4 acres)

E. Biogas-Project

Based on the number of dairy animals available, a biogas plant can be constructed near the farm shed itself. The dung collected from 3-4 milk animals is enough to generate 2 m^3 capacity of biogas. This can be used for the preparation of gruel to the dairy unit, lighting two lamps in the farm-stand, mushroom and spawn production.

F. Spawn and Mushroom Project

From the cropping component, sorghum grain could be obtained as per the schedule, which is the basic material for spawn production. From the mother spawn, we can inoculate the sorghum seeds to produce spawn bottles by recycling the biogas plant. Mushroom cultivation is a viable proposition in the mixed farming system and the production technology is also simple. Small and marginal farm families with thatched shed and less expenditure can do the mushroom cultivation (Chelliah and Chandrasekaran, 1989). *Pleurotus sp*., the commonly grown edible type of mushroom, can be grown easily. Paddy straw is cheap and a good substrate for cultivation of oyster mushroom. It is clear that in a small project with a production of 3 kg of mushroom the farm family can get a net income of Rs. 6,400/ha and a per day income of Rs. 35.00.

G. Improved Technology Available

Efforts to develop improved technologies for stepping up the production of dry lands are being made since long in the country, as a result the technologies capable of increasing productivity and income appreciably have been generated. A critical examination of different technologies reveals that the dry land technology should consist of soil and water conservation, crop management, efficient cropping system and Integrated Farming System (Rajendra, 1994).

(A) Existing Cropping System

The normal cropping programme of the tract is sorghum and cowpea and was raised for grain purpose during Northeast monsoon period.

(B) Improved Cropping System

In this system, field crops and both tree and forage crops were included. All the improved technologies available for ensuring a successful crop production and income of the farm in dry lands were applied.

A. Crop Components

(i) Sorghum + Cowpea (grain) : 0.20 ha
(ii) Sorghum + Cowpea (fodder) : 0.20 ha
(iii) Subabul + Cenchrus ciliaris : 0.20 ha
(iv) Acacia senegal - tree fodder : 0.20 ha
(v) Prosophis cineraria - tree fodder : 0.20 ha

B. Goat Rearing

Six Tellichery goats (5 female and 1 male)

(C) Technology Integrated

Taking advantage of the rain received during March and April, summer ploughing was resorted for better absorption of rainwater and eradicates the perennial weeds. Since the slope of the land is less than one per cent and the rainfall is comparatively low, broad bed furrows 1.5 m interval was formed to increase the intake of rainwater. Much care was taken in selecting the improved crop varieties. The selected varieties have high yield potential, high harvest index, flexibility in sowing time; better response to management, compatibility for crop mixture and resistance to pests and diseases. The sorghum seeds were treated with 10 per cent potassium dihydrogen phosphate and the cowpea seeds with zinc sulphate solution for drought tolerance.

These field crops were sown during third week of September, (Pre-monsoon sowing) before the onset of North East Monsoon. Optimum seed rate and depth of sowing were ensured for getting sufficient population. Application of organic manure in the form of enriched farmyard manure (750 kg/ha) and composted coir pith waste and tank silt would augment the moisture holding capacity of the soil. Sorghum seeds were treated with Azospirillium and cowpea. with Rhizobium to increase the root growth to absorb moisture from deeper layers of soil at times of soil drought. Foliar spraying of antitranspirants like Kaolin and also potassium chloride (0.5 per cent) at critical stages when crops suffered due to acute moisture reduced the yield loss by 10-15 per cent.

(I) Tree Farming

The tree crops subabul, *Acacia senegal* and *Prosophis cineraria* were planted for fodder, timber and fuel. These perennial fodder tree crops are highly drought resistant hardy plants. In deep vertisol, fruit trees like pomegranate, sapota and guava were planted. Annual crops were raised in the interspaces of the perennial crops for about 4 or 5 years.

(II) Water Harvesting

In the rainfed area, about 50-60 per cent of the total rainfall is received during a short period of 3 months. About 50 per cent of the rain goes as run off. Having farm pond could collect this water. A farm pond was dug in an area of 7-8 cents (280-320 m^2), which could store about 240 m^3 ($1m^3$ = 1000 lit) of water during monsoon period. The water so collected was used for pot watering the tree crops at least during the critical stages. The water could be stored in the farm pond for about 30-35 days under Coimbatore condition. For grown up trees, slopy basins can be formed so as to collect rainwater at the root zones.

(III) Goat Rearing

Tellicherry goats was integrated with the improved cropping under dry land in Coimbatore. These goats are medium size animals with average weight of 23-25 kg capable of yielding 1.5 lit. milk per day. The goats yield more meat, exhibiting grazing habit similar to sheep. They thrive on several types of grasses and leaves and fairly resistant to diseases. Initially five female and one male goats were maintained and in course of time they multiplied to 20:1 in a area of one hectare land. These goats were stall fed under deep litter system.

(IV) Organic Recycling

Organic waste was effectively recycled in Integrated Farming System. The goats were fed with annual crops like (sorghum and cowpea) and perennial fodder trees (*Acacia senegal* arid *Prosopis cineraria*) at the rate of 3 kg of dry fodder, 3 kg of green fodder and 100 gm of concentrate per adult animals. The animals were grown in deep litter system and the manure so collected was applied to the field. We could get an average of 3 tones of organic manure valued Rs. 750.00 to Rs. 1,000, which is an additional income by way of recycling of the organic waste.

(V) Economics

The system of integrating crops, goat and trees recorded increased net income compared to the conventional cropping system in all the five years of experimentation. The above mentioned IFS gave a mean net income of Rs. 5666/ha/yr where as the conventional cropping system gave a net income of Rs. 1919/ha/yr. (Rajendran, 1994). This resulted in an additional net income of Rs. 3747/ha/yr. The net profit per day worked out was Rs. 15.52/ha in the Integrated Farming System whereas it was only Rs. 5.26 under conventional cropping system. The average net income from IFS under crop and animal component were Rs. 3085 and Rs. 2584 respectively. In IFS, 54.4 per cent of net income was obtained from crop component and remaining 45.6 per cent was from the animal component.

The average income from the crop component in IFS exceeded the average income of crop in the conventional cropping system. This was due to inclusion of various crops and trees and improvement in soil fertility due to effective recycling. This also provides employment opportunity to family labour round the year. The average labour requirement for maintenance of goat works out to 52 man days per year and it was 101 man day /ha/yr for the crop component under integrated farming system. The total employment generated under IFS was 153 man days/ha/yr whereas it was 40 days/ha/yr under conventional cropping system thus generating an additional employment opportunity of 113 man days/ha/yr.

The different farming systems developed independently, without any consideration for utilizing the wastes arising out of the industrial, agricultural and other activities of the communities are not viable. Hence, in course of time, these wastes have assumed staggering dimensions. Thus detailed investigations were undertaken for recycling these wastes more profitably under the integrated farming systems. The recent scientific advances have pave the way to blend our traditional practices of integrating various farming systems according to the present day needs. The adoption of this farming system brings down the operational cost considerably and increases the yield per unit by recycling the wastes or the byproducts of the various farming systems.

A Model Study at TNAU, Coimbatore, Tamil Nadu

One hectare of dry land was selected in Tamil Nadu Agricultural University, Coimbatore. The mean annual rainfall of this tract is 648 mm. The mean maximum temperature ranges from 29 °*C* to 35 °*C* with the minimum temperature range of 17 °*C* to 23 °*C*. The soil type is clay loam. Six goats (5 female and one male) were taken up for the study with the following cropping programme.

Cropping System

IFS

1. Sorghum + Cowpea (grain) : 0.20 ha
2. Sorghum + Cowpea (fodder) : 0.20 ha
3. Subabul + *Cenchrus cilliaris* grass : 0.20 ha
4. *Acacia Senegal* (tree fodder) : 0.20 ha
5. *Prosophis cineraria* (tree fodder) : 0.20 ha

Control

Sorghum + Cowpea (grain) : 0.20 ha

Feed for the Goat

The goats were allowed in the developed pasture viz., *Cenchrus cilliaris* grass + *Clitoria ternata* (leguminous plant) grown in the pasture land. Dry fodder was fed to the animal as stall feeding at 2 kg/day.

Crop Yield

The straw and green fodder yield from the experimental area was recorded and income calculated. The animal yield was recorded in terms of meat, the weight gained by the six animals for one year. The results of the one year experiment reveals that the net income under Integrated Farming System may be increased by increasing the herd size/ha that can be fed with the biomass produced from one hectare of land, besides increasing the employment opportunity round the year.

Model IV (Coconut Based Integrated Farming System)

Coconut is mainly grown as a pure crop under rainfed conditions. In India, ten million people depend on directly or indirectly on coconut cultivation, processing and marketing. Being a perennial crop, it starts yielding from fifth year extending up to sixty years providing insufficient farm income to sustain even small families. One of the feasible way to step up production and to enhance the farm income is to grow compatible intercrops in association with coconut. In addition, part of the garden can be entitled to grow fodder crops which in turn help in rearing milch animals, generating conventional energy and satisfies the family requirement.

Coconut utilizes only 40 per cent of the solar radiation and 23 per cent of the soil space and remaining area is found wasteful and can be efficiently tapped through companion crops. An endeavor was undertaken in the existing coconut plantation aged six years planted with normal spacing 7.5 x 7.5 m at a density of 180 palms/ha to assess the suitability in intercrops. Out of one acre taken for study, 60 cents were alloted to crop components and remaining 40 cents left for rearing milch animals.

Crop Components

tier : Coconut
tier : Fruit crops (Papaya, anona, pomegranate, banana) Vegetables: Bitter grourd, ribbed gourd, lab lab.
tier : Spices and condiments: ginger, turmeric.
tier : Coriander, mentha (Pudhina).
tier : Grass: Bajra Napier.
Trellis : Betel vine and pepper.

Fertilizers. were applied as per recommendations of each crops. Irrigations were scheduled once in a week except betel vine which was irrigated in alternate days. Water requirement of coconut (120 litres/day) was apportioned and shared by all crops, monitored through a water meter. Income derived is furnished in Table 14.16. The results revealed that growing compatible intercrops in coconut realized the income of Rs. 9,477.25/ha for a period of ten months over the pure coconut plantation. Additionally there was an employment generation of 400 man days.

TABLE 14.16: Economics of coconut based farming system

Sl. No.	Crop	Quantity (kg)	Amount (Rs.)
1.	Ribbed gourd	42	126.00
2.	Bitter gourd	15	60.66
3.	Coriander	-	615.75
4.	Ginger	10.2	307.50
5.	Turmeric	806	6448.00
6.	Grass	1320	1320.00
7.	Sunnhemp	600	600.00
Total			**9477.25**

INTEGRATED NUTRIENT MANAGEMENT (INM) THROUGH IFS

India would need to produce additional food grains of 100 and 160 m t. by 2030 and 2050 AD respectively, to feed its projected population. More than 85 million, out of 105 million operational holdings in India are with less than 2 ha farm size, necessiating intensive cultivation involving the modern concept of time and space by adopting heavy dose of chemical fertilizers leading to accumulation of chemical residues in the soil (Swaminathan, 1990). As agriculture becomes intensive, soil toxicity and deficiencies of various kinds threaten the sustainable production, yields start declining in high production areas, if sustainability concepts are to be integrated in different farming situations. One of the main objectives in IFS is recycling of produces/wastes among the components involved. The end product of such recycled materials is used as organic manure for crops. In this venture, some of the products are improved in their quality on recycling and thus the program provides double benefit to the farmer. A judicious mix of one or more enterprises with cropping complements each enterprise through effective recycling of waste/residues and in addition, when carefully chosen keeping in view the soil and environmental conditions.

Integrated farming system approaches are not only a reliable way of obtaining fairly high productivity with substantial fertilizer economy, but also a concept of ecological soundness leading to sustainable agriculture. The need for system approach of nutrient supply to the system as a whole to increase the fertilizer use efficiency in economizing the costly mineral fertilizers by accounting residual effect of applied fertilizers and preceding crops was greatly emphasized (Rangasamy, 1994).

Crop and Animal Waste

The productivity of livestock mainly depends on the availability of quality feed and fodder in requisite quantity. It is observed that most of the livestock population in the country in underfed. The residues of different agricultural crops constitute the major source of fodder for livestock. In India, till late 1980's the main source of livestock feed consisted of conventional agricultural residues like rice straw, wheat, sorghum and maize stalk, sugarcane trash and remains from pulses. Based on the grain : straw ratio, approximately 334.5 million tones of agricultural crop residues are available in India.

The area under green fodder is not increasing to meet the increased demand. The area under green fodder in the country is estimated to be approximately 4 per cent of the total cultivated area. Since the farmers feel that animal power is one of the costliest inputs, they switched over to mechanical power for most of the operations. Moreover, the animals meant for milch and meat are replaced by cross-bred and improved varieties, which need to be supplemented with quality concentrated feeds. Thus, the estimated total production of crop residues in 2020 AD to the tune of 447.0 m t is to be effectively utilized. The details of the crop residues in India as given by FAO (1985) projected to 2020 AD. are furnished in Table 14.17.

TABLE 14.17: Projected availability of crop residues in 2020 A.D.

Crop	Quantity of residue (m.t.)	
	FAO (1985)	Projected for 2020 A.D.
Rice	118.9	160.0
Wheat	57.5	80.0
Maize	21.0	35.0
Millets	40.0	56.0
Sorghum	41.0	52.0
Sugarcane	43.0	64.0
Total	**321.0**	**447.0**

In addition to crop residues, there is a possibility for collecting the following quantities of animal voids in India (FAO, 1985).

Cow dung	:	750 m.t.
Buffalo dung	:	250 m.t.
Voids of small ruminants	:	130 m.t.
Total	:	1130 m.t.

At present, the voids are being utilized for fuel and as FYM/compost. The same objectives can very well be attained with boosted value through recycling is found through experimentation. When recycled through biogas unit, there is good possibility to improve the organic source of nutrients apart from generation of fuel energy, when fossil fuel is getting depleted. Apart from the major nutrient, there is good amount of increased availability of secondary and trace elements required for the excellent performance of crop communities. By virtue of adoption of one of the modem agricultural technology viz., integrated farming system (IFS) in the farm activity, there is a possibility of improving untapped potential of each and every produce by recycling with dual benefits.

The possibility of improving the quality of products/waste materials of different components involved in the integrated farming system trials were conducted at Coimbatore for lowland, upland with supplemental irrigation and rainfed land through effective recycling and integrated nutrient supply system and their results are presented.

Lowland Farming

In the lowlands IFS, cropping, poultry, fishery and mushroom enterprises are involved with a view to recycle the residue and byproducts of one component over the other. The trial was conducted for a period of five year (1992-1996) in an area of 0.4 ha (1.0 acre) considering the average holding size of marginal and small category farmers. In one acre farm, 90 cents were assigned for crop activity and the remaining 10 cents allotted to fish pond. Twenty fowls sheltered over 10 cents of fish pond to feed 400 polyculture fingerlings gave about 700 kg poultry droppings in a cropping year. This could yield about 33.7, 21.4, 10.1 kg of N, P and K respectively. In IFS, at the end of one year after the harvest of grown up fish about 4500 kg of silt was obtained from the pond with a nutrient content of 3.52, 1.38 and 1.06 per cent, N, P and K respectively. This could relieve the burden of applying equal quantity of inorganic fertilizer to the crop component of the IFS. Even after accounting the nutrient value of the voids of fowls as manure, additional benefit of 124.7 kg, 40.7 kg and 37.6 kg of N, P_2O_5 and K_2O respectively, can be achieved by recycling (Table 14.18) (Rangaswamy, 1996).

TABLE 14.18: Nutrient value of recycled poultry manure

Particulars	Content (per cent)			Nutrient added (kg)		
	N	P	K	N	P	K
Raw poultry manure	4.81	3.06	1.44	33.7	21.4	10.1
Settled silt of the pond in which poultry dropping is used as source of fish feed.	3.52	1.38	1.06	158.4	62.1	47.7
Additional benefit through recycling				124.7	40.7	37.6

Similarly, inclusion of mushroom with the production capacity of 2 kg/ day as one of the components in IFS utilizes about 1800 kg of paddy straw and could yield about 2340 kg of mushroom spent after the harvest of edible mushroom at the end of one year. The enhancement in weight in the mushroom spent is due to unharvested mycelial growth. The nutrient value of mushroom spent in IFS is given in Table 14.19.

TABLE 14.19: Nutrient value of mushroom spent in IFS

Particulars	Content (per cent)			Nutrient added (kg)		
	N	P	K	N	P	K
Paddy straw	0.62	0.24	1.72	11.2	4.3	31.0
Mushroom spent	0.71	0.29	1.74	16.6	6.8	34.6
Additional benefit through recycling				5.4	2.5	3.6

Field experiments were conducted at Tamil Nadu Agricultural University, Coimbatore to identify and optimize the component linkage for lowlands of Tamil Nadu and to sustain the productivity through effective recycling of wastes from the linked components associated in the system during September, 1993-August, 1995. The components of integrated farming systems in lowlands involve crop + fish + mushroom, crop + poultry + fish + mushroom and crop + pigeon + fish + mushroom. The efficiency of the component linkages was evaluated predominantly on the basis of productivity, its income and employment generation with the possibility of utilizing recycled organic wastes as nutrient to enrich the soil fertility.

The results on productivity, economics, employment generation, energy budgeting and resource recycling revealed the advantage of appropriate component linkage for lowland integrated farming systems (Jayanthi, 1995). Experiment on enterprises linkage for lowland farming systems revealed that rice-soybean-sunflower and rice-sesame-maize cropping systems each in 0.18 ha with pigeon (40 pairs), fish (400 polyculture fingerlings in 0.04 ha of ponded water) and mushroom (2 kg/ day) was the best in obtaining higher rice grain equivalents than the conventional cropping system of rice-green gram-maize (0.20 ha) and rice-sunnhemp-maize (0.20 ha) cropping systems (Mode. I). Cropping + pigeon + fish + mushroom integration earned the highest gross and net return with better per day return and benefit: cost ratio. Integration of cropping with pigeon + fish + mushroom generated the highest employment of 798 man days/ha/year with 1.16 man days from the allied enterprises linked with equi-distribution through out the year. Integration of poultry + fish + mushroom + cropping applied with recycled poultry manure sustained the productivity of soil through the addition of bio-resource residue with better NPK nutrient supply potential (Model 2).

The growth of fish cultures included was comparable among artificial feeding and the treatment with poultry and pigeon dropping feeding. Fish production was higher with artificial feeding than with other two methods (Model 3). However, the net returns and benefit: cost ratio were higher in the fish fed with poultry dropping followed by pigeon dropping.

Rice-Fish-Azolla Farming System

Field experiment was conducted at Agricultural Research Station, Bhavanisagar, Tamil Nadu to develop an integrated N management practice for rice-fish-azolla farming in wetland. Farming systems consisted of rice+rice+fish and rice-rice+azolla+fish and two levels of nitrogen (100 and 75 per cent of recommended) with and without green leaf manure (*Sesbania rostrata*) applications. In the rice-fish system, rice and fish crop were raised together (synchronous system) in rice field. Field trenches were provided with 1.0 m depth and 1.5 m width occupying 10 per cent of the rice area, for sheltering the fish. *Azolia microphylia* was grown in rice field throughout the cropping period. The dual culture method of growing azolla with rice has gained widespread adaptability, because standing water is available in rice field from seedling to panicle maturing in lowland rice fields and is effectively used as biofertilizer for rice. Azolla cultivation in rice field can improve the fish food organisms. Fish culture in the decomposition of organic matter and promote release of nutrient from soil. The excreta of fish directly fertilize the water in rice fields leading to increase in utilizable source of N to the rice crop. Integration of allied components like azolla + fish with rice in lowland farming could provide wider scope for bio-resource recycling.

Rice-rice + azolla + fish farming with 75 per cent recommended N as well as incorporation of green leaf manure resulted in higher productivity with increased economic returns and improved the soil fertility through recycling of organic residues. The quantum of organic residue addition and N added through recycling were higher in rice-rice + azolla + fish farming with *Sesbania rostrata* incorporation (Table 14.20). The unutilized fish feed, decayed azolla and fish excreta settled at the fish trench bottom had a higher nutrient-value, which can be recycled to enrich the soil (Balusamy, 1996).

TABLE 14.20: Organic residue recycling and N addition through residues

Systems	Crop residue	Organic residue added (kg/ha/yr)		Total N addition (kg/ha/yr)
		Fish trench residue	Azolla biomass	
Rice-Rice	3620	-	-	19.85
Rice-Rice + Azolla + Fish	3860	952	49800	136.29

Farming in Upland with Supplemental Irrigation

Bearing in mind the advantage of recycling of cowdung, a waste from the animal component (3 adult + 2 calves), biogas unit was linked in the IFS of upland with supplemental irrigation instead of directly utilizing it as FYM. It gave a good proportion of methane gas, a fuel supplement, apart from its enhanced manurial value of the slurry that comes out of the biogas chamber. Moreover, by this recycling, some of the weed seeds present in the raw cowdung also get killed during digestion process, thus improving the quality of the slurry over its raw materials used. The quality improvement by way of recycling the cowdung through biogas chamber was studied by analyzing the NPK content of slurry and FYM prepared utilizing the cowdung from the system. The total quantity secured from the unit over the period of 365 days has been taken on equal weight basis and the analyzed data is furnished in the Table 14.21 and Table 14.22. The net income and employment generated through this system were Rs. 20,702/ha/year and 620 additional man days/ha/year respectively. Recycling of cowdung also led to the production of 730 m^3 of biogas with the possibility of enhancing the nutrient value of NPK to the tune of 44.5 kg, 65.9 kg and 28 kg respectively in a year. Trace elements like Fe, Mn, Zn and Cu are also present in an enhanced level over FYM (Rangasamy and Premsekhar, 1994).

TABLE 14.21: Nutrient content in biogas slurry and FYM

Particulars	Nutrient content						
	Major element (per cent)			Trace element (ppm)			
	N	P	K	Fe	Mn	Zn	Cu
Biogas slurry	1.43	1.21	1.01	4200	550	150	52
Farm Yard Manure	0.94	0.56	0.72	4000	490	100	45

TABLE 14.22: Nutrient saving by recycling (in kg)

Particulars	N	P	K	Fe	Mn	Zn	Cu
1. Quantum of nutrient in slurry obtained (11.0 t on dry weight basis)	157.3	133.1	144.4	46.2	6.05	1.65	0.57
2. Quantum of nutrient in FYM obtained (11.6 t) in an year	112.8	67.2	86.4	44.0	5.39	1.10	0.49
3. Additional nutrient realized by way of recycling	44.5	65.9	28.0	2.2	0.66	0.55	0.08

Rainfed Farming

The model integrating Tellichery goat as a component with crop activity in rainfed condition has been tried for a period of five years. The data collected on the availability of organic source of crop nutrients is furnished in Table 14.23. The organic manure like litter from the goat unit can readily be used for soil application, and thus will help in enriching the soil. Goat droppings are found to be a good energy source, which can also be linked with biogas unit before it is utilized as manure. This will generate good value of gas (22 kg of goat dropping will generate one cubic meter of gas as against 30 kg of cattle dung) as well as enhance nutrient availability. Thus, through recycling in the farming systems approach, the potential of each produce can be exploited to a greater extent.

TABLE 14.23: Nutrient gain from 20+1 productive tellichery goats unit Under deep litter system

1. No. of kidding/year			1.5
2. No. of kids/kidding			1.5
3. No. kids/year/adult			2.25
4. No. of kids from 20 adults/year			45
5. Mean wt. of dropping recorded/day (excludes the pellets dropped during open grazing)	**Adult**	**Kid**	
	900 g	110 g	
6. Dropping received in an year	6800+1800	=	8600 kg
7. Deep litter waste obtained/year	8600+2600	=	11200 kg
8. Wt. of coir waste used for the stall/year			2600 kg
9. Analytical value of deep litter was obtained from the unit (per cent)	**N**	**P**	**K**
	1.79	0.95	0.82
10. Total nutrient available from the goat unit/year (kg)	200.5	106.4	91.8

The experimental evidences indicated above, dearly show that the recycled organic residues/wastes of crops and allied activities could supplement the chemical fertilizers. Moreover, this will also provide opportunity to reduce the environmental pollution by way of burning the fossil fuel in the process of manufacturing the chemicals. The soil health problem by way of dumping more and more quantities of inorganic fertilizers can also be solved to greater extent by proper utilization of organic sources as indicated above under different field situations. This programme, if properly implemented, by involving small and marginal farmers who occupy more than 83 per cent of the total holdings in this country, will certainly pave way for increasing the productivity, profitability and also will help in sustaining the potential of the production base viz., soil.

"Everything can wait but not agriculture"

Jawaharlal Nehru

15 Crop and Land Management Practices

"If you give a fish to a person he will be contented for the day and if you teach him how to fish he can maintain himself throughout his life and his generations"
"Nature does the Business and Man is merely the Manager"

Farmers get less time for sowing of planting of different crops due to introduction of multiple cropping systems. Delayed crop operation severely affects crop growth and yield. Sub-standard tollage with a field full of clods, stubbles and weeds affects crop establishment stand and encourages weed growth and creates other problems. Human labour and animal power are the main sources of energy for the farmers of the state. The wooden plough used by most of the farmers is insufficient. Acute shortage of animal power during early cultivation and human labour during transplanting and harvesting is being felt by the farmers. Timeliness of agricultural operations is one of the most important factors for successful crop production. The use of improved implements and machinery is important for completing the farm operations in time.

Considering the above circumstances, the recent approach for promotion of farm implements and machinery has been reoriented and systematized.

The most important factors in dry farming are the conservation of moisture for proper and timely utilization by the crop. In order to suggest suitable dry farming methods and practices for an area, a thorough study of the climatic conditions, adaptable crops, nature of the soil and topography of the area is essential. The recommended practices may be broadly grouped as: (1) Crop management practices, and (2) Land management practices.

CROP MANAGEMENT PRACTICES

Selection of crops, tillage practices, proper sowing method, control of plant population, weed control, and application of the appropriate quantity of manures and fertilizers are the practices that are to be considered.

Selection of crops will involve not only the crop itself but also the particular strain of the crop suitable for dry farming conditions. Plant breeding has taken great strides and varieties of several crops have now been evolved that are drought resistant. The duration of the crop is another aspect and relatively short duration crops are more suited under dry farming conditions.

TILLAGE PRACTICES

Tillage is the mechanical manipulation of soil to provide environment suited for crop growth, maintenance of infiltration capacity, aeration, and to control weeds. Tillage plays an important role in seed germination and crop establishment. It has an influence on soil structure which in turn influences infiltration, aeration and soil erosion.

Traditionally tillage consisted of ploughing and harrowing so that the soil is loosened, clods are broken and the crop residues are buried. The objective was the preparation of a good seedbed. A number of modified tillage operations are available now, with the objective of providing improved soil-water-plant relations, reducing soil erosion and runoff, helping in moisture conservation and reducing the time and cost of tillage operations. The term minimum tillage is used to describe the preparation of the seedbed with minimum soil disturbance. The practice consists of opening the land just to plant the seed and use of chemical herbicides to control weeds. The major type of tillage operations and their purpose are as follows:

Conventional Tillage

This refers to the general seedbed preparation consisting of ploughing, harrowing and surface planting.

Strip or Zone Tillage

In this method of tillage the seedbed is prepared by cultivating the soil in a narrow strip in the adjacent area to the proposed planting row. The area in-between the rows are either left untilled or tilled in a different manner.

The area around the planting zone, also referred to as **seedling environment zone** is prepared in such a way that favourable conditions for sowing and seedling emergence are created. These conditions include proper soil structure, aeration and soil temperature. The area in between the crop rows is known as the **water management zone**. The operation in this zone influences the infiltration, surface detention and erosion. Tillage operation for the area is designed such that these objectives are achieved for the particular conditions. In the broad-bed and furrow system, zone tillage is adopted. The broad beds and furrows have been found to be suitable for managing the deep black soils in India where surface drainage during the monsoon period is a problem (Kampen, 1979). The beds function as "minibunds" at a grade and they help in reducing the velocity of surface runoff and increase the infiltration opportunity time. The excess water is removed through the furrows. A permanent bed and furrow system provides waster control for *in situ* soil and water conservation.

Mulch Tillage

Any tillage operation that will leave a substantial part of the residual vegetative material like leaves, stalks etc. on or near the surface as a protective cover is known as mulch tillage. The residual material of the preceding crop is the one that is left as the mulch. This practice is sometimes referred to as stubble mulching. The major benefits of these mulches are: (1) mulches reduce the beating action of raindrops and reduce splash erosion, (2) they reduce sheet erosion by retarding the surface flow, (3) they help in infiltration through maintenance of an open soil structure, and (4) they help in controlling soil temperature.

A layer of dry soil formed at the top of the profile is known as soil mulch. In the post rainy season both these type of mulches, viz., soil mulch or organic mulch help in moisture conservation which is beneficial to crop growth. Keeping vegetative stubbles in trenches (40 cm deep and 15 cm wide and 4 m to 8 m apart) is known as vertical mulching could enhance more retention of rainfall and making it available as soil moisture.

Tillage practices adopted have a great bearing on moisture conservation. Both the tillage methods and tillage implements have an influence on the preparation of seed bed and moisture conservation. There is considerable difference of opinion about the usefulness of ploughing as preparatory tillage in dry farming areas. In majority investigations in low rainfall areas in India, shallow ploughing as well as only harrowing has produced as good yields as deep ploughing. Shallow cultivation was adequate to kill weeds and increase rainfall infiltration. Fine pulverization of soil at times may lead to excessive soil erosion. In high rainfall regions (rainfall over 1000 mm annually) good effects have been obtained by deep ploughing.

Deep tillage is generally found to be beneficial in (1) areas with hard pans, (2) areas with persistent weeds, and (3) areas with specific textural profiles like alfisols and related soil where increase of infiltration into the soil is beneficial. In general yield increases were clear in deep soils, crops with deeper root systems and crops grown during monsoon season.

Deep cultivation apart from hard pan developed in soil, if any, opens up soil profile for better penetration and retention of incident rain and also helps in root development. Tillage practices influence the post monsoon crops also. A combination of tillage practices suited for soil and climatic conditions of the region needs to be developed.

Seeding methods are important as these help in the germination and initial establishment of the crops. These methods are to be worked out depending upon the crops and soil conditions. For example, in case of rice grown under rainfed conditions, higher yield were obtained when it was sown with a seed-drill with pass wheels. Maize sowing with a planter is helpful in obtaining good germination and also to keep the desired row spacing. In case of wheat, again sowing with a seed drill helps in better emergence and higher yields as compared to broadcasting or sowing behind the plough.

Dry sowing is a method that has been used for deep black soils (vertisols) in India (Kampen, 1979). In this approach, crops are sown shortly ahead of the rains in areas where the precipitation commences fairly reliably and where there is good probability of follow-up rains to ensure establishment of the germinating crop. Dry sowing is required as these soils become unworkable with the monsoon rains. For dry sowing, tillage of the soil is somewhat difficult. However, suitable equipment and one or two preseason rains help in the land preparation.

Regulation of plant population and weed control are important so that the limited moisture available in the soil is efficiently utilized. Application of fertilizers, when done at the appropriate time, helps in increased crop yields under dry farming conditions.

Other crops will have different fertilizer requirements. Bulky organic manures were also found to be useful both nutritionally and in improving the physical conditions of soils.

LAND MANAGEMENT PRACTICES

These include all the soil erosion control and water conservation practices. Land management practices that are of significance in dryland agriculture are water harvesting and runoff recycling, tied ridging and sub-soiling.

Water Harvesting

Water harvesting is the process of collecting and storing water from an area that has been modified or treated to increase runoff from precipitation. A water harvesting system usually consist of a catchment or water collecting area, a water storage facility and auxiliary components like conveyance and sediment control structures.

Water harvesting systems are used not only for collecting water for plant growth but also for collecting water for domestic consumption. In some parts of the world, rain water from roof tops is collected in containers and later used for domestic purposes.

Catchment Area Treatments

The catchment area treatments designed to increase runoff from rainfall can be grouped into three general categories:

1. Topography modifications,
2. Soil modifications, and
3. Impermeable coverings or membranes.

Topography modifications consists of practices like brush clearance, land grading, construction of diversion channels etc., which help in increasing the runoff from land areas. Soil modification treatments consist of chemicals applied to the soil surface by spraying or mixing to reduce infiltration. One such chemical used is a sodium based salt (NaCl) which is mixed in the top 2 cm of soil and the area is wetted and compacted to a firm smooth surface. The sodium salt disperses the clay, plugs the soil pores and reduces the hydraulic conductivity. Bitumen or asphalt has also been tested as soil modification treatments. These are generally suited for use on fine sandy soils and have a projected effective life of 2 to 5 years.

Impermeable or water proof sheeting or membranes are used when the areas to be covered are relatively small. These materials consist of concrete, sheet metal and rubber covering. Several types of plastic and other thin sheeting have been tried as soil coverings. Most of these thin film coverings were found to be susceptible to mechanical damage and deterioration due to sunlight. Their use therefore, is limited to small areas as well as for relatively shorter durations.

Runoff Farming Systems

Runoff farming is a water harvesting system specially designed to provide water for growing plants.

These could be broadly grouped in two types. One is the direct water application system by which the runoff is stored in the soil profile of the crop growing area. The other is the supplemental water system by which runoff water is stored off-site and applied to the crop as needed.

Inter-plot water harvesting is used under very low rainfall conditions. The surface runoff from the upper areas gets collected in the lower part and thus more water is available in the lower part. In case enough rainfall is expected in the area, a crop like maize requiring good drainage may be grown in the contributing area and crops requiring more water can be put in the receiving area. Chemical sprays, compaction etc., are sometime used in the contributing area to increase runoff.

Micro-catchments are land areas surrounded by small earth bunds with infiltration pit in the lowest corner. The diamond shaped basins are known as Negarim micro-catchments and widely used in Israel (Critchley and Siegert, 1991). Runoff is collected from within the basin and stored in the infiltration pit wherein suitable species of trees or bushes are planted. The layout and design depends on the local conditions.

Runoff Collection and Recycling

Even in semi-arid regions, some runoff occurs from the agricultural lands, particularly during the rainy season. In the supplemental water system, the runoff is collected and stored in a pond at a convenient location. The stored water can be used for providing supplemental irrigation to crops located near the pond. If the areas to be irrigated are upstream to the pond (in the donor catchment) water lifting devices are to used to lift the water. In case of down stream areas, gravity flows from the pond are possible.

The stored water in the pond is usually limited and may not be enough to give full irrigation to the crops. The following issues need to be considered.

1. Identification of critical periods of crops when the effects of moisture stress would be maximized.
2. Reducing penetration losses from the ponds through sealing or lining procedures.
3. Efficient use of the harvested runoff water.

The stored water may be used to give one or two irrigations to the crops at the critical periods so that the crop survives and produces maximum possible yield. Optimization procedures like linear programming and dynamic programming together with crop water production functions can be used optimal irrigation schedules. Efficient water application methods are also desirable. Instead of flooding furrow irrigation and alternate furrows could help in better use of the water.

Tied Ridging

Tied ridging of covering the land surface with closely spaced ridges in two directions at right angles so that a series of rectangular basins are formed. The purpose of such basins is to retain the rain water till it infiltrates into the soil. Tied ridging can be successful when carefully designed and constructed. The system can be successful on level ground or when the amount of water which can be stored in the basins, plus the amount infiltrating during the storm, is more than the worst storm likely to occur. Failure of a ridge, particularly on sloping land, can cause a series of failures of other ridges. In general tied ridging has been found to be successful on permeable soils rather than on shallow soils. To prevent failure of the ridges, they are constructed on grade with ties lower than the ridges so that the failure and runoff will be along each ridge and not down the slope. It is also advisable to back up the system with other measures like terraces or bunds.

Tied ridging is also referred to as basin listing. Special implements are used capable of forming the basins i.e., the ridges as well as the ties in one operation. Basin lister is such as implement. A basin lister consists of one or two shovels like bodies mounted on a shaft and wheels. When the implement is drawn these shovels make interrupted basins on the land surface. Rain water gets collected in these basins and is slowly absorbed by the soil.

Subsoiling

This practice consists of deep ploughing or chiseling using special equipment known as chisel plough or sub-soiler. The objectives in sub-soiling are:

1. Break and shatter plough soles or other impermeable layers in the soil profile,
2. Loosen soil profile to considerable depths to permit deep leaching of accumulated salts in the upper layers,
3. In some cases bring the clod-forming subsoil to the surface for wind erosion control, and
4. To deepen the effective plough zone depth for crop growth.

The effectiveness of the subsoiling practice for achieving any of the objectives and ultimately helping crop production is dependent upon the particular site conditions. The effects of subsoiling do not last long and the operation has to be repeated after some years. To adopt this practice a study of the local condition and the results obtained under these conditions should be known.

Watershed Based Soil and Water Conservation

Water is the limiting natural factor in crop production in the arid and semi-arid areas. In most of these areas, the only water available is the rain that falls on a given area. The watershed (or catchment) being an independent hydrological unit, the rainwater management in relation to conservation and utilization should be planned on a watershed basis.

All the land treatments like contour cultivation, bunding, and bed and furrow system are executed on an individual plot basis. The runoff from these areas is conveyed through grassed waterways to prevent erosion. The water at a convenient place is collected in farm ponds to be subsequently used for irrigation.

All the above outlined practices are location specific and the package of practices to suit the particular area taking into consideration the climate, soil and crops need to be developed.

SMALL WATERSHED DEVELOPMENT FOR INCREASED FOOD PRODUCTION

To some people the term watershed is new; to others it is familiar but not well understood. The increasing need for improved water and soil conservation and utilization in order to meet the needs of a rapidly extending population is bringing the term into more common usage. A watershed (also called a catchment) is a land area from which surface water drains to a given point.

Where do you find a watershed? Everywhere! Every farmer's field, city lot, or factory site is a micro-watershed. The next time it rains you can easily observe this. Micro-watersheds are parts of mini-watersheds of small streams, and finally all of these form the watershed of a large river which empties into a bay or ocean. In other words, thousands of small watersheds make up larger ones.

Watershed Management

Watershed management involves management of the land surface and vegetation so as to conserve and utilize the water that falls on the watershed and to conserve the soil for immediate and long-term benefits to the farmers, his community and society. In the past most watershed management studies have been oriented towards forest or grassland areas in the upper reaches of a river system, with the runoff generally being stored for use of at locations far downstream from the area being managed. Such work is important and should be continued. However, in countries of the Semi-

Arid Tropics (SAT), especially India, a very high proportion of the land is cropped (more than 80% in some districts). Thus, there is an urgent need to investigate better land and water management systems including, where possible, collection, storage, and use of water within small cropped watersheds. There is an urgent need for information to make the most effective utilization of supplement water from runoff or ground water to **"backstop"** rainfed agriculture and to reduce the risk involved in using inputs such as improved seeds, fertilizers, and implements.

At ICRISAT Center, natural watersheds are being used not only for water-balance investigations, but also for resource conservation and utilization research on an operational scale, using annual drawn equipment and human labour. Similar studies are being planned at several states experiment stations in India and in one area of Africa.

The major aim of research on small agricultural watersheds is to improve the utilization of every drop of water that falls on farmers' fields. This can be accomplished by improving rainfall productivity, which is expressed in kilo-grams of food, feed, fibre, or monetary value per centimeter of rainfall. The final aim is increased and more stable crop production on the small farms of the SAT.

Improved watershed management systems for rainfed cropping in the SAT can:

1. Provide year-round erosion protection by a graded bed-and-furrow system.
2. Move water slowly in graded furrows so as to increase infiltration, provide crop drainage, and reduce erosion.
3. Avoid concentration of water except in grassed water ways.
4. Provide by established grasses and legumes in waterways, forage and protection against erosion.
5. Collect and store excess water from surface runoff for supplemental irrigation, especially for shallow soils.
6. Decrease risk through optimum use of collected runoff and ground water.
7. Reduce downstream flooding of cities and crops and sedimentation of large reservoirs and tanks.

The Watershed-Based System of Soil and Water Conservation

Since water is the first limiting natural factor to crop production in the Semi-Arid Tropics (SAT), improving techniques for management and conservation of water and soil for increased crop production becomes the primary aim of farming systems research. In rainfed agriculture, the only water available is the rain that falls. Thus, the watershed is the natural focus for agricultural development involving land and water management in relation to improved crop production and soil and water conservation in the SAT.

Contour bunding, in contrast to watershed management, employs distinctly different concepts of water conservation and management. In contour bunding (which is a very common practice of present day India), runoff may flow in a concentrated manner, causing erosion between bunds. The runoff collects at the bund and is forced to flow across one or more small watersheds and then is finally disposed of in roadside drains or gullies. In cropped watersheds cultivated in graded (150 cm) beds and furrows, excess water is allowed to flow slowly through small furrows to grassed drainage ways and is then safely conducted to a tank and/or outlet. The velocity of flow of the water is controlled by the direction and slope of the furrows. Since the bed-and-furrow system can be a permanent land feature, it can provide considerable protection against soil erosion on a year-round basis, even during the prolonged hot and dry non-cropping season when occasional high-intensity rains may occur. Experience at ICRISAT indicates that the soil in the broad bed (plant zone) becomes more friable due to lack of traffic and compaction, and because of this tillage and land preparation costs are less than in the flat cultivated system.

The graded broad bed system is an improved soil and water conservation practice which controls runoff and erosion and facilitates infiltration of water into the soil uniformly over the watershed area. This is especially important for seedling establishment early in the rainy season. The system provides opportunity for surface water collection tanks and groundwater recharge for wells which can then provide water for year-round domestic use as well as supplemental irrigation. The broad ridged system also facilitates the application of supplemental irrigation where needed. Results of ICRISAT Center indicate that the optimum slopes for the bed-and-furrow system are 0.3 to 0.6 per cent on Alfisol (red soil) and 0.4 to 0.8 per cent on Vertisols (black cotton soils).

Benefits

Improved watershed development and management in the upper reaches of catchments has many potential benefits for the individual, the community, and for society.

1. For the Individual
   a) Improved crop drainage and rainfall utilization.
   b) Increased and more stabilized agricultural production.
2. For the Community
   a) Reduced flooding and water logging.
   b) Reduced soil erosion; better conservation of natural resources for the future.
   c) Increased agricultural productivity
   d) A more dependable clean water supply for domestic and industrial use.
3. For Society
   a) Reduced sedimentation of reservoirs.
   b) Reduced flooding of down stream homes, cities, and agricultural lands.
   c) Opportunities for reforestation, and beautification of non-cultivated land for improvement of the environment.
   d) Increased productivity and improved quality of life in rural area.

Since upstream watershed development benefits society, society should provide assistance for this development. This could be done in several forms:

1. Provide necessary technical assistance in watershed management by soil conservation and extension service technicians.
2. Provide low-interest development loans, and where appropriate, partial subsides.

SOIL AND WATER CONSERVATION MEASURES

Water and soil conservation practices for agricultural lands includes, *in situ* or inter-terrace rainwater conservation practices, conservation of rain water at terrace level through bunds and guiding the excess runoff for safe disposal through grassed waterways to the farm ponds/tanks/dams for its storage and recycling to the agricultural lands. These are called hard ware measures, which are of permanent type provided for improvement of relief, physiographic and drainage features. These are executed with major Government support with the purpose to check soil erosion, regulate overland flow and reduce peak flow. The present approach to reduce runoff by adopting suitable *in situ* management practices includes, tillage practices comprising primary tillage operations, i.e., summer or deep ploughing either every year or once in three years depending upon the soil type, land smoothening to avoid local depressions, frequent harrowing and secondary tillage practices with frequent inter-cultivations. The other rainwater conservation practices include, adoption of small section bunds or vegetative barriers on contour, contour sowing, opening ridges and furrows across the slope, tied ridging, zing terracing, scooping, compartmental bunding, broad bed and furrows, broad bed and ridge, vertical mulching and dead furrow formation in every row or at 3 m interval on contour depending upon the rainfall situation in black and red soils. In addition, evaporation control/*in situ* rainwater conservation measures using mulches i.e., soil, sand and vegetative mulches have resulted in greater benefits especially in winter (*rabi*) crops in deep black soils.

Apart from these management practices, increasing the infiltration rate and moisture retention capacity of soil by improving their physical conditions with application of amendments and organic materials is of greater importance in the integrated approach for rainwater conservation in rainfed/dry land areas. Adoption of these options depending upon the rainfall, soil type, and land topography/slope would reduce runoff and increase *in situ* rainwater conservation in rainfed and dry land areas and reduce the ill effects of occasional dry spells (Mishra *et al*, 1999). All these measures are software measures, which are mostly responsible for initiation rainwater conservation and management and easily and voluntarily adopted by the farmers as they can be made integral part of the agronomic measures/package of practices. *In situ* rain water conservation is a vital component of dry land crop management practices. Earlier efforts were mainly concentrated on strengthening and formation of bunds across the slope. This resulted in reducing soil erosion rather than achieving uniform rainwater distribution in the soil profile. Present emphasis is mainly concentrated in increasing the opportunity time of water penetration to soil through land configurations, and applications of amendments and organic materials. With appropriate demonstration and action learning exercise the *in situ* rainwater conservation measures can be easily popularized. Several **Indigenous Technical Knowledge (ITKs)** relating to *in situ* rainwater conservation measures are in practice, befitting the agro-ecological settings. These ITKs can be converted to **Medium Technical Knowledge (MTK)** by addressing the researchable and extension issues (Misrea, 2002).

The Semi-Arid tracts are mostly characterized with red and black soils. The problems associated with different soil types are different in physical, chemical and hydrologic terms. The red soils have low water holding capacity, higher infiltration and crusting tendency. In black soils even though the water holding capacity is high, low infiltration rate results in greater loss of soil and rainwater. In medium deep black soils the crack formation results in wetting of subsoil with first showers. In the years of low rainfall, the soil profile wetting is not uniform and results in a dry layer in the profile. This ultimately results in lower crop yields. Major part of the country's rainfed agriculture is fed by the Southwest monsoon in addition to the Northeast monsoon especially for the black soils region of Deccan plateau of Karnataka, Andhra Pradesh and Tamil Nadu. Hence, its onset, continuity, intensity, volume and withdrawal patterns have a tremendous influence on the agricultural production. High intensity rains produce volumes of water beyond the intake capacity of the soil and may leave the soil dry at lower depths. With intermittent long dry spells this situation affects rainfed crops adversely, even in areas with moderate to high rainfall. Thus, improving soil surface conditions to increase infiltration and improving water-holding capacity are two basic requirements in dry lands. The inter-terrace management practices for *in situ* conservation of rainwater and ensuing its uniform distribution within the field and throughout the crop growth period assume paramount importance in dry land crop production. The research efforts on rainwater management have resulted in identification of several useful technologies for *in situ* rainwater conservation. However, the choice of the most appropriate practices is a function of the soil type rainfall characteristics, and topographic features. Hence, *in situ* rainwater conservation plays a greater role for stabilized/sustained crop yields in the Semi-Arid Tropics of India. This can be achieved with appropriate tillage and *in situ* rainwater conservation practices at the individual farm level.

The research results at both the research stations and in the farmers' fields indicate that the *in situ* rainwater conservation practices reduce the runoff, soil and nutrient losses and recharge the profile both during rainy and post rainy season and increase the yields of different crops especially during drought years. The suitable *in situ* moisture conservation practices for different crops at some dry land centres of India are mentioned in Table 12.1.

TABLE 15.1: Data of the experiments/demonstrations in the Farmers' fields at different Dry land Centres/ locations on *in situ* moisture conservation practices

Location	Crop	Suitable inter-terrace land treatments
Bijapur	Safflower	Compartmental bunding
	Chickpea	Compartmental bunding, Ridges and furrows
	Rabi Sorghum	Tied ridging
Akola	Pigeon pea	Opening furrow at 30 DAS after every 2 weeks
Bellary	Rabi Sorghum	Compartmental bunding, Ridges and furrows
	Chickpea	Compartmental bunding
	Safflower	Bedding system
Kovilpatti	Rabi Sorghum	Compartmental bunding
Sholapur	Chickpea	Compartmental bunding
Bangalore	Pigeon pea	Ridges and furrows, Furrow at 3 m interval
Anantapur	Groundnut	Dead furrow at 3 m interval Contour cultivation

INDIGENOUS TECHNICAL KNOWLEDGE (ITKS) ON SOIL MOISTURE CONSERVATION

The present farming situations warranted the consideration of the ITKs in formulating projects with detail analysis of the missing links in research. Rainwater conservation begins from seedbed preparation. Although farmers practice many indigenous technologies in soil and rainwater conservation, the documentation and refinement of these technologies are the major thrust areas in research for greater crop productivity that ultimately improves the economic conditions of the farming community. Some of the simple ITKs documented on the *in situ* soil moisture conservation adopted in different parts of India are presented in Table 12.2.

The potential ITKs on *in situ* water conservation was reviewed by Mishra *et al.*, 2002.

The specific researchable issues pertaining to different ITKs adopted in different rainfed regions of India are presented in Table 12.2. A systematic scientific study may change the indigenous Technical Knowledge (MTK). The research results will benefit both farming community as well as the extension agencies i.e., the Government or non-Government organizations in up scaling the technology.

TABLE 15.2: Researchable issues in Potential ITKs on *in situ* Moisture Conservation Measures

Indigenous *in situ* water	Researchable Issues
Furrow opening in standing crops For Rainwater conservation	1. Modification of implement with different serrated blades and introducing additional tines
	2. Effectiveness in conserving soil moisture
Wider row spacing in pearl millet for rainwater conservation and weed control	1. Plant geometry and population research in different rainfall situation
Crop residue management for Improving soil organic matter and Water holding capacity	1. Quantification soil and water conservation and yield advantage.
	2. Better or improved implements for crop residue Incorporation
	3. Alternate ways of composting and application
Mulching in turmeric cultivation For rainwater conservation	1. Quantification of soil loss, improvement of soil quality and water availability
	2. Use of alternative organic material to Sal leaves as mulch
Broad bed and furrow practice for Rainwater conservation and runoff Disposal	1. Width of broad bed needs to be evaluated for different crops and rainfed situations
	2. Identification of suitable low cast tractor/bullock Drawn implement for layout of BBF
Set-row cultivation for soil and rainwater conservation and improvement of soil properties	1. Quantification of rainwater conservation and water use efficiency (WUE) of the crops
	2. Improvement in soil health and crop yield over years
Summer/pre-monsoon tillage for Harvesting early rainfall, weed Control and initiate timely seeding	1. Identification of appropriate tillage implements for soil and water conservation
	2. Evaluation of root: shoot ratio and quantifiacation Of WUE of crops
Formation of *Gurr* for rainwater and soil conservation	1. Effect of bullock and tractor drawn *Gurr* on runoff reduction, soil water conservation and crop productivity
Geen manuring practice for water conservation and soil health improvement	1. Growing of green manure crop and its management in improving soil health and crop productivity
	2. Economic evaluation of the system by addressing issues of sustainability
Tank silt application to improve Soil fertility and water holding Capacity tank silt application and its effect on crop	1. Method and quantity of tank silt application in different soils
	2. Improvement in soil water and fertility with productivity with Government's programme of tank desiltation
	3. Cost effectiveness of silt application especially

Up-scaling of ITKs

Prevailing ITKs should invariably be given priority. All the projects on resource conservation and management should focus on the viable and applicable ITKs relating to soil and water conservation. The stakeholders such as farmers, NGOs, extension officers, scientists, administrators, policy makers and people's representatives may popularize the ITKs through different programmes for improving soil and crop productivity on sustainable basis. The ITks on *in situ* soil and moisture conservation are not up scaled and are attributed to the constraints in adoption and unawareness of the effectiveness of such practices indifferent agro-ecological settings. The present documentation process has definite bearing on the future course of action in farming new projects.

- The documentation exercise should be in-built in extension and research especially in natural resource management.
- The suitable ITKs may be adopted and validated other Agro-ecological regions.
- The documented ITKs should be translated in all regional languages and published for the benefit of farming communities.
- Suitable modifications pf the ITKs through on-farm research would help developing appropriate and acceptable technologies in different Agro-ecological environments.

WATER RESOURCES AND THEIR DEVELOPMENT

The estimated increase in world population, currently at an average rate of 2 percent per annum, signifies that the world water-demand will double henceforth about every 35 years, at least for the time being. The increase in water requirements must be far more considerable in the arid regions than in the humid regions, for two main reasons: (1) that water is the limiting resource in the dry regions and therefore, determines the extent to which other resources can be developed, and (2) that an increase in water requirements is concomitant with a rise in the standard of living. It is in the drier regions of the world that the standard of living tends to be the lowest.

The existing shortage of water supply in the dry regions does not necessarily imply lack of water resources. In many of the dry regions there are potential water resources which, with the necessary know-how and funds could be developed. Much additional water can also be made available by improving water conservation, by increasing the efficiency of use of existing water resources, and by a more rational allocation of water among competing demands.

WATER CONSERVATION PRACTICES

The soil as a moisture reservoir: The soil, to the bottom of the root zone, constitutes a moisture reservoir of vital importance to agriculture. Water that infiltrates into this reservoir can be stored with relatively little loss for fairly long periods-longer than generally realized. Unlike humid climate where heavy soils are more effective in holding water, in arid climate sandy soils are more effective in holding water, in arid climate sandy soils are better since the precipitation is the limiting factor. In heavy soils percolation is slower, the depth wetted is less and thus, the proportion of precipitation lost by runoff and evaporation is greater.

In the Mediterranean region, crops such as sorghum, melons and sesame can be grown during the hot summer months, without receiving a drop of rainfall from sowing to harvesting; they can produce high yields on the moisture stored in the soil during the winter months.

The amount of precipitation taken in by the soil depends on runoff and infiltration. By reducing runoff and increasing infiltration, the amount of water stored in the soil can be increased with beneficial effects on crop production, provided the total amounts of rainfall are sufficient.

PAST AND FUTURE TILLAGE

Tillage is hard work and energy expensive. Since people began to cultivate crops, they did only what tillage was necessary to plant and to control weeds. As cultivation became more sophisticated, tillage operations and equipment were altered and specialized. Powered machinery and cheap fossil fuels brought in the age of "maximum" tillage –loose fine seedbeds, weed less trash less fields, and extensive mixing soils, lime, plant residues, and sometimes fertilizers.

Now, it is realized that often this extensive tillage is not good. Compaction of soils, the need for soil conservation, and the increased costs of energy and labour have caused a new appraisal of our concepts about tillage.

Purpose of Tillage

The major purposes for tillage are to prepare an adequate seedbed and to control weeds. To these are added several other purposes, such as improving aeration, increasing water infiltration, to make furrows for irrigation, and to bury crop residues. The important of the various reasons for tillage varies with geographic location, soil differences, crop grown, and the climate. A detailed list of reasons for tillage is given in the following items:

Seedbed Preparation

Tillage to prepare the seedbed is most needed when planting small and/or high-cost seed (lettuce, tomatoes, sugar beets, alfalfa, clovers). It is less critical for vigorous large seed ed plants (corn, small grains, soybeans, dry beans, and sorghum). The soil covering depth for a seed is about 4-5 times the seed diameter for large seeds and 1-3 times its length for a small seed. A planted seed needs to be firmly set into moist soil, where it can get adequate water to germinate. The soil must be loose or aggregated enough to allow good root penetration and to not form a thick hard crust that hinders seedling emergence.

Special beds are being tried for various purposes. For example, a special 152 cm wide bed for strawberries in the Santa Ana, California, area has a contour of gentle slopes on each ridge-half from the ridge center and a shallow irrigation furrow in the center. Two rows of strawberries are planted on each of the sloping sides. The raised center furrow gives good water penetration to the plant roots, the sloping and raised plant sites keep the berries well drained and dry above any runoff or storm water that might be at the normal ground level.

Weed Control

One crop seeds germinate, weed control is essential. Weeds compete with the crop for water, nutrients, and light. Until herbicides became available in the year 1940s, tillage for weed control was an integral part of crop production. Weeds were ploughed under in land preparation and killed by cultivation once or several times during early stages of the crop growth. But cultivation was only partially effective, many weeds were in the plant row where they were not removed except by hand, also the cultivator always cut some shallow crop roots. Herbicides to kill weeds are more effective and eliminate extra trips over the field, which results in both a fuel (energy) and labour savings. Tillage for weed control is still effective and sometimes used. Certain crop systems need weed control other than herbicides. Crop rotation is one solution. A recent suggestion recommends a rotation in tillage (deep plough every several years, less tillage other years) as an improvement in management. Throughout the world, particularly in developing counties, weed control by tillage and by hand is still common. Often the area cultivated is limited by the ability to control weeds.

Incorporation of Surface Trash

The 'clean' field has been the 'ideal' for many land users. Burial of plant and weed residues and seeds has reduced plugging of planters and has reduced new weed and insect infestations.

Loosening the Soil

Equipment compacts the soil, often developing pans, which may reduce or inhibit both root and water penetration. The heavier the machinery and the finer and more moist the soil, the greater will be the compaction problem. In one soil in the southern Great Plains, 35 years of cropping to grain sorghum reduced infiltration rates from 23 cm /h above the compaction zone to 0.5 cm/h in the compacted area, thus was a 46-fold greater infiltration in the non-compacted soil.

Corrective tillage may be useful on several types of natural soil pans. These **soil pans** (hard, compacted, or impermeable layers) often develop from natural soil processes or parent materials stratification. These include **clay pans, silt pans** (some are **fragi pans,** a particular type of silty surface pan), and **cemented** horizons. The latter pans were formed, mostly in the geologic past, as soluble cements, including carbonates, sulfates, silica, and iron and aluminum forms, moved down and precipitated in layers of soil. These pans can frequently be ripped and opened by tillage, allowing water and root penetration. Deep chiseling is an energy expensive operation but may be justified by the greater production.

Shaping the Soil

Surface flow irrigation or ridges to allow drainage of excess water, are often essential. About 70 percent of irrigated land is surface irrigated, and most of this is by furrow irrigation. Tillage is needed to form the ridges and furrows and to smooth the land surface to avoid "hills and valleys" in the field, over which irrigation would otherwise be difficult or impossible.

Incorporation of Lime and Fertilizers

Lime, to be most efficient, needs to be mixed with the soil. If it is added at the soil surface, it will move down slowly. The soil acidity will be corrected slowly with depth. The "over liming" at the soil surface could favour some fungi diseases. Many management schemes involve broadcast followed by incorporation methods of fertilizer additions. All of these require tillage for incorporation.

Control of Insects and Diseases

The timely ploughing under of crop residues is an effective means of controlling certain insects. The Hessian fly, which is a serious infestation in many wheat fields, can be controlled by ploughing under infested wheat stubble and volunteer wheat. The wheat joint worm is held in check by destroying all volunteer grain. Ploughing under corn stalks reduces next year's crop of European corn borers. Timely cultivation aids in reducing grasshopper infestations by drying out their eggs. The cotton boll weevil and the pink ball worm in cotton are held in check by the early destruction and ploughing under of cotton stalks.

Improving Water Relations

Tillage destroys soil crusts, loosens soil, and generally increases water intake when water is next applied after tillage. Tillage should be minimal and the land left as rough as feasible to still have good seedling establishment. Crop residues help keep cultivated soil porous. In Minnesota, when residues were partially incorporated to 15 cm with a chisel cultivator in the fall, they provided eight times more infiltration before runoff started and four times more infiltration during runoff than occurred on spring ploughed, disked, and harrowed land.

The higher water storage in the root zone resulting from deep tillage permits irrigation before planting and less irrigation during crop growth. Tillage on a clay loam to a depth of 41 cm was still beneficial after 3 years.

When no crop is grown for a year for the purposes of increasing the total soil stored water, it is called **fallow.** Weeds and volunteer growth are controlled by two to four weed control tillage passes (cutting blades slide along below soil surface) during fallow. Water storage is often low, but the storage of even in inch or two of water is essential for increasing yields in areas of limited rainfall. In the northern Great Plains in the United States 2.5 cm of water produces on the "average" an extra 2.4 bushels per acre of wheat (161 kg/ha). A gain of 5-15 cm of soil water can increase grain yields appreciably. Waster gains from fallow very considerably. Except in low rainfall areas of less than 20-30 cm, where a profitable yield is impossible without fallow, the value of allowing land to be idle a full crop year is now being questioned. One problem in annual planting is that a fallow year is sometimes necessary to obtain enough moisture in the fall to **germinate** the seed.

Aerating Soil and Speeding Warming

Compacted soil may have inadequate air exchange. The lowered aeration can be growth limiting in some clayey soils, but it is unlikely to be a problem in sandy soils. Any soil cover reduces the amount of heat radiation reaching the soil. Thus, plus keeping the soil wetter, slows warming of the soil surface in a cool spring. Tillage that exposes soil to the sun, especially as drained ridges, will speed warming of the soil where seed will be planted.

Ease of Planting

Anyone who has attempted to operate a planter in soil covered by bulky plant residues has experienced the frustration of constantly removing accumulated trash from the equipment and trying to achieve uniform planting depths. Tillage to incorporate portions of this residue greatly facilitates planting.

REASONS FOR REDUCED TILLAGE

The reasons for going to reduced tillage systems are primarily to reduce erosion of soil, save time in operations, and save on costs of fuel. Fuel costs are about 35 per cent of the energy used in production with conventional tillage. Fertilizers are about 20-25 per cent, machinery is about 20 per cent, and other chemicals are about 5-8 per cent of production energy costs. Reducing tillage operations should reduce costs of both fuel and equipment, it may, however, increase costs for pesticides and nitrogen fertilizers. Studies done suggest that the total reduction in tillage costs with reduced tillage are about 5-10 per cent lower than with conventional mould board or disk plough tillage.

Erosion of soil has finally been recognized by decision makers as the bankrupting policy of soil management that it is. Top soil must be kept in place. Pressure from national and state soil conservation groups and controls by the U.S. Environmental Protection Agency (EPA) to prevent erosion of soil sediments into waters have encouraged the use of reduced tillage. Less erosion occurs with reduced tillage.

The U.S. Department of Agriculture started in 1988 to implement a plan to allow government payment only to those farmers (1) who have a farm plan to limit soil erosion, and (2) who follow the plan to reduce erosion so that it does not exceed the rate of soil formation. The allowable erosion is known as the "T" (tolerance) rate.

Other benefits or factors contributing to the increased use of reduced tillage were the following:

1. More area can be planted in a short time, important if weather is bad.
2. No-tillage soil dries less than with conventional tillage.
3. Double cropping (two crops per growing year) is possible with no-till by reducing time needed for seedbed preparation in many temperate areas usually producing one crop per year.
4. Steeper land can be farmed because of better erosion control.
5. More water usually is stored in the root zone.
6. Equipment requirements are lower.
7. The lower costs permit farming marginal lands.
8. Less tillage reduces soil compaction.

GOAL FOR THE FUTURE

The goal for the future of agriculture is to develop a permanent agriculture. That goal has not changed over the years, but the choices we make to accomplish it may now be somewhat different than they were previously. For the foreseeable future, agriculture must supply the food and much of the fiber to an increasing world population. This production should be as economical as possible, both for the producers and to consumers. While doing this, the soil must not be permitted to be degraded or destroyed. To accomplish this goal, the following important factors are needed:

1. Reduce erosion to low, replaceable rates of loss. This will maintain what soil we have, perhaps improve it.
2. Minimize the costs of production by lowering energy consumption. Use less fuel and less equipment, use chemicals of their highest efficiency but lowest possible amounts.
3. Select management systems which do not cause land degradation or uncontrollable ecological upsets.
4. Select management systems that will accomplish items 1, 2, and 3 but will still be highly productive.

"Success is where preparation and opportunity meet"

16 Bioconservation of the Environment

> *"Organic materials were practically the only external source of nutrients to crops before introduction of inorganic fertilizers. The recycling of various forms of residues has the advantage of converting the surplus farm waste into useful product for meeting nutrient requirement of crops, besides maintaining the soil condition and improving the overall ecological balance".*
>
> *Dr. M. S. Swaminathan*

Sustainable rural development is the management and conservation of the natural resource base and the orientation of technological and institutional change in such a manner as to assure the attainment and continued satisfaction of human needs for the present and future generations. Such sustainable development, in the agriculture, forestry and fishers sectors, conserves land, water, plant and animal genetic resources, is environmentally non-degrading, technically appropriate, economically viable and socially acceptable.

ORGANIC CYCLE

Organic Farming is a production system, which avoids or largely excludes the use of synthetically compounded fertilizers, pesticides, growth regulators, and livestock feed activities. To the maximum extent feasible, organic farming systems rely upon crop rotations, animal manures, legumes, green manures, off-farm organic wastes, mechanical cultivation, mineral-bearing rocks, and aspects of biological pest control to maintain soil productivity and tilth, to supply plant nutrients, and to control insects, weeds, and other pests.

Biological Farming

Biological Farming has become synonymous with farmers using the Reams fertility system as the basis for crop production. ***Eco-agriculture*** is the term use to describe this system by the monthly Acres, U.S.A. The Reams system is based on the LaMotte-Morgan soil test and the use of rock phosphate, calcium carbonate, and compost to achieve nutrient ratios of 7:1 calcium to magnesium, 2:1 phosphorus to potassium, and so on. Biological farming allows the use of selected chemical fertilizers (avoiding disruptive materials such as anhydrous ammonia and potassium chloride) and adopts low-inputs approaches to use of herbicides and insecticides.

Nature Farming

Nature Farming was developed in Japan in the 1930s by Mokichi, who later formed the Mokichi Okada Association (MOA). Nature farming parallels organic farming in many ways but includes special emphasis on soil health through composts rather than organic fertilizer. When possible, Kyusei Nature Farming, a branch group, emphasizes use of microbial preparations in addition to traditional Nature Farming is most active in the Pacific rim, California and Hawaii.

Regenerative Agriculture

In **regenerative agriculture** bunds on nature's own inherent capacity to cope with pests, enhance soil fertility, and increase productivity. It implies a continuing ability to re-create the resources that the system requires. In practice, regenerative agriculture uses low-input and organic farming systems as a framework to achieve these goals.

Permaculture

Permaculture is a contraction of permanent agriculture. The term was coined by Bill Mollison, an Australian Forest Ecologist, in 1978. Permaculture is concerned with designing ecological human habitates and food production systems, and follows specific guidelines and principles in the design of these systems. To the extent that permaculture is not a production system, per se, but rather a land use planning philosophy, it is not a specific method of production. Thus, practically, any site-specific ecological farming system is amenable to permaculture.

Mixed Farming

This is for appropriate diversification in organic Agriculture Systems, when ideally means mixed farming, or the integration of crop and livestock production on the farm. In this way, cyclic processes and interactions in the agro-ecosystem can be optimized, for example, using crop residue in animal husbandry and manure for crop production. This does not mean that some specialization is not possible; there is often a focus on a certain activity. This enables the farmer to acquire a high level farming is not a dogma because it is often not feasible for every individual farm. In such situation, one should look for an appropriate scale in the relevant region or watershed to connect animal with plant production. Various types of cooperation offer a longer-term solution for building organic agro-ecosystems between several farmers, each with appropriate expertise and specialization.

Diversification of species, biotopes and land use as a means to optimize the stability of the agro-ecosystem is another way to indicate the mixed farming concept. The synergistic interactions between plants, animals, soils, and the biosphere support this idea.

Ecological Farming Systems

The shift from chemical to ecological agriculture should, however, be gradual. A sudden switchover can spell disaster and discourage farmers from taking to this course. At least, two to three seasons will be needed for the transition, and during the interim years the farmers could build up a sufficient organic base to fertilize the fields and improve the productivity of the soil. Once, a state of ecological balance is attained and there are a good number of beneficial organisms to check the explosion of pests and pathogens, crop yields of a high order can be ensured. The cost of cultivation can be brought down substantially as many farm-grown inputs can be integrated efficiently in the farming system.

Ecological farming can be sustainable and profitable too. It is a self-reliant method and has to be an integrated system. Since it is a knowledge-intensive practice, one has to keep pace with the dynamics of nature to increase the biological productivity of the soil.

Objectives of Ecological Farming

Bringing to match the crop, soil and climate of a region the ecology and farming and gaining from the economy and efficiency of inputs are the objectives of eco-farming. According to Sankaram (1996), the technology may be grouped into four categories.

1. Those that reduce the environmental burden of green house gases (CO_2 and CH_4), global warming, ozone depletion. All attention to promote renewable sources of energy (drought animal power), electrical energy from garbage disposal and biogas from organic wastes.
2. Those that reduce the demand on land, water and biodiversity without adverse effect on agricultural production and nutritive value of food. Nurse the soil back to health, change cropping patterns to maximize ecological productive efficiency, improve water use efficiency through conjunctive use of rain, tank, underground, well and river waters; reduce conveyance losses; phase out subsidies.
3. Those that continue to improve crop productivity, under shrinking land resources, (genetic and agronomic) hybrid vigour, gene pyramiding; multiple cropping patterns, integrated nutrient management and integrated pest management.
4. Reduce hunger and poverty, adopt, cost effective farming to bring equity of food price and wage, encourage job promoted growth right in the village to arrest migration as ecological refugees.

Integrated Intensive Farming System (IIFS)

The Integrated Intensive Farming Systems (IIFS) methodology provides the pathway to achieving an evergreen revolution in agriculture. According to Prof. M.S. Swaminathan, Chairman, M.S. Swaminathan Research Foundation (MSSRF), Madras, the IFS involves agricultural intensification, diversification and value-addition. It helps improve physical and economic access to food, thereby fostering sustainable food security at the level of each individual in a house hold. Emphasis on food security at the individual level is important in view of the growing feminization of poverty, according to him.

As the same indicates, IFS involves intensive use of farm resources. To be ecologically sustainable, such intensification should be based on techniques which are knowledge-intensive rather than capital-intensive, and which replace, to the extent possible, market-purchased chemical inputs with farm-grown biological inputs.

Such a shift in the nature of inputs used is brought about through integrated farming involving animal husbandry, fishery and agro-forestry. Such integrated farming practices which constitute the second element of the IIFS strategy provide scope for organic recycling. The third element of the strategy is value-addition to every part of the plant and animal biomass through the establishment of bio-refineries. On-farm and Off-farm employment can then be linked in a symbiotic manner. In household, IIFS aims to reduce the number of hours of work and add economic value to each hour of their work according to Dr. Swaminathan.

With an ever-expending urban population, the demand for a side range of processed and pre-cooked foods will expand, as urbanization leads to diversification of foods habits. The IIFS with its emphasis on crop-livestock-fish integration responds to this challenge. Also, the emphasis placed on post-harvest technology in IIFS helps both producers and consumers. To sum up, IIFS involves a pro-nature, pro-employment, pro-women and pro-poor orientation to technology and dissemination. It leads to labour diversification and not displacement. It leads to resource-based, in contrast to the widely prevalent commodity-based, agricultural development planning, according to him.

Dr. Swaminathan has prescribed ***seven pillars of IIFS*** for successful implementation of such systems. Soil health care is fundamental to sustainable intensification. In addition, vermin-culture constitutes essential components of IIFS. IIFS farmers should maintain a ***soul heath card*** to monitor the impact of farming systems on the physical, chemical and microbiological components of the soil.

Water harvesting and management is the next crucial component. Crop and pest management including integrated nutrient supply (INS) and integrated pest management (IPM) systems form important components as well. Energy management and post-harvest component, choice of crops, other components of farming system and information technology are also important. These will help to ensure economic viability, environmental sustainability, and social and gender equity in IIFS villages.

It should be emphasized that IIFS will succeed only it is a human-centered rather than a mere technology-drive programme. It essence is the symbiotic partnership between farming families and their natural resources endowments of land, water, forests, flora, fauna, and sunlight. Without appropriate public policy support in areas such as, land reform, security of tenure, rural infrastructure. Input and output pricing and marketing small farming will find it difficult to adopt IIFS according to him. The eco-technologies and public policy measures needed to make IIFS a mass movement should receive concurrent attention. The programme will fail if is based solely on a technical quick-fix approach.

On the other hand, it can trigger an evergreen revolution, if mutually reinforcing packages of technology, training, techno-infrastructure and trade are introduced, according to Dr. Swaminathan.

Ecological agriculture that effectively combines use efficiency of input and economic yield maximization will begin to prevent the abuse of natural resources. On its progress depends the achievement of food security and sustainability of farming systems. Ecological agriculture will reactivate the fatigued green revolution to move forward through promotion of soil health, package of agronomic practices with each component a blend of tradition and modernity, and a policy to value each input based on its intrinsic worth free from subsidies.

There is the need to recapture local knowledge about farming systems. A synthesis of traditional wisdom and ecological prudence of topical agriculture with modern advances in technology will help higher use efficiency of inputs and sustainable crop yield maximization.

This symbolic of low input sustainable agriculture (LISA) and sustainable agriculture research and education programme (SAREP) are widely promoted in the U.S.

Low External Input Supply Agriculture (LEISA)

Low-Input Supply Agriculture

The term low-input agriculture has been defined as a production activity that uses synthetic fertilizers or pesticides below rates commonly recommended by the Extension Service. It does not mean elimination of these materials. Yields are maintained through greater emphasis on cultural practices, IPM and utilization of on-farm resources and management.

Although, the term ***low-input farming*** has often been used to describe any system of alternative agriculture, it can be seen that it is distinctly different from organic farming etc. Nevertheless, any system that reduces purchased chemical inputs can be called **Low-input Farming,** some examples are:

1. Poultry litter can replace nitrogen fertilizers in the production of watermelons.
2. Legume cover crops can supply the total nitrogen requirements of pecan trees.
3. Compost amended potting mixes produce superior vegetable transplants than traditional soil less mixes.
4. No-till vegetable systems are feasible using reduced herbicide rates to kill over crops.
5. Subterranean clover living mulches supply nitrogen and weed control in peach orchards.

Important characteristics of LEISA systems are that they are based on a preventive approach whereby the problem is tackled at its roots, as its roots, as opposed to the more symptom-curing nature of modern agriculture. Ecological and biological principles are the basis of the farm system. The main ecological criteria for LEISA are:

1. Balanced use of nutrients and organic matter,
2. Efficient use of water resources,
3. Diversity of genetic resources,
4. Efficient genetic resources,
5. Efficient use of energy resources,
6. Minimal negative environmental effects, and
7. Minimal use of external inputs.

Municipal / Industrial / Hazardous Wastes and Wastewater

Biological treatment systems contain living organisms and require carbon, nutrients and water as a vital part of the process. Such systems are affected by temperature, light, movement and chemical conditions of pH, oxygen, salinity, and metals and the interrelations of the organisms present. Microorganisms perform best under steady state environment. Environmental changes if at suitable rates will allow the biomass to acclimate.

Bio-conservation and biotechnological research for the future is striving to develop new strains of microorganisms and processes for destruction of organic toxics and hazardous materials heretofore immune to treatment and to shorten process and retention time. The need is present in the market place and biotechnology will respond.

Municipal Treatment Plant Sludge

Wastewater sludge is being generated in enormous quantizes at sewage treatment plants, particularly at activated sludge facilities. Regulations have mandated both the end of ocean dumping of sludge and provisions for full secondary treatment. These regulations result in increased production of sludge and the necessity to treat and dispose of it an acceptable manner. Anaerobic digestion is one of the processes employed in the stabilization of this sludge, to remove from the raw sludge its ordour, pathogens, putrescibility and other offensive characteristics (Table 13.1).

Using the methane produced by the anaerobic digestion of sludge to supply power for the sewage treatment plant is not new. This practice was used in the 1940s and 1950s by municipalities, but was generally abandoned in the 1960s when electricity became inexpensive. Aerobic digesters were chosen in place of anaerobic ones because of their relative ease of operation and resistance to upset. However, they are energy users. Energy costs have resulted in a renewed interest in anaerobic sludge digestion since the methane produced can often heat the digester and supply the bulk of the power needed by the entire treatment plant.

Desulfurization

The bacteria, expert at mineral leaching can be applied to water decontamination with other microorganisms. In many mining operations, water is pumped out of the mines to prevent flooding. Water used in million processes becomes laden with soluble inorganic ions. Mine drainage from abandoned mines loaded with a variety of metal salts. The practice has been to evaporate this water in holding ponds or to neutralize and acid flow and precipitate the metals with time.

Many organisms have the ability to concentrate, accumulate or precipitate metals allowing the recovery of elements of economic importance. Many bacteria are known to concentrate potassium, magnesium, manganese, iron, calcium, nickel and cobalt. Other bacteria produce complexing agents which selectively extract metals from dilute solutions. Algae's concentration silica, green-brown algae and fungi concentrate zinc and other heavy metals. Mosses and higher plants concentrate mercury, nickel, zinc, uranium, cesium and strontium.

TABLE 16.1: Sludge treatment processes and their functions

Unit Processes	Functions
Thickening (Blending)	– Water removal
	– Volume reduction
	– Post process efficiencies
	– Blending
Stabilization (Reduction)	– Pathogen destruction
	– Volume and weight reduction
	– Putrescibility control
	– Gas production
Conditioning (Stabilization)	– Improve dewatering/thickening rate
	– Improve solids capture
	– Improve compactability
	– Stabilization
Dewatering	– Water removal
	– Volume and weight reduction
	– Change to damp cake
	– Reduces fuel requirements for incineration/drying
Heat drying	– Water removal

Unit Processes	Functions
	– Sterilization
	– Utilization
Reduction (Stabilization)	– Destruction of solids
	– Water removal
	– Conversion
	– Sterilization
Final disposal	– Utilization (Cropland)
	– Utilization (Energy)
	– Utilization (Land reclamation)
	– Disposal (Landfill)
	– Disposal (Ocean)

Sulfite reducing bacteria of the Desulvovibrio, Desulfotomachulum, Desulfobacter, Desulfococcus, Desulfonema, and Desulfosarcina general are especially adept *at* metal removal from water by producing hydrogen members of these groups embrance a wide range of salinity or osmotic pressure, temperature, hydrostatic pressure, pH, Eh and other environmental conditions.

These organisms have been put to work in mien wastewater cleanup operations. Setting ponds inoculated with sulphate reducing bacteria (desulfovibrio, desulfotomaculum), and sulphur oxidizingthiobacili (*Thiobasil thiooxidants*, T. *thiopiasu, T. denitrifican),* and algae (*Chara, Spirogyra, Oscillatoria)* have been effective in lowering the concentration of uranium, selenium, and molybdenum in wastewater. A variety of metallurgical effluents containing high concentrations of sulphate ions, a number of microorganisms can utilize this sulphate and convert it to an insoluble, stable non-leachable form.

Desulfovibrio reduces sulphate to sulphide. Chlorobium and Chlrobium photosynthetically oxidize to elemental sulphur. A mutualism between these bectaria proposed.

$$SO_4 = \xrightarrow{\text{Desulfovibo}} H_2S \xrightarrow[\text{Chromatium}]{\text{Chlorobium 6}} S$$

A series of tests applied to solvent extraction raffinate demonstrate that a gas purged mutualistic system of *Desulfovibro* and *Chlorobium* can be used for the efficient conversion of sulfate to elemental sulphur. The extraction of minerals from ores, its benefication to a high quantity material and its fabrication into a useful product are all sources of very toxic materials. Many industrial wastes contain valuable metals diluted in a large mass or volume. Processes must be developed to simultaneously extract these valuable metals and reduce the attack on the environment.

Tests have been conducted using *T. ferroxidans* and *T. thiooxidants* to extract economically interesting metals from wastes.

1. Jarosite – residue which accumulates during zinc production;
2. Sulidic dust concentrates from copper processing;
3. Fly ash from apyrite-roasting process;
4. Slag from a lead somelting process.

The results indicate that it is possible to stimulate bacteria already to work on these waste dumps to leach into solution economically valuable metals.

Flotation is probably the most common unit operation in metallurgical operations. One of the problems is that the reject water is laden with flotation chemical agents. Laboratory experiments have tested the ability of *Escherichia coli, Proteus retigerii, Klebsiella pneumoniae and Pseudomonas aerucinosa* to biograde sodium hexadecyl-sulfate and dodecylamine acetate. *Klebsiella* and *Proteus* appear to be the most efficient organisms. They handled very well the sodium hexadecyl suphate, manage with the sodium oleate but had difficulty with the amine. *Pseudomonas fluoresccens* has been used to remove floatation agents from wastewater.

Nitrification / Denitrification

The nitrogen cycle is complex like all bio-geological cycles. Its essential components are:

Nitrogen Fixation

$$N_2 + H^+ \xrightarrow{\text{Rhizobium}} NH_3$$

Nitrification

$$NH_3 + O_2^+ \xrightarrow{\text{Nitrosomas}} NO_2^-$$
$$NO_2^- + O_2^+ \rightarrow NO_3^=$$

Denitrification

$$NH_3 + NO_3^+ \xrightarrow{\text{Pseudomonas}} H_2O \text{ or } N_2$$

During the biological denitrification, bacteria reduce nitrite or nitrite to nitrogen gas. The heterotrophic bacteria require a source of carbon. In denitrification reactions, the supplemental source is often provided by methanol. The nitrogen reaction is a respiratory process encompassing the following reactions.

$$NO_3 + 1/3\ CH_3OH \rightarrow NO_2 + CO_2 + 2/3\ H_2O$$
$$NO_2 + \tfrac{1}{2}\ CH_3CH \rightarrow \tfrac{1}{2}\ N_2 + \tfrac{1}{2}\ CO_2 + \tfrac{1}{2}\ H_2O + OH$$
$$NO_3 + 5/6\ CH_3OH \rightarrow \tfrac{1}{2}\ N_2 + 5/6\ CO_2 + 7/8\ H_2O + OH$$

Biological reactions have been used in the conversion of ammonia-nitrogen to nitrate-nitrogen with attendant reduction in chemical oxygen demand and total organic carbon in coal gasifier effluents and municipal wastewaters.

Nitrogen, in its various forms, can deplete dissolved oxygen levels is receiving waters, stimulate aquatic growth, exhibit toxicity toward aquatic life, affect chlorine disinfection efficiency, present public health hazards, and affect the suitability of wastewater residue. Nitrogeneous materials enter the aquatic environment from natural or man-caused sources. Natural sources include precipitation, dustfall, non-urban runoff, and biological fixation. Activities that may increase quantities of nitrogen added to the aquatic environment from fertilization of agricultural land and combustion of fossil fuels. Other man-related sources include urban and livestock feedlot runoff, municipal wastewater effluents, subsurface drainage wastes. The average concentration of nitrogen from natural sources is difficult to estimate but range from 0.02 mg/l to 0.2 mg/l. Nitrogen concentration in raw municipal wastewaters are well documented, and values range from 15 to 50 mg/l of which approximately 60% is ammonia nitrogen, 40% is organic nitrogen, and negligible amount (1%) is nitrite nitrate nitrogen. Nitrogen concentrations of other man-related sources vary widely depending on the source. Treatment of these and other nonpoint sources is difficult if not impossible to treat or industrial sources of nitrogen and their treatment. For the purpose of this control methods can be divided into three broad categories:

1. Biological methods or removal.
2. Chemical/physical methods of removal,
3. Other methods of removal.

Biological methods include nitrification in suspended growth, and attached growth systems, using trickling filters rotating biological contractors, and packed bed reaction. Biological denitrification in both suspended and attached growth reactors is a developing method. Chemical physical methods for removal include breakpoint chlorination and ozone treatment, selective ion exchange, and ammonia stripping. In addition, other methods such as aquatics have been discussed.

The term nitrification is applied to the reaction in nature of the biological oxidation of ammonium (NH_4) first to the nitrite (NO_2), then to the nitrate form. The conversion of ammonium to nitrate is caused by the bacteria Nitrosomonas as follows:

$$2NH_4 + 3O_2 \xrightarrow{\text{Bacteria}} 2NO_2 + 4H\ 2H_2O$$

The nitrate is turn are oxidized by the bacteria nitrobacter according to the following:

$$2NO_2 + O_2 \xrightarrow{\text{Bacteria}} 2NO_2$$

The overall nitrification reaction is as follows:

$$NH_4 + 2O_2 \xrightarrow{\text{Bacteria}} NO_3 + 2N\ H_2O$$

For this reaction to go to completion about 4.6 mg/l of oxygen are required per gram of ammonia nitrogen. This oxygen demand is often identified as the **nitrogenous oxygen demand (NOD).**

The problem with nitrogen in wastewater is related to the NOD if ammonia nitrogen is discharged into the environment. This problem can be mitigated by either converting the ammonia and organic nitrogen to nitrate or by eliminating the nitrogen in the wastewater. Biological methods of nitrogen removal are classified under two broad categories, biological nitrification and denitrification.

Nitrification

The application of biological nitrification in municipal wastewater treatment is applicable to cases where ammonia removal requirements exist, without the need for complete nitrogen removal. Nitrification systems are broken up into combined carbon oxidation-nitrification and separate stage nitrification process and each can be divided further into suspended and attached growth processes. Suspended growth processes are those which suspend the biological solids in mixed liquor by mixing mechanism. A subsequent clarification stage is required for returning these solids to the nitrification stage. Attached growth processes, on the other hand, retain the bulk of the biomass on the media and, therefore, do not require a solids separation stage for returning the solids to the nitrification reactor.

Attached Growth Processes

The trickling filter, rotating biological contactor and the packed bed reactor is the principal biological methods of nitrification. The trickling filter and rotating biological contactor can be considered for combined carbon oxidation-nitrification and separate stage nitrification, whereas the packed bed reactor is only considered as a separate stage process.

Trickling Filter: (combined oxidation – nitrification): The tickling filter consists of a bed of highly permeable media to which micro-organisms are attached and waste water is percolated. The filter media usually consists of rock, packing etc. one to four inches in diameter with a 32 depth of three to eight feet. More recently, some filters 30 have been built with plastic pot redwood media and depths of 15 to 25 feet.

The development and maintenance of nitrifying organisms on a trickling filter depends on factors such as organic loading and temperature. Kinetic theory for combined carbon oxidation nitrification has not been developed, therefore, the approach applied to date is largely empirical and realist mostly on specification of an organic loading rate suitable for application to each media. A typical loading rate for rock media to attain 75% nitrification is 10 to 20 lb, $BOD_5/1000^3$/day. At higher loading rates, such as that above 30 to 40 lb $BOD_5/1000^3$/day, very little nitrification occurs. The specific media surface selected has an effect on the allowable organic loading. Greater specific surface in the media allows greater biological film development and a greater concentration of organisms within a unit volume.

Trickling Filter (Separate Stage Nitrification): The development of the two stage suspended growth system for nitrification. Later was observed that nitrification occurred as a side affect of the original use of the two-stage trickling filters. Initially, the two-stage trickling filtration process was developed to increase the removal of organics in the effluent of high rate trickling filters. In separate stage nitrification applications, the rate of nitrification is proportional to the surface area exposed to the liquid being nitrified. Plastic media is commercially available with specific surface available from 27 ft^2/ft^3 and as such is the best alternative for a two-stage suspended growth system because of its high specific area. Pilot plant studies indicate that with temperatures of 13 to 19°C and influent NH_4-N of between 8 to 18 mg/l NH_4-N can be reduced to 2.5 mg/l for 1000 ft^2/lb NH_4-N oxidized/day. Whereas with temperatures of 7 to 11°C and the same influent ranges, removal rates only averaged 3.0 mg/l per 4000 ft^2/lb NH_4-N oxidized /day. When effluent levels of less than 2/5 mg/l NH_4-N are desired consideration should be given to breakpoint chlorination for reducing the ammonia residual rather than increasing the surface area of the filter since a surface area of 10,000 to 12,000 ft^2/lb NH_4-N oxidized/day will be required to reduce ammonia effluents to between 1 to 2.5 mg/l. Tests using rock media have shown the NH_4-Noxidation is only surface area. A 1:1 recirculation ratio is normally considered adequate for this type of system. Effluent clarification is not required in separate stage trickling filter systems since organisms are attached to the media and solid levels are low. In pilot plant studies, it was found that the same low influent solids of 9 to 28 mg/l were evident in the effluent.

Rotating Biological Contractors (combined Carbon Oxidation – Nitrification): The rotating biological contractor or disk (RBD) has seen limited use in the United States in combined carbon-oxidation-nitrification. The RBD process consists of a series of closely spaced circular disks of polystyrene or polyvinyl chloride. The disks are slowly rotated and submerged in the wastewater. The disks are slowly rotated and submerged in the wastewater. In operation, biological organisms become attached to the surface of the disks and in about 10 days in two weeks, from a slime layer of aerobic biomass. In rotation, the discs biomass contact wastewater and then the atmosphere for absorption of oxygen. Disks rotation is the mechanism used for the removal of the excess solids from the disks by the shearing forces as the disks passes through the liquid. The treated effluent is mixed and in suspension, and the wastewater flow carries the suspension out of the disk section into a secondary clarifier for separation and disposal. Disk units are normally housed to avoid temperature drops, prevent algae growth on the disk surfaces, and to protect from weather.

As a RBD system operates, in series, organic matter is removed in the first disk stages and the subsequent disk stages are used for nitrification. Nitrification only occurs after the bulk of the BOD_5 has been oxidized. After the low levels of BOD_5 have been reached, the disk is no longer dominated by heterotrophy, and nitrification can proceed.

Rotating Biological Contractors (Separate Stage Nitrification): In order to obtain separate stage nitrification, the wastewater influent must be pretreated to remove organic carbon. As such separate stage nitrification is very adaptable to upgrading existing secondary sewage treatment systems to meet low ammonia nitrification effluents. The BOD_5/TKN must be sufficiently low to ensure a significant fraction of nitrogen to biomass. Some alternative pretreatment systems used are chemical treatment in activated sludge, roughing filters, and trickling filters. In addition to the separate stage trickling filter. Rotating Biological Contractors are very adaptable to separate stage nitrification.

The purpose for RBD combined carbon oxidation – nitrification may be applied to separate stage nitrification. The only difference is if the secondary effluent being treated has a BOD_5 or suspended solids content of less than 20 mg/l, the secondary clarifier may be eliminated. The most common RBD systems used is a four-stage system having four rows of disks. But the more is stages, the greater the nitrification.

Packed Bed Reactors (PBR) (Separate Stage): A packed bed reactor consists of a bed of media where nitrifying biological mass is developed, overlaying a support media and inlet chamber to which wastewater influent is introduced through an up flow configuration. Media bed types that have been used are anthracite coal, silica sand and, stones, gravel, and a plastic packing media. Both air and high purity oxygen have been employed as a means of supplying the necessary oxygen for nitrification. Air can either be injected directly into the inlet chamber of distributed across the PBR floor. Oxygen can be either bubbled directly into the PBR. Only monthly backwashing of light density anthractice or plastic media is considered necessary due to turbulence developed in the bed. Experience has shown that backwashing of a gravel media reactor is required on a more frequent basis, perhaps as much as once per day. Oxidation rates for a typical reactor range from 4 to 27 lb NH_4-N oxidized per 1000 ft^2/day. A typical reactor using crushed coal and loosely packed plastic can reduce average ammonia nitrogen concentration from 15 mg/l to less than 1 mg/l. This is at an ammonia loading of 6.9 lb NH_3-N/day/1000 ft^3 of media volume, detention time to 3.25 hours, operation temperature of 10°C. Temperature has a strong affect on the PBR process by increasing the temperature of liquid from 10°C, the detention can be cut to less than two hours to obtain the same effluent characteristics.

Suspended Growth Process: The activated sludge process is the main suspended growth biological method nitrification used to treat sewage. Basically, there are two methods considered for nitrification by activated sludge systems: combined – carbon – nitrification, in the same tank where carbonaceous removal is accomplished; and separate stage nitrification which has developed to isolate the carbonaceous removal and the nitrification processes so that each could be separately controlled and optimized.

Not all of the modifications to the activated (Combined Carbon -Oxidation – Nitrification) process are appropriate for nitrification applications. Some of the more successful applications are the complete mix plant, conventional plant, extended aeration plant, contact stabilization plant, and step aeration plants.

Conventional Plant: Conventional plant consists of a series of rectangular tanks with a length to width ratio of 5 to 50. Tanks hydraulics is plug flow. In the system, sludge from the aeration tank is returned to be mixed with influent coming into the tank. When the safety factor for a conventional plant is compared to that of a complete mix plant, it is determined that the continental plant is more efficient. However, the conventional plant has the advantages that the carbonaceous oxygen demand is concentrated at the head of the tank, making it more difficult to supply enough air in the front of the tank to supply both carbonaceous and nitrogenous oxygen demands. Air diffusion systems for conventional plants must be specifically designed to handle this concentrated load, otherwise, the first portion of the tank is not available for nitrification and the nitrification volume is reduced.

Complete Mix Plant: A complete mix plant is designed so the effluent entering the plants aeration tank is evenly distributed. This design provides uniformity of load to all points and elimination of the oxygen transfer problems of the conventional plant. Due to the increased amount of short-circulating of the influent to effluent, ammonia concentrations may be slightly higher than a conventional plant. Load characteristics are similar to a conventional plant. Temperature plays a critical roll in nitrification. At the minimum temperature of 10°C, the hydraulic relation is about 14 hours whereas at a 20°C, the retention time is decreased to 6 hours. Therefore, location and temperature of the plant must be determined before sizing.

Extended Aeration Plant: Extended aeration plants are similar to complete mix plants except that hydraulic retention times are 24 to 48 hours instead of 2 to 14 hours used in the complete mix plants. Endogenous respiration is maximized in this type of plant, consequently 25 to 35 day solids retention times are not uncommon. Because of the long aeration time, they do suffer heat losses because of this long net growth, extended aeration plants can be expected to nitrify unless the temperature is less than 10°C.

Contact Stabilization Plants: Contact stabilization plants are not well suited for complete nitrification because insufficient biological mass is present in the contact tank for complete nitrification of ammonia. Since, the ammonia is not absorbed on the biological floc, ammonia will bleed through to the effluent. In contact stabilization, instead of mixing the influent wastewater with return sludge is re-aerated in a sludge re-aeration tank prior to mixing with the influent wastewater. Back-mixing between the contact and sludge re-aeration tanks is prevented by the use of weirs. BOD_5 removal, not nitrification, is fairly high in this system because the bulk of the organics in domestic wastewater are particulate and can be absorbed by biological solids for later oxidation in the sludge re-aeration tank.

Step Aeration Plant: In the step aeration plant, the influent wastewater is introduced at several points along the aeration tank even though the return sludge is introduced at the head of he plant like a conventional plant, this influent distribution reduces the initial oxygen demand usually experienced in contact stabilization plants, is avoided in a step aeration plant because of the longer contact time. In any case, some NH_4 bleed-through that is experienced in contact stabilization plants is avoided in a step aeration plant because of the longer contact time. In any case, some NH_4 bleed through will occur from short circulating of the influent to the effluent and insufficient contact time for complete organic nitrogen hydrolysis and oxidation of ammonia.

Advantages and Disadvantages

The combined systems single stage is more compact than the separate stage system, but at the same time the chances for toxic upset are greater. Suspended growth systems (both combined and separate stages) have the advantage over the attached growth systems in that effluent ammonia concentration are lower, but this is overcome in attached growth systems, stability is not linked to a secondary clarifier as in the suspended growth system. Separate stage attached growth system is less sensitive to cold weather operations than the combined attached growth systems.

Denitrification

Biological denitrification is used to remove nitrogen from wastewater when it is in the nitrite or nitrate form. In the process, ammonia-nitrogen in wastewater is first converted to nitrite and subsequently nitrate. The nitrate nitrogen is then converted to a gaseous nitrogen, primarily nitrogen gas. This is accomplished through suspended and attached growth systems using methanol as the carbon source and combined carbon oxidation – nitrification – denitrification system, using wastewater as an endogenous carbon source.

Pseudomonas, Micrococcus, Achromobacter, and Cacillus are principal nitrate-reducing bacteria that accomplish denitrification. These bacteria accomplish nitrate reduction by nitrate dissimilation whereby nitrate or nitrite or nitrite replace dissimilation whereby nitrate or nitrite or nitrite replace oxygen in the reparatory processes of the organism under anoxic conditions. Using methanol as the carbon source, the energy reaction is represented as follows:

Step 1: Nitrate to Nitrite

$$6NO_3 + 2CH_3OH \longrightarrow 6\,NO_2 + 2\,CO_2 + 2\,H_2O$$

Step 2: Nitrite to Nitrogen Gas

$$6NO_3 + 3CH_3OH \longrightarrow 3N_2 + 3CO_2 + 3H_2O + 6OH$$

Overall Energy Reaction

$$6NO_2 + 3CH_3OH \longrightarrow 3N_2 + 3CO_2 + 3H_2O + 6OH$$

A Typical Synthesis Reaction is

$$3NO_2 + 14CH_3OH + CO_2 + 3H \longrightarrow 3C_5H_7O_2N + 4H_2O$$

Overall Nitrate Removal

$$NO_3 + 1.08CH_3OH + H \longrightarrow 0.065C_5H_7O_2N + 0.47N_2 + 0.76CO_2 + 2.44H_2O$$

Where nitrate, nitrite and dissolved oxygen are present, the methanol requirement (Cn) can be computed using the following:

$$C_n = 2.47\,(NO_3\text{-}N) + 1.53\,(NO_2\text{-}N) + 0.87\,(DO)$$

Typically, methanol feed ratio three parts methanol per one part nitrogen by weight.

Suspended Growth Systems

Suspended growth denitrification is usually carried out in plug-flow type activated – sludge systems. The anaerobic bacteria obtain energy for growth from the conversion of nitrate or nitrite to nitrogen gas but require an external source of carbon for cell syntheses. Methanol is commonly used as this carbon source, it is nitrite or nitrate that is the pollutant to be removed, where as in conventional treatment, carbon is the pollutant. The methanol added is a carbon source but should not be added in excess, if left in the effluent, the methanol will exert a BOD_5. Because the nitrogen gas formed in the denitrification reaction hinders the setting of the mixed liquors, a nitrogen gas stripping reactor precedes the denitrification clarifier. Removal of residual methanol induced BOD_5 is an added benefit of the stripping tank.

Attached Growth System

Denitrification in attached growth systems, are accomplished by a variety of column configurations. Column systems are nitrogen gas packed beds; submerged high porosity packed bed; and submerged high porosity fluidized bed systems. Since there is a variety of media used to measure reactions, general design criteria for these systems are not readily available. In order to size the reactor volume, one approach is to estimate the level of biomass on the media surface and then the measured reaction rates per unit biomass to obtain the nitrogen removal capacity of the column. This is the best done through pilot plant experiments. Based on the measured rates, surface nitrate removal rates are classified as high surface removal rates reflecting the formation of extensive biological film, and low surface removal rates, reflecting minimal surface film development.

Nitrogen Gas Filled Packed Bed Reactor: The nitrogen gas filled packed bed reactor is covered and filled with nitrogen gas. This eliminates the necessity of having to submerge the medium to maintain anoxic conditions. The column media consists of plastic (corrugated sheet modules) similar to those used in trickling filters.

Aquatics

The use of aquatic plants are nutrient removal agent, is not a concept. The water hyacinth (*Eichlornia crassipest*) is the most researched plant in this area to date. The water hyacinth is a floating plant that covers vast area of water surface in the southern United States. The plant interferes with navigation, causes flood control problems, and restricts recreational activities such as fishing, boating, and water skiing. The water hyacinth is a native of South America and was introduced to the United States in the late 1800's. Any constructive use of this plant is most welcomed.

In 1968, 61% removal of PO_4-P was accomplished by growing water hyacinths after a five day detention time. However, most tests to date have shown that after 25 to 30 days of continuous operations, phosphate removal efficiency declines until only 5 to 8% removal efficiency is observed. Also, hyacinth removal efficiency is much less during colder months. Research done at Gainsville, Florida, found that the nutrient removal capacity of water hyacinth was directly related to point surface area in order to remove 44% of the phosphorus; a one million gallon pond with 5.1 acres of water hyacinths was needed. For the small ponds used in the Gainsville test, influent values ranged from 3.37 to 3.44 mg/l with effluent values of 1.82 to 1.86 mg/l. Effluent had a four-day detention time and the pond a depth of one foot. Tests were also done with a deeper pond and it was found the shallower pond, had better the PO_4-P removal efficiency. Nutrient uptake by the hyacinth was good during the area growth phase and vertical growth phase, but if the hyacinths growth was good during the area growth was not limited, lesser efficiencies in PO_4-P removal was noted when the pond became overgrown with hyacinths. This could lead to anaerobic conditions in the pond. This experiment did not show that nitrogen removal efficiency is better from water hyacinths than PO_4-P removal efficiency; however, some phosphorus removal is obtained.

BIOCONSERVATION AND OF BIOLOGICAL DEGRADATION OF HAZARDOUS WASTES

The bulk of potentially hazardous waste is generated from four industries: primary metals, organic chemicals, and explosives; electroplating and metal finishing, inorganic chemicals. The hazardous waste management industry often deals with input materials in the five groupings of: metals/metal; finishing: plaint/ solvent/ coatings; organic; petroleum and inorganics.

Industrial Waste Problem, especially those involving hazardous waste were in ongoing problem. They have a high profile, and as such are reported regularly in the media.

The hazardous waste does not consist of one chemical or attribute; the definition of it is a wide one. The Environmental Protection Agency (EPR) stated in its report to Congress in 1977 that:

The term "**hazardous waste**" means any waste or combination of wastes which pose a substantial present or potential hazard to human health or living organisms because such wastes are lethal, non-degradable, persistent in nature, biologically magnified, or otherwise cause or trend to cause detrimental cumulative effects. General categories of hazardous waste are toxic chemical, flammable, radioactive, explosive, and biological. These wastes can take the form of solids, sludge, liquids, or gases. So, the range of hazardous wastes and their effects is large and diverse.

Attributes of Hazardous Wastes are: irritation, corrosively and bio-concentration. The observable effects are a function of exposure, the longer a person is exposed, the more serious the effects. Nearly all industry produces some waste and no one is likely to independently shut down an industry that makes the waste, we need some guidelines and regulations.

Bio-conservation and biological treatment are expanding fields, which have shown success in remediation of hazardous waste problems. These are not the only treatment technologies available and do not have to be used in isolation, nor is it without limitations.

EPA Listed Five Most Common Strategies

1. Reduction in the overall amount of waste that is produced.
2. Reuse of the processing chemicals or solvents with little or no treatment, sometimes within the same company.
3. Recycling the processing chemicals through a treatment process.
4. Research, development, and demonstration (RD&D) of new waste-reduction techniques. RD&D can range from basic research, to demonstrating the feasibility of tailoring an existing technology to the process of a particular business.
5. Replacement of the entire industrial process.

Determination of Biological Degradability

Some terms are important in understanding the fate of hazardous wastes in biological systems. They are *biodegradable*, *persistent*, *recalcitrant* and *mineralization*.

A compound that is biodegradable can be changed by the action of microorganisms to another compound. This does not necessarily mean that the product is less toxic than the parent compound, or that intermediate toxic compounds are formed that may inhibit the degraders, or that the product will not be toxic to the next degraders or to man himself.

The bacterial degradation of 7 - 12 dimethylbenzanthracene in non-acclimated soil is degraded aerobically, the per cent 14 C DMBA had a half life of 17 days. There is a decrease in the parent compound accompanied by increase metabolites, 4-hydroxy, 5-hydroxy and 10-hydroxy DMBA and 7.12 dihydro-12-methyl 7-mrthylene benz (a) anthracene. But there was some associated Mutagenecity that was thought to be due to the polarity of the compounds, with those that are low and moderately polar having more mutagenic potential than the highly polar compounds.

The pH of the soil also plays a role since at low pH fungi dominate the ecosystem and they are more likely to form epoxides (associated with Mutagenecity) than bacteria which are dominant at higher pH.

Recalcitrant means that the compound cannot be biodegraded under any circumstances i.e., it is the compound itself that is inherently resistant and not the treatment system that has failed to account for some vital fact. Persistence is a **conditional** property of the biodegradable compound in that it may be biodegraded if the correct circumstances favouring biodegradation are present.

Mineralization is the complete conversion of an organic compound to the end products of carbon dioxide and water. Primary biodegradation is the single transformation of a compound and partial biodegradation is somewhere between primary degradation and mineralization.

Reasons for Degradation of Some Compounds by Microorganisms

The degradation is done not as a favour to us but because the organisms gain energy needed for growth, repair, reproduction and other biological functions needed for survival. Food sources contain in hydroxyl (-OH) or carboxylic acid (-COOH) groups. Oxidation reactions take place where electrons are transferred along an electron chain with compounds accepting and passing on electrons to a terminal electron accepter. The coenzymes may be NADH and NAD which get reduced and then coupled with the electron transport chain to produce high energy bonds in ATP. Many compounds are biochemically inert such as alkanes, saturated ring structures and unsaturated benzene. They are devoid of oxygen and not subject to dehydrogenation reactions. The ability of bacteria to utilize these compounds lies in the fact that they can catalyze oxidation using oxygen. Other bacteria have enzymes which work without oxygen and coenzymes are needed.

Studies have been done that attempt to relate the physical and/or chemical properties of a chemical to its recalcitrance; however, experience has shown that many chemicals which are though not to the removed by abiotic mechanisms actually undergo transformation in model treatment systems.

Genetics of Bacteria

Bacteria can only do things for which they have a genetic capability. They must be able to produce the right enzymes to do the job, and the right environment must exist for them to be able to produce the right enzyme. If a chemical is present in concentrations either too low or too high, then it may not be biodegraded: too low and the enzymes will not be induced, too high and the compound may be toxic. One reason why bacteria are robust at biodegradation is that they may still be able to degrade a different compound to the one that they normally utilize if the active site of the molecule has not been altered.

The xenobiotic still need to be able to induce the production enzymes or the reaction will stop. The bacteria will not be able to metabolize it and will die. The bacteria will not be able to metabolize it and will die. This use of a different substrate is fortunate for us and has been termed **gratuitous** metabolism.

Complicating biodegradation further is **cometabolism** where two substrates are needed i.e., one compound can not fulfill all the bacterial needs and act as the sole carbon and energy substrate. A second compound is needed as growth substrate. Biological degradation treatment systems have to be designed to take source for growth and yet if too much substrate is added, initially the bacteria are unable to metabolize it. They need a period of **acclimation** where they are growing and strengthening and perhaps even undergoing genetic changes. If the intermediate is toxic then the bacteria may be killed or the formation of the next strain that metabolizes it may be stopped so that it accumulates and destroys the system.

Abiotic Treatment Techniques

Wastewater Treatment

Selection of a treatment process is dependent on the nature of the wastewater and the quality of the effluent desired. Hazardous components of the wastewater may be either separated or converted to non-hazardous forms in order to permit the disposal of the wastewater effluent by conventional methods, which are separated from the wastewater must be disposed of. This may take additional steps e.g., sludge dewatering.

Liquids-Solids Separation

Separation of suspended matter from wastewater can be accomplished by a number of different processes. Large heavy solids are easier to remove than finely divided light solids.

Screening devices are used to remove large pieces of solid matter that would interfere with screening devices may consist of bars, rods or wires, perforated plates, gratings or wire mesh.

Gravity sedimentation, this process involves the containment of wastewater for a sufficient period of time to allow some or all of the suspended materials to either settle out or float to the surface of the wastewater. In its simplest form as a batch process, a given volume of wastewater to transferred to a vessel and held there until nearly all the settle able and floatable matter separates. The floating matter can be skimmed off and the wastewater decanted for discharge or further treatment. Sludge may be allowed to collect until several batches of wastewater have been processed. Then, it is removed. The vessel may have a conical bottom so that the sludge can be removed via a valve. Large setting ponds may be constructed which are drained periodically to permit sludge removal. Solids-contact or sludge-blanket clarifiers are useful for where sludge are flocculent flocculent and of low density. They are designed with large mixing and reaction zones that coupled with the sludge blanket account for greater efficiency on solids removal. Gravity sedimentation works like the clarifiers but more time is taken for the settling.

Dissolved-Air Floatation is useful for suspended matter that does not sink or float in a reasonable period of time. Separation is brought about by the introduction of finely divided gas bubbles, which become attached to the particulate matter, causing it to float to the surface where it is removed by skimming. Introduction of the gas bubbles is usually accomplished by reducing the pressure of the wastewater causing dissolved gases to be released. This is commonly used to separate greasy or oily matter from industrial wastes.

Granular-Media Filters or Deep Bed Filtration is a polishing step that removes small amounts of suspended solids and produces highly clarified water. Chemical coagulation and sedimentation typically precede this stage. Graded sand and pulverized coal is commonly used in the filter beds. Conventional operation is usually by down-flow. The ability of the granular media filter beds to produce a clear effluent results from the straining action and adhesion, which removes particles finer than the pore space.

Surface Filters make use of a fine medium such as a cloth or close mesh screen. In the rotary vacuum filter, the medium is in the form of a continuous belt and it rotates over a perforated drum that is partially submerged in the slurry to be filtered.

Water is pulled through the filter cake that forms on the belt to the inside of the drum, where it is transferred to the vacuum system.

Centrifugation is a useful alternative to filtration for sticky sludge that do not dewater rapidly on a filter. They operate by a rapid rotation of a liquid suspension which induces a much greater force than gravity to hasten the separation of the suspended matter.

Chemical Treatment

Chemical treatment is a widely used process for the destruction or separation of hazardous constituent in wastewater. This can be done by neutralization of acidic or alkaline wastewater until a suitable pH is obtained. Precipitation/Coagulation/Flocculation is use for the removal of heavy metals. Precipitation refers to the formation of a solid phase, coagulation is where the containment is trapped by the formation of a precipitate and flocculation is the agglomeration pf a coagulating chemical.

Oxidation-Reduction or the redox processes are used for converting toxic pollutants to harmless or less toxic materials into other forms that are more easily removed. These processes involve the addition of chemical reagents to wastewater, causing changes in the oxidation states of substances both in the reagents and in the wastewaters. In order for one substance to be oxidized, another must be reduced. Ozone is a powerful oxidizing agent that is usually used to eliminate traces of organics. Wet oxidation utilizes oxygen at temperatures and pressures up to 350°C and 180 atmospheres to treat organic wastes.

Ion Exchange involves a change in the chemical form of a compound, the exchange of ions in solution with other ions held by mixed anionic or cationic groups or charges. Typically, a waste solution is percolated through a granular bed of the ion exchanger, where certain ions in solution are replaced by ions contained in the ion exchanger. If the exchange involves cations, the exchanger is called a cation exchanger and correspondingly anion exchanger is one that involves anion exchange.

Physical Methods

There are several methods used for separating pollutants from wastewater activated carbon, steam stripping, evaporation, reverse osmosis and solvent extraction. The chemical and physical characteristics of the pollutant are important in the selection of the physical removal method. Steam stripping is effective for substances that have an appreciable vapour pressure at the boiling point of water, whereas evaporation is effective for those chemicals that will not volatilize. Soluble, small organic molecules are absorbed by activated carbon, large ions are separated by reverse osmosis.

Activated Carbon Adsorption: Here the inorganic and organic chemicals are absorbed into activated carbon. Usually, hydrophobic chemicals are more likely to be removed. The degree of adsorption is linked to the molecular weight, methanol-water coefficient or solubility (these are also linked to the recalcitrance and or toxicity). The smaller the size of the grain, the more surface area is available and so equilibrium is reached quicker with powdered activated carbon compared to the granular form. But then the powdered form needs more pumping to get the wastewater through and hence the costs are increased. There are two principle systems, one is **downward flow** through the bed (pressure or gravity flow) and the other is **upflow** through a packed or expanded bed. Activated carbon adsorption is applicable to the treatment of dilute aqueous wastes, but they should be treated to remove suspended solids, oil and grease.

Temperature and pH are also important for the different compounds to be treated. The carbon is either disposed of or regenerated. Carbon has also been added directly to biological treatment effluent in a contacting basin. The advantages of this are that the sludge toxicity is reduced by selectivity removing the toxic organics from solution and that the carbon adsorption capacity is extended by bio-regeneration of the '**biocompatible**' species adsorbed on the surface, for aqueous solvent waste containing contaminants in concentrations up to 10,000 mg/L, the activated applicable treatment. However, these concentrations may be toxic to the sludge or they may be easily stripped to the atmosphere, thereby creating another hazard. The sludge may also contain recalcitrant waste, due to sorption of the contaminants and be difficult to dispose it.

Evaporation is the process that heats the liquid, venting the vapours to the atmosphere and concentrating the pollutants into slurry.

Reverse Osmosis: Osmosis is the process where a solvent (e.g., water) moves from an area of low concentration to high across a semi permeable membrane, which does not allow the dissolved solids to pass. In reverse osmosis, a pressure is greater than the osmotic pressure is applied so the flow is reversed. Pure water will then flow through the membrane from the concentrated solution.

Solvent Extraction is a process whereby a dissolved or adsorbed substance is transferred from a liquid or solid phase to a solvent must be immiscible in the liquid and differ in density so that gravity separation is possible and there is minimal contamination of the raffinate with the solvent. The hydrophobic solutes are more likely to be extracted. Solvent extraction can be performed as a batch process, or by the contact of the solvent with the feed in staged or continuous equipment.

Steam stripping is where water vapor at elevated temperatures is used to remove volatile components of a liquid. Countercurrent flow is generally used to promote gas-liquid content, thus allowing soluble gaseous organics from the liquid waste to be continuously exchanged with molecules within the stripping gas. Again, this is useful only for waste with low water solubility.

Incineration

Incineration is a high temperature oxidation process that converts the principal elements (carbon, hydrogen and oxygen) in most organic compounds to carbon dioxide and water. Given with the problems of disposing on land, incineration may take on a lead role in waste treatment. However, it is not without its problems, there is fear among the general public about the nature of the stack emission but it is an efficient method. The destruction of the molecular structure usually eliminates the toxicity of the chemical. But the existence of other elements in a waste may result in the production of particulate pollutants that require removal in off-gas treatment systems. There are several types of incineration available.

The **liquid injection incinerators** operate by spraying waste mix with air into a chamber where oxidation lakes place flame. The purpose of spraying is to atomize the waste into small droplets, which present a large surface area for rapid heat transfer, thereby increasing the rate of vaporization and mixing with air to promote combustion. Air is supplied to provide the necessary mixing and turbulence. These incinerators are widely used for destruction of liquid organic wastes.

Rotary-Kiln Incinerators are designed to process solids and tars can not be processed in the liquid incinerator. The rotary-kiln is a cylindrical shell lined with refractory material that is horizontally mounted at a slight incline. It is rotated from 5 to 25 times an hour at high temperatures, 1500 to 3000°F with excess air, the residence time varied depending on the nature of the waste. The rotation causes a tumbling action that mixes the waste with air. The primary function is converting, through partial burning and volatilization, solid wastes to gases and ash/residues. If the ash is free of dangerous levels hazardous wastes it is put in a landfill.

Fume Incineration: Large quantities of organic vapour fumes are produced by many industries, including fat rendering, metal painting and varnishing and various types of printing. These vapours are generally mixes of hydrocarbons, alcohols and acetates. The mixes may not be actually toxic but they do cause odour problems. An integrated heat-recovery system demonstration at Case-Hoyt Company, near Rochester, captures waste heat and simultaneously reduces plant air emissions by oxidizing solvents into harmless gases as the heat is recovered.

The Multiple Hearth Incinerator is used for waste that is difficult to burn or that contain valuable metals that can be recovered. It consists of a refractory –lined circular steel shell, with refractory hearths located one above the other. Solid waste or partially dewatered sludge is fed to the top of the unit, where a rotating plow rake plows it across the hearth to drop holes, the un-combusted material falls to the next hearth the process is repeated until the combustion is complete.

Fluidized Bed Incinerators (FBI) is applicable to the destruction of halogenated organic waste streams. This type of incinerator consists of a vessel in which inert granular particles are fluidized by a low velocity air stream, which is passed through a distributor plate below the bed. An FBI consists of a wind box, through which combustion air in introduced to the reactor zone, containing a bed of sand, waste injection and removal ports. Temperatures are in the range of 1,300 to 2,100°F, gas residence times usually a few seconds. They have been used to treat municipal sewage sludge, low quality fuels, pulp and paper effluents, and food processing waste, refinery waste, radioactive waste and miscellaneous chemical waste.

A Modern Salt Incinerator uses a molten salt such as sodium carbonate as a heat transfer and reaction medium. In the process, waste material along with air is added below the surface of the bed so that any gases formed during combustion are forced to pass though the melt. Reaction temperatures in the bed range from 1,800 to 2,000°F and residence times are less than a second. Any acidic gases formed are neutralized by the alkalinity of the bed. This can change the fluidity of the bed so it needs replacement frequently.

Plasma Incineration is based on the concept of reducing or pyrolyzing waste molecules to the atomic state using a thermal plasma field. The system uses very high energy at temperatures near to 10,000°C to break bonds of hazardous waste chemical molecules down to the atomic state. An electrode assembly ionizes air molecules, which create a plasma field. Hazardous waste mixtures interact with the field, forming simple molecule such as carbon dioxide, hydrogen chloride, and other minor matrix compounds such as acetylene are formed. Westinghouse Electronic has a mobile plasma are unit called **Pyroplasma** that reportedly treats liquid wastes at the rate of a 3 gal/minute. The high temperatures decompose PCB and other wastes in an oxygen-deficient atmosphere Hydrogen chloride are treated with sodium to form water and salt.

Lime or Cement Kiln Incineration: A cement kiln is basically a large rotary kiln in which raw materials are fed countercurrent to combustion gas flow. The wet process kiln uses a 30% water slurry feed are the most suitable for hazardous waste destruction. The products formed are alkaline and so act as a scrubber, removing acid gases formed during combustion. This system operates at 2,800°F resulting in very efficient removal of wastes. A project is in operation at Blue Circle Atlantic Inc., a cement company in Ravena where it is evaluating the incorporation of hazardous wastes into an existing cement-kiln operation. The facility is not yet in operation but should be able to handle large amounts of waste and alleviate some problems locally.

Wet Air Oxidation

Wet air oxidation involves the aqueous phase oxidation of organic materials at high temperature and pressure. A major advantage over other incineration methods is that the water in the waste stream is kept in the liquid state. Water is pumped into the reactor along with oxygen which is heated by the hog effluent. Two types of reactor are used, a bubble tower reactor and a stirred tank cascade reactor. This process is good for wastes that are too dilute to incinerate but too toxic for biological methods. The products are usually acetic acid and carbon dioxide.

Solidification Techniques

There are several innovative non thermal processes that have been developed under the SITE program that immobilize wastes vitrification or other types of solidification. The researchers at Battelle Pacific Northwest Laboratories have developed an in-situ vitrification process (which was (which was originally designed for the containment of nuclear wastes) in which electrodes are sunk into a contaminated area and attached to a diesel powered generator. The current produced temperatures of about 3,600°F which is much higher than the fusion temperature of soil. An exhaust hood is placed over the site to collect and treat any combustion products. The result is a massive glass like product consisting of completely immobilized organics, inorganic, steel drums and other components that are essentially looked up and inert. The time taken to complete the process depends on the electrode depth and frequently. Another solidification process uses a reagent called **Urrichem** that immobilizes slurred hazardous components. The contaminated soil is excavated and intimately mixed with the Urrichem off-site. After blending, the slurry is pumped out of the mixer and hardens into concrete like mass within 24 hours. Chemfix is a process developed by chemfix Technologies (Metaire LA). In this technique, a proprietary blend of soluble silicates and organic and inorganic slurries into a cross-linked, clay like matrix.

Alternative methods have also included landfills, deep well injection disposal and ocean dumping.

1. **Landfills** were developed because of it was believed that by placing waste in designated ground areas, there would be a neutral decomposition over time. Unfortunately the water table rises in a landfill and this mounding effect means that we get a leaching of the water containing toxics since the water flow is always from areas of high to low head, i.e., there is a gradient set up that favours water movement originating in the landfill and away from it thus contaminating our ground water supplies.
2. **Deep Injection Wells** have similar problems. When the aquifers in which the deep well sit are pumped, the contaminants are drawn into the well.
3. **Ocean Dumping** has unset the delicate ecosystem balance in some areas. Thus, there is a need for effective techniques that industry can safety use to try and combat the nations hazardous waste problems. Biological treatment may be such.

Biological Control Methods

Biological treatment of hazardous wastes is becoming a realistic option in the treatment strategies for hazardous wastes. As we research more about specific bacteria and fungi that do the work and know more about what conditions favour degradation, we are finding more uses for them. A major advantage is that the cost is much less than incineration. A limitation is the length of time involved, since biological treatment can take months. There are different mechanisms that we can use to degrade a compound. Selection is based on the type of pollutant, type of degradation (i.e., aerobic or anaerobic), and various other parameters.

These are suspended growth systems such as activated sludge, aerated legoons and aerobic digester where the microorganisms are suspended and fixed film processes which include filers and rotating biological contactors where the microorganisms grow on a fixed surface. Since toxin degradation is a slow process, the fixed film processes may be best since the microbial retention is a very high. In aerobic degradation, the major portion of the cost is the aeration. Wastes in any form – can be treated biologically. The treatment can be done several ways: by applying the wastes in situ, using soil filter gases, and with wastewater treatment systems. Care must be taken in assessing biological treatment since non degradable pesticides and other organics may be unaffected by the treatment or toxic intermediates may be formed and therefore, a finishing process such as carbon may be needed.

Anaerobic digestion of complex organic matter to methane consists of following four successive steps:

1. **Hydrolysis:** Where organic polymers are hydrolyzed to their individual monomers as a result of enzymatic attack.
2. **Acid Formation:** The hydrolyzed compounds are converted intracellular by a group of acid forming bacteria to simple compounds, such as volatile fatty acids, alcohols, CO_2, NH_2, and H_2.
3. **Acetogenesis:** The products of the acid formation are transformed to acetic acid, CO_2, NH_2 and H_2.
4. **Methane Formation:** Acetic acid, CO_2 + H_2 and methanol are transformed to CH_4 and CO_2 by the strictly anaerobic methane forming bacteria.

Depending on the composition of the wastewater, each of the four steps can be rate limiting. When the wastewater contains a considerable fraction complex (undissolved particles), the hydrolysis can be rate limiting, when the pollution consists mainly of acetic acid and other volatile fatty acids, the methane formation can be the rate limiting step.

TABLE 16.2: Historical Benefits and Limitation of Anaerobic Treatment of Wastewaters

Benefits	Limitations
– Low production of waste biological solids	– Anaerobic digestion is a rather sensitive process, e.g., the presence of specific compounds, such as $CHCl_2$, CCl_4 and CN-
– Waste biological sludge is highly stabilized product that can be easily dewatered	– Relatively long period of time are required to start up the process when using digested sludge.
– Low nutrient requirements	– Anaerobic digestion is essentially a pretreatment method; an adequate post treatment is usually required before the effluent can be discharged into receiving waters
– No energy requirement for action	– Anaerobic treatment only seems economically feasible when the influence COD-concentration is high and the temperature is 33 to 37°C

Benefits	Limitations
– Production of methane, which is useful end product	– Little practical experience has been gained with the application of process to the direct treatment of wastewater
– Very high loading rates can be applied under favourable conditions	
– Active anaerobic sludge can be preserved unfed for many months	

The anaerobic formation is also helpful in managing hazardous wastes because it results from three main following needs:

1. To utilize renewable resources for chemicals and fuels.
2. To improve energy conservation in waste.
3. To balance economic and political dependence on petroleum, this should be conserved for synthesis of higher molecular weight products that can only be produced from renewable resources of too high cost.

Land Treatment

Land treatment is the biological option most widely used to treat hazardous wastes. Many of the industrial waste treatment sites using biological treatment are at petroleum refineries. The steps in the process typically are: spread the waste on the land, let it dry, till the soil to mix the waste in control moisture, and if necessary and nutrients so that the waste-destroying bacteria will grow.

In a land treatment system, bottom liners under the treatment zone are generally not needed physical, chemical and biological processes in the soil are relied on to degrade, immobilize or transform the wastes to the environmentally acceptable forms. Most pollutants are captured and transformed in the top 6-12 inches (15-30 cm) in the soil, though the treatment zone may extend down 5 feet.

Land treatment is suitable with wastes such as petroleum refinery sludge, creosote sludge and wastewaters, and processing sludge from wood, paper and textile manufacture. It is also widely used for municipal sludge and wastewaters, and food processing sludge and wastewaters. Oil, metals and other constituents of environmental concern are successfully controlled by land treatment.

The process has been encounteringly effective, as shown by polycyclic aromatic hydrocarbon removals of 90% after 193 days of treatment. Total cost, including excavation of the waste and construction and operation of the facility, was approximately $ 80/cu.yd. This cost is much less than incineration, the only other in situ treatment technique that completely eliminated the hazardous compound.

Composting

Composting uses less space than land treatment and it controls gases and leach ate. In composting, piles of the waste 3-6 ft high are treated. Aeration is provided either b turning the piles mechanically or through a forced aeration system. Bulking agents such as wood clips are sometimes added to facilitate mixing and oxygen transfer. These bulking agents are typically screened out at completion of composting, and mixed in with the next batch.

Composting has been used mostly in treating municipal solid wastes. These include industrial wastewater treatment sludge, food processing wastes and some industrial wastes containing low levels of pesticides.

A pilot scale bacteria based composting system has successfully been used for treating diesel fuel contaminated soils. Aeration is provided by an air compressor that draws a vacuum at the bottom of the pile and returns the air added as necessary. Nutrients and water are provided through an irrigation system located at the top of the pile. The system is a fully enclosed vessel that is batch loaded. Studies to date have shown that the diesel contaminants have half-lives in the range of 30-60 days.

Liquid / Solids Treatment System (LSTS)

These are hybrids, intermediate between land treatment and conventional water-suspended biological systems. The waste is not in a solid form as in land treatment, nor in water is in conventional municipal wastewater treatment, but half way between. The wastes are in a suspended solid, slurry or sludge form.

LSTS systems reduce the level of contaminants in a waste primarily by:

1. **Dissolving the organics:** The key step, dissolving the wastes from solid to liquid, can result from microbial action and/or physical/chemical action.
2. **Biodegrading** the dissolved organics into less toxic and less environmentally significant forms.
3. **Releasing the products as gases** to the atmosphere.

There are two basic LSTS process types:

Single-Reactor and Two-Stage: In either case it is vital to have sufficient mixing of the liquid/solid mixture to achieve effective mass transfer of both organics in the liquid phase, adequate oxygen transfer to the microorganisms, and to keep the suspension of solids in the solution. The single batch reactor is the simpler process and the most widely used. Here both the dissolving and biodegradation steps occur together in a single tank, pit or lagoon. Air is introduced to provide mixing for the extraction process and oxygen for biological reasons. Aeration and mixing may be provided by submerged aerators, floating aerators or compressor/Spurger (lance) systems.

At completion of treatment, the aerators and mixing are stopped and the solids allowed settling. The treated liquid is decanted off the top, while the treated solids are treated further, by conventional land treatment or by disposal on-site. If the process is done in a lagoon or surface impoundment, it may be possible to decant the liquid and leave the solids in place.

LSTS have the potential to achieve the treatment goals in a shorter time using less land area than other on-site biological management approaches, such as land treatment and composting. The process uses tank reactors, and thus may eliminate regulatory obstacles or concerns related to the possible land-treatment ban or to contaminant migration in the groundwater.

Soil Bio-Filters

Hazardous wastes in gaseous form can be biologically treated too. On-site bio-filter can reduce nuisance odours, volatile organic compounds and particulates in waste gases. The soil bio-filter involves two mechanisms: mechanical filtration and treatment by bacteria on the soil particles.

Bio-filter consists of a distributing pipe network underneath soil or gravel. The gas to be treated is pumped through the media and escape through holes in the pipe wall. It rises through the media bed which is usually 3-1-0 feet deep. Pollutants captured on the bio-filter can be removed by filtration, physical and chemical absorption, chemical oxidation and structural treatment.

The technique is not a new one, filters have been used for many years to remove ordours from agricultural, food and wastewater processing operations, and emissions from paper and chemical manufacturing plants. The soil filters have successfully removed low concentrations of H_2S, SO_2, NH_3 and NO_2 as well as particulars in these gases. Other potential applications include treatment of exhaust gases from air stripping units for wastewater and solvents, from mechanical technologies such as belt presses for oily sludge, from aerated waste treatment systems and form in situ soil stripping of **volatile organic compounds (VOCs)** using vacuum and forced aeration methods. A soil bio-filter is being planed for the treatment of VOCs being removed from contaminated soil and groundwater in New Jersey. Natural gas will be introduced with the contaminated air stream to stimulate growth of the needed methane oxidizing bacteria.

Alternative Efforts for Bio-Conservation of Environment

The Use of Biogas in Diesel Engines

The technology of the treatment of organic wastes has lately had an almost explosive growth. This has been initially due to the increase in the environmental conscience (late 1960s and early 1970s), and in a second stage due to the energy crisis.

On the other hand, the biotechnology itself has had a very important development, partly because of the very same reasons, but also because of the better knowledge of the biochemistry and genetics of microorganisms. This way, new processes have been developed and old processes have been improved. Special importance has been acquiring the anaerobic processes, such as sewage water treatments, biological treatments of industrial wastes, industrialized landfills, anaerobic fermentation of agricultural wastes, and others. Their particular advantage is that an energetic (biogas) is produced and its use as a fuel may become in many cases the main reason for the implementation of the process. Depending on the particular application of the process, a wide range of technological complexity of plants exists. Some are ready for or are already in production, while others are still in the developing stage.

Biogas has use as an energetic for direct combustion in heating applications and also for the running of internal combustion engines in replacement of traditional fuels. In the latter sense, the firing of Otto engines with gas, whether it were biogas, natural or producer gas has had a long history and has been already developed application, under the name of Gas Engines. However, when it comes to the running of diesel engines with methane or directly with biogas, there has been very little done due to the ignition problems which are involved. This then, is the main subject of this work.

Methane Production by Continuous Digestion of Farm Wastes

Bacterial methanogenesis is ubiquitous in most anaerobic environments. The biogenesis of methane has been known since the 18th century. The first attempts of production and utilization of this gas evolving from decaying organic matter are also very old. On other hand, the fundamental investigations concerning this phenomenon are, in this case, relatively new.

The development of biogas as an energy supply has only been conducted within special circumstances. For example, it has been done when for economical, historical, or geographical reasons, when other energy sources were lacking. In the same way it is a conjunction of such factors which triggered a rising interest for biogas at the beginning of the 1970s.

Realistic evaluations based on economic and scientific data greatly reduce the enthusiasm shown more than a decade ago. They also allowed the determination of the main application fields for the different processes of anaerobic digestion producing methane rich biogas.

The spreading of these processes is slowed by some of their common features:

1. The energy produced has a low density.
2. The substrates are geographically dispersed.
3. Plants and equipments are relatively expensive, (investment/price of the energy produced ratio is high).
4. Difficult checking proper to every high scale biological process (great number of parameters to be undertaken).

For these reasons anaerobic digestion tends to limited to the valorization of various types of wastes and to the improvement of existing processes of pollution treatment.

Thermophilic anaerobic digestion of agricultural and urban wastes is possible. The aim is to find out the gradual setup of a process working at relatively high temperatures (55 to 60°C) with high solids substrates.

Agricultural and urban wastes have high solids contents (20 to 40%) because most of cellulosic wastes have a dry content high above 20%. The digestion of such wastes in conventional systems requires a previous dilution to reach values around 10 per cent. This will induce an increase in the size of plants treating a given bulk of wastes and will setup an important problem for the handling and disposal of the resulting wastewater. Also the equipments and techniques for the handling of semisolid wastes are very similar to those utilized for raw urban wastes. So it would be possible to avoid the purchases of some expensive equipment and to shorten the personnel training when implementing an anaerobic digestion unit within an existing treatment plant (receiving urban wastes for example).

The benefits of on-farm application of the anaerobic digestion of livestock manure and crop residues are possible, however, there is a high initial investment for construction coupled with a long pay back period and the need for skilled operator expertise to prevent reactor failure due to the process variability inherent in a farming situation. The technique is based on methane production from the fermentation of organic materials requires the concerted action of a symbiotic population of facultative and obligate bacteria. The symbiotic bacterial population could be conveniently divided into two major groups **acid forming** bacteria and **methane forming** bacteria. The group that decomposed the initial organic substrate to produce a mixture of volatile fatty acids, hydrogen, ammonia, carbon dioxide, and less significant quantities of other metabolic by-product were classified as acid forming bacteria. Methane forming bacterial referred to the groups that converted substrates such as acetic acid, hydrogen, and carbon dioxide to methane.

There are three major groups of symbiotic bacteria:

1. **The First Group,** called fermentative or **acidogenic bacteria,** degrade carbohydrates (largely cellulose) lipids, and proteins to form alcohols, organic acids (lactate, succinate, etc.), and long chain fatty acids.
2. **The Second Group,** called **acetogenic bacteria,** produce acetate, carbon dioxide, and hydrogen from the end products of the first stage.
3. **Methanogenic bacteria** produce methane and carbon methane and carbon dioxide via mainly the following two reactions. Acetic acid fermentation.

$$CH_3COOH \rightarrow CH_4 + 2H_2O$$

And CO_2 reduction:

$$CO_2 + 4H_2 \rightarrow CH_4 + 2H_2O$$

The acidogenic and acetogenic bacterial groups are the most prolific, while the methanogene have the slowest growth rate. A healthy population of methanogenic bacteria prevents the accumulation of volatile acids that could lead to digester failure. A balanced state of dynamic equilibrium among the anaerobic groups can be most easily maintained within a controlled environment. A wide variety of physical and biochemical parameters influence the types and members of bacteria that constitute stable methane fermentation. Environmental control for the continuous generation of methane gas required that parameters such as the concentration of microbial nutrients, pH, substrate concentration, the numbers of viable microorganisms, and the total volatile acid content be maintaining within acceptable limits.

Anaerobic Digestion Applied to Livestock Wastes

The benefits of anaerobic digestion for on-farm applications are as follows:

1. The effluent form anaerobic digestion has greatly reduced solids content, less offensive odours, and fewer handling problems.
2. Fermentation results in an effluent with lower solids content than aerobic digestion since more of the organic material in converted to gas and less to cell mass.
3. Approximately 5% of the degraded organic waste material is converted to biological solids during anaerobic digestion, while most of the remainder is converted to biogas.
4. In comparison, the biomass yield from aerobic digestion is in the range of 50% of the mass of converted organic substrate.
5. It has been frequently demonstrated that biological oxygen demand (BOD) reductions greater than 80% are achievable from anaerobic digestion of livestock manures.

The value of anaerobic digester effluent compares favourably with raw manure as a crop fertilizer because the inorganic nutrients (minerals) pass through the digester un-depleted. The amount of nitrogen available for plant assimilation is increased by the conversion of protein nitrogen to ammonia. However, if the effluent is stored before use, the ammonia may be subject to volatilization to the atmosphere. Pathogenic bacteria such as *Salmonella sp.* and *Brucella sp.* are destroyed during anaerobic fermentation, and livestock can be grazed on pastures that have been spread with digester sooner than would be acceptable if raw manure was used. A major advantage of anaerobic digestion is the production of biogas that is typically about 60% methane with a fuel value of 2.24×10^7 l/m^3. For example, a 450 kg dairy produces approximately 3.9 kg of volatile solids as waste per day. From this, a methane yield in the range of 0.36 to 0.8 l m^3, equivalent to the fuel value of 0.2 to 0.51 of diesel oil, could be expected.

Anaerobic vs. *Aerobic Biotechnology*

Anaerobic digestion is becoming an important and attractive process for waste treatment as a result of a large research effort devoted to the development of the understanding of its fundamentals in the last few years, supported by significative advances in the process engineering aspects of anaerobic reactors. The field of potential applications of anaerobic biotechnology in waste treatment is spreading fast, in many cases it has become an economic alternative to conventional aerobic treatment process. The fact that almost any organic substance that can be aerobically degraded can also be so anaerobically, has been important in providing such interest in anaerobic processes.

Some of the salient feature of anaerobic process may be listed as follows:

Advantages of Anaerobic Treatment

1. Low production of stabilized sludge (0.04 to 0.14 kg sludge/kg COD removed, as compared 0.3 to 0.5 sludge/kg COD removed in aerobic process).
2. Possibility of recovering 80 to 90 per cent of the energy present in the metabolized waste as methane gas (which may be used as fuel).
3. Low consumption of nutrients.
4. Lower energy requirements than aerobic treatment (no need for aeration and smaller amounts of sludge for dewatering and disposal).
5. High degree of waste stabilization (about 90%) compatible with very high organic loadings.
6. Well acclimated anaerobic reactors can be preserved unfed for periods up to 1 year without any appreciable deterioration in sludge properties.
7. Anaerobiosis sharply enhances public health quality of the wastes by inactivating pathogenic bacteria, as well as some viruses and parasites, when properly conducted no nutrient odor is produced.

Limitations of Anaerobic Treatment

1. Great sensibility to some specific compounds such as $CHCl_3$, CCl_4 and CN-, which have been found to be toxic.
2. Due to slow growth rate of the anaerobic bacteria the first set up may take several months, although a period of 8 to 12 weeks is possible when inoculation with digested sewage is carried out.
3. A rather high temperature (30 to 40°C) is desirable to reach high conversion rates.
4. Anaerobic digestion is essentially a pretreatment method.
5. It has been estimated that anaerobic digestion offers a potential net energy saving of about 2.3 x 104 kJ/kg.

Fixed-Film* vs. *Suspended Growth Reactor

The first reactors used in anaerobic treatment of domestic sewage sludge were of the completely mixed or the CSTR type. For these reactors, the hydraulic retention time is similar to the solids retention time (or sludge age). As the anaerobic bacteria growth rates are slow, these reactors required high hydraulic retention times and consequently high reactor volumes for anaerobic digestion of low strength wastes, which resulted in an economically unfeasible reactor for large quantities of wastes.

One way to overcome this problem is to increase biomass retention time inside the reactor, either by creating inside the reactor condition promoting natural biomass tendency to originate "granulating" flocs with enough size to be speared by setting and recycled to the reactor, or by introducing in the reactor a solid support where on the bacteria could attach and create a film that could be kept inside the reactor.

In either case, hydraulic retention time can now be much smaller then the solids retentions time in these fixed film reactors. Using fixed film reactor waste may be effectively treated in hours, as compared with several days for suspended growth.

The increased biomass retention time not only digester volume and corresponding capital costs, but also allows methane production rates per unit of reactor volume 2 to 10 times greater than those reached with suspended growth reactors, thus increasing their utilization for energy production.

Other advantages of fixed compared to suspended growth reactors include susceptibility to washout, higher tolerance to shock loading, capacity to sustain active microbial culture even after periods of starvation and facility to withstand considerably higher transient and chronic concentrations of toxics.

As in the past, suspended growth reactors are mainly used for concentrated wastes with an appreciable suspended solids content, while fixed-film reactors are more appropriate for the digestion of soluble low and high strength wastes; the introduction of leachate types of digesters for solid wastes of high solid contents complicates the distinction made above but increases the potential use of anaerobic digestion.

Biogas from Sanitary Landfill Technique in Egypt

The Cairo area generates about 4300 ton/day (tpd) municipal solid waste (MSW), about 1500 tpd are collected and recycled by private garbage collectors for feeding pigs and preparing organic manure at very severe conditions.

The garbage was collected from three big vegetable markets of Cairo, dried and ground. The chemical composition of the ground garbage was as follows: moisture content, 10,02%; volatile solids, 78.27%; organic carbon, 45.397%; total nitrogen content, 0.945%; and C?N ratio, 49.61. Similarly sewage sludge was collected from the biggest sewage treatment plant of Cairo at Zenin. The chemical analysis was as follows: total solids, 2.269%; volatile solids, 68.089%; organic carbon, 39.492%; total nitrogen, 1.981%; and C/N ratio, 19.935. The remaining 2800 tpd were collected by governmental systems to be burned in several dumping places thus losing energy, fertilizers and soil conditioners as well. Air pollution is also one of the threats. The sanitary landfills combined with production of high Btu gas and soil conditioners could be a good solution for the increasing MSW accumulated in Cairo and other cities of Egypt. The fermentation of city garbage in combination with water or with sewage sludge at different levels – 10.25 and 35% TS (Total solids) were conducted in Cairo. Fermentation took place in 1.25 and 3.5 l fermentors. Some of the fermenters were kept at 35 and 55°C. The experiments lasted for 4 to 6 months. Gas volume, methane content, pH, soluble-N, and contents of total and volatile solids were estimated regularly. Coliform and Salmonella groups (pathogenic bacteria) are anaerobic acid former and cellulose decomposers and help in fermentation of garbage and sewage in cities of Egypt.

Reactor Treating Paper and Board Mill Effluent

During the last decade great progress has been made on anaerobic wastewater treatment. Different reactor types have been developed and compared to aerobic methods, the advantages are numerous. Until recently almost all applications and principle for anaerobic wastewater treatment were used for food processing industries. In 1983 reactor technique was started for paper industry.

Anaerobic vs. Aerobic Treatment: The anaerobic method of waste treatment offers a number of significant advantages with few serious or limitations of the process. At the present state of process technology, there are few drawbacks of the anaerobic process left and all of its benefits are still valid, as well be shown in this paper and others.

Anaerobic digestion of complex organic matter to methane consists of four successive steps:

1. **Hydrolysis:** Where organic polymers are hydrolyzed to their individual monomers as a result of enzymatic attack.
2. **Acid formation:** The hydrolyzed compounds are converted intra-cellulary by a group of acid forming bacteria to simple compounds, such as volatile fatty acids, alcohols, CO_2, NH_3, and H_2.
3. **Acetogenesis:** The products of the acid formation are transformed to acetic acid, CO_2, NH_2, and H_2.

4. **Methane formation:** Acetic acid, CO_2 + H_2 and methanol to CH_4 and CO_2 by the strictly anaerobic methane forming bacteria.

Depending on the composition of the wastewater, each of the four steps can be rate limiting.

When the wastewater contains a considerable fraction complex (un-dissolved particles), the hydrolysis can be rate limiting; when the pollution consists mainly of acetic acid and other volatile fatty acids, the methane formation can be the rate limiting step.

The essential feature of the **anaerobic contact process** is that the washout of the active anaerobic bacteria mass from the reactor is controlled by a sludge separation and recycles system. The major problem in the practical application of the contact process has always been the separation and concentration of the sludge from the purified effluent. Several methods have been used for this purpose: i.e., **plain sedimentation, setting combined with chemical flocculation or vacuum degasification, floatation,** and **centrifugation**. The digester should be mixed thoroughly by gas recirculation, or mechanical agitation.

Bio-Energy Recovery and Conservation in Israeli Agriculture – Anaerobic Digestion of Cotton Stalks

Israeli agriculture is probably one of the most energy intensive industries in Israel and in World agriculture. Its increasing productivity and development has involved increased energy use. Despite its energy intensity, Israel has managed to utilize its natural resources for agriculture. For instance, the NEFAH project – the utilization of agricultural wastes and biomass for production of energy (biogas), animal food supplement, substrates for crops and plants and industrial products (fibre boards) by the anaerobic digestion process – is unique in its comprehensive approach and is beginning to be widely used in Israeli Kibbutzim.

In Israel over 10% of the total national energy consumption is used for agricultural production, and only about 2.5% in the form of direct energy input. The profitability of many crops depends very much on the prices of fuel and its availability. The high energy intensity of Israeli agriculture is mainly due to its water needs and the constant growing of the agricultural production for export, mainly to the European market. Therefore, reduction in energy consumption and use of alternative energy sources are in very intensive sources are in very intensive development in Israel.

Two ways to ensure that agriculture has the energy that is needs are:

1. Substitution of plentiful energy agricultural users with energy sources.
2. Conservation and more efficient energy management.

Conservation of Energy in Greenhouses

Air shipment, greenhouses heating, artificial lighting, and fertilizers makes this exported crop very energy dependent. More than half of the production cost of cut flowers is for energy. Therefore, reduction in energy consumption and use of alternative cheaper energy sources are of main goals in the Israeli agriculture.

The energetic aspects of growing plants under covering (**greenhouse and plastic covers**) are important. It is clear that the plant does not need energy of night. Greenhouses need energy to maintain temperature in winter. During the summer days greenhouses need energy to lower the temperature. Therefore, most new technologies for new structures of greenhouse or new ways of heating are based on insulating the greenhouse, collecting the surplus heat during the day, and releasing it during the night or even during some of the winter days.

In Israel, anaerobic digestion of cotton stalks and several new technologies have been examined, including closed greenhouse in which air chances are limited, uses of hygroscopic salts, developing heat exchanges light/air on the roof, and the hydro-solaric greenhouse system.

Conservation of Energy in Animal Husbandry

Until a few years ago not much was done by Israeli agricultural section to reduce the negative impact of energy soaring prices on the farming community. Because of the special nature of the energy consumption pattern of Israeli animal husbandry, not much direct energy can be saved. Most energy saving will be achieved through better use of inputs. The following are some examples of energy saving:

1. Instruments for better control of the temperature inside poultry houses, and using the heat of the chickens themselves in a more efficient way.
2. Reducing energy use by better design of equipment.
3. Reducing energy demands for hot water by using waste heat from compressors of the milking system.
4. Reducing energy use of water pumping which is the main sector consuming energy in Israeli agriculture.
5. Reducing energy consumption in fish ponds by reducing aeration.
6. Reducing indirect energy of fertilizers, pesticides and irrigation system operations.

Alternatives to Current Pesticides Uses

One can avoid using toxic pesticides or lessen their environmental and human health impact. In many cases, improved management programmes can cut pesticide use between 50 and 90 per cent without reducing crop production or creating new diseases. Some of these techniques are relatively simple and save money while maintaining diseases control and yielding crops with just as high quality and quantity as we get with current methods.

Behavioural Changes

Crop rotation keeps pest populations from building up For instance, a soybean-corn rotation is effective and economical against white-fringed weevils. Mechanical cultivation can substrate for herbicides. Flooding fields before planting or burning crop residues and replanting with a cover crop can suppress both weeds and insect pests. Habitat diversification, such as restoring wind breaks, hedgerows and ground cover on water courses, not only presents soil erosion but also provides perch areas and nesting space for birds and other predators that eat insect pests. Growing crops in areas where pests are absent makes good sense. Adjusting planting times can avoid per outbreaks, while switching from huge monoculture fields to mixed polyculture (many crops grown together) makes it more difficult for pests to find the crops they like. Tillage at the night time can greatly reduce pest populations. For instances, spring or fall plowing can help control over wintering com earthworms.

An important behavioural adjustment can occur in our attitudes and performances. Farmers can be persuaded that the bare earth look in row crops made possible by powerful herbicides is not good for the soil or for themselves. Allowing a few weeds to creep in or planting a cover crop between rows may make more sense in the long run. Consumers may have to learn to accept fruits and vegetables are less than perfect.

Biological Controls

Biological controls such as predators (wasps, ladybugs, praying mantises); or pathogens (viruses, bacteria, fungi) can control many pests more cheaply and safely than broad spectrum, synthetic chemicals. *Bacillus thuringirnsis* or Bt. For example, is a naturally occurring bacterium that kills the larvae of Lepidoptera (butterfly and moth) species but is harmless to mammals. A number of important insect pests such as tomato hornworm, corn rootworm, cabbage loopers, and others can be controlled by spraying bacteria on crops. Larger species are effective as well. Ducks, chickens and geese, among other species, are used to rid fields of both insect pests and weeds. These biological organisms are self reproducing and often have wide prey tolerance. A few mantises or ladybugs released in your garden in the spring well keep well keep producing offspring and protect your fruits and vegetables against a multitude of pests for the whole growing season. Plants too, can be effective weapons against insect pests. Home gardeners often plant a border of insect repelling plants, such as garlic or marigolds, around their crops. On a larger scale, farmers might also benefit from this approach.

Herbivorous insects also have been used to control weeds. For example, the prickly pear cactus was introduced to Australia about 150 years ago as an ornamental plant. This hardy cactus escaped from garden and found an ideal home in the dry soils of the outbreak. It quickly established huge, dense stands that dominated 25 million ha (more than 60 million ac) or grazing land. A natural predator from South America, the cactoblastis moth, was introduced into Australia in 1935 to combat the prickly pear. Within a few years, cactoblastis larvae had eaten so many prickly pears that the cactus has become rare and are no longer economically important.

Genetics and bioengineering can help in our war against pests. Traditional farmer have long known to save seeds of disease resistant crop plants or to breed livestock that tolerate pest well. Modern science can speed up this process through selection regimes or by using biotechnology to transfer genes between closely related or even totally unrelated species. There are concerns about the ecological dangers of moving too fast in this new area. Releasing sterile males to interfere with insect pest reproduction has been used successfully in some cases. Screwworms, for example, are the flash –eating larvae of flies that lay their eggs in scratches or skin wounds of livestock. They were a terrible problem for ranchers in Taxes and Florida in the 1950s, but release of massive numbers of radiation-sterilized males reproduction and eliminated this pest in Florida. In Taxes, where flies continue to cross the border from Mexico, control has been more difficult, but continual vigilance keeps the problems manageable.

Other promising approaches are to use hormones that upset development or sex attractions to bait traps containing toxic pesticides. Many municipalities control mosquitoes with these techniques rather than aerial spraying of insecticides because of worries about effects on human health. Briquettes saturated with insect juvenile hormone are scattered in wetlands where mosquitoes juvenile hormone are scattered in wetlands where mosquitoes breed. The presence of even minute amounts of this hormone prevents larvae from ever turning into biting adults. Unfortunately, many beneficial insects as well as noxious ones are affected by the hormone, and birds or amphibians that eat insects may be adversely affected when their food supply is reduced. Some communities that formerly controlled mosquitoes have abandoned these programs, believing that having naturally wetlands is worth getting a few bites in the summer.

Integrated Pest Management (IPM)

IPM is a flexible, ecologically based pest control strategy that uses a combination to techniques applied at specific times, aimed at specific crops and pests. It often used mechanical cultivation and techniques such as vacuuming bugs off crops as an alternative to chemical application. IPM does not give up chemical pest controls entirely but rather tries to use the minimum amount necessary only as a last resort and avoids broad spectrum, ecologically disruptive products. IPM relies on preventive practices that encourage growth and diversity of beneficial organisms and enhance plant defenses and vigor. Careful scientific monitoring of pest populations to determine economic thresholds, the point at which potential economic damage justifies pest control expenditures, and the precise time, type, and method of pesticide application is critical in IPM.

The trap crops, small areas planted a week ot two earlier than the main crop, are also useful. This plot matures before the rest of the field and attracts pests away from other plants. The trap crop then is sprayed heavily with enough pesticides so that no pests are likely to escape. The trap crop is destroyed so that workers will not be exposed to the pesticide and consumers will not be at risk. The rest of the field should be mostly free of both pests and diseases.

A successful IPM program that could serve as a model for other countries is found in Indonesia, where brown plant hoppers developed resistance to virtually every insecticide and threatened the country's hard-won self-sufficiency in rice.

Reducing Pesticide Exposure

The total numbers and amounts of different chemicals to which we are potentially exposed is overwhelming. The one-half million metric tons of pesticides used in the United States every year contain about 600 active ingredients combined with more than 1200 presumably inactive carriers, solvents, preservatives, and other ingredients in about 25,000 commercial products. Less than 10 per cent of the active pesticide ingredients have started only recently. Since 1972, only forty pesticides have been banned. Critics change that this slow pace puts all of us at risk and the assessment and regulation must proceed more quickly.

Regulating Pesticide

Three federal agencies share responsibility for regulating pesticides, used in food production in the United States; the Environment Protection Agency EPA), the Food and Drug Administration (FDA) and the Department of Agriculture (USDA). The EPA regulates the sale an use of pesticides under the Federal Insecticide, Fungicide, and Rodenticide Act (FIFRA), which mandates the "registration" (licensing) of all pesticide products. Based on scientific studies, the EPA determines which pesticides will not pose significant risks to human health or the environment when used according to label directions. Under the Federal Food, Drug, and Cosmetic Act (FFDCA), the EPA also sets "tolerance levels" or limits for the amount of pesticide reduces that lawfully may remain in or on foods marketed in the United States, whether grown domestically or imported from abroad. The FDA and USDA enforce pesticide use and tolerance levels set by the EPA. These agencies can size and destroy food shipments found to contain pesticide residues in violation of limits set by the EPA.

Non-chemical pest control methods, on-farm education, and biotechnology and the development of pest resistant crop varieties are all planned to play a role in this program. Savings on chemical costs are expected to be about $ 100 million per year.

We need special standards for children because of their higher susceptibility to toxins and special dietary needs. Up to a tenfold margin of safety (i.e., a one-in-ten standard) will be applied to foods making up a major part of children's diets. We should also study and consider endocrine hormone-disrupting effects of pesticides in setting tolerance standards and should publish consumer information pamphlets explaining the risks and benefits of pesticides food that may exceed a one-in-million risk, and tips for reducing pesticide exposure.

Personal Safety

What can we do personally to reduce pesticide use and to protect ourselves from pesticide exposure? Following practices are suggested for alternative methods of pest control and can be used at home too.

1. Plant ground cover in your yard that can compete successfully with weeds. Learn to appreciate a mixture of different plants rather than striving for a perfect, lush carpet of grass. What is wrong with a few dandelions, anyway?
2. Install or repair screens on doors and windows to keep out insects. Caulk around window, doors, and baseboards to seal routes of entry for ants, roaches, mice, and other pests. Put a few bay leaves in containers of flour or cereal to discourage flour weevils.
3. Clean up spilled food and empty garbage regularly to eliminate food for ants or roaches. Keep food in tightly sealed containers. Sprinkle borax around drains and ducts where pests may be lurking. Use sticky flypaper to catch flies to catch flies rather than spray dangerous chemically inside your house of apartment.
4. Get rid of aphids, scales, mites, and other houseplant pests by washing leaves and stems individually with rubbing alcohol, or spray plants with a dilute solution of dish soap in water. Simply washing vegetation with pure water will remove many pests and improve the health and appearance of you plants.
5. A saucer of stale beer in your garden will attract and drown sludge. If you have only a few worms or bugs, pick them off by hand. Purchase some predators such as ladybugs or praying mantises to protect garden plants. If you have a lot of bugs in your house or apartment, a praying mantis can make an interesting house pet.
6. Drain stagnant water in or near your yard that might serve as breeding sites for mosquitoes. Pick up dead branches, fallen leaves and other liter that provide a haven for termites and other pests. Put up houses for bug-cating purple martins or bats.
7. Accept slightly blemished fruits and vegetables. If your garden, trim away bad parts or give up a portion of your crop rather than saturate the environment with dangerous chemicals. Interplant insect-repellent plants such as marigolds, basil, peppermint or single together with your sensitive garden plants.
8. If you must toxic chemical pesticides, use them in the smallest possible amounts and apply them only where and when it is necessary. Rather than spray chemicals, use trap or application that limit dispersal into the environment.
9. Read Organic Gardening and Farming reference books which to find out alternative approaches to pest control. It should be noted that "natural" or organic does not necessarily make a chemical good, no does being synthetic necessarily make one bad. Some natural products such as nicotine are highly toxic. Others such as mercury and lead are just as likely to bio-concentrate and persist in the environment, as are synthetic compounds.

"We must not wait for things to come. If we want it, we must do something about it"

Erwin Schrodinger

17 Systems of Crop Production

"What we to cure is not crisis but a chronic malady and the remedies for it are of routine rather than of a spectacular nature".
Professor M. L. Dantwala

The relationship between man and his environment is hardly ever a static or 'closed' system and if so only for a limited time, because sooner or later human ingenuity or mistakes, natural disasters or chance discoveries such as rice as a food crop will bring about changes from within the ecosystem. It may be altered by external influences, by new ideas, techniques and implements or by exotic plants such as potato and groundnut.

Food is the basic requirement of man. The successful domestication of plants, animals, soil and water by settled human communities has been the starting point of culture. Tropical Asia provides a typical example. Here the predominant pattern has been one of almost uninhabitable uplands side by side with heavily populated lowlands where the cultivation of rice as a staple food crop has developed over the past five millennia supported by irrigation under monsoonal, seasonally dry conditions.

HUMAN SETTLEMENT AND LAND USE

The basic spatial pattern of population and land use as described in early colonial times in tropical Asia must have existed for a long time over five thousand years depending on the geography and climate of a location. For example in peninsular India, three main spatial units had evolved whose general features have persisted until date.

The most densely populated and cultivated regions were and still are the coastal and riverine alluvial plains, mainly on account of their high potential of continuous irrigated rice cultivation. In the Western Ghats, abundant rainfall during the major part of the year and unfailing water supply along numerous mountain streams even during the drought period seemed to intensify multiple cropping of rice, the staple food crop. In the adjoining latertitic high lands, luxuriant growth of coconut and areca nut, palms, tapioca, banana, spices and condiments are seen. In the East, where the rainfall is low, thousands of tanks have been constructed to store the rain water for extensive stretches of rice fields. Whenever irrigation is not possible as in the case of uplands, a variety of millets, pulses, oilseeds and fibre crops are grown under rainfed conditions.

In the hilly areas, the widespread land use of shifting cultivation of the past has been given place to permanent intensive agriculture of plantation crops such as tea, coffee, rubber, cardamom, etc., and the growing of profitable vegetable crops like potato, cabbage, cauliflower, carrot, pea and beans apart from fruit trees, like pears and plums. At present different cropping systems are intensively practiced in many parts of the world.

SYSTEMS OF CROPPING

Sun light, soil, water, plants, animals and human beings constitute the basic resources of agriculture. In tropics, due to sun light available throughout the year, more than two crops can be easily raised if other resources are available. Growing of more than two crops in the same field in sequence in a year is called **multiple cropping.**

In many instances, the seeds/plants of second crop are sown /planted before the harvest of the first crop. Similarly, the seeds/plants of third crop are sown/planted before the harvest of the second crop. Growing of more than two such crops in the same field in a year is called **relay cropping**.

It is a common practice, to raise two or more crops in the same field in the same season particularly under rainfed agriculture. The seeds may be mixed sown or sown separately either in paired rows or alternate rows or in different proportions of rows depending on the need of the farmers and successful growth of the mixed crops. This practice is called **mixed cropping** if seeds are sown mixed and **inter-cropping** if raised in rows.

A combination of these different cropping methods is called **crop-cafeteria** wherein every available space is used for growing crops in the same field over time.

In case of plantation crops, three or more crops are raised simultaneously in a mixed pattern with different heights and canopy to take advantage of the sunlight. Growing together of coconut, areca nut, pepper, banana, coco, pine-apple, fodder grasses etc. are a common practice. This is called **multi-storied cropping** or **multi-tier cropping.**

MULTIPLE AND RELAY CROPPING

The cultivation of three or four crops during a year has become possible in irrigated areas of tropics. The development of high yielding varieties in tuber crops namely potato, sweet potato and tapioca has added a further dimension to yield possibilities. Hundreds of different cropping patterns to suit different soils, climate and locations have been tried in many parts of the country particularly in India. The general approach has been the standardization of various alternative rotation patterns all of which are scientifically sound from the view point of maintaining the long term productivity of the soil but from which the farmer can choose according to his capacity to mobilize inputs, market conditions and seasonal behaviour.

New systems of cropping are rapidly coming into existence in many parts of irrigated areas. In the states of Punjab, Haryana and parts of Uttar Pradesh, rice during *kharif* and wheat during *rabi* are now popular. In Bengal, wheat is popular as a *rabi* crop in rice fields. In the rice fallows of South India, sorghum, bajra (*Pennisetum*) and cotton are raised during the off-season. The following data (Table 17.1) indicate the possibilities.

TABLE 17.1: Yields and total production of three crop annual sequence at various centres during 1970-71

State	Centre	Soil type	Crop sequence			Yield in kg/ha			Grain produced per day Kg/ ha/day
			Kharif	Rabi	Summer	Kharif	Rabi	Summer	
Andhra Pradesh	Tirupathi	Red sandy	Rice (IR 8)	Rice (IR 8)	Rice (Hamsa)	4438	6410	3090	42.9
	Maruteru	Medium Block	Rice (Pankaj)	Rice (Padma)	Mung (Kaveri)	4666	5583	910	34.1
U.P.	Banaras	Alluvial	Rice	Wheat	Rice	3732	5042	4309	39.4
Karnataka	Mangalore	Laterite	Rice (MTU 20)	Rice (Padma)	Rice (Padma)	2450	1770	4570	34.3
Tamil Nadu	Bhabanisagar	Red loam	Rice (Adt 27)	Rice (IR 8)	Rice (Co 29)	3643	3908	5106	35.6
Orissa	Bhubaneswar	Laterite	Rice (Padma)	Rice (T 141)	Rice (Padma)	2648	1970	3127	28.8
			Jute (JRO 212)	Rice (T 141)	Rice (Padma)	1738	2055	4485	33.7

Source: Annual Progress Report of All India Co-ordinated Agronomy Experiment (1969-71).

Potentials

Experiments conducted in various research stations have indicated that more than 20 tonnes per hectare of grain per year can be produced. The continuous rice cropping at Tamil Nadu Agricultural University, Coimbatore had produced 22 tonnes per hectare with rice IR 8 over a period of one year with three crops. Similar results have been obtained at Inter National Rice Research Institute, Manila. The principles used are the improvement of harvest index (proportion of economic yield to the total biological yield) as well as the ability of crops to respond to good management through restructuring of plant hybrid vigour and the enhancement of per day productivity through the introduction of photo-insensitive varieties thereby facilitating multiple cropping.

Multiple cropping provides an effective agricultural method to harvest solar energy. Green plants can be made to harness sunlight throughout the year, wherever there is water to sustain crop growth. There is vast potential for using solar energy in arid and semi-arid areas for household purposes thereby saving valuable organic matter to soil.

Advantages

Multiple and relay cropping is a very potent tool for improving the economic position of a farmer with a small holding and for banishing unemployment and underemployment. Some of the beneficial effects of the technology are the following:

1. Increase of yield and income.
2. The employment potential of agriculture can be greatly enhanced.
3. There is scope for better utilization of resources such as animals and machineries, irrigation water etc.
4. Better advantage of market particularly in case of horticultural crops.
5. There is scope for self-sufficiency in meeting the needs of animals for feeds and fodders.
6. There is appreciable increase in real income among all classes of farmers big, medium and small as well as landless labour.
7. If properly handled, this may become a powerful method of minimizing the incidence of pests and diseases.

Scope of Multiple Cropping

When crops are grown continuously in sequence and with relay and mixed crop patterns, there is greater scope for reducing the cost of cultivation of crops without adversely affecting yield.

Tillage: When single or double crops are grown in the same field in a year, the fallow period available in between the crops help in soil hardening, weed growth and loss of nutrients due to erosion. This necessitates more number of tillage operations than needed. Experiments on tillage have indicated that the concepts of minimum and zero tillage have reduced the cost of production without affecting the yield. The case of relay cropping such as sowing cotton in the stubbles of finger millet, has clearly demonstrated that there is no need for preparatory cultivation for cotton for successful growth. Similarly in the paired row system of mixed crops, the inter-space is covered by the second crop and helps in controlling weeds and reducing the time needed for after cultural operation. Thus, this technology helps in reducing the cost if tillage operations considerably.

Manures and fertilizers: In the package of practices formulated for any crop, the cost on manures and fertilizers is considerable. Many experiments conducted on cropping system, instead of individual crop, has shown considerable saving on the manorial budget particularly with P and K without affecting the yield. The usage of a leguminous crop in the sequence helps in the reduction of N requirement for the following next crops.

Water: In the irrigated garden land mixed crops, the water loss by evaporation is considerably reduced. The design of irrigation method is changed due to the paired row system and considerable water saved.

Plant Protection: The opportunity for greater weed growth in the multiple, relay and mixed cropping system is considerably reduced due to lesser area and time for its proliferation. Again due to the mixing of crops, the spread of pests and diseases are reduced. These help in the reduction of chemicals needed for their control.

Some Ground Rules for Introduction

1. No two crops sharing in common, the same pests and diseases should be grown either mixed or succession.
2. One crop should be deep rooted and another with more shallow roots so that different layers also differ in rooting pattern and if a deep rooted crop such as cotton is grown during *kharif*, a wheat variety such as Sonalika or Sharbathi sonara may suffer less from micronutrient deficiency than Kalyan sona which also taps nutrients from the lower layers of the soil.
3. For the restoration of soil fertility, at least one leguminous crop should find a place in the sequence. If the stalks of the soil would help to improve soil structure.
4. The requirement of different nutrients very among the crop. So also their capacity to absorb them. Hence, it is better to grow crops of different families in sequence.

Multiple Cropping in Farmers Fields

In India up to mid 1960s wheat was grown in the *rabi* season while rice was confined to *kharif* in the northern regions as a single crop in their respective areas depending on soil and rainfall pattern. Even the irrigated wheat was cultivated as a mono-crop after *kharif* fallow. But with the development of high yielding, photo-insensitive and short duration varieties and hybrids of maize, sorghum, pearl millet, wheat and rice, an entirely new era in crop management has been ushered. It was established that there is no need to keep the land fallow. Double and triple cropping has become common in most of the irrigated area where water supply is assured.

Major states growing irrigated wheat in India are Punjab, Haryana, Uttar Pradesh, Bihar and Rajasthan. The new varieties of rice fitted very well in rotations with wheat in Punjab and Haryana replacing maize, pearl millet or pulses, the normal *kharif* season crops (Table 17.2).

TABLE 17.2: Statistics in the growing importance of rice in the traditional wheat growing states

	Haryana		Punjab	
	1971-72	1978-79	1971-72	1978-79
Area (000 ha)	291	463	450	1053
Production (000 t)	536	1241	920	3073
Yield (kg/ha)	1842	2680	2044	2918

Research in multiple cropping has resulted in the development of several crop rotations of different cropping intensity as shown in Table 17.3.

TABLE 17.3: Crop rotations developed

Location	Crop sequence			Yield in q/ha			Total yield per year q/ha
Ludhiana (Punjab)	Rice	Wheat	Mung	67.8	36.3	7.4	111.5
	Maize	Wheat	Mung	34.1	38.6	9.4	82.1
Bichpuri (U.P.)	Pearl millet	Wheat	Mung	32.4	57.6	10.2	100.2
Pantnagar (U.P.)	Maize	Wheat	Mung	43.5	40.7	5.9	90.1
Delhi	Rice	Wheat		50.0	38.0		88.0

Among these 'rice-wheat' double cropping and rice–wheat-green gram triple cropping to a limited extent have become popular with the farmers.

Constraints for Adoption

Although the experiments conducted at research stations have indicated the high profitability of cropping pattern of 300 per cent and above, these are not much popular with the farmers due to the inherent constraints for adoption.

1. The land, water and other farm resources are decidedly inferior to those of Experimental Stations and farms in which experiments are conducted.
2. The farms, being small ranging between 0.5 to 3.0 hectares, needs fencing and protection unless the adjoining farmers also practice the same pattern.
3. Many farmers are share-tenants with limited assets. Hence, high profit may not be an inducement.
4. The family labour and cash flow are limited and as such the farmer may not be in a position to tackle the needs of multiple cropping.
5. The non-farm and other farm activities competed for labour at the critical times.
6. In most of the villages, the contract able labour pools and credit sources are limited.
7. Inputs that can not be purchased locally at farm level costs may not be applied as suggested.
8. Crops that local markets can not absorb at profitable prices may not be adopted.
9. In areas depending on onset of rainfall and water resources, the field operation may not be started on time. In these areas, the risks are very high because of likely water shortage.
10. While quantitative and qualitative resource deficiencies account for part of the gap, socio-economic factors hinder adoption of modern techniques and efficient management.

These constraints indicate the need for systematic cropping systems research both in the experimental stations and the farmers holding as on site research.

CROPPING SYSTEMS RESEARCH

Although the evolution of different cropping patterns is concerned with the agronomists, the adoption by farmers needs a co-ordinated approach involving crop management research and farm management economics research. Besides there is the need for farmer involvement and scrutiny so as to evoke an alternate cropping pattern with economic viability. There is need to consider the following aspects in the research activities.

1. The resource base of farm family and community that governs adoption potential.

2. The research findings should have greater variability in physical and socio-economic environment for wide scale adoption.
3. The indigenous technical knowledge evolved in a location on practical experiences over centuries should be given its due importance when new methodologies are tried.
4. As the research needs in all these activities require a large number of human resources, the limited supply of qualified research personnel have to be taken into account.

NEED FOR ON-SITE RESEARCH

The data generated in the experimental farms may not take into account the variability found in the physical and socio-economic environment of the large number of farmers,. Soil, physiographic and climate variability are more pronounced in tropics. Homogenous, agro-ecological zones are smaller. Hence, there is greater need for land evaluation to assess variability. Similarly, in the socio-economic environment, the farming community can be differentiated into small, marginal, medium and large farmers, illiterate and semi-literate, owner and tenant cultivators etc. To cater to needs of such a clientele in the physical environment, there is need for on-site research, the results of which may provide a basis for judging the following.

1. Technology that is within the management and resource limitation of small and marginal farmer and can be adopted without Government programme.
2. Technology that faces specific resource constraint such as weak market, managerial deficiency, etc. have to be modified to meet the needs of the farmers of the locations.
3. Technology that encounters problems which can not be solved or that can not be over come by a programme of resource augmentation, market development or support or extension activities may not be any use to the farmers.
4. The technology suggested should be such that it should be flexible and provide alternate plans in carrying the farm operations.
5. The technology suggested should have taken into account all the environmental and socio-economic resources of the farmer and due importance given to the indigenous technical knowledge developed in the locality.

Cropping systems research methods that evaluate limitations by the environment and farm resource base is a continuous process and the farmers should be given alternate plans indicating their benefits and drawbacks. Where components of pattern are economically viable; the farmers adopt those components rapidly. When components are not compatible with physical, biotic or socio-economic factors, cooperative research with farmers in their field helps recognize field level constraints. For large scale following by farmers, the studies on the cropping pattern may help in suggesting to Government for its input wherever necessary and for to the researchers to correct defects on the technology proposed.

CROPPING SYSTEM

A system consists of several components which are closely related and interacting among themselves. In agriculture, management practices are usually formulated for individual crops. However, farmers are cultivating different crops in different seasons based on their adaptability to a particular season, domestic needs and profitability. Therefore, production technology or management practices should be developed keeping in view all the crops grown in a year or more than one year if any sequence or rotation extends beyond one year. Such a package of management practices for all the crops leads to efficient use of costly inputs, besides reduction in production cost. For instance, residual effect of manures and fertilizers applied and nitrogen fixed by legumes can considerably bring down the production cost if all the crops are considered than individual crops. In this context, cropping systems approach is gaining importance.

SYSTEM

System means an arrangement of components which process inputs into outputs. Each system consists of boundaries, components, interactions between components, inputs and outputs.

Crop System

Crop system is an arrangement of crop populations that transform solar energy, nutrients, water and other inputs into useful biomass. The crop can be of different species and variety, but they only constitute one crop system if they are managed as a single unit. The crop system is a sub-system of a cropping system. For example, in the maize crop system, maize is the dominant crop which is grown in association with other crops.

Cropping System

Cropping system is a land-use unit comprising soils, crop, weed, pathogen and insect subsystems that transform solar energy, water nutrients, labour and other inputs into food, feed, fuel and fibre. The cropping system is a sub-system of the farming system.

The cropping patterns used on a farm and their interaction with farm resources, other farm enterprises, available technology and environment (physical, biological and socio-economic) which determine their make up, constitute the cropping system.

Livestock System

Livestock system is a land-use unit comprising of animals and auxiliary feed sources transforming plant biomass into animal products.

Farming System

Farming system is a decision making and land-use unit comprising the farm household, cropping and livestock systems that produces crop and animal products for consumption and sale. The farming system is a sub-system of a higher level land-use system, such as village or water-shed that, in turn, forms a component of the agricultural sector of the regional system.

Entire complex of development, management and allocation of resources as well as decisions and activities which, within an operation form unit or combination of units, results in agricultural production and the processing and marketing of the products.

A farming system is a complicated interwoven mesh of soils, plants, animals, implements, workers, inputs and environmental, physical, biological and social influence with the strands held and manipulated by a person called the farmer who, given his preferences and aspiration, attempts to produce outputs from the inputs and technology available to him.

Farming system is a complex inter-related matrix of soil, plants, animals, implements, power, labour, capital and other inputs controlled in part by families and influenced to varying degrees by political, economic, institutional and social forces that operate at many levels.

Farming system is the scientific integration of different interdependent and interacting farm enterprises for the efficient use of land, labour and other resources of a farm family which provides year round income to the farmers specially located in the handicapped zone.

Integration of various farm enterprises in a farm ensures growth and stability in overall productivity and profitability. It also ensures recycling of residues, optimization of resources, and minimization of risk and generation of employment. Various enterprises that could be included in farming system are crops, vegetables, fruits, flower cultivation, dairy, poultry, fish, goat, pig, sericulture, mushroom cultivation, agro-forestry, bee keeping, silviculture, agro-based industries and food processing. A judicious mix of enterprises complementary to cropping and suited to the given farm situation and farmer's preference would bring overall prosperity.

Under farming system, the farm is viewed in a holistic manner. Farmers are subjected to many socio-economic, bio-physical, institutional, administrative and technological constraints. Farming system conceptually is a set of elements or components that are inter-related which interact among themselves. At the centre of the interaction is the farmer exercising control and choice regarding the type and results of interaction.

Farming system is a resource management strategy to achieve economic and sustained production to meet diverse requirement of farm household while preserving resource base and maintaining a high level environmental quality (Lal and Millar, 1990).

Farm Household System

Farm household system is a group of usually related people who, individually or jointly, provide management, labour, capital, land and other inputs for the production of crops and livestock, and who consume at least part of the farm produce.

Agro-Ecosystem

Agro-ecosystems are ecological systems modified by human beings to produce food, fibre or other agricultural products. Like the ecological system they replace, agro-ecosystems are often structurally and dynamically complex, but their complexity arises primarily from the interaction between socio-economic and ecological processes.

Agro-Ecological Zone

Agro-ecological zone is an area of similar soil, vegetation, and population density characteristics, resulting in a similar (sub) type of cropping system. Because temperature and rainfall do not vary significantly through out the region, climate is not included as a variable in the definition of agro-ecological zones.

Efficient Cropping Zone

The efficient cropping zone is the zone where both yield and spread indices are maximum.

$$\text{Yield index of a crop} = \frac{\text{Average yield 6 state/zone}}{\text{Average yield 6 country}} \times 100$$

$$\text{Spread index of a crop} = \frac{\text{Area in state/zone}}{\text{Area in country}} \times 100$$

The most efficient zone of a crop is where both the indices are above 100. in case the indices are below 100 that zone is the least efficient zone of a crop.

Cropping System and Patterns

Cropping system varies widely from the simplest system of two crops a year in sequence to complex intercropping with many crops. Multiple cropped lands can be broadly grouped into lowlands, irrigated uplands, and rainfed uplands.

The cropping system and patterns are the yearly sequence and spatial arrangement of crops or of crops and fallow on a given area.

SHIFTING CULTIVATION / LAND ROTATION / JHUMMING

1. Forest land is cleared and cultivated. Due to cultivation of the same crop on the same cleared forest land year after year; soil productivity is lost and the crop is shifted to other slashed and burnt land.
2. Here same crop is grown year after year. In this case land is rotated but crop is fixed. Therefore it may also be called land rotation.
3. Shifting of land hence called shifting cultivation. Also called **Jhum** cultivation.
4. It causes soil erosion.
5. Practised in north-eastern states of India, Chhotanagpur Plateau of Jharkhand, M.P. and in hilly areas.

Crop Rotation

1. Crop rotation is the reverse of land rotation. Here land is fixed but crop is rotated year after year.
2. On a certain land, repeated cultivation of crop or crop fallow land in a certain sequence is called crop rotation.
   a) Rice – Wheat – Mung (Green gram):
   b) Rice – Mustard – Chilli
3. Here wheat is rotated by mustard and mung by chilli in the next agriculture year.
4. Maintains and even improves soil fertility and stabilises income.
5. It checks the soil erosion and conserves moisture.
6. Ensures a balanced programme of work throughout the year.
7. Suitable crop rotation is the key of modern scientific agriculture which aims to produce maximum yield by maintaining soil productivity.

LEY FARMING

To define, "leys" are temporary, short term: two to five years specially sown pastures consisting of grasses and legumes. These are used for grazing in succession with cultivated crop (Sample, 1951). The leys are established with intention of plough the land again after a predetermined period (Davies, 1960). Ahlgren (1956) defined it as a system of farming in which grasses and legumes are included in proper rotation for hay, silage and pasture to care for maximum livestock needs and to improve and conserve the soil fertility.

Thus, ley farming is a dynamic and integral system of agriculture which encompasses soil, plants and livestock together. It denotes a system of farming whereby emphasis is placed on the value of grass-legume mixtures to get fodder for livestock and build up soil fertility for obtaining higher yields of subsequent arable crops.

The production of a grass-legume mixture for fodder to provide adequate amounts of proteins and carbohydrates in cattle feed would also be the basis of an improved livestock industry to provide more milk for better nutrition of our primarily vegetarian people and better draught power for the cultivation of land. A cut and carried ley crop would, therefore, be helpful in producing more food for animals and enhancing soil fertility.

The practice of ley farming can be employed to introduce a measure of stability in our agriculture. It would contribute enormously for increasing productivity or agriculture under conditions of traditional farming where farm inputs are low. Ley farming has proved to be a beneficial practice in western countries. In Europe, major part of land is under grasses e.g. acreage under grasslands in United Kingdom alone amounted to 73.8% in 1949 which increased to 76.7% in 1958 most of this increase being in temporary leys (Davis, 1960).

SUSTAINABLE AGRICULTURE / ECO-FARMING

The concept of sustainable agriculture has come up because yield from modem farming technique (modern commercial agriculture) reaching a plateau and the environmental problems due to excessive use of chemicals, fertilizers and pesticides in food chain.

Types of Agriculture	Objective	Quantity of Input	Output
Subsistence Agriculture i.e. low level equilibrium	To sustain life and family need	Low	Low
Commercial Farming i.e. high level equilibrium	To obtain high Income	High	High
Sustainable Agriculture i.e. Natural equilibrium or ecological equilibrium	Ecological balance	Low	High

Sustainable agriculture is that form of farming which produces sufficient food to meet the needs of the present generation without eroding the ecological assets and productivity of life supporting systems of future generations. Natural farming is an excellent illustration of sustainable agriculture. It is also known as ecological farming/ eco-farming or organic farming or perm culture. It is called eco farming because ecological balance is given importance and organic farming because organic matter is the main source for nutrient management. It is a system of cultivation with use of manures, crop rotation and minimal tillage. Sustainable agriculture also involves agro-forestry and multi-level cultivation and integrated animal husbandry. The term sustainability denotes the characteristic of a process that can be maintained indefinitely and sustainable use of the eco system refers to making use of the system without impairing its capacity for renewal or regeneration.

CROPPING SYSTEMS

System means a set of elements/components that are inter-related and interacting among themselves. The objective of any cropping system is efficient utilization of all resources *viz.* land, water and solar radiation, maintaining stability in production and obtaining higher net returns. The efficiency is measured by the quantity of produce obtained per unit resource in a unit time.

The objective of traditional agriculture was to increase the production by two means:

1. by increasing the area of cultivation.
2. by increasing the productivity per unit area of the crop. But two more dimensions are added to modern agriculture.
3. to increase the production per unit time.
4. to increase the production per unit space.

In the traditional cropping systems, mixtures and rotation were developed by the farmers over years of experience by trial and error to suit specific ecological and socioeconomic conditions to attain yield stability, whereas modern scientific cropping system has three pillars:

(i) Genotype, (ii) Geometry of planting and (iii) Management practices.

1. Genotype means genetic make up of seed.
2. Geometry of planting means:
   i) Shape of planting pattern on the land surface.
   ii) Space of the area available for the individual plant. Geometry of planting may be circular, rectangular, square type or cubical. It is indirectly related to plant population. Cubical pattern of planting has maximum plant population. Plant population may be defined as
   iii) Size of area available to the individual plant.
   iv) Number of plants per unit area.
3. Management practices: include all the practices of crop production. For the cropping systems, management includes.
   i) Type and arrangement of crops in time and space i.e. cropping pattern.
   ii) Choice of variety.
   iii) Method of stand establishment.
   iv) Pest management and harvest.

DIFFERENCE BETWEEN CROPPING PATTERN AND CROPPING SYSTEM

Cropping Pattern means:	Cropping System means:
– Crop rotation practiced by a majority of farmers in a given area or locality.	– Cropping patterns and its management to derive benefits from a given resource base under specific environmental condition.
– Type and arrangement of crops in time and space.	– The cropping patterns used on a farm and their interaction with farm resources, other farm enterprises and available technology which determine their make up.
– Yearly sequence and spatial arrangement of crops or of crops and fallow on a given area.	– Pattern of crops taken up for a given piece of land, or order in which the crops are cultivated on a piece of land over a fixed period, associated with soil management practices such as tillaqe, manuring and irrigation.
– The proportion of area under various crops at a point of time in a unit area.	

FARMING SYSTEMS

It represents an appropriate combination of farm enterprises *viz.* cropping system, livestock, poultry, fisheries, forestry and the means available to the farmer to raise them for increasing profitability. The term 'farming systems' and' mixed farming' are used interchangeably but have some subtle differences.

MIXED FARMING / INTEGRATED FARMING SYSTEM

Defined as a system of farming on a particular farm which includes crop production, raising livestock, poultry, fisheries, bee keeping etc. to sustain and satisfy as many needs of the farmer as possible. The objective of mixed farming is subsistence while higher profitability without altering ecological balance is important in farming systems. Cropping system is an important component of a farming system.

SOLE CROPPING / SOLID PLANTING

It is defined as the cultivation of one crop variety alone in pure stand at normal density in a certain time and place.

CROPPING SCHEME

The plan according to which, crops are raised on individual plots of a farm with an object of getting the maximum return from each crop without impairing the soil fertility. Soil fertility and crop yields are important factors in determining which cropping programme to follow, but at the same time, the amount of labour and capital required and prices in relation to the resources influence the adoption of a particular cropping scheme.

Monoculture

Monoculture is the repetitive growing of the same crop on the same land.

Or

Monoculture is also called single cropping or mono-cropping. A system of growing the same crop on the same land year after year is called mono-cropping or monoculture.

Cropping Intensity / Cropping Index

CI = No of crops grown in a year in a piece land x 100

Cropping intensity in India is about 127%

TYPES OF CROPPING SYSTEMS

1. **Mono Cropping:** C I is always 100%
2. **Multiple cropping:** Growing of two or more crops on same field in a year. It is the intensification of cropping in **time** and **space** dimensions i.e. more number of crops within a year and more no. of crops on the same piece of land at any given period. The principal limiting factors in adoption of multiple cropping are low intensity of irrigation and preponderance of long duration varieties.

Recently one more dimension i.e. to maintain the soil fertility, is added in the definition of multiple cropping. Thus multiple cropping is defined as the cultivation of two or more crops on the same field in a year without deteriorating the soil fertility.

Intensive Cropping

Cropping systems based on climate, soil and water availability have to be evolved for realizing the potential production levels through efficient use of available resources. The cropping system should provide enough food for the family, fodder to the cattle and generate sufficient cash income for domestic and cultivation expenses. These objectives could be achieved by adopting intensive cropping. Methods of intensive cropping include multiple cropping and intercropping. Intensive cropping may pose some practical difficulties such as shorter turn-around time lapse for land preparation before the succeeding crop and labour shortage at periods of agricultural activities. These practical handicaps can easily be overcome by making modifications in the cropping techniques. Alteration of crop geometry may help to accommodate intercrops without loosing the base crop population.

Multiple Cropping

The intensification of cropping in time and space dimensions is growing two or more crops on the same field in a year.

- Inter Cropping
- Mixed Cropping
- Sequential or sequence cropping or non-overlapping Cropping.
- Relay cropping or overlapping Cropping.
- Double, Triple and Quadruple Cropping
- Ratoon Cropping
- Multi-storey / Multilevel/Multi-tier Cropping
- Paira / Utera Cropping

Inter-cropping

It refers to growing of two or more crops simultaneously, on the same piece of land, base crop necessarily in distinct row arrangement. Crop intensification is in both time and space dimensions. There is intercrop competition during all or parts of crop growth.

Or

Intercropping is the cultivation of two or more crops simultaneously on the same field, with or without a row arrangement (row intercropping or mixed cropping). Relay intercropping is the growing of tow or more crops on the same field with planting of the second crop after the first one has already flowered. The crops grown in intercropping are called a crop association, and are denoted with a + sign. The crops involved can be annuals only, a combination of annual and perennials or perennials only. Important intercropping systems for various regions are given in Table 17.4 and 17.5.

TABLE 17.4: Some recommended intercropping systems for Indo Gangetic alluvial plains

S. No.	System	Ratio
1.	Wheat + Raya	8 : 1 or 12 : 4
2.	Winter maize + Peas (green pods)	1 : 1
3.	Toria + Gobhi Sarson	1 : 1
4.	Potato + Mustard	3:1
5.	Potato + Wheat + Summer moong	1 : 2 : 2
6.	Mustard + Gram	4 : 1
7.	Autumn cane + Wheat	1 : 2
8.	Autumn cane + Potato	1 : 1
9.	Autumn cane + Raya	1 : 2
10.	Autumn cane + Toria	1 : 2
11.	Autumn cane + Winter Maize	1 : 1
12.	Maize + Green/Black Gram	1 : 1
13.	Maize + Soybean	1 : 1

S. No.	System	Ratio
14.	Pigeon pea + Green/Black gram	1 : 1

TABLE 17.5: Intercropping systems for region other than the Indo-Gangetic alluvial plains

S. No.	System		Ratio
1.	Semi-arid Alfisol of Southern Telengana	(i) Sorghum + Pigeonpea	2 : 2
2.	Vidharbha Region of Maharashtra	(i) Sorghum + Mungbean	1 : 1
		(ii) Sorghum + Pigeonpea	2 : 1
3.	Malwa plateau of Madhya Pradesh	(i) Sorghum + Pigeonpea	2 : 1
4.	Vetisols of south cast Rajasthan	(i) Sorghum + Mungbean	1 : 1
		(ii) Sorghum + Pigeonpea	2 : 1
5.	Deccan region of Maharashtra	(i) Pearl millet + Pigeonpea	2 : 1
6.	Deccan region of Karnataka	(i) Pearl millet + Pigeonpea	2 : 1
7.	Vertic inceptisols of Rajkot region	(i) Pearl millet + Pigeonpea	4 : 1
		(ii) Peanut + Pigeonpea	6 : 1
8.	Sub-humid Alfisols region of Orissa	(i) Rice + Pigeonpea	5 : 2
9.	Sub-humid Alfisol region of Chotanagpur	(i) Rice + Pigeonpea	4 : 1
10.	Semi-arid Alfisols Ralayaseema region	(i) Peanut + Pigeonpea	5 : 1

Here cropping intensity in space dimension is achieved. Such crops are usually grown along with those crops which have larger row interval. Intercropping includes alley cropping, strip cropping, contour cropping, paired row cropping, skip cropping, parallel cropping, companion cropping, multi-storied cropping, and synergetic cropping. But the major intercropping systems are

Parallel Cropping

Cultivation of such crops which have different natural habit and zero competition e.g., Black gram/Green gram + Maize

The peak nutrient demand period for black gram or green gram is around 30-35 days after sowing while it is 50 days after sowing for Maize.

Companion Cropping

Such intercropping where the production of both intercrops is equal to that of its solid planting. e.g., Mustard/Potato/Onion + Sugarcane.

By growing of the above three intercrops individually with sugarcane, the production of none is hardly affected.

Multi-Storey / Multi Tiered / Multi Level

Cultivation of two or more then two crops of different heights simultaneously on a certain piece of land in any certain period. e.g., Sugarcane + Mustard + Onion/ Potato

Synergetic Cropping

The yields of both crops are higher than of their pure crop on unit area basis e.g. Sugarcane + Potato. On the basis of per cent of plant population used for each crop in intercropping system; intercropping is of two types viz, additive series and replacement series.

Additive Series Intercropping

In such type of intercropping, one crop is main crop or base crop and another crop is intercrop. Intercrop is introduced into the base crop by adjusting or changing crop geometry i.e. addition of intercrop to the base crop. Here plant population of base crop is same to what recommended population in pure stand whereas that of intercrop is less. Cropping husbandry is according to the base crop. This type of intercropping is prevalent in India and its main objective is to get additional income and to cover risk. e.g. sowing of potato in the field of sugarcane in between the rows of cane at a 90 cm.

Replacement Series Intercropping

Both the crops are component crops. Neither is the base crop nor is the intercrop. It means the plant population of both component crops is less than their recommended population in pure stand. Such type is widely practiced in western countries. Her 10th row of wheat crop is replaced by mustard and row to row distance of wheat crop is somewhat reduced to adjust the mustard crop.

Mixed Intercropping

Growing two or more crops simultaneously with no distinct row arrangement is also referred to as mixed cropping.

Row Intercropping

Growing two or more crops simultaneously where one or more crops are planted in rows is often simply referred to as intercropping.

Patch Cropping

The component crops are planted in patches.

Strip Intercropping

Growing two or more crops simultaneously in strips wide enough to permit independent cultivation but narrow enough for the crops to interact agronomically.

Relay Intercropping

Growing two or more crops simultaneously, during the part of life cycle of each is called relay intercropping. A second crop is planted after the first crop has reached its reproductive stage of growth but before it is ready for harvest. Often it is simply referred to as relay cropping.

Main Crop / Base Crop

This is the one which is planted at its optimum sole crop population in an intercrop situation and the second crop is planted in between rows of the main or base crop with a view to obtain some extra intercrop yield without sacrificing the main or base crop yield.

Component Crop

The component crop is used to refer to either of the individual crops making up the inter-cropping situation.

Pre-requisites of Successful Intercropping

The main objective of intercropping is higher productivity per unit area in addition to stability in production. Intercropping utilizes resources efficiently. There are certain important pre-requisites of intercropping:

1. The time of peak nutrient demands of component crops should not overlap.

Maize + Urd/Mung.

2. The peak nutrient demand is 30-35 days after sowing in case of urd/mung where as it is 50-55 days after sowing of maize.
3. Competition for light should be minimum among the component crops.
4. Complimentarily should exist between the component crops.
5. The differences in maturity of component crops should be at least 30 days.
6. Competition for CO_2 and water should also be minimum among component crops.

The advantage and disadvantage of intercropping are presented below:

Advantage and Disadvantage of Intercropping

Advantages of Intercropping	Disadvantages of intercropping
– Better use of growth resources including light, nutrients and water.	– Labour intensive
– Suppression of weeds	
– Reduced plant and disease incidence yield stability	– Control of pests and diseases or chemical weed control may be difficult
– Ecological stability i.e. improvement of soil health and agro-ecosystem	– Mechanisation is difficult to realize
– Other e.g. spreading of labour, physical support of one crop to another and home gardening leading to a more pluriform food supply is a good example of realization of intercropping advantages.	– Disadvantages caused by adverse competitive effects or by allelopathy (can of course be avoided).

Mixed Cropping

Mixed cropping is the cultivation of two or more than two crops simultaneously on the same land without definite row pattern or fixed ratio. Sowing of seeds is generally by broadcasting method commonly practised in dryland area of India. It is subsistence farming. Main objective is to lessen the risk of total crop failures, and to satisfy the farmers in food and fodder. Intercropping is the new concept of mixed cropping. Scientific study of mixed cropping was firstly done by La-Flitze (1928)

Sequence / Sequential Cropping or Non-Overlapping Cropping

Growing of two or more crops in quick succession, on the same piece of land in a farming year is called sequential cropping. The sowing of the succeeding crop and harvesting of the preceding crop may be done simultaneously or in a quick succession. e.g. just after the harvesting of maize, potato is sown and just after digging of potato, chilli is sown. Sequential cropping is also called non-over lapping cropping because there is no overlap between the two or more cropping. Ratooning or ratoon cropping may also be classified under sequential cropping.

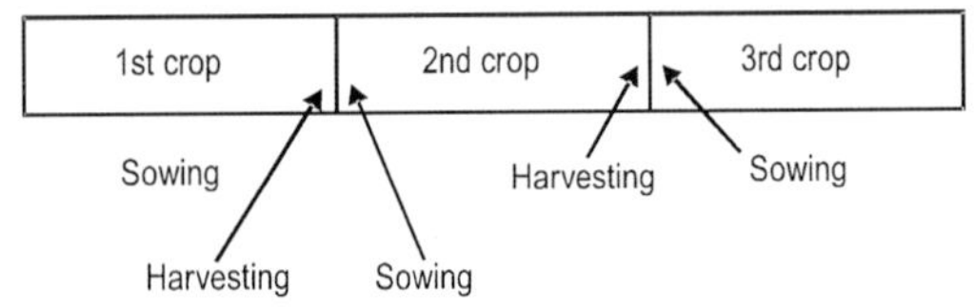

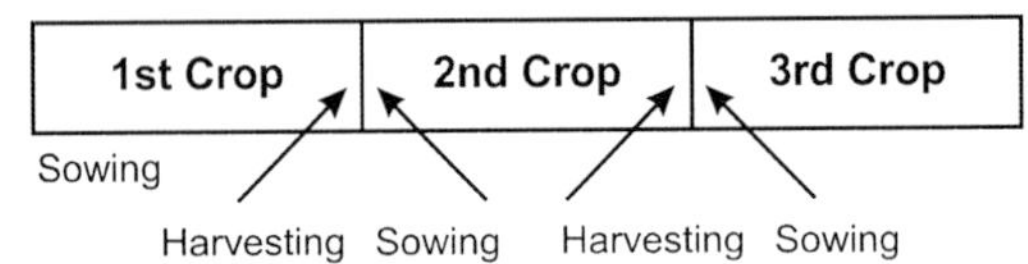

Relay Cropping or Overlapping Cropping

Its concept has been derived form relay race in which four runners run in the field having in their hand. The first runner passes on its flag to the succeeding partner and 2nd to 3rd and 3rd to the 4th runner. Relay cropping can be defined as growing two or more crops simultaneously during the part of the life cycle of each. Succeeding crops are planted before harvesting the preceding crop. Generally 2nd crop is planted after the first crop has reached its reproductive stage of growth but before it is ready for harvest e.g. Potato is planted before the harvesting of Maize and Radish is sown before harvesting of potato. Paira (Bihar & W.B.) or Utera (M.P) cropping may be example of relay cropping. Paira or utera cropping means sowing of lathyrus or lentil (generally pulses) before the harvesting of rice in lowland area. Planting of nursery crop may also be an example. Dr. Sher Singh Bains has suggested the following relay cropping pattern for north-west part of India

Mung – Maize – Potato – Wheat

1st crop		IInd crop		IIIrd Crop		IVth Crop
Sowing of 1st crop	Sowing of 2nd crop	Harvesting of 1st crop	Sowing of 3rd crop	Harvesting of 2nd crop	Sowing of 4th crop	Harvesting of 3rd crop

Double, Triple and Quadruple Cropping

- **Double cropping:** Growing two crops a year in sequence.
- **Triple cropping:** Growing three crops a year in sequence.
- **Quadruple cropping:** Growing four crops a year in sequence.

Ratoon Cropping

The cultivation of crop for regrowth after harvest, although not necessarily for grain, is called ratoon cropping.

Multistorey / Multilevel / Multi-tier

Intercropping is mostly prevalent in plantation crops like coconut and arecanut. There is scope for intercropping in coconut garden upto the age of 8 years and after 25 years. During this period, there is adequate light transmission to the ground which permits cultivation of intercrops. The practice of growing different crops of varying height, rooting pattern and duration is called 'multi-tier cropping'. The objective of this system of cropping is to utilize the vertical space more effectively. In this system, the leaf canopies of intercrop components occupy different vertical layers. The tallest components have foliage tolerant of strong light and high evaporative demand and the shorter component(s) with foliage requiring shade and or relatively high humidity. e.g.,

Integrated Farming System

Coconut + black pepper + cocoa + pineapple. The schematic representation of planting of these crops is shown in Fig 17.1

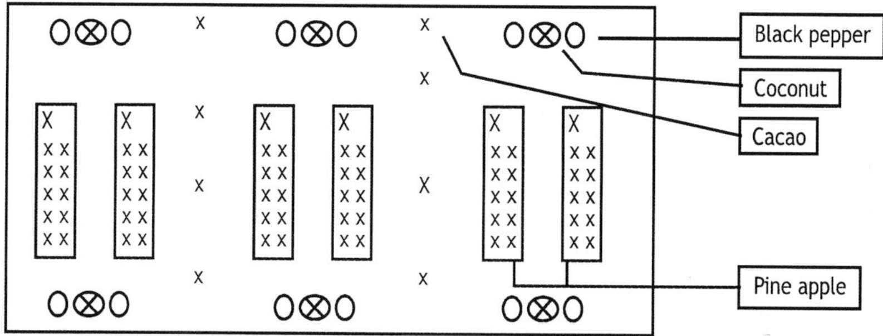

FIG. 17.1: Multistoried cropping.

In this system, coconut is planted with a spacing of 7.5 m. Rooted cuttings of black pepper are planted on either side of coconut about 75 cm away from the base. On the coconut trunk at a height of about one metre from the ground level, the vines of pepper are trailed. A single row of cacao is planted at the centre of space between coconut rows. Pine apple is planted in the interspace.

Coconut growing to a height of more than 10 m occupies the top floor. Black pepper is growing to about 6-8 m height forms the second floor. Cacao with its pruned canopy of about 2.5 m height and pine apple growing to about 1 m height form the first and ground floors respectively.

In another multi-tier system in coconut, ginger or turmeric and partial shade loving vegetables form the first tier, banana in second tier, pepper in third tier and coconut or arecanut in the final tier.

In the arecanut plantation, tuber crops are predominantly intercropped. Elephant yam, tapioca, greater yam and sweet potato are common intercrops in humid tropics. Banana and pine apple are also cultivated as intercrops in arecanut gardens.

In coffee based multi-tier cropping system, first tier is with pine apple, second tier with coffee, third tier with cacao/mandarin orange and final tier with fast growing shade trees necessary for coffee plantation. (e.g., dadaps and silver oak).

Paira / Utera Cropping

Growing of such crops sown a few days or weeks before harvesting of standing mature crops is called paira/utera cropping and the sown crop is called paira/utera crop e.g. Lathyrus in rice. Paira cropping in succession may constitute relay cropping.

Pulses like mung, black gram, gram, lentil and lathyrus come up well as paira cropping in the *rabi* season. Where the land after harvest of paddy remains slaucy making it unsuitable for tillage, these crops may be broadcast atleast 15 - 20 days before harvest using 1.50 times the seeds required for normal sowing.

CHOICE OF CROPS FOR MULTIPLE INTENSIVE CROPPING

The crops should be arranged in such a way that Allelopathic effect, temporary immobilisation of nutrients and depletion of nutrients from the same layer of soil do not occur.

Legumes have a definite place in any cropping system because of their limited demand on resources like water, nutrients and light, their adaptability to varying environmental conditions and capacity to fix atmospheric N in root nodules.

Vegetables deserve their due place in intensive cropping because of high cash and nutritional values, most vegetables can be raised as seedlings and planted, thus reducing main field duration and competition with the associated crop. They can be grown in interspaces in orchards and plantations, they can be interred or relay cropped in cereal base crops. Short duration, photoperiod insensitive genotypes are suited for intensive cropping system.

RELATED TERMINOLOGY USED IN MULTIPLE CROPPING SYSTEM

- **Cropping index:** The number of crops grown per annum on a given area of land multiplied by hundred.
- **Cropping pattern:** The yearly sequence and spatial arrangement of crops or crops and fallow on a given area.
- **Cropping system:** The cropping patterns used on a farm and their interaction with farm resources, other farm enterprises and available technology which determine their makeup.
- **Monoculture:** Repetitive growing of the same sole crop in the same land.
- **Sole cropping:** One crop variety grown alone in pure stand at normal density.
- **Rotation:** Repetitive cultivation of an ordered succession of crops (or crops and fallow) on the same land. One cycle often takes several years to complete.
- **Mixed farming:** Cropping systems which involve the raising of crops, animals and/or trees.
- **Land equivalent ratio (LER):** Ratio of the area needed under sole cropping to one of intercropping at the same management level to give an equal amount of yield. **LER** is the sum of the fractions of the yield of the intercrops, relative to their sole crop yields.
- **Income equivalent ratio (IER):** The ratio of the area needed under sole cropping to produce the same gross income as one ha of intercropping at the same management level. **IER** is the conversion of **LER** into economic terms.

ADDITIONAL TERMS USED TO DESCRIBE VARIOUS FORMS OF MULTIPLE CROPPING

- **Agrisilviculture:** Growing of trees for timber, but with cultivated crops beneath.
- **Area-time equivalency ratio (ATER):** The ratio of number of hectare-days required in monoculture to the number of hectare-days used in intercropping to produce identical quantities of each of the components.
- **Calorie index:** A single productivity index which incorporates caloric equivalents by all crops in a system.
- **Competition effect:** Competition of intercropped species for light, nutrients, water, carbon dioxide and other growth factors.

- **Complementary effect:** Effect of one component on another which enhances growth and productivity, as compared to competition above.
- **Component crops:** Individual crop species which are a part of the multiple cropping system.
- **Component technology:** The procedure for growing each component crop.
- **Farm enterprise:** An individual crop or animal production function within a farming system which is the smallest unit for which resource use and cost-return analysis is normally carried out.
- **Farming systems research (FSR):** The study of whole farm system, which includes all the enterprises on the farm, their biological, economic and cultural operations and usually implies some involvement of the farmer in the research process.
- **Interplanting:** All types of seeding or planting a crop into a growing stand. It is used especially for annual crops grown under stands of perennial crops.
- **Maximum cropping:** Attainment of highest possible production per unit area per unit time without regard to cost or net return.
- **Monetary index:** A single productivity index which incorporates monetary value of all crops in a system.
- **Over yielding:** Production of component crop in an intercrop which is higher than the sum of appropriate monoculture crops. This is indicated by an **LER** greater than unity.
- **Residual effect:** Effect of the previous crop in a sequential cropping pattern on the productivity of the current crop.
- **Simultaneous poly culture:** Simultaneous growth of two or more useful plants on the same plot. It includes mixed cropping, intercropping, interculture, interplanting and relay planting.
- **Spatial arrangement:** The physical or spatial organisation of component crops in a multiple cropping system.

PLANT INTERACTIONS

Interaction between Component Crops

In intensive cropping, crops are grown in association (intercropping) or in sequence (sequential cropping). In such situations there is possibility of interaction between the component crops. The interaction is mainly due to response of one species to the environment as modified by the presence of another species. Interaction may be competitive or non-competitive or complementary. If the crops are grown in association and the growth of either of the concerned species is not affected, such type of interaction is called as non-competitive interaction or interference. If one species is able to help the other it is known as complementary interaction. e.g. supply of N fixed by the legumes to the associated non-legumes. (This is otherwise known as annidation). One species may have greater ability to use the limiting factor and will gain at the expense of the other. This is called as competitive interaction or interference. Interaction may also occur in some other manner by way of producing toxic substances and affecting the establishment and growth of the associated species. This is called allelopathy.

Interactions in Intercropping

Factors such as light, water, nutrient, oxygen and CO_2 are required for plant growth. In mixed or intercropping situations, the component species compete for the growth factors. The close proximity of species causes sub-optimal utilization of the growth factors and hence there is inequitable distribution of resources among the plants. Generally competition will develop between two components or within the components. The nature of such competition may vary depending on the density and proportion of each component and planting pattern.

Light

Photo-synthetically active radiation (light) is directly related to photosynthetic rates of component crops. Intercropping can increase light interception by as much as 30-40%. When one component is taller than the other in an intercropping system, the taller components intercept most of the solar radiation. The pattern of light interception also varies according to the age of the crop. Light transmission by coconut is only about 20% at the age of 8-10 years and this remains almost constant till about 25 years of age. Subsequently the per cent light transmission increases progressively as the canopy coverage of ground decreases. When the palms are at about 40 years of age, light transmission increases to about 50%. In an intercropping situation where the component crops have different growth durations, the peak demand for light would occur at different times. For example, in maize + green gram intercropping, the peak light demand for maize occurs at 60 days when green gram is ready for harvest. In such combinations, competition for light is less among the component crops and there is greater light use in intercropping than in pure stands. In general, the component crops under intercropping situations are grown in such a way that competition for light is minimized. Proper choice of crops and varieties, adjustment of planting density and pattern are the techniques to reduce competition and increase the light use efficiency.

Moisture and Nutrients

Competition for moisture and nutrients may result in two types of effects on the less successful or suppressed component. First, the roots to this component may grow less on the sides towards plants of aggressive component. Secondly, plants affected by competition for soil factors may have increased root/shoot ratio. The aggressive component generally absorbs greater quantity of nutrients and soil moisture. In legume and non-legume combination, the latter takes up large amount of P, K and S. As a result, the legume may show deficiency of these nutrients. Such effects, however, may be mitigated by appropriate fertilizer application. Among intercrops, sorghum and pearl millet are more competitive in extracting nutrients. Generally the intercropped stands remove greater amount of nutrients than sole crop stands.

Allelopathy

Allelopathy is any direct or indirect harmful effect that one plant has on another through the production of chemical substances that escape into the environment. Some crops may be unsuitable to be grown as intercrops because they may produce and excrete toxins into the soil which are harmful to the associated components. Allelochemicals produced from the leaves of Eucalyptus globules drastically reduce the germination of mustard sown underneath.

Annidation

Annidation refers to complementary interaction which occurs both in space and time.

- **Annidation in space:** The canopies of component crops may occupy different vertical layers with taller component tolerant to strong light and high evaporative demand and the shorter component favouring shade and high relative humidity Multi-storeyed cropping in coconut and planting shade trees in cocoa and tea plantations use this principle. Similarly, root systems of component crops may exploit the nutrients in different layers of soil and hence utilize the resources in a better way with much less competition.
- **Annidation in time:** when component crops of widely varying duration are planted, their peak demand for light and nutrients are likely to occur at different periods, thus reducing competition. In a combination having early and late maturing crops (sorghum + redgram), when early maturing crops are harvested, conditions become favourable for the late maturing crop (redgram) to putforth its full vigour.

Other Complementary Effects

In an intercropping system involving a legume and a non-legume, part of the N fixed in the root nodule of the legume may become available to the non-legume component. With the presence of rhizosphere microflora and mycorrhiza, one species may lead to mobilization and greater availability of nutrients not only to the species concerned but also to the associated species. Provision of physical support by one species to the intercropped climbing species may improve the yield of the climber. Examples are coconut + pepper and maize + beans. The taller component acts as wind barrier protecting the shorter components from lodging.

Interactions in Sequential Cropping

In sequential cropping, sole crops are grown in succession. Hence, competition for light, water and nutrients as found in mixed cropping does not occur. In relay cropping, a short span of overlapping occurs between two crops in sequence. Hence the relay sown crop at the seedling stage may have competition for light from the standing crop. Such type of competition may be minimized by proper choice of crops and varieties and adjustment of time and method of planting.

In intensive multiple cropping involving two or more crops in sequence, the main adjective is to harvest as much solar energy per unit area per unit time as possible. In rice based cropping system, the solar energy use efficiency ranged from 1.58 to 2.02% of PAR in Uttar Pradesh. In Coimbatore, the efficiency ranged from 0.82 to 1.77% of PAR for the whole year in cotton based system. Inclusion of a C_4 plant in the summer increased the efficiency.

In sequential cropping, the preceding crop has considerable influence on the succeeding crop. This includes the complementary effects such as release of N from the residues of the previous crop, particularly legume, to the following crop and carry-over effects of fertilizer applied to the preceding crops. The adverse effects include allelopathy, temporary immobilization of N due to wide C/N ratio of the residues and carry over effect of pest and diseases. Introduction of pulses as summer crops results in improved yield of rice in *kharif* season.

SEQUENTIAL CROPPING SYSTEMS

In India, food crop is predominantly grown in most suitable seasons and thus particular food crop is basic to the cropping system followed by the farmers. Accordingly the cropping systems are usually referred to as:

- Rice-based cropping system

- Sorghum-based cropping system
- Pearl millet-based cropping system
- Wheat and gram-based cropping system

Some of the cropping systems based on commercial crops are (i) cotton-based, (ii) groundnut-based (iii) sugarcane-based, (iv) plantation crop-based, and (v) vegetable-based cropping system.

The grain production potential in different regions of the country ranges from 11 to 18 t/ha. In maize-potato or toria -wheat- mung system followed at IARI, New Delhi, it was possible to produce 14-15 tonnes of food per ha per annum without impairing the soil health. The results of multiple cropping demonstrations under irrigated conditions showed that production potential can be as high as 19.8 t/ha in cereal-based cropping system of rice – rice – rice. The yield potential of multiple cropping varies from region to region depending up to the physical and socio-economic conditions.

RICE BASED CROPPING SYSTEM

Sequential cropping is the most suitable cropping pattern of rice-rice in the rice areas. Traditionally single crop of rice with long duration cultivars is planted during the rainy period. However, due to the advent of short duration, photo-insensitive varieties, multiple cropping is being now practiced. The cropping system adopted under lowland condition can be grouped into two broad categories such as all rice and mixed rice and upland crop sequential cropping systems. These can be further grouped as below.

1. All rice cropping pattern
   i) Rice- Rice (R-R)
   ii) Rice- Rice – Rice (R –R –R)
2. Mixed rice and upland crop sequence
   i) Rice –upland crop (R-UC)
   ii) Upland crop – Rice –upland crop (UC-R-UC)

ALL RICE SEQUENTIAL CROPPING SYSTEM

Rice crop is repeatedly grown wherein two or three crops are raised. With three crops of rice, the land is cropped at least ten months in year. This is possible only in irrigated areas where water is adequate even during the dry periods. Crop failure due to water stress normally does not exist. Among the three crops, productive rice crop is the one which coincides with the dry months. During this period, solar radiation is high and the occurrence of there calamities such as cyclone or pest infestation is very low.

Two rice crops are mostly cultivated in many parts of the rice growing areas in India. The first crop is a short duration crop of rice which is cultivated during the *Kharif* season. This is followed by the second crop of rice during *rabi* season with medium duration varieties. Single crop of rice with the medium or long duration varieties is most prevalent in the command areas. Cultivation is restricted to the period of water availability in the canal. All rice sequence is followed in some pockets of the river command areas of Orissa, Andhra Pradesh, Karnataka and Tamil Nadu.

MIXED RICE AND UPLAND CROPS

Only one rice crop is grown during the peak rainy period. In such situations, there is enough time left to grow a fast- maturing annual crop between the first monsoon rain and the time of transplanting. Such crops must mature early. The crops uniquely suited for these conditions are vegetables, such as green corn, cowpea, green soybean or a green manure. The crop can easily be harvested within 80 days since only cobs are harvested.

In most occasions immediately after the harvest of the main rice crops, it is possible to plant another crop (rice fallow crop). Because rice crop is usually harvested at the end of the rainy season when the soil is still wet, there is usually enough moisture in the soil to support the succeeding crop. Some of the crops that are suited to this condition are green gram, black gram, cowpea, soybean, etc. Rice–upland crop is the commonly used system. Green gram is the popular choice as upland crop after rice.

SYSTEMS WITH MORE THAN THREE CROPS

With ratooning or relay cropping and special cultural practices that avoid puddling in lowland rice, up to five crops can be harvested in the same rice field over 12 month period (Bradfield,1970). In rice fields with adequate irrigation and good water control, Bradfield experimented at IRRI, Philippines, with various cropping patterns that could maximize productivity per unit area. The most successful pattern was a sequence of five crops consisting of rice, sweet potato, soybean, sweet corn and green soybeans. The tillage pattern varied for each crop and number of days to harvest were 102, 100, 85, 66 and 60 respectively. This pattern gave three times higher gross income than normally obtained in a rice- rice pattern.

PROMISING RICE–BASED CROPPING SYSTEMS IN INDIA

Promising rice-based cropping systems with high productivity and intensive land use for diverse agro-climatic situations of India have been identified in multi- locational experiments under Cropping Systems Research Project (Table 17.6).

TABLE 17.6: Promising rice based cropping systems in India

System	Region
Rice – Indian mustard	Kashmir valley
Rice – wheat – sorghum (Fodder)	Jammu region
Rice – wheat – green gram	Punjab, Haryana, U.P.
Rice – mustard	North Konkan
Rice – field bean	South Gujarat
Rice – groundnut	South Karnataka
Rice – black gram/sesame/cotton	Thanjavur, Tamil Nadu (Old Cauvery delta)
Rice – groundnut	Thanjavur, Tamil Nadu (New Cauvery delta)
Rice – mustard	Assam
Rice – wheat – jute	West Bengal
Rice – groundnut	Coastal Andhra Pradesh and Orissa
Rice – green gram	Eastern Madhya Pradesh

CROP SEQUENCES UNDER IRRIGATED UPLAND CONDITIONS

Under irrigated uplands (garden lands) depending upon the water availability or irrigation potential, more than two crops can be grown in a year. Crop sequences and their production potential have been tested under different agro climatic conditions. At Delhi, four crop sequence of green gram – maize – potato- wheat gave the highest production of 13.6 t/ha. However, three crop sequence of green gram – maize – wheat was found to be the most ideal for the small and marginal farmers (Table 17.7).

TABLE 17.7: Crop sequences and their production potential, New Delhi, India

Crop sequences	Total production (q/ha)
Two crops	
Maize – wheat	90.1
Three crops	
Green gram – maize – wheat	
Green gram – pigeon pea – wheat	68.8
Cowpea – pearl millet – wheat	92.9
Four crops	
Green gram – maize – potato – wheat	135.9
Green gram – maize – toria – wheat	112.8

Under All India Co-ordinated Agronomic Experiments Project, Production potentials of the crop sequences under irrigated condition were studied. Some of the important sequences and their yield potential are given in Table 17.8.

TABLE 17.8: Crop sequences, duration and yield levels in different states of India

State	Crop sequences	Duration (days)	Yield (q/ha)		
			Grain	Fodder	Vegetable
Himachal Pradesh	Soybean – wheat	323	77	-	-
	Soybean – potato	283	34	-	72
	Maize – wheat	299	65	-	-
Haryana	Pearl millet – wheat – cowpea (f)	323	83	63	-
	Maize – wheat – cluster bean (f)	323	82	103	-
	Pearl millet – wheat – maize (f)	232	81	126	-
Uttar Pradesh	Maize – wheat – pearl millet (f)	301	108	316	-
	Sorghum – wheat – green gram	306	63	-	-
	Sorghum – wheat – cowpea	313	59	-	-
Rajasthan	Pearl millet – wheat – green gram	339	35	-	-
	Green gram – wheat – green gram	295	22	-	-
Madhya Pradesh	Maize – wheat – cowpea	323	70	-	-
	Maize – wheat – green gram	325	70	-	-

f-fodder

In multi – location experiments, under Cropping Systems Research Project, promising cropping systems with high productivity as well as intensive land use for diverse agro climatic situations have been identified (Table 17.9).

TABLE 17.9: Potential cropping systems for different agro climatic situations

Region	System
Western Haryana	Cotton – wheat
Rajasthan	Groundnut – wheat

Region	System
Madhya Pradesh	Soybean – wheat
North Gujarat	Pearl millet – mustard
Sourashtra region – Gujarat	Cotton – groundnut
Western Maharashtra	Sorghum – wheat
Western Maharashtra	Groundnut – wheat
Marathwada region of Vidarba region	Cotton – groundnut
North Karnataka	Sorghum – bengalgram
South Karnataka	Cotton – groundnut

CROPPING SYSTEMS IN DRY LANDS

Normally only one crop is grown under dryland condition and cultivation is restricted during the rainy season. However, the intensity of cropping can be increased through sequential or intercropping, depending on the rainfall and moisture storage capacity of the soils (Table 17.10).

TABLE 17.10: Cropping pattern with varying rainfall and soil moisture storage capacity

Rainfall (mm)	Storage capacity of soil (mm)	Cropping pattern
350 – 625	100	Single crop in *kharif*
650 – 750	100	Intercropping can be attempted
780 – 900	150	Sequential cropping is possible
900 – above	200	Sequential cropping is assured

Source: Sharma, 1980

Choice of crops: Selection of a suitable crop and genotype is important to get higher yield in dry farming conditions. Deep rooted crops extract moisture from the deeper layers and they also have slow rate of transpiration. Leguminous crops are very well adopted for rainfed conditions because of their root system. Among the cereal, barley has better ability to harvest moisture from deeper layers than wheat. Moisture loss from the plant surface is less in barley. Similarly linseed has greater ability to trap soil moisture under water scarce situation.

CROPPING SYSTEMS FOR DRY LANDS

Important factors such as total rainfall, soil type and water availability have to be considered while practicing a particular cropping pattern in a place/ region. Cropping system for different regions as suggested from the results of AICARP (1970) are given in Table 17.11.

TABLE 17.11: Cropping systems for different region of India under dry lands

Regions/soil type	Crops	
	Kharif	Rabi
Northern Region		
Samba (Jammu)	Maize	Wheat
	Green gram	Barley
	Cowpea	Wheat
	Sunflower	Wheat/barley

Regions/soil type	Crops	
	Kharif	Rabi
	Groundnut	Barley
	Pearl millet	Bengal gram
Punjab	Sorghum	Wheat
	Maize	Wheat / Potato / Barley
Hisar (Haryana)	Pear millet	Cluster bean / Chickpea
Arid soils (Sierozemic soils)	(Mono-cropping is the rule)	
Central Region		
Dehra Dun (Uttar Pradesh)	Maize	Wheat
	Rice (upland)	Wheat/Bengal gram
Agra (Uttar Pradesh)	Gram	Mustard
Entisols (Alluvial soils)	Pearl millet	Chickpea / Barley
Varanasi (Uttar Pradesh)	Sesame	Chickpea / Mustard
Entisols (Alluvial soils)	Bengal gram	Barley / Mustard
	Rice	Chickpea / Mustard
	Pearl millet	Chickpea / Mustard
Eastern Region		
Ranchi (Bihar)	Rice	Chickpea / Linseed / Barley
Alfisols and related red soils	Maize	Rapeseed / Bengal gram
Bhubaneswar (Orissa)	Rice	Linseed / Mustard
Alfisols and related red soils	Finger millet	Horse gram
	Maize	Horse gram
Rewa region (M.P.)	Rice	Chickpea / Lentil
Vertisols and related black soils	Sorghum	Bengal gram / wheat
	Green gram	Wheat / Chickpea
Jhansi region (M.P.)	Sorghum	Bengal gram
Vertisols and related black soils	Pearl millet	Bengal gram
	Black gram	Safflower / Chickpea
	Sesame	Bengal gram
Indore (M.P.)	Maize	Chickpea
Vertisols and related black soils	Soybean	Safflower
	Soybean	Wheat
Western Region		
Udaipur (Rajasthan)	Green gram	Safflower
(Vertisols and related black soils)	Sorghum	Bengal gram
	Pearl millet	Bengal gram
	Maize	Wheat
	Black gram	Wheat
	Sunflower	-
Anand (Gujarat)	Pearl millet	Wheat
	Pigeon pea	-
	Castor	-
	Sunflower	-

Regions/soil type	Crops	
	Kharif	Rabi
Akola (Maharashtra)	Green gram	Safflower
Vertisols and related black soil	Sorghum	Safflower
Sholapur (Maharashtra)	Green gram	Safflower
Vertisols	Pearl millet	Bengal gram
	Black gram	Sorghum
Southern regions		
Anantapur (AP)	Sorghum	Safflower / Horse gram
	Pearl millet	Cowpea / Black gram
	Black gram	Sorghum
	Green gram	Pearl millet
	Cowpea	Pearl millet / Sorghum
Mono-cropping is generally adopted		
Hyderabad	Pearl millet	Cowpea
Alfisol	Sorghum	Horse gram
Bijapur	Sorghum	Green gram / Black gram
Vertisols	Green gram	Sorghum / Safflower
Bangalore (Karnataka)	Cowpea	Finger millet
Alfisols	Cowpea	Chilli
Mysore (Karnataka)	Groundnut	Sorghum
	Setaria	Safflower
	Hybrid	Cotton
	Pearl millet	
Bellary (Karnataka)	Cotton	Sorghum
	Black gram	Bengal gram
		Safflower
Kovilpatti (TN)	-	Cotton, Sorghum, Pearl millet, Cowpea, Horse Gram, Chilli

Intercropping System

Multiple cropping in the form of intercropping is predominant in the regions of dry, humid and semi-arid tropics. Farmers normally grow two or more crops together on the same piece of land. Different crops are cultivated in mixed cropping: simple mixture of only two or more crops is the most common.

The objectives of this practice are:

1. Insurance against total crop failure under aberrant weather conditions or pest epidemics.
2. Increase in total productivity per unit land area.
3. Judicious utilization of resources such as land, labour and inputs.

When two crops are to be grown together, they are chosen in such a way that there is variation in their growth duration. The peak periods of growth of the two crop species should not coincide. In such arrangements, a quick maturing crop completes its life cycle before the other crop starts. Willey (1979) described the concept as temporal complimentarity. Greater the differences in maturity and growth demands of the crop components, more opportunity is provided for greater exploitation of growth factors and over yielding. This will be achieved either by genetic difference in crop species or manipulation of planting dates. Normally short and long duration crops are grown together. Food crops are usually mixed with cash crops to ensure both sustenance and cash income.

Combination of species under intercropping depends on climate, location preferences and other site specific factors. Some of the important intercropping systems are described below:

Intercropping in Cereals

Intercropping with cereal is an excellent way of improving the resource utilization because the cereal utilizes the rainy season resources while late maturing crop exploits the post-rainy season resources such as residual moisture. Sorghum is most commonly inter cropped with pigeon pea on a variety of soils. Sorghum is harvested after 3½ to 4½ months and pigeon pea matures in about 6-9 months depending on the genotype. Grain legumes such as cowpea, green gram, black gram and groundnut are the other recommended intercrops with sorghum.

Maize + bean (*Phaseolus vulgaris* L.) is a predominant intercropping system in Central and South America and parts of East Africa, where rainfall ranges from 700 to 1500 mm spreading over 4 - 5 months. Soybean is also a good compatible companion crop with maize.

Pearl millet is a quick tillering and fast growing crop that attains full canopy development within 20 - 30 days of seedling establishment. It can be intercropped with groundnut, black gram or castor.

Finger millet is widely intercropped with field beans (*Dolichos lab- lab*) in Karnataka State, India, where 4 – 12 rows of the millet are alternated with one row of beans.

Intercropping in Pulses

For pigeon pea, short duration grain legumes such as black gram and soybean are the best companion crops in peninsular India. Groundnut is also a suitable intercrop. Sorghum is commonly intercropped with pigeon pea.

Intercropping in Cotton

Cotton is initially a slow growing crop. Any short duration and fast growing crops such as green gram, black gram, groundnut or cluster bean are the compatible companion crops. Onion is also a preferred intercrop in cotton.

Intercropping in Sugarcane

Sugarcane is slow growing upto 80 – 90 days. Since the crop is planted in rows in 0.8 – 1.0 m apart, considerable space is available for intercropping. Short duration crops maturing in 80 – 90 days can be advocated as intercrops. Black gram and soybean are found suitable. The green manure, dhaincha *(Sesbania aculeata)* can be sown in the inters pace and incorporated at about two months of its growth.

Intercropping for Dry Lands

Intercropping recommended for different regions of India under drylands conditions are given in Table 17.12.

TABLE 17.12: Intercropping for different regions in India

Region	Intercropping
Samba (Jammu)	– Castor + black gram / sesame / cowpea / sunflower
	– Maize + cowpea (fodder)
	– Maize + green gram
Hissar (Haryana)	– Pearl millet + cluster been / green gram
Dehra Dun (U.P.)	– Pigeon pea + soybean
	– Maize + pigeon pea
Agra (U.P.)	– Pearl millet + black gram / green gram
	– Pigeon pea + green gram / pearl millet
Varanasi (U.P.)	– Pigeon pea + maize / black gram / millet
Ranchi (Bihar)	– Barley + rapeseed
	– Wheat + Bengal gram / rapeseed / lentil
Bhubaneswar (Orissa)	– Pigeon pea + black gram / sorghum / groundnut
Rewa region (M.P.)	– Pigeon pea + sorghum + finger millet black gram / sunflower
Jhansi (M.P.)	– Sorghum + black gram / green gram / soybean
	– Wheat + Bengal gram / mustard
	– Barley + Bengal gram / mustard
	– Bengal gram + linseed / mustard
Indore (M.P.)	– Pigeon pea + sorghum
	– Cotton + soybean
Udaipur (Rajasthan)	– Sorghum + black gram
	– Pearl millet + black gram
	– Sunflower + black gram
Anand (Gujarat)	– Pearl millet + red gram / castor / sunflower
Akola (Maharashtra)	– Pigeon pea + black gram / green gram / sorghum / groundnut
	– Cotton + sunflower / black gram / groundnut
Sholapur (Maharashtra)	– Pigeon pea + pearl millet / sorghum / black gram
	– Cotton + sunflower / groundnut
Anantapur (A.P.)	– Pigeon pea + green gram / pearl mill / sorghum / foxtail millet / groundnut / sunflower
Bijapur (A.P.)	– Cotton + groundnut / sorghum / pearl millet / green gram
Mysore region (Karnataka)	– Pigeon pea + pearl millet
	– Cotton + pear l millet
	– Groundnut + cotton
Kovilpatti (T.N.)	– Cotton + black gram / green gram
	– Sorghum + black gram / cowpea / Dolichos lab lab
	– Pearl millet + green gram / cluster bean / cowpea / Dolichos lab lab

CROP ROTATION

Crop rotation is a process of growing different crops in succession on a piece of land in a specific period of time, with an object to get maximum profit from least investment without impairing the soil fertility. One cycle may take one or more farming years to complete. Farming year is 12 months for irrigated areas and is limited to the period of adequate soil-water availability for crop growth in rain fed areas.

PRINCIPLES OF CROP ROTATION

- The crops with tap roots should be followed by those which have a fibrous root system. This helps in proper and uniform use of nutrients from the soil.
- The leguminous crops should be grown after non-leguminous crops because legumes fix atmospheric N into the soil and add more organic matter to the soil.
- More exhaustive crops should be followed by less exhaustive crops because crops like potato, sugarcane, maize etc. need more inputs such as better tillage, more fertilizers, greater number of irrigations etc.
- Selection of the crop should be demand based.
- The crop of the same family should not be grown in succession because they act like alternate hosts for insects, pests and diseases.
- An ideal crop rotation is one which provides maximum employment to the family and far in labour, the machines and equipments are efficiently used and all the agricultural operations are done timely, simultaneously maintaining soil productivity.
- The selection of the crops should be problem based
- On sloppy lands which are prone to erosion an alternate cropping of erosion promoting and erosion resisting crops like legumes should be adopted.
- Under dry faming the selection of crops should be such which can tolerate the drought.
- In low-lying and flood prone areas the crops should be such which can tolerate water stagnation.
- The selection of crops should suit farmer's financial conditions.
- The crop selected should also suit the soil and climatic conditions.

ADVANTAGES OF CROP ROTATION

- Agricultural operations can be done timely for all the crops because of less competition.
- Soil fertility is restored by fixing atmospheric nitrogen, encountering microbial activity, avoiding accumulation of toxins (HCN etc.) and maintaining physical properties of soil.
- An ideal crop rotation helps in controlling insects, pests and diseases. It also controls the weeds in the fields.
- Proper utilization of all the resources and inputs could be made by following crop rotation.
- The farmer gets a better price for his produce because of its higher demands in the locality or market.

ROTATIONAL INTENSITY

This is calculated by counting the number of crops grown in a rotation and is multiplied by 100 and then divided by the duration of the rotation.

$$\text{Rotation intensity (RI)} = \frac{\text{No. of crops grown in a rotation}}{\text{Duration of rotation}} \times 100$$

e.g., Maize – barley – late potato – green gram: **First year**
Paddy – Potato – Wheat – Black gram: **Second year**

$$RI = \frac{8}{2} \times 100 = 400\%$$

CROPPING SCHEME

The cropping scheme is a plan according to which crops are grown on individual plots of a farm during a given period of time with the object of obtaining maximum return from each crop without impairing the soil fertility. Thus a cropping scheme is related to the most profitable use of resources, land, labour, capital and management.

PRINCIPLES OF CROPPING SCHEME

- *Area under building and layout.* A proportionate area is allocated for buildings and layout. If the farm area is 50 hectares or less, then the area under buildings and layout allocated is 8 -10 per cent of the total area, if the farm area is more then 50 hectares then the area allocated is only 5 per cent.
- *Number of plots.* The number of plots should be either equal to the duration of rotation or a multiple of that e.g., if the total duration of rotations in a cropping scheme is 5 years, then the number of plots must be either 5, 10, 15, or 20 etc
- *Selection of the crops.* Selection of crops depends upon the situation of the farm as discussed below.
- *Near a city.* If the farm is near a city, the farmer should grow vegetables on at least 60% of his net cultivated area.
- *Near a sugar factory.* If the farm is situated near a sugar factory then the farmer should put at least 60% of his net cultivated area under sugarcane or sugar beet.
- *Near a canning factory.* In such a locality, the farmer should grow vegetables, e.g., peas, tomato etc on at least 60% of his net cultivated area.
- *Near a vanaspati factory.* The farmer should grow oil seed crops, e. g., groundnut, sunflower etc. in such a locality.
- *Near a dairy farm.* If the farm is located near a dairy farm, the owner should go for fodder cultivation. Apart from fodder he can raise oil seeds and pulses on 10 – 15% of his net cultivated area.
- *Near a cold storage.* In such areas, the farmer should grow potatoes, onions etc.
- *Near a highway or railway station.* Then farmer can go for perishable crops like vegetables etc. In such cases, the farm is considered to be within the proximity of a city area.
- *Facilities available on the farm.* Selection of crops also depends upon the facilities for power, irrigation, inputs, labour, transport etc.
- *Area of individual plots.* The area of individual plots should be equal unless topography, soil conditions or any other person compels the farmer to layout.
- *Area allotted to crops for calculation of cropping intensity.* If the farmer takes crop mixtures then the relative area under different crops has to be considered so that none of them are adversely affected due to their mutual competition.
- *A crop planning* which can utilize inputs and other resources available on the farm without wastage should be adopted by the farmer.

CONCEPTUAL BASES OF MULTIPLE CROPPING

A cropping system is the kind and sequence of crops grown on a given area of soil over a period of time. It may be a regular rotation of different crops in which the crops follow definite order of appearance on the land or it may consist of only one crop grown year after year on the same area (monoculture). Other cropping systems may include different crops but lack definite or planned order in which crops follow one another or growing of two or several crops mixed together. Multiple cropping is thus a one year cropping system in which two (double cropping) or more crops are grown in succession within a year. If all these crops are cereals, multiple cropping will resemble monoculture in its advantages and requirements. If legumes are also included it will have certain features of traditional rotations.

A soil management system includes the cropping system plus the other associated practices in raising the crops, such as frequency and method of tillage, use of fertilizers, irrigation, drainage etc. thus effect of cropping system on soil productivity depends on other practices used with it in making up the systems of soil management.

Recent developments in the field of soil management (frequency and method of tillage, residue management and use of fertilizer, herbicides, insecticides and fungicides) have enabled a multiple cropping system to be feasible, desirable and commendable in tarai where the temperature and the sun shine are such that one or the other crop could be grown throughout the year provided water is available in plenty for irrigation.

If the system of soil management is attuned to the long term maintenance of soil productivity with the newer tools then the maximization of yield over a unit area in a year will involve in bio-energetic terms (Energy is the only common scientific denominator of the many sciences with which agriculture is concerned) the conversion of optimum amount of electromagnetic light energy received from the sun into useful yield of chemical energy through the photosynthetic mechanism of crop plants. This chemical energy has to be digestable, palatable, movable and storable. The wastage of the converted energy is to be minimized through the control of pests and diseases which convert chemical energy in the form which is not usable by man. The output of a crop will thus depend on the rate of working (net assimilation rate), size (leaf area) and the duration of the photosynthetic system. Recently we have varieties and hybrids of cereals available in the country which have a morphological frame to withstand high fertilizer doses. The disposition of their leaf area in space and time (dating, size and geometry of leaf area in its duration) can make full use of the growing condition and in particular of the available light energy. Fortunately these varieties have shorter duration than the traditional ones an attribute which fit them very well in a multiple cropping system. In paddy the newer varieties are not only short duration but insensitive to photoperiod and weekly thermo sensitive. These attributes enable their cropping throughout the year in some parts of the country.

Vegetable growers of our country have a long record of multiple cropping. They have developed soil management systems which maintain the productivity of their soil at a higher level than those of their neighbours who have been growing grain crops. The system of their soil management included, heavy dressing of organic manures, more tillage and inter culture, manual removal of weeds and following a rotation of vegetable crops which kept the diseases at minimum. With the availability of the short duration high yielding varieties of cereals the multiple cropping system is now to be extended to cereals but with a quite different soil management system. Users of this system will be mostly dependent on fertilizer use, chemical herbicides and insecticides, disease resistant varieties of plants, soil fumigation etc.

Two sets of forces determine the most profitable cropping system for a farm. One set involves the effect on an individual field and is measured in long term crop yield possibilities. The other involves effect on the whole farm and therefore, includes not only yield but also the economic and management factors of investment in implements and machinery, seasonal distribution of labour and prices and prices and of the produce:

Let us examine the first set in retrospect and prospect. Cropping systems followed in major parts of the country are:

1. Continuous cropping in which on conscientious effort is made to maintain productivity. This is one year cropping system in which the main *kharif* crop alternates with *rabi* fallow or *kharif* fallow alternates with *rabi* crop;
2. Rotation of two or more than two years involving occasional fallowing, legumes for fodder, seed or green manure;
3. Mixed cropping of one year either during *kharif* or *rabi* or during both the seasons;
4. Cropping systems with a lack of definite planned order, and
5. In certain pockets shifting cultivation in which soil is cleared, cropped for three or four years and then abandoned.

This classification is, of course, too generalized and is not exhaustive. There is considerable and necessary overlap between these systems. Some of these will continue in the dry farming area but wherever water has been made available either in dry farming area or humid areas all will give way to multiple cropping systems. During the recent years a renewed emphasis has been placed on minor irrigation programmes in almost all states. All these minor irrigation sources are destined to become nucleus around which a programme of multiple cropping will develop.

For these traditional cropping systems soil management programmes were based on occasional fallowing, green manuring, frequent and cultivation during fallow period to remove weeds and hot weather cultivation to speed up the burning of organic matter for releasing nitrogen. Experimentally, continuous wheat with *kharif* fallow has given steady yield of nearly 900 kg/ha. In North India, mono-culture of paddy over centuries has given similar yield without manuring. Thus wheat and paddy yields have stabilized on lower level and maintained through the operation of natural nitrogen fixing process. The year to year variation in yield around these averages is due to seasonal variation. Green manuring-wheat rotations have shown no long term effect on soil productivity. In short term they have given response equal to the application of 20-30 kg N/ha. Old rotational trials have shown the value and utility of rotations under the management followed. Long term trials at Pusa, Coimbatore, Cuttack have shown that plots receiving only balanced inorganic fertilizers have maintained yield and organic matter content of the soil. Since nitrogen was almost universally deficient for maximum yields of non-legumes there was an urgent need of including a legume in rotation and rotations like sorghum-wheat were utter failure without manuring. The whole picture has changed since nitrogen has become more and more available. Often the farmers and some agronomists raised the question 'How would the soil physical factors be affected under intensive cereal cropping in multiple cropping systems?'

The importance of good tilth for high yield has been generally accepted but seldom found in research results to be as important as it appears. On sandy soils the tilth is seldom a problem. On medium textured soil (silt loams and loams) an acceptable level of soil tilth can be maintained in the plough layer without bringing legumes in the cropping system with minimum tillage. On fine textured soil (silty clays, clay and clayey loams) favourable tilth is more necessary than in coarse and medium textured soils. On such soils problem of structure may become more severe when one grows the crops which residues to the soil. Farmers, who have such soils, should watch carefully the water retention in their soil. If seed bed is cloddy or field dries slowly after rains or water stands between the rows in low places then the soil management system used is not maintaining soil structure. More fertilizer to grow more residues or less tillage or both is needed. If paddy is included in such a cropping system to alternate with upland crop, it should be grown like upland crop without puddling. One agronomic fact has always to be remembered i.e. any damage from over cropping is temporary and can be corrected by changing the soil management system which provide greater return of organic residues, less frequent tillage and perhaps more deep rooted crops.

As far as the economic factors are concerned, the cost benefit ratios are more favourabie now for following multiple cropping. Improved fertilizer technology, higher analysis of fertilizers to reduce handling and transportation costs, more efficient application techniques including bulk handling of dry mixture etc. are expected further to reduce cost. The peak labour requirement for harvesting of crops and sowing of *rabi* crops is generating a demand for labour saving machines. Thus the multiple cropping programme will not only accelerate the chemicalization of the India agriculture, but it would also lead to mechanization. Double cropping with no or inadequate manuring or with inclusion of a legume has been practiced for long in the country. What is needed is to provide more purchased inputs. The economics is in favour.

Although written in another context the basis of multiple cropping according to Melsted (1954) is, it is a philosophy of maximum crop production per acre of land with a minimum of soil deterioration. The philosophy is based on the concept that high production is good for soil that minimum tillage promotes soil tilth and conserves soil organic matter, that high fertility promotes high yield, and lessens the losses of soil humus, that large amount of decomposable organic matter in soil is essential to good soil tilth and good soil physical conditions, and cover in the form of living mulch is good protection against all forms of erosion.

Many of the concepts on which multiple cropping is based are only tentative. The preliminary results which have been achieved in national demonstrations, by progressive farmers and in experimental plots are encouraging. They seem to indicate that the concepts are sound and future will support them. When multiple cropping will be adopted on large scale many soil management problems are likely to come up. For their solution integrated research done by plant breeders, agronomists, soil scientists and agricultural engineers will be absolutely necessary.

PRE-REQUISITES FOR MULTIPLE CROPPING

There are a number of essential features for developing a successful multiple cropping programme.

Even in less intensive cropping, success is the end result of the timeliness of the following operations: Choosing the proper crop, selecting the correct variety, providing high quality vigorous seed, treating the seed before planting, preparing the soil properly, planting the seed and correct setting of the plants, cultivation for stimulating plant growth, controlling weed and insect pests, preventing and controlling diseases and proper and timely harvesting. The timeliness of cultural operations becomes much more critical factor in multiple and relay cropping. A few of the conditions considered to be pre-requisite to successful multiple cropping are enumerated here.

1. There must be adequate supply of available water and it must be available at the time it is needed. Since intensively conducted multiple cropping will involve irrigated farming and because the timeliness of the operations is extremely critical, it is absolutely essential that a dependable source of water be available.

Fortunately tube wells and pumping sets are becoming available in greater and greater numbers year after year. As three to four crops will be taken from every acre the area that will be commanded by a well will be only 5 – 10 acres especially during the summer period from April to June because the consumptive use of water will be very high due to the higher evaporative demand during this period. With the available supply of under ground water it is possible to follow relay cropping over large areas. It is also equally important that the water available is well controlled in its distribution. For this purpose land levelling and provision of adequate drainage is absolutely necessary.

2. The second pre-requisite is an adequate and timely supply of seed, fertilizer and plant protection materials. These basic inputs have to be available when required and in sufficient quantities if multiple cropping is to be successful. The time available between two crops is very short and this period will be the peak period of farm activities, hence all these inputs should be available to the farmer without wasting much of his time.
3. The use of short duration varieties and hybrids. Crops of short duration are generally more useful taking into consideration the yield per unit area per unit time rather than the longer duration ones. Fortunately, now such varieties of paddy and wheat, jowar and bajra and maize are available and the more are in the nursery of the breeders for release. For ensuring continuous increase in the yield of crops extensive research on various aspects of crop production is necessary because new problems of soil management always crop up.
4. A fourth pre-requisite, is the development of new market, processing and new uses of the crops which are not traditionally grown in the locality but have a comparative advantage during some period. For example, crops like soybean, cheena and zaid legumes fit very will in the multiple cropping. New uses of these crops have to be developed for generating a higher demand for them.
5. Another important aspect of relay cropping is a high man-hour reguirement which can be reduced by the availability of machinery for tilling the land, harvesting of crop, and pumping irrigation water. In a multiple cropping system, the time available between two crops is extremely short and time in this system literally means money. Time available for many of the farming operations under the multiple cropping system will be reduced. This implies higher degree of mechanization. When relay cropping becomes more and more widespread, there will likely be increasing demand for machinery to work the land and harvest the crops.
6. A sixth pre-requisite is a better under standing of scientific agriculture, or in other words, a need for better farm managers. The farmers must learn to be efficient in irrigation water application; efficient in harvesting techniques; and he must learn to be judicious in the selection of crops and varieties which he grows. There are many skilled farmers capable of undertaking such scientific agriculture, but they are dependent on research information on which to base decisions. There is a need for production-oriented research, including this area of intensive planning, to lead the way so that the farming community can invest and expand along an already explored and proven path.

SYSTEM OF RICE INTENSIFICATION (SRI)

SRI involves the use of certain management practices, which together provide better growing conditions for rice plants, particularly in the root zone, than those plants grown under traditional practices. SRI was developed in Modagascar in the early 1980s by Father Henri de Lauhanie, a Jesuit priest. In 1990, Association Tefy Saina

(ATS) was formed as a Malagasy NGO to promote SRI. It has since been tested in China, India, Indonesia, Philippines, and Bangladesh with positive results.

Most rice farmers plant fairly mature seedlings (20-30 days old), in clumps, fairly close together with standing water maintained in the field for as much of the season as possible. These practices seem to reduce the risk of crop failure. It seems logical that more mature plants should survive better, that planting in clumps will ensure, that some seedlings should result in more yield, and that planting in standing water means the plants will never lack water and weeds will have little opportunity to grow.

System of Rice Intensification (SRI) is emerging as an important agro-ecological alternative for food security of small and marginal farmers in Orissa and India.

SIX PRACTICES IN SRI

1. Young Seedlings

Seedlings are transplanted early. Rice seedlings are transplanted when only the first two leaves have emerged from the initial tiller or stalk, usually when they are between 8 and 15 days old. Seedlings should be grown in a nursery in which the soil is kept moist but not flooded. The seed sac should be kept attached to the infant root, because it is important energy source for the young seedlings. The young seedlings should be planted so carefully that the root tip is not left pointing upward. The effect of young seedlings (14 days old) over normal seedling (21 days old) was studied on growth and yield of rice and the yield of rice in various treatments was furnished in Table 17.13.

TABLE 17.13: Effect of spacing and age of seedlings on the yield of rice (2003)

Number of seedlings with spacing	Young seedling (kg/ha)		Normal seedling (kg/ha)	
	Kuruvai (Kharif)	Late Thaladi (Rabi)	Kuruvai (Kharif)	Late Thaladi (Rabi)
One seedling at 20.0 x 20.0 cm	6951	5377	6329	4859
Two seedling at 20.0 x 20.0 cm	6553	4868	5142	4572
Three seedling at 20.0 x 20.0 cm	6355	4402	5802	3905
One seedling at 22.5 x 22.5 cm	7444	6138	7714	5160
Two seedling at 22.5 x 22.5 cm	7745	5019	7384	4710
Three seedling at 22.5 x 22.5 cm	7641	4538	7120	4230
One seedling at 25.0 x 25.0 cm	7149	4978	6725	4142
Two seedling at 25.0 x 25.0 cm	6951	4286	6329	4450
Three seedling at 25.0 x 25.0 cm	6752	4093	6329	3567

2. Single Seedling

Seedlings are planted singly rather than clumps of two or three or more. This means that individual plants have room to spread and to send down roots. They do not compete as much with other rice plants for space, light and nutrients in the soil. The effect of number of seedlings/hill was studied was furnished in Table 22.1.

3. Wider Spacing

Seedlings are planted in a square pattern with plenty of space between them in all directions. Usually, they are spaced at least 22.5 cm x 22.5 cm. It helps for vigorous root growth and more tillering. The square patter facilitates *in situ* incorporation of weeds by cono weeder. Roking marker for square planting of rice seedlings at 25 x 25 cm^2 spacing in a 2 M width bed area is also used for new women laborourers.

4. Lesser Seed Rate

SRI method requires much lower seed rate (5-8 kg/ha) than traditional methods (75-100 kg/ha).

5. Moist but Un-flooded Soil Conditions

Rice is traditionally been grown in water under submerged condition. In SRI method, soil is kept moist but not flooded during the vegetative period, ensuring that more oxygen is available in the soil for the roots, occasionally the soil should be allowed to dry to a point of cracking except in saline soils. This will allow oxygen to enter the soil induce the roots to grow and search for water. In SRI method, un-flooded condition is only maintained during vegetative period and from flowering to harvest, 1-3 cm of water is kept in the field as is done in the traditional method.

6. Cono Weeding

Weeding can be done by cono weeder. The cost of cono weeder will be around Rs. 300 to 1000/' depending upon the type and materials used. First weeding should be done on fourteen days after transplanting and this should be continued up to 40 days at 7-10 days at 7-10 days interval. At least two or three weedings are recommended. This practice of churning of soil seems to improve soil structure and increases the aeration of the soil. Thus the incorporation of weed biomass into the soil results in enrichment of CO_2 near to root zone, increases the biological activities, increases soil microbes population and activities, results in better nutrient availability in soil and uptake by plants. If cono weeder is not available, rotary weeder is recommended.

> ***"Nothing in life is to be feared. It is only to be understood"***
> ***Marie Curie***

18 Concept of Ideal Plant Type

"What we observe is not nature itself, but nature exposed to our method of questioning"

INTRODUCTION

Of many factors responsible for low yield of crops in India, the poor genetic yield potential of the varieties of different crops is most important. Under ideal conditions when other production inputs are adequately available, the genetic make up of a variety is the most significant factor determining the yield. In recent years, due to all around effort on the part of agricultural scientists, it has been possible to cultivate high yielding varieties of cereal crops which have often been termed a **"New Plant Type".** This terminology has been given because of their altered morphological attributes responsible for many advantages associated with them. Since these new plant types posses the trait of high responsiveness towards heavy fertilizer application, they are also called as fertilizer responsive varieties. However, keeping in view their morphological frame work conducive to heavy fertilizer application, some workers have also used the term "**Adaptable Varieties**". Adaptable here means mainly the physiological attributes of a variety responsible for controlling the assimilation of absorbed nitrogen in plant body, translocation and storage of photosynthetic products and processing more activity of roots under heavy application of fertilizers along with the ability of resistance to lodging and diseases. But the term **"New Plant Type"** seems to be more appropriate and reasonable as it can very easily express the extent of improvement incorporated over the old type varieties. The so called improved tall varieties cultivated by the farmers, generally grow very tall and process low yielding potentiality due to:

1. Weak and tall straw, susceptible to lodging under heavy fertilizer application.
2. Inefficient leaf arrangement responsible for poor photosynthetic activity and less utilization of solar energy.
3. Many associated attributes like un-synchronous tillering, long growth duration, and susceptibility towards the attack of diseases and insect pests.

New plant types on the other hand, do not posses these defects and have been further improved to increase their production efficiency where the morphological frame work has been genetically linked with other yield contributing characters.

RECENT DEVELOPMENT

Recent developments in plant breeding have made significant contribution towards the concept of new plant types. The successful efforts of altering the morphological architecture of crop plants and making them suitable for cultivation under high fertility status of soils, hqave opened a new vista in developing the varieties suitable for good agronomy.

New plant types, therefore, should morphologically be dwarf in growth habit with hard and stiff straw, erect and dark green leaves remaining active for a larger duration. Agronomically, they should be highly responsible to heavy fertilizer application. Such varieties should physiologically be well equipped for more dry matter production and high yields. Besides this, they should be adaptable under different agro-climatic conditions and be of short growing duration.

NEW PLANT TYPES

During the last four decades relatively much improvement towards developing varieties with **new plant type concept**, have been made in crops like wheat, rice, Jowar, Bajra, and maize. Important features of such new plant types of cereals and grain crops are as follows:

1. Dwarfness

The new plant types are generally dwarfed in height with certain exceptions. This dwarfness has been introduced in these varieties through dwarfing gene-Norin. The basic architecture of these plants is different and they are short and stiff, not more than knee-high but could take heavy amount of fertilizers without lodging or falling over.

2. Efficient Leaf Arrangement

The leaves of the new plant types are relatively narrow, thick, erect and dark green with optimum leaf area index composed of properly arranged leaves, which remain active for longer period after flowering. Physiologically such leaves are well equipped for dry matter production even after flowering because of high sunlight interception. Since leaves of new plant types, remain active and green even up to the maturity stage, they play prominent role in supporting the grain formation resulting into more number of fertile grains per ear head.

3. Synchronous Tillering

The growth and development of these varieties are more or less rhythmic. The proper germination, formation of all the tillers during a specific period and timely maturity of all the tillers are the important features of this rhythmic growth. They have the highest synchronization co-efficient as regards the development and maturity of grains. This applies tp different ears of a plant which are developed simultaneously because of synchronous tillering behavior. However, synchronization of tillering has been found to be dependent on other factors like moisture, proper secondary root growth and adequate nutrient availability during the period.

4. Low Florate Sterility

Under heavy fertilizer application, varieties other than new plant types produce more sterile florates. Whereas the new plant types have a very low percentage of florate sterility which is largely because of the synchronized tillering resulting into uniform ear head formation supported by longer physiological activity of leaves at maturity. Low florate sterility in these varieties has also been attributed due to increased activities of roots at grain formation stage.

5. Shorter Duration of Growth

The growth duration of new plant types is generally, shorter than other varieties. The optimum growth duration of a variety is more important for scheduling its irrigation and manuring for obtaining higher yield. High correlation with nitrogen application and growth duration has been obtained in several studies. At low nitrogen level, longer growth duration, correlates with satisfactory grain yield. Whereas under heavy nitrogen application, a short duration variety is preferred for obtaining higher dry matter production as well as more grain production efficiency i.e., the grain yield per unit area per unit time. Additionally with shorter growth duration, these varieties can very well fit in under high cropping intensity programme like multiple cropping and relay cropping.

6. Adaptability of Different Crop Seasons

Almost all the new plant types are photo-insensitive and thus completely resistant to the fluctuations in the day length. Therefore, these varieties can be grown under all crop seasons provided inputs like irrigation and fertilizer are adequately made available. With this quality, higher yields can be obtained with these varieties under care and management in wide range of day length. However, some of the varieties of new types are thermo-sensitive and hence are affected by variation in temperature during growing season.

7. Absence of Seed Dormancy

The seeds of new plant type are of not dormant type i.e., they do not require any rest period, often called as dormancy period. Freshly harvested seed may be used for sowing after obtaining a desired moisture percentage. This quality associated with new plant types has been found quite helpful in the seed multiplication programme of high yielding varieties. The absence of seed dormancy renders the multiplication of seed possible within short span of time. This along with photo-insensitivity, makes these varieties quite suitable for adaptation under high intensity cropping programme.

8. Effective Translocation of Food Material from Plants to Grain

The new plant type varieties have more potentiality to absorb and assimilate nutrients from soil throughout the growth duration which in combination with higher photosynthetic activities enables them for higher dry matter production. The built-in-efficient plant mechanism in these varieties coupled with fewer organs respiring at flowering stage permit more efficient use of respiration for growth and grain production. After flowering this enables effective translocation of accumulated food materials of straw for grain formation.

9. Responsive to Heavy Fertilizer Application

The most important feature of new plant types is that they possess the trait of high responsiveness towards fertilizer application. Their optimum requirement for nitrogen is two and half to three times more than the requirement of so called improved local varieties. Similarly, the phosphorus and potash requirement of new plant types are also one and half to two times more in comparison to local types.

Under low fertility status, their yielding ability is not fully utilized and very often they give quite poor yield under sub-optimum conditions. It is, therefore, necessary to supply adequate quantity of fertilizers specially, N. P. K. in order to exploit their high yielding potentiality to the maximum. Experiments carried out throughout the country have clearly indicated that on an average they require about 100 to 120 kg nitrogen, 50 to 70 kg phosphorus and 40 to 60 kg potash per hectare under optimum condition of soil moisture status.

10. Lodging Resistance

High yielding new plant types are generally dwarf in growth habit with strong and stiff stem. This provides considerable resistance against plant lodging. Under heavy fertilizer application the tall varieties are bound to lodge resulting in to substantial decrease in yield. Contrary to this, dwarf high yielding varieties seldom lodge unless too heavy fertilizer application has been made coinciding excessive water application. Because of incorporation of dwarfing genes in these high yielding varieties, they possess the trait of high responsiveness towards heavy fertilizer application without lodging. These characteristics coupled with their altered morphology results into very efficient utilization of nutrients and water applied.

11. Yield Potentiality

The new plant types are known for their high yield potential. These varieties are capable of yielding 2 to 3 times more grain yield in comparison to the local tall improved types. This is probably because of their altered morphology which results into efficient utilization of water, nutrient and radiation and increased metabolic activities with high dry matter production. Their grain to straw yield ratio is roughly about 1:1. However, these strains are more susceptible to any degree of variation in manageable inputs. Inadequate and untimely supply of nutrients, irrigation water and plant protection measures may result into partial or complete failure of crop. One the other hand, under optimum conditions and proper care and management the new plant types of cereals have given 6 to 7 tonnes of grain yield per hectare.

12. Disease Susceptible

The only biatus associated with the new plant types is the disease susceptibility. With luxuriant vegetative growth, the varieties offer scope for insect pests and disease attack. However, attempts are being made to develop disease resistant new plant types.

FACTORS RESPONSIBLE FOR SUCCESSFUL CULTIVATION OF NEW PLANT TYPES

It is now well known that in our drive for higher food production and to attain self sufficiency, the cultivation of high yielding varieties have a decisive role to play. These new plant types are not only superior for lucrative income because of their high production potential, but have also proved to be quite suitable under different agro-climatic conditions of India. Demonstrations conducted in the recent past throughout the country have shown that these varieties possess the high trait of responsiveness towards various manageable inputs. However, to exploit their yielding ability to the maximum an understanding of the cultivation practices and other requirement is not only necessary but a must. Such a study must be extensive and thorough in itself. This has already been pointed out in preceding chapter that any inadequacy and untimeliness in meeting the various requirements may cause tremendous variations in the expected quantum of return. This apparently is a disadvantage associated with these plant types.

In general, for any new plant type an optimum application of 100 kg N, 70 kg P_2O_5 and 60 kg K_2O are recommended in almost all the locations. This is a very high requirement as against improved local varieties. Thus, therefore, is an aspect of great importance particularly in view of the present high cost of fertilizers and their unavailability. Chances of crop failure due to prolonged drought, floods, diseases or insect pests have also to be reckoned with. Most of these new plant types are reported to be susceptible to many diseases. As such the successful cultivation of these varieties is much more complex, requiring special attention at every successful cultivation of new plant types are:

1. Proper Nutrition

Nutrition requirement of these new plant types are 2 to 3 times greater than for the local improved varieties. To have profitable production, therefore, the yield should also increase proportionately. In spite of the very favourable yield obtained from these varieties, the yield and fertilizer response data reported in the past are not very convincing leading to a controversy between agronomist and economist. Keeping in view the limited supply of fertilizers in the country coupled with high cost, the possibility of taking up ambitious programmes of cultivation of these varieties is limited at present. On the other hand, realizing the enthusiasm of the farmers and high yielding potentiality of varieties, it is desirable that these varieties extensively be cultivated. One has to be very considerate while selecting these varieties for cultivation. Once it is adopted, its all requirements must be made available in appropriate amount and at proper time. Area to be cultivated under these varieties must be fixed keeping inn mind the resources available. Without proper application of all the nutrient one cannot obtain the desired and economic return. At the same time the amount of fertilizers should be decided only after assessing the nutrient status of the field. There is no sense of applying higher doses of nitrogen, beyond what is required. Right type of fertilizers is also equally important and should be decided according to the pH and type of soil. While adopting heavy nitrogen application appropriate amount of P and K must be applied for obtaining maximum efficiency of applied N. In most of the cases cultivators use heavy nitrogen application but neglect the application of P and more particularly K. This condition sometimes produces various intricate metabolic problems resulting into physiological diseases. For economizing the fertilizer use, foliar application of fertilizer along with soil application be adopted. The proper manorial programmes have been discussed in length for various crops separately.

2. Proper Irrigation

Consideration of various aspects of irrigation of high yielding varieties assumes greater importance. The irrigation requirement of these varieties is slightly more than the local varieties but its timely application synchronized with the critical stages is of much significance. Irrigation ignored at any critical stage may cause greater variation in yielding ability of these types. Therefore, these varieties should be adopted only under irrigated conditions. The need of adequate and timely irrigation is imperative when heavy nutritional supply is advocated. It may be conceded that these initial difficulties in the cultivation of high yielding new plant types are unavoidable and constant efforts are necessary. to overcome them. However, for successful cultivation these varieties should be irrigated as and when desired.

3. Plant Protection

The new development in agriculture, involving new crop varieties, cropping pattern and use of heavy amount of fertilizers, often provides favourable condition for the multiplication of diseases and pests and for their appearance more frequently and in more virulent forms. The newly introduced high yielding varieties of cereals are comparatively more susceptible to disease and insect pests because of less resistance accompanied with profuse vegetative growth. Therefore, for the successful cultivation, the plant protection measures assume greater importance. Any negligence of this part may result in the total failure of the crop. Constant supervision in this regard and frequent preventive measures are thus a must for these types. However, in recent years efforts are being made to increase the resistance potential of these varieties towards many diseases and pests.

Conclusively this can be said that for the successful and economic cultivation of new plant types almost all the inputs should be adequately made available. Besides the inputs the scientific technical Know-how regarding their agronomy is equally important. These aspects have been described. But it should be born in mind that unless these factors are easily and adequately available there is no use of adopting these varieties for higher production. In as much as these varieties have given even poorer performance as compared with improved local varieties under sub-normal condition. For exploiting their yielding potentiality to the maximum these pre-requisites are essential or some improved agronomic techniques should be involved so as to make their cultivation possible and economical even under sub-normal conditions.

RECENT APPROACHES TOWARDS RECONSTRUCTING "NEW PLANT TYPES"

Cultivation of new plant types with high genetic yield potentiality and which have a built incapacity to make efficient use of available resource, have proved one of the best methods for increasing productivity. With successful attempts in the recent past, some of the new plant type varieties in different crops have been introduced which have the capacity to with stand various hazards like cold or drought, salinity-alkalinity, water logging and disease and pests. However, some of the important considerations in this direction are must while breeding such strains in future, which are as follows:

1. Genetic Amelioration for High Yields

The main approach towards this direction is to improve the yield level through genetic amelioration of yield components. A desirable plant type in this context, may be defined as the required plant structure (morphology) function and developmental sequences which can best suit a particular environment for producing maximum economic returns. Several alternations in reconstructing the plant type are thus desired to bring about a striking improvement in yield. For example, in crop like arhar and cotton a major reconstruction of the plant type requires above all a reduction in the spreading and bushy growth habit in order to accommodate greater plant population per unit area. This reconstruction entails an improvement to the index (economic/biological yield) which is obviously of very great value in a crop plant. It is, therefore, imperative in future breeding programme to incorporate some major advances through genetic manipulation in the direction of high harvest index. Already, the dwarf wheat and rice pointed out the possibilities of progress in these directions. Even in the case of Sugarcane, average yield has improved to the extent of 25% and sugar output has also increased consequently by almost 18% due to advent of hybrid varieties with desirable plant types. Break through is in sight in respect of development of early maturing varieties giving high cane yield. Recently, attempts have been made to develop umbrella like plant type in gram which has restricted number of main basal branches but has high number of fruiting branches which are expected to exhibit minimum competition in its own plant community.

2. Stability of Yield Potentiality

Strains with desirable plant type concept introduced in the very beginning, have shown to some extent the genetical break down of this yielding potentiality after a period of continuous cropping. Variability of soil types, agro climatic conditions and sub-optimum supply of required inputs, are some of the causes for such deterioration. Stabilizing the yield, therefore, is of significant importance. Attempts thus be made to stabilize the genetic yield potentiality of cultivars, which may not be affected by climatic, soil and cultural variability.

3. Stability of Production

Most of the new plant types of wheat and rice are relatively photo and thermo-insensitive which can cover very large areas in several parts of the world. Introduction of photo-insensitivity has proved to be very useful. One result of such a change in the physiological make up of the plants is the extensive cultivation of wheat both in time and space. The optimum time of sowing wheat in India has been the last week of October. However, the new strains can be sown as late as the end of December. This has also made possible of taking successful cultivation of wheat in southern and eastern parts of the country where earlier it had limited acreage.

4. The Concept of Per Day Productivity

With growing demand for increasing the agriculture production, apart from the introduction of high yielding varieties, cropping patterns with high cropping intensity like multiple and relay cropping are also being adopted. However, such efforts are limited to areas having assured irrigation facilities. Notwithstanding, with the availability of new plant strains known for their ability to withstand adverse condition and hazards, double cropping system is also being adopted under rainfed or dry land conditions. The evolution of early maturing varieties is providing a solution to this problem. This aspect has also been found useful for diversified type of production in pulses, oilseeds and fodder crops in addition to cereals. The major objective in developing such short duration varieties has been to generate crops which complement each other rather than compete. In an intensive cropping scheme of the kind, the par day, rather than the absolute yield pf a variety, assumes greater significance.

5. Up-grading of Cultivars for Better Nutrition

In developing country like India, malnutrition to children is very common. Efforts, therefore, are essential to upgrade the nutritive value of agricultural product particularly cereals. Similarly, high HCN content in sorghum during early stage is fatal to the cattle, high tannin content in sorghum and Bajra adversely affects Fe absorption and crucid acid in mustard oil results in Atherosclerosis of heart. While increasing the genetic yield potentiality of food grains, significant improvement in their nutritional quality is also must.

6. Achievement in the Evolution of New High Yielding Varieties

In recent years number of varieties has been developed in various crops suitable for different agro-climatic conditions with high genetic yield potentiality and nutritive value. For example two gram varieties i.e., G. 214 and G. 130 have been bred for rainfed and adequate rainfall or limited irrigation conditions respectively. The release of cowpea variety, FS- 68, which matures in 63 days and Varsha of mung maturing in 55 to 60 days are notable achievements. G.304 of Hooka Tobacco yields about 20% higher with far superior quality. Prakash in the case of raya and Sangam in Toria have been found much promising. Apart from these, two gram varieties H. 208, H. 355, cowpeas variety HFC. 42-1, sweet Sudan grass 59-3, lucern T. 9 is imminent. In the case of wheat outstanding achievements has been the evolution of variety WH.147. Similar achievements have also been made in crop like barley, oat, sunflower, castor etc. in different parts of the country which have been discussed in the chapters dealing with different crops in the book entitled **"Crop management and Integrated Farming"** written by Prof. S>C>Panda.

FUTURE PLANT BREEDING STRATEGY FOR DEVELOPING NEW PLANT TYPES

In the process of Green Revolution, role of new plant types or idio-type needs no emphasis The high yielding varietal programme has been pivotal to this revolution Now with the availability of sufficient hybrids and new plant types in several crops, efforts are therefore, necessary to exploit the maximum energy resources either solar or nuclear. While breaking the present day yield barriers in crop plant we will have to approach for a plateauning of yield levels. Better plant type, use of bio-regulators, formation of new production system and up-grading yield level through manipulation of photosynthetic efficiency will be main objectives for such efforts.

In above context, the end point in all such efforts, is the formulation of the appropriate plant type for each crop for given environment. In the area of advance agronomy the main objective is to engineer for massive yield. Under such circumstances, the plant type should be of closed or compact habit to facilitate dense planting, fertilizer responsive and tolerant to disease and pests. On the other hand, in areas of low productivity, the immediate objective is to stabilize yields at moderate levels. Thus plant type for such areas should be of open habit to facilitate sparse planting and characteristically high stability of production over a wide fluctuating environmental conditions of soil and season. Apart from this, genetic resistance to pests and diseases and drought are essential. The plant breeding end point therefore, be a hormogeneous population like pure line or hybrid for areas of advanced agronomy and heterogeneous population like synthetic, composite and varietal blends for areas of low productivity.

In addition to the reconstructing morphological frame-work of plants, their production physiology is also very important. Improved varieties are much more efficient in per day Calories production along with efficient potentiality of dry matter production. Essential point is to consider how much of the energy and dry matter produced by the plant is stored in the grain. The day productivity and grain and straw ratio of some varieties in different crops are presented in the following Tables No. 18.1 and 18.2 respectively.

It is evident from the Tables that the improved varieties have more productive potential and less straw: grain ratio indicating that their photosynthetic efficiency is sufficiently high. However, we have been able to increase their yielding potentiality only through altering straw and grain ratio than productivity *per se.* In such varieties it is the distribution of dry matter which has been exploited rather than increasing photosynthetic efficiency. Thus, exploiting the photosynthetic efficiency of plants should be considered in future breeding programmes.

TABLE 18.1: Productivity of Some Varieties

Crop	Varieties	Yield in t/ha.	Growing Period	Cal/ha per day x 10^3
Rice	ADT. 27	4.2	105	141
	IR.1	4.9	140	121
	Jaya	5.6	130	151
Wheat	NP 710, 824	2.9	135	74
	Sharbati Sonara	4.9	120	140
Maize	Ganga 101	1.7	95	65
	Ranjit	1.2	110	39
Deccan		3.0	110	98
Sorghum	CSH-1	3.0	100	106
Local		2.0	150	47

Source: N. Parthasarthy S.S. Raynjan IRC News Letter Vol. XXI 4, 1972.

TABLE 18.2: Grain : Straw Ratio in SOME varieties

Crop	Varieties	Total dry Matter (g/m^2)	Straw : Grain Ratio	Yield Q/ha
Rice				
	T.N.!	750	0.93	50.10
	Tainan. 3	697	0.91	45.30
	MTU.15	–	1.20	36.55
Wheat		(g/Plant)		
	NP. 824	5.32 (g/Plant)	1.48	33.4
	Kalyan Sona	6.24	1.23	37.3
	HD. 1155	4.24	1.14	31.2
Sorghum		(g/Plant)		
	CSH. 1	225.5	1.14	55.30
	CSH. 2	383.8	1.63	94.30

Source and Sink Ratio

High yields from a variety can be obtained only when enough photosynthetic products are first produced and then accumulated in greater extent in form of grain. The photosynthetic efficiency of variety thus becomes **'Source'** as product of photosynthetic products and efficient distribution, of this product is the **'Sink'.** For exploiting yielding potentiality of a variety the relative size of source and sink is required to be increased because high yielding potential could only be realized when a balance increase both in the Source and Sink has been achieved. High yielding varieties introduced in different crops have high Source and Sink ratios. These varieties can manifest their full capacity when enough size of producer (Source) and receiver (Sink) of photosynthetic products are formulated and then allowed to function.

NEW PLANT TYPES AND DRY LAND FARMING

In India, out of 137 million hectares cultivated land, only 27 million hectares are irrigated. The rest 110 million hectares receive no irrigation and crop cultivation in such areas is relatively difficult which depends mainly on intensity frequencies of rainfall. Crop cultivation in these areas is called as **rainfed farming.** The extent of total precipitation in rainfed zones may vary from 400 mm to 1000 mm. In certain localities the rainfall do not exceed above 500 mm annually. Such areas in many sense are more problematic than the rainfed areas as far as crop production is concerned. These areas are characterized by inadequate and uneven distribution of rainfall. Such areas that receive annual precipitation ranging from 400 to 1000 mm with erratic distribution are called as **"dry area"** and crop husbandry pertaining to such area is known as **"dry land farming".** About 47 million hectares of land in the country falls under this category and of the total rainfed area of 110 million hectares amounting to nearly 42.5 per cent. These areas are characterized by, besides inadequate and ill distributed rains, high evaporation rate (200 to 350 cm. annually), high summer temperatures up to 49°C at some places, desiccating winds of high velocity and poorly fertile soils, often subjected to erosion. Apart from the uncertain and inadequate rainfall, long dry intervals between rains, low humidity and run off contributes significantly in partial or complete failure of crops in these areas.

In spite of our effort to increase irrigation facilities, crop cultivation in large areas in Tamil Nadu, Mysore, Maharashtra, Gujrat, Rajasthan, Haryana, U.P. and Bihar depends on vagaries of rainfall. The main problems in these tract are the control of soil erosion and shortage of moisture. In arid and semi arid regions animal husbandry has been more important than crop husbandry. However, due to continuously decreasing productivity, on account of over grazing, farmers have been compelled to adopt crop cultivation. The cultivators of these areas are still adopting crop husbandry on the basis of traditional approaches i.e., (1) low intensity cropping, (2) little or no use of fertilizer or manure, (3) excessive tillage operations and (4) raising low value crops. Generally, long duration and slow growing varieties are grown in the hope that small change in the behavior of rainfall may not affect the yield much. In as much as, depending up on the rainfall alone, larger areas in these tracts have become one crop area. In places where rainfall is both deficient and unreliable only a catch crop is practiced. The single crop farming usually means the poor productivity and precarious standard of living

The course of increasing the agricultural productivity in such areas by introducing high yielding varieties is limited unless suitable techniques are developed. In fact the main criticism against high yielding varieties programme is that it serves in relatively better off areas and neglects dry areas. This criticism may not be well informed but nonetheless it underlines the urgency of a major strategy for technological transformation of unirrigated farming. The area under dry land farming being vast, a small percentage of increase in yield will mean large addition to the total agricultural production. It is, therefore, imperative to find out some ways and means for maximizing the production in these areas, per unit time per unit water stored in the soil. In the past many efforts have been made to increase the production by adopting various improved techniques but without much success. The only limitation in such efforts has been a variety suitable to fit in with the rainfall distribution pattern in dry areas with high genetic yield potentiality. The **concept of a new plant type** in this concern will altogether be different. A variety possessing wide adaptability, short in growth duration, drought tolerant with high genetic yield potentiality will be called as New Plant Rype for dry land farming.

Improving Plant Materials for Dry Land Farming

Attempts have been made in India to breed plant type of short stature and high yielders in recent years and considerable success have been achieved in improving the productivity of crops both under irrigated and dry farming conditions. Hybrid vigour has also been utilized to stabilize the production under rainfed condition. In terms of modern technology such plant material are called as **"Ideo Types"** instead of new plant types. Ideo types, are the strains of suitable dry land crops characterized by short growth duration, effective root system, drought tolerance, high yielding potentiality with altered morphology of the plants conducive to dry conditions. Efforts have been made by the plant breeders to incorporate the desired characters on the basis of findings obtained in the investigations carried out in recent years. Some of the outstanding results obtained are given below:

1. Effective growth of roots with deep penetration is desired under dry land condition for proper utilization of nutrient and moisture.
2. Correlation between root penetration and ear emergence in wheat has been observed. Cultivars earing between 90 to 100 days had deep root system than earing in 70 to 90 days (Asana and Singh, 1967).
3. Plant with few leaves just sufficient to maintain the photosynthetic output and growth, will economize the water expenditure.
4. Under dry conditions, leaves horizontally disposed have been found better for light interception contrary to vertically disposed leaves leaves most effective under irrigated conditions.
5. Under dry condition, maturity characters like number of grains per ear head and test weight of 1000 grains are most effective and significantly after the total output. Whereas with adequate supply of water number of ear heads per plant is the potent contributory factor towards yield.

It is fortunate that almost all the high yielding varieties are of short duration. This is itself makes it possible for the crop to evade drought. In high yielding new plant types the root-shoot ratio is (also very favourable) narrow making them more tolerant to drought. With these qualities in new plant types, efforts have been made to screen out some promising suitable strains for dry land farming. The details of such selected strains have been discussed else where. Although the tall varieties to certain extent have provided strains more suitable for dry farming, yet these varieties in an arid climate are an effective sink for adverted energy. As the aridity increases, the difference in the water needs between a tall and short plant are accentuated to such degree that they render the survival of tall vegetation in arid areas most difficult. In addition to this, yield potentiality of tall varieties is very low. Therefore, to have successful and profitable cultivation in dry areas, besides evolving techniques of soil and crop management, so as to make best use of stored water, an ideotype is a must. However, a detailed study regarding the critical periods in the life span of these ideotypesis also important from crop management point of view. Critical period has been defined as the stage at which the crop suffers most due to a given adverse factor, particularly water stress. If during the critical period there is moisture stress, the crop will be severely affected, even if the condition has been favourable throughout the remaining portion of growing period. Ascertaining the critical stage for moisture stress in these types, will prove much useful in scheduling the proper time of their sowing in accordance with the rainfall distribution pattern of the locality so as to avoid drought during critical periods. In case of wheat, this period for moisture occurs 15 days before earing and in maize, 15 days after appearance of male inflorescences. On the results mentioned above Asana (1968) has suggested a model of **ideotype** in wheat for dry land farming which are as follows

Root Growth: must have deep root system at least up to the depth of four feet and should be related to time of ear emergence.

Leaves: should be horizontally disposed and be about seven in number with a large size flag leaf.

Peduncle: it should be long enough to cover substantial part of stem.

Ear: should be long with more number of grains per ear. Fertile branched ears are preferred.

Time of earing: ear must emerge at such time so that grains may develop for at least five weeks at a mean time maximum temperature around 25°C.

Ideotype in Rice

The area under rainfed upland rice (700-1100 mm rainfall zone) is nearly 5 million hectares and forms 15-20 per cent of total area under rice. These areas have low production potentiality because of poor soil fertility status and lack of suitable varieties. Since the rice plants are very much susceptible to moisture stress at any stage, a suitable drought resistant variety is therefore, very necessary to stabilize the yield in dry areas. The **ideal plant type** for rainfed uplands rice should possess the following characteristics (Rao et al., 1972).

Ideal Plant Type for Rainfed Uplands Rice

1. Early in growth duration (65-100 days)
2. Deep root system with maximum ramification of roots at deeper zone (25 cm).
3. Dwarf (below 100 cm) in height with erect leaves and thick stem.
4. Early vigour with moderate tollering (6.4) and synchronous flowering.
5. Good expression of the panicle at the maximum density (about 400 panicles/ m^2).
6. Panicle weight type with high grain number per panicle.
7. Moderate seed dormancy.
8. Low requirement for iron and manganese.
9. Resistant to diseases like Helminthosporium.

HIGH INTENSITY CROPPING PROGRAMMES AND NEW PLANT TYPES

The necessity for increasing the food production is one of the major problems of the whole world. The world to-day is faced with a serious ever-increasing food shortage. Particularly, the developing countries are increasingly finding it difficult to feed their fast growing population. In India, there is a current shortage of about 16 million tonnes of food grains which will go up to 45 million tones by the end of 2010. This shortage may still go higher with the present rate of increase in population. The problem of agricultural production and rate of growth in agriculture, keeping in view the ever increasing population, have been worrying all since long. Despite considerable investment of resources in farming, irrigation and agro-industries in Five Year Plans, the output has not been up to the expectation. During the first three plans we had about 2.7 per cent increase in production of food grains. But to satisfy the required food grain demand, the rate of increase in production must average 8.2 per cent or more per year.

In the past, the increase in agricultural production has been obtained by increasing the area under cultivation to certain extent. In most of the developing countries to-day the frontier of available cultivable land has already been reached. In

India, the increase in area is hardly one per cent per annum. Presently 43 per cent of the total area is under cultivation in India and is the highest in the world. The scope for further increase in this direction is very limited. Even the marginal lands are being cultivated at present. Therefore, the production has to be increased within the limitation inflexible land area and the only course left is to increase the agricultural productivity. The agricultural productivity of the developing countries is very low as compared to the developed countries, but the possibility of increasing it is immense. By developing new varieties responsive to heavy fertilizer application with high yield potentiality and making the supply of desired inputs of production, the productivity can be stepped up substantially. In the last 15 years whatever the success has been achieved in agricultural production in the country, is only due to the introduction of high yielding varieties of crops. These varieties have opened vast avenues for increasing the productivity of agriculture per unit area per unit time by increasing the intensity of cropping. In the battle for greater food production, therefore, one of the soundest strategies for success in the existing conditions of India is the maximum utilization of cultivable lands by increasing the intensity of cropping i.e., cultivating more than two crops in a year. The pre-requisites for such crop plan will be the short growing, high yielding and fertilizer responsive varieties that would fit into an annual cycle of operation for three crops grown in one year. But above all these, the adequate supply of fertilizers, assured irrigation and technical know-how are the must for successful implementation of the such programme often called as **"high intensity cropping programme".** Under high intensity cropping programme two systems have been developed, namely the **Multiple Cropping** and **Relay Cropping**.

Multiple Cropping

The main problem in Indian Agriculture is the low average yield per unit area unit time. For low yield, poor soil fertility and low yield potentiality of varieties have been accounted as the main factors, whereas the low productivity per unit of time results from the fact that most of the land are cultivated only during a part of the year. Foe several months in the year land remains devoid of any crop. The long duration varieties of most of the important crops have been partly responsible for a state of affairs.

With the introduction of quick maturing, photo-insensitive, fertilizer responsive dwarf varieties of cereals, coupled with the recent concept of minimum tillage in crop production, it has become practically possible to increase the intensity of cropping under irrigated condition. The cultivation of two or more than two crops in succession in order to increase the intensity of cropping is called **Multiple Cropping.** It may consist of mono-culture when only cereals are cultivated or it may be double cropping nature when legumes are also included. **The concept of multiple cropping programme** envisages that the field is not kept fallow for more than a few days in the year for cultural operation. The high yielding varieties programme is thus, one of the methods for maximizing yield per hectare. The important factor in agriculture is is time factor. The multiple cropping programme which is the second major plank of new agriculture strategy aims at maximizing yield per day and the total yield in the whole year. The recent developments in the field of soil management have made multiple cropping quite feasible in tropic where the temperature and sunshine are such that one or the other crops could be grown throughout the year provided water is available in plenty for irrigation. In this context Melsted has mentioned that **"Multiple cropping is a philosophy of maximum crop production per hectare of land with a minimum of soil deterioration.** The philosophy is based on the concept that high production is good for soil, that minimum tillage promotes soil tilth and conserves soil organic matter; that high fertility promotes high yield and lessens the loss of soil humus; that large amount of decompoble organic matter in soil is essential to good soil tilth and good soil physical conditions and that the cover in the form of living mulch is good protection against all forms of erosion." As a matter of fact all these concepts are the essential features of multiple cropping. For successful implementation of multiple cropping the concept of minimum tillage and its proper utilization and application of adequate quantity of manures and fertilizers are very essential. It is also obvious that growing of more crops will add, substantial quantity of organic matter in form of crop residues in soil which helps in proper maintenance of physical properties of soil. As such multiple cropping programme, apart from increasing the production per unit time per unit area, is also helpful in increasing the efficiency of farm and efficient utilization of available resources.

In India, the index of multiple cropping is very low as compared to other countries. The **index of multiple cropping** may be defined as **the ratio between gross and net cropped area.** This index is only 1.5 in India as compared to about 2.5 in many agriculturally advanced countries. Such low index is primarily because of lack of irrigation facilities. Apart from this mono-culturing, unskilled management of soil and lack of fertilizers are other major factors for low multiple cropping index.

The multiple cropping is, essentially a programme for irrigated area with adequate fertilizers, and plant protection measures. With these facilities, it is not only possible to increase the intensity of cropping but the yield of crops can also be increased and stabilized. It has now become possible to obtain more than 100 quintals of grain in addition to more than 25 quintals of pulses and oil seed from one hectarein one year with successful implementation of multiple cropping by increasing the cropping intensity up to 400 per cent. However, the success and the achievements of such cropping programmes, depends to a large extent on the high yielding strains of cereals and other crop plants tailored to fit into such system. The increase in per hectare yield is being made possible by a number of multiple cropping patterns which the new varieties are generating. In recent years, spectacular progress in this direction has been made in the various parts of the country. For example, the division of Agronomy, I.A.R.I., New Delhi, has successfully demonstrated the technique of adopting multiple cropping and obtained 125 quintals of total grain production in one hectare. This is possible with a one year cropping pattern, which involves the cultivation of wheat, mung, maize and mustard in succession. The early maturing wheat variety Sharbati Sonara with yield potential of about 45 quintals per hectare was planted in January and harvested by 15th of April. The wheat crop was followed by a crop of mung which was grown from 20th of April to 30th June giving an yield of 4.5 quintals/ha. The third crop raised was of hybrid maize from 7th July to 30th September giving a yield of 60 quintals per hectare. Lastly, a crop of Brassica (toria) was taken from October to December with an yield of of nearly 15 quintals/ha. Similar demonstrations have also conducted in other areas such as:

1. Jute-Rice-Wheat
2. Hybrid Maize-Caulif lower-/Potato-Onion
3. Rice-Mustard-Rice
4. Rice-Ragi-Rice/Groundnut.
5. Rice-Wheat-Mung
6. Maize-Rice-Rice
7. Rice-Wheat-Rice
8. Groundnut-Jute-Rice and so on.

In order to increase the intensity of cropping and to promote the cultivation of more profitable crops, production efforts will be diversified. Efforts will be made to obtain the maximum possible efficiency of available inputs particularly water and land. In addition to this, guidelines on the inclusion of crops, selection of areas, strengthening of plant protection measures, supply of timely and labour saving implements and machinery etc. were also formulated for profitable multiple cropping.

Relay Cropping

It is also a high intensity cropping programme in which at least four crops are taken in one year. In objective, multiple cropping programme as well as relay cropping are almost the same, however, the pattern of growing crops differ. Relay cropping is new system of cropping analogous to **relay race** where a crop hands over the land to next crop in quick succession. Sometimes the next crop is sown in standing field of previous crop when it is almost matured or ready for harvest within a week or two. This is done usually to accommodate more crops in this system in one year. Commonly this system is called **"Para cropping".**

The system of relay cropping has been quite popular since long to the vegetable growers of India. Most of the vegetable growers are still cultivating 4 to 5 crops per year. However, this system could not be adopted with grain crops, due to lack of short duration varieties. Now with the introduction of short duration new plant types of cereals it has become possible to adopt **relay system of cropping** in case of grain crops. In the recent past much efforts have been made to popularize this system with the cultivators of irrigated areas and it is gratifying that considerable success have been achieved. One of the limitations in this system is short time for tillage operation. Since the time lag between the harvesting of one crop and sowing of the next, is very short, adequate tillage is not possible. The **success of relay cropping** therefore, depends on the effective management of land based on minimum tillage concept. Usually under such cropping programmes, the lack of the tillage is supplemented by increasing the frequency of inter-cultural operations. Inclusion of legume crop also helps in this direction in addition to maintain the soil productivity. Thus, short duration varieties of cereals, suitable legume crops, effective tillage implements and availability of other inputs are the pre-requisites for successful implementation of relay cropping. According to Dr. Bains **"the relay cropping aims** at increasing the efficiency of farm inputs to maximize outputs through a judicious **selection** of crops and their varieties tailored to fit into a tight schedule farm operations and opening vast avenues of employment opportunities even on small farms. It is designated to batoned in such a way that three food crops viz., mung, maize and wheat and one cash crop toria (*Brassica compestris*) are grown on the same piece of land within one crop calendar year".

The crop calendar, pre-sowing tillage requirements and fertilizer schedules recommended are as fallow:

CROP CALENDER

Crop	Variety	Sowing time	Harvesting time	Duration (days)
1. Mung	Pusa Baisakhi	3rd week of April	End of June	65-70 in rows 50 cm apart
2. Maize	Hybrid Ganga-3	End of June in rows 30 cm a part	First week of October	85-90
3. Mustard (Toria) selection)	Toria T. 22 (Punjab	3rd week of September in rows 50 cm apart	Last week of December	100-105

Crop	Variety	Sowing time	Harvesting time	Duration (days)
4. Wheat	Sharbati Sonara	End of December	3rd week of April	105-110

PRE SOWING TILLAGE EQUIPMENT

Crop	Tillage
Mung	One cultivation after irrigating the wheat stubble land.
Maize	Plant in standing crop of Mung
Toria	Plant in standing maize crop.
Wheat	Three cultivation after irrigating toria stubble land.

SEED RATE AND FERTILIZER SCHEDULE

Crop	Seed rate (kg/ha)	Fertilizer kg/ha		
		N	P_2O_5	K_2O
Wheat	150	100	50	50
Maize	15	100	75	50
Toria	5	49	-	-
Mung	50	-	-	-

In this plan sowing of crops has been recommended in lines 50 cm apart for wheat. This spacing facilitate the sowing of next crop in standing condition of previous crop as well as in the application of fertilizers as top dressing and inter-cultural operations. The required quantity of fertilizers should be drilled between the rows either by hand operated fertilizer drill or by opening by country plough. Care should be taken in selecting the varieties particularly of wheat. Since the sowing of wheat usually delayed by two weeks than the proper time, varieties suitable for late sowing should only be selected.

The concept of minimum tillage with main emphasis on weed control has been fully exploited in these programmes. The main objectives of tillage operation is control of weeds, removal of stubbles and providing suitable tilth for proper germination, is very well obtained by the schedule of tillage operations recommended.

While preparing the relay cropping programme for any area care should be taken to select suitable short duration crops. As a matter of principle, out of the crops selected one should be restorative legume crop, one neither exhaustive nor restorative as toria and two should be exhaustive crops (cereals). Similarly, the exhaustive crops should be followed alternately by legume and oil seed crop.

The four crops in **one year relay cropping plan** developed at IARI, New Delhi has 400 per cent cropping intensity and have given balanced production of 90 quintals of cereal, 10 quintals of high protein pulse grain in addition to 150 quintals of nutritious legume green fodder and 8 quintals of oil seeds containing 33 per cent edible oil extensively asa cooking medium. Although this recommendation has been made good under the agro-climatic conditions of Delhi region, has been found quite remunerative, but it may not hold good for other parts of the country. However, keeping its main principle in view, minor alterations in selection of crops and their varieties can be made by farmers as per the locality and available resources, for example, as an alternative to toria. Graig's Defiance or on 1202 varieties of potato be planted after maize and harvested by the end of December. Similarly, IR 8 or any other dwarf short duration rice variety can be taken in place of maize in the areas having adequate irrigation facilities.

Adoption of relay cropping, therefore, is a mean for higher production in limited time. It provides ample opportunities for maximum utilization of available farm resources and thus increases the output and improves economic status of the cultivators.

Pre-Requisites of Multiple and Relay Croppings

In addition to the technical know-how, efficient management and care, certain other requirements are very much essential for successful implementation of multiple and relay cropping. These requirements, in true sense, are the pre-requisites of such crop planning, which are as follows:

1. New Plant Types of Crops,
2. Maintenance of Soil Fertility,
3. Availability of Required Inputs,
4. Effective Tillage Implements,
5. Proper Crop Residue disposal and
6. Effective Plant Protection measures.

1. New Plant Types

The high yielding new plant types of cereals are not only known to increase per hectare yield but also the per day productivity of land. The important features of these varieties have already been discussed earlier. The short growing duration, photo-insensitivity and altered morphology make them much suited under high intensity cropping programmes. Only with such varieties, it is possible to increase the yield per hectare per unit time. These new plant types are high yielding and short in growing duration and their per day productivity is also high. They are very much responsive to manageable inputs particularly fertilizer and irrigation, and as such their yield potential can be stabilized by proper application of these inputs even under intensive cultivation.

2. Maintenance of Soil Fertility

Continuous cropping in a particular land is found to cause soil fertility problems and therefore, success multiple cropping solely depends on maintenance of soil fertility. In case the multiple cropping programme includes mono-culturing (cereal after cereal) the problem of maintenance of organic content is also important. Under this system of cropping judicious application of chemical fertilizers will be needed. Unless this input is adequately made available, growing of two or more crops will not be possible. Experimentally, it has been proved that it is always better to include one legume crop in this system. Legume crops have a remarkable power to improve physical character of soils, particularly structural status of sandy loam soil. Such structure building properties of legumes is further enhanced by the application of phosphorus. The organic and the total phosphorus content of soils have been found to increase. Legumes fertilized with phosphorus significantly contribute to the soil aggregate formation and an improved aggregation status is further manifested by higher available water and water retention characteristics of soil. Since nitrogen is universally deficient in our soils, for maximum yields of non legumes, there is an urgent need of including a legume in this system. Besides this balance inorganic fertilizer application coupled with once or twice application of organic manure every alternate years is very necessary for the maintenance of desired productive status of the soil. Long terms trials at Pusa, Coimbatore, and Cuttack have shown that plot receiving only balanced inorganic fertilizer have maintained yield and organic matter content of the soil. The four crops in a year system therefore, be followed only when the fertilizers are adequately available.

3. Availability of Required Inputs

Beside fertilizers, other inputs particularly irrigation is the most important for the successful implementation of multiple cropping system. In most of the irrigated conditions, where inputs like fertilizers and improved seeds are easily available, this type of cropping system has given quite a lucrative remuneration to the cultivators. However, this system can also be implemented even in rainfed (unirrigated) conditions with a little modification in the cropping pattern. For this, concerned efforts are required to find suitable crops, their varieties and agronomic practices which can fit in **relay sequence of multiple cropping.**

4. Tillage Implements

The importance of good tilth for high yield has been generally accepted but seldom found in research results to be as important as it appears. On sandy soils the tilth is seldom a problem. Even on medium textured soils (silt loam and loams) an acceptable level of soil tilth can be obtained with minimum tillage. Buy on fine textured heavy soils (silt clay, clay and clayey loams) favourable tilth is more necessary than the coarse and medium textured soil. Since in multiple cropping, crops are taken in a relay sequence, very important to develop suitable implements for the mechanical manipulation of soils in the plough layer within the time gap available. Sometime the deficiencies of tillage operations due to paucity of time are also remedied by increasing the intercultural operations. The main objectives of tillage operation on irrigated soils are to burry weeds, uproot stubbles and incorporate organic matter besides preparing the seed bed into good tilth as also to increase the microbial activities by aeration. These objectives can also be achieved by increasing the internal operation by effective implements. The need therefore, is to device suitable, handy and efficient tillage implements for this system of cropping.

5. Crop Residues

The **multiple cropping or relay cropping systems** are often confronted with the problems of crop residues. The crop residues left by the crop must be cleared or be well incorporated in soil after decomposition. Due to high intensity of cropping the possibilities of residual nitrogen effect in this system is also not possible and as such the crop residues which C/N ratio is liable to remain in field for longer duration. Efforts is thus desired to work out suitable agronomical measures to overcome this problem. To certain extent the decomposition of crop residues can be enhanced by the application of some nitrogen (depending upon the crop residues, soil and soil moisture) in the soil. This nitrogen should be given immediately after the harvest of crop fallowed by tilling with soil inverting plough and planking.

6. Plant Protection Measures

Effective plant protection measures are very essential under high intensity cropping programmes. High yielding new plant types of cereals are often attacked various diseases, and insect pests. In addition, under such crop plan, if plant residue remains undecomposed, it provides ample opportunities for attack of termites and other insects. It is therefore, necessary to take every possible care to prevent the crops from attack of diseases and insect pests. It has also been observed, particularly when mono-culturing is adopted under high intensity programmes that some physiological diseases occur due to imbalanced soil fertility status or due to the deficiency of certain microelements. Such diseases have been reported in case of rice due to K and Zn deficiencies (Akagore and khairs) and in wheat due to certain micro element deficiency. Correct identification of such diseases is essential at the initial stage of their appearance and measures should be taken to overcome them.

> ***"The important thing in science is not so much to obtain New facts as to discover new ways of thinking about them"***
> ***William Bragg***

19 Production and Package of Practices

> *"The secret of rapid agricultural progress in the under developed countries is to be found much more in agricultural extension, in fertilizers, in new seeds, in pesticides and in water supplies than in altering the size of the farm, in introducing machinery, or getting rid of middlemen in the marketing process".*
>
> *W. A. Lewis*

INTRODUCTION

Production is a process whereby certain goods and/or services are used to create goods and/or services of a different nature. Production is the name given to the process of conversion of certain inputs (resources) into a consumable form. Farm production likewise refers to producing of food, fibre and livestock by using several different kinds of inputs. Land is used by the farmers as a factory which helps them producing the desired crop. To this manufacturing plant (land), labour and capital are added to cultivate, plants and harvest the crop. When considered necessary, fertilizer is also added by the management. Water may either be provided by rainfall or by artificial irrigation methods. Application of all these inputs results in the desired crop (out put). The crops so produced are, in turn, fed to animals which produce meat, milk, eggs and many other livestock and poultry products through complex biological processes.

Agricultural production economics is, thus concerned with the quantitative relationships which are basic to production processes in agriculture. "These relationships take the form of input-output patterns, and the various types of interactions among the individual inputs themselves and among the products which contribute to the output. It is also concerned with levels of factor costs and product prices and with the nature of production patterns which allow the attainment of certain desired optima, like profit maximization or cost minimization". Subject matter of production economics covers all agricultural problems that fall under the scope of resource allocation and marginal productivity analysis. The farmer as a production economist is hence concerned with any phenomena which have a bearing on economic efficiency in the use of agricultural resources.

Products are the result of the use of resources or services of resources, examples are wheat, maize, milk, eggs, etc. Production is a process of transformation of certain resources or inputs like land, labour (human and bullock), seeds, fertilizer, irrigation water, into products like wheat, rice, milk. Farm production decisions involve application of the principles of production economics. It is, therefore, essential for a farm operator or a planner to understand basic concepts, and relationships pertaining to the economics of agricultural production.

PRODUCTION ECONOMICS

Farm production economics is a broad division or field of specialization within the subject of agricultural economics. It is concerned with the choice of production patterns and resource use in order to maximize the objective function of the farm operator, their families, the society or the nation, within a framework of limited resources. It is concerned with choosing of available alternatives or their combinations with a view to maximizing the returns and/or minimizing the costs.

Production economics is concerned with two broad categories of decisions in the production process:

1. How to organize resources in order to maximize the production of a single commodity i.e., to make choices from among various alternatives ways of using resources.
2. What combination of different commodities to produce.

   The goals of agricultural production economics are two fold:

   a) To provide guidelines in using the farm resources most efficiently and
   b) To facilitate the most efficient use of resources from the stand point of the economy.

FACTORS OF PRODUCTION

There are three main factors of production, viz., land, labour and capital at the disposal of the farmer. It is the job the farmer to use these three inputs in combination on the farm. In doing so, the farmer has to perform two distinct functions. In the first place, he has to act entrepreneur or proprietor of the farm business and is responsible for farming the general policy or plan on which his business and his system of farming is based. In the second place, he has to act as chief executive or farm manager and is responsible for the administration of that plan. These both functions of the farmer are concerned with the fundamental problem of ascertaining and adopting the optimum combination of the land, labour and capital at the disposal of the farmer. Under any given set of circumstances, the optimum combination is that which will enable the farmer to obtain the maximum financial input from his farm as a whole.

Production decisions facing the individual farmers are of three types:

1. Farmer may wish to produce a certain amount of a given product having certain amount of resources to use. He will have to decide the most profitable amount of resource to use in the production of a commodity.
2. Having a certain combinations of resources, he may have to decide most profitable combination of resources to use to produce a specific amount of given output.
3. Having a certain combination of resources he is also faced with problem of choosing the most profitable mix of products to produce.

PACKAGE OF PRACTICES

In large areas of the world, agricultural production is mostly a private business carried on in million of farms. Efforts to promote agricultural progress to meet rising demand for its products, generated by the rapidly growing population, therefore, must be persuasive and must work through market forces. Price policy, therefore, is an important instrument for providing boost to stagnant agriculture.

The amount of change in agricultural production which is brought about by the farmer as a sequence of change in prices of agricultural products is known as supply response. Thus the function of the factual change in per acre production of agricultural crop with respect to the change in its market price is known as the crop acreage response. With a view to optimizing the use of farm resources on an individual farm level and to rationalize the use of agricultural resources from a national angle, the Package of Practices for Field Crops, Fruit Crops, and Plantation Crops and Tips of Horticultural Crops of the State of Orissa are given below for increasing technical knowledge and use for agricultural production.

PACKAGE OF PRACTICES FOR FIELD CROPS

Cereals

Unit – One acre (0.4 ha)

Sl. No.	Item	HYV rice (*Kharif*)		Improved rice (*Kharif*)	HYV rice (*Rabi*)
1.	Land type	High medium, medium and low		Medium and Low	Medium and Low
2.	Seed variety	Khandagiri, Parijat, Vandana, Annapurna, Subhadra, IR 36, Naveen, Kanark, Lalat, Bhoi, Padmini, Surendra, Swama, Pooja, Sarala, Savitri, Gaytri, Ramchandi, Indravati, Mahanadi, Jagabandhu, Tulasi, Panidhan, Durga, CR-1014		T-141, BAM-6, T-90, T-242	Khandagiri, Parijat, Vandana, Lalat, Konark, Naveen, bhoi, Kharvel, Satabdi
3.	Seed rate (kg)	25		25	25
4.	Seed treating chemical (g/kg)	2.5 to 3		2.5 to 3	2.5 to 3
5.	Sowing/Planting time	June-July		June-July	Dec.-Jan.
6.	Spacing (cm)	15 x 10, 20 x 10		20 x 10	15 x 10
7.	Compost/ FYM (Tonnes)	2		2	2
8.	Chemical fertilizer (kg)				
	– N	24	32	20	32
	– P_2O_5	12	16	10	16
	– K_2O	12	16	10	16
9.	Interculture (nos)	2		2	2
10.	Irrigation (nos)	-		-	20
11.	PP measures (nos)	6		3	6
12.	Bullock labour (days)	24		19	21

Sl. No.	Item	HYV rice (*Kharif*)	Improved rice (*Kharif*)	HYV rice (*Rabi*)
13	Human labour (days)	82	60	80
14.	Yield in q (approx)	16-22	12	25

Cereals

Unit – One are (0.4 ha)

Sl. No.	Item	HYV Wheat (Irrigated)	Maize (Kharif/Summer)	HYV Jowar (Kharif)
1.	Land type	High and medium	High	High and medium
2.	Seed variety	Sonalika, Kalyansona, UP-272, Sagarika, UP-262	Vijaya, Jawahar, Navjot, Madhuri, Naveen, Decan Hybrid, Ganga-5, Kargil-633	CSH-1, CSH-2, CSH-5, Swarna, CSV-13, Varsha, CSV-15
3.	Seed rate (kg)	40-50	6	4-5
4.	Seed treating chemical (g/kg)	2.5-3	2.5	2.5
5.	Sowing/planting time	2nd FN of Nov.	Jan. – Feb.	May – June
6.	Spacing (cm)	Row to row 15-20	60 x 45	45 x 15
7.	Compost/FYM (t)	2	2	2
8.	Chemical fertilizer (kg)			
	– N	24-32	32	24
	– P_2O_5	16-20	16	12
	– K_2O	12-16	16	12
9.	Inter-culture (nos)	2	2	2
10.	Irrigation (nos)	6	-	-
11.	PP measures (nos)	2	2	2
12.	Bullock labour (days)	16	11	9
13.	Human labour (days)	45	84	55
14.	Yield in q (Approx)	10-12	12-15	8-10

Cereals

Unit – One acre (0.4 ha)

Sl. No.	Item	Barley (Rabi) (Irrigated)	Ragi (Kharif) (Rainfed)	HYV Bajra (Kharif)
1.	Land type	Medium	High	High
2.	Seed variety	Ratna	AKP-2, AKP-3, AKP-7, Neelachal, Dibyasinha, Godavari, Bhairabi, Chilika	HB-1, HB-2, HB-4
3.	Seed rate	40	4	4

Sl. No.	Item		Barley (Rabi) (Irrigated)	Ragi (Kharif) (Rainfed)	HYV Bajra (Kharif)
4.	Seed treating chemical (g/kg)		2.5	2.5	2.5
5.	Sowing/ Planting time		2nd FN of Nov	June, July & Sept	June-July
6.	Spacing (cm)		20 x 7	15 – 20 x 10	45 x 10
7.	Compost/FYM (t)		2	2	5
8.	Chemical fertilizer (kg)				
	– N	16	16	24	24
	– P_2O_5	8	8	12	12
	– K_2O	8	8	12	12
9.	Inter-culture (nos)		2	1	2
10.	Irrigation (nos)		2	1	2
11.	PP measures (nos)		2	1	2
12.	Bullock labour (days)		16	9	9
13.	Human labour (days)		45	57	55
14.	Yield in q (approx)		12	8-12	10

Pulses

Unit – One acre (0.4 ha)

Sl. No.	Item	Green gram (Kharif, Pre-Rabi, Summer)	Black gram (Kharif, Pre-Rabi, Summer)	Bengal gram (Pre-Rabi)
1.	Land type	High & High Medium	High & High Medium	Medium
2.	Seed variety	Pusabaisakhi, Jyoti, Kopergaon, T-44, Sujata, K-851, PDM-11, PDM-54, Dhauli, Pusa-105	Pant U-26, T-9, Sarala, Pant U-30, Pant U-35, Co-1, Pragati, PDU-1, KU-301	JG-62, Radhey, JG-74, H-208, C-235, ICC-4, Annegiri, L-155, K-4 (Kabuli), BDN 9-3, BG-203
3.	Seed rate (kg)	10	10	20-25
4.	Seed treating chemical (g/kg)	2.5	2.5	2.5
5.	Sowing/Planting time	June-July/ Sept-Oct/ Feb-March	June-July/ Sept-Oct/ Feb-March	Oct-Nov
6.	Spacing (cm)	25 – 30 x 10	25 – 30 x 10	30 x 10
7.	Compost/FYM (t)	2	2	2
8.	Chemical fertilizer (kg)			
	N	8	8	8
	P_2O_5	16	16	16
	K_2O	8	8	-
9.	Inter-culture (nos)	1	1	1
10.	Irrigation (nos)	-	-	-

Sl. No.	Item	Green gram (Kharif, Pre-Rabi, Summer)	Black gram (Kharif, Pre-Rabi, Summer)	Bengal gram (Pre-Rabi)
11.	PP measures (nos)	1	1	1
12.	Bullock labour (days)	7	7	7
13.	Human labour (days)	30	30	27
14.	Yield in q (approx)	4	4	6-8

Pulses

Unit – One acre (0.4 ha)

Sl. No.	Item	Cowpea (Kharif, Rabi, Summer)	Horse gram (Rainfed)	Field Pea (Rabi-irrigated)
1.	Land type	High, Medium	High, Medium	Medium, Low
2.	Seed variety	FS-68, SEB-2, Swarna	DS-1-2, Local	EC-33866, GC-66 T-163, Hans, Rachana, Dhusar Swamarekha
3.	Seed rate (kg)	8-10	20	20-25
4.	Seed treating chemical 9 g/kg	2.5	2.5	2.5
5.	Sowing/Planting time	June-July, Sept-Oct., Feb-March	Sept – Oct.	Oct. – Nov.
6.	Spacing (cm)	45 x 15	30 cm in line	30 x 10
7.	Compost/FYM (t)	2	2	2
8.	Chemical fertilizer (kg)			
	– N	8-10	5	10
	– P_2O_5	16-20	10	20
	– K_2O	8-10	-	-
9.	Inter-culture (nos)	2	1	-
10.	Irrigation (nos)	2	-	1
11.	PP measures (nos)	2	-	1
12.	Bullock labour (days)	9	7	-
13.	Human labour (days)	40	27	10
14.	Yield in q (approx)	7	6	8

Pulses

Unit – One acre (0.4 ha)

Sl. No.	Item		Arhar (Kharif)	Arhar (Rabi)	Lentil
1.	Land type		High	High	Medium
2.	Seed variety		Prabhat, Pragati, Pusa Ageti, UPAS-120, Jagruti, Sweta, Mukta, Laxmi, Sarada, Basanta, Bahar	UPAS-120, C-11, Pusa Ageti, ICPH-8, Jagruti	B-77, Hyb. 1, Pusa-8, Pant L-15 K-76, DPL-15
3.	Seed rate (kg)		8	12	16

Sl. No.	Item		Arhar (Kharif)	Arhar (Rabi)	Lentil
4.	Seed treatment chemical (g/kg)		2.5	2.5	2.5
5.	Sowing/Planting time		June-July	Sept. – Oct.	Oct. – Nov.
6.	Spacing (cm)		45-60 x 20	30 x 15	20 x 5
7.	Compost/FYM (t)		2	2	2
8.	Chemical fertilizer (kg)	Early	Late		
	– N	8	8	8	8
	– P_2O_5	16	24	24	16
	– K_2O	8	8	8	8
9.	Inter-culture (nos)		2	2	1
10.	Irrigation (nos)		1	2	-
11.	PP measure (nos)		3	3	1
12.	Bullock labour (days)		8	8	7
13.	Human labour (days)		40	40	30
14.	Yield in q (approx)		8	10	5-6

Oil Seeds

Unit – One acre (0.4 ha)

Sl. No.	Item	Groundnut (Kharif – Rainfed)	Groundnut (Rabi – Irrigated)	Mustard (Rabi – Irrigated)	Mustard (Rabi – Rainfed)
1.	Land type	High, Medium	Light loamy soil	High, Medium	High medium, Medium
2.	Seed variety	AK 12-24, Kisan, JL 24, ICGS-11, ICGS-44, GG-2, Kadri-3	AK 12-24, Kisan, JL 24, Jawan, Smruti	PT-303, PT-507, Vlnaya, Pravati, Anuradha, Varuna, Pusa bahar, Pusa basant, Pusabold, B-85, YSB-6 Pravati	TS-29, M-27, T-9, PT-303, PT-507, Anuradha
3.	Seed rate (kg)	50 (Kernel)	50 (Kernel)	4	4
4.	Seed treating chemical (g/kg)	2.5	2.5	2.5	2.5
5.	Sowing/Planting time	June	Sept-Oct., Jan-Feb.	Oct.	Oct.
6.	Spacing (cm)	25 x 10	20 x 10	20-30 x 10	20 x 8-10
7.	Compost/FYM (t)	2	5	5	2
8.	Chemical fertilizer (kg)				
	– N	8	8	24	12
	– P_2O_5	16	16	12	6
	– K_2O	16	16	12	6

Sl. No.	Item	Groundnut (Kharif – Rainfed)	Groundnut (Rabi – Irrigated)	Mustard (Rabi – Irrigated)	Mustard (Rabi – Rainfed)
	Lime (q)	5 (One month before sowing)	5 (One month before sowing)	-	-
9.	Inter-culture (nos)	2	2	2	1
10.	Irrigation (nos)	-	8	3	-
11.	PP measures (nos)	2	3	3	2
12.	Bullock labour (days)	12	12	9	9
13.	Human labour (days)	70	70	40	36
14.	Yield in q (approx)	6	8	6	3

Oil Seeds

Unit – One acre (0.4 ha)

Sl. No.	Item		Sesame (Irrigated wd)	Sunflower (Rabi/Summer)	Safflower (Rabi – Rainfed)
1.	Land type		Sandy loam, High land	High, Medium	High, Medium
2.	Seed variety		Vinayak, Kanak, Kalika, Usha, Uma, Nirmala, Prachi	Sunrise, Morden, EC-68415, EC-68414, MSFH-8, MSFH-17, MSFH-1, KBSH-1, PAC-36, PAC-308	C-437, A-300, A-62-8, Manjira
3.	Seed rate (kg)		4	4 (2 for hybrids)	8
4.	Seed treating chemical (g/kg)		2.5	2.5	2.5
5.	Sowing/Planting time		Feb.	Oct-Nov., Feb.-March	2nd FN Sept. to Mid. Oct.
6.	Spacing (cm)		20 x 3	45 x 30, 60 x 30 (Hybrids)	30 x 5
7.	Compost/FYM (t)		2	2	2
8.	Chemical fertilizer (kg)		HYV	Hybrid	
	– N	12	16	32	10
	– P_2O_5	6	16	32	10
	– K_2O	6	12	24	10
	Boron (kg)	-	4	4	-
9.	Inter-culture (nos)		1	2	2
10.	Irrigation (nos)		3	3	-
11.	PP measures (nos)		2	2	2

Sl. No.	Item		Sesame (Irrigated wd)	Sunflower (Rabi/Summer)	Safflower (Rabi – Rainfed)
12.	Bullock labour (days)		9	9	9
13.	Human labour (days)		36	45	30
14.	Yield in q (approx)		4	4-8	3

Oil Seeds

Unit – One acre (0.4 ha)

Sl. No.	Item		Castor (Kharif)	Castor (Rabi/Summer)	Niger (Rainfed)
1.	Land type		High	High, Medium	High
2.	Seed variety		NPH-1 (Aruna), SA-2 (Suitable for Dry land cropping and can withstand drought for 14-20 days), Bhagya, Subhagya	NPH-1 (Aruna), SA-2 (Drought tolerant), Clipping encourages more branching in both the varieties.	Keonjhar-1, Alsi-1, IGP-72, Deomali, IGP-76
3.	Seed rate (kg)		8	8	4
4.	Seed treating chemical (g/kg)		2.5	2.5	2.5
5.	Sowing/Planting time		June	Sept-Oct/Jan.	Aug. – Oct.
6.	Spacing (cm)		60 x 45	60 x 45	30 cm in line
7.	Compost/FYM (t)		2	2	2
8.	Chemical fertilizer (kg)		Unirrig	Irrig	
	– N	20	20	30	16
	– P_2O_5	30	12	20	8
	– K_2O	20	12	20	8
	Borax (kg)	2	2		-
9.	Inter-culture (nos)		2	2	2
10.	Irrigation (nos)		-	6-8	-
11.	PP measures (nos)		2	2	1
12.	Bullock labour (days)		9	9	9
13.	Human labour (days)		30	30	30
14.	Yield in q		4-5	5-6	4

Fibres

Unit – One acre (0.4 ha)

Sl. No.	Item		Cotton (Kharif-Rainfed Improved Hybrid		Cotton (Rabi-Irrigated) Improved Hybrid		Jute	Mesta
1.	Land type		High land with well drained loamy, black cotton soil		High and High-medium (Mal)		High and Medium land	High land
2.	Seed variety		SRT-1, H-777, MCU-5, LRA-516	Banni, Savitha	MCU-5, SRT-1, H-77, LRA-5166	Banni Savitha	JRC-212, JRC-7447, JRO-878, 7835, 524, Mahadev, Rabati, Subdariffa Rabati, Baladev (C), Jayadev (C)	Cannabinus, HC 538, HC 867, AMC-108, AMV-1, AMV-3, AMV-4, HS 7910, HS 4288
3.	Seed rate (kg)		2	2	5	2	3 (for Caps) 2 (Olitorious)	6
4.	Seed treating chemical (g/kg)		2.5	2.5	2.5	2.5	2.5	2.5
5.	Sowing/ Planting time		June	June	Sept-Oct January	Sept-Oct January	March-April	May-June
6.	Spacing (cm)		90 x 45, 60 x 30	120 – 150 x 60-75	75 x 25, 45 x 25	90 x 45	30 x 5-7	40 x 10
7.	Compost/FYM (t)		2	2	2	2	2	2
8.	Chemical fertilizer (kg)					Caps.	Olit.	
	– N	32	48	32	48	24	16	10
	– P_2O_5	16	24	16	24	12	8	8
	– K_2O	16	24	24	24	12	8	8
9.	Inter-culture (nos)		5	5	3	3	3	2
10.	Irrigation (nos)		-	-	10-12	10-12	-	-
11.	PP measures (nos)		8	8	8	8	3	3
12.	Bullock labou (days)		10	10	10	10	11	11
13.	Human labour (days)		109	109	109	109	86	62
14.	Yield in q (approx)		6	7	8	8	10-12	7

Other Crops

Unit – One acre (0.4 ha)

Sl. No.	Item	Tobacco	Sugarcane (Irrigated)
1.	Land type	High and High Medium	High and Medium land
2.	Seed variety	Bidi type: Ananda-3, Gajapati, Pyruvithunum	Early: Co-6907, Co-7208, Co-87263 (Surya), Co-7704, Co-7805
		Chewing type: Bhagyalaxmi Meenakshi	Med Late: Co-62175, Co-87044, Co-6304, Co-86249 (Bhabani), Co-8201, Co-84044 (Uttara)
3.	Seed rate (kg)	0.2-0.3	3000-5000
4.	Seed treating chemical (g/kg)	2.5	Dipsetts in 0.5% Emisan or 0.25% Aretan for 10 minutes
5.	Sowing/Planting time	Sept-Nov	Jan-March
6.	Spacing (cm)	60 x 40	Early-60, Med. Late 75-90
7.	Compost/FYM (t)	2	10
8.	Chemical fertilizer (kg)		
	– N	24	80
	– P_2O_5	16	24
	– K_2O	16	40
9.	Inter-culture (nos)	2	3
10.	Irrigation (nos)	5-6	20
11.	PP measures (nos)	3	6
12.	Bullock labour (days)	11	13
13.	Human labour (days)	114	262
14.	Yield in q (approx)	5	350-400

PACKAGE OF PRACTICES FOR FRUIT CROPS

Papaya (*Carica papaya*)

It produces throughout the year and requires less area for tree, easy to cultivate

Varieties: Coorg Honey Dew, Co-1, Co-2, Pusa Delicious, Sunrise solo, Pusa Majesty, Pusa Nanha, Pusa Dwarf.

Soil: Loamy soil, Well drained- aerated soil with high pH (8.0) and low pH (5.0) should be avoided.

Planting distance: 1.4m x 1.6m

Pit size: 0.5m x 0.5m x 0.5m

Planting time: July – August

Planting material: Seedlings

Manure and fertilizer: (per pit)

FYM – 2 baskets			Top dressing- FYM- 2 baskets	
Basal	N	- 50 g	N	- 50 g
	P_2O_5	- 40 g	P_2O_5	- 40 g
	K_2O	- 90 g	K_2O	- 90 g

Spraying of 0.5% Zinc Sulphate (Twice) and One spray of Borax (0.1%)

Time of application: June-July and October – November.

Inter culture: One in two months.

Weed Control: Deep hoeing is recommended during first year to check weed growth. Hoeing should not be done in rainy season or after fruiting since the plants are shallow rooted.

Application of Fluchloralin or Alachlorin (0.8 kg/ac or 0.4 ha as pre-emergence, 2 months after transplanting

Irrigation: During dry season, once in a week (from October tll May)

Over irrigation may cause root-rot disease

Proper care should be taken to save the seedlings in the field especially against heavy rainfall in early stage.

Plant Protection

Insect pests: Red spider – Sulphur dust.

Diseases: Anthracnose – Dithane M-45 @ 3%.

Papaya mosaic – Check insect vactor (aphids) by spraying Rogor or Metasystox.

N.B.: Plant 2-3 seedlings pee pit and remove the male plants as soon as their sex is known. However to ensure pollination, it is desirable to retain ten male plants for a group of hundred plants. Solo, Pusa Magesty, and Co-3 produce both hermaphrodite and female plants and hence, gynodioecious lines. So every plant is capable of producing fruits.

Banana (*Musa* sps.)

It is an import crop for subsistence farmers and has multifaceted uses:

Varieties: Poovan (Champa), Robusta (Semi tall), Dwarf vavendish (dwarf), Monthan, Bontha

Soil: Loamy, well drained Sandy loam soil is the best soil.

Soils having pH upto 8.5 with suitable amendments it can be grown.

Optimum pH is 6.5 to 7.5.

Planting distance: 2m x 2m (Tall), 1.8m x 1.8m (Dwarf)

Pit Size: 0.6m x 0.6m x 0.6m

Planting time: June - July

Trench planting is practiced in wet land cultivation of coastal region of Puri district. Land is prepared like rice using plenty of water gauge wheel. Water is drained from the field allowing to set a day. Planting is done by simple pressing the suckers into the wet field. After a week, 15 cm deep trenches are opened both ways maintaining 4-6 plants in each block. Deepening of trenches by 25 cm is taken up every month till suckers put forth 1-3 leaves.

Planting Material: Healthy sword suckers (about 3-4 months old)

75 cm height, 15 cm base width, 2.5 – 3 kg weight

Manure and Ferilizer

Per pit

Basal - FYM – 2 baskets (10 kg FYM)

N	- 25 g
P_2O_5	- 30 g
K_2O	- 90 g

Top dressing - (Days after planting)

	60 days	**90 days**	**120 days**
N	25 g	30 g	-

	P_2O_5	30 g	-	25 g
	K_2O	90 g	-	90 g

Potassium is indispensable in Banana nutrition due to its role in vital function.

Inter-culture: Twice in a year.
Regular hand weeding is essential as weeds reduce yields upto 40-50% depending upon the variety.

Weeding: Chemical weed control by using 0.4% Glycel spray is also effective.

Irrigation: During dry season, once in a week. Temporary bunds are made around a block of 10-20 plants and irrigation is carried out block wise.
Drip system of irrigation saves 40 – 50% water

Mulching: With sugarcane trash @ 4 tons/ ac or 0.4 ha provides effecting mulching for conservation of moisture.

Plant protection

Insect pests:
1. Aphid – Rogor 0.05% or Metasystox 0.05%
2. Rhizome weevil – Phorate granules @ 1 g/plant.

Disease:
1. Sigatoka leaf spot – Dithane M-45 @ 25%
2. Anthracnose – Dithane M-45 @ 3%
3. Bunchy top – Metasystox 0.05% or Rogor 0.05%
4. Panama wilt – Carbendazim (1 g/litr), dipping before planting.

NB: For good yield a minimum of 10-12 leaves should be maintained on the mother plant.

Guava (*Psidium guajava*)

It is a popular fruit, available throughout the year except during the summer season, hardy crop, requires little care. Best quality guavas are obtained where low light temperature (10°C) prevails during winter season.

Varieties: Allahabad Safeda, Lucknow-49, Chittidar, Arka Mridula, Allahabad Surkha.

Soil: Loamy, Well drained, pH 4.5 – 8.2, Soils of river basin.

Planting distance: 7m x 7m

Pit size: 0.75m x 0.75m x 0.75m

Planting time: July – August

Planting material: Grafts or gooties

Prunning: Fruiting trees should be pruned to check over crowding and to control plant height, the terminal shoots on the periphery may be headed back at about 40 cm level in alternate years after harvesting or in pring.

Manure & fertilizer:

Basal -	FYM	- 2 baskets
	N	- 60 g
	P_2O_5	- 30 g
	K_2O	- 60 g
Top dressing	N	- 60 g
	P_2O_5	- 30 g
	K_2O	- 60 g

Time of application: June – July and October

Inter-culture: Twice in a year. Guava plants do not require much care after

planting.
1. Pre-monsoon.
2. Postmonsoon.

Irrigation: During dry season, ones in a week.
The fruit quality is adversely affected by high soil moisture content during harvest.

Plant Protection

Insect pests:
1. Fruit fly – Sevin-2% + 17 molases
2. Mealy bug – Thimet

Diseases:
1. Fruit rot – Dithane-Z-78 @ 2%
2. Wilt – Bavistin 0.1%
3. Canker – Dithan- Z-78 @ 2%.

NB: Both rainy and winter season crops are heavy compared with spring crop. For fruit quality, the winter crop is the best. The rainy season crop can be avoided by spraying 10% urea on Allahabad Safeda and 20% on Lucknow- 49 at the time of peak flowering in Summer Season.:

Citrus (*Citrus* sps.)

Varieties:
Loose skinned: Nagpur orange, Coorg orange (Mandarin), Khasi Orange, Kinnow.
Tight skinned: Mosambi, Malta (Mosambi), Sathgudi, Lemon, Eureka, Lisbon, Lucknow seedless, Kagzikalan,
Lime: Kagzi, Pramalni, Vikram, Chakradhar, PKM-1,.

Soil: Loamy, alluvial well drained.

Planting distance: Orange- 6m x 6m, Lime- 5m x 5m

Pit size: 0.75m x 0.75m x 0.75m

Planting time: With the onset of monsoon (June – August)

Planting material: Healthy budded plants/Air layering (gootee)

Manure & fertilizer: (per pit)

Basal -	FYM	- 3 baskets
	N	- 125 g
	Neem oil cake	- 200 g
Top dressing -	FYM	- 3 baskets
	N	- 125 g
	P_2O_5	- 80 g
	K_2O	- 145 g

Age of tree (in year)	FYM (kg)	Gram (g)/tree		
	N	**P_2O_5**	**K_2O**	
1	20	50	40	120
2	25	100	40	120
3	30	200	80	240
4	40	300	160	360
5	50	400	320	480
6 & above	60	500	400	720

Applied in 2 equal splits doses, 1st before January & rest after fruit set,. followed by irrigation. Micro nutrient spray is done after the major flush of leaves which as expended 2/3rd of normal size. Time of application is during June and October in one to two

	does.	
	Zinc Sulphate	2.5 kg
	Boron	500 g
	Copper Sulphate	1.5 kg
	Lime	4.5 kg
	Manganese Sulphate	1 kg
	Urea	5 kg
	Magnesium Sulphate	1 kg
	Warer	450 litres
	Ferrous Sulphate	1 kg

Inter-culture: Twice in a year (Soil tillage, Mulching and Weed control)
1. Pre-monsoon.
2. Post-monsoon.

Irrigation: 10-15 days interval during winter
5-7 days interval during summer

Plant Protection

Insect pests: 1. Caterpillar, 2. Aphids, 3. White fly – Spray Endosulfan

Diseases: 1. Fruit rot - Dithane Z – 78 @ 0.2 %
2. Canker - Streptomycin - 500 ppm.
3. Powdery Mildew - Sulfex @ 0.3 %.

Pomegranate (*Punica granatum*)

It is a crop of arid and semi-arid regions.

Varieties: Paper shell, Spanish ruby, Dholka, ganesh, Vadki, Kandhani, Ruby, Jyoti, Mridula

Soil: Loamy well drained, Tolerates Salinity up to 9.0 EC/mm and sodicity 6.78 ESP.

Planting Distance: 4 x 5 m (200 plants/ac or 0.4 ha)

Pit size: 0.75m x 0.75m x 0.75m. 50g of Carbaryl dust/pit as precaution against termites.

Planting time: July – August

Planting material: Seedlings and cuttings/Air layering or gootee

Training: A multistem training with 4-5 stems/hill is beneficial. In general application of 600 -700g N,
200-250 g P_2O_5 and 200 – 250g K_2O/tree/year is optimum.

Manure & fertilizer:

Basal-	FYM	-	2 baskets
	N	-	320 g
	P_2O_5	-	100 g
	K_2O	-	230 g
Top dressing-	FYM	-	2 baskets
	N	-	320 g
	P_2O_5	-	100 g
	K_2O	-	230 g

Pruning: Removal of ground suckers, water shoots.
Light thinning and pruning of old spurs to encourage new growth is required.

Time of application: July – August and October – December.

Inter culture: Pruning and training frequently done in order to maintain shape, size and number of branches in a tree. Intercropping is beneficial.

Mulching: Use of soil cover to conserve soil moisture is quite essential. It is done through the use of soil cover with black polythene. Banana trashing, paddy husk groundnut shells are good mulch materials.

Irrigation: During dry season, once in a week.

Fruit cracking: Pinolene (5%0 as vapour guard and GA (15 ppm) and applying Boron (0.2%) reduce fruit cracking.

Plant Protection: Caterpillar – Spray Endosulfan 0.2%.

It flowers and provide fruits throughout the year.

N.B.: Flowering in June is advantageous. Therefore, a rest period is given during April-May so as to enable prolific harvest of a given time. For this, the 1st irrigation should be given during Mid May followed by regular irrigation until the onset of monsoon.

Mango (*Mangifera indica*)

It is the king of the fruits, grown in almost all the districts of the State.

Varieties: Langra, Banganapalli, Suvarnarekha, Himsagar, Amrapalli, Golabkhas, Bombay gree.

Soil: Loamy well drained. < 7.5 soil pH (Slightly acidic soil).

Planting distance: 10m x 10m

For dwarf variety, Amrapalli 2.5m x 2m

Pit size: 1m x 1m x 1m.

Planting time: July-August

Planting material: Healthy grafts (about one year old)

Manure & Fertilizer:

Basal	FYM	-	4 baskets
	N	-	60 g
	P_2O_5	-	20 g
	K_2O	-	60 g
Top dressing	FYM	-	4-6 baskets
	N	-	60 g
	P_2O_5	-	20 g
	K_2O		60 g

Organic manure and phosphatic fertilizers should be applied immediately after harvest, whereas ammonium sulphate should be given before flowering.

$ZnSO_4$ (0.3%) during February, March and May

Borax (0.5%) spray after fruit set twice at monthly intervals.

0.5% $MnSO_4$, after blooming.

Time of application: July and September.

Inter-culture: Twice in a year. Intercropping helps in checking weed growth and reduces nutrient losses.

1. Pre-monsoon.
2. Post-monsoon.

Irrigation: Irrigating grown up trees after fruit set at 10 days interval.

Plant Protection

Insect pests: 1. Hopper, 2. Mealy bug, 3. Fruit fly – Sevin 0.2% + 0.1% Molases.

Diseases: 1. Powdery mildew – Bavistin @ 0.1% at first flowering and subsequent spray at fortnight interval + Sulfex @ 0.2%.

2. Anthracnose – Control measures as recommended for

powdery mildew.

3. Die back – Spraying of Bordeaux mixture (5:5;50) or M-45.

Sapota (*Archas sapota*)

It is popularly known as Chiku, cultivated for its delicious sweet fruits.

Varieties: Cricket Ball, Kalipalti, Kirtibarthi, Dwarapudi, PKM-1, Calcutta special, Pala.

Soil: Loamy, well drained.

Planting distance: 7m x 7m or 10m x 10m

Pit size: 0.75m x 0.75m x 0.75m

Planting time: July – August

Planting material: Healthy grafts or gooties

Weed control: 0.8 kg Bromacil + 0.8 kg Diuron/ac as pre emergence spray.

Manure & fertilizer:

	FYM	N	P_2O_5	K_2O
Basal	50 kg	45 g	145 g	60 g
Top dressing -	45 g	145 g	60 g	

Time of application: June – July and November – December

Inter-culture: Twice in a year, Pre-monsoon and Post-monsoon

Irrigation: During dry season, once in a week.

Adoption of drip irrigation system is beneficial, saving 40 % water with 70-75% higher net income.

Plant Protection

Insect pests: 1. White ants - Seven @ 2%

Disease: 1. Leaf spot – Dithane-z-78 @ 2%

NB: It gives only one crop from summer flowering in April and May. Under moisture stress also, it produces crop only.

Litchi (*Litchi chinensis*)

It is famous for its excellent quality, pleasant flavour, Juice pulp with attractive colour.

Varieties: China, Calcutta, Muzafarpur, Early Badana, Late bedana, Deshi, Purbi, Gulabi,

Soil: Loamy, well drained, pH – 5.5 – 7.0 is ideal.

Planting distance: 8m x 8m

Pit Size: 0.75m x 0.75m x 0.75m

Planting time: July-August

Planting material: Gooties

Manure & fertilizer: (per pit)

	FYM	N	P_2O_5	K_2O
Basal	2 baskets	60 g	20 g	60 g
Top dressing -	75 g	20 g	90 g	

Plants grown under deficiency to NPK can flower but do not set fruits. Plants grown under Mg deficiency do not even bloom.

Time of application: June – July and October- November

Inter-culture: Twice in a year- Pre-monsoon and Post-monsoon

Weeding: Apply pre-emergence herbicides - Diuron or Atrazine @ 2 kg/ac or 0.4 ha at one month interval to keep weeds under control.

Irrigation: During dry season, once in a week in summer. No irrigation is required during winter in fruiting trees before fruit set.

Plant Protection:

Insect pest: Mite – Rogor @ 0.05% or Metasystox @ 0.05%

NB: Maintenance of good sanitary condition is a must to keep litchi orchards healthy and disease free.

Jack Fruit (*Artocarpus heterophyllus*)

Varieties: Singapore or Ceylon jack,

Soil: Loamy, well drained.

Planting distance: 10m x 10m/ 12m x 12m

Pit size: 1m x 1m x 1m

Planting time: June- August

Planting material: Seeds and seedlings

Manure & fertilizer: (per pit)

	FYM	N	P_2O_5	K_2O
Basal	3-4 baskets	60 g	80 g	90 g
Top dressing	3 baskets	125 g	80 g	290 g

Time of application: July and September

Inter-culture: Twice in a year, Pre-monsoon and Post-monsoon

Irrigation: Dring dry season at 12-15 days interval

Plant Protection:

Disease: Fruit rot – Use Blitox or Dithane M-45.

Pineapple (*Ananas sativa*)

Varieties: Kew, Queen, Giant Kew, Mauritious, Jaildhup

Soil: Loamy, well drained pH 5.0 – 6.0 without hard pan or stones.

Planting distance: Ridge and furrow system of planting

Plant population **Per acre (0.4ha)**	**Plant to plant (cm)**	**Row to row (cm)**	**Trench to trench (cm)**
14,000	30	60	20
17,000	30	60	90
21,000	25	60	90

Pit size: 10 cm deep holes

Planting time: July – August/ October – November

Planting material: Healthy suckers of 300 g weight

Manure & fertilizer: (per plant)
N – 12 g in six splits in a year
P_2O_5 – 4 g
K_2O - 12 g in two splits
FYM - 25-30 Cart loads/acre or 0.4 ha as basal dose

Time of application: First split 2 months after planting

Inter-culture Hoeing, weeding and earthing up after each application of fertilizer

Weed control: Diuron (1.2 kg/ac) or Bromacil + Diuron @ 0.8 kg/ac each

Irrigation: During dry season, once in 15-20 days

Mulching: Black polythene or saw dust results better growth

Use of growth regulator: NAA based compound – Planofix and Celemone @ 10-20 ppm induces flowering – NAA (200 -300) ppm after 2-3 months of fruit set increases the size of fruits by 15 – 20%

Plant Protection:

Diseases: Heart rot or Stem rot

Dip the suckers in Boardeaux mixture (1:1:3) before planting

PACKAGE OF PRACTICES FOR PLANTATION CROPS (UNIT – ONE ACRE OR 0.4 HA)

Coffee

Land and Soil type: Coffee grows well at height from 750 to 1050 m. Soils with high organic matter is the most suitable. The most favourable rainfall is 1500 -1875 mm per annum.

Seed variety: *Coffea Arabica*, *Coffia sanaramon*

Seed rate: 2.5 to 3.0 kg.

Seed treatment/ Management: Mulching with 5 cm thickness of paddy straw to ensure even temperature for germination in nursery.

Apply cow-dung slurry and urea solution once in a month to maintain young seedlings

Sowing/Planting time: Sowing in nursery bed may be done in December-January. Planting in main field is in July -August.

Spacing and Planting Method: 180 cm x 180 cm (1000 plants/ac. Pit size – 45 x 45 x 45 cm.

6-7 months seedlings with 3-4 pairs of leaf should be planted under temporary shade with Enthrina plant. Silver Oak, Mango, Jackfruit etc. 650 nos. of shade plants be planted per acre at 3.60m x3.60 m distance.

For Hybrids, spacing – 120 x 120 cm

Manure & fertilizer: Compost/FYM – 20 Cart load

Fertilizer in split Doses (in kg)

	1st	2nd	3rd	4th
N	45	60	60	80
P_2O_5	30	45	45	60
K_2O	45	60	60	80

Inter-culture: Weeding, Hoeing and Mulching be done three times.

Plant Protection: Three spraying with urea, insecticide and fungicide

Man days: 250, 640, 433, 450, and 525 for successive years including shade plants

Yield/acre (approx): 675 kg/year (5th to 30th year)

Chashew

Land and soil type: Shallow to deep sandy loam, laterite and coastal sandy soil

Seed variety: M-44/3, WBDCV, BLA-39-4, BLA-139-1, T-56

Local varieties collected from mother trees of 10-15 years old and yielding 10-15 kg/plant per year

Seed rate: 2.5 kg (150 nuts/kg, or less)

Seed & Seedling Treatment & Pot planting: Sun dry for 2-3 days. Soaked in water for 24 hours and sown at a depth of 5-7 cm in polythene bags filled with a mixture of sand, soil and compost in 20:40:40 proportion, Keep stalk end up.

Sowing/Planting time: 20th May to 10th July

Spacing: 8m x 8m (70plants per acre), Pit size – 45 x 45 x 45 cm

Compost or FYM: 8-9 Cart load

Fertilizer (g/plant)

	1stYr	2ndYr	3rdYr	4thYr
N	200	200	250	300
P_2O_5	60	100	125	150
K_2O	-	60	125	150

Inter-culture: 2-3 weeding, hoeing and mulching

Plant Protection: Foliar spray with urea 1-2 % or more with insecticide and fungicide at pre-monsoon, post-monsoon, pre blossom and post blossom times

Man-days (Labour): 110, 50, 60, 40 respectively from 1st Yr to 4th Yr.

Yield/acre or 0.4 ha (approx): 20, 60, 160, 240 kg per annum respectively from 4th year to 7th year 8th year onwards 240 kg.

Sisal

Land and soil type:	Well drained loamy soil	Well drained red-loamy soil
Seed variety:	*Agava sisalana* (seedling variety)	Local slips of Chitrada Bankisol for propagation
Seed rate:	2200 Bulbils/Suckers	8000 Slips (8 q)
Seed or Seedling Treatment and Pot planting:	Bulbils kept in nurseries for one year to attain a height of 22.5 to 30 cm and weight about a kg. Treat the roots with Dithane-Z-78 @ 1.2 kg	Roots be treated with 0.2% Chlorrpyriphos solution
Sowing/Planting time:	June – July	June – July
Spacing:	1m x 1m leaving 3m strip in Between two successive double rows.	90 cm x 60 cm Pit size – 15 x 15 x 15 cm.

Compost/FYM:

Fertilizer (g/plant):

	4 – 5 Cart load		1 Cart load		
1st Yr.	**2nd Yr**	**3rd Yr.**	**1st Yr**	**2nd Yr**	**3rd Yr**
N	3	4.5	4.5	8	8 8
P_2O_5	3	4.5	4.5	4	4 4
K_2O	3	4.5	4.5	4	4 4

Inter-culture:	2-3 weeding, hoeing and	2 weeding and hoeing. mulching.
Plant Protection:	Spray 4:4:50 Bordeaux Mixture @ 200 litres and Apply Aldrin dust @ 3 to	Intercropping of green gram/ black gram/cowpea (1:2) 5 kg twice a year
Man-days (Labour):	62, 46, 20 respectively 50 From 1st to 3rd year	
Yield per acre or 0.4 ha (approx):	180.224, 270 kg of fibre year).	5-6 q (2nd year onwards till 12th

TIPS ON HORTICULTURAL CROPS

Sl. Common name	Oriya name	Botanical name	Spacing	Population	Mode of population
Fruit Crops					
1. Grape	Angur	Vitis vinifera	3x2	1600	Grafting
			4x3	830	Hybrid varieties
2. Aonla	Aonla	Phyllanthus embica	9x9	124	Graft/Seeding
3. Papaya	Amrutbhanda	Carica papaya	2.5x2.5	250	Seeding
4. Mango	Amba	Mangifera indica	9x9	124	Grafting
			5x5	400	Hybrid varieties

Sl. Common name	Oriya name	Botanical name	Spacing	Population	Mode of population
5. Elephant apple	Oau	Dillenia indica	8x4	300	Seed & layering
6. Banana	Kadali	Musa spp	2.1x2.1	2000	(Tall variety) Sucker
7. Carambola	Karamanga	Averrhoa carambola	5x5	400	Seedling
8. Orange	Kamala	Citrus reticulate, Citrus sinensis	6x6	275	Grafting
9. Sour Orange	Kandhia	Citrus aurantium	6x6	275	Graft/seeding
10. Date palm	Khajuri	Phoenix dactylifera	5x5	400	Seedling
11. Rose apple	Golapjamu	Syzyguim jasbos	6x6	275	Seed layering
12. Cherry	Cherry	Malphisia punicifolia	2x2	2500	Seedling, graft
13. Jamun	Jamu	Eugenia jambolana	10x10	100	Seedling
14. Jamroll	Jamulola	Syzygium melaccensis	7x7	200	Seed layering
15. Pomegranate	Dalimba	Punica granatum	5x5	400	Seedling, layering
16. Fig	Dimiri	Ficus carica	5x5	400	Layering, graft
17. Mulberry	Tul	Morus nigra	7x7	200	Seedling, Budding
18. Coconut	Nadia	Cocos nucifera	7.5x7	5 175	Seedling
19. Pear	Naspati	Pyrus communis	5x5	400	Shield budding
20. Rangpur lime	Naranga	Citrus limonia	6x6	275	Seedling/Budding
21. Jack fruit	Panasa	Artocarpus heterophyllus	10x10	100	Seedling
22. Guava	Pijuli	Psidium guajava	6x6	275	Graft
23. Peach	Peach	Prunus persica	4x4	625	Graft
24. Plum	Plum	Prunus salicima	4x4	625	Budding
25. Ber	Barakoli	Zigyphus mauntiana	5x5	400	Eye budding
26. Custard apple	Badhial	Annona squamosa	5x5	400	Seedling, Seed
27. Bael	Bela	Aegle marmelos	7x7	200	Patch budding
28. Sapota	Sapota	Achras sapota	9x9	124	Graft
29. Pineapple	Sapuri	Ananas sativa	1x0.3	35000	Suckers
29. Apple	Seo	Malus sylvestris	4x4	625	Graft
30. Strawberry	Strawberry	Fragaria vesca	0.4x0.25	1 Lac	Stem cuttings
32. Litchi	Lichu	Litchi chinensis	7x7	200	Layering/Gooti
33. K. Lime	Lembu	Citrus aurentifolia	5x5	400	Seedling
34. Avocado	Dalchini Phal	Cinnamomum zylanicum	5x5	400	Seed, Stem cutting
		Cinnamomum camphora	10x7	140	Seed/Gooti
35. Loquat	Japanplum	Mespilus japonica			Seedling/Gooti
36. Phalsa	Phalsa	Grewia asiapice	25x2.5	1100	Seed/graft/Stem
Plantation Crops					
1. Coffee	Coffee	Coffea Arabica	2.5x2	2000	Seedling

Sl. Common name	Oriya name	Botanical name	Spacing	Population	Mode of population
		Coffea canephore	3x2.5	1300	Seedling
2. Cashewnut	Kajubadam	Anacardium accidentale	7x7	200	Graft
3. Cocooa	Coco	Theobroma cacao	4x4	625	Graft, Seed
4. Arekanut	Gua	Areca catechu	3x3	1100	Seedling
5. Tea	Cha	Camellia sinensis	1.2x0.75	10890	Branch
6. Oil palm	Tala Tela	Eaeis guineenis	8.5x8.5	137	Seedling
7. Rubber	Rubber	Hevea brasiliensis	5x4.5	450	Seed Budding
Vegetable Crops					
1. Brinlal	Bagan	Solanum melongena	0.6x0.45		Seedling
2. Tomato	Tamat	Lycopercicon escu	0.9x0.6		(kharif) Seedling(winter)
3. Chilly	Lanka	Capsicum annum	0.6x0.3		Seedling
4. Okra	Bhendi	Abelmoschus esculentus	0.6x0.45 (K)		Seedling
			0.6x0.3 (summer)		
5. Cabbage	Bandha kobi	Brassica oleracea	0.45x0.3 (Early)		Seedling
	Var; Capitata		0.6x0.4 (Late)		
6. Cauliflower	Fool kobi	Brassica oleracea	0.45x0.3		Seedling
	Var. Botrytis				
7. Knolkhol	Ull kobi	Brassica caulorapa	0.45x0.3		Seedling
8. Pea	Matar	Pisum sativum	0.3x0.1		Seedling
9. Capsicum	Capsicum	Capsiicum spp	0.6x0.45		Seedling
10. Bean	Bin	Phaseolus vulgaris	0.3x0.15		Seed
11. Country bean	Simba	Dolichos lablab	1.2x1.2		Seed
12. Cowpea	Jhudunga	Vigna sinensis	0.3x0.1		Seed
13. Cluster bean	Guanr	Cyamopsis tetragonoloba	0.45x0.15		Seed
14. Pointed gourd	Potal	Trichosanthes dioica	1.5x1.5		Seed
15. Radish	Mula	Raphanus sativus	0.3x0.1		Seed
16. Beet	Bit	Beta vulgaris	0.45x0.15		Seed
17. Carrot	Gajar	Daucus carota	0.37x0.7		Seed
18. Turnip	Salgam	Brassica campestris	0.3x0.15		Seed
19. Onion	Piaj	Alium cepa	0.2x0.1 (K)		Seedling
			0.2x0.15 (Rabi)		
20. Potato	Alu	Solanum tuberosum	0.45x0.25		Seed potato
21. Sweet potao	Kandamula	Ipomoea batatas	0.6x0.25		Stem

Sl. Common name	Oriya name	Botanical name	Spacing	Population	Mode of population
22. Colocasia	Saru	Colocasia antiquorum	0.6x0.45		Rhizom
23. Yam	Desi Alu	Dioscorea spp	0.75x0.75		Tuber
			0.9x0.9		
24. Tapioca	Gaya Alu	Manihot esculenta	0.9x0.75		Stem
25. Pumpkin	Kakharu	Cucurbita moschata	1.5x1.5		Seed
26. Snake gourd	Chhachindra	Trichosanthes anguina	1.5x1.5		Seed
27. Ridgegourd	Janhi	Luffa accutangula	1.2x1.2		Seed
28. Bettergourd	Kalara	Momordica charantia	1.2x1.2		Seed
29. Cucumber	Kakudi	Cucumis sativus	1.2x1.2		Seed
30. Bottle gourd	Lau	Lagenaria siceraria	1.2x1.2		Seed
31. Pound gourd	Tinda	Citrullus vulgaris	1.5x1.5		Seed
32. Small gourd	Kunduri	Coccinia indica	2x2		Stem, Bulb
33. Spine gourd	Kankada	Momordica dioica	1x1		Stem, Bulb
34. Water melon	Tarbhuj	Citrullus lanatus	1.2x1.2		Seed
36. Musk melon	Kharbhuj	Cumis melo	1.2x1.2		Seed
37. Lettuce	Lettuce	Latuca sativa	0.3x0.2		Seed
37. Poi	Poi	Basella alba	1.5x0.6		Seed, Stem
38. Greens	Saga	Amaranthus spp			Seed
39. Palak	Palanga	Spinacia oleracea	0.1x0.1		Seed
40. Moringa	Sajana	Maringa oleifera	2.5x2.5		Seed, Seedling
Flower Crops					
1. Aparajita	Aparagita	Clitorda ternate			Seed
2. Iswar Jata	Iswarjata	Celasia plumose			Seed, Layering, Seedling
3. Crossandra	Itamal	Crossandra infundibuliformis	0.6x0.3		Seedling, Layering
4. Nymphea	Kain	Nympaea lotus			Root
5. Yellow Oleander	Kaniar	Thevetia nerifolia			Graft, Stem
6. Cosmos	Cosmos	Cosmos bipinnatus			Seed
7. Bougain villea	Kagaja Phool	Bougainvillea spp	2.5x1.5		Stem
8. Keoda	Kia	Pandanus odoratissimus			Root
9. Karbira	Karabira	Nerium oleander			Stem
10. Celosia	Kukudachula	Celosia cristata	0.3x0.3		Seed
11. Barbades	Krushnachuda	Delonix regia			Seed, graft
12. Kadamb	Kadamba	Anthocephalus cadamba			Seed
13. Kamini	Kamine	Murraya exotica			Stem
14. Rose	Golap	Rosa spp	1x1		Eye budding

Sl. Common name	Oriya name	Botanical name	Spacing	Population	Mode of population
15. Gomphrena	Godibana	Gomphrena globosa			Seed
16. Hybrid rose	Golap	Rosa hybrida	0.75x0.75		Seed/Eye budding
17. Gladiolus	Gladioli	Gladiolus spp	0.3x0.3		Seed, corm
18. Marigold (African)	Gandu	Tagetes erecta	0.4x0.3		Seedling
	Marigold (French)	Tagetes patula	0.2x0.2		Stem
19. Champak	Champa	Michellia champaca			Graft, Seed
20. Chhentpeela	Chhenaphool	Lagerstroemia indica			Stem
21. Jasmine	Jui	*Jasminum grandiflorun*	1.2x1.2		Stem
22. Jasmine	Jui	Jasminum auriculatum	2x1.5		Stem
23. Zinnia	Zinnia	Zinnia elegans	0.3x0.15		Seed
24. Crops jasmine	Tagara	Tabernaer Montana coronaria			Stem
25. Dahalia	Dahalia	Dahlia spp			Stem
26. Bottle brush	Bottle brush	Callistemon lanceolatus			Graft, Seed
27. Madhumalati	Madhumalati	Quisqualis indica			Stem
28. Hibiscus	Mandar	Hibuscus rosasinensis			Graft, Stem, Seedling
29. Tuberose	Rajanigandha	Polyanthes tuberose	0.3x0.2		Corm
30. Radhatamala	Radhatamala	Passiflora coerulea			Stem
31. Braziletto	Radhachuda	Caesalpinia pulcherrima			Seed, Graft
32. Gardenia	Sugandharaj	Gardenia florida			Graft
33. Sthalapadma	Sthalapadma	Hibiscus mutabilis			Stem, Seedling
34. Chrysanthemum	Sebati	Chrysanthemum indicum			Corm
35. Balasam	Haragoura	Impatiens balsamine			Seed
36. Henna	Henna	Cestrum noeturrnum			Stem
37. Nelumbium	Padma	Nelumbium nelumbo			Stem
Spice Crops					
1. Ginger	Ada	Zingber officinale	0.2x0.15		Rhizome
2. Large Cardamom	Aleicha	Amon suboletus	2x2	2000	Seedling
3. Elaichi/ Cardamom	Gujurati	Elettana cardamomomum	2x2	2500	Seed, Seedling

Sl. Common name	Oriya name	Botanical name	Spacing	Population	Mode of population
4. Black pepper	Golmaricha	Piper nigrum	2.5x2.5	2500	Seedling
5. Nutmeg	Jaiphala	Myristrica fragrans	8x8	156	Seeding Graft
6. Cumin	Jira	Cuminum syminum	0.3x0.15		Seed
7. Dillseed	Juani	Anethum graveolens			Seed
8. Cinnamon	Dalchini	Cinnamomum zeylanicum	2x2	2500	Seedling
9. Bayleaf	Tejpatra	Cinnamomum tamalla	2x3	1600	Seedling
10. Tamarind	Tentuli	Tamanindus indicus	10x10	100	Graft
11. Coriander	Dhania	Coriandrum sativum	0.5x10		Seed
12. Onion	Piaja	Allium cepa	0.15x0.10		Seed/clore
13. Long peper	Pippali	Piper longum	2.5x2.5	2500	Seedling
14. Fennel	Panamaduri	Foeniculum vulgare	0.3x0.15		Seed
15. Poppy	Postak	Papaver somniferum			Seed
16. Mint	Pudina	Menta viridis	0.15x0.15		Seed
17. Almond	Pestabadam	Prunus amygdalus			Seed
18. Curry leaf	Bhrisangapatra	Murraya koenigil	1.5x1.5	4000	Seedling
19. Fenugreek	Methi	Trigonella foenumgraecum	0.3x0.1		Seed
20. Garlic	Rasuna	*Allium sativum*	0.25x0.1		Corm
21. Mustard	Sorisa	Brassica juncea	0.45x0.3		Seed
22. Clove	Labanga	*Eugenia caryophyilus*	6x6	275	Seedling
24. Chilli	Lanka	Capsicum spp	0.6x0.3		Seedling, seed
Medicinal and Aromatic Plants					
1. Sarsaparilla	Anantamula	Hemides musindicus			Root
		Hemidesmus indicus			
2. Ipecae	Ipecac	Caphaelis ipecauanha	.2x0.2		Seed
3. Isobgol	Isabgol	Plantago oata			Seed
4. Eucalyptus	Eucalyptus	Eucalyptus citriodora			Seed
5. Acalypha	Indramarisha	Acalypha indica			Leaf, root
6. Kantakari	Kantabaigana	Solanum khasianum			Seed
		Solanum incanum	0.5x0.5		Seed
7. Chirayeta	Chirayeta	Swertia chirayeta			
8. Liquorica	Jastimadhu	Glycyrrhiz glabra			
9. Davana	Davana	Artemisia allens	0.2x0.15		

Sl. Common name	Oriya name	Botanical name	Spacing	Population	Mode of population
10. Foxglove	Tilapuspi	Digitalis lanata	0.6x0.3		
11. Tulsi	Tulsi	Ocimum sanctum			Seed
12. Dil/Sowa	Dil/Sowa	Anethum graveolens	0.6x0.3		Seed
13. Butea	Palas	Butea monosperna/			
		Butea frondosa			
14. Palmarasa	Palmarasa	Sympopogon martine	0.6x0.6		Seed, Seedling
15, Betelvine	Pana	Piper bitel			Stem Cutting
16. Bellerica	Bahada	Terminaliabellerica			
17. Diascorea	Banasaru	Diascorea floribunda	0.45x0.3		Rhizome
18. Belladona	Belladona	Atropa belladone	0.6x0.45		Seed
19. Brahmi	Brahmi	Bacopa monniera			
20. Calamus	Bacha	Aocrus calamus	0.3x0.3		Peak branch, corm
21. Markingnut	Bhalia	Semecarpus anacardium			
22. Mulhat/	Bhringaraj	Glycyrrhiza glabra			Stem
	Liquo rice				
23. Mahua	Mahula	Bassialangifolia			Graft
24. Citronella grass	Citronella	Cymbopogan winterianus	0.6x0.6		Root
25. Rauwolfia	Sarpagandha	Rauwolfia serpentine	0.6x0.3		Seed, Stem, Root
	or Patalgaruda	R. tetraphylla			
26. Senna	Senna	Cassia angustifolia	0.3x0.3		Seed
27. Cinchona	Cinchona	Cinchona ledgeriana	2x2	2500	Seed
28. Chebulic	Harida	Terminalia chebula			Seed
29. Hengu	Hengu	Ferula assafoetida	0.5x0.5		Seedling
30. Henna/	Henna/	Lawsonia inermis	0.3x0.3		Seed, Stem
Mehendi	Manjuati				
31. Lemongrass	Lemongrass	Cymbopogan citrates	0.6x0.3		Seed, Root
		C. flexuosus			
32. Linaloe	Linaloe	Brusera depechianum	7.5x4	300	Stem
33. Lahangal		Gloriosa superrba			Corm, Seed
34. Pasaruni		Paederia foetida			
35. Satabar	Satabari	Asparagus rasimosus			Rhizome
36. Gruta Kumari	Gruta Kumari	Aloe vera			
37. Sweta Chitrac		Plumbage zylanica			Seed, Stem

Sl. Common name	Oriya name	Botanical name	Spacing	Population	Mode of population
38. Sadabihar	Sadabihari	Vinka resia			Root
39. Asgandha	Aswagandha	Withenia semnifera			Root, Seedling
40. Suna mukhi	Sanamukhi	Casia angusthifolia			Seedling
Forest Plants					
1. Ashoka	Ashoka	Saraca indica			
2. Tomaentosa	Asana	Terminallia tomentosa			
3. Arjuna	Arjuna	Terminallia arjun			
4. Peepal	Aswatha	Ficus religiosa			
5. Hog plum	Ambada	Spondias pinnate			
6. Pongamia	Karanja	Pongamia glabra			
7. Wood apple	Kaitha	Peronia elephantum			
8. Camphor	Karpura	Cirrameomum comphora			
9. Yellow teak	Kuruma	Adina cardifolia			
10. Persimmon	Kendu	Diospyrus empryopteris			
11. Nuxvomica	Kochila	Strychnous nuosvomica			
12. Malay bush beach	Gambhari	Gmelina arborea			
13. Tree of sorrow	Gangasiuli	Nyctyanthes arborpirisis			
14. Sandal Wood	Chandan	Sandain album			
15. Classia	Chakunda	Cassia spp			
16. Beef Wood	Jhaun	Casuarina equisetidia	3x15	2220	
17. Palm	Tala	Borassus **flabellifera**			
18. Mast tree	Devdaru	Polyathia longifolia			
19. Datura	Dudura	Datura stramonium			
20. Neem/ Margosa	Neem	Azadirachta indica			
21. Connal ball	Nagachampa	Cauropita guianensis			
22. Pine	Pine	Pinus longifolia			
23. India Coral	Paladhua	Erythrina indica			
24. Indian Kinoo	Polanga	Petrocarpus marisapium			
25. Fasi	Fasi	Anageissus patifoula			
26. Banyan Tree	Bara	Ficus benghalensis			
27. Grape fruit	Batapi	Citrus bengalensis			
28. Indian Oak	Mahalimba	Mellingonia hortensis			

Sl. Common name	Oriya name	Botanical name	Spacing	Population	Mode of population
29. Red Sendal	Raktachandan	Pterocarpus santalinus			
30. Bullock heart	Ramphala	Annona reticulate			
31. Rudrakshya	Rudrakshya	*Elaecarpus ganitrus*			
32. Sal	Sal	Shorea rebuste			
33. Teak	Saguan	Tactona irandis	2x2	2500	
34. Sisoo	Sisoo	Dalbergia sisoo			
35. Womans tongue	Sirisa	Albizzia lebbek			
36. Simili	Simili	Salmalia malabarica			

ADOPTION OF NEW TECHNOLOGY

New farm technology can not produce results of its own unless it is adopted by the farm population. Adoption of new technology shall further depend upon the rate of technical progress. Thus, in order for new technology to become acceptable, technical change must occur side by side. The development and utilization of technological knowledge greatly varies with the economic and social development of a country. Tradition ridden societies are less receptive of new ideas and innovations compared to those societies that are more educated and socially emancipated. Economically developed societies offer greater incentives for creativity than those of the less developed ones. Naturally, therefore, greater chances of the technical progress exist in the developed countries compared to the less developed societies. This, in its turn enables an advanced society to take better advantage of new technology and reach new heights of economic progress.

"We must also initiate positive measures to reduce regional imbalances as otherwise the tension caused by such balances will inhibit the very process of development"

Indira Gandhi

20 Natural Disasters and its Management

"If agriculture in this country fails, we fail,
The Government fails and the Nation fails,
There is no help for us but to succeed in Agriculture".

Nehru

WEATHER RELATED DISASTERS

World Watch Institute (WW) at Washington reported climate changes and the latest dossier on weather-related disasters in 1998. Preliminary estimates of weather related economic looses worldwide during 1998 total $ 89 billion which is 48% over the previous record of $ 60 billion in 1996 and far exceeds the $ 55 billion total for the entire decade of the 1980s. Among the most severe disasters of 1998 were the hurricane havoc in Central America and the Floods in China and Bangladesh. Hurricane Mitch, the deadliest Atlantic storm on 200 years, caused an estimated 11,000 deaths in Central America. The flooding in China's Yangtze River resulted in 3700 deaths, displaced 223 million people, inundated 25 million ha of crop land and cost $ 30 billion. The Yangtze basin has lost 85% of its forest cover to logging, agriculture and in recent years, wetlands have been drained and the river heavily dammed. As against flooding of the basin taking place once in 20 years in the previous centuries, the last few decades have seen flooding occurring in nine out of 10 years. Bangladesh suffered its most extensive flood of the century in the summer of 1998. Rice harvest was reduced by 2 million tones. Damage estimates exceed $ 4.3 billion. An ice storm in Canada and New England in the U.S. in January 1998 cost $ 2.5 billion. Floods in Turkey in June 1998 caused $ 2 billion in damages and floods in Argentine and Para guay cost $ 2.5 billion.

Rice cultivation has been hit in about 20% of the total area in Punjab. In Andhra Pradesh, rain and flood have hit 8% of the area covered under groundnut, sunflower, pigeon pea and Sorghum. In Rajasthan, about 20% of the mustard cultivation area was hit by the rain.

WEATHER ABERRATIONS

Many types of weather aberrations occur which include tropical storms, tornadoes, dust storms, hail storms, etc. Agricultural production is affected due to prolonged dry spells, heat waves, cold waves and frost.

Tropical Storms

The violent and destructive types of tropical cyclones originate over the tropical oceans. They are named differently according to the places of occurrence.

Hurricanes

They occur in the Caribbean and off the Pacific of Mexico during late summer or early autumn. The important feature of the hurricane is its wind force. Winds are only moderate at the outer margins of the system, but velocities increase rapidly toward the centre, reaching a maximum along the outer edge of the central eye. Wind speed of 65 knots (118.8 km/hour) to 130 knots has been recorded. Heavy rainfall and flooding may cause extensive damage.

Typhoons

They occur in the sea of China, the Philippines, Japan and Islands of the Western Pacific; in the Indian Ocean, they are normally called cyclones. However, this should not be confused with cyclones in general.

Thunderstorms

Thunderstorms occur regularly in the tropics. Thunder and lightening accompany the typical mature thunderstorm. Intensive precipitation occurs from the mature storm. Raindrops are large and it is literally a **"cloud burst".** They cause large proportion of rainfall and have considerable climatological importance. Thunderstorms are most common in summer due to surface heating of landmass.

Tornadoes

Tornadoes are the most violent storms of lower troposphere. They mostly occur during spring and early summer. They have been reported at widely scattered locations in the middle in the latitudes and tropics. Crop losses are heavy due to this event.

Dust Storms

Dust storms take place during summer in the months of April, May and June. During this period, high temperature and low humidity prevail. Air gets heated, expands and blows with high velocity. It carries dust particles into the atmosphere. When it becomes violent, it acquires the form of dust storm. Dust and sand particles may deposit on the crops.

Frost

A single hardy frost may wipe out entire orchards. The frost hazard is in the greatest intensity during the critical stages of crop growth. Young seedlings are killed unless frost hardly varieties are chosen. Flowering stage is the critical period for most of the field crops.

Fog

The entire hilly region of North India is under the influence of foggy days throughout the year. Non-irrigated *rabi* crops get benefited from fog. The negative aspect of fog manifests itself when it persists for several consecutive days blocking the sunlight.

MICROCLIMATE

Microclimate is the climate in which plants and animals live. It differs from the macroclimate which prevails above the first few meters over the ground. Temperature changes drastically in the first few centimeters from the surface into the soil or into the air. Changes in humidity with elevation are the greatest near the surface. Very large quantities of energy are exchanged at the surface in the process of evaporation and condensation. Wind speed decreases markedly as the surface is approached. Micro-climate is largely a function of energy (heat and light) and moisture transfers within the plant environment.

AMELIORATING MICROCLIMATE

Soil

Agricultural practices are aimed at achieving the best possible arrangement of chemical, biochemical and physical factors in the microenvironment of plants. Considerable research efforts have been directed towards altering soil temperature under field conditions by cultural practices such as mulching, irrigation and tillage.

Opaque mulches act as thermal insulators and considerably dump the diurnal wave of soil temperature. Mulch material increases the minimum temperature. Lower temperature and high vapour pressure gradients in mulched crop result in restricted evapo-transpiration losses. In the early morning hours, the soil temperature is slightly higher than in non-mulched plots.

Successful production of winter wheat in Canadian prairies requires no-till seeding into standing stubble in order to trap snow. The snow cover acts as an insulating blanker that protects the wheat crowns from exposure to low temperature (-10°C or lower), thereby the risk of winter kill and/or low temperature injury.

Warm water may be useful for irrigation Temperature changes in soil due to addition of warm (or cold) water are small and quite transitory. The effect may last for less than a day. Because of its great heat capacity, soil temperature will not markedly change by heat contained in water.

Canopy Climate

Canopy temperature is altered by arrangement of rows. Air temperature patterns are relatively more in wide row arrangements. Horizontal temperature gradients are larger and more persistent in the wide row fields. Planting pattern also influences the distribution of radiation in the canopy as well as the total amount of incident radiation intercepted by the crop. East-west rows have higher net radiation than north-south rows. In arid areas with low relative humidity, a shade cloth windscreen would decrease the radiation load on crops. This helps in conserving water both by the decreased energy load and by functioning as a diffusion barrier and creating humidity blanket under the windscreen.

Wind Breaks and Shelter Belts

Crop plants can be protected from excessively strong winds with the beneficial use of wind breaks and shelter belts. The use of windbreaks goes far back in the history of pastoral and agricultural civilization. Wind problems have been of major importance in determining the characteristics agriculture of the regions. Plants are subject to damage caused by excessive chilling, high temperature, desiccation or direct mechanical injury. Wind breaks and shelter belts can reduce these stresses and have been found to be profoundly beneficial to the growth of plants.

In northern latitudes, a significant amount of the annual precipitation comes in the form of snow. The wind breaks help in snow trap and uniform distribution of snow and thus improves the soil moisture supply to the spring grown crops. Snow trapped and retained by the shelter belts also provides an insulating layer to prevent winter kill of sensitive crops such as winter wheat.

The primary effect of any wind break system is the reduction in wind velocity. Air flow modification depends upon the factors such as barrier height, porosity, physical characteristics of the barrier material (Leafiness), barrier spacing, wind speed and wind direction. Protecting crops from excessive wind and from sand blasting can modify the crop yields. In arid regions, direct evaporation of moisture from the soil is reduced with the reduction of wind speed. Yield increases from the sheltered crops have been reported for both dry land (rainfed) and irrigated conditions.

NATURAL DISASTERS

Earthquake

We all know what an earthquake is. When the surface of the earth trembles, we call it an earthquake. Sometimes it trembles mildly, but sometimes the shaking is very violent.

Causes: The inside of the earth is a very hot burning mass of liquid matter. When the heated matter cools, it contracts. Its result is that the upper part of the earth is drawn down. Then this produces cracks on the surface of the earth. The water of the ocean flows down through these cracks and comes in touch with the hot liquid. Sometimes the rain water also finds its way down. It comes in contact with the heated matter. Steam is thus produced, and it wants to come out and the earth is violently shaken. Thus the earthquake is caused.

Effect: If the earthquake is mild, it does not do much harm. But if it is violent, its effect is dangerous. It causes a great loss of life and property. If the earth trembles for two to three minutes violently, many brick-built houses fall down and the inmates are buried alive. It is difficult for men to remain standing. If they can not run away to an open place, they may be killed by the fall of houses. Due to severe shocks, trees are uprooted and the earth of one place may go down and that place may be filled with water from below. Sometimes ponds and rivers swell all on a sudden and sometimes the water of a river or oceanic dried up, and land appears there. Thousands of lives are lost. Many people become wounded. Many people may save their lives but their properties are lost. Thus they become poor. Crops are destroyed. As a result a famine may break out. People from other parts of the country help these people with sufficient money to make good their losses.

Some notable earthquakes: There was a violent earthquake in India in 1897. A terrible earthquake visited Japan in the year 1925-26. It caused a great loss to that country. Even the Indians helped Japan at that time. In 1934, there was a great earthquake in Bengal also, but its effect in North Bihar and Nepal was terrible. Several towns in North Bihar were destroyed as a result of this earthquake. Many people lost their lives. Many were rendered homeless. Indians as well as Englishmen helped the people of Bihar. In 1951, there was a terrible earthquake in upper Assam. It causes a great loss.

A July 1976 (circa 8.0 magnitude) earthquake that destroyed the Chinese city of Tangshan officially claimed around 255,000 lives, although per United States Geological Survey ("USGS") published material, unofficially, the death toll could have as high as 655,000 people. It is understood that the earthquake occurred during the night when many people were in bed.

A January 1995 (circa 7.2 magnitude) earthquake divested the Japanese city of Kobe – claiming at least 5,500 lives. Per a World Health Organization ("WHO") press release, this earthquake promoted a four-day international symposium on earthquakes and health with the intention of developing recommendations that could be used by countries to improve preparedness and reduce the vulnerability of their populations to earthquake disasters, as well as to introduce rehabilitation measures to cope with their aftermath. As the WHO pointed out, deaths and material losses are often just the beginning of hardships that attack communities ... earthquake disasters leave thousands injured and permanently disabled, putting enormous strain on medical and health care systems that may already be stretched to the limit ... the psychological and social damage they cause can not be measured ... common among victims of man-made natural disasters, past-traumatic stress disorder (PTSD) persists long after the event and, in certain causes, interferes with an individual's functioning.

As per the information (as on 15.9.2002) an earthquake was occurred at 0359 hours on 14th September, 2002 of moderate intensity with a magnitude of 6.0 on Richter Scale. The tremor was experienced in South, Middle and North Andaman's in the Andaman and Nicobar Islands. As per information received from the Government of Andaman & Nicobar Administration, the Aerial Bay Jetty at Dighpur suffered one major crack and three minor cracks. Some residential buildings and Tehsil Office, DFO Office, Telephone Exchange etc. at Aerial Bay and Keralpuram suffed some minor damage/developed some cracks. There was also damage to Diglipur Hospital and water supply mains at Klighat and Dighpur. None of the Government building has been severely affected. A survey is being taken up to assess the damage to the private houses and Government buildings. **No loss of life reported.**

India's Days of Sorrow: Just imagine for a while that, one day before you get up from the bed in the morning, you find that your house/apartment is falling. You run to save your life. Somehow you save yourself but within seconds you find that you have lost everything. Everything means including your family members your belongings your house.

On January 26, 2001 India celebrated Republic Day. As the colourful parades got under way the continent reverberated to the worst earthquake in India's history, measuring 7.9.

The epicenter in the northern province of Gujarat was a scene of devastation. The city of Bhujwhere 150,000 people lived was turned into rubble with hardly a building left standing. Businesses were ruined, infrastructure torn apart and basic services were rendered now non-existent. Total death toll is likely to cross 1,00,000 and total damage in terms of value is likely to cross 30,000 crore Indian rupees. The physical damage is apparent but the human cost is yet to be counted with tens of thousands of people killed and injured and whole communities wiped out. India's national day will long be remembered as a day of sorrow.

Conclusion: The earthquake is a terrible enemy of mankind. Men can not control it. They do not know when it will take place. They can prevent its occurrence. So they have only to depend upon God to save them from this great calamity. Wooden houses should be built where earthquakes occur very often.

Flood

In the rainy season, rivers and canals are full to the brim. Sometimes water rises so high that it overflows the banks. Fields, and sometimes, houses lie under water. This rise of water is called a **flood**.

Cause: The flood is caused in many ways. Sometimes rain falls heavily for days together. The rain water of mountains flows down the river. Sometimes the river can not hold so much water. Then the water overflows its banks and floods the adjacent lands. Sometimes snow melts in mountains. The water flows down the river and floods the country. A flood is also caused by the breaking of the embankment of a river. The water of the river rushes through the breach and floods a vast tract of land. Government has many dams to store the surplus water of some rivers. This helps to prevent flood. But these dams sometimes become full to the brim. It is then necessary to allow the stored up water to flow out. As a result a sudden rush of water overflows the banks of rivers and canals and causes flood.

The Scene of the Flooded Area and its Effect: During a flood, the water of the rivers rises very high. It sweeps away everything that comes in its way. Crops are destroyed. Many houses are ruined. Many people are rendered homeless. Their belongings are washed away. Many people and cattle are drowned. The sufferings of people know no bounds. Railway lines remain under water, and sometimes these are also washed away. The whole area under a flood presents a very horrible sight. The miseries of the people can not be fully described. Famine and epidemic diseases break out as the effect of a flood.

Measure to help the sufferers: People try their utmost to help the flood-stricken people. They start relief committees. Money, food and cloth are collected. Volunteers are sent to the flooded area to help the people. Food, cloth and money are distributed to them. Medicine also is distributed free to them. Government also grants them help. They grant peasants loans of seeds and money and provide people with work.

Some Notable Floods: Floods are common to all counties. Floods occur every year in some parts of India. In Bengal, there was a terrible flood in 1913. It is known as Damodar flood. Large areas of land in the districts of Burdwan, Hooghly and Midnapu were flooded. The North Bengal flood of 1922 also caused a great loss of life and property. A terrible flood visited North Bengal and East Bengal in 1931. In 1934, a terrible flood caused a great loss to the people of Assam. Their condition became extremely miserable. Even the people of other parts of India helped the sufferers. In recent years terrible floods have caused great loss to Assam, North Bengal, Bihar and Orissa. In South India also there was a terrible flood by the rise of water in the Cauvery and other rivers.

Floods have been an integral part of the human experience ever since the start of the agricultural revolution when people built the first permanent settlements on the great riverbanks of Asia and Africa. Seasonal floods deliver valuable topsoil and nutrients to farmland and bring life to otherwise infertile regions of the world such as the Nile River valley. Flash flood and large 100-years floods, on the other hand, are responsible for more deaths than tornadoes or hurricanes. In the United States along over the past 60 years, an average of 127 people has lost their lives due to floods.

Latest Floods Events

Flooding in Indochina: Seasonal rains in Indochina in throughout August and September 2002 gave rise to deadly floods across the peninsula.

1. **Isidore Dumps Heavy Rains on Southern U.S.:** On September 26, 2002, Tropical storm Isadora came ashore in Louisiana and moved northward, bringing high winds and heavy rains inland, and causing flash flooding in some locations. This TRMM image and animation shows rain fall accumulations on the ground over a 5-day period.
2. **Isidore Brings Heavy Rains to Cuba:** On September 23, 2002, Hurricane Isidore tracked south of Cuba and produced up to 40 inches of rain over western Cuba, the Isle of Youth, and adjacent waters before passing into the gulf of Mexico. Thus false-colour TRMM image, and animation, shows rain accumulation over time.
3. **Flooding in Southern France:** Torrential rains led to severe flooding in southern France that killed as many as 26 people and forced hundreds from their homes.
4. **Flooding in Central China:** Heavy, frequent rains in central China during the summer 2002 gave rise to heavy flooding throughout the country.
5. **Flooding in Algeria:** Heavy rains in Algeria over the past have led to deadly flooding.
6. **Flooding on Elbe River:** Heavy rains in Central Europe over the past have led to some of the worst flooding along the river in over 100 years.
7. **Floods; flooding along Danube River:** Heavy rains in Central and Eastern Europe have led to some of the worst flooding the region has witnessed in over 100 years.
8. **Floods in Bangladesh and India:** For the first two week of July 2002 floods ravaged eastern India and Northern Bangladesh.
9. **Foods Devastate Southern Texas:** A week of rains in early July 2002 brought flooding to San Antonio and other parts of southeastern Texas.

Floods in India

The National Remote Sensing Agency (NRSA) sends regular flood maps along with data obtained through Rader satellite/microwave satellite for assessing the intensity of flood situation in various States.

Neal Real Time Flood Mapping using Satellite remote Sansing RADARST data of 26th and 28th July 2002 was procured and analyzed. Flood inundation map using and data transmitted to the **Assam** Remote Space Application Centre (ARSAC), Guwahti and NESAC, Shillong for dissemination to user Departments. As per the NRSA inundation impact was estimated from the satellite data of 31st July 2002, the inundation in Dibrugarh, Dhemaji, Sibsagar, Jorhat, Lakhimpur, Golaghat, Nowgong, Marigaon, Darrang, Sonitpur, Cachar, Karimganj, and Hailakhandi Districts was 1,66,493 hectares. 51298 hectares crop area submerged and 974 villages marooned in 5 districts.

Flood inundation map using 8th August 2002 and data transmitted to the **Bihar** Remote Sensing Application Centre (BRSAC), for dissemination to user Departments. As per the NRSA, inundation impact was estimated from the satellite data of 8th August, 2002, the inundation in Darbhanga, Madhubani, Muzaffarpur, Sitamarhi, Supaul, Samastipur, Begusarai, Madhepura, Katihar, Purnia, Khagariya, Bhagalpur and Saharsa Districts was 2,71,143 hectares.

Assam

According to the report **(as on 19.8.2002)** of the State Government, 22 districts were affected by heavy rains and floods, these are Kamrup, Nalbari, Barpeta, Darrang, Dhemaji, Dhubri, Golpara, Jorhat, Morigaon, Sonitpur, Tinsukia, Dibrugarh, Sibsagar, Golaghat, Nagaon, Karimganj, Kokrajhar, Lakhimpur, Bongaigaon, Cachar, Hailkandi and Arbi Anglong.

Arunachal Pradesh

As per report **(as on 27.8.2002)** received from the State Government, out of 15, 14 districts affected by flash flood/landslide during June and July, 2002. About 20,000 populations affected and 7 houses damaged. 11 human lives lost including 7 Army personals and 20 cattle lost. Estimated value of damage to public properties is Rs. 34.66 crore.

Bihar

As per information **(as on 20.9.2002)** received from the State Government, 25 Districts *viz.*, Sitamarhi, Nalanda, Muzaffarpur, Darbhanga, Patna, Katihar, E. Champaran, W. Champaran, Sheohar, Gopalganj, Smastipur, Madhubani, Saharsa Supaul, Shekhpura, Siwan, Araria, Kishanganj, Purnea, Bengusarai, Madhepura, Saran, Khagaria, Lakhisarai, and Bhagalpur were affected by flood. The State Government has deployed 7119 country boats and distributed essential commodities like wheat, readymade food, matchboxes, candles, cash dole of Rs. 1147.10 lakh, Rs. 111.93 lakh cash for chana etc., and 92493 liters kerosene oil among affected persons. 560 Health Centres, 281 Veterinary Centres and 766 Relief camps were functioning. 287459 meters polythene sheets distributed. 29397 packets have been distributed/air dropped by army helicopters; 154744 packets by military boats and 1918.15 q other materials in Madhubani, Sheohar and Darbhanga.

Gujarat

There was a report **(as on 7.9.2002)** about the floods in some parts of the State. The office of the Commissioner for relief, Government of Gujarat was contacted for seeking factual position. It was informed that 19 human have lost their lives due to recent heavy rains and floods in the State. The State Government is taking necessary relief operations. Detailed report is awaited from the State Government.

Kerala

As per information **(as on 20.9.2002)** 14 districts has been affected due to heavy rains. 23245 population affected of 776 villages and 2184 families affected, 27 humen lives lost. 42 persons injured and 8 persons missing. 275 houses fully damaged and 2060 houses partially damaged. 214 families evacuated and 21 number of relief camps were opened.

Manipur

Manipur **(as on 27.8.2002)** was affected with heavy rains since 8th August, 2002 large areas in the four valley districts have been seriously affected by floods from 11th August, 2002. Due to floods approximately 42,000 ha of rice crop and 6552 ha under pisciculture have been severely affected. 30, 024 houses, 1, 09,615 families affected by the flood. 2 persons had lost their lives. The State Government is proposing to provide immediate relief to the affected families by way of distributing of rice, dal, salt, drinking water and provision of medical coverage.

Maharashtra

There was a media report **(as on 6.9.2002)** about the heavy rain and floods in some parts of the State. According to the report there was heavy rainfall from 1-3rd September 2002 causing floods in some parts of the 7 districts *viz.*, Nanded, Parbhani, HIngoli, Yavantmal, Akola, Buldhana and Washim, taking a toll of 35 human lives and damaging 7000 houses. About 25 persons were surrounded by flood water in Nanded were rescued by sending helicopter.

Uttar Pradesh

As per information **(as on 31.8.2002)** about 443 villages of 8 districts of the State have been affected due to heavy rains, lighting etc. About 2.58 lakh population affected. 326 fully and 1289 partially houses damaged. 6 human lives and 15 cattle lost. Financial assistance was also given to affected peoples. 255 country boats deployed.

Delhi

As per information from the Control Room of Flood Control Department of Delhi, the water level of the river Yamuna was 204.71 meters at 1700 hours on the 16th September 2002 against the danger level of 204.80 meters. The level of the river at 1500 hours of the same day was 204.80 meters. There was receding trend of the water level on 16th September, 2002.

Preventive Measures

We can not fully check the occurrence of flood. We should try to dig a large number of passages for water. We should see that water may flow away easily. Railway bridges should be made in large numbers. Some of the river banks should be raised and embankment must be strong, Dams should be made tostore the water of these rivers.

The flood is a great evil. But it has some good effects. It makes the soil fertile. It also washes away all impurities of land. Still we never welcome a flood in a terrible form.

Famines, Droughts and Heatwaves

Famines

A general scarcity of food-grains of extreme nature in a country is called **famine.** A famine may be due to various causes. It is also caused when food crops are sent to other countries. It is sometimes caused by the ravages of locusts and other insects.

Effect: When a famine breaks out in a country, the condition of the people becomes miserable. People go without food for a number of days. Sometimes they live only on leaves and roots of trees. They sell their cattle and all household articles for money to buy food with. Many people sell their children for food. For want of food people are reduced to skeletons. Theft and robbery become common. Famine presents a horrible scene in a country. Many people die for want of food. Epidemic diseases break out, because people eat whatever they get.

In India, famines break out very often. Crops are destroyed by floods. Sometimes crops fail for want of rain. So, there is then shortage of food. People are generally poor. Their condition becomes pitiable. Many great famines broke out in India. Of these, the Bengal famines of 1770 and 1943, the Orissa famine of 1869, the Deccan famine of 1879 and the Gujarat famine of 1900 were terrible. It is impossible to describe fully the sufferings of the famine-stricken people.

How to Prevent Famines: The Government of a country should take steps to prevent the out break of famines. Farmers should be trained to cultivate their lands on a scientific basis. Improved machineries should be used by farmers, Some rivers cause flood. These should be controlled. Dams should be made to store up their water. A large number of canals should be dug so that fields may be watered easily. In this way, food crops will grow in plenty. There should be good roads and a net work of railways. Thus, if crops fail in one part of the country, and a famine breaks out, other parts may supply it with sufficient food grains.

How to help the Sufferers: When a famine breaks out, it is the duty of the Government and the people to help the sufferers. Government should open relief funds. People should raise money and help the famine-stricken people. Government should also open relief works to give employment to the sufferers. Loan should be given to them. Revenue should be remitted. Rice and other grains should be sent frm all parts of the country to the famine-stricken area. It should be given free or at a low price to the people. In West Bengal, the Ramakrishna Mission, Bharat Sevasram Sangho and other charitable institutions help the sufferers with money, food and cloth whenever there is a famine.

It is beyond the power of men to prevent famines altogether. They can, however, lessen their suffering if steps are taken beforehand.

Droughts

A monsoon is defined as a seasonal shift in wind direction, being derived from the Arabic word "mausim", meaning season. The word itself does not mean heavy rain, although the misnomer is not baseless. In a true monsoon climate, seasonal wind shifts typically cause a drastrict change in the general precipitation and temperature patterns. However, the monsoon may also be associated with dry weather as well, since the "wet" monsoon phase of warm, moist air as seasonally replaced by a "dry" monsoon of cool, dry air. This phenomenon is the dominant feature of low-latitude climates stretching from West Africa to the western Pacific Ocean. To understand why these are favoured areas, we need to discuss some of the driving forces behind monsoons and the Earth's weather in general.

Droughts, famines and heat waves are much harder to quality. Drought is defined in three ways: (1) an extended period of below normal rainfall, (2) a long term depletion of ground water or (3) the stunting of vegetative growth due to a lack of water. Famines/drought has occurred in the Sahel of Africa in 1910-1914, 1940-44, 1970-1985. The drought in the Sahel claimed over 600,000 in 1972-1975 and again in 184-1985. The effects are devastating and impacts cab span just a couple months or stretch to a decade or more. Some historical droughts/famines with loss of life include.

Asian Droughts

As per the (United States) National Oceanic and Atmospheric Administration ("NOAA"), 20th Century drought related natural disasters claiming significant numbers of lives. Numerous droughts related disasters have occurred over the Asian mainland this country. The most notable include:

1. Indian Drought of 1900-1/4 to 31/4 million die due to drought, starvation and disease.
2. Chinese Famine of 1907-Over 24 million perishes from starvation.
3. Chinese Famine of 1928-1930-Over 3 million perish in northwest china.
4. Chinese Famine of 1935-5 million Chinese die in water is called the **"New Famine".**
5. Chinese Drought 1941-1942- Over 3 million perish from starvation.
6. Indian Drought of 1965-1967-Over 1.5 million perish in India.
7. Soviet Union (Ulkraine and Volga regions) 1921-1922-1/4 to 5 million perish from drought.

A severe drought will nevertheless upset growth and other macroeconomic estimates of the government. A growth target of even 6-6.5 per cent, as estimated by the Reserve Bank of India, looks hard to achieve, with agriculture likely to record negative growth in 2002-2003. When agriculture output fell 0.2 per cent in 2000-

2001), GDP growth was just 4 per cent, notwithstanding a 6.7 per cent growth in manufacturing.

In 2001-2002, the economy grew 5.4 per cent, with agriculture up 5.7 per cent on a low base and manufacturing up just 2.8 per cent. Manufacturing clearly suffered for ac rop in farm output in the preceding fiscal. Likewise, industry is showing signs of pick-up in 2002-2003, growing 3.8 per cent in April-May, but the trend may not sustain itself in the later part of this fiscal or in 2003-2004 if rural demand flags.

A drought also increases pressure in the environment, in terms of falling ground water tables and reduced tree cover. Industry would have to contend with a power crisis, due to the expected shortfall in hydel generation. Similarly, even the prosperous irrigated regions would be hit by a fall in dam reservoir levels.

Drought to Cut Indian Growth

The Indian economy is heavily dependent on agriculture, which contributes about one-quarter of GDP. About 70 per cent of the workforce works in agricultural related activities and the prosperity of the sector play a key role in determining consumer demand. SSB said the drought's impact will also have a knock-on effect on industry and services sectors.

Indian's growth is expected to be lower than forecast for the current financial year because of the worst drought in more than a decade. Investment bank Salomon Smith Berney (SSB) has sharply lowered its economic growth estimate for the financial year to 4.8 per cent, from 5.5 per cent projected earlier "Rain has been inadequate, down 24 per cent from normal levels as of 24 July 2002. The water level in reservoir has fallen considerably and is 52% less than a year ago", the firm said in a research note. It is believed that this will affect growth from 3% to zero, resulting in lower full year 2003. Gross Domestic Product (GDP) at 4.8%, at 4.8%, compared with 5.5% "SSB sais Agriculture Minister Ajit Singh told to BBC's World Business Report that it was the worst drought since 1987.

Heat Waves

Just as cold waves are injurious to crops in winter, heat waves are also injurious in summer. Deccan and Central parts of the country experience hot waves during March, April and May, during this period, temperatures go beyond 43°C. Blowing of hot winds leads to pollen and premature fruit drop. Most of the water bodies dry up in a short time leading to water shortage.

Heat evasion by shading of plants appears to be effective to minimize the adverse effect of heat wave. A number of shade structures from wood or fibre are used, especially. To protect vegetable crops grown on sandy soils during summer. Wind beaks also can reduce the heat wave effect by decreasing the velocity of hot winds. For the details the chapter 2, Abnormalities in Weather, and Weather Hazards and their Mitigation, may be referred in this context.

Cyclone

A cyclone is a violent storm. It never comes alone. Heavy showers of rain, thunder and lightening are its companions. When a cyclone blows, it moves round and round in the form of small circle. It always changes its course when it blows. It does not move in the same direction like an ordinary storm. It occurs mostly in warm parts of the world. India is in this region of the world. Hence, cyclones often blow over her. In cold countries there are no cyclones.

The signs of a cyclone: There are some signs of the blowing of a cyclone. The weather becomes very hot. There is not a breath of wind. Patches of dark clouds gradually spread in the sky. At least, the sky becomes dark. A storm sets in. The wind then begins to blow violently. From these we can understand that a cyclone will begin.

The actual cyclone: Then it rains very heavily. Flashes of lightening are seen. Thunders begin to peal loudly. The wind begins to blow furiously. There is a roaring noise. The wind continues to blow for some hours. Sometimes it continues for a day or two.

Effect: The effect of the blowing of a cyclone is terrible. Big trees fall down. Roofs of houses are blown away. Many a house is destroyed. Many a man is killed. The water of the river rises. There are huge waves in the river. Many boats sink down. Many men are drowned. Sometimes steamers are also sinking. Everywhere dead bodies of men and beasts are found. Crops are destroyed. People suffer much. They become homeless. There is scarcity of food. Famine breaks out. The miserable condition of the people can not be fully described. Other people then try to help them with money, food and clothes. The cyclone is a terrible evil. But man has no power to prevent it.

Wild Fire

Wild fire is a natural part of Earth's environment; sparked in most cases by lightening strikes and occasionally by volcanic eruption. Wild fires consume a million or more square kilometers per year. Humane also start many fires – sometimes accidentally, but mostly deliberately. Fire is a tool used in agriculture to clear crop lands and help return nutrients to the soil. Prescribed fires clear away dead and dying vegetation to help rejuvenate forests and reduce the risk of larger, uncontrolled wild fires. Fire is also a tool people use to help clear away forests for human developments. Annually humans' burn anywhere from 750,000 to 8.2 million squares km of forest and grassland around the world.

Latest Fires Events

1. **Fires and Deforestation in Brazil:** Scores of fires were burning in the Amazonian Rainforest in Mato Grosso State, Brazil, from Mid May well into late summer 2002.
2. **Fires in Central and Southern Africa:** Hundreds of fires were burning in several central and southern African countries during the annual burning season. These images show fires scattered across Democratic Republic of Congo, Tanzania, Angola, and Zambia from late July and into September 2002.
3. **Fires: Willium Fire Northeast of Los Angeles:** The fast-moving Willims Fire northeast of Los Angeles is affecting several thousand acres of Angeles National Forest in the San Gabriel Mountains. Dozens of homes in nearby suburbs are threatened.
4. **Fires and Smock in Eastern Russia:** In mid-September 2002, numerous fires in eastern Russia created a smoke plume that stretched for eastward over the Sea of Okhotsk. These true colour images were captured by the Moderate Resolution Imaging Spectroradiometer (MODIS) instruments on the Terra and Aqua satellites.
5. **Fires in Western Russia:** There were many fires burning in Western Russia throughout the summer of 2002. Beginning in April and continuing into September, an unusually active fire season created smoky skies all summer long.
6. **Fires in Argentina:** This image shows fires burning throughout southern Bolivia, Argentina, and Uruguay on September 8, 2002.

7. **Widespread Burning across South Central Africa:** The southern African fire season is under way in summer of 2002.
8. **Wild fires and Haze cover Borneo:** Haze from fires in thick over the Indonesian island of Borneo on August 19, 2002. Scores of wild fires (red dots) were detected by the Moderate Resolution Imaging Spectroradiometer on the Terra satellite.
9. **Fires Scorch Oregon:** Several large wildfires burned out of control in Oregon through most of July and into August 2002, while there was also fire activity in Washington and California.
10. **Trick Fire near Williams, Arizona:** Drought conditions in late August 2002 are forcing firefighters to haul in water from Grand Canyon National Park to Williams, Arizona via railroad car in order to combat the Trick Fire.

Volcanic

Volcanic eruptions are among nature's most spectacular displays, at once both awesome and deadly. Volcanoes can present a major hazard to those who live near them for a variety of reasons: (1) Pyroclastic eruptions can smother large areas of land scope with hot ash, dust and smoke within a span of minutes to hours; (2) red hot rocks spewed from the mouth of a volcano can ignite fires in nearby forests and towns, while rivers of molten lava can consume almost anything in their path as they reshape the land scape; (3) heavy rains or a rapidly melting summit snow pack can trigger lahars-sluices of mud that can flow for miles, overrunning roads and villages; and (4) large plumes of ash and gas ejected high into the atmosphere can influence climate, sometimes on a global scale. Check out our related links for volcanoes.

The August 79 A.D. eruption of Italy's Vesuvius volcano buried the nearby towns of Herculaneum and Pompeii – it is estimated that at least 3,500 people died. It is further estimated that up to 18,000 people died as a subsequent December 1631 eruption into: the Vesuvius volcano lies on Italy's western coast in the Compania district – about 7.5 miles from Naples – it is understood that around 2 million or so peple live in the volcano's immediate vicinity.

The April 1815 eruption of Indonesia's, Tembora volcano, claimed around 92,000 lives – many as a result of starvation after crops were destroyed.

The August 1883 eruption of Indonesia's, Krakatoa volcano, claimed around 32,000 lives – the majority as a result of consequent tsnamis (which also resulted in the destruction of many villages) – the explosion was reportedly heard in Australlia.

The (United States) Mount St. Helens eruption during May 1980 – according to the United State Geological Survey ("USGS"), the eruption was triggered by an 5.1 magnitute earthquake that occurred about 1 mile beneath the volcano. At least 57 people died as a result of the eruption, damage to property/infrastructure (per the "USGS) included; "27 bridges, nearly 200 homes. Blast and lahats destroyed more than 185 miles of highways and roads and 15 miles of railways".

The July 1995 eruption of Montserrat's Soufriere Hills volcano - to be updated/ finalized.

Latest Volcanoes Events

1. **Ash fall around Shiveluch Volcano:** Unrest at Sheveluch Volcano on Russia's Kamchatka Peninsula continues. This image shows ash fall from recent small eruptions.
2. **Chikurachki Volcano:** Chikura chiki volcano located on Paramushir Island in the North Kuriles began erupting on January 25, 2002.
3. **Chiliques Volcano, Chile:** The Chiliques volcano, which has not erupted in at least 10,000 years, is now showing signs of life.

4. **Eruption of Pago:** Mount Pago on the island of New Britain, New Guinea, has been erupting since July 2002. About 8,000 people have been evacuated from villages surrounding the volcano.
5. **Indonesia's Ruang Volcano Erupts:** Mount ruang, a stratovolcano in Indonesian Sulawest Islands, erupted on September 25, sending a large plume of ash streaming westward toward Bomeo and Sumatra.
6. **Mauna Kea and Mauna Loa, Hawaii:** Thereia snow atop the Mauna Loa and Mauna Kea volcanoes in Hawii meanwhile the onging eruption of Kilauea continues.
7. **Miyake**-Jima Eruptions: Japan's Miyake-Jima continues to erupt and discharge 26, 2002, spewing lava quantites of sulphur dioxide gas from the summit crater.
8. **Nyamuragira volcano erupts:** Nyamuagira volcano erupted on July 26, 2002, spewing lava high into the air along plume of stream, ash, and sulphur dioxide.
9. **Smoke Plume from Mount Etna:** A smoke plume from Mount Etna that a streming almost directly to the East.
10. **Three Sisters Volcanoes:** The ground near one of the long-dormant. Three SistersVplcanoes in Oregon's Cascade Mountains has risen approximately 10 centimeters since 1996, meaning that underground magma is slowly flowing into the area.

Severe Storms

Earth's atmosphere is a fluid in motion, constantly working towards equilibrium by attempting to smooth out temperature irregularities between the poles and the equator. Thunderstorms, tornadoes, flash floods, hurricanes, typhoons, winter storms and nor'easters are all powerful examples of how the atmosphere attempts to reach that balance. All to often severe storms come with a price – as severe storms worldwide have caused billion dollar losses.

Latest Severe Storms Events

1. **Hurrican Hernan:** Hurricane Hernan was moving northwest in the Pacific Ocean just off the coast of Baja California on September 3, 2002.
2. **Tropical Storm Edouard:** Tropical Storm Edouard creates a bank of clouds off the southeast Atlantic Coast on September 2, 2002.
3. **Tropical Storm Fay:** Tropical Storm Fay formed on September 5, 2002, in the Gult of Mexico before coming ashore in Taxas.
4. **Tropical Storm Gustav:** Tropical storm Gustav brought high surf and heavy rains to the North Carolina coast before veening northeast on September 10, 2002.
5. **Tropical Storm Hagupit:** Tropical storm Hagupit was closing in on Hong Kong on September 11, 2002.
6. **Tropical Storm Isidore:** With sustained winds of 65 miles per hour and heavy rains, the ruminant of Hurricane Isadore was already battering the coast of Louisiana on the evening of September 25, 2002, hours before the eye of the storm was expected to make landfall. As much as 20 inches of rain is expected in some areas.
7. **Tropical Storm Makkhala Hits China:** Packing sustained winds of 55 knots and gusts of more than 70 knots. Tropical Storm Mekkhala hit China on September 25, 2002.
8. **Typhoon Ele:** Typhoon Ele was moving north in the central Pacific on September 25, 2002.
9. **Typhoon Higos:** Typhoon Higos churning in Eastern Pacific Ocean.

10. **Typhoon Sinlaku:** On September 7,2002, Typhoon Sinlaku made landfall over eastern China.

Land Slides

These are sudden, short lived geomorphic events that involve the rapid-to-slow descent of soil or rock in sloping terrains. They occur worldwide, often in conjunction with natural hazards like earthquakes, floods, or volcanic eruptions. Land slides can also be caused by excessive precipitation or human activities, such as deforestation or development, that disturb natural slope stability.

Summer of 2001 was a season of devastating landslides. During June and July, landslides and floods throughout China's southern provinces left thousands dead or homeless. At least 35 people died in a landslide west of Katmandu, Nepal, and intense rain triggered severe flooding and landslides in 20 countries in West Virginia. In August, a landslide caused Ecuador's Sote oil pipeline to break, spilling some 1,000 barrels of crude oil and halting oil transport for a week.

Mass wasting, the ground ward gravitation of rock, debris, and dirt, continually sculpts Earth's landscapes. Research indicates that on most slopes, a certain amount of downhill movement is occurring constantly, although it may often be imperceptible.

On average, landslides in the Himalayan United States cause $ 1 to $ 2 billion in property damage and more than 25 fatalities per year. Posing threats to settlements and structures, landslides often result in catastrophic damage to highways, railways, waterways and pipelines. According to the U.S. Transportation Research Board, annual costs for the repair of minor slope failure by state departments of transportation exceed $ 100 million.

To determine where protective measures are necessary, scientists produce landslide inventory and risk assessment maps for many areas in the United States and Canada. But because landslides do not display a clear relationship between magnitude and frequency, as do earthquakes and floods. Landside studies are challenging to scientists. "It is very difficult to represent landslide hazards in quantitative terms over large areas", said Vemon Singh Roy, Senior Research Scientist at the Canada Centre for Remote Sensing.

Diseases

The so-called **"Black Death"** affected Europe during the mid-14th Century- fatality estimates vary with some commentators suggesting that up to 50% of Europe's population succumbed to the disease. The influenza ("Spanish Flu") pandemic of 19/18/19 claimed up to 40 million lives world wide. It is questionable whether or not this item falls within this category but at least 4,000 people died in London (England) during December 1952 as a result of smog (smoke-laden fog)- that being said, it is suggested (per the Met Office) that "but exactly how many people perished as a direct result of the fog will never be known many who died already suffered from chronic respiratory or cardiovascular complaints without the fog, they might not have died when they did".

Locusts

Per the United Nations, Food and Agriculture Organization (FAO) there are a number of species of locusts – some are identified below:

1. The African Migratory Locust – found in Africa.
2. The Australian Plague Locust – found in Australia.

3. The Bombay Locust – ranging through south-west to south-east Asia.
4. The Brown Locust – found in Southern Africa.
5. The Desert Locust – ranging through Africa and Asia.
6. The Moroccan Locust – ranging through north-west Africa to Asia.
7. The Oriental migratory Locust – found in south-east Asia.
8. The Red Locust – found in eastern Africa.
9. Tree Locust – ranging through Africa, the Mediterranean and Near East regions.

According to the FAO, the last Desert Locust plague occurred during the period 1986 to 1989. The FAO identify the following countries as being "historically been a source of many locust outbreaks and plagues: Djibouti, Egypt, Eritrea, Ethiopia, Oman, Saudi Arabia, Somalia, Sudan and Yemen.

DISASTER MANAGEMENT (FLOODS, EARTHQUAKE, CYCLONE AND LANDSLIDES)

Attempts at predicting when and where natural disaster will occur have met with some success in recent years. At present, China, Japan, Russia, and the United States are the countries most actively supporting such research. In 1975 the Chinese predicted the magnitude 7.3 quake at Haicheng, evacuating 90,000 residents only two days before the quake destroyed or damaged 90 per cent of the city's buildings. One of the clues that led to this prediction was a chain of low-magnitude tremors, called foreshocks that had begun about five years earlier in the area. Other potential clues being investigated are tilting or bulging of the land surface and changes in the Earth's magnetic field, in the water levels of wells, and even in animal behaviour.

Both physical solutions and institutional strategies should be developed to manage natural disasters. For flood disaster control the water flow should be done. Institutional strategies do not rely on physical barriers and include floodplain management, flood- and disaster-warning systems, education and preparedness programmes.

Natural disasters impart lessons at a very high cost of life and property. But if those lessons do not lead to learning and knowledge generation then it is a very heavy cost to bear. This lack of learning from the past hurts most at the recurrence of disasters. The earthquake in Gujrat (*26th January 2001, an earthquake of 7.9-magnitude on Richter scale struck Gujrat, India, with its epicenter in Kutch, causing destruction of three towns and death of more than 20,000 people*) and the subsequent chaos was an indicator of how crucial disaster planning is to manage relief and rehabilitation during disasters.

Structural Solutions

The major physical method used to mitigate water disasters in India are to build river and sea dykes. Dykes protect crops and other properties against floods.

In Viet Nam the Red River Delta in the north, people have built 3,000 km of river dykes and 1,500 km of sea and estuary dykes to protect against flooding. Many of these dykes are old and built of poor materials using inadequate manual construction technology. This results in the dykes suffering distress due to local slumping, under-seepage and piping. Further, the construction of dykes has gradually reduced over bank areas which used to be available for excess flood flows, with the result that river levels have become higher and higher. There are now many areas where the floodways are as much as five or six metres higher than the land protected by the dykes. Along the coastal plains in the central part of Viet Nam, the people have built sea dykes to protect themselves against extreme events such as typhoons. These dykes are essential for agriculture in the central provinces. When these dykes are overtopped or breached, the fields are flooded by saltwater and rendered unproductive for years. If this happens frequently, there are not sufficient resources in the community to repair or upgrade the dykes, and the people become impoverished and prone to malnutrition. In some parts, the attacks on coastal lands by the sea have been aggravated by water-resource structures, particularly dams, have interrupted the natural supply of sediment to the coast. This has sometimes caused the coastline to recede by as much as one kilometer in 20 years. Houses and villages have been destroyed and people have been killed.

In Delta area, a system of drainage channels and pumping stations should be used to make agriculture possible. However, drainage and construction work in areas of the Delta with pyrite-rich marine sediments have produced acid-sulphate soils containing high concentrations of sulphuric acid. This has limited the development of agriculture in the Delta.

Institutional Arrangements

Strategies for flood-disaster mitigation which do not rely on structural measures are gaining greater emphasis in India. For example, there has been a problem of population pressure causing unauthorized settlement in annually inundated floodways and in hazardous coastal areas.

Land management authorities still need to develop the ultimate power to enforce appropriate land-care practices, and to specify matters such as minimum floor levels and structural requirements for houses in disaster-prone areas.

Flood-Warning Systems

Flood-warning systems could do much to mitigate water disasters in India. What facilities exist are often unreliable, and the procedures for issuing warning suffer from a lack of modern communication technology.

Water-disaster mitigation is further complicated by a lack of expertise to assess damage of ability to assess requirements for emergency relief, and by the need for improved awareness of the threat of water disasters at all levels. Apart from those communities living near or having to care for the dykes, public awareness of the threat of water disasters is low even though the risk of flooding is very high.

Government of India Efforts

Government of India has constituted various committees to meet out the problems of natural disasters.

Standing National Crisis Management Committee

Standing National Crisis Management Committee is for effective coordination and implementation of response and relief measures in the wake of disasters. The chairman of the committee is the Cabinet Secretary and the members are the Secretaries of Ministry of Home Affairs, Ministry of Defence, Ministry of Power, Chairman, Railway Board; Department of Urban Employment and Poverty Alleviation, Ministry Petroleum and Natural Gas, Department of Shipping, Department of Roads Transport and Highways, Department of Telecom Services, Department of Rural Development, Department of Agriculture & Cooperation, Department of Drinking Water Supply, Department of Food & Public Distribution, Ministry of Water Resources and Joint Secretary, Cabinet Secretariat. The Ministry of Home Affairs, Government of India will provide necessary secretariat assistance.

Task Force

The compositions of Task Force on drought management have been constituted under the chairmanship of Deputy Prime Minister and the members are the Minister of Finance and Company Affairs, Minister of Agriculture, Minister of Consumer Affairs, Food & Public Distribution, Minister of Rural Development, and Dy. Chairman, Planning Commission. The Task Force will meet as frequently as necessary and initiate appropriate measures for effective drought management. Whenever, necessary, the Task Force may interact with the State Governments. The Task Force will be serviced by the Ministry of Agriculture.

National Crisis Management Committee (NCMC)

The composition of Standing National Crisis Management Committee for effective coordination and implementation of response and relief measures in the wake of disasters is constituted. The Chairman is the Secretary of Cabinets rank. The members are Secretary, Ministry of Home fairs, Secretary, Ministry of Defence, Secretary, Ministry of Power, Chairman, Railway Board, Secretary, Department of Urban Employment & Poverty Alleviation, Secretary, Ministry petroleum & Natural Gas, Secretary, Department of Shipping, Secretary, Department of Roads Transport and Highways, Secretary, Department of Telecom Services, Secretary, Department of Rural Development, Secretary, Department of Agriculture & Cooperation, Secretary, Department of Drinking Water Supply, Secretary, Department of Food & Public Distribution, Secretary, Ministry of Water Resources, Joint Secretary, Cabinet Secretariat.

The Ministry of Home Affairs, Government of India will provide necessary secretariat assistance.

Inter-Ministerial Group (IMG)

Composition of the Reconstituted Inter-Ministerial Group (IMC) constituted vides Ministry of Home Affairs, Government of India OM No 31-2/2002-NDM (T) Dated the 24th July, 2002. The Chairman of the group is Secretary, Ministry of Home Affairs and the members are Central Relief Commissioner (CRC), Secretary, Planning Commission, Secretary, Department of Expenditure, Ministry of Finance, Secretary, Department of Agriculture & Cooperation, Secretary, Ministry of Water Resources, Secretary, Department of Rural Development, Secretary, Ministry of Power, Secretary, Ministry of Petroleum & Natural Gas, Secretary, Department of Urban Employment & Poverty Alleviation, Secretary, Ministry of Road Transport& Highways, Secretary, Department of Health, Secretary, Department of Food & Public Distribution, Secretary, Department of Drinking Water Supply, Joint Secretary (NDM).

Policy for Acceptance of External Assistance

The present policy of not issuing a formal appeal on behalf of the Government, either directly or through any other agency to attract relief will continue. However, relief donated on a voluntary basis will be accepted and acknowledged as a sign of international solidarity.

There is no objection to NGO's issuing appeals for donations provided it is clear that the appeals are not at the instance of the Government of India. In the case of UN organizations and agencies (like OCHA) such appeals would imply endorsement by member countries and they are advised against appeals for international assistance.

Disaster Management Facility

The mission of the World Bank is to fight poverty. An important part of this mission is providing assistance to prepare for and recover from natural or man-made disasters that can result in great human and economic losses. Indeed, developing countries suffer the greatest costs when disaster hits: more than 95 per cent of all deaths caused by disasters occur in developing countries; and losses due to natural disasters are 20 times greater (as per cent of GDP in developing countries than in industrial countries.

Moreover, poorly planned development can turn a recurring natural phenomenon into a human and economic disaster. Allowing dense population on a floodplain or permitting poor or un-enforced building codes in earthquake zones is as likely as a natural event to cause casualties and losses. Similarly, allowing the degradation of natural resources increases the risk of disaster.

In this connection, the Disaster Management Facility (DMF) aims to reduce human suffering and economic losses caused by natural and technological disasters. We do this by helping the World Bank provide a more strategic and rapid response to disasters, and promoting the integration of disaster prevention and mitigation efforts into the range of development activities.

Disaster Operations

Disaster prevention	Flood insurance, hurricane awareness, hazardous weather awareness, family disaster plan, secure your home, disaster supply kit, pet plan, blueprint for safety, building codes.
Evacuation	On-line evacuation and shelter maps, evacuation maps, shelter status, have a place to go.

Disaster Management Information System (DIMS)

The Disaster Management Information System, is a civil society initiative supported by socially conscious institutions and individuals, companies and organizations. The system was launched by SRISTI, (Society for Research and Initiative for Sustainable Technologies and Institutions) on the 18th of January 2002 at Indian Institute of Management, Ahmedabad, and Gujarat, India.

SRISTI initiated an effort to build a ***"Disaster Management Information System".*** Though this initiative we are trying to develop a database driven information system for Disaster Management Authorities (DMA) in various states, NGOs and other organizations. We appealed to NGOs, relief workers, DMAs and individuals to share their experiences and volunteer services and resources to the online database maintained at our website. The database currently contains more than a thousand volunteers who have offered to volunteer their services and resources in time of emergency. About 700 organizations and institutions are also listed on the site, besides other resources and web links.

The DMIS is a wholly voluntary activity run with contributions in terms of time and services by SRISTI volunteers, NGOs and above all civil society institutions across the world. All the information shared with us is accessible to all, except where the volunteer has chosen to limit accessibility only to the relevant authorities.

SRISTI participated in the relief and rehabilitation work in Kutch. But the relief work suffered immensely due to lack of information and proper planning. When we tried to get answers to important questions that were cropping up – for intense, whether there exists a database on the distribution of available resources and expertise with individuals, institutions and corporations – all we got in response was a blank. This pointed to the urgent necessity of building a system for disaster mitigation and for documenting experiences of individuals and organization, which might act as a knowledge resources and help in better coordination in the case of future disasters.

A Guideline for Disaster Management

Database

On various resources, skills, and services required for relief at short notice. It will have information on safety equipments, oxygen cylinders and various other equipments, sills and other information required to deal with emergency.

Logistics

One of the most difficult problems to be handled is the organization of supply chain for relief. Several colleagues at IIMA also developed logistics system for Collectorate at Bhuj. This system could not be integrated with the GIS system so that supplies could be tracked right up to the village. This needs to be done. Likewise, other elements of logistics need to be put in place.

Technological Needs

Whole range of technical questions regarding buildings, cutting concrete slabs, rescue and relief emerged which needed to be solved on the spot. The best practices have to be put in use.

Self Reliance

The community self-reliance, lot of aid led to excessive inventory at the household level leading to reduced incentives for work and self-help. Likewise, there were areas where communities came together to help each other. The lessons of community self-help need to be put together. A database of volunteers who can move at short notice will need to be developed.

Communication Infrastructure

This is a serious problem and has to be resolved once for all. It will require network of ham radios, use of All India Radio, setting up help lines, etc. We will also have to create information dissemination system and develop mechanism for capacity building.

Emergency Preparedness

Drills will have to be organized to keep society prepared for dealing with such emergencies. One will have to learn from the experience of other similar drills.

Forecasting

Whenever possible, disasters which can be anticipated over time or space need to be looked into. For instance, buildings erected on land fill areas which were wet lands or low lying areas were more likely to get damaged, as was borne out by the recent evidence.

CAREER IN DISASTER MANAGEMENT

The term '**Disaster**' owes its origin to the French word "Desastre" which a combination of two words 'des' meaning 'bad' and 'astre' meaning star. Thus, the term 'disaster' refers to 'bad' or 'evil star'.

A disaster may be defined as "serious disruption of the functioning of society, causing widespread human material and environmental losses which exceed the ability of the affected society to cope using its own 'resources'. Disasters pose a serious threat to the normal life as well as the process of development and strike with sudden violence destroying lives and structures and throwing family apart.

With burgeoning population pressures, urban industrial growth deforestation and cultivation of marginal lands the human induced hazards have also increased. The impact of disasters is multidimensional affecting in all aspects – domestic, social economic and environmental, etc. A disaster is a product of hazards such as earthquake, flood or windstorm coinciding with a vulnerable situation which may include communities, cities or villages.

Among the top ten countries in terms of average annual loss of lives due to natural disasters, all except Australia are developing countries. India, amongst those relations is the most vulnerable to natural hazards. Due to its geographical position, climate and geological setting, India has been experiencing natural disasters every year. In India, natural disasters claim a significant toll of population, destroy/damage more than two million houses and convert many thousand hectares of land into infertile land annually. About 60 per cent of the total country is vulnerable to earthquake, 8 per cent area prone to cyclones and 12 per cent vulnerable to floods. The 5700 km long coastline is vulnerable to cyclones and 68 per cent of Net Sown Area (118 districts) is vulnerable to drought.

Types of Disasters

Disasters can be categorized as of three types:

i) Natural disaster
ii) Man-made disaster
iii) Hybrid disaster

1. **Natural disasters** are the result of natural phenomenon for instance, earthquake, tsunami, volcanic eruption, hurricane, tornado, avalanche or flood.
2. **Man-made disasters** are of an anthropogenic origin, and exemplify some of the terrible accidents that have resulted from human beings interaction with artificial environment, which they themselves have created. Bhopal gas tragedy, Chemobyl accident in Russia, etc. are some of the examples of such disasters.
3. **Hybrid disasters** arise from a linkage of anthropogenic events and natural events. For example – spread of disease from a community in which the disease is endemic to a community in which has no natural immunity, destruction of forests and resulting increased floods, large scale deaths due to smog, etc. Hybrid disasters can also be categorized and termed complex disasters or compound disasters in many parts of the world one kind of disaster can trigger another type of disaster which in turn triggers subsequent kind of disasters resulting in huge loss of life and property.

What is Disaster Management?

Disaster management is a multidisciplinary area covering a wide range of issues such as rescue, relief, reconstruction and rehabilitation. It also requires multi-sectoral governance, scientists, planners volunteers and communities all having important roles to play. These roles and activities span the pre-disaster, during the disaster and the post-disaster phases. Since their activities are complementary as well as supplementary to each others, there is a critical need to for coordination in the activities.

Disaster management is nothing but skilful ways, means and methods of controlling a disaster. When awareness and education about disasters are provided to the people, disaster management becomes a simpler task. India being a poor and developing country, lack of awareness and illiteracy plays an important role in the vulnerability of the masses. Disaster management techniques or methods are based on the economic status of nay country and hence it varies from country to country. Developed countries manage disasters better and quicker than developing countries.

Climate change is already changing the geographic distribution, frequency and intensity of weather-related hazards and threatens to undermine the resilience of poorer countries and their citizens to absorb loss and recover from disaster events. Climate change therefore, magnifies the impact of disasters on people and assets in developing countries. With changing climate globally the role of disaster management experts is also growing day by day. According to the UN findings global disaster risk increased between 1990 and 2007 by 13 per cent in terms of economic loss in the case of floods, due to rapid world population and GDP growth in disaster prone areas in relative terms the trends are stable and may even be falling. Global disaster risk is highly concentrated in poorer countries with weaker governance. The most intensive risk is found in a very small portion of the earth's surface. Just three countries - Bangladesh, China and India (all heavily populated) – account for 75 per cent of the mortality risk from floods only. In such a scenario, the job of disaster management experts is also in great demand in these countries.

Importance of Disaster Management Experts

Disaster Management Experts are trained to manage emergency situations and respond to the needs of the people and areas affected in a quick and efficient manner. They may be responsible for evacuating the people from a disaster/calamity affected town, managing the distribution of flood to the people in crisis, or providing/ overseeing the medical care to the injured. Disaster Management Experts are always in demand for large and small emergencies like hurricane, earthquake, terrorist attacks or chemical disasters.

Disaster management, as an emerging field needs the qualified personnel in the field. Disaster management graduates will find opportunities in the development sector with government and non-government organizations, with consultancy firms, or could go on to pursuing research, training and teaching careers.

Response to Disaster Management

The Ministry of Home Affairs is the nodal Ministry for coordinating relief, response and overall natural disasters management, and the department of Agriculture and Cooperation is the nodal Ministry for drought management. Other Ministries responsible for disaster of other kind are Ministry of Civil Aviation for air accidents, Ministry of Railways for rail accidents, Ministry of Environment and Forests for Chemical disasters and forest fire, Ministry of health for biological disasters (controlling spread of epidemics) and Department of Atomic Energy for Nuclear disasters.

Disasters – Taining and Education

The Ministry of Human Resources and Development (HRD) in the tenth Five Year Plan included Disaster Management in the curriculum of schools and professional education.

Central Board for Secondary Education (CBSE) also introduced Disaster Management subject in Social Sciences in Classes VIII in the year 2003 and for Class IX in 2004 and later on in Class X.

Career Opportunities / Prospects

Disaster Management personals can mostly find jobs in Government Organizations, Emergency Services, Law Enforcement Departments etc. They can also be employed in relief agencies, Non-governmental Organizations (NGOs) and International Agencies such as the United Nations etc. Private sectors also employ Disaster Management Experts to fulfil their Corporate Social responsibilities.

Administrators/Civil servants, civil engineers, police and defence personals, fire fighters and other protective service workers also undergo training and attend workshops to augment their skills in disaster management.

In the field of Information Technology also System or Network Administrators database Analysts, Security Administration or Operation Analysts are also offered positions to participate in the organization's disaster recovery, planning and management.

Social workers, engineers, medical health experts, environmental experts, scientists, rehabilitation workers can also have careers in disaster management, disaster mitigation and rehabilitation.

Courses

In the field of disaster Management the following courses are offered:

1. **M.A./M.Sc. in Disaster Management:** The Master's Programme in Disaster Management offered by the Jamsetji Tata Centre for Disaster Management, aims at enhancing knowledge capacities and skills, and perspective on disasters. When enabling an understanding of disaster from scientific point of view and technology in the prediction, mitigation and response, the programme also foster a critical and reflective appreciation of current debates in disaster management within the framework of social and environmental justice, state and civil society dynamics, development, conflict and displacement, and globalization
   Spread over four semesters, this Master's programme is a pioneering initiative within the country and has certain unique features which will enable mid-career professionals to combine their work with study. Research is a compulsory component of this programme. The Programme also has a multi level entry/exit option where student coming for a semester can obtain a certificate in Disaster Management on completion of the stipulated requirements. Students who wish to continue through the next semester would obtain a Diploma Alternately, a student on completion of the Certificate programme could come back after a break (not exceeding five years from the data of completion of the Certificate Programme, and join in for the second semester in order to obtain the Master's degree a student can complete the four semesters withut a break or in an incremental manner after the credits required for a diploma.
2. **Certificate Programme in Disaster Management.**
3. **Post Graduate Diploma Programme in Disaster Management.**
4. **National Institution of Disaster Management (NIDM)'s Online Course in Disaster Management.**

Institutes / Universities Offering Courses in Disaster Management

1. Tat Institution of Social Science, Mumbai: M.A./M.Sc. in Disaster Management
2. IGNOU
3. University of North Bengal, P.O. North Bengal University, District. Darjeeling-734430.
4. International Centre of Madras University, University of Madras, Chepauk, Chennai-600005 offers one year Post-Graduate Diploma in Disaster Management.

5. Sikkim Manipal University of Health, Medical and Technological Sciences, Tadong, Gangtok, Sikkim in association with Indian Institutue of Ecology and Environment (A-15, Paryavaran Complex, Saket Maidan Gandhi Marg, New Delhi-110030) offers Post-Graduate Diploma in Disaster Management.
6. Disaster Management Institute (DMI), E-5, Paryavaran Parisar Arera Colony, Bhopal-462016 (organizes training for working managers and Government officials in the management of natural disasters.

Nature does the Business and Man is merely the Manager"
"It you are late in doing one thing in
agriculture, you are late in all things'
Cato, 2nd Century B.C.

21 Role of Sustainable Agriculture

> *"What we have to cure is not a crisis but a chronic malady and the remedies for it are of a routine rather than of a spectacular nature."*
> ***Professor M. L. Dantwala***

SUSTAINABLE AGRICULTURE

In the colonial India, progress in agriculture was through exploitation of resources to serve the interests of rulers in power. In the post independence era, we were faced with the daunting task of feeding an increasing population and were burdened by recurring import of food grains. Attaining self-sufficiency in food grain production was the challenge and a goal before the agricultural scientists. The elements of the strategy to achieve self-sufficiency in food production within a short time were to expand the area under cropping, make large investments in development of water resources, input supplies and market infrastructure. Development and promotion of high yielding crop cultivars responsive to high inputs was the cornerstone for green revolution of the 1970s.

We are self sufficient in food grain production. Now the question is whether to follow the same strategies which resulted in green revolution or to redefine them. Consideration of our past achievements, new challenges and the related developments world wide can answer the question.

- Although we are now self sufficient in food grain production, the per capita availability has not increased significantly and a large fraction of population are still below poverty line due to lack of purchasing capacity,
- At the present growth rate of 2 per cent per annum, we will be about 1.3 billion by 2025 and 1.6 billion by 2060. Hence, there is need to produce more from shrinking resource base,
- Major gains in productivity and production in the past three decades have been from areas which had no serious limitations to production (adequate irrigation, productive soils, ideal climate, etc.). Productivity and production increase was moderate in eastern and central India and Deccan plateau, and
- Even in areas with no serious limitations for production, technology adopted for enhancing productivity has also simultaneously weakened the resource base resulting in a series of second generation problems.

- Small landholders in the tropics are mainly fed up with rain fed farming and it is being carried out with high risk. In a constant struggle to survive, farm communities have developed numerous ways of obtaining food and fibre from plants and animals. (TAC/ CGAIR, 1988). A wide range of different farming systems have been developed, each adapted to the local ecological conditions (Okigbo, 1978; Richards, 1988; Dupre, 1990). A closer look at these traditional farming systems reveals that they are not static; they have changed over the generations and particularly quickly over the last few decades, primarily as a result of the research and development activities of the local people. (Wieskel, 1989; Owasu, 1990). However, rapid changes in economic, technological and demographic conditions demand adjustments in smallholder farming systems. New market opportunities, promotion of chemical inputs and financial constraints may lead farmers to seek short term profits and pay less attention to keeping their agriculture in balance with the ecological conditions. In recent years, the negative environmental and soil impacts of high external input agriculture (HEIA) have become increasingly obvious (Wali, 1992; NRC, 1993). At the same time, many disadvantaged communities of small holders are being forced to exploit the resources available to them so intensively that, environmental degradation is setting in. Hence it is important to seek new approaches to agricultural development, which will benefit small farmers, half degradation of natural resources and restore degraded soils and ecosystems.

I. DEFINITION AND CONCEPT OF SUSTAINABLE AGRICULTURE

(A) Sustainable Agriculture

In 1987 the World Commission on Environment and Development (WCED, 1987) called attention to the immense problems and challenges facing world agriculture for meeting present and future food needs and to the need for a new approach to agricultural development.

The agricultural systems that have been developed over the past few decades have contributed greatly to the alleviation of hunger and the raising of standard of living of poor people. (Dora, 1983; Wilken, 1987) have served their purposes up to a point. But they were developed for the purposes of a smaller, more fragmented world. However, new realities reveal their inherent contradictions realities while require agricultural systems that focus as much attention on people as they do on technology, as much on resources as on production, as much on the long term as on the short term. Only such systems can meet the challenges of the future (WCED, 1987).

The production scenario is beset with ecological, technological and demographic problems. In contrast, the demand scenario features a high growth rate in food requirements. Such a situation will lead to dwindling global food reserves and an escalation in the cost of food grains and other agricultural commodities. In this context, the need for redoubling efforts in enhancing agricultural production and in promoting agrarian prosperity is obvious. Fortunately, there is growing interest in promoting sustainable agriculture which is also referred to by other names such as alternative farming, regenerative agriculture, and natural or organic farming, eco-farming and permeaculture.

(B) Concept and Meaning of Sustainable Agriculture

The concept of sustainable agriculture has been conceived differently by different authorities/authors. In economic terms, the use of resources today should not reduce real incomes in the future. Sustainable development encompasses several aspects or dimensions. The Word Conservation Union, in canning for the earth 1991, defines sustainable development as improving the quality of human life while living within the carrying capacity of supporting ecosystems. Earlier, the World Commission on Environment (1988) defined sustainable development as development that meets the needs of the present without compromising the ability of future generations to meet their own needs.

In essence, sustainable development means protecting the natural resources needed for food production and cooking fuels, while expanding production to meet the needs of growing population. In other words, sustainable development means more efficient use of arable lands and water supplies as well as development and adoption of improved agricultural practices and technologies to increase yields (World Resources, 1992-93).

Sustainable agricultural systems must maintain or enhance (i) biological and economic productivity of crops, (ii) enhance the efficiency of use of input, (iii) lesser adverse environmental impacts both on and off the farm, (iv) minimize adverse environmental impacts on adjacent and down stream environments, (v) minimize the magnitude and rate of soil degradation and to enhance soil quality and resilience so that the crop productivity can be sustained with minimum adverse impact on soils and environment, and (vi) enhance compatibility with social and political conditions.

The word 'sustainability' is now widely used in development circles. But what does it really mean? According to a dictionary definition, 'sustainability' refers to 'keeping an effort going continuously, the ability to last out and keep from falling. In the context of agriculture, 'sustainability' basically refers to the capacity to remain productive while maintaining the resource base. For example, the Technical Advisory Committee of the Consultative Group on International Agricultural Research (TAC/CGIAR 1988) states: "sustainable agriculture is the successful management of resources for agriculture to satisfy changing human needs while maintaining or enhancing the quality of the environment and conserving natural resources".

However, many people use a wider definition, judging agriculture to be sustainable if it is (after Gips 1986);

- Ecologically sound, which means that the quality of natural resources is maintained and the vitality of the entire agro-ecosystem from humans, crop and animals to soil organisms is enhanced. This is best ensured when the soil is managed and the health of crops, animals and people is maintained through biological processes (self-regulation). Local resources are used in a way that minimizes losses of nutrients, biomass and energy, and avoids pollution. Emphasis is on the use of renewable resources.
- Economically viable, which means that farmers can produce enough for self-sufficiency and/ or income, and gain sufficient returns to warrant the labour and costs involved. Economic viability is measured not only in terms of direct farm produce (yield) but also in terms of functions such as conserving resources and minimizes risks.
- Socially just, which means that resources and power are distributed in such a way that the basic needs of all members of society are met and their rights to land use, adequate capital, technical assistance and market opportunities are assured. All people have the opportunity to participate in decision-making, in

the field and in the society. Social unrest can threaten the entire social system, including its agriculture.

- Humane, which means that all forms of life (plant, animal, human) are respected. The fundamental dignity of all human beings is recognized, and relationships and institutions incorporate such basic human values as trust, honesty, self-respect, cooperation and compassion. The cultural and spiritual integrity of the society is preserved and nurtured.
- Adaptable means that rural communities are capable of adjusting to the constantly changing conditions for farming: population growth, policies, market demand etc. This involves not only the development of new, appropriate technologies but also innovations in social and cultural terms.
- These different criteria of sustainability may conflict and can be seen from different view points; those of the farmers, the community, the nation and the world. There may be conflicts between present and future needs; between satisfying immediate needs and conserving the resource base. The farmer may seek high income through high prices for farm products; the national government may give priority to sufficient food at prices, which the urban population can afford. Choices must continually be made in a never-ending search for balance between the conflicting interests. Therefore, well-functioning institutions and well deliberated policies are needed on all levels-from village to global- in order to ensure sustainable development.

In agricultural development, raising production is often given primary attention. But there is an upper limit to the productivity of ecosystems. If this is exceeded, an ecosystem will degrade and may eventually collapse, and fewer people will be able to survive on the remaining resources than before. This implies that, when the limits on the supply side are reached, something has to be done on the demand side, e.g. other sources of income, emigration, lower consumption level, and population control. Production and consumption have to be brought into balance on an ecologically sustainable level. Although sustainability must be seen as a dynamic concept which allows for the changing needs of an increasing global population (TAC/CGIAR, 1988), basic ecological principles oblige us to recognize that agricultural productivity has finite limits. Why has the concept of sustainability gained increasing importance with reference to agricultural development? This becomes evident if we take a look at the present situation of world agriculture.

(C) Sustainability Defined

Sustainable agriculture can be defined as that form of agriculture aimed at meeting the food and fuel needs of the present generation without endangering the resource base for the future generations. It is an efficient management system of renewable resources including soil, forests, crops, fish, livestock, biodiversity and ecosystems without degradation, to provide adequate food and other needs for the current and future generations.

The Technical Advisory Committee of the Consultative Group on International Agricultural Research (CGIAR) defined sustainable agriculture as successful management of resources for agriculture to satisfy changing human needs while maintaining or enhancing the quality of the environment and conserving natural resources (TCA, 1989).

It is frequently argued that sustainability is not a concern of developing world. What we need is productivity increase. Concerns of environment are largely those of the developed world. Are the goals of enhanced productivity and sustainability in conflict? The answer is simply no. In fact, there is a strong nexus between the two. In a majority of cases, further improvement in productivity not likely to come about in the absence of resource conserving strategies. By implication, the strategies to enhance productivity should not simultaneously reduce the quality of resource base. We simply cannot afford productivity increases at the cost of resource degradation. The recovery of degraded resources may be, too costly, if not impossible.

Sustainable agriculture should not be viewed as a system of maintaining the present high productivity levels of crops without endangering the resource base for future generations. It aims at sustaining the productivity of the natural resource base by conserving and improving ecological foundations essential for continuous advances in crop and animal productivity. It is a continuous process for increased productivity of crops and animals without weakening the resource base.

(D) High and Towards Low External Input and Sustainable Agriculture (LEISA)

In view of the limited access of most farmers to artificial external inputs, the limited value of these inputs under LEIA conditions, the ecological and social threats of 'green revolution' technology and the dangers of basing production on nonrenewable energy sources, the strong emphasis on High External Input Agriculture (HEIA) in agricultural development must be questioned. However, it is also open to question whether it will be possible to raise world food production sufficiently without the use of such external inputs. Besides, natural; as opposed to artificial inputs can, also have detrimental environmental effects.

LEISA is an option which is feasible for a large number of farmers and, which can complement other forms of agricultural production. As most fanners are not in a position to use artificial inputs or can use them only in small quantities, it is necessary to concentrate on technologies that make efficient use of local resources. Also, those farmers who now practice HEIA could reduce contamination and costs and increase the efficiency of the external inputs by applying some LEISA techniques. It is important that the agro-ecological knowledge of both scientists and farmers be applied, so that internal and external inputs can be combined in such a way that the natural resources are conserved and enhanced, productivity and security are increased and negative environmental effects are avoided.

LEISA Refers to those Forms of Agriculture that:

- Seek to optimize the use of locally available resources by combining the different components of the farm system, i.e., plants, animals, soil, water, climate and people, so that they complement each other and have the greatest possible synergistic effects.
- Seek ways of using external inputs only to the extent that they are needed to provide elements that are deficient in the ecosystem and to enhance available biological, physical and human resources. In using external inputs, attention is given mainly to maximum recycling and minimum detrimental impact on the environment.
- LEISA does not aim at maximum production of short term but rather at a stable and adequate production level over the long term. LEISA seeks to maintain and, where possible, enhance the natural resources and make maximum use of natural processes. Where part of the production is marketed, opportunities are sought to regain the nutrients brought to the market.

Numerous developing countries are now implementing so-called structural adjustment programs that involve policies such as devaluation of exchange rates, reduction of government spending and intervention, reduction of subsidies and removal of price controls. In this way, the demand for imports is to be curtailed and the purchase of local goods stimulated, so as to reduce the balance of payment and government deficits and to promote national economic growth. LEISA appears to fit within this context, as it is less demanding on imports and credits than the conventional approach to agricultural development. At farm, regional and national level, LEISA implies the need for closely monitoring and carefully managing flows of nutrients, water and energy in order to achieve a balance at. a high level of production. Management principles include harvesting water and nutrients from the watershed, recycling nutrients within the farm, managing nutrient flow from farm to consumers and back again, using aquifer water judiciously, and using renewable sources of energy. As these flows are not confined by farm boundaries, LEISA requires management not only at farm level but also at district, regional, national and even international levels. At each level, technologies are sought to make the flow cycle as short as possible and. to balance the flows. In this book, the focus is on practices that can be applied at farm level. Questions related to techniques and system at village level and above are equally important, but should be addressed in a separate study.

LEISA incorporates the best components of indigenous farmers' knowledge and practices; ecologically sound agriculture developed elsewhere, conventional science and new approaches in science (e.g. systems approach, agro-ecology, biotechnology). Thus far, conventional science has served mainly HEIA, but the contributions it could make to LEIA should be explored to the full. LEISA practices must be developed within each ecological and socio-economic system. The specific strategies and techniques will vary accordingly and will be innumerable. The experience thus far of developing LEISA systems cannot provide universal, ready-made answers for the problems of farmers in other areas, but can provide some indication of principles and promising possibilities.

The process of combining local farmers' knowledge and skills with those of external agents to develop site-specific and socio-economically adapted farming techniques has been given the name 'Participatory Technology Development' (PTD). Farmers work together with professionals from outside their community (e.g., extension workers, researchers etc.), in identifying, generating, testing and applying new techniques. OTD seeks to strengthen the existing experimental capacity of farmers and to encourage continuation of the innovation process under local control (Haverkort *et al.*, 1988). The experience of combining indigenous and scientific knowledge through a process of PTD indicates strongly that it is indeed possible to transform LEIA to LEISA: Low External-Input and Sustainable agriculture. This approach to agricultural development appears to be better adapted to the needs and opportunities of LEIA farmers and to fit better into their cultural context than the conventional approach.

(E) Sustainable Agroecosystems

An alternative to the chemical dependence is to maximize the contributions of bio-diversity to pest control and nutrient cycling and to attain optimal productivity with minimal inputs. Edwards and Grove (1991) proposed an analogous term for management of nutrients, integrated nutrient management. This approach capitalizes on the adaptive features of traditional systems and incorporates additional advantages of conventional and innovative technology. It is important to recognize a strong link between the availability of organic matter and both bio-diversity and nutrient cycling (Palm *et al.*, 1987). The practice in many developing countries of removing organic matter from the land for fuel and other purposes is a serious constraint to long-term sustainability (Oram, 1988). The most sustainable farming practices and components of the man managed bio-diversity can be developed only by understanding the functions of the agro-ecosystem and low social and economic conditions of the farmers and their climatic and adapted environments impact upon overall crop and animal productivity. No matter how well the agro-ecosystem functions biologically, it is sustainable only if it is socially and economically sounds (Altieri, 1987).

(F) Indices of Sustainability

Quantification of sustainability is essential to objectively assess the impact of management systems on actual and potential productivity, and on environment. One can assess sustainability or several indices (Lal, 1994). Indices may be simple involving one parameter or complex involving several parameters. Although general principles may be the same, these indices must be fine-tuned and adapted under local environments. Some indices of Sustainability includes the following:

1. **Productivity (P).** Production per unit of resource used can be assessed by,

$$P = P/R$$

Where, P is productivity, P is total production and R is resource used.

2. **Total Factor Productivity (TFP),** It is defined as productivity per unit cost of all factors involved (Herdt, 1993).

$$TFP = \frac{p}{\sum_{i=1}^{n} (Ri \ X \ Ci)}$$

where, P is total production, R is resource used and C is cost of the resource, and n is the number of resources used in achieving total production.

3. **Coefficient of sustainability (Cs).** It is measure of change in soil properties in relation to production under specific management system (Lal, 1991).

$$Cs = F\ (oi, ad, Om)\ t,$$

where, Cs is coefficient of sustainability, Oi is output per unit that maximizes per capita productivity or profit, ad is output per unit decline in the most limiting or non-renewable resource, Om is the minimum assured output, and t is the time. The time scale is important and must be carefully selected.

4. **Index of sustainability (Is).** It is a measure of sustainability relating productivity to change in soil and environmental characteristics (Lal, 1993; Lal and Miller, 1993).

$$Is = f\ (Pi*Si*Wi*Ci)\ t,$$

where, Is index of sustainability, Si is alteration in soil properties, Wi is change in water resources and quality, Ci is modification in climatic factor and t is time.

5. Agriculture Sustainability (Ag.). It is a board based index based on several parameters associated with agricultural production (Lal, 1993).

$$As = d\ (Pt * Sp * Wt * Ct *)\ dt,$$

where, As is agricultural sustainability, Pt, is productivity per unit input of, the limited or non-renewable resource, Sp is critical soil property of rooting depth, soil organic matter content, Wt is available water capacity including water quality, and Ct is climatic factor such as gaseous flux from agricultural activity, and t is time.

Sustainability Coefficient (SC)

It is a complex and a multipurpose index based on a range of parameters, and is similar to As. It is defined as:

$$Sc = (Pt*Pd*Pm)\ t$$
$$Sc = d\ (Pi*Wt*Ct)\ dt$$

Where, Pt is productivity per unit input of the limited resource, Pd is productivity per unit decline in soil property, Sc is critical level of soil property, Wt is soil water regime and quality, Ct is climatic factor, and t is time.

Crop Productivity as an Indicator of Sustainability

A measure of crop productivity is a good integrator of all soil, water, climatic and biotic factors. It is important to assess potential vis-à-vis actual productivity. In a science based management system, actual production exceeds potential production in soils of low inherent fertility and in harsh environments. The potential productivity, soils' productive potential within a biome, can be estimated by several models e.g., GERES (Richie et al., 1989) and Tropical Soil Productivity calculator (Aune and Lal, 1994). If land availability -is a limiting factor, appropriate indices of productivity are Land use factor (L), Land Equivalent Ratio (LER), and Area Time Equivalent Ratio (ATER). etc.

The Land use factor (L) is defined as the ratio of cropping period C plus fallow period F to cropping period C (Okigbo, 1978).

$$L = C + F/C$$

The factor L is generally high for low intensity system e.g., shifting cultivation.

The LER is calculated as follows (Willey and Osiru, 1972):

$$LER = \sum_{i=1}^{n} \left(\frac{Yi}{Ym} \right)$$

where, Yi and Ym are yields of component crops in the inter crop and monoculture system, respectively, and n is the number of crops involved.

Because crops involved vary widely in their maturity period, ATER index considers the crop duration (Hiebsch and Mc Collum 1987).

$$ATER = \frac{1}{t} \sum_{i=1}^{n} \left(\frac{d\ Yi}{Ym} \right)$$

where, d is the growth period of the crop in days and t is the time in days for which the field remained occupied i.e., the growth period of the longest duration crop. Numerical values of ATER approaches that of LER for a mixture consisting of crops of approximately identical growth periods i.e.; when t = d in comparison, productivity can also be expressed terms of the resources use efficiency of the most limiting resource e.g., wafer, nutrients, energy or labour.

II. COMPONENTS OF SUSTAINABLE SYSTEMS

A wide array of critical environmental threats is posing a challenge to the mankind all over the world. Some of these threats include degradation of soil, water, forest and marine resources essential to increased production of food, fodder, fibre, timber, etc. Degradation of natural resources is the main issue threatening sustainable development of agriculture.

1 Cause for Resource Base Degradation

Three main reasons attributed by FAO (1989) for degradation of natural resources are:

1. Lack of access to land, inputs and other production resources by the majority of rural households and communities,
2. Lack of awareness among policy makers of the economic costs involved, and
3. Lack of environmentally sound alternative technologies acceptable to farmers.

Subsistence farmers are forced to over exploit natural resources by way of mining soil nutrients, cultivating steep slopes, overgrazing rangeland, and excessive collection of fuel wood for their survival. It must be realized that the cost and level of inputs needed to restore the productivity of degraded lands is much greater and returns much lower in several cases.

2 Major Components of Sustainable Systems

Major components of the sustainable agricultural systems are:

1. Soil and water conservation to prevent degradation of soil productivity and lengthening crop growing season for optimum crop productivity,
2. Efficient use of limited irrigation water without leading to problems of soil salinity, alkalinity and high ground water table,
3. Crop rotations that mitigate weed, disease and insect problems, increase soil productivity and minimize soil erosion,
4. Integrated nutrient management that reduces the need for chemical fertilizers, improves the soil health and minimize environmental pollution by conjunctive use of organics, inorganic and bio-fertilizers,
5. Integrated pest management that reduces the need for agrochemicals by crop rotation, weather monitoring, use of resistant cultivars, planting time and biological pest control, and
6. Management systems to control weeds by preventive measures, tillage, timely inter-cultivation and crop rotation which improve plant health.

All such ecologically sound practices required more information, trained labour and management skills per unit of production than current input intensive agriculture.

III. EMERGING ISSUES FROM PRESENT TECHNOLOGY

Technology generated and implemented for increasing productivity of crops during the past three decades resulted in depletion of resource base besides creating several other crop production problems.

1. Biodiversity

Development of high yielding cultivars has been possible due to wide range of genetic source material provided by land races. Apparently, the vast genetic diversity within the same crop species is being forced out by a narrow genetic range of economically advantageous cultivars. The very success now threatens the source of genetic diversity on which further progress depends. Widespread use of modern cultivars raises question about the stability of production and threatens the endemic breakout of pest and disease attack. As the biodiversity is reduced, natural processes that control and influence habitat quality and genetic expression decline. As the biodiversity is reduced, internal and natural controls must be replaced by more externally applied artificial controls in the form of management and inputs. Similar to crops, there is need to conserve the biodiversity among animals to sustain the overall improvement in their production capacity witnessed in recent years.

2. Pests and Diseases

Pest and disease problems are now more serious than ever before. This situation is the result of monoculture of selected crops leading to buildup of pests and diseases. Carryover of stem borer (pink borer) and buildup of soil born pathogens in continuous cereal systems and *Phalaris minor* weed problem in rice-wheat system are example of such serious problems. Farmers have to incur huge amounts on plant protection.

3. Agrochemicals

Modern cultivars are responsive not only to fertilizers but also to chemicals used in plant protection. In general, pest, disease and weed problems are more in modern improved cultivars compared to traditional varieties. Keeping in view the productivity of high yielding cultivars, farmers used the chemicals indiscriminately. The natural enemies thus eliminated with chemicals resulted in pest and disease incidence. Some of the pests and weeds developed resistance to the pesticides and herbicides, thus complicating the plant protection. Pesticide, fungicide and herbicide residues entering food chain has serious health hazards.

4. Accelerated Soil Erosion

The most deleterious impact of agriculture on environment is man induced accelerated soil erosion. Continuous cultivation has replaced longer fallows and extended further into areas where either the climate and/or soil are marginal for cultivation. Continuous cultivation and decline in soil organic matter content has resulted in blowing of lighter drier soils and the sheet washing, rilling and gullying of heavier soils. When once the top soil is lost, the parent material will not respond to any inputs of management practices for improving the crop productivity. Of the soil lost by water erosion, high proportion find its way finally into oceans. It reduces the efficiency of reservoirs and irrigation channels.

5. Irrigation Related Problems

Inadequacies in irrigation water management resulted in serious resource degradation problems. In areas of canal irrigation without adequate drainage, salinity, alkalinity and high water table are the serious constraints in enhancing the productivity. There are other areas where unrestricted installation of tube wells has resulted in continuous lowering of ground water tables leading to increasing costs in lifting water for irrigation. Many productive soils have become unproductive under canal irrigation due to misuse of irrigation water. Most water management problems are system related and defy solutions arrived at in piece meal fashion.

6. Depletion of Soil Fertility

With increasing harvests, rates of fertilizer application have increased several folds in intensive cropped areas. In general, more quantities of nutrients are removed from soil than restored through fertilizers. Availability of FYM is decreasing as tractors replace draught animals in many cases and large quantities used as fuel. Hence, use of organics is not to the desired level. There is imbalance in the use of nutrients resulting in low use efficiency. Decline in quality and quantity of organic matter in most soils is adversely affecting soil biodiversity and biological regulation of soil process. Strategies to reverse the process of soil fertility decline or enhancement of soil productivity require a long term perspective. These and other related issues are by no means amenable to simple solutions-solutions that are economically, socially and environmentally viable. This calls for systems approaches in finding solutions through multidisciplinary and multi-institutional efforts to understand interdependence of different components of ecosystem. This prompted the necessity for recasting the production technology in use for over three decades. Sustainable production systems appear to be the way out of emerging problems from present production technology.

IV. FACTORS AFFECTING ECOLOGICAL BALANCE

Major factors affecting ecological balance are deforestation and overgrazing rangelands, accelerated soil erosion, irrigation related problems, over exploitation of ground water and indiscriminate use of agrochemicals.

1. Deforestation and Overgrazing of Range Lands

Perennial vegetation such as trees and grasses successfully prevent soil erosion and runoff from fallows. Forests influence climate of a region due to their effect on wind direction and hence the rainfall. Deforestation and overgrazing modifies the climate and the biodiversity besides loss of valuable genetic resources used in breeding programme for developing high yielding cultivars. While some plant and animal species can adapt to the direct or indirect impact of particular agro-ecosystem, others have been unable to do so. Habitat reduction or destruction is inevitably accompanied by the disappearance of plant and animal species.

2. Accelerated Soil Erosion

Accelerated soil erosion is currently a major environmental problem in tropical and subtropical areas of the world as a consequence of population growth and demand for food. When once the vegetative cover is lost, the bare soil is exposed to the vagaries of wind and intensive rains leading to accelerated soil erosion. The productive soil is lost, making the soil unsuitable for crop production. Runoff from arable land contributes to nutrient enrichment (eutrophication) of the water into which it drains. The result is an increased rate of algae growth, increase in turbidity and depletion of oxygen. Aerobic plants and animals die and the rate of decomposition and recycling of nutrients can decline to a level at which a fresh water lake may be regarded as virtually dead.

3. Irrigation Related Problems

Poor quality of water is one of the main factors turning good soils into saline or sodic soils. Provision of irrigation, without adequate drainage, leads to the same problems, as that with poor quality water. Many canal irrigated lands have become unproductive due to salt problems and high ground water table. Total area is suffering from water logging ranges between 6 and 8.5 M ha while that affected by salinity is around 9 M ha. It is the misuse and the associated poor water use efficiency which lead to depriving the tail-enders of the valuable water resources and diminishing the potential of otherwise prime land which would have otherwise benefited from the created wealth of irrigation water.

4. Over Exploitation of Ground Water

Water table depth is one of the main factors determining the economics of irrigation. Rapid increase in number of wells has resulted in a situation in which a large number of well owners competing to extract water from the limited aquifers, with a concomitant lowering of the ground water table. Deepening of water tables in areas where less rain water is captured than necessary to offset the withdrawals is likely to lead ultimately to a permanent water deficit situation. Further, overdraft of ground water aggravates surfacing of harmful fluorides and salts. The areas which face prospects of this type abound specifically in peninsular India. Sustainable agriculture aims at maximization of the ground water recharge through appropriate interventions favouring minimization of surface runoff.

5. Indiscriminate Use of Agrochemicals

Among the agricultural impacts on environment that have attracted much attention are those resulting from rapid escalation in amount of chemical fertilizers and pesticides.

Chemical Fertilizers

Surplus soluble inorganic fertilizers, particularly nitrogen, which have not taken up by the plants are leached out of the system. Others such as phosphorus and potassium are not so susceptible to loss by leaching as is nitrogen, except under abnormal condition. Overuse and abuse of chemical fertilizers harm the biological power of the soil. This must be prevented as all nutrient transformations are negotiated by soil microflora. Organic matter is the source of energy to soil microflora and its content is the index of soil health.

There is mounting concern about the increasing levels of nitrate in river and ground water. In some areas it has exceeded the recommended 11.3 mg l-1 set by the World Health Organization. Intensive arable agriculture is undoubtedly one of the main sources of nitrate pollution. Nitrates have been identified as one of the causes of fatal condition, methemoglobinemia in young babies. In addition they are a source of carcinogenic nitroso compounds.

Pesticides

Use of pesticides to control animal pests (insecticides), plant diseases (fungicides) and weeds (herbicides) to cope up with crop protection opened the doors for several problems. Exclusive reliance on chemical pesticides has resulted in problems such as pesticide resistance, resurgence, residues and environmental pollution. Despite large scale use of pesticides, more pest outbreaks of bollworms and whitefly in cotton, BPH and leaf folder in rice, pyrilla and borers in sugarcane and pod borer in pigeon pea are mainly due to destruction of natural enemies. Application efficiency is usually low, with less than 20 per cent chemical applied reaching the target. Pesticide residues are disposable into the air, soil and water and other nonagricultural ecosystems, where they are subjected to transformations. Most pesticides enter the agro-ecosystem via green plants either by direct foliar absorption or by uptake from water and soil. If the pesticide taken by animals through the feed is not voided from the animal system, these are stored in certain organs such as liver. They are highly toxic at relatively low levels. This residual toxicity contributes to elimination of vulnerable species.

V. MANAGEMENT PRACTICES FOR SUSTAINABLE AGRICULTURE

The integrated concept of sustainable agriculture owes its origin to the philosophy of 'holism' which enunciates that all things are connected and their interactions in nature are complex. Stimulus creates response not only in that but in the system as a whole. An ancient Tamil proverb "No fodder, no cattle; no cattle, no manure; no manure, no crop" indicates importance of integrated concept. Although our flora and fauna are rich and diverse, limitations and constraints in other resources (land, water and climate) make it imperative that we make the best optimal use of these resources based on scientific principles and practices.

1 Soil Management Practices

The intrinsic capacity of natural resources becomes suboptimal due to unintended introduction of degradation processes (irrigation mediated salinisation). Damage can also be caused by overstraining natural resources beyond their carrying capacity. Arable farming in soils of land capability IV and above or the bid to produce super yields artificially through high intensity farming are typical examples of overtaking natural resources. Indiscriminate addition of restorer inputs (fertilizers) and their-inefficient use can also lead to non-sustainability of a farming enterprise.

Soil Degradation

Soil degradation is decline in the productive capacity of land due to processes induced mainly by human activities. According to National Bureau of Soil Survey and Land Use Planning, out of a total geographical area of about 328 M ha, 187 M ha (57%) are suffering from different soil degradation problems. Water erosion is the major problem causing loss of soil in 132 M ha and terrain deformation in 16.4 M ha. Wind erosion is dominant in western region in 13.5 M ha. Human induced chemical degradation is observed in 14 M ha, causing salinisation in 10 M ha and loss of nutrients and organic matter in 4 M ha. Unless short and long term measures are taken to assess our basic resources in order to arrest degradation and restore productivity, we may not be able to fulfill our obligation to sustainable agriculture.

Soil and Water Conservation Technology

It has been estimated that about 5334 mt of soil is lost through erosion annually from agricultural lands besides 8.4 mt of plant nutrients. Soil and water conservation practices for agricultural lands include contour farming, agronomic, cultural and biotic conservation measures, and runoff. Cycling and mechanical measures such as contour bunding, graded bunding, contour ditches and bench terracing on slopes. Alternate land use systems such as pastures, fuel-fodder-timber, etc. are to be followed on non- agricultural lands falling under land capability classes V, VI, VII and VIII.

Watershed Based Approach

The major abiotic constraints that cause un-sustaining rainfed agriculture include:

- Low and variable rainfall,
- Removal of vegetative cover and its non-replacement,
- Declining soil fertility,
- Extension of cultivation to marginal lands,
- Accelerated land degradation, and
- Socioeconomic factors.

For the development of rainfed agriculture, the watershed model conceptually provides an envelope through which many of the steps for sustaining agriculture can be implemented. The landscape watershed units can be effectively subdivided into discrete hydrological units. Since, the watersheds are spatially laid from ridge to valley, they most effectively conserve land and water resources and help secure water availability throughout the crop growing season. The land area of the watershed drains into a common point. Hence, the drainage water can be easily stored in above ground storage structures for recycling during droughts or for growing additional crop. There is no alternative to capturing the runoff water and recycling it for stabilizing and sustaining the productivity of rainfed crops.

The impact and effectiveness of watershed initiatives such as Sukhomajri, the Mysore Resettlement and Development Agency (MYRADA) and the Ralegoan Sidhi have clearly shown that it is possible to embed ecologically sound rainfed agriculture in economically sustainable development. Watershed management has been discussed in Chapter 8.

Soil Fertility Management

There is a net negative balance of soil nutrients in relation to the total uptake of nutrients by crops and the nutrients added to the soil. Progressive soil fertility depletion will assume as one of the serious limitations in future for sustainable agriculture.

As early as 1974 the need for integrated nutrient management (INM) was elucidated. The INM philosophy combines economic and efficient traditional and improved technologies to grain from the symbiosis and synergy of crop-soil environment bio-interactions. The approach is flexible and minimizes use of chemicals but maximizes use efficiency. The concept is for optimization of the effects of all available sources of plant nutrients to improve soil fertility availing nature's gifts. Agriculture is the most important sources of renewable wealth. Development of an integrated plant nutrient system (IPNS) involving an approximate mix of organics, biological nitrogen fixation, phosphate solubilizing microbes and need based chemical fertilizers would be crucial for the sustainability of production and soil as the resource base for it. At the farm level, integrated nutrient management systems (INMS) should be able to maximize the use of residues from the farm, wastes from the livestock and the household before supplementing them with chemical fertilizers. At the watershed level, this should take into account an inventory of organic resources available from plants and livestock, agro-forestry systems, bio-fertilizers and availability of need based chemical fertilizers. INMS include:

- **Recycling of organic wastes** is such as crop residues, dung and urine from domesticated animals and wastes from slaughter houses, human excreta and sewage, biomass of weeds, organic wastes from fruit and vegetable production and household wastes, sugarcane trash, oil cakes, press mud and fly ash from thermal power plants. Materials not suitable for direct application can be better applied by composting and vermi composting.
- **Biological nitrogen fixation** is through blue-green algae (BGA) and *azolla* for rice, *rhizobium* for legumes, *azotobacter* and *azospirillum* for others. *Bacillus subtiles*, *B. circulans* and *Aspergillus niger* (fungus) available as cultures can solubilize the unavailable phosphorus and increase its availability to plants, and
- **Green manuring** is an age old practice for improving soil fertility and supplying part of nutrient requirement of crops. The stem nodulating green manure plant, *Susbania rostrata* can fix 100-250 kg N/ha in 45 to 55 days has great scope in rice culture. Availability of noncompetitive season in cropping sequence and irrigation water are the limitations to the spread of green manures. There is also greater scope for green leaf manuring for rice and other crops from the looping of various multipurpose trees popularized through afforestation and agroforestry systems.

Popularization of biogas plants, encouraging legumes in crop rotations and intercropping systems and use of sewage, sludge and effluents for agriculture can also be the components of INMS. Components of INMS and their approximate contributions are indicated in Table 21.1.

TABLE 21.1 : Components of INMS and their contribution

Component	Scope	Equivalence under favourable conditions	Factors
Fertilizer	All crops	One unit of 60-30-30=10-15 units of wheat, 10-15 units of rice	Variety, weather and management.
FYM	To the extent Available	One unit of FYM = 3 kg fertilizer for single crop and 5 kg for double cropping	Composition and quality
Compost	Where available	1 t = 3.5 kg N for rice.	Composition and Quality
Crop residues	Where available	1 t = 3.5 kg N in *kharif* rice, 2 kg N In *rabi* rice.	Composition, quality and interval between Applications
Green manure	Where available	45 to 50 days sesbania = 30-60 kg N/ha for rice	P application and economics
Rhizobium	Legumes	Equivalent to 10-40 kg N/ha	Quality of inoculums, P application and time Of incorporation
Blue green Algae	Rice	10 kg BGA = 15-25 kg N/ha	Suitable culture, temperature, moisture and P application
Azolla	Lowland rice	10 t fresh azolla = 20-25 kg N/ha.	P, moisture, pH, temp, incorporation.

Reclamation of Salt Affected Soils

Introduction of canal irrigation accelerated the appearance of salts in those places where they had existed at some depth in the soil profile. The gestation period between canal irrigation and appearance of water-logging and salinity was rather small. While canal irrigation may salinise the soil through rise in water table, well irrigation usually causes salt buildup due to poor quality water. While incomplete weathering of constituent soil minerals ensures a steady source of salts, the inadequate rainfall favours their deposition in the surface layers in dry lands. Recent estimates show that about 10M ha is affected by salinity and alkalinity.

Removal of excess salts from saline soils is accomplished by simple leaching provided good quality water is available for the purpose and water table is either reasonably deep or there is provision of subsurface drainage and disposal of drainage effluents. In lands with shallow but good quality ground aquifers, vertical drainage with shallow irrigation pumps has been successful.

In dry lands with saline ground water or where salinity is associated with water logging, immediate reclamation may not be possible and the solution narrows down to living with salinity. The best alternative is to plant such lands with salt tolerant plant species like *Atriplex*, *Leptochloa*, etc. The other plant well established in saline/ alkaline soils is *Prosopis juliflora* which can produce fuel and small timber.

The reclamation technology for alkali lands has been refined by CSSRI, Kamal. Its main components are grading and bunding the fields, application of amendments, judicious fertilizer use and growing rice and wheat in rotation. Installation of shallow irrigation pumps lower the water table as ground aquifers in most alkali areas are of good quality. Gypsum is mixed with 10 to 15 cm soil depth. Only small amounts of additional irrigation water can complete exchange reaction and leach down reaction products below the root zone of rice.

Management of Coastal Saline Soils

Technologies for management of these soils need both Government and community participation. The technologies-include provision of embankments with one way sluice gates to prevent the ingress of sea water into land, at the same time draining the excess rain water from inland into the sea, provision for the rain water storage from late monsoon rains for irrigation in the winter season and provision of surface and subsurface open drains with pumps, wherever necessary. The acid sulphate soils which occur along the west coast can also be reclaimed by the above measures along with liming.

Desertification and Arid Soils

The arid zone of India occupies about 12 per cent of the geographical area and is beset with desertification hazards due to very high human and livestock pressure. The CAZRI, Jodhpur developed suitable technologies for rangeland improvement, afforestation, establishment of shelterbelts and stabilization of sand dunes.

- Contour furrows (60 cm wide and 22 cm deep) on rangelands at a distance of 8 to 10 m across the slope for moisture conservation,
- Inter-row water harvesting system in which -30 cm wide ditch alternated with 70 cm wide raised bed,
- Improved grasses like *Lasiurus sindicus*, *Cenchrus ciliaris*, *Dicanthium annulatum* and *Panicum antidotale* for high forage yield on sandy soils with 250 mm annual rainfall. A mixture of grass seeds is more advantageous than single species for pasture development,
- Introduction of Stylosanthes hamata after light soil working in the established pasture of *Dicanthium annulatum* for increasing the productivity of grassland,
- Agri-pasture systems in which crops such as *Vigna radiata* are grown between perennial protective strips of forage *Lasiurus sindicus*, and
- Introduction of forage trees such as *Prosopis cineraria*, *Calligonum polygonoides*, Alcacia seinegal, *A. nilotica*, etc. in the pasturelands for additional fodder advantage without any detrimental effect on grass.

2. Management of Water Resources

Considering the fact that 80 to 90 per cent of the developed water resources are diverted towards irrigated agriculture, its efficient management is of crucial importance. However, in practice, it is hardly realized to any appreciable extent. The efficiencies of different irrigation projects vary from 14 to 40 per cent. Conveyance losses vary from 22 to 50 per cent depending upon the location and management level. Sustainable recommendations have been evolved for efficient use of irrigation water.

- On-farm water management technology such as land shaping, land levelling and efficient design and layout of irrigation methods,
- Conveyance of water through pipes/prefabricated channels to prevent seepage losses,. Scientific scheduling of irrigation both for adequate and deficit water supplies,
- Furrow/alternate furrow/skip furrow irrigation for all row crops,

- Introduction of sprinkler and drip Irrigation systems in water scarcity areas,
- Conjunctive use of surface and ground water and canal and tank water to solve drainage problems,
- Provision of drainage facilities in canal and in tank command area for increasing the productivity of crops,
- Adopting proper cropping systems in place of continuous lowland rice,
- Adoption of recommended production technology for crops and cropping systems,
- Integrated watershed development to use rain water, soil water, ground water and runoff water efficiently, and
- Recharging ground water during rainy season through basins, recharge wells, diversion of water to store on lands and by construction of check dams and percolation ponds.

It is a better strategy to increase irrigated area than to increase irrigation intensity. Inadequate control over irrigation water, poor timing and allocation of irrigation water are serious limitations in efficient management of water from irrigation projects. A close coordination and dialogue between irrigation engineers and agronomists will result in improvements.

3. Management of Rain Water

There is wide variability in rainfall occurrence over space and time. About 72 per cent of the cultivated area of the country depends entirely upon rainfall. Even after full irrigation potential is realized, about half of the cropped area would continue to be under rainfed agriculture. Hence, efficient rain water management is of prime importance for ensuring stability and sustainability in agricultural production for protecting the soil resource and minimizing drought and flood hazards. Options for sustainable dryland agriculture are:

- Agro-climatic analysis that gives an index of probability of rainfall,
- Adoption of efficient cropping systems based on agro-climatic analysis that give an index of rainfall probability, evapo-transpiration for a given location and period of water stress and sufficiency during the season,
- Contingent crop planning for aberrant weather conditions,
- Adoption of improved soil and water management practices,
- In-situ soil moisture conservation practices,
- Runoff harvesting and recycling, and
- Improved package of management practices for different cropping systems

On a sustainable basis, management of rain water holds the key to ride out the weather and make the dryland agriculture a reasonable success.

4. Integrated Pest Management

Plant protection determines the effectiveness of other inputs in crop production. Exclusive reliance on chemical pesticides resulted in problems, such as pesticide resistance, resurgence, residues and environmental pollution. This led to the development of integrated pest management (IPM) strategies. The IPM should be understood as an ideal combination of agronomical, biological, chemical, physical and other methods of plant protection against entire complex of pests, diseases and weeds in a specified farming ecosystem, with the object of bringing down the population of harmful organisms to economically insignificant levels with minimum interference on the activity of natural beneficial organisms. Bioagents are an integral part of IPM. The IPM has been developed for cotton bollworm, sugarcane internode borer, grapevine mealy bug, coffee berry borer, coconut rhinoceros beetle, wilt of oilseeds, pulses, banana, pepper, rhizome rot of ginger, turmeric and damping-off of vegetables.

Though several parasitoids, predators and pathogens of pests, antagonistic microorganisms were known to be effective for several decades, they were not commercially exploited because of quick knock down effect and easy availability of chemical pesticides instead of biopesticides and IPM. Consequently predators like Cryptolaemus, parasitoids like Trichogramma, insect pathogens like Bacillus thuringiensis and nuclear polyhedrosis virus (NPV) for pest management were commercialized during the past decade.

Agronomic Components in IPM

Agronomic adjustments, necessary for higher yield, are at the same time are directed at preventing mass multiplication and spread of pests, diseases and weeds by modifying the crop microclimate. As such, their creative application is one of the potential levers of phyto-sanitation aimed at creating optimal conditions for crop production. Agronomic components of IPM include land preparation (tillage), cultivar selection, time of sowing, plant population, intercultural, manures and fertilizers, water management, intercropping and crop rotation.

Land Preparation

Ploughing brings about unfavourable conditions for multiplication of pests, diseases and weeds inhabiting the soil. Quiescent stages (pupae) of harmful organisms will be exposed to dehydration or to predation by birds and other stages may be mechanically damaged or buried deep in the soil. Summer ploughing reduces the reserve of several pests.

Cultivar Selection

Cultivars with high yield potential and quality without resistance to pests and diseases are the main cause of frequent epidemics and mass multiplication of pests and diseases. A number of cultivars resistant/tolerant to pests and diseases have been developed. Selection of such cultivars can bring down the pest and disease incidence considerably.

Time of Sowing

As weather influence developmental rhythm of plants as well as growth and survival of pests and diseases, serious setback occurs when the weather conditions are such as to bring about coincidence of the susceptible crop stages with highest incidence of pests and diseases. Therefore, adjustments in sowing dates are often resorted to as an agronomic strategy to escape the crop loss. In general, early sowing in the season considerably reduces the pest and disease incidence.

Plant Population

Plant population per unit area influences crop microclimate. Dense population restricts wind movement within the plant canopy leading to higher humidity buildup. This creates congenial conditions for pest and disease multiplication. Keeping the total plant population constant, inter and intra row plant population can be adjusted to minimize the humidity buildup within the crop canopy.

Inter-Cultivation

Mechanical or manual inter-cultivation suppresses the pests, diseases and weeds similar to that of preparatory tillage. Weeds that could serve as alternate hosts to insects and pathogens can be effectively controlled with inter-cultivation. Microclimate conducive to pest and disease buildup will be disturbed by inter-cultivation.

Manures and Fertilizers

Excessive nitrogen application increases susceptibility of crop to sucking and leaf eating pests because of the succulence of the crop conferred by nitrogen. High rates of nitrogen than the recommended rate to hybrids without corresponding increase in phosphorus and potassium is the main factor for heavy pest and disease incidence. It is known that balanced fertilizers application (NPK) help the crop to tolerate pest and diseases considerably.

Irrigation and Drainage

Modification of natural water supply changes the biological equilibriums between crop and its pests and diseases. Irrigation can reduce the soil inhabiting pests by suffocation or exposing them to soil surface to be preyed upon by birds. Granular soil applied insecticides and herbicides do not become available until dissolved in water. Systemic chemicals are not absorbed by plant in the absence of adequate soil moisture. Waterlogged or saturated soils create microclimate of the crop conducive to buildup pests and diseases. Provision of adequate drainage during periods of heavy rains is a must for crops in heavy soils.

Intercropping

Many pests prefer feeding on a particular plant to others. This preference may be exploited to reduce pest load on crop. Groundnut as component crop in cotton reduces pest load on cotton. Based on the pests infesting a particular crop, companion or border crops can be grown to minimize the incidence of pests and diseases.

Crop Rotation

Advantages of crop rotation have been well documented in literature. It serves the purpose of plant protection by reducing the nutrition of pest in every subsequent year or season or totally deprive them of food. It reduces the amount of initial infection by disease and ensures unfavourable stratum for its development. Although, dramatic production gains have kept pace with population growth, there is hardly any room for complacency, as the demand for food will continue to rise in the absence of any let up in the population surge. Agriculture growth combined with income diversification through more profitable enterprises by shifting from a commodity approach towards a natural resource management centered approach will be essential to solve the problem of poverty and build a sustainable food security system. As natural resource management has multiple dimensions, it requires integrated initiatives. To cope with this challenge, future research and development agenda will have to be multidisciplinary and trans-institutional in function and holistic in content.

5. Input Management for Sustainable Agricultural Systems

The concept of two global commonalities - biological diversity and nutrient cycling among agro-ecosystems is supported by the literature on ecosystems and their management and the rapidly emerging literature on agro-ecology. Organic matter is the basis of all bio-geochemical cycles. The fundamental issues concerning efficient use of organic matter are leakage of nutrient from agro-ecosystems and the rates of decomposition. Organic matter and the nutrients if contains are lost from soils by run off and mineralization (Tiuy, 1990), both of which can be controlled by appropriate tillage practices (Campbell *et al.*, 1995; Lal *et al.*, 1994). Loss of nutrients to mineralization is also controlled by assuring sufficient inputs of plant or animal material to maintain the soil organic matter (SOM) reserves (Woodmansee, 1984). Legumes are important in maintaining SOM and increasing soil N suffer. In addition, they prefect the soil from run off water and wind erosion and improve infiltration, agro-forestry systems use leguminous and other trees to provide alternative crops (Steppler and Lundgren, 1988), produce animal forage and fuel, recycle nutrients for crop use, and protect soil from wind and water erosion (Altieri, 1987).

Plant biodiversity plays an important role in pest, disease and weed management. Crop rotations are effective in controlling pests, diseases and weeds (Altieri, 1987). Living mulches control weeds and minimize the need for herbicides (Regnion and Jahnke, 1990); Increases in structural diversity within the crop canopy leads to greater diversity in insects and less damage from insect pests (Stinner and Blair, 1990). Integration of animals into agro-ecosystems offers further diversity and stability. Mc Infire and Cryseels (1987) summarized and potential benefits of integration of crops and animals. Integration of animals facilitate nutrients movement and increase the opportunities for efficient nutrient management across the whole farm system. Animals increase overall net productivity of the farm and reduce environmental degradation by serving as alternatives to crops on the marginal areas of farms by utilizing crop residues as feed.

6. Optimizing Nutrient Availability

A very important condition for good plant growth and health and, indirectly, for good animal and human health is the timely provision of sufficient and balanced quantities of nutrients that can be taken up by the plant roots. Nutrient deficiencies and imbalances are main constraints to crop production, especially in regions with poor and very poor or alkaline soils. There is a constant flow a nutrient through the farm. Some of the nutrients are lost of exported, by export of products, erosion, leaching and volatilization. For example, it has been estimated that in Africa nutrient losses through soil erosions and other processes exceed application of artificial fertilizers (Stocking, 1986). If the farm is to remain productive, it must be ensured that the amount of nutrients leaving the farm does not exceed the amount returned to it. In other words, over time, there must be a positive nutrient balance.

7. Micronutrient Deficiencies

Due to intensive cropping the micronutrients are removed to a considerable extent, which control various aspects of plant growth. A study at Ranchi, India revealed that applying NPK 10:25:25 kg per ha led to depletion of Zinc by 0.619/ha and copper by 0.49/ha this can depress yields by up to 4 t/ha in rice, 2 t/ha in wheat and 3.4 t/ha in maize. Also iron is a limiting factor in rice production in. the new rice-wheat rotation evolved in the nontraditional rice growing areas of Punjab. One of the solutions to correct these micronutrient deficiencies is greater use of organic manures and multiple cropping with legumes. At Punjab Agricultural University, Ludhiana, field experimental results proved that application of poultry manure; pig manure and farmyard manure were effective in meeting zinc, requirements in a maize-wheat rotation. Also cultural practices such as prolonged submerging of the field can be used to tackle iron and manganese deficiencies (Sharma, 1985).

8. Limiting Nutrient Losses

Nutrient losses can be limited by:

- Recycling organic wastes by returning them to the field, either directly or treated (composted, fermented etc).
- Applying organic and artificial fertilizers in such a way that nutrients are not leached by excessive rain or volatilized by high temperature or solar radiation.
- Reducing losses due to run-off and soil erosion.
- Minimizing nutrient losses due to biomass burning.
- Reducing volatilization of nitrogen by denitrification under wet soil
- conditions.
- A voiding leaching by using organic and, artificial fertilizers which release, nutrients slowly, maintaining a high, humus content in the soil and intercropping plant species with different rooting depth.
- Limiting nutrient export in products by producing crops with relatively high economic value relative to nutrient content.

9. Use of Chemical Fertilizers

The use of chemical fertilizers is essential for obtaining high crop yields. However, many small landholders and resource-poor farmers cannot offer costly fertilizers. Most soils in the tropics are so deficient in primary nutrients that it is imperative that strategies to be developed for adding them from outside the ecosystem. There is some potential for enhancing N supply by biological N fixation. Additional N and other nutrient must be supplied. The requirements for chemical fertilizers, however, can be reduced considerably by decreasing losses, recycling nutrients and through biological N 'fixation'.

10. Nutrient Recycling

Nutrient recycling is an important strategy for sustainable crop production. It involves returning nutrients removed by crops to the soil for further use. In addition, soil fauna (e.g., earth worms, termites) also play an important role in recycling of plant nutrients. Growing deep-rooted crops is important in order to recycle nutrients from the sub soil by returning them through crop residue to the surface where the succeeding shallow rooted crops can use them. Use of mulches, incorporation of crop residues and animal wastes, growing legumes as intercrops in cereals etc., can substantially reduce chemical fertilizer requirements.

11. Use of Crop Residues

Crop residues contain substantial quantities of plant nutrients. The beneficial effects of returning crop residues as mulch on crop yields are well known (Akimbo and Lal, 1980, Kang, 1993). These benefits are not only to the recycling of plant nutrients but also to improvement in soil moisture and temperature, enhancement of soil structure and soil erosion control. The nutrient composition of the crop residues of some of the important crops is given in Table 21.2.

TABLE 21.2 : Nutrient Composition (per cent) of crop residues of major crops grown in the tropics.

Crop/Species	N	P	K
Cowpea stem	1.07	1.14	2.54
Cowpea leaves	1.99	0.19	2.20
Rice	0.58	0.10	1.38
Maize	0.59	0.31	1.31
Oil palm (Processed fiber)	1.24	0.10	0.36
Sesbania leaves	4.00	0.19	2.00
Cratolaria spp	2.89	0.29	0.72
Tephrosia spp	3.73	0.28	1.78
Azalia spp	3.68	0.20	0.11
Typha spp	1.37	0.21	2.38
Water hyacinth	2.04	0.37	3.40

Source: FAO, 1990.

12. Biological Nitrogen Fixation

Augmenting the nitrogen supply to crops through biological nitrogen fixation is a viable offer for resource - poor farmers pf the tropics. The amount of N fixed by legumes can range from 20 to 250 kg/ha/yr depending on the species, soil type, climate and agro-eco-region. Some common legumes that can be grown as cover crops and the quantity of N fixed by these crops are listed in Table 21.3.

TABLE 21.3: Reported quantities of N field by various legume crops

Crop/Species	N fixed (kg/ha/yr.)	Reference
Alfalfa	78 - 222	Hrichel, 1987
Peanut	87 - 222	Retner et al., 1979
Cowpea	65 - 130	Alexander, 1977
Field peas	174 - 195	Heichel, 1987
Soybean	170 - 217	Thurlow and Hiltboid, 1985
Birds foot	49 - 112	Heichel, 1987
Chickpea	24 - 84	Rennie and Kemp, 1984
Common bean	70 - 124	Heichel, 1987
Faba bean	77 - 250	Heichel, 1987
Vetch	111	Heichel, 1987
Ladino clover	164 - 187	Heichel, 1987
Lentil	167 - 188	Larson et al., 1989
White lupin	193 - 247	Rinaudo et al., 1983
Sesbania spp.	267	

13. Use of Biofertilizers

Biofertilizers have been recognized as important inputs in integrated plant nutrition systems. The use of legume green manure (Meelu and Morris, 1988; Abrol and Palaniappan, 1987; Palaniappan and Siddeswaran, 1990), blue green algae and Azolla (Singh, 1989; Singh and Singh, 1992; Adil and Katre, 1992; Venkataraman and Shanmugasundaram, 1992) for rice: Azotobacter and Azospirillum for wheat; millets and vegetable crops (Kavimandan et at., 1987 Agarwal and Tilak, 1988; Wani *et at.*, 1987, Tilak and Dwivedi, 1989, Wani, 1990, Negi *et at.*, 1991; Shende, 1987; Bhandari and Somani, 1990; Subbian and Palaniappan, 1992); Rhizobium for pulses and oil legume crops (Somani and Bhandari, 1990; Verma and Bhattacharyya, 1992; Baldev et at., 1988; Patra, 1990), Phosphate solubilizers (Gaur 1990; Guar and Gaind, 1992) and Vesicular Arbuscular Mycorrhizae) (Baguaraj *et al.*, 1979; Tilak, 1988) for various crops is well reported on an average these biofertilizers can minimize the use of inorganic N by 25 to 50 kg/ha.

14. Green Manuring

The green manure crops when applied improve the physical and chemical properties of the soil (Singh and Paroda, 1994; Mahapatra *et al.*, 1995). Green manures also increase the fertilizer use efficiency of crops when applied in combination with inorganic fertilizers (Karuppiah and Thangamuthu, 1986; Abrol and Palaniappan, 1987; Subbian and Palaniappan, 1992; Thakur *et al.*, 1995). Among the green manure crops, special attention is being given to *Sesbania rostrata*, which bears stem nodules in addition to the root nodules. The amount of N contributed in terms of fertilizer N equivalence ranges from 80 to 120 kg/ha. In a field trial comprising different green manure crops, it was found out that *Sesbania rostrata* produced the highest biomass (20-25 tons/ha) and accumulated a maximum of 150-220 kg N/ha. (Palaniappan and Siddeswaran, 1990). The common leguminous green manure crops used in tropics and their N content are given in Table 21.4.

TABLE 21.4 : Common green manure crops and their N content

Crop Species	Scientific Name	Biomass	N per cent (Moist)
Sunnhemp	*Crotalaria juncea*	21.2	0.43
Dhaincha	*Sesbania aculeate*	20.0	0.43
Dhaincha	*Sesbania rostrata*	19.6	-
Pillipersara	*Phaseolus trilobus*	18.3	1.10
Mungbean	*Phaseolus arvensis*	8.0	0.53
Cowpea	*Vigna sinesls*	15.0	0.49
Guar	*Cyamposis tetragonoloba*	20.0	0.34
Senji	*Melilotus alba*	28.6	0.57
Khesari	*Lathyrus sativus*	12.3	0.54

Source: Abrol and Palaniappan, 1987.

15. Development of Integrated Farming Systems

The challenge is to design practical integrated farming systems that can be adapted to regions, and the sustainable design and implementation of integrated farming systems require collaboration among agricultural, social, economic, and ecological disciplines. They also depend upon the full participation of farmers. It may be probably impossible to develop efficient integrated farming systems without interdisciplinary planning and implementation. A major challenge to agricultural research and development is the need for modifying institutional structures and research and development methodologies to allow such collaboration.

SUSTAINABLE DRY LAND AGRICULTURE

A wide array of critical environmental threats is posing a challenge to mankind all the world over. Some of these threats include degradation of soil, water, forest and marine resources essential to increased production of food, fodder, fibre and timber etc. The problem of degradation is compounded in dry areas because of fragile ecosystem. Common to all areas is the health-threatening pollution which is wide spread. Stratospheric ozone depletion, global climate change, and loss of biodiversity, are some of the concerns receiving the attention of all developed and developing countries alike.

Deterioration of natural resources is the main issue threatening sustainable development of dryland agriculture, more so in the Third World Countries. Three main reasons are attributed to this state of affairs (FAO 1989).

1. Lack of access to land, inputs and other productive resources by the majority of rural households and communities,
2. Lack of awareness among policy makers of the economic costs involved, and
3. Lack of environmentally sound alternative technologies acceptable to farmers.

Subsistence dryland farmers are forced to over exploit natural resources by way of mining soil nutrients, cultivating steep slopes, over grazing rangeland, excessive collection of fuel wood, in order to survive and feed themselves. The government policies should, therefore, strive to create favourable circumstances, whereby the farmers conserve natural resources, and are not forced to destroy them. It must be realized that productivity of degraded lands are much greater, and returns much lower, in harsh environment of semi-arid tropics than in the humid temperate climates (Parr *et al.*, 1990).

Desertification of Dry lands

The United Nations Conference on Desertification held in Nairobi, Kenya, in 1977 defined desertification as the "diminution or destruction of the biological potential of the land, and can lead ultimately to desert like conditions". Drought was considered as just one cause of desertification. More importantly, human misuse of land or mismanagement practices like overgrazing, tree cutting, improper tillage methods, poor water distribution systems, and over exploitation, are now regarded as the direct cause of desertification.

The principal desertification processes are: (1) degradation of vegetative cover, (2) water erosion, (3) wind erosion, (4) salinization. (5) soil crusting and compaction, (6) reduction in soil organic matter, and (7) excess of toxic substances in the soil (Dregne, 1990). The list suggests that the word land degradation in an arid land context is nothing but desertification. The dominant problem varies from over grazing in rangelands, to water and wind erosion in rainfed crop lands, to salinization to irrigated land. In drylands, however, the most serious long term land degradation problem is water and wind erosion.

"Degradation on drylands (arid, semi-arid, and sub-humid) is called **desertification** because desert-like conditions appear where none existed before. Desertified drylands are characterized by low primary productivity (less than 400 kilograms of dry matter per hectare per year) and poor rain use efficiency (usually less than 1 kilogram of dry matter produced per hectare per year for each millimeter of rainfall." (World Resources -1988-89).

Estimates of soil erosion world wide are rough. Crop lands lose far more top soil each year than is replaced naturally. About six to seven million hectares of crop land are rendered unproductive because of soil erosion each year. Sodication (water logging, salinization and alkalinization) is responsible for reducing the productivity of upto 1.5 million hectares of crop land each year.

Recovering Degraded Lands

The International Union for Conservation of Nature and Natural Resources (IVEN) adopted guidelines for rehabilitating or restoring degraded lands. The recommendations were developed at a workshop on restoration ecology held in Varanasi, India, in December, 1987. The source of important guidelines are:

- "Critical sites – such as those especially liable to erosion, source of saline water in the upper reaches of a watershed, or habitats of threatened and endangered species – should be rehabilitated or restored first.
- For the initial re-vegetation, plants should come from different successional stages to lessen the chances of natural succession being thwarted and to let slow-dispersing species be included.
- Plants that facilitate succession or maintain nutrient and water cycles should be included. Nitrogen fixes are an important example, as are plants whose roots can penetrate beyond infertile upper soil layers.
- Local people should be able to contribute traditional knowledge about local plants and soils.
- Rehabilitation programmes, should provide economic benefits to the participants.
- Laws and customs affecting the area under recovery should be reviewed periodically.
- Tree population should be established on graded lands rather than on undisturbed or otherwise productive tracts.
- As coroilary to recovery projects, protected areas should be set aside to conserve plant and animal wild life species that may be needed for future rehabilitations or restorations."

Role of Rainfed Agriculture

Althiugh irrigation is known to boost yields dramatically; India could enhance food production with more efficient farming. Rainfed Agriculture, which accounts for roughly two thirds (68 per cent) of the total cultivated area in India is crucial to its food production. Managing the small, fragmented holdings where this type of farming is practiced calls for a cooperative approach if economics of scale are to be utilized (Bottral, 1991; Singh, 1988).

Areas depending on rainfed agriculture are diverse both with regard to soils and rainfall availability. 'Improved farming practices in rainfed areas could enhance production otherwise by poor soils, in particular the black and red types found in Cenytral and South India'.

Environmentally Sustainable Dry Land Farming Systems

Principles or relevant components of environmentally sustainable farming systems should include the following (Lal *et al.*, 1988).

1. Reducing soil erosion and improving water conservation.
2. Inclusion of legumes and cover crops in crop rotations.
3. Agroforestry as an alternate land use system.
4. Judicious use of organic wastes.

Environmentally sustainable dryland farming systems emphasize on conservation and efficient utilization of natural resources. "Agronomic practices of conservation, tillage and mulch farming, rotational cropping, use of legumes and cover-crops to grow nitrogen and suppress weeds, and efficiently utilizing cattle manure and sewage sludge" (Lal *et al.*,1988), are some of the ingredients of sustainable farming systems source of the experimental truths are: (1) No till with residue mulch is an effective erosion control measures even on steep slops (Lal, 1984), (2) Adverse effects of crust can be minimized by mulching, (3) One of the several useful measures for water conservation is the practice of tied-ridging, (4) Ridge cropping is yet another practice widely adopted in rainfed farming, (5) In addition to supplying nitrogen, legume cover-crops suppress weeds-*Stylosanthes* sp. Suppressing perennial weeds spear grass-*Imperata cylindrica*, in the tropics and, (6) Cover and grain crops can also be grown together as live mulches – a desirable practice so long as the host crop does not compete with the grain crop.

As compared with monoculture systems, perennials used in the agroforestry systems impart ecological stability, besides adding biologically fixed nitrogen. "Perennial hedge rows established on contour may reduce run-off and erosion and provide additional mulch with readily mineralizable N".

Scientifically proven components that go into developing sustainable farming systems, and sustainable farming systems for different ecological regions, are given in Tables 21.5 and 21.6 respectively.

TABLE 21.5: Scientifically proven components for developing sustainable farming systems

Objective	Recommended components/such systems
A. Controlling Erosion	Conservation tillage, strip cropping, cover crops, mixed cropping and agro-forestry, mulch farming, terracing, tied ridging, shelter belts, wind breakes.
B. Conserving Water	Mulch farming, eco-fallow, weed control, tied ridges, irrigation.
C. Alleviating Compaction	Sowing deep legumes, chisel ploughing, reducing vehicular traffic, guided traffic.
D. Enhancing Soil fertility	Legume based rotations, cover crop, organic amendments, chemical fertilizer
E. Improving Soil fauna maintaining soil tilth	Mulch farming, crop diversification, reduced tillage, organic amendment, planted fallows.
F. Post Management	Rotations, crop diversification residue management, pesticides, use based on threshold needs.

Source: Lal and Miller, 1990.

TABLE 21.6: Sustainable farming systems for different ecological regions

Soils		Sustainable Systems
Structurally Active soils	(a)	1. Grain crop-cover crop rotation using improved cultivar 2. No till or minimum tillage methods. 3. Residue mulch, 4. Chemical amendments, 5. Planted fallow.
	(b)	'a' with agro-forestry system using suitable woody perennial or shrubs 'b' in association with growing alternate strips of woody shrubs/cover crops with grain crops.

Soils		Sustainable Systems
Structurally	(a)	1. Mixed cropping with compatible mixtures 2. Tied ridges, following ploughing at the end of rains. 3. Organic compost or Farm Yard Manure. 4. Chemical amendments, 5. Planted fallow 6. Irrigation.
	(b)	'a' above with contour hedge rows of woody perennials at 1 to 2 m vertical interval.
	(c)	'b' above and integration with livestock, based on light grazing.

Source: Lal and Miller, 1990.

Policy Issues

Sustainability related issues are many. The technical aspects of sustainability, sustainable agriculture and sustainable development of dryland agriculture, have been discussed in the preceding paragraphs. The ill of all ills is the problem of land degradation. In the context of India, demographic pressure is yet another issue causing concern.

In order to tackle the problem of deteriorating natural resources, in particular land and use policy is called for. Land use according to land capability, and putting human and livestock population pressure on land in accordance with its carrying capacity, should form the central theme of land use policy. The mismanagement of land should be halted through people support and strong legislative measures. Agriculture on steep slopes and marginal lands should be given a go-bye. Alternate land use systems, is place of arable cropping, should be given incentives necessary to popularize them among small and marginal farmers. The technologies for rehabilitating marginal and wasted lands are available. Concerted efforts on the part of the Government are needed to push through programmes of rehabilitation and restoration of such lands. A strong political will and commitment is needed to combat forces leading to land degradation.

Farming systems that could ensure sustainability need to be developed for each of the agro-ecological regions, so as to conserve soil, water, vegetation, wild life, etc. Farming systems that include multi-purpose nitrogen fixing tree species, animals (livestock), leguminous crops which are efficient utilizes of land and water should be given a premium in the scheme of things. Minimum tillage and use of crop residues as mulch should be the guiding principles for farming systems.

In arid regions, introduction of tractors has done irreversible damage to the ecosystems. Whereas carrying out timely operations it extremely important in dry regions, is should not be at the cost of soil structure and stability. As a policy, tractorization should be prohibited in fragile ecosystem, particularly in arid regions.

Human resources development is vital. In fact, environment development programme and human resource development programmes should go together. An education policy which emphasizes on conservation and management of natural resources, and family planning, will go a long way in achieving sustainability.

"Increasing yields and input use efficiency will require new farming concepts that focus on fine-tuning of seed, nutrients, water, pesticides, energy and labour inputs for smaller management units".

P. C. Robert, 1995

22 Soil Amendments

> ***"Organic materials were practically the only external source of nutrients to crops before introduction of inorganic fertilizers. The recycling of various forms of residues has the advantage of converting the surplus farm waste into useful product for meeting nutrient requirement of crops, besides maintaining the soil condition and improving the overall ecological balance".***
>
> ***Dr. M. S. Swaminathan***

Soil amendments are those materials or chemical substances which can amend or correct the soil reaction to make it suitable for crop cultivation at profitable levels. By correcting the soil reaction, these chemicals increase the availability of most of the plant food elements which had been fixed in the problem soils besides supplying Ca, Mg and S etc. to plants after their application. They also act as flocculants and correct toxicity due to excess availability of elements in untreated soils.

LIMING IN ACID SOILS

The dominance of hydrogen ions over OH is called acidic condition of soil because pH of soil is based on hydrogen ion concentration. Therefore, according to pH soil may be classified into following acidic groups.

Soil group	pH range
Extreme acidic	Below 4.5
Very strongly acidic	4.5 to 5.0
Strongly acidic	5.1 to 5.5
Medium acidic	5.6 to 6.0
Slightly acidic	6.0 to 6.5
Neutral	6.6 to 7.3

The hydrogen ions are closely attached with colloidal particles of soil. Therefore, liming is required to replace hydrogen ions from soil colloids. Liming materials, when added to acidic soils, releases Ca which is ionized. This Ca replaces Mg, Na, K, NH_4 or H from the colloidal complex of the soil. This process is scientifically termed as ionic exchange.

When liming material is added to soils after ionization calcium becomes free. This calcium, being very active, immediately replaces H ions from the exchange surface of soil colloidal complex. The replaced hydrogen ions are also very active and these ions react with OH ions resulting into water. However, this correction of acidity is not permanent because after sometimes the calcium of soil micelle reacts with carbonic acid, which is produced by decomposition of organic matter, and forms $CaCO_3$ and $Ca(HCO_3)$:

$$CaCO_3 + H_2CO_3 \longrightarrow Ca(HCO_3)_2$$

Ca $(HCO_3)_2$ is not water soluble, therefore, Ca becomes inactive and hydrogen again dominates on soil micelle which converts the soil into acidic again.

LIME REQUIREMENT OF THE SOIL

Lime requirement depends upon the pH and exchange capacity of the soil. The exchange capacity is governed by organic matter content and colloidal clay content of soil. These values can be calculated by the help of following equations:

$$\text{Organic matter} = \%\ \text{nitrogen} \times 20$$
$$\text{Exchange capacity} = \%\ \text{Clay} \times 0.57 \times \%\ \text{nitrogen} \times 38$$
$$(\text{m.e./100 g of soil})$$

According to soil type and pH condition of the soil following quantities of lime has been recommended.

TABLE 22.1: Lime Requirement at Different pH Levels

Soil pH	Lime requirement in kg/ha for different soil types		
	Sandy soils	Loam	Clay loam
5.00	1125	1687	2625
5.20	975	1462	2275
5.40	825	1237	1925
5.60	675	1012	1575
5.80	525	787	1225
6.00	375	562	875
6.20	225	337	525

Source: Dr. J. S. Kanwar, Lime requirement of different soils under varied pH, Ind. Farm. 9(2): 27-28, 1959.

Lime requirement is also governed by following factors:

1. Extent of acidity.
2. Purity of liming material.
3. Fineness of material.
4. Texture of soil – lighter the soil lesser will be the lime requirement.
5. Types of crops to be grown – Legumes require more lime than non-legumes, potato needs less lime because of attack of scab disease under alkaline reaction.
6. Cation exchange capacity of the soil.
7. Organic matter content of the soil.
8. Exchangeable Hydrogen.
9. Clay content of the soil.

ADVANTAGE OF LIME

1. Liming materials supply Ca which helps in flocculation of soil. Thus it improves soil structure.
2. The Ca, supplied by liming material, helps profuse bacterial growth as they die under strong acidic conditions of soil. As a result it improves the decomposition of organic matter which improves the soil aeration and other physical properties of soil.

3. Calcium, being very active, replaces other cations from the soil micelle; therefore, it helps in increasing availability of K by releasing K from absorbed surface of the soil.
4. Liming materials supply Ca and Mg to soil which the secondary essential elements for growth and development of plants.
5. Under strong acidic reactions the phosphate gets fixed in the form of Al and Fe phosphate. Liming increases soil pH due to which Al and Fe are released and monocalcium phosphate is formed (most available form to crop plants).
6. In every case liming releases CO_2 from carbonic acid which is used as food by microbes. Thus, CO_2 finally increases the extent of nitrogen fixation in the soil.

LIMING MATERIALS

There are several liming materials which are used for liming purposes. These liming materials include carbonate or ground limestone, burnt limestone hydrated limestone, blast furnace slag and ashes. Of these, only types of liming materials are commonly used in India which is ground limestone, burnt limestone and hydrated limestone.

Carbonates or Ground Limestone

This group consists of limestone and $CaCO_3$. Sometimes limestone consists of $MgCO_3$ in the extent of 45 per cent than that material is technically called as dolomite. According to relative proportion of $MgCO_3$ and $CaCO_3$ the materials are termed as dolomite or high magnesium limestone containing higher quantity of $MgCO_3$ than $CaCO_3$ whereas high calcium limestone contains higher $CaCO_3$ than $MgCO_3$ contents.

Grading

Grading of limestone is done based on purity percentage and fineness. The grades are given as under:

1. **According to Purity Percentage:** Based on purity percentage the liming materials may be classified as under:
   a) **Poor grade:** When limestone contains 80% calcium and $MgCO_3$, the limestone is said to be of poor grade.
   b) **Good quality grade**: When limestone has Ca and Mg carbonates up to 90 per cent, the grade is known as of good grade.
   c) **Best quality grade**: When limestone consists of 95 percent or even more of Ca and Mg carbonates, the grade is called as of best quality.
2. **Fineness:** The efficiency of limestone always depends upon fineness of materials because fineness offers more exposed area for chemical reactions and becomes highly efficient. According to fineness the best quality limestone should pass through 100 mesh sieve and 90 per cent should pass through a 20 mesh sieve while all should pass through a 10 mesh sieve.

Line of Action of Liming Material

Limestone, when applied under acidic reaction, undergoes following reactions to neutralize acidity:

1. Firstly the limestone neutralizes the acidity of soil solution which is mainly due to dominance of carbonic acid obtained by decomposition of soil organic matter. Calcium carbonate converts into calcium bicarbonate after reacting with carbonic acid as given under:

$$CaCO_3 + H_2CO_3 \rightarrow Ca\ (HCO_3)_2$$

$$\text{Soil}^{H}_{H} + Ca\ (HCO_3)_2 \rightarrow \text{Soil}^{Ca} + 2H_2O + 2CO_2$$

2. Secondly the calcium carbonate itself reacts directly with soil micelle and the calcium replaces two hydrogen ions from micelle as shown below:

$$\text{Soil}^{H}_{H} + CaCO_3 \rightarrow \text{Soil}^{Ca} + H_2O + CO_2$$

Thus carbonate, when added to the soil, raises its pH and corrects soil acidity.

Burnt Lime

The burnt lime is produced by heating limestone at a temperature over 800°C during which carbon dioxide is left behind. The product is called burnt lime, lump lime or oxide of lime. Following equation takes place:

$$\underset{(100\text{-}44)}{CaCO_3 - CO_2} = \underset{(56)}{CaO}$$

Thus burning (heating) of 100 kg of calcium carbonate 56 kg of CaO or burnt lime is obtained. Similarly, 84.3 kg of magnesium carbonate produces only 40.3 kg of MgO.

The quality of lime depends upon purity percentage and fineness as given above. The CaO, to be used for liming or agricultural purposes, must have 85 to 98 per cent purity and fineness should be like that of carbonate.

Line of Action: Burnt lime, after its application in soil, reacts with water and forms calcium hydroxide. Now this calcium hydroxide reacts with organic acids, particularly with carbonic acids, particularly with carbonic acid, and forms calcium bicarbonate. This calcium bicarbonate reacts with soil micelle and each calcium ion replaces two hydrogen ions from soil micelle resulting into rise in soil pH as given below:

$$CaO + H_2O \rightarrow Ca\ (OH)_2$$
$$Ca\ (OH)_2 + 2\ H_2CO_2 \rightarrow Ca\ (HCO_3)_2 + H_2O$$

$$\text{Soil}^{H}_{H} + Ca\ (HCO_3)_2 \rightarrow \text{Soil}^{Ca} + 2\ H_2O + CO_2$$

Hydrated Lime

Hydrated lime is also called as agricultural hydrate, water slaked lime, quick lime slaked or calcium hydroxide, Ca $(OH)_2$. This can be manufactured by adding water to calcium oxide as explained below:

$$CaO + H_2O \rightarrow Ca\ (OH)_2$$

Sometimes, calcium oxide forms calcium carbonate when it reacts with CO_2 in soil as shown below:

$$CaO + H_2O \rightarrow Ca\ (OH)_2$$
$$Ca\ CO_2 + CO_2 \rightarrow Ca\ CO_3 + H_2O$$

The $CacO_3$, thus formed, is washed off away with runoff water.

Line of Action: Hydrated lime also undergoes the same chemical reaction in the soil as that of ground limestone and burnt lime.

$$Ca(OH)_2 + H_2CO_3 \rightarrow Ca(HCO_3)_2 + 2H_2O$$

$$Soil\genfrac{}{}{0pt}{}{H}{H} + Ca(HCO_3)_2 \rightarrow Soil\,Ca + 2H_2O + 2CO_2$$

Blast Furnace Slag

Slag is a by-product of steel factories which can be used as liming material under acidic soil reaction. Calcium present in the slag helps in neutralizing acidity by replacing hydrogen ions from the soil micelle. Release of Ca neutralizes acidity, raises soil pH and the phosphate present in the slag as citrate soluble phosphate becomes water soluble and hence available it becomes available to plants due to conversion from tricalcium phosphate to monocalcium phosphate. The efficiency of liming material depends upon fineness, organic matter content of the soil etc.

Relative Efficiency of Liming Materials

The efficiency of liming materials to correct acidity depends entirely upon calcium content of the material. Thus higher the calcium content greater will be the efficiency of the material. Considering this they may be compared as under:

1 kg of CaO is equivalent to 1.784 kg of $CaCO_3$ and
1 kg of Ca $(OH)_2$ is equivalent to 1.351 kg of $CaCO_3$

By the help of the following conversion factors we find out the required quantity of any liming material available in the market for use:

TABLE 22.2: Conversion Factors for Liming Materials

From	To	Multiply by
CaO	$CaCO_3$	1.784
$CaCO_3$	CaO	0.560
Ca $(OH)_2$	CaO	0.757
MgO	$MgCO_3$	2.091
$MgCO_3$	MgO	0.478
Mg $(OH)_2$	MgO	0.691
$Mg(OH)_2$	CaO	0.960
$MgCO_3$	CaO	0.664
MgO	CaO	1.390
$MgCO_3$	$CaCO_3$	1.186
MgO	$CaCO_3$	2.480
CaO	Ca	0.751
Ca	CaO	0.399
MgO	Mg	0.608
Mg	MgO	1.658
Ca	$CaCO_3$	2.496
Mg	$CaCO_3$	4.112

Calcium Oxide Content of Liming Materials

The relative efficiency can be measured by determining the CaO content of various liming materials which is given below:

TABLE 22.3: CaO Content of Limes

Types of liming materials	CaO %
Limestone and chalk	54
Burnt lime/Quick lime	95
Ground limestone	54
Hydrated lime	78

Besides purity, fineness is also very important factor which controls the efficiency of liming materials. According to U.S. (American) studies made at Ohio the efficiency rating is given in Table 22.4.

TABLE 22.4: Efficiency Rating Based on Fineness of Liming Materials

Fineness	Efficiency Rating (%)
Materials passing through 60 to 100 mesh sieve	100
Materials passing through 20 to 60 mesh sieve but not a 60 m.s.	60
Materials passing through 8 to 20 mesh sieve	20

The efficiency of liming materials also depends upon neutralizing values of various materials as explained below:

Materials	Neutralizing power
CaO	179
$Ca(OH)_2$	136
$CaCO_3$	100

Over liming

Sometimes farmers apply liming materials without knowledge of required quantity which may be either much lower than the required quantity and fails to correct the acidity or it may be much higher which is called as over liming and may have following harmful effects:

1. Sudden change in soil reaction or rise in soil pH above neutral range due to over liming causes death of beneficial micro-organisms which deteriorates soil fertility.
2. Over liming is harmful for many plants because several species of plants fail to grow under alkaline soil reaction.
3. Rise in soil reaction or soil pH due to over liming reduces nutrients like Al, Fe and Mn is drastically reduced. Such elements become deficient and need correction.

Light Liming

Using lower than the required quantity of liming material is termed as light liming which increases cost of liming but the soil use to have all ill effects of acidic soil reaction.

TIME OF APPLICATION OF LIMING MATERIALS

The liming materials should be used in non-cropped or fallow lands because its application as top dressing produces heat during chemical reaction in soil which is very injurious or harmful for growing plants and also for useful micro-organisms. Therefore, they must be applied at least a fortnight before sowing or transplanting of crops except burnt lime and hydrated lime which are fast reacting chemicals and should be applied only one week before sowing or transplanting of crops.

METHOD OF LIME APPLICATION

The material may be broadcasted or drilled according to convenience and it should be well mixed into the surface soil for initiating a rapid chemical action. If very high quantity of liming materials (6 to 8 tonnes/ha) has to be applied then for better use the material must be applied in 2 to 3 splits or supplementary doses because sometimes apart of the material is washed off due to heavy rains.

TABLE 22.5: Quantity of Lime (Total) Removed by Principal Crops

Crops	Removal of Ca and Mg in kg/ha	
	Ca	Mg
Wheat	18	12
Potato	130	32
Sugarbeet	55	56
Tobacco	165	22
Rapeseed	110	19
Carrot	63	30
Radish	40	30

Source: Soil fertility – theory and practices by J. S. Kanwar, 1976, an ICAR publication page 209.

SOIL AMENDMENTS USED UNDER ALKALINE SOIL REACTIONS

Salinization process results in very heavy accumulation of basic salts at the soil surface and soon after salinization or even simultaneously alkalization process begins in which Na dominates on the soil micelle by replacing other cations. This sodium is strongest deflocculent of soil particles. Gradual increase in Na^+ and other anion like OH^- causes rise in soil pH which results in alkalinity and requires reclamation for bringing back to normal or neutral reaction. The alkalinity may be classified into following groups based on soil pH; however, pH is not the only criteria:

Soil reactions	Soil pH
Mid alkaline	7.4 – 7.8
Moderately alkaline	7.9 – 8.4
Strongly alkaline	8.5 – 9.0
Very strongly alkaline	9.1 – and above

Exchangeable sodium percentage, E.C. and pH give three major groups of soil types as mentioned below:

Saline Soils: These soils contain exchangeable sodium percentage less than 15 E.C., more than 4 m.mhos/cm at 25°C and pH less than 8.5.

Saline-alkali Soils: These soils contain exchangeable sodium percentage more than 15 E.C., more than 4 m. mhos/cm at 25°C and pH more than 8.5.

Alkaline Soils or Non-saline Alkali Soils: These soils contain E.S.P. more than 15 E.C., 4 m mhos/cm at 25°C and pH within 8 to 10.

ALKALINITY CAUSES FOLLOWING HARMFUL EFFECTS

1. Presence of sodium, a highly deflocculating agent, causes deflocculating of soil particles which destroys the soil structure.
2. Higher concentration of soluble salts cause physiological drought to crop plants.

3. Assimilation of P, K and Ca in crop plants is hindered.
4. Accumulation of some salts becomes toxic to crop plants.
5. Higher contents of Ca and Mg under high pH reverts the soluble B, Zn and Cu into carbonates, hydroxides and oxides of these elements which are neither soluble nor available to crop plants.

AMMENDMENTS

There are certain chemicals which can be either strong acidic in nature for form acidic chemicals neutralize they contain more Ca. Thus the acidic chemicals neutralize alkaline salts directly whereas Ca replaces exchangeable sodium from the soil micelle. This way the ill effects of alkalinity may be countered. These chemicals are given below:

Sulphur

Elemental sulphur having low price may be economically applied for reclaiming alkali soils. This contains lime ($CaCO_3$) which provides Ca for replacing sodium from the soil micelle. Sulphur is most effective reclaiming agent as one atom of sulphur replaces 4 exchangeable sodium ions. One tonne of sulphur is equivalent to 5.38 tonnes of gypsum in reclaiming alkali soils. Sulphur is slow acting amendments and takes 2-3 weeks time for action towards reclamation under favourable moisture and temperature conditions because oxidation process is very slow. The sulphur must be well mixed into the soil after its application and so far it is possible only fine materials must be applied. It has found that application of sulphur at least one fortnight before sowing or planting of the crop gives better results.

Line of Action

Sulphur when applied, goes under microbial oxidation which completes in slightly more than a week's time. After oxidation it forms sulphur trioxide which reacts with soil water and forms sulphuric acid (H_2SO_4). The sulphuric acid reacts with $CaCO_3$ and forms $CaSO_4$. The Ca present in $CaSO_4$ replaces sodium from the soil micelle and forms sodium sulphate which either leaches down or is washed away from the field. The reactions are given below:

$$2\,S + 3\,O_2 \Leftrightarrow 2\,SO_3$$

$$SO_3 + H_2O \Leftrightarrow H_2SO_4$$

$$H_2SO_4 + CaCO_3 \Leftrightarrow CaSO_4 + CO_2 + H_2O$$

Or

$$H_2SO_4 + 2\,CaCO_3 \Leftrightarrow CaSO_4 + Ca(HCO_3)_2$$

(According to Kelley, 1951)

$$\text{Micelle}\begin{matrix}Na\\Na\end{matrix} + CaSO_4 \Longleftrightarrow \text{Micelle}\,Ca + Na_2SO \longrightarrow \text{Washing}$$

↓Leaching

Also sometimes

$$\text{Micelle}\begin{matrix}Na\\Na\end{matrix} + Ca(HCO_4)_2 \Longleftrightarrow \text{Micelle}\,Ca + H_2O + Na_2CO_3$$

(According to Kelley, 1951)

Amount of Sulphur Required

The amount of sulphur required for reclaiming alkali soils may be calculated in two ways- First is pH and second exchangeable sodium percentage in meq/100 g of which may be given as under.

Table 22.6: Amount of Sulphur in kg/ha (when 95% pure) in bring down Soils pH to 6.5

pH of soil	Soil type	
	Sandy soils	Clayey Soils
7.5	400 – 600	800 – 1000
8.0	1000 – 1500	1500 -2000
8.5	1500 – 2000	1500 – 2000
9.0	2000 -3000	2000 – 3000

Whereas the amount of sulphur required to replace indicated amounts of exchangeable sodium may be given below:

Assuming one acre foot to weigh approximately 4,000,000 lbs and 6 acre inches about 2,000,000 lbs.

TABLE 22.7: Amount of Sulphur to replace Indicated Exchangeable Sodium from Micelle

Exch. Sodium in meq/100 g of soil	Amount of sulphur required in tones per	
	Acre foot	6 acre inches
1	0.32	0.16
2	0.64	0.32
3	0.96	0.48
4	1.28	0.64
5	1.60	0.80
6	1.92	0.96
7	2.24	1.12
8	2.56	1.28
9	2.88	1.44
10	3.20	1.60

Iron Sulphate

Iron sulphate is normally used to reclaim alkali and saline-alkali soils having a pH upto 9.0 The sulphate present init reacts with water and forms sulphuric acid directly which helps in reclamation undergoing following equation:

$$FeSO_4 + H_2O \Leftrightarrow H_2SO_4 + FeO$$

$$H_2SO_4 + CaCO_3 \Leftrightarrow CaSO_4 + CO_2 + H_2O \textit{ Or}$$

$$H_2SO_4 + 2\,CaCO_3 \Leftrightarrow CaSO_4 + Ca\,(HCO_3)_2$$

$$\text{Micelle}\begin{matrix}Na\\Na\end{matrix} + CaSO_4 \Longleftrightarrow \text{Micelle}\,Ca + Na_2SO \rightarrow \downarrow$$

And also

$$\text{Micelle}\begin{matrix}Na\\Na\end{matrix} + Ca(HCO_4)_2 \Longleftrightarrow \text{Micelle}\,Ca + H_2O + Na_2CO_3$$

The amount to be applied always depends upon sulphur content of the material. The equivalent quantity of the amendments is given below:

One tonne of sulphur is equivalent to = 3.06 tonnes of H_2SO_4
5.30 tonnes of Gypsum
8.69 tonnes of iron sulphate
3.13 tonnes of limestone ($CaCO_3$)

Gypsum

It is white coloured common calcium salt found as natural deposit in various parts of Rajasthan, Tamil Nadu and Himachal Pradesh which is applied to reclaim alkali soils. Its calcium replaces Na from soil micelle and improves physical properties of saline-alkali soil and alkali soils. In solubility is found up to 0.25 % which means that under field conditions. 3 to 4 acres foot water is needed to dissolve 4 to 5 tonnes of gypsum. It is the cheapest source of sulphur for reclaiming alkali soils as elemental sulphur is very costly and is mostly used for manufacture of H_2SO_4. Presently the high grade gypsum is being used for manufacture of ammonium sulphate and only lower grade gypsum having 60 to 70% $CaSO_4$ is made available for agricultural use as soil amendment.

Scientists working with gypsum in saline and alkali soils reported that use of gypsum at the rate of 4 to 6 tonnes with 10 tonnes of well rotten FYM gave best response. The residual effect of gypsum has been found up to 4 to 5 years and the yield of few crops like barley, rice and wheat was found to increase many folds.

Gypsum Requirement of Soil

Saline-alkali or alkali soils are found to have abundance of sodium salts. To replace all the exchangeable sodium from the soil micelle, a fixed quantity of calcium salt or gypsum is needed so that physico-chemical properties of the soil can be improved and better crops can be grown. According to Schoonover one milli-equivalent of gypsum per 100 g of soil corresponds to 1.72 tonnes of gypsum/acre foot of soil.

Method of Gypsum Application

There are two methods of gypsum application. In the first method gypsum is applied as broadcasting and then it is thoroughly mixed into the soil. In the second method it is applied along with irrigation water (Fullmer, 1950) where gypsum bag is kept in the irrigation channel and irrigation water is slowly flown in the channel. Thus the gypsum gets dissolved in water and uniformly distributed throughout the field. It should be applied about 2 to 4 weeks before sowing or transplanting of the crop.

Line of Action

$$\text{Micelle}^{\text{Na}}_{\text{Na}} + CaSO_4 \Longleftrightarrow \text{Micelle}\,\text{Ca} + Na_2SO \rightarrow \downarrow$$

The sodium sulphate is leached down by impounding water in the field.

Rate of Application

General recommended rate of application of gypsum is 3 to 5 tonnes/ha, however, it depends upon exchangeable sodium percentage and the pH of the soil. Moreover, following rates of gypsum application has been in practice.

It is not as efficient as sulphur and 5.38 tonnes of gypsum is equivalent to one tonne of sulphur. Its equivalent ratio of other amendment is given below:

One tonne of gypsum = 0.186 tonne of sulphur
0.570 tonne of H_2SO_4
1.620 tonnes of ferrous sulphate
0.580 tonne of limestone
0.850 tonne of calcium chloride
1.290 tonnes of ammonium sulphate

TABLE 22.8: Amount of gypsum required to replace Given Exchangeable Sodium

Exchangeable sodium Meq/100 g of soil	Gypsum in tones per acre foot	Gypsum in tones per 6 acre inches
1	1.72	0.9
2	3.40	1.7
3.	5.20	2.60
4	6.90	3.40
5.	8.60	4.30
6	10.30	5.20
7	12.00	6.00
8	13.70	6.90
9.	15.50	7.70
10	17.20	8.60

Limestone

It is useful and suitable primarily for acidic soils or slightly saline soils having pH below 8.0, because its solubility decreases with increase in soil pH. When it is applied it dissolves in water and calcium replaces sodium from the soil micelle which brings the soil reaction towards neutral or near neutral. There are two possible reactions when limestone is applied into the soil:

$$\overset{Na}{\underset{Na}{\text{Micelle}}} + CaCO_3 \Longleftrightarrow \overset{Ca}{\text{Micelle}} + Na_2CO_3$$

And

$$\overset{Na}{\underset{Na}{\text{Micelle}}} + HOH \Longleftrightarrow \overset{H}{NaOH} + \text{Micelle}$$

$$\overset{Na}{\underset{Na}{\text{Micelle}}} + CaCO_3 \Longleftrightarrow \overset{Ca}{\text{Micelle}} + CO_2 + H_2O$$

The solubility of lime stone depends upon pH values and presence of exchangeable hydrogen ions of soil and accordingly it reacts very slowly in alkali soils. The finely ground material is more effective than coarse materials. It should be broadcasted and well mixed into soil for distribution. 3.13 tonnes of lime stone is as effective as one tonne of sulphur for reclaiming alkali soils.

Pyrite

Use of pyrite as an amendment is a recent development in chemical amelioration and reclamation of alkali soils. Pyrite is a natural deposit at Amjhore in Rohtas district of Bihar to the estimate extent of 385 million tones containing about 40% sulphur and 1500 million tones of 10% sulphur. The overall reserves of low grade pyrite of less than 20% sulphur are about 2000 million tones. This pyrite is highly pyrophoric and contains reactive mineral 'Marcasite' (FeS_2). The agricultural grade pyrite has following analysis:

TABLE 22.9: Chemical Composition of Amjhore Pyrite used as Agriculture Grade Pyrite

Constituents	Per cent
Sulphur	22.0
Iron	20.0
Magnesium (MgO)	0.5 – 0.8
Calcium (CaO)	0.1 – 0.3
Alumina (Al_2O_3)	6.0 – 8.0
Silica (SiO_2)	45.0 - 50.0
Carbon	2.0 – 5.0
Zinc	0.02
Copper	0.05
Manganese	0.01

Source: P. K. Awasthi, Promotion and marketing of Pyrites, TSI-FAI symposium, Sulphur in Indian Agriculture.

Presently Agriculture Grade Pyrite containing 22 % sulphur are distributed mainly as soil amendment for reclamation of alkali soils and Govt. of U.P. and Bihar are subsidizing the cost of materials by 50 to 75%.

Method of Application

In case of pyrite application the soil is first irrigated, ploughed after attaining working condition then pyrite is broadcasted and thoroughly mixed into soil where it goes under both chemical and microbial oxidation in presence of moisture and air. After oxidation it forms sulphuric acid or sulphate of sulphur.

Mode of Action

Following chemical reactions take place:

$$FeS_2 + 2\,H_2O + 7\,O_2 \xrightarrow[\text{Thiobacillus Ferro - Oxidance}]{\text{Oxidation}} 2\,FeSO_4 + H_2SO_4$$

$$4\,FeSO_4 + O_2 + H_2SO_4 \rightarrow 2\,Fe_2\,(SO_4)_3 + H_2O$$

$$Fe_2\,(SO_4)_3 + FeS_2 \rightarrow 3\,Fe\,SO_4 + 2\,S$$

$$2\,S + 3\,O_2 + 2\,H_2O \xrightarrow[\text{Thio – oxidance}]{\text{Thiobacills}} 2\,H_2\,SO_4$$

Under calcareous conditions following reaction takes place:

$$CaSO_3 + H_2SO_4 \longrightarrow CaSO_4 + H_2O + CO_2$$

$$\begin{matrix} Na \\ Micelle \\ Na \end{matrix} + CaSO_4 \longrightarrow Micelle + Na_2SO_4\ (\text{Leaching}) \downarrow$$

Under non-calcareous soil:

$$\text{Micelle}\begin{matrix}Na\\ Na\end{matrix} + H_2SO_4 \longrightarrow \text{Micelle}\begin{matrix}H\\ H\end{matrix} + Na_2SO_4 \text{ (Leaching)} \downarrow$$

Soil Conditioners

These are such materials which are used to bring about required physical properties of soil or it is used to improve and maintain the physical conditions of the soils. Crop residues, organic manures and other organic materials are the organic soil conditioners. Other synthetic organic materials which are used as soil conditioners are – Polyvinyl alcohol (PVA), Carboxymethyl cellulose (CMC) and Krilliium conditioners. These materials use to form soil aggregates or they use to stabilize soil aggregate formed by mechanical manipulations. However, its application is found restricted to green house, glass house or in growing high value crops like vegetables, ornamental plants or spices and condiments etc.

CONVERSION FACTORS FOR FINDING OUT ELEMENT OR COMMERCIAL MATERIAL

To find out the equivalent of one material or element A in terms of another B multiply the amount A (kg) by the factor in column A to B. To calculate a reversed amount multiplies kg of B by factor in column B to A.

Multiply by			
A	**B**	**A to B**	**B to A**
Free lime (CaO)	Calcium sulphate	2.4277	0.4119
Calcium sulphate ($CaSO_4$)	Gypsum ($CaSO_4$ 2 H_2O)	1.2647	0.7907
Free lime (CaO)	Gypsum	3.0702	0.3257
Free lime	Carbonated lime ($CaCO_3$)	1.7848	0.5604
Free lime	Calcium (Ca)	0.7147	1.3942
Free lime	Tricalcium phosphate Ca_3 $(PO_4)_2$	1.8437	0.5424
Calcium carbonate ($CaCO_3$)	Calcium (Ca)	0.4004	2.4973
Calcium carbonate	Calcium hydroxide $Ca(OH)_2$	0.7403	1.3508
Calcium hydroxide	Calcium	0.5409	1.8487
Calcium hydroxide	Calcium oxide (CaO)	0.7508	1.3212
Sulphur trioxide (SO_3)	Sulphur (S)	0.4005	2.4972
Sulphuric acid (H_2SO_4)	Sulphur	0.3269	3.0587
Magnesia (MgO)	Magnesium (Mg)	0.6032	1.6579
Magnesium sulphate ($MgSO_4$)	Magnesium	0.2020	4.9501
Magnesium sulphate	Magnesia (MgO)	0.3349	2.9858
Epsom salts ($MgSO_4.7H_2O$)	Magnesia	0.1639	6.1135
Epsom salts	Magnesium (Mg)	0.0987	10.1356

"There is only one way to recommend Nature, and that is obeying it."

23 Biofertilizers

> ***"Natural resources determine the course of development and constitute the challenge which may or may not be accepted by the human mind"***
> ***W. Arthur Lewis***

During post green revolution period the Indian Agriculture has become a chemical agriculture in which numerous chemicals like insecticides, nematicides, herbicides and commercial fertilizers are being used in producing crops. This has resulted into a major shift in microbial population and whole rhizosphere is getting polluted. Nutrient dis-balance, fast depletion in soil fertility and continuous deterioration in physical properties of soil are some of the added disadvantages of chemical agriculture. Bio-fertilizers play an important role in improving soil fertility and thereby boosting crop yields. Bio-fertilizers are microbial inoculants which are capable in nitrogen fixation, phosphate solubilising and decomposing organic matter at a faster rate. Besides, the Vesicullar Arbuscular Mycorrhizae (VAM) and Plant Growth Promoting Rhizobacteria (PGPR) are also very important microbes which help improving the soil fertility and boosting crop productivity.

CLASSIFICATION OF MICROBIAL INOCULANTS

These inoculants may be classified into following groups based on their mode of action:

1. Nitrogen Fixers
   a) Symbiotic: Those found in symbiosis or in complementarily with legumes i.e., Rhizobium inoculants.
   b) Non-symbiotic: Those which are found associated with cereals, millets and vegetable like:
      i) Bacteria
         (1) Aerobic: These which function in presence of oxygen including Azotobacter, Azomonas, Azospirillum, Mycobacterium.
         (2) Anaerobic: Closterdium, Chlorbium, Chromatium. They are capable of functioning in absence of oxygen.
         (3) Facultative anaerobes: This may also function in absence in absence of oxygen which includes Bacillus, Enterobacter, Escherichia, Klebsiella, Rhodospirillum
      ii) Blue green algae: These algae are capable of fixing atmospheric nitrogen besides adding organic matter to the soil. This includes Anabaena, Nostac, Tolypothrix, Anabaenopsis

2. Azolla
3. Phosphate Solubilizing Microbes
4. Vesicular-Arbuscular Mycorrhyza (VAM)
5. Plant Growth Promoting Rhizobacteria (PGPR)
6. Sulphur Solubilizing Microbes.

NITROGEN FIXERS

It has been estimated that the atmosphere contains 78% nitrogen amongst its gaseous components but most of it remains unutilized due to the fact that majority of plants can not fix atmospheric nitrogen. A group of micro-organisms is identified capable of fixing this nitrogen, though a negligible amount, into their roots through the process of symbiosis with legumes which is left into the soil after completion of life cycle of these plants. These inoculants are discussed below:

Symbiotic Bacteria

It is observed that every hectare of land surface area contains about 78,000 tonnes of atmospheric nitrogen over it in the form of inert N_2 gas. The biologically fixed nitrogen by a few prokaryotic organisms directly from atmosphere is hardly about 175 million tones out of which approximately 79% amount is fixed under terrestrial fixation depending upon- nitrogenase enzyme complex, ATP content and source of strong reductant. Rhizobium has been one of the most important and widely used inoculants symbiotic bacteria. Which use to fix atmospheric nitrogen through the process of symbiosis however, the extent of nitrogen fixed by Rhizobium bacteria varies from species to species of host plant. The extent of nitrogen fixed by some legume crops is mentioned in the following Table 23.1.

TABLE 23.1: Amount of Nitrogen Fixed by Important Legume Crops

Crops	N fixed (kg/ha)	Crops	N fixed (kg/ha
Alfalfa	100-300	Green gram	50-55
Chickpea	26-63	Lentil	35-100
Clover	100-150	Peas	46
Cluster bean	37-196	Pigeon pea	68-200
Cowpea	53-85	Soybean	46-130
Groundnut	112-152	Fenugreek	44

Source: Role of biofertilizers in upland crop production by Lee and Wani (1989).

Rhizobium Species

There are seven Rhizobium species so far as given below:

TABLE 23.2: Classification of Rhizobium species with their Respective Host Crops

Rhizobium species	Cross-inoculation group	Host genera	Crops
R. japonicum	Soybean group	Glycine	Soybean
R. leguminusarum	Pea group	Pisium	Pea
		Vicia	Vicia
		Lathyrus	Lathyrus
		Lens	Lentil
R. lupine	Lupine group	Lupinus	Lupines

Rhizobium species	Cross-inoculation group	Host genera	Crops
		Ornithopus	Serrdella
R. meliloti	Alfalfa group	Medicago	Lucerne
		Melilotus	Melilotus
		Trigonella	Fenugreek
R. phaseoli	Bean group	Phascolus	Beans
R. species	Cowpea group	Vigna	Cowpea
		Cajanus	Pigeon pea
		Cicer	Chick pea
		Crotolaria	Sunhemp
		Arachis, etc.	Groundnut
			Green gram
			Black gram
			Cluster bean
			Dhaincha
			Acasia
			Prosopis
			Acasia
			Prosopis
			Dalbergia
			Albizzia
			Indigofera
			Teprosia
			Atylosia
			Stylo, etc.

Biology of Nodulation and Process of Nitrogen Fixation

Nodulation or nodule formation on a root is a result of interaction between a particular *Rhizobium species* and a specific host root system. The Rhizobia enter the roots of legumes either through root hairs as in case of clovers and Lucerne or directly as a point of origin of lateral roots as in case of groundnut. The entry point appears to be swollen infected part as 'infection thread' which later produces nodules. It is believed that lectins (phytohaemaglutinins) in clover roots function as molecular bridges between the common antigens on the bacterial cell and the root surfaces. The infection thread carries Rhizobium bacteria and enters the root cortex from the root hairs or epidermal cell of the roots. It branches and traverses in the root cortex till release of the contents of hairs into a tetraploid cell. The cell forms a nodule meristems which grows out of the cortex region as a lateral growth. The meristematic tissue is soon organized into a central bacteroid zone surrounded by a cortex. The cells, present in the bacteroid tissues, contain stationary rhizobial cells which undergo transformation into non-motile bacteroid are the ultimate stages in the life cycle of Rhizobium which starts from motile cocci in rhizosphere and changes into motile bacilli in the nodule (merely a protective tissue over bacilli). The shape, size and number of nodules depend upon the bacterial strains and host plant cultivars. Biological nitrogen fixation is an anaerobic process in which nodule develops a mechanism to exclude oxygen from the bacteroid which is the site of fixation. The bacteroid is surrounded by membranous envelope containing leghaemoglobin. The leghaemoglobin and the extent of bacteroid tissue of the nodule have direct bearing on extent of nitrogen fixation (Bergersen and Briggs, 1958; Chopra and Subba Rao, 1967). Nitrogenase, an enzyme, mediates the reduction of Nitrogen (N_2) to ammonia (NH_3). This process requires a source of ATP and reductiant which are provided by photosynthates (products of photosynthesis). It is understood that 12 molecules of ATP are needed to fix one molecule of N_2 by reducing to NH_3. However, according to some thoughts about 28 molecules of ATP are required to fix one molecule of N_2. The ammonium ions (NH_4^+ is assimilated by the GS-GOGAT (glutamine synthatase glutamate synthase) pathway.

Significance of Inoculants in Agriculture

It is believed that the crops which are native to India or the indigenous crops do not need inoculation due to presence of specific bacteria but it has been observed that inoculation helps in increasing nodulation and the nitrogen fixing efficiency is found to be significantly increased. Moreover, abundance of nodulation does not necessarily mean that all the nodules are effective ones and they would efficiently fix atmospheric nitrogen because the nitrogen fixing ability of the nodules depends upon the symbiotic capacity between rhizobium and host plants. Kulkarni and Joshi, 1988 reported that out of 87 groundnut rhizobial strains isolated from different parts of India, only 5 strains were found to be effective.

Therefore, it makes essential to go for inoculation with improved strains so that the highest crop yields may be obtained with an increased fixation of nitrogen by the rhizobium bacteria.

Factors Affecting Efficiency of Inoculants

The factors which are found responsible to affect the extent of nitrogen fixation are discussed below in brief:

1. **Soil Reaction:** Acute acidity or alkalinity both is detrimental for nodulation and the inoculants are found to perform very poorly under such conditions. Saline soils mostly contain sulphates and chlorides of sodium, calcium, magnesium and small amounts of carbonates and bicarbonates which are found to have adverse effects on nodulation either individually or collectively. Thus, the extent of salt tolerance of host legume and rhizobium species decides their survival and efficiency. In a study conducted by Subba Rao *et a*l, 1972; Singh *et al.*, 1972; Subba Rao *et al.*, 1974 at IARI, New Delhi with Lucerne it was found that *R. meliloti* could tolerate NaCL level up to 3% but seeds of Lucerne could not germinate beyond 1.5% salt. Beyond 0.7% NaCL nodulation in Lucerne was inhibited under aseptic condition. Similar results were observed with KCL and $MgCL_2$. The tolerance limit for carbonates and bicarbonates was much lower than chlorides. Nodulation was inhibited at 0.1 to 0.2 % levels although rhizobia could tolerate up to 0.5% and Lucerne seedlings up to 0.5 to 0.75 per cent of carbonate/bicarbonate.
2. **Toxicity and deficiency of Nutrient Elements:** It has been observed that manganese and aluminium cause severe toxicity to inoculants resulting into very poor to no nodulation at all. These elements are found in abundance under acidic conditions.

   On the other hands, deficiency of certain elements like B, Co, Mo and Zn in soils cause drastic reduction in nodulation (both size and number of nodules were reduced). Application of these elements in groundnut resulted into higher nodulation and more umber of effective nodules/plant.

3. **Soil Type:** Rhizobium bacteria prefer aerobic conditions therefore; soils having better aeration promote nodulation as the extent of damage of nodules is found to be much higher in pigeon pea when sown in vertisols (up to 86%) as compared to that planted in Alfisols (only up to 20 %). Thus, soils with higher organic carbon/humus, appropriate moisture regime, free from accumulation salts, etc. are always better for an increased nodulation and its efficiency.
4. **Poor Water Management:** The efficiency of nodule is drastically reduced by either prolonged water logging or excessive moisture stress in the soil as both the conditions cause a marked reduction in size, count and efficiency of nodules. The rhizobium bacteria are found to perform at the best of their capacity when provided with field capacity of moisture regime.
5. **Relative Performance of Various Inoculant Strains:** The new strains which are going to be introduced must be competitive and more effective compared to the native rhizobium strains and their populations; otherwise they will be suppressed by the existing rhizobium population. Thus, the introduced rhizobium will not have any beneficial effect.
6. **Plant Health and Extent of Bacterial Predators:** Attack of insect-pest and diseases results into poor growth, health and development of host plants which invariably use to have very poor nodulation due to poor nodulation due to poor symbiosis. Bacterial predators or insect damage can seriously affect the nitrogen fixing ability of nodules. Sithanathan *et al.*, 1981 reported a marked nodule damage of red gram by Dipteran larva (Rivellia angulata).
7. **Nutrient Status of Soil:** It has been noticed and scientifically observed that the inoculation response was much less pronounced in the presence of high soil mineral nitrogen than under conditions where such nitrogen was immobilized and unavailable. Similarly, P rich soils or soils fertilized with D.A.P. promote nodulation of crops and an increase in nitrogen fixation capacity has been observed.
8. **Effect of Pesticides:** Effect of various pesticides have been studied, the important of which are mentioned here in the text. Dalapon has been found to inhibit the rhizobial growth while Amitrole delays nodulation and reduces nodule count. Many other agricultural chemicals have shown variable effects on symbiosis. However, seed treatment with fingicides like thiram, captan, agronol, hermal, brassicol and agroson did not affect nodulation adversely and has encouraged nodulation. Application of Bavistin, thiram, dithane M-45, defolatan stimulated nodulation.
9. **Antagonistic Effect in Soil:** It includes the effects of seed coat diffusates and inhibitory effects of other soil micro-organisms or rhizobia. Some legume seeds like that of chickpea, pea, soybean, green gram, cowpea, cluster bean, lablab, horse gram and groundnut contain water soluble toxins which inhibit rhizobial activity but these toxins can be removed by pre-soaking in plain water (Dadarwal and Sen, 1971; Sen *et al.*, 1972; Dadarwal and Sen, 1973; Kandaswami *et al.*, 1974). Some seed-borne bacterial and fungi are also found to be antagonistic to *rhizobia* (Jain and Rewari, 1974). A part from these, some *rhizosphere microflore* may also inhibit activity of the growth of *rhizobia*. The growth of *rhizobia* is found to be adversely affected by bacteria like *Bacillus, Alkaligenes, Erwinia, Aerobacter, Corynybacterium, Aithobactor, Brevibacterium, Agrobacterium, Sarcina, Enterobacter* and *Micrococcus.* The important fungi which inhibit growth of r*hizobia* include *Alternaria, Aspergillus, Penicillium, Rhizopus, Acrothecium, Eusarium, Rhizoctonia, Curvularia, Pythium, Trichoderma, Cladosporium, Mortierella, Scolecobasidium, Sordaria, Thielavia* and *Mucor.* Contractly to this,

species of *Pacilomyces, Phoma* and *Rhizopus* are found to stimulate the growth of *Rhizobium*.

10. **Synergistic Effects of Combined Nodulation:** Seed inoculation with *Beijerinckia indica* and *Azotobacter chroococcum* resulted into better nodulation in soybean. Similarly nodulation improved when soybean seed was treated with *R. japonicum* and *Azospirillum brisilense.*

Method of Seed Inoculation

Carrier based inoculants are mixed with little water for preparing slurry in which small quantity of sugar or jaggary or gum is added so that the inoculants may get energy for their prolonged survival. The slurry is poured over the seeds which should be kept in a container. The seed is mixed well with slurry of inoculant by pouring the seed and slurry into another container. Thus, by pouring fourth and back into both the containers the seed is nicely mixed with inoculant. Now the treated seed should be dried in cool and dry shady place and sown immediately in the field. In case of acidic soils pelleted seed with inoculants should be sown.

Stem-nodulating Legumes: *Aeschynomene, Sesbania* and *Neptunia* are three genera of legumes whose species bear nodules on stem and roots. In studies it has been found that *S. rostrata* fixes 83 to 109 kg N/ha whereas S. *sesban* (which does not possess stem nodules) fixes only 7 to 18 kg N per hectare. Thus the plants having stem nodules are best green plants. The species of *Rhizobium* inducing nodulation on both stem and root in *S. rostrate* has been designated as *Azorhizobium caulinodans* (Drefus and Dommergues, 1981; Ndoye and Drefus, 1988).

Effects of Nodulate Non-Legumes

Recently white clover seedlings were treated with cellulose and pectolyase followed by treatment with polyethylene glycol (PEG) and calcium chloride with a view to break the cell wall burrier in the root hairs for successful nodulation. These treatments resulted into exposure of protoplasts of root-hairs for active reception to rhizobia which normally do not invade white clover root hairs due to cell wall burrier. The nodules, thus formed, on white clover roots, induced by heterologus *R. loti* fixed nitrogen. Similarly, pre-treatment of seedlings of rice (*Oryza sative*) and wheat *Triticum aestivum),* both monocots also resulted in root nodulation when inoculation when inoculated with *R. trifolii* and *R. loti* or their mixture. The nodules were formed. Similar procedure adopted to nodulate rape seedlings by inoculating with *R. leguminosarum and Bradyrhizobium* gave successful results. These nodules were similar to normal nodules and had appreciable amount of nitrogenase activity (Cocking *et al.,* 1990). The rape seedlings were found to have nodules, when inoculated with *R. parasponium,* even without the enzyme and PEG treatment.

AZOTOBACTER INOCULANTS

Azotobacter is a non-symbiotic heterotrophic nitrogen fixing bacterium found in Indian soils which was first isolated from root *rhizosphere* of cereals by *Beijerinck* in the early part of this country (nineteenth century). The *Azotobacter spp.* can use a variety of carbon and energy sources for their proper growth and nitrogen fixing efficiency. Soil samples collected from over 200 fields used for growing sorghum and millet of Northern Western India showed the number of nitrogen fixers ranging between 100 to 1, 00,000 per gram of soil. *Rhizosphere* study of pearl millet conducted at ICRISAT, Hyderabad showed that out of 546 isolated only 17% isolates had nitrogenase activity in vitro and only 11% *Azotobacters* were found while a maximum population of 72% was that of *Azospirilla*.

Classification of Azotobacter

Several workers have grouped *Azotobacter spp.* under *Azotobacteraceae,* whereas others group them under *Bacteriaceae.* The *Azotobacteraceae* is recognized to have three genera *viz.,, Azotobacter, Baijerinckia* and *Derxia.* DNA analysis of these three groups showed that *Baijerinckia* and *Derxia* are genetically distinct from genera *Azotobacter* and they recognized 3 groups within genus. Azotobacter: (1) *A. chroococcum, A. beijerinckia* and *A. vinclandii,* (2) *A. insignis* and *A. macrocytogenes* and (3) *A. agilis.* Accordingly, the first group is called as genus *Azotobacter*, group second as *Azotomonas* and group third is known as *Azotococcus*. However, the most commonly known *Azotobacter spp.* in Indian soils are: *A. chroococcum, A. vinelandii, A. paspalli, A. beijinckii, A. macrocytogenea, A. insignis, A. agilis* but of these the most common spp. found in majority of Indian soils is *A. chroococcum.*

Crop Response to Azotobacter Inoculation

This bacteria promotes growth and development of crop plants by helping in synthesis of auxin, vitamins, growth substances and antifungal antibiotics. These factors help increasing crop yields in the range of 7 to 12% when crops like wheat, rice, barley, oats, maize, sugarcane and vegetables like potato, brinjal, tomato, cabbage and onions were inoculated with *Azotobacter chroococcum*. In another study pearl millet seed was treated with *Azotobacter chroococcum.* In another study pearl millet seed was treated with *Azotobacter chroococcum* and *A. Lipoferum,* which resulted into an increase in yield of grain to the extent of 8 and 11 per cent respectively (Wani *et al.*, 1988). Again Wani *et al.*, (1990) while compiling the response pattern of Azotobacter concluded that in about 60 % of the field trials conducted in Russia, Israel and India the inoculation has boosted the crop yields, however, the extent of increase in yield varied with crops, varieties, places seasons, agro-techniques, bacterial strains, fertility status, capability of inoculant with native soil microflora, etc. Increase in crop yields has also been reported by Pandey and Kumar (1989). They reported that an increase of 2 to 45% in vegetables, 7 to 28% in cotton and 9 to 24% in sugarcane had been obtained by inoculation of seeds of these crops with *Azotobacter*. It has been observed that the highest response may be obtained by application of moderate dose of nitrogen.

How to Inoculate the Seed with Azotobacter

The crop seeds are treated with slurry prepared from inoculation as explained in case of rhizobium inoculant. In case of transplanted crops, the seedling roots are dipped in the slurry for 10-30 minutes following which they must be transplanted immediately. Long duration crops as sugarcane may be given secondary inoculation with slurry near root zone in early crop growth stages or the inoculant may be mixed well with organic manures which should be applied near root zone during early growth period.

AZOSPIRILLUM INOCULANTS

The crops, grown in tropics like India, use to have nitrogen fixing *Azospirillum spp.* inside their roots and in rhizosphere. The pearl millet rhizosphere, studied at ICRISAT showed about 72% *Azospirillum* out of total nitrogen fixing isolates. It increases crop yields by acting similar to *Azotobacter*. It promotes uptake of NO_3, NH_4, H_2PO_4, K and

Fe besides producing antibacterial and antifungal compounds of inducing resistance against various plant diseases. It has been observed that sorghum plants grown by *Azospirillum* treated seeds had greater resistance to shoot-fly due to an increased level phenol content in the shoots of treated plants.

Classification of Azospirillum

Beijerinck found out that *Spirillum lipoferum* may fix atmospheric nitrogen in 1925 which was confirmed by Backing in 1963 by using $^{15}N_2$. Dobereiner (1975), Dobereiner and Day (1975) and Dobereiner *et al.*, (1976) used this species commercially in Brazil. The associative symbiosis has been found in case of sorghum and millets, C_4 plant types which are characteristics with high rate of photosynthesis and lowest rate of photo-respiration.

Two species of Azospirillum have been so far recognized viz., *Azospirillum lipoferum* (Beijerincki) Comb. nov. and *A. brasillense* sp-nov.

Crop Response to Inoculation with Azospirillum

Inoculation of pearl millet with *A. brasilense* and *A. lipoferum* resulted into an increase of about 17 % and 11% in grain yield respectively over untreated control. Sorghum when treated with *A. brasilense* gave about 31% higher grain yield compared to untreated control. Seeing the performance of pearl millet and sorghum under *Azospirillum* treatment, the ICAR workshop on sorghum and pearl millets has recommended the use of *Azospirillum* in cultivation of these crops by farmers.

Method of Azospirillum Inoculation

The seeds of sorghum and pearl millet may be treated or inoculated with slurry prepared by mixing by mixing inoculum with water sicker (500 ml water + 6 g of carboxy methyl cellulose). The method of inoculation should be same as that of rhizobium inoculant. In case of transplanted crops, the seedling roots should be dipped in the slurry for about 10-30 minutes and immediately after this they must be transplanted.

BLUE GREEN ALGAE (CYNOBACTERIA OR BGA)

Cynobacteria are found in many diss-similar ecosystems which make being green, are capable of performing photosynthetic activity as well as they fix the atmospheric nitrogen. Thus the energy required for nitrogen fixation is met with photosynthesis or the photosynthates. Frank (1889) and Prantl (1889) were the first scientists to explain possibility of these algae to fix atmospheric nitrogen under transplanted water logged/rice ecosystems. De (1939) and Singh (1942) were the first persons in India to explain the agronomic potential of BGA in fixing atmospheric nitrogen and attributing natural fertility in tropical flooded rice fields. The nitrogen fixing activity was confirmed by stimulating through acetylene reduction activity (ARA).

Biology of Nitrogen Fixation by BGA

The actual mechanism of nitrogen fixation has not yet been demonstrated but it is believed that following steps help in fixation of atmospheric nitrogen by blue green algae:

Nitrogenase Synthesis

This is the most important enzyme responsible for fixation of atmospheric nitrogen into cells of blue green algae. These algae use to synthesis nitrogenase which is very similar to that found in other nitrogen fixing organisms. The nitrogenase catalyses are the reduction of N_2 for which it requires ATP, reductant and Mg^{2+}. Nitrogenase has two oxygen-sensitive components which fix the atmospheric nitrogen in combination. The protein found in Fe use to bind Mg ATP and transfers electrons to the MoFe protein which use to bind and reduce in N_2. Thus, the nitrogen gets fixed in the form of protein. Of these blue green algae are the most efficient ones.

Oxygen Protection

The nitrogenase enzyme, being very sensitive to oxygen, needs O_2- protection mechanism in algal cells for an efficient nitrogen fixation. The BGA is capable of evolving special mechanism of structural modification of vegetative cell into heterocyst which provides oxygen protection for the enzyme so that it may perform best nitrogen fixation activity within the algal cells.

In case of non- heterocystous forms of algae, a temporal separation of photosynthesis and nitrogen fixation has been reported by Weare and Benemann, 1973. Gallon *et al.*, 1974 reported that in the unicellular Glococapsa, the O_2 protection is achieved either by a temporal separation of photosynthetic O_2 evolution and nitrogen fixation, an O_2 linked hydrogen uptake by the presence of tocopherols serving as an antioxidant or by certain internal membranes which may represent some intracellular protective compartment, though this is yet to be confirmed.

Reductant and ATP

The nitrogenase can get reductant from various sources like photosystem I-dependent reaction in presence of light, from oxidative pentose pathway in light or dark and from glycolysis or Kerbs cycle or from both. The heterocystous BGA may fix N for a limited period in dark which generates reduced ferredoxin by non-photosynthetic means. The reductant required for nitrogen fixation in dark may also be supplied by pyruvate, isocitrate, glucose-6-phosphate and/or 6-phosphogluconate.

The heterocystous algae get ATP for support of N fixation through cyclic photosphorylation in light under adequate supply of fixed carbon. Under aerobic conditions, in dark, the ATP may come from oxidative phosphorylation. Thus, nitrogen fixation is not hindered.

NH_4 Assimilation

In most of the algae the NH_4 assimilation takes place through glutamine synthetase along with ferrodoxin linked glutamate synthase which are present in vegetative cells while glutamate synthase and aminotransferases are present in heterocysts. Thus, it is assimilated in both the groups of algae having vegetative cells as well as heterocysts.

During the process of nitrogen fixation, NH_4^- is the first stable product which is assimilated and produces glutamine and glutamate as first organic product from NH_4^+. Aspartate is formed from glutamate.

Classification of BGA

Over 100 strains or types are found to have nitrogenase of which the heterocystous forms use to fix atmospheric nitrogen aerobically whereas the non-heterocystous ones fixes N_2 anaerobically or micro- aerobically except unicellular Glococapsa which fixes N_2 aerobically. According to Gallon (1980), Stewart (1980) and Vekataraman (1981) the blue green algae can be classified as under.

TABLE 23.3: Classification of BGA

Group	Genera
Unicellular	Synechococcus
	Glococapsa (symbiotic with other organisms)
	Aphanotheca
	Dermocarpa
	Xenococcus
	Myxosarcina
	Chroococcidiopsis
	Pleurocapsa group (a symbiotic type)
Filamentous, non-heterocystous	Plectonema baryanum
	Lyngbya
	Trichodesmium
	Oscillatoria
	Pseudanabaena
	Microcoleus chthanoplastes
	Schizothrix
	LPP group
Filamentous, heterocystous	*Anabacna* (a symbiotic type)
	Nostoc
	Nodularia
	Cylinderospermum
	Scytonema (a symbiotic type)
	Calothris (a symbiotic type)
	Anabacnopsis
	Mastigoeladus
	Fischerella (a symbiotic type)
	Tolypothrix
	Aulosira
	Stigonema
	Hapalosiphon
	Chloroglocopsis
	Camptylonema
	Glocotrichia
	Nostochopsis
	Rivularia
	Scytonematopsis
	Westiclla
	Westicllopsis
	Wollea
	Chlorogloca

Source: Blue Green Algae, Biological nitrogen fixation by Venkataraman, G. S. an ICAR publication, page 47.

Preparation and Multiplication of BGA

The algal cultures can be prepared in galvanized iron trays or cemented pits of 4.8 m x 0.3 m x 0.25 m dimension lined with polythene sheets consisting of 10 kg soil and 200 g superphosphate having about 5 to 15 cm deep water in the pits. After settling of the soil, sawdust mixed soil based starter culture of BGA is spread over the water surface of the pits or trays. One week later the algae forms a thick scum on the water surface. The water is now drained out or dried, the algal flakes are collected and stored in polythene bags for use.

Application of Algal Culture

The algal culture is applied at the rate of 10 kg/ha after one week of rice transplanting and the field is flooded with 2.5 to 5 cm deep water. The irrigation is stopped after the algae has formed a thick scum and the algal mat is incorporated into the soil which adds about 25 kg of N/ha.

Factors Affecting Efficiency of BGA

The efficiency of blue green algae is found to be adversely affected by several factors and some of which are mentioned below:

Water Level: It requires standing water for formation of balls in unicellular BGA or mat in case of scum and filamentous BGA, however, abundant growth has been observed on sufficiently moist soil without standing water. It has been found that wetting and drying encourages growth of BGA, controls weeds and fixes more atmospheric nitrogen.

Light: BGA are phototrophic, hence their growth and nitrogen fixing efficiency are positively related with solar radiation, however, the rice crop canopy of fast growing varieties has a depressing effect on the growth and nitrogen fixing ability of BGA as the canopy has shading effects over growing BGA. Roger and Reynand (1982) observed and reported that a reduction in light availability to BGA by 50% after 15 days, 85% after one month and 95% after two months of transplanting has been noticed. Therefore, the BGA inoculum must be applied soon after establishment of rice seedlings.

Temperature: The most appropriate temperature for an enhanced nitrogen fixing efficiency of BGA has been found between 30 and 35°C, and both higher and lower temperature have their depressing effects over growth and efficiency of N fixation of BGA. Management of suitable water level can regulate the temperature for an efficient working of BGA.

Soil Reaction: A neutral to alkaline soil reaction having a pH range of 6.5 to 10 has been found to be most preferential for an accelerated growth, development and nitrogen fixing fixation by BGA. Thus, growing BGA does not fix atmospheric nitrogen only but it helps reclaiming alkalinity by adding organic matter to the soil.

Mineral Nutrients: Phosphorus is the key element responsible for boosting growth and biomass of BGA which also increases extent of nitrogen fixation by them, however, all the plant nutrients are required for their better growth. BGA has been found to consist about 0.05 to 0.18% P on dry weight basis and prefers to have available soil P in the range of 0.2 to 6 ppm. Other important nutrients for an increased growth of BGA are Mo (0.2 ppm.), Fe, Mg, Ca, and nitrogen. Ammoniacal form of nitrogen is not congenial for better growth; hence nitrate or amide form of nitrogen should be applied in split doses for achieving better growth, higher biomass and more nitrogen fixation.

Agronomic Practices: These practices include tillage operations, nutrient management, method of rice planting, incidence attack of insect-pest and diseases in the crop. Little or surface tilling is better for boosting growth of BGA. Split application of P and N helps increasing growth, development and nitrogen fixing ability of BGA. Results of studies indicate higher performance of BGA under transplanting of rice seedlings in rows at wider spacing than direct seedling and closer spacing.

Predators and pathogens like bacteria and fungi often cause rapid death of BGA. The most harmful of these are zooplankton, mosquito larvae, chadocerans, copepods, ostracodes and snails which use to damage the whole population of BGA.

Pesticide Tolerance: The response of BGA to pesticide application shows greater variations, although their biological activity is little affected by application of recommended doses of these chemicals. Self-contradictory reports are available as according to some workers certain chemicals have stimulatory effects while others have found them inhibitory. Some chemicals like Stam F-34 and DGMU are found to inhibit both the growth and heterocyst differentiation, while a few like Michete has shown a mutagenic action, however, developing pesticide resistant or tolerant strains offers greater possibility of use of BGA.

Response of Rice to BGA: Experiments conducted with BGA at CRRI, Cuttack and All India BGA trials have shown an increase in rice yields to the extent of 6 to 35% along with addition of about 25 to 35 kg N/ha per crop season and good amount of organic matter and available P content of the soil. However, the crop response depends upon several factors as dry season, shallow depth of water; clean water and longer duration of crop varieties give better results. Beneficial effects of algal inoculation have also been observed on a number of other crops like barley, oats, tomato, radish, cotton, sugarcane, maize and capsicum.

Recovery of Algal Nitrogen: During growing period, BGA liberate relatively large quantities of combined nitrogen mainly in the form of polypeptides along with lesser amount of free amino acids, vitamins and auxin like substances. Approximately, 40% of total nitrogen fixed by algae has been found to be recovered by rice plants. The nitrogen uptake takes place by either active as well as passive assimilation processes.

AZOLLA

Azolla, an aquatic fern capable of fixing atmospheric nitrogen in symbiotic process having symbiotic association with a heterocystous BGA, Anabaena azolla under temperate as well as tropical rice growing areas. Azolla is found to have 0.2 to 0.3% nitrogen on fresh weight and 3 to 5% on dry weight basis. Azolla has been in use very widely in countries like China and Vietnam for centuries but it is of a recent introduction in India.

Classification of Azolla

Azolla pinnata has two morphologically different types: (i) *A. pinnata* var. *pinnata,* and (ii) *A. pinnata* var. *imbricate.* The imbricate forms are found to be distributed on the west coast of Africa and Madagascar and in Australia, whereas the pinnata forms occur in South East Asia from India to Japan and from Indonesia to Papua, New Guinea, Variety innata is being cultivated for agricultural purposes in Vietnam, China, Thailand, Indonesia, Sri Lanka and India.

Presently there are seven Azolla species which are under cultivation in India *viz., Azolla filiculoides* and *Azolla rubra* which may be grown under temperate conditions as they are cold tolerant, and *A. maxicana, A. microphylla, A. caroliniana, A. nilotica* and *A. pinnata* may be grown under tropical conditions as they are tolerant to high temperatures.

Nitrogen Fixation: Azolla is found to have nitrogenase activity through which they use to fix atmospheric nitrogen as in case of BGA.

Application of Azolla Culture

The purpose of application governs the rate of Azolla inoculum application and accordingly higher quantities (0.1 to 0.3 kg/sq. m or 1 to 3 tonnes/ha) in case of multiplication plots while a relatively lower rate of 0.05 to 0.1 kg/sq. m. or 0.5 to 1.0 tonne/ha as used in case of dual cropping. In case of green manuring Azolla culture is inoculated in the main field about 15-20 days before transplanting of rice seedlings so that the fern may be incorporated in the field at the time of puddling. In case of dual cropping it is inoculated about one week after transplanting of rice seedlings. For this purpose i.e., for rice-azolla dual cropping, the nicely prepared field is divided into smaller plots of 100 sq. m by raising bunds and 5-10 cm deep water level is maintained up to 20-25 days after inoculation so that a thick azolla mat may be formed. After about 25 days of inoculation when thick mat is formed the water level is reduced and azolla mat is mixed into the soil while weeding.

Effect of Azolla on Soil Fertility

Azolla green manuring done at CRRI is reported to add about 15-20 tonnes of green matter per hectare consisting of about 20-40 kg nitrogen/ha while rice-azolla dual cropping adds about 20-30 kg nitrogen/ha. Experiments conducted with azolla at CRRI, Cuttack indicate that azolla green manuring followed by azolla-rice dual cropping can add 60 to 70 kg nitrogen/ha whereas azolla green manuring followed by two dual crops can fix about 90 kg nitrogen/ha.

Pattern of nitrogen mineralization indicates that azolla releases nitrogen after its decomposition which completes within about 8-10 days of incorporation of azolla into the soil. It has been further observed that about 67% of fixed nitrogen by azolla is released for crop-use within 35 days of inoculation only. The studies indicate that unincorporated azolla release nitrogen very slowly for the use of rice plants and the rate of release s of nitrogen is much slower than that of nitrogenous fertilizers. The availability of azolla-nitrogen to rice plants is about 70% of ammonium sulphate-nitrogen, however, the rate of release of nitrogen in case of azolla is faster than of BGA.

Response of Rice to Azolla Application

Trials conducted with azolla show that azolla green manuring increases 0.5 to 2.0 tonnes of grain yield per hectare while azolla dual cropping gives an additional yield of 0.5 to 1.5 tonnes of grain/ha over control. Patel *et al.*, (1980) have reported, while summarizing the response of rice to azolla inoculation in Model Agronomic Trials conducted at four locations that rice yields obtained from azolla inoculated plots are comparable to that of 60 kg N/ha as given below:

TABLE 23.4: Effect of Azolla on Rice Yields in Model Agronomic Trials

Treatments	Rice yield in tones per hectare at				
	Cuttack	Bhubaneswar	Kharagpur	Titabar	Mean
Uninoculated (Control)	2.75	2.94	2.48	2.92	2.77
Azolla (10 t/ha) base	3.29	4.00	3.75	3.63	3.67
Azolla dual cropping*	3.08	3.60	3.05	3.12	3.21
Fertilizer 30 kg N/ha	3.05	3.38	3.74	3.00	3.29
Fertilizer 60 kg N/ha	3.25	4.20	4.19	3.25	3.27

*Fresh azolla @ 1 t/ha was incorporated 20 days after transplanting of rice seedlings.

Pillaiet al., (1982) compiled the rice yield response to Azolla from the experiments conducted at various locations of All India Co-ordinated Rice Improvement Project (AICRIP) and concluded that incorporation of 6 tonnes Azolla/ha into the soil as green manuring was more effective than dual cropping (rice+Azolla). The findings further, indicate that green manuring with 6 tonnes Azolla/ha with 25 kg N applied through urea produced 94% of the rice yield obtained from application of urea 50 kg N/ha. The results are presented below:

TABLE 23.5: Response of Rice to Azolla Manuring as Compared to Fertilizer N under AICRIP

State	No. of trials	Rice yields t/ha (average of 4 seasons)				
		Control	25 kg N	50 kg N	Azolla green manuring (6 t/ha)	Azolla green manuring (6 t/ha + 25 kg N/ha)
A.P.	3	3.60	4.58	5.01	4.15	4.79
Assam	10	2.89	3.16	3.15	3.15	3.21
Bihar	2	3.58	4.18	4.33	4.49	4.41
Karnataka	2	3.88	4.35	4.55	4.39	4.47
Kerala	3	2.89	3.36	3.55	3.00	3.28
M.P.	4	2.18	2.65	3.10	1.16	1.39
Maharashtra	3	2.81	3.69	4.05	3.50	4.24
Manipur	3	3.01	3.16	3.45	3.34	3.80
Punjab	2	2.34	3.50	3.91	2.61	3.66
Tamil Nadu	10	3.89	4.62	4.81	4.50	4.69
Uttar Pradesh	10	2.06	2.77	3.29	2.52	2.92
West Bengal	3	4.10	4.96	5.15	4.45	4.93
Mean	**55**	**3.10**	**3.75**	**4.07**	**3.44**	**3.82**

Factors Affecting Azolla Efficiency

There are numerous factors which affect the efficiency of azolla. Some of the important factors are described below:

Water: Azolla is a free floating fern, therefore, it can not tolerate drying except *A. caroliniana* which may survive on the moist soil for a longer period. Therefore, a standing water of 5 to 10 cm depth is most desired level of water in the fields where it is to be inoculated however, the growth of fern is not adversely affected up to a water depth of 30 cm The water should be clean for an accelerated activity of azolla.

Light: The azolla being green performs photosynthetic activity, therefore, light is the most essential for better growth, development and nitrogen fixation, however, during summer it prefers partial shading so that it may escape scorching due to desiccating heat. The azolla being multiplied in multiplication pond for use of green manuring, should have bright sunlight for an accelerated biomass production. Dual cropping (Azolla + rice culture) has been found to be adversely affected by shading effects of rice canopy created with advancement of rice growth, specially in case of dwarf rice cultures which grow very fast. It prefers longer days over shorter days.

Temperature: A temperature ranging between 25 and 30°C has been found to be the most ideal for better growth and development of azolla however, *A. pinnata* has been observed to grow successfully at a temperature between 14 and 35°C at Cuttack, Orissa. The response of azolla to variation in temperature depends upon the species of azolla *viz.,*, *A. filiculoides* and *A. rubra* are cold tolerant and may be grown under temperate climatic conditions while species like *A. mexicana, A. microphylla,*

A. caroliniana, A. nilotica and *A. pinnata* are tolerant to high temperature due to which they are the most ideal species for tropical and sub-tropical regions of the country.

Phosphorus Level of the Soil: Phosphorus is known to play a vital role in the growth and multiplication of azolla. Phosphorus level of 20 ppm produced more biomass than at lower levels and *A. filiculoides, A. mexicana* and *A. caroliniana* are the most sensitive species to phosphorus levels. It has been found that application of 1 kg of phosphorus to azolla could fix nitrogen to the tune of 1.9 to 2.5 kg. The importance of phosphorus nutrition for azolla was stressed by Talley *et al.,* (1977) and Kulasooriya *et al.,* (1980) for an accelerated nitrogen fixation.

Compatibility of Azolla Species: The compatibility was studied in pot experiments, where *A. filiculoides* and *A. caroliniana* were inoculated separately and also in mutual combinations in standard N-free medium. The mixed inoculation of *A. pinnata* and *A. filiculoides* resulted in the maximum biomass growth followed by *A. pinnata* and *A. caroliniana* combination whereas the growth rate was reduced when *A. filiculoides* and *A. caroliniana* were inoculated in combination. Combination of all the three species together also gave a comparable biomass yield. This has greater agronomic significance because individual azolla culture inoculations may sometimes be badly affected by pests or diseases.

Effect of Neem-cake on Nitrogen Fixation by Azolla: Neem cake has been found to stimulate nitrogenase activity for an accelerated nitrogen fixation and increases biomass production significantly. This increase in rate of multiplication of azolla and increase in biomass yield is due to insecticidal and fungicidal properties of neem-cake.

Effect of Insecticides on the Growth and Nitrogen Fixation: Field trials were conducted to see the effect of insecticides like furadan, phorate, monocrotophos, thiodan, carbosulfan and chlorpyriphos on nitrogenase activity and ammonia assimilating enzymes. It was observed that both the nitrogenase activity as well as ammonia assimilation was stimulated by insecticidal treatment of azolla.

Effect of Herbicides on Ammonia Assimilating Enzyme: Most commonly used herbicide butachlor was used to assess its effects over ammonia assimilating enzymes and growth of azolla. The azolla samples were drawn on third day of butachlor application where upon it was observed that the ammonia assimilating enzymes were inhibited considerably with an increase in the concentration of butachlor and the growth of azolla was drastically reduced, by butachlor application.

Effect of Fungicides: Effect of Benomyl was studied on nitrogenase activity of azolla and it was found that the nitrogenase activity was drastically inhibited at higher rate of Benomyl application.

Soil Reaction: A pH range between 5 and 8 has been found to be most appropriate for an accelerated growth of azolla and best nitrogenase activity however; it may survive in a pH range between 3.5 and 10. Higher efficiency and better growth of azolla may be attributed at a pH between 5 and 8 due to higher nutrient availability as under both the extreme pH conditions the nutrient availability is reduced and many micro-nutrients become deficient. Acute acidity causes toxicity of Al and Fe and P gets fixed. Similarly under extremely high pH most of the micro-nutrients and P become less available. A salt concentration of 90 to 150 mg/litre of water is found to be most optimum for better growth of azolla and that makes a yardstick to measure salinity.

Agro-techniques

Use of azolla inoculum at recommended rate is always better for a profused matting of the fern. Its inoculation at an early rice growth stages is better because of least shading effect of rice canopy, however, its inoculation before establishment of rice seedlings may damage them, specially under very high water level which may cover the rice seedlings and they may start rotting.

Rice planting in conventional method at regular spacing or planting at irregular spacing in haphazard manner results into maximum shading over azolla and hence the growth and nitrogenase activity are found to be adversely affected. Paired row planting at 40 cm between two paired rows with a 15 cm spacing between the hills within the rows gives better growth of azolla. The gap with the paired rows or between two narrow rows should be 7 cm.

Planting of rice seedlings in paired rows reduced shading effect and also facilitates incorporation of azolla biomass as green manures or even dual cropping. Planting rice in double rows in east-west direction favours the growth and efficiency of azolla.

The fern is found to be attacked by certain insects like Pyralis, Nymphula and Chironomus species besides numerous species of snails which damage azolla under hot and humid conditions. Application of any insecticide may control these insects but use of biological or natural pesticides like neem oil, karanj or neem cake may boost the azolla growth and efficiency.

Under poor field preparations and drying of fields, the rice gets infested by a variety of weeds which are often controlled chemically but some of the herbicides are found toxic to azolla. The order of toxicity is found to be butachlor, benthiocarb, 2,4-DEE, 2,4-DNa. Thus, 2,4-D sodium salt can be used safely, however, this can not control most of the associated weeds like *Echinocloa spp.* etc. Therefore, application of butachlor and benthiocarb becomes unavoidable. Azolla inoculation after 3 weeks of benthiocarb or butachlor application has no toxic effects.

Azolla is also attacked by fungal and bacterial pathogens which may reduce the population and the efficiency of the fern hence application of 0.5 to 1.5 % benlate solution or 10 kg of copper sulphate is beneficial.

PHOSPHATE SOLUBILIZING MICRO-ORGANISMS

They are heterotrophic in nature which has ability to solubilize inorganic phosphorus present in soluble sources. The group includes bacteria like *Bacillus magaterium, B. circullans, B. subtilis, Pscudomonas straita, P. rathoms*; Fungi like *Aspergillus awamori, Penicillium digitatum, Trichoderma spp.* and yeast like *Pchwamiomyces occidentails.*

Mode of Action

These micro-organisms excrete organic acids like oxalic, lactic, succinic, citric, glutamic, glyoxalic, maleic, fumeric, tartaric and L-Ketobytyric which help in weathering rock phosphate and tricalcium phosphate by reducing the particle size to nearly amorphous forms. The organic acids produced by these micro-organisms form stable complex of Al, Ca, Fe and Mg. Besides, they can also mineralize organic phosphate into soluble forms. The micro organisms synthesize fungistatic and growth promoting substances found beneficial in boosting plant growth.

Factors Affecting Efficiency of Micro-organisms

The efficiency of these micro-organisms depends largely upon carbon content and sources in soil, extent of phosphate present in the organic and inorganic sources, particle size of rock phosphate, moisture percentage and temperature of the rhizosphere. Their efficiency is accelerated when organic manures like FYM, leaf mould, green manure etc. are applied.

Crop Response to Inoculation

The seeds of crops like cereals, vegetables and forages were inoculated with phosphate solubilising microbes in soil well supplied with organic manures, rock phosphate and bone meal. The crops inoculated with *B. megaterium* var. phosphaticum (phosphobacteria) gave 5 to 10 per cent higher yield. Potato seed tubers, when treated with *A. awamori, P. striata,* and *B. polymyxa* gave 50 to 60 per cent higher tuber yield than untreated ones.

VESICULAR ARBUSOCULAR MYCORRHIZAE (VAM)

Mycorrhiza is found in symbiotic association between plant roots and fungal mycelia which is common with majority of agricultural crops under temperate and tropical regions. VAM are found in a varied ecological conditions ranging from aquatic to desert environment however, they are not cultivated on nutrient media.

Mode of Action

The fungi use to infect plants root and spread inside them consisting of two structures – first arbuscules which use to help in the transfer of nutrients from the fungus to plant roots and second vesicles (a sac like structure) which store phosphorus in forms of phospholipids. Besides, phosphorus they also accumulate Zn, Cu and S. VAM promote the plant growth and development by stimulating nutrient uptake and by inducing disease resistance in plants.

Classification of VAM

The important VAM, so far identified, belong to four families like – *Endogonaceae* (Endogone), *Glomaceae* (Glomus, Selerocysis); *Acanlosporaceae* (Entrophosphora, Acanlospora) and *Gigasporaceae* (Gigaspora, Scutellispora).

Factor Affecting Efficiency of VAM

The performance of VAM is governed by several factors. It has been observed that the cereals, when rotated with legumes had higher root associated with higher number of VAM propagules than cereals in mono-cropping. Addition of FYM while prolong and continuous application of chemical fertilizers and long fallows reduce mycorrhizal colonization in crop roots. Mustard reduces VAM colonization when grown mixed with wheat. Similarly, conditions congenial for better growth and development of crop plants encourage mycorrhizal colonization however, application of fungicides like benomyl has depressing effect but use of nematicides stimulates colonization. Deep summer ploughing, prolonged water logging, use of soil sterilents and drastic fumigation reduce the number of VAM propagules in soil.

Crop Response to VAM Inoculation

VAM inoculation is not very popular compared to *Rhizobium, Azotobacter* or *Azospirillum* because of inability to produce clean pure inoculum on commercial basis as they have to be produced on living host plants. Results of the field trials indicate increase in yield and phosphate content of the soil. Increase in yield ranges between 8 and 46% over control.

PLANT GROWTH PROMOTING RHIZOBACTERIA (PGPR)

There is a group of bacteria (Rhizobacteria) found in rhizosphere which promotes plant growth due to which they are called as rhizobacteria. The group includes several genera *viz.,*, *Arthobacter, Azotobacter, Cellulomonas, Enterobacter, Pseudomonas, Bacillus, Erwinia, Flavobacterium, Bradyrhizobium, Streptomyces,* and *Xanthomonas.* Besides promoting plant growth, they also induce resistance against several plant diseases in crops.

Mode of Action

These bacteria suppress the growth of harmful bacteria in the rhizosphere or on the root surface and infect the root system where they form colonies later. The mechanism through which they suppress multiplication or even growth of harmful bacteria includes substrate competition and niche exclusion, production of siderophores and antibiotics. Which cause resistance in plants against dexterous rhizophore micro organisms and improves either growth or yield as observed by increase in potato tuber yield by suppressing cyanide producing rhizosphere micro-organisms.

Crop Response to PGPR Inoculation

Wheat, potato, sugarcane, oat, tomato, and groundnut were tested PGPR inoculum *B. subtilis* (A-13) for their productivity and results showed a significant increase in yield ranging between 7 to 33 per cent. However, the performance was found to be greatly influenced by host spp., inoculum spp. microbes present in rhizosphere, environment and soil factors. Detailed information are lacking because of very little work is done on the aspect.

Sulphur Solubilizing Microbes: They are heterotrophic types of bacteria which help in transformation of unavailable organic sulphur compounds present in organic matter or organic manures into sulphate and sulphate is observed by higher plants. These bacteria are known as sulphur oxidizing bacteria which work as follows:

$$\text{- (HS)} + O_2 \xrightarrow[\text{Oxidation}]{\text{Enzymic}} SO_4^{=} + H^+ + \text{Energy (taken by microbes)}$$

Organic Combinations

Effects are being made to isolate these organisms and use them for rapid transformation as Indian agriculture is going to face a threat severe deficiency in near future.

"Perfection comes from practice"

24 Land Utilization and Production Pattern

"Corruption is the mother of vegetation".
"Nature does the business and man is merely the manager."

During the course of any development programme of national resources, land utilization occupies a special attention. At any particular time, it is determined by temperate moisture, topography, soil and physical structure. Obviously, land has the characteristics of its fixity in supply and scarcity. Therefore, land use pattern is directly concerned with the problem arising to the process of deciding upon and carrying out into action the optimum use. In a dynamic world, certain modifications can occur in the existing pattern of land utilization. It deals with the process of putting various types of land to the optimum use. In brief, a close look of the present land use pattern and its trends will go a long way to recommend the scope of systematic and planned shift in the pattern.

CLASSIFICATION OF LAND

It is a process which "assigns each tract of land in an area to its proper class in a system of classes". The classes in the system are defined in terms of the qualities or characteristics with which the classification is concerned. In India, the classification of land has had its roots in agricultural statistics. The collection of such statistics in 19th century when in view of the famines and local shortages of food that were then confronting the country and also with a view to furthering the economic exploitation of the country's resources, it became necessary to know how the available land was utilized. Till 1950, land was classified into the following five broad classes:

1. Area under forests,
2. Area not available for cultivation,
3. Uncultivated lands excluding current fallows,
4. Area under current fallows, and
5. Net area sown.

This classification continued almost intact all these years, the definitional changes made in one or two states being minor, later on, it was realized that the above classification afforded only a broad outline of land utilization in the country, and did not give a very clear picture of the actual area under different categories of land use. Moreover, the data given in the respect of different States of India were also not comparable in certain cases due to lack of uniformity in the methods of classification. The Government of India, therefore, appointed a committee on the co-ordination of Agricultural Statistics in India to work out the details of the annual and periodical inquiries to be organized in the pursuance of the recommendations of the committee. In March, 1950, the following classification of areas for each district of the country was given:

1. Forests,
2. Barren and uncultivable land,
3. Land put to non-agricultural uses,
4. Permanent pastures and other grazing lands,
5. Culturable wastes,
6. Miscellaneous tree crops and groves not included in the net area sown,
7. Current fallows,
8. Other fallows, and
9. Net area sown.

The total of these nine classes would be equal to the geographical area according to village records.

LAND-USE PATTERN BEFORE INDEPENDENCE

According to General Survey of India, the total geographical area of the country was 80.63 crore acres but the land utilization statistics, i.e., the distribution of land among its important uses like cultivation, grazing, forestry, settlement, etc., was available for 72.63 crore acres only, i.e., about 90 per cent of the total area, the rest being of mountains, hills and other unaccessible tracts. In short, till 1946-47 main trends were as follows:

1. The net area sown did not increase appreciably except in Uttar Pradesh. The area growing more than one crop increased by about 2 per cent and the total; cropped area, therefore, showed some increase which however, lagged for behind the rapidly increasing population.
2. Irrigated area increased by about 10 per cent mainly through the extension of canals. The area irrigated from minor irrigation works remained almost static over this long period. It indicates that the new construction at best kept pace with those works which were going out of use for want of repairs or otherwise, e.g., through the extension of canal irrigation.
3. The area under current fallow remained at the level of 1919-20 till the early forties and thereafter it showed some increase, particularly in the cotton growing tracts, possibly because of a sudden decline in the cotton area which was left partly fallow. Hyderabad was the only State which showed a continuous increase in fallows. In the opinion of special committee set up by the Hyderabad Government the apparent increase in the area under fallows was some what an exaggeration.

Land-Use Pattern in India (2000-01)

The land use classification according to 2001-02 agricultural census in India and for 2002-03 is given in the following Table 24.1.

TABLE 24.1: Land Utilization Pattern of India

	1960-61		2000-01		2002-03	
	m ha	%	m ha	%	m ha	%
Total geographical area	328.73		328.73		328.73	
Reporting area	298.46	100	306.08	100	306.06	100
1. Forest	54.04	18.1	69.22	22.6	69.07	22.6
2. Land not available for cultivation	50.75	17.0	42.88	14.0	43.50	14.2
(a) Land put to non-agricultural uses	14.84	5.0	23.66	7.7	24.25	7.9
(b) Barren and uncultivated land	35.91	12.6	19.22	6.3	19.25	6.3
3. Other uncultivated land excluding fallow land						
(a) Permanent pasture and other grazing land	13.97	4.7	10.90	3.6	10.57	3.5
(b) Land under tree crops and groves not included in area sown	4.46	1.5	3.35	1.1	3.36	1.1
(c) Culturable waste	19.21	6.4	13.64	4.5	13.49	4.4
4. Fallow lands	22.82	7.7	25.01	8.2	33.21	10.8
(a) Fallow lands other than current fallow	11.8	3.8	10.11	3.3	11.68	3.8
(b) Current fallows	11.64	3.9	14.90	4.9	21.53	7.0
5. Net sown area	133.20	44.6	141.08	46.1	132.86	43.3
Total cropped area	152.77		186.36		175.65	
Cropping Intensity (%)	114.7		132.10		132.2	
Net Irrigated area	24.66		54.83		53.07	

(Source: Agricultural Statistics at a Glance 2005)

The total geographical area of the country was 328.73 million hectares of which 22.6 per cent was under forest. The land not available for cultivation was 14 per cent of the total area. Barren land unfit cultivation was 6.3 per cent. The uncultivated land excluding fallow land was 9 per cent which included permanent pastures, land under miscellaneous trees and culturable waste lands. The fallow lands accounted for 10.8 per cent of the total area. Only 46.1 per cent of the total geographical area was under cultivation (net area sown). Double cropped area is 44 million hectares resulting in cropping intensity (Total cropped area/net area sown x 100) as 132 during 2002-03 compared to 114.7 per cent in 1960-61. Therefore, over time, net sown area has gone up from 133.20 million hectares in 1960-61 to 141.10 million hectares in 2000-01. This is because more and more of barren lands have been reclaimed. Similarly, the area under forests has increased as the land not available for cultivation went down.

TYPES OF LAND USE

There are nine types of land use such as:

1.	Cultivated land	a. Land used for cultivation of field crops.
		b. Orchard.
		c. Land fit for forage crops.
		d. Fallow.
2.	Grass land or pasture	
3.	Wood land or area devoted to forest.	
4.	Idle land or fit for wild life preservation.	

5.	Public services:	High ways, Temple, Air port, Cremation ground, or burial place
6.	Mining area.	
7.	Industrial site.	
8.	Recreation ground:	Park, garden etc.
9.	Protected area:	Reservoirs, Watershed etc.

LAND-USE PLANNING

Now, it is evident that these resources have been used and misused by a trial and error method and a revision of land and water use methods in these areas in imperative if the hopes to take care of an ever increasing population are to materialize. Therefore, the rational use of land and water resources is slowly becoming a basic policy of the States. Generally, land use planning is simply the process of deciding the suitability of land for various uses on the basis of agro-economic social and climatic conditions including, such things as what crops to grow, in what quantity and in what order. But such a system should be flexible and adjustable to developments in technology such as the evolution of drought or disease resistant crops, labour saving devices, etc.

STEPS IN LAND-USE PLANNING

There are a number of ways in which the various steps in planning might be taken up. In any case, it is necessary at times to proceed from one step to another and then back to the first as adjustments in one phase are required by adjustments in another. The main steps may be summarized as follows.

1. Develop farm management information to show the highest alternative use for the different types and grades of land and water resources, and the size and organization of operating units in which such use or combination of uses can be carried out.
2. Develop the information as to present land use and type and size of existing operating units, mapped in place.
3. Integrate soil survey information into types (farming, grazing, etc.) and grades of those types, showing the potential crop yield or other productivity of such grades, and map this classification material in total detail.
4. Develop detailed map information showing tenure and ownership class, land tax, status, public facilities, local government's financial conditions, and other pertinent data showing the institutional and social pattern which has developed from the present utilization of the land resources and farm operating units carrying on such use.
5. Delimit major areas of 'homogeneity of characteristics' by association and comparison of all the detailed map data, and compile the statistical analysis, for such areas, from the sources from which the maps were developed.
6. By bringing together the analysis of information and the ideas of individuals, groups, and agencies involved, the adjustment procedures and measures which will best fit in the needs of area are developed.
7. For areas where adjustments are indicated, prepare a quantitative summation on types and degrees of such changes. For example, the land classification map may show an area consisting largely of such grades of farm land. The present land use map may show a large part of such land now in cash grain production. The operating units map and other map data on present situation reflects certain aspects of these circumstances.

Thus, it is necessary to appraise inherent characteristics of land e.g., the distribution of soil types, degree of slopes, minerals, and similar surface and sub-surface features including the native vegetation, water resources and features of the climate. This process involves field observations. Since every human activity can be regarded as involving some use of land and characteristics of the present uses of land is also necessary to know why some particular land has been in some particular use. Only then the ideal land-use can be recommended in terms of capabilities.

CHANGE IN LAND UTILIZATION PATTERN

Total geographical area of the country is 328.73 million hectares according to professional survey by Surveyor General of India. But according to village papers prepared by the revenue staff, reporting area came out to 306.25 in 2000-01.

1. **Area under Forests:** It includes actually forested class or administrated as forests under any legal enactment dealing with forests whether state owned or private. Therefore, the figures presented here do not tally with those of forest department because certain lands though no wooded are taken as forest by the Forest Department while these are not treated as such by the Director of land records. The area under forests has increased from 54.05 million hectares in 1960-61 to 69 million hectares in 2002-03 accounting for 22.6 per cent of the total geographical area. Therefore, there has been increase in the area under forests over time but it is still less than the requirements to have ecological balance, since the effective area under forests is only about 23 per cent of the geographical area.
2. **Land put to Non-Agricultural Use:** It includes (a) Barren and uncultivable land and (b) Land put to non agricultural uses. The first category includes absolutely barren and uncultivable land like mountains, deserts etc. which can not be brought under plough except at exorbitant cost. The (b) category includes land covered by buildings, roads, canals and railways or other appropriated for non-agricultural uses. As compared to 35.91 million hectares of barren land in 1960-61, it was only 19.25 million hectares in 2002-03. Therefore, overtime some such land have been put under forests. Due to increasing pressure of population, urbanization and industrialization, the land put to non-agricultural uses has constantly been increasing. It was 5 per cent in 1960-61 and has gone up to 7 per cent in 2002-03.
3. **Other Uncultivated Land excluding Fallow Lands:** This category of land classification includes land available for cultivation either not taken up for cultivation or abandoned later on for one reason on the other and includes.
   a) Permanent pastures and other grazing land.
   b) Land under miscellaneous trees and groves.
   c) Cultivable wastes.

The land under all these sub categories is decreasing over time mainly due to more area being brought under cultivation.

1. **Fallow Lands:** It denotes cultivable land which after abandonment remains uncultivated **over long period called old fallow** and those kept uncultivated during the current year are called **current fallows.** The current fallow land was 11.64 million hectares in 2002-03. Regarding fallow land other than current fallows was 11.18 million hectares in 1960-61 which declined to 10.11 million hectares in 2000-01.
2. **Net Area Sown:** The area under cultivation of various crops was 133.20 million hectares in 1960-61 which increased to 140 million hectares in 1980-81 and

141.10 million hectares in 2000-01 accounting for about 46.1 per cent of the total geographical area. The year 2002-03 was a drought year so there was decline in net area sown.

3. **Gross Cropped Area:** The gross cropped area includes total area under crops during the year. In case different crops are raised on the same land during the year, the same area is counted more than once. The gross cropped increased faster than the net area sown because area sown because area sown more than once has increased from 19.57 million hectares in 1960-61 to 48.74 million hectares in 2001-02. However, gross cropped area was 152.77 million hectares in 1960-61 which rose to 186.36 million hectares in 2002-03.
4. **Net Area Irrigated:** The net area irrigated has shown marked increase overtime touching a level of 54.83 million hectares in 2000-01 against 24.66 million hectares in 1960-61.
5. **Gross Irrigated Area:** In 1960-61, gross irrigated area was 27.98 million hectares which rose to 49.78 million hectares in 1980-81 and further 75.14 million hectares in 2002-03.

OPTIMAL LAND USE PATTERN

In order to optimal land use pattern the following suggestions are made.

1. The **cropping pattern** may be in agreement with the soil and water requirements of the area. Crops standing longer on the land and needing more water could be easily substituted by leguminous crops, Lucerne, pulses, oilseeds, which are ready within shorter span.
2. The areas like West Bengal, Bihar, Orissa, Eastern U.P. and Assam and the eastern coastal regions, which practice **mono-culture,** should resort **to multiple cropping** practices due to irrigation facilities in the area. In area producing jute, efforts should be made for double cropping and particularly introducing high value crops.
3. In the irrigated command areas of North and South India, water consuming crops like **rice and sugarcane;** Cotton, Wheat, Vegetable crops and Fruit could be planted.
4. **Dry farming** should be undertaken in low rainfall area with proper conservation of moisture, and reducing loss of moisture by evaporation through deep ploughing and harrowing. Crops like millets, pulses, oilseeds, and forage crops may be raised. In these regions, drought resistant varieties may be cultivated. Sunflower is another crop which can be successfully adopted.
5. **Hill Slopes** in Northern part of the country may be developed with the cultivation of temperate fruits like peach, plum, apricot, apple, walnut, pears, as they are in great demand. In Southern parts, vegetables like pea, cabbage, cauliflower, etc. may be grown.
6. In arid and semi-arid regions, **fast growing fuel trees** may be planted along road/railway lines and along the canal bunds, and pastures raised for livestock farming could be a useful vocation for the local people.
7. **The Hilly Areas** may be utilized for **mixed farming and raising of sheep and goat herds**. Water harvest technology could be appropriate for this region.

CROPPING PATTERN

Cropping pattern refers to the proportion of area under different crops at a point of time. A change in the cropping pattern means a change in the proportion of area under different crops. Cropping pattern in India is determined by natural factors like climate, soil conditions and rainfall etc. Although, the impact of green revolution is uneven, yet it has led to new pattern. At the beginning of the century, more than 83 per cent of land was put under food crops and about 17 per cent under non-food crops. But, by 1950-51, area under food crops had declined to 74 per cent and that of non food crops it had increased by 26 per cent. This shift in food grains was mainly due to the higher profitability of non-food grains.

Area under Food Grains and Non-Food Grains

A brief description of the area under different crops has been presented in the Table 24.2.

Table 24.2 reveals that in 1950-51, 76.7 per cent area was put under food grains only 23.3 per cent was under non-food grains. In 1970-71, the area under food grains declined to 75.4 per cent while the area under non-food grains went up to 24.6 per cent. Further, the area under food grains declined to 68.9 per cent in 1990-91 and that of non-food grains it went up to 31.1 per cent. Further, in 2001-02, area under food grains was 60.3 per cent while it was 39.7 per cent for non-food grains. This shift in the allocation of area under food crops to non-food crops reflect a change from subsistence cropping to commercial cropping. No doubt, the area under non-food crops as a proportion to total cropped area is increasing, but still there is dominance of food crops over non-food crops in the present cropping pattern of India.

TABLE 24.2: Allocation of Area under Food grains and Non-food grains

Year	Area under Food grains (In percentage)	Area under Non-food grains (in percentage)
1950-51	76.7	23.3
1960-61	75.7	24.3
1970-71	75.4	24.6
1980-81	75.0	25.0
1990-91	68.9	31.1
2000-01	62.8	37.2
2001-02	60.3	39.7

Source: Economic Survey 2003-04.

CROPPING PATTERN AND PRODUCTION TRENDS

Among the major crops, rice, wheat, pulses and oilseeds occupy significant proportion of total cropped area in the country (Table 24.3). The area under rice increased from 34 million hectares in 1960-61 to about 45 million hectares in 2000-01. Similarly, the area under wheat almost doubled during this period. Maize crop showed fluctuation from year to year with positive trend. Bajra crop declined sharply owing to low profitability relative to alternative crops. The area under pulses have remained more or less stable thus necessitating large scale imports to meet the requirements of increasing population. The oil seed crops covered more area overtime but again due to increase in area under irrigation, food grains particularly rice and wheat did not allow fast increase in area under oilseeds. Sugarcane and cotton crops observed significant fluctuations from year to year.

TABLE 24.3: Cropping Pattern showing Area under Major Crops (Million ha)

Crops	1960-61	1970-71	1980-81	1990-91	2000-01	2004-05
Rice	34.1	37.6	40.1	42.7	44.7	42.12
Wheat	12.9	18.2	22.3	24.2	25.7	26.5
Maize	4.4	5.8	6.0	5.9	6.6	7.5
Bajra	11.5	12.9	11.7	10.5	9.8	9.7
Jowar	18.4	17.4	15.8	14.4	9.9	9.1
Gram	9.3	7.8	6.6	7.5	5.2	6.7
Pulses	23.6	22.6	22.5	24.7	20.3	22.5
Oilseeds	13.8	16.6	17.6	24.1	22.8	27.0
Sugarcane	2.4	2.6	2.7	3.7	4.3	8.9
Cotton	7.6	7.6	7.8	7.4	8.5	8.9
Total cropped area	**131.9**	**152.8**	**165.8**	**172.6**	**185.7**	**187.9**

Source: Economic Survey 2003-04.

1. Rice

Rice is the main *kharif* crop of the country. It is transplanted in June and July and matures between September and November. Due to high water requirements, it is grown in the high rainfall areas. Moreover, if it has to depend barely on rainfall, it requires not less than 30 cm rainfall per month. In India, only 9 per cent of the area enjoys 30 cm rainfall per month in the months of July and August.

In view of this the states like West Bengal, Punjab, U.P., Coastal Andhra Pradesh, Bihar, Orissa, Tamil Nadu and Assam have become the most important states producing rice.

Area, Yield and Production Trends

The area, yield and production trends in respect of rice have been given in Table 24.4. Rice is the most important food grain crop in India. In 1950-51, it was grown on 30.8 million hectare which in 2000-01 increased to 44.7 million hectares and to 42.1 million hectares in 2004-05. Thus, it is accounted for about 30 per cent of the total area. It reflects that area under rice has been going up, the rate of increase was fast in the early period and has slowed in the later period.

TABLE 24.4: Production Trends of Rice

Year	Area (million hectares)	Production (million tones)	Yield (kg pr hectare)
1950-51	30.8	20.6	668
1960-61	34.1	34.6	1013
1970-71	3.7.6	42.2	1123
1980-81	40.1	53.6	1336
1990-91	42.7	74.3	1740
2000-01	44.7	85.0	1901
2004-05	42.1	85.3	2026

Source: Economic Survey, Various Nos.

Total production of rice which was 20.6 million tones in 1950-51 increased to 34.6 million tones in 1960-61 and further to 42.2 million tones in 1970-71. Again in 1980-81, production has increased to 53.6 million tones and 93.3 million tones in 2001-02. As regards the yield, it has increased from 668 kg per hectare in 1950-51 to 1123 kg per hectare in 1970-71 and 1740 kg per hectare in 1990-91. In 2004-05, production stood at 2026 kg per hectare. It is expected to rise to 2051 kg per hectare.

Future Strategy

In view of increasing production of rice, the following measures may be taken up:

1. To increase the area under HYV, emphasis needs to be given to develop hybrid rice varieties.
2. To increase the stress on export of rice.
3. To take intensive steps to raise the yield of upland rice by waterharvesting with the aid of package practices.
4. To make proper use of rice straw and avoid air pollution by its burning.

II. Wheat

Wheat is the staple food crop of India. Warm and cold climate is required at the time of its sowing. Hot climate is needed for its ripening. The wheat region is responsible for 70 per cent of the area and 76 per cent of the production. It includes Punjab, Uttar Pradesh, Haryana, flanked by Rajasthan and Gujarat in the West region and West Bengal and Bihar in east region. This area enjoys extensive irrigation system ranging from 97 per cent in Punjab to 70 per cent in Madhya Pradesh.

Area, Yield and Production Trends

Table 24.5 highlights the trends of area, yield and production for wheat. Wheat was grown on only 9.8 million hectares in 1950-51. After the Green Revolution, it has consistently improved its position. In 1980-81, the area under wheat was 22.3 million hectares which steadily increased to the extent of 24.2 million hectares at 1990-91 and further to 26.5 million hectares in 2004-05. The production of wheat has increased from 6.5 million tones in 1950-51 to 11.0 million tones in 1960-61 and again to 36.3 million tones in 1980-81. Further, it has increased to 55.1 million tones in 1990-91 and expected to reach to 72 million tones in 2004-05. Considering the yield per hectare it has increased from 663 kg per hectare in 1950-51 to 1307 kg per hectare in 1970-71 and to 1630 kg per hectare in 1980-81. Further, in 1990-91, it was recorded to be 2281 kg per hectare and is expected to increase to the level of 2718 kg per hectare in 2004-05. During the last few years, it appears that the yield of wheat has stagnated at around 2700 kg per hectare.

TABLE 24.5: Production Trends of Wheat

Year	Area (million hectares)	Production (million tones)	Yield (kg per hectare)
1950-51	9.8	6.5	663
1960-61	12.9	11.0	651
1970-71	18.2	23.8	1307
1980-81	22.3	36.3	1630
1990-91	24.2	55.1	2281
2000-01	25.7	69.7	2718
2004-05	26.5	72.0	2718

Source: Economic Survey, Various Nos.

Following steps may be taken to increase the production of wheat:

1. To raise the area under HYV with assured irrigation facilities.

2. To use more chemical fertilizers.
3. To use better transfer of technology with adequate extension service.
4. To boost export of wheat from the country.

III. Coarse Cereals

The coarse cereals include maize, jowar, bajra, etc. These crops are generally sown in June-July and harvested in September-October. The coarse cereals are important in areas where irrigation facilities are not well developed, for example, Maharashtra, Karnataka, Rajasthan, Gujurat, Bihar, Madhya Pradesh, U.P. and Andhra Pradesh. The area under coarse cereals has declined overtime. It was 44.96 million hectares in 1960-61 which declined in 2002-03 to 26.3 million hectares. The detailed information of production of coarse cereals has been given in Table 24.6

The total production has increased from 15.38 million tones in 1950-51 to 32.70 million tones in 1990-91. However, in 1995-96 it has slide down to 29 million tones in 2004-05. The yield per hectare has increased from 433 kg per hectare in 1950-51 to 900 kg/ha in 1990-91. During 2004-05, it rose to 1168 kg/ha. Thus, it is clear from the above analysis that the increase in production is mainly due to increase in yield. Because these are mostly rainfed crops and yield is a gamble of monsoon.

Since coarse cereals are less profitable as compared to wheat and rice under irrigated conditions, so with the increase in area under irrigation, the area under coarse cereals is being shifted to rice and wheat crops.

TABLE 24.6: Production Trends of Coarse Cereals

Year	Area (million hectares)	Production (million tones)	Yield (kg per hectare)
1950-51	37.67	11.38	433
1960-61	44.96	23.74	528
1970-71	45.95	30.55	665
1980-81	41.78	29.02	695
1990-91	36.32	32.70	900
2000-01	30.26	31.08	1027
2004-05	29.08	33.92	1168

Source: Economic Survey, Various Nos.

IV. Oilseeds

The nine important oilseed crops include groundnut, rape seed and mustard, sesame, caster seed, linseed, soybean, sunflower and safflower. Heavy expenditure has to be incurred by the country for import of edible oils, which amounted to Rs 11974 crores in 20003-4.

Area, Production and Yield

Table 24.7 brings out the trends of area, production and yield of oil seeds. The area under oil seeds was 10.73 million hectares in 1950-51 increased to 24.15 million hectares in 1990-91. The area under this crop has rosen to 27.00 million hectares during 2004-05.

TABLE 24.7: Production Trends of Oilseeds

Year	Area (million hectares)	Production (million tones)	Yield (kg per hectare)
1950-51	10.73	5.16	481
1960-61	13.77	6.98	507
1970-71	16.64	9.63	579
1980-81	17.60	9.37	532
1990-91	24.15	18.61	771
2000-01	22.77	18.44	810
2004-05	27.00	26.10	967

Source: Agri. Stat. at a glance, 2005)

The production has increased from 5.16 million tones in 1950-51 to 18.61 million tones in 1990-91 which further declined to 15.06 million tones in 2002-03 but increased to 26.10 million tones in 2004-05. The yield per hectare has increased from 481 kg per hectare in 1990-91 and 967/ha in 2004-05. Thus, year to year fluctuation in yield add risk in these crops.

V. Pulses

Pulses deserve the special attention on the food front in the country. The area under pulses has varied between 19.1 and 24.7 million hectares and production between 8.4 and 15.2 million tones. Generally, these are grown in the rainfed conditions. But due to their poor returns to the cultivators in comparison to the competing crops, the pulses in the country are grown on marginal lands. However, the severe shortage of pulses in the country has resulted in large scale imports which was worth of Rs 2251 crores in 2003-04.

Area, Production and Yield

Table 24.8 shows the trends of area, production and yield of pulses. The area under pulses increased from 19.1 million hectares in 1950-51 to 23.6 million hectares in 1960-61. However, in 1970-71 it declined to 22.6 million hectares and again to 22.5 million hectares in 1980-81. But, in 1990-91, the area under pulses again stood at 22.5 million hectares in 2004-05.

TABLE 24.8: Production Trends of Pulses

Year	Area (in million hectares)	Production (in million tones)	Yield (kg pe hectare)
1950-51	19.1	8.4	441
1960-61	23.6	12.7	539
1970-71	22.6	11.8	524
1980-81	22.5	10.6	473
1990-91	24.7	14.3	578
2000-01	20.3	11.0	544
2004-05	22.5	15.4	595

(Source: Agri. Stat. at a Glance, 2005)

Regarding the production of pulses, it was recorded to be 12.7 million tones in 1960-61 as against 8.4 million tones in 1950-51. Thereafter, it remained at low ebb, for two consecutive decades, i.e., 11.8 million tones in 1970-71 and 10.6 million tones in 1980-81. In 1990-91, the production of pulses was 14.3 million tones. But the production of pulses increased to 15.4 million tones in 2004-05. The yield per hectare of pulses was 441 kg /ha in 1950-51 which in 1960-61 increased to 539 kg/ha in 1960-61. In 1970-71, it was reduced to 524 kg/ha and further to 595 kg/ha in 2004-05.

VI. Cotton

Cotton is sown from mid April to mid May and Picking starts in the month of September. It occupies an area about 9 million hectares. The state of Maharashtra alone occupies about 36 per cent of the total area under cotton in the country. Gujarat shares 21 per cent, Karnataka 13 per cent and Madhya Pradesh 10 per cent of the total cotton growing area. In a sense, these four states are taken together account for 80 per cent area under cotton. Moreover, most of the cotton growing areas in the country are under the high to medium rainfall zone.

Area, Production and Yield

The picture of area, production and yield in respect of cotton is given in Table 24.9. The area under cotton has increased to 7.6 million hectares in 1970-71 from 5.9 million hectares in 1950-51. However, in 1990-91, it has declined to 7.4 million hectares in 1980-81 and further; in 2004-05 it touched 8.9 million hectares. Regarding the production of cotton it was recorded to be 5.6 million bates (1 bale= 170 kg of cotton lint) in 1960-61 as against 3.0 million bales in 1950-51. Moreover, in 1990-91, it was 9.8 million bales and 12.9 million bales in 1995-96. In 2000-01, it declined to 9.5 million bales due to decline in area and yield. The production showed a record production of 17 million bales in 2004-05. The yield per hectare of cotton was 88 kg of lint in 1950-51 which in 1960-61 increased to 125 kg. In 1995-96, it was 242 kg and showed a fall to 190 kg per hectare in 2000-01. The decline in cotton yield is due to severe attack of pests. Now it has gone up to 324 kg/ha mostly due to introduction of Bt cotton varieties.

TABLE 24.9: Production Trends of Cotton

Year	Area (In million hectares)	Production (in million bales)	Yield (Lint kg/ha)
1950-51	5.9	3.0	86
1960-61	7.6	5.6	125
1970-71	7.6	7.0	106
1980-81	7.8	7.0	152
1990-91	7.4	9.8	225
2000-01	8.5	9.5	190
2004-05	8.9	17.0	324

(Source: Agri. Stat. at a Glance, 2005)

VII. Sugarcane

Unlike other crops, sugarcane is an annual crop. It is grown mostly in the states of Uttar Pradesh, Bihar, Punjab, Haryana, Karnataka, Tamil Nadu, and Andhra Pradesh. The state of Uttar Pradesh alone shares more than 50 per cent area under sugarcane. The production of sugarcane has to be associated with the processing through sugarcane mills apart from suitable agro climatic conditions.

Area, Yield and Production

Production trends for sugarcane have been shown in Table 24.10. The areas under sugarcane increased from 1.7 million hectares in 1950-51 to 2.6 million hectares in

1970-71 and again to 3.7 million hectares in 1990-91. By 2000-01, it reached to the level of 4.3 million hectares and in 2004-05 it fell to 3.6 million hectares but again marginally rose to 4.4 million hectares in 2001-02.

The production of sugarcane was 57.1 million tones in 1950-51 increased to 126.4 million tones in 1970-71. It has again increased to 241.0 million tones in 1990-91 and 297 million tones in 2001-002. In term of yield per hectare of sugarcane it was 33 tones in 1950-51 which almost doubled in 1990-91 and was recorded to be 69 tones per hectare in 1970-71 but declined to 64 tones per hectare in 2004-05.

TABLE 24.10: Production Trends of Sugarcane

Year	Area (in million hectares)	Production (in million tonnes)	Yield (tones/ha)
1950-51	1.7	57.1	33
1960-61	2.4	110.0	46
1970-71	2.6	126.4	48
1980-81	27	154.2	58
1990-91	3.7	241.0	65
2000-01	4.3	296.0	69
2004-05	3.6	232.3	64

(Source: Agri. Stat. at a Glance, 2005)

DETERMINANTS OF CROPPING PATTERN / CROP SYSTEM

The cropping pattern in any country is determined by a number of factors like natural, economic, historical etc. In addition to this, Government Policy can also affect changes in the cropping pattern. In our country, factors responsible for cropping pattern are under stated.

1. **Natural Factors:** Natural factors are the most important determinants of cropping pattern. These factors pertain to the physical characteristics and the natural endowments of a region. For instance, in the areas having adequate rainfall, rice is grown while the areas having low rainfall, choice will naturally fall upon to produce jowar and bajra. Similarly, the soil of the Indo-Gangetic plains is suitable for the growth of rice and wheat. The temperature, humidity, soil etc. are other factors due to southern states are more suitable for rice, jute etc. while the northern states are more suitable for wheat crop.
2. **Economic Factors:** Economic factors are playing the significant role in determining the cropping pattern in a country. The following are some of the economic factors influencing the cropping pattern in India.

    a) ***Comparative yield level and stability in yield:*** Some crops have higher yield potential in specific areas and yields are also stable year after year and thus less risky. So such crops are preferred over other crops e.g. tea in Assam.

    b) ***Relative price and price stability:*** The higher and remunerative price of some commodities, less seasonal variation and assured prices of some commodities cause increase in area under such crops. For example, sugarcane is preferred in sugar mill area.

    c) ***Availability of farm inputs:*** Cropping pattern also depends upon the farm inputs available *viz.*,, seeds, fertilizer, controlled and assured water supply through irrigation etc. and among these irrigation is the most important.

3. **Tenancy System**: Some particular crops are grown due to historical regions. If the land is divided into a number of small plots with ownership vested in numerous small and marginal farmers, the tendency will be to grow food crops to meet domestic requirements. Contrary to it, if the ownership is vested in landowners, the tendency will be to produce more food crops.
4. **Technical Factors:** The cropping pattern also influenced by the technical factors such as nature and capacity of irrigation facilities, the entire method of cultivation being followed from the traditional period is bound to change. With this, new crop rotation system and better crops can be followed. In India, due to the extension of irrigation facilities, the cultivation of sugarcane, tobacco, oilseeds etc. have increased substantially. Moreover, with the availability of irrigation facilities, even double or triple cropping is also successfully done. Again, in the absence of irrigation facilities in some other parts of the country, the concept of "dry land farming" is also gaining its importance in recent years.
5. **Social Factors:** The prevalent social environment also influences the cropping pattern. For instance, in the pre-independence, farmers were illiterate and had narrow out look. They were bound to traditional culture and the same cropping pattern was continued by the successive generations. But after interdependence due social awareness, farmers became more and more responsible to price changes. The new techniques of production have changed to the outlook of the cultivators.
6. **Irrigation Facilities and Soil:** Irrigation facilities and type of soil in a particular region affects the cropping pattern. A farmer can not grow a crop ignoring climate, soil and irrigation facilities.
7. **Size of the Farm:** The farm size in India is the smallest than that in most tropical countries. Thus, our cropping pattern is also different from those countries. In a way, cropping pattern on small farms is more in form of food grain crops while the large farmers prefer commercial crops.
8. **Religious Factors:** Tobacco growing is not preferred by Sikh farmers. Raising of pigs is not preferred by Muslim farmers due to religious factors.
9. **Government Policy:** Cropping pattern may also be influenced by government policy undertaken in the form of administrative and legislative measure. Supply of inputs by the government, intensive scheme for various crops, government campaign like grow more food or any legislative provision by the government, transportation and marketing provision also help to finalize the cropping pattern in the country.
   a) Like U.S.A., U.A.E., and other countries, the legislative compulsion for structuring a desired cropping pattern in different regions of the country are suggested.
   b) Government may appoint District Planning Officers who introduce crop planning in every season considering various seasonal changes and changes in other factors such as yields, prices and demand etc.
   c) The Government should also provide adequate transport and marketing facilities and also make provision for consolidation of holdings.
10. **Pest and Diseases:** Generally, farmer tries to grow only those crops which are ecologically practicable and they are not subject to pests and diseases in the region. Therefore, certain crops subject to serious pests and diseases in a specific area are normally discouraged.
11. **Provision of Infra-structure:** Another factor responsible for choice of cropping pattern of a region is the availability of infra-structure like processing, transport and communication facilities, credit facilities etc. It is seen that these facilities

are commonly available in towns and adjoining to these towns, thus they intend to cash crops as they give highest returns.

12. **Influence of Technology:** Cropping pattern is also influenced by the latest and modern techniques of production. Since the inception of green revolution, cropping pattern has radically been changed. In a sense, high-yielding varieties of seeds, irrigation, chemical fertilizers and mechanization has provided a base for multiple cropping.
13. **Export and Import Possibilities:** The choice of cropping pattern largely depends upon the comparative advantage of these crops. The products which can be improved at cheap rates are discouraged in the cropping pattern. On the other hand, the possibilities of export increases the scope of such crops

The National Commission on Agriculture quote, "A wide range of factors *viz.,*, nature of soil, climatic conditions, size of holdings, availability of various inputs like irrigation, fertilizers, manures and pesticides and processing facilities, comparative economics of alternative crops etc. have a bearing on the individual farmer's decision to grow a particular crop. These decisions can be influenced and modified to a considerable extent through deliberate policy measures aimed at including changes in physical, economic and institutional environments of agriculture, or offering various incentives by way of input supplies, processing and marketing facilities". It means that, these are factors which affect the cropping pattern or the farmer's decision to grow a particular crop.

Similarly, NCAER has given many crucial suggestions for improving the cropping pattern. They are stated as under:

1. The present way of haphazard allocation of area to different crops by individual farmers is not in the best interests of the farmers or of the country. Thus, the crop plans drawn out for different irrigation projects are not strictly adhered to, because individual farmers may not like to change their existing crop pattern either due to inertia or to ignorance, resulting in a wastage of capital, land and irrigation resources. As persuasion is slow remedy, it favours legislative compulsion (as in U.A.R. arid USA) to bring about a cropping pattern in different areas.
2. District Planning Officers may be appointed who could plan the crops for each season according to changes in the seasonal and other factors, like prices, yields, input supply etc. The farmers should work in close cooperation with departments of agriculture, irrigation, etc. so that chalked out programmers may be implemented.
3. An Agriculture Mechanization Corporation should be set up where the size of holdings is too large and where the average farmer is unable to manage with hired labour. The corporation should advance the farmers the requisite machinery, and
4. The Government should give the greatest importance to the promotion of transport, marketing facilities and consolidation of holdings.

FUTURE OF CROPPING PATTERN

With the advent of new agricultural strategy and green revolution, extensive cropping pattern is giving a way to intensive cropping. It has provided an ample opportunities to the Indian farmers to raise manifold production. This is possible by making easy availability of chemical fertilizers, plant protection, improved techniques of production, assured irrigation with appropriate price policy.

With the rapid increase in population, size of holding has been decreased and expected to decline continuously in future also. This is a serious challenge in years to come. Therefore, following factors are greatly responsible for the present form of cropping pattern and land use pattern.

1. Comparative advantage of crops.
2. Increased irrigation potential.
3. Increased use of fertilizers and pesticides.
4. Better crop varieties and quality of seeds.
5. Higher level of production technology for major cereals, cotton and sugarcane, etc.

Besides, the production trends and the cropping pattern of our country would be influenced by the following decisions of the Government of India.

1. A steady growth of food grain production to meet the growing needs and a substantial increase in pulse production is to improve the nutritional quality of the diet of the people.
2. To aim at self sufficiency in oilseed and pulses production to eliminate import of edible oils and pulses.
3. To increase production of export oriented crops like tea, coffee, tobacco, cashewnut, species, jute, cotton, sugarcane, flowers, basmati rice, durum wheat, fruits and vegetables.
4. Efforts may be made to use land in optimum way. This can only possible when such crops are grown towards which land is more responsive.
5. Various facilities like irrigation, crop insurance, subsidies for input, marketing, finance demonstration of the profitability of different crops, and storage may be provided to the cultivators. The future changes of cropping pattern requires basic conditions to be fulfilled as:
   a) Output of all existing crops should be increased so as to meet the consumption needs of the people.
   b) There should be a shift towards high value crops, which in turn will obviously add to the total value of output. Thus commercial crops may be preferred.
   c) Efforts should be made to shift the area from coarse food grains to superior quality of food grains such as basmati rice.
   d) The production of these crops should be encouraged which have world demand.
   e) Land use plans for plain and hilly areas, salt affected, water logged lands and dry area may be properly drawn.
   f) Research on soil fertility, water use, dry farming may be done continuously.
   g) Necessary provision of financial help may be made available.
   h) The processing units should be supplied with regular raw material.

DRY LAND FARMING

Dry land farming means the introduction of agricultural operation in the arid and semi arid land. Therefore, steps have been taken to start special type of agricultural operation in those dry lands which remained barren at the earlier stage.

Dry land farming in India is thus defined broadly to cover rainfed agricultural operation dominated by low water requiring crops in those arid and semi-arid tropical regions of the country. To start dry land farming, it requires alternate farm seasons of cropping season and fallow season. In order to catch early showers, first the ploughing and harrowing operations are done. The second ploughing and harrowing operations are done during the rainy season to open the soil to the relatively heavier rain. The ploughing and harrowing operations are done for preventing loss of land moisture by evaporation. The ploughing operation uproots weeds which absorb moisture from the soil and the harrowing operation usually prepares a dry, dusty soil on top level which will act as blanket for preventing evaporation. In this type of farming, special efforts are made by the farmers for conserving soil moisture and also for using limited rain water to a maximum extent.

Future of Dry Farming

The main features of dry farming are:

1. Low agricultural productivity,
2. High degree of vulnerability to vagaries of nature,
3. Instability in output,
4. Complete dependence on rains,
5. Application of traditional agricultural methods and
6. Requirements of low assets level.

Thus, the dry farming is basically considered as a subsistence cultivation which has total dependency on rains.

Significance

The importance of dry farming is increasing year by year in our country. With the continuous growth in the size of population, the gap between the requirements and the supply of agricultural output is increasing gradually leading to a crisis in respect of both food and non-food crops. The World Bank in its recent study observes that at the turn of the country, the total requirements of food would increase to about 235 million tones which would again require an additional production of food grains to the level of 65 million tones over the present level of production. With the present initiative, the production of food grains would increase at best by another 35 million tones, leaving a gap of 30 million tones. Under this situation, the only way open to the country is to introduce and promote better dry farming which has a huge potential throughout the country.

The development of technology for dry land farming has remained neglected. Considering the potential and also its importance, the Indian Council of Agricultural Research (ICAR) evolved dry farming technology for its application in those areas which are experiencing inadequate rainfall even in the rainy season. In India, a huge area of land remained fallow for want of assured water supply required for agriculture. The Planning Commission has also made an estimate of such arid lands where dry land farming technology can be adopted. As per this estimate, a total of about 37 million hectares of such arid lands, spread over 92 districts in different regions of the country can be brought under the dry land farming technology which could serve the twin purpose of controlling soil erosion, as well as increasing agricultural production and productivity.

"Your aspirations are your possibilities"
"In every progressive economy, there has been a steady shift of employment and investment from the essential 'Primary' activities to secondary activities of all kinds and to a still greater extent into tertiary production."
A. G. B. Fisher

25 Modern Concepts of Agricultural Research

"Pure science, undertaken without regard to application to human needs, is usually ultimately of direct benefit to mankind"

Irving Langmuir

"A country which wants to progress, wants to progress in a hundred ways We have, therefore, to take into consideration the order of preference what is first, what is second and what is third. There are so many things we want to do in India, and we want to do them quickly and passionately. The question of finding the proper way to reach a certain goal becomes important".

Jawaharlal Nehru

Modern agriculture is **scientific**. Man imitated nature, maneuvered the soil fertility and productivity, crop architecture, developed artificial source of nutrients with large concentrations in small quantities and readily available form to the crops, pesticides, to protect the crops from pests and rodenticides and herbicides to turn down the competition of other vegetation with the crop. Several equipment and machineries were developed to till the land, sow the seed, plant the seedlings and apply fertilizers, spray the chemicals, harvest the crop and trench the produce. The water-bodies were tapped to irrigate larger area of the crops. Dry land technologies wre developed to utilize the natural resources efficiently, conserve the soil and reduce the runoff losses and minimize the risk of crop failures due to insufficient and irregular distribution of rainfall. Crops and cropping systems were indentified for site specific situations. Thus, the past extensive agriculture was transformed to the exploitative intensive agriculture.

The ancient agriculture was not sustainable. India encountered 14 famines from 11 to 17 century. There were 12 famines and 4 severe scarcities from 1765 to 1855. Several famines and scarcities again prevailed between 1860 and 1908. Hunger and malnutrition was destined even in the first half of the 20th century. The human population was low. A total of 238.4 million(m) people inhabited in 1901. The population increased exponential in the second half of the century and crossed an ever record of 1,000 m by 2000.

TRANSFORMATION FROM OLD TO NEW CONCEPTS OF AGRICULTURE IN INDIA

The modern concepts of agriculture in India commenced during 1950s. The food grain crops were cultivated on an area of 97 m ha producing 50.8 m tones (t) in 1950. The per caput availability of food grains was only 395 g/day. To sustain the need of food for the increasing population, their cultivation was extended to 124 m ha by 1970. Increase in irrigation was another effort to meet the increasing food requirement. The gross irrigated area increased from 22.5 to 37.0 m ha by this time. The fertilizers were introduced and 69,000 t were applied to the crops in 1950. Their consumption increased to 2.18 m t in 2 decades. The high yielding fertilizer responsive wheat varieties were introduced in 1963 from Mexico and rice varieties from IRRI in 1964. These interventions more than doubled the food grain production to 108 m t in 1970. Thus, the population increased by 21.79 %, while food grain production increased by 115 % during this 20 year period. This revolutionary achievement relaxed the country from the clutches of dependence on other countries for import.

Thomas Malthus (1798) predicted disaster due to adverse imbalance between human number and their capacity to produce food, when global population was only 360 m. The paddock brothers predicted doom's day for India due to recurrent famines. Several foreign experts in 1960s concluded that India has no option except to become the home of serious and recurrent famines. The famous economist Paul eldritch in 1966 predicted that sometimes between 1970 and 1985 the world will undergo vast famines and millions of people will starve to death.

The modern concepts of agriculture disproved all these forewarnings and India achieved the **Green Revolution.** Indian food grain production increased progressively despite the cultivated area remained constant since 1970. This crucial limitation was fairly overcome by the modern technologies to increase the productivity/unit area and thereby the total food grain production. New ideotypes with canopy architecture to capture more sunlight and CO_2, the root mass with high CEC-effective to respond to the artificial nutrients through fertilizers of partitioning the more dry matter of the plant into grains vis-à-vis the increasing irrigation potential which made intensive cropping feasible added to increased level of food production to match the need for the increasing population until the turn of the century. The per caput availability of food grains was raised to 460 g/day. The high yielding varieties covered 97 % of the area by 1997. The fertilizer consumption increased steadily to 20.94 m t and gross irrigated area to 82.68 m ha by 2005.

NEW CHALLENGES OF MODERN AGRICULTURE

Currently, decline and deterioration of natural resources emerged as a serious concern for the sustainability of food production. The yield plateaus and several new generation problems of soils are considered as the silent ill effect of over exploiting the resources through the Green Revolution technologies. The soil have become sick for sustained crop production and the environment unsafe for human health Some of the widely recognized problems are the low nutrient status of the soils due to intensive cultivation. The crops are unhealthy with prominent symptoms of multi nutrient deficiencies due to the indiscriminate quantities of fertilizers applied. The continuous and massive application of fertilizer without the organic manures has created lifeless and dusty soils which are more prone to erosion. The use of heavy machinery created the problem of soil compaction which reduces the ability of the soil to absorb water and reduce the crop yield. The water runoff the compact soil and contribute to erosion. The faulty irrigation practices led to the development of barren lands by accumulating salts to toxic level and creating water-logged conditions. The over exploitation of irrigation by shifting from centrifugal pumps to submersible tube well to extract water lowered the depth of water table. The excessive tilth eroded the soils leading to loss of nutrients, reduced the organic carbon through the evolution of CO_2 destructed the stable aggregates and disrupted the void proportion for proper root growth and development. High nitrate content in ground water has been detected in intensively cultivated rice-wheat system. The indiscriminate use of insecticides created resurgence with no satisfactory control of pests viz., *Helicoverpa armigera* on cotton and red gram, pod borer in rice etc. and tolerance to chemical toxicity of the herbicide *Isoproturon* to the weed *Phalaris* minor in wheat and butachlor to wrinkle grass in rice. The balance between the harmful insects and predators has disrupted. The pests continued to evolve and through the process of natural selection build up resistance to the chemicals. As a result yesterday' pests have become super-pests to withstand large doses to 5 or more pesticides and are less likely to be troubled by the predators. Chemical residues in food increased the incidence of many types of cancer particularly among the farmers. Over 40,000 Third World farm workers are killed each year and up to one million made ill or permanently injured due to exposure or mishandling of the agro chemicals. Rodent control in the fields by the regular use of zinc phosphide @ 2 % in cereal baits has become ineffective in Punjab and Haryana. These are some of the classical examples to cause severe yield losses in crops. The eutrophication of water bodies due to the accumulation of excesssalts in soil due to heavy dose of fertilizer applications rendered soils less productive for cultivation. The factor productivity for 1 kg fertilizer was 35 kg food grain during Green Revolution. This has come down to < 12 kg. Therefore, to get the same response to fertilizers as in the past, currently higher fertilizer dose as needed. Surekha (2007) reported that the soil fatigue in intensively cultivated irrigated rice lands led to the use of 40 more nitrogen to produce the same quantity of rice as 10 years ago. The cultivation of Bt cotton was over whelming to the farmers. The *Helicoverpa* larvae were fairly controlled until 90 days after sowing the crop. This technology saved the farmers from committing suicides. But, in 2007-2008, the pest became virulent early at 80 days in Andhra Pradesh. This threat appears to be due to the silencing effects of the genes. A hitherto major pest of custard apple, mango, and red gram – the mealy bug infestation emerged as a new and serious pest on Bt or non-Bt cotton. The application of confidor lost its toxic effect on the sucking pests. There are several such examples that the modern technology is challenged and demand a revision of human intelligence to search alternative options.

SYSTEMS RICE INTENSIFICATION (SRI)

The resource conservation technology for rice cultivation was evolved by Fr. Henrie Laulanie at Madagascar during 1983. He migrated from France in 1961 and settled at Madagascar to practice and profess agriculture. The technology showed a dramatic improvement in rice yield to harvest 2-fold more grain with just half the water. Also, it conserved the seed. This yield improvement was possible with only 5-10 kg seed against convention of sowing 50-100 kg/ha. The technique is also eco-friendly. Since, the fields are not flooded but kept moist 'methane' the major green-house gas is not produced. This was not known outside Madagascar until 1997 as it sounded too good to be true.

The farmers harvesting 2 t grain realized even 8 t/ha. Its potential was tested in China, Indonesia, Cambodia, Thailand, Bangladesh and Sri Lanka. The technology spread successfully in 18 districts of Sri Lanka. The first International conference on SRI was held in China in April 2002. It was known in India during 2000. In Andhra Pradesh, it was introduced in 2002. Several large scale demonstrations were conducted in the farmers' fields. Systematic research was then conducted in the research farms. There was no appreciable increase in yield compared to the conventional method. However, water conservation was the advantage. To produce 1 kg grain 2,710 litres of water was needed through this novel approach in contrast to the 3,720 litres through the conventional method. The labour do not have the skill to pull the 8-12 day-old tiny and delicate seedlings from the nursery with the seed intact and soil adhering to the roots without damaging the root system. It was not possible to be extremely cautious to irrigate the field frequently and shallow to keep the soil moist without flooding or allowing it to dry to develop hair cracks. The high intensity rains in the *kharif* season may submerge the seedlings, uproot and wash them away easily. The frequent failure of electricity is not uncommon. This leads to drying of soil. The depressions and elevations due to imperfect leveling of field create water logged and dry patches. Most of the farmers with initially tested this concept shirked it off since 2006.

"Rice plays a unique role in providing food to the majority of the world population. Timely harvesting ensures optimum grain quality, higher market and consumer acceptance, since the grain is less likely to break when milled. When harvesting is not done in time, grain may be lost due to shattering, lodging, damage by rats, birds and insects etc."

ORGANIC FARMING

A network of experiments initiated in India to study the feasibility of reverting to organic from the chemical agriculture. The results of experiments conducted at three different ecological regions of Andhra Pradesh on maize-onion, turmeric-sesame and rice-rice cropping systems are discussed.

The experiment on maize-onion cropping system at Rajendranagar (Table 25.1) indicated that the recommended dose NPK to maize (120:60:60 kg/ha) and NPKS to onion (150:60:60:40 kg/ha) produced more yield than the organic treatment by the application of FYM, vermin compost and neem cake each applied at 1/3 recommended rate of N. The integrated supply of nutrients through 50 % RDF with 50 % N equivalent through FYM in general produced yield similar to RDF. The soil recorded considerable improvement in OC, available P and K due to the external input of nutrients compared to the initial level after 4 years. The OC increased from the initial 0.36 to 0.49 % due to the continuous application of fertilizers and also by the integrated supply of nutrients. But, high value 0.70 % OC was recorded by the organic source of nutrient supply.

The yield responses and the effect on soil fertility after 5 years of turmeric – sesame cropping system at Rudrur in response to different sources of nutrients are presented in Table 25.2. The rhizome yield of turmeric and seed yield of sesame reduced due to the continuous addition of organic source of nutrients in all the 5 years. The application of 50 % RDF with 50 % N equivalent through FYM largely compensated the yield loss due to the organic nutrients alone. Maximum yield was realized by continuous application of fertilizers @ 180:75:120 kg/ha NPK to turmeric and 30:60:40 kg/ha NPK to sesame. The organic carbon content was maintained to the initial level even after 5 years of continuous fertilizer additions while, it reduced to the organic farming.

TABLE 25.1: Influence of source of nutrients on yield (T/ha) in maize-onion Cropping system & soil chemical properties after 4 years at R'nagar

Treatment	2004-2005		2005-2006		2006-2007		2007-2008	
	Maize	Onion	Maize	Onion	Maize	Onion	Maize	Onion
50% NPK-50% FYM	3.15	14.37	2.20	15.14	4.17	9.33	4.36	18.48
1/3 N FYM + VC+NC	1.90	8.77	1.62	9.20	1.94	6.31	2.18	9.02
RDF Initial	4.29	10.86	2.23	14.41	3.39	10.23	4.78	19.58

pH	EC dS/m	OC%	Av. P (kg/ha)	Av. K. (kg/ha)
7.4	0.35	4.49	13.3	349
7.4	0.36	0.70	34.6	377
7.2	0.39	0.49	34.6	359

TABLE 25.2: Influence of source of nutrients on yield (t/ha) in turmeric- sesame system and soil chemical properties at Rudrur

Treatment	2003-2004		2004-2005		2005-2006		2006-2007		2007-2008	
	Termeric	Sesame	Termeric	Sesame	Termeric	Sesame	Termeric	Sesame	Termeric	Sesame
50% NPK-50% FYM	5.02	0.79	2.70	4.8	4.31	5.0	4.05	0.51	4.10	0.52
1/3 N FYM + VC+NC	4.52	0.65	2.31	3.9	4.02	4.1	3.28	0.45	3.42	0.46
RDF Initial	5.43	0.76	3.05	7.6	4.43	7.6	4.23	0.58	4.57	0.60

pH	EC dS/m	OC%	Av. P. (kg/ha)	Av. K (kg/ha)
7.8	0.34	0.41	35.8	230
7.7	0.30	0.30	45.6	242
7.8	0.47	0.47	41.4	246
7.3	0.48	0.46	41.0	225

The production trends in rice-rice cropping system in the coastal ecosystem at Maruteru and the effect on soil fertility are given in Table 25.3.

TABLE 25.3: Influence of source of nutrients on yield (t/ha) of rice-rice cropping system at Maruteru

Treatment	2003-2004		2004-2005		2005-2006		2006-2007		2007-2008	
	Kharif	Rabi	Kharif	Rabi	Kharif	Rabi	Kharif	Rabi	Kharif	Rabi
50% NPK-50% FYM	4.50	3.92	5.39	5.41	4.36	6.40	3.86	4.80	5.09	5.55
1/3 N FYM + VC+NC	4.46	3.16	5.13	5.08	4.01	5.01	3.45	3.77	4.21	4.37
RDF Initial	4.87	3.42	5.26	5.83	4.78	6.52	4.23	5.30	5.30	5.30

pH	EC dS/m	OC%	Av. P. (kg/ha)	Av. K (kg/ha)
5.8	0.76	0.80	340	179
6.6	0.41	1.03	357	171
5.7	0.29	1.06	382	170
6.6	0.37	1.00	310	184

*VC: Vermi compost, NC = Neem cake, RDF = Recommended dose of fertilizers.

The organic treatment did not support yield similar to those due to continuous application of fertilizers @ 60:40:40 kg/ha NPK in *kharif* and 120:60:60 kg/ha during *rabi* in any one of the 5 years. But the integrated supply of nutrients by the application of 50 % RDF and 50 % N equivalent through FYM was almost equally productive due to the application of fertilizers alone. The organic carbon content was improved by the cultivation of crops through the external input of nutrients. But, there were no distinct variations due to the source of nutrients.

By and large the experiments conducted elsewhere in the country do not provide evidence that the organic farming is more efficient to increase the production or the soil organic carbon, available P and K compared to the chemical fertilizers. It is not possible for all the farmers to raise organics on the farm in sufficient quantities to supply the needed nutrients to the crops. If purchased, the costs are prohibitive. Nonetheless, organic farming is highly profitable despite severe reduction in yield, if the produce is certified for export after cumbersome processes of international standards and heavy investments. It is not applicable to all the farmers who cannot afford to invest high nor ready to lose the yield substantially. It is estimated that about 30 m t of nutrients are removed by the crops annually in India. The nutrient addition has reached 22 m t by 2006. Still there is a gap of 8 m t. All the organic sources of nutrients pooled do not exceed 6-8 m t. They will be sufficient to supply 1/5 nutrient requirement of crops in the country, if agriculture is to be shifted from fertilizers to organic. Therefore, this concept has no scope to flourish in future. Awareness to combine every organic source with the fertilizers will help in the physical improvement of efficiency for high yield.

INTEGRATED PLANT NUTRIENT SUPPLY

The decline in productivity of rice-wheat cropping system due to over exploitation of natural and indiscriminate use of chemical inputs in South Asian countries was the turning point to integrate the plant nutrient supply system and conserve the soil-health by improving physical, chemical and biological properties for sustained production of crops. The appropriate combination of mineral fertilizers with organic manures, crop residues, green manures, composts, rotation with N-fixing crops, bacterial cultures that fix –nitrogen, solubilize the fixed P in the soil and microrrhizae etc. varies with the system of land use, ecological, social and economic conditions. Therefore, long-term experiments were initiated in rice based and other cropping systems throughout India in the early 1980s. The concept help in avoiding soil degradation and deterioration of water and environmental quality by promoting carbon sequestration and checking the losses of nutrients to water bodies and atmosphere. The organic source of nutrients act as slow release fertilizer as it synchronizes the nutrient demand set by the plants (Gill and Brar, 2005). The 20 years search in this direction from a rice-rice system in the alfisol at Rajendranagar in Andhra Pradesh showed that the green manuring with *Glyricidia maculate* which is local available in plenty is the best source to sustain the crop productivity and conserve the soil health. The incorporation of *Glyricidia* @ 60 kg N equivalent on dry weight basis in the soil about 2 weeks before planting of rice in *kharif* and fertilizing the crop only with 60:30:30 kg/ha NPK, produced grain and straw yield similar to the effect of the recommended dose of kg/ha NPK (120:60:60). The soil organic carbon was more than the initial value and as much due to inorganic cultivation. The substitution of 50 % NPK fertilizers with Farm Yard Manure improved the organic carbon to a maximum of 0.70% compared to the initial value of 0.54% (Table 25.4).

TABLE 25.4: Influence of integrated nutrient management treatments On soil fertility after 20 years at R'nagar

Treatment	pH	EC (sD/m)	OC%	Av. P (kg/ha)	Av. K (kg/ha)
No fertilizer	8.06	0.45	0.55	30.07	159
RDF	8.01	0.47	0.67	48.56	166
50% *N Glyricidia* +50% N fertilizers	8.03	0.55	0.57	36.71	165
50% N FYM + 50% N fertilizers	7.67	0.46	0.70	40.69	158
50% N rice straw + 50% N fertilizers	8.02	0.47	0.59	32.97	171
Initial	8.50	0.24	0.54	34.00	224

In the tank irrigated medium deep soils at Rudrur, the rice-mustard cropping system also confirmed the usefulness of integrated supply of nutrients. The substitution of 50% N fertilizer dose to *kharif* rice with FYM or *Glyricidia* green leaf manure incorporation sustained the yield with the recommended dose of NPK (120:60:60 kg/ha) to rice and to mustard (80:40:30 kg/ha) supplied sole through the fertilizers during 14 years of study. The soil organic carbon improved even in the unfertilized crop after 5 years. This was probably due to the incorporation of organic matter through harvested stubbles and the fallen leaves that were incorporated into the soil every season. The organic carbon was more due to the integrated nutrient management treatments over the continuously fertilized soil (Table 25.5). The soil available P and K increased after 14 years of cultivation of rice-mustard cropping system with different treatments.

TABLE 25.5: Influence of source of organic nutrients combined with Fertilizers on soil fertility after 14 years of rice cropping system

Treatment	pH	EC (sD/m)	OC%	Av. P (kg/ha)	Av. K (kg/ha)
No fertilizer	7.63	0.58	0.59	25.47	328
RDF	7.77	0.64	0.59	26.73	338
50% *N Glyricidia* +50% N fertilizers	7.60	0.65	0.64	28.10	316
50% N FYM + 50% N fertilizers	7.73	0.62	0.64	26.80	320
50% N rice straw + 50% N fertilizers	7.67	0.64	0.67	26.83	326
Initial	8.10	0.37	0.52	18.00	304

In Maruteru, the integration of green leaf manure with *Sesbania aculeate* or FYM replacing 50% NPK fertilizer dose during *kharif* in rice-rice cropping system sustained yield on par with the continuous application of fertilizers. The organic carbon content of the soil did not change substantially either by the cultivation of crops with fertilizers or through integrated nutrient supply after 15 years (Table 25.6). There was a substantial reduction to the initial value.

TABLE 25.6: Influence of source of integrated nutrient management on Soil fertility after 15 years rice-rice cropping system at Maruteru

Treatment	pH	EC (sD/m)	OC%	Av. P (kg/ha)	Av. K (kg/ha)
No fertilizer	6.05	0.34	0.83	32.17	220
RDF	5.98	0.36	0.89	34.47	174
50% *N Glyricidia* +50% N fertilizers	5.96	0.33	1.01	35.23	169
50% N FYM + 50% N fertilizers	6.15	0.35	0.89	33.07	168
50% N rice straw + 50% N fertilizers	6.11	0.32	0.86	32.65	214
Initial	5.10	0.38	0.90	38.00	344

The results obtained from a wide net work of experiments in the country from several long-term experiments has act a sound confidence that the soil can be conserved from degradation of its physical and chemical properties without serious yield reduction in rice-based cropping systems by adopting the integrated nutrient approach. These results indicated that should be a surge of renewed interest to decompose every farm waste and incorporate into the soil and conserve its structural properties and improve the water and nutrient use efficiency for better productivity.

TRANSGENICS

The transfer of an unrelated desired gene from one organism to another or non-crossable species is called **transgeneic.** The transgenic plants or animals so derived are called **genetically modified organisms.** Such variation of crop plants are now emerging as strong tools for yield security because of their promise to resist the damage due to pests and diseases, tolerance to the herbicides, high level of salinity, alkalinity and aridity etc.

In India, cotton is grown on 9 m ha i.e., 20 % of the area in the world. But, its tyranny is that the boll worms and sucking pests cause severe economic damage. The cotton growers often commit suicides. The pest *Helicoverpa armigera* has developed a high level of resistance to several pesticides. The transgenic Bt cotton were introduced for commercial cultivation since 2002 in India amidst several skepticisms of tolerance of the insect even to this bacterial gene effect. It was also known that *Helicoverpa armigera* has the capacity to develop resistance to the protein Cry 1 AC in 7-8 generations. The conventional varieties progressively made a way for the Bt varieties. The farmers were comfortable. The larvae of *Helicoverpa* were fairly untraced until 90 days after sowing. But by 2007, the Bt cotton is infested with pest by 60 days. It is also severely infested by hitherto a negligible pest – the mealy bugs. The story of miseries turned around. The risks of this tremendous and potential technology are now seen in 5 years. Several research workers already cautioned that the gene silencing may occur in transgenic plants and the transgenic cease to function. The **emerging biotechnology** can erode the biodiversity increasing uniformity in production and imposing intellectual property right to turn life forms into private property (Rethinam and Sreedhar,2000. Desirable new phenotypes created by introducing foreign DNA into plants are frequently unstable leading to loss of newly acquired traits. This genetic instability is due to the inactivation of transgene referred as gene silencing. Resistance problems are likely to be developed in Bt crops. The continuous expression of toxin creates a strong selection pressure. They are also likely to affect non-target organisms and neighbouring fields that share similar pest complexes ending up with resistant insect population colonizing their fields. **Genetic engineering in agriculture** will do nothing to stave off world starvation. On the other hand, it is likely to do much to increase such problems (Anuradha, 2000). To overcome this challenge integrated gene management along with precision farming is suggested. **Cysgenesis** is now considered to open new option for second Green Revolution in India. It can be used directly for the improvement of existing varieties which already proved to be safe for use in the market (Jacobsen and Nataraja, 2008).

CROP SUBSTITUTION

It is possible in the near future that the traditional food crops can be substituted with factory produced foods viz., starch, proteins and fats grown in fermentation tanks by the use of modern techniques of cell and tissue culture. This will replace some of the agricultural crops such as vanilla, sugar or oilseeds to be grown at current production levels.

Nano Technology

It is the manipulation or self assembly of individual items, molecules or molecular clusters into structures to create materials and devices with new or vastly different properties. The prefix **'nano'** is the Greek word meaning **dwarf**. One nm corresponds to the million part of a mm. **Nanotechnology** utilizes structures of 0.1 to 100 nm in size. The application of this technology to agriculture and food industries was first addressed by a United States Department of Agriculture road map in September, 2003. It has a potential to revolutionize the agricultural and food industry with new tools for the molecular treatment of decreases, detection and enhance the ability of plants to absorb nutrients etc. **Nano-structured catalysts will be** available to increase the efficiency of pesticides and herbicides with lower doses. Smart sensors and delivery systems will help agricultural industry combat viruses and other pathogens (Joseph and Morrison, 2006). The low-cost sensors are developed to detect the presence of food borne pathogens, filters to remove undesirable compounds from foods and beverages and nano-particles to store flavours and nutrients inside foods and release them at designated organs in the body when they are needed. **Sensors** also detect low levels of molecular signals of spoilage and food borne pathogens within minutes of exposure. The packaging materials for food products can also be structured at the molecular level. Plastics can be manufactured with different nano-structures to gain various gas and moisture permeabilities to fit the requirements of specific products, viz., fruits, vegetables, beverages and wine. This will improve the shelf-life, flavor and colour preservation of the products. **Food safety is also ensured by nano-structured films and packaging materials that prevent the invasion of pathogens and other micro-organisms.**

A range of smart agricultural inputs and products are developed viz., nano-seed varieties with inbuilt pesticides that release under certain environmental conditions. Molecular food manufacturing appears to be the future achievement for food security and safety. **The nano-capsules** make it possible to alter the nutritional composition, flavor and other attributes of food to many consumers personal testes, psychological requirements and smart food for packaging able to detect the presence of pathogens. These capsules respond to the body's requirement. They remain dormant in the body and release the nutrients to the cells more efficiently and when needed. The additions of nano-particles to the existing food increase the absorption of nutrients. Tuna fish oil, a source of omega 3 fatty acids has been successfully encapsuled to incorporate in the bread in Australia. These capsules break open only in the stomach and thereby avoid unpleasant teste of the fish oil.

Nano-scale devices with novel properties could be used to make agricultural system smart. They can be used to identify plant health issues before they become visible to the farmer. They will be capable of responding to different situations by taking appropriate remedial actions. If not, they will alert the farmer to the problem. Thus, the smart devices act as both a preventive and an early warming system. They deliver the chemicals in a controlled and targeted manner. The encapsulation and controlled release methods have revolutionized the use of pesticides and herbicides. The formulation certain nano-particles within 100-250 nm size range that dissolve in water more effectively than the existing ones thereby increasing their activity. The manufacture of nano-emulsions – suspensions of nano-scale particles are: water or oil-based containing uniform suspensions of pestcidal or herbicidal nano-particles in the range of 200-400 nm. They can be easily incorporated in various media like gels, creams and liquids. They have multiple applications for preventive measures, treatment or preservation of the harvested product.

A few examples of nano-technology are: The application of growth regulator **'Primo MAXX'**, if applied prior to the on-set of stress viz., heat, drought, disease or traffic, strengthen the physical structure and turf grass and allow it to withstand ongoing stresses throughout the growing season. The encapsulated product '**Karate ZEON' is a broad control spectrum on primary and secondary insect pests of cotton, rice, peanuts, and soybeans.** It contains the active compound – cyhalothrin - a synthetic insecticide-based on the structure of natural pyrethrins. It breaks open to release its contents when it comes into contact with alkaline environment, viz., the stomach of certain insects. **Nano-products** are also under preparation to tailor them in such a way that they release the chemicals slowly or quickly in response to different signals e.g., magnetic fields, heat, ultra sound and moisture etc. **Nano-technology** can also be used to clean ground water. The aluminium oxide nano fibres of 2 nm diameter are used as water purifiers. Filters made from these fibres remove viruses, bacteria and protozoan cysts from water.

Conservation Agriculture

The concept of conservation agriculture was developed in USA during 1974. This was due to the necessity. The price of oil increased steep. To cut down the exorbitant expenses on tillage machinery, Government imposed minimum disturbance of soil. Essentially, it has 3 components viz., minimum disturbance of soil, surface cover with crop residues and suitable crop rotations.

Impact of Conservation Tillage in Andhra Pradesh

Intensive tillage is necessary after the harvest of rice. The primary tillage to open the soil and secondary tillage to break the clods and make a soft seed bed has become common practice. This is followed by a secondary tillage to prepare a congenial seed bed in the conventional system. This is illustrated with the data on performance of rice-groundnut cropping system to planting technique of rice and tillage requirement of groundnut in Andhra Pradesh (Table 25.7).

TABLE 25.7: Influence of planting techniques of rice on seed yield (t/ha)

Treatments	Year			
	2003	2004	2005	2006
Dry seeding	2.83	2.08	1.52	1.72
Transplanting	3.92	3.87	4.38	3.51
In-situ green manuring and transplanting	3.81	4.02	4.55	4.06

The results showed that in all the 4 years transplanted rice produced more yield than the direct dry seeded crop. The data on yield of groundnut in response to different tillage practices in rice field sown in puddle and unpuddled soil Table 25.8 showed that results are clear that the groundnut require intensive land preparation by ploughing twice followed by running the rotavator to prepare fine tilth. The performance was better after dry than the transplanted rice continuously for 3 years. The sub-surface compact layer in the puddle soil was not optimal for better root development and was perhaps more compact for the growth and development of pegs and pods. However, due to the intensive cultivation for 4 years a better tilth was obtained in the previously puddle than the unpuddled field.

TABLE 25.8: Influence of tillage practices in rice fields on pod yield of groundnut (t/ha)

Treatment	Year			
	2003	2004	2005	2006
Dry seeded rice-ploughing once and harrowing twice	1.921	1.940	2.150	1.199
Ploughing twice and harrowing thrice	2.299	2.075	2.275	1.267
Ploughing twice and rotavator twice	3.240	2.210	2.362	1.460
Transplanted rice-ploughing once and harrowing twice	1.723	1.505	1.780	1.068
Ploughing twice and harrowing thrice	2.200	1.680	1.676	1.271
Ploughing twice and rotavator twice	2.884	1.775	1.984	1.335
Green manure and Trans-ploughing once and harrowing twice	1.809	1.730	1.950	1.468
Ploughing twice and harrowing thrice	2.283	1.865	2.073	1.566
Ploughing twice and rotavator twice	3.033	1.990	2.174	1.619

In Andhra Pradesh, rice is cultivated on an area of 4 m hectares. Out of this 3.8 m ha is transplanted in the puddle while rest is sown direct. In the coastal regions black gram is sown as a relay crop before the harvest of rice on the residual soil moisture. Sesame and mustard are also relay cropped to a limited extent. This is **a zero till cultivation.** It saves expenditure on land preparation and yield about 2 t/ha of black gram seed. The fodder is highly valued for its high protein and palatability to the cattle. The yellow mosaic virus infestation has become a serious yield limiting disease since 2005-06. It devastated the entire crop during 2007-2008. This challenge compelled the farmers to search for alternate cropping system. In 2005-2006, farmers attempted exploratory efforts by sowing maize between the stubbles of harvested rice at optimum moisture for good germination.

The canal water is usually released in July and withdrawn in February. Rice nurseries are raised in June-July for about a month and transplanted. The crop is harvested in October-November. Soil moisture is retained for a long time in the heavy black soils. The evaporation is slow due to cold weather with low temperature, high humidity and low wind velocity. Therefore, land preparation becomes difficult. About 2-3 weeks time is needed for the implements to work. There is little time for this operation. If land preparation is delayed, the soil become hard and develops cracks. If ploughed under over moist conditions heavy clods are exposed. Maize matures in 4 months unlike black gram in 80-90 days. Therefore, time of sowing for maize is critical. Farmers recognized the need to sow the crop early to match with the availability of water until the grain filling stage. They skipped of the tillage practices to create a soft seed bed. Seeds were dibbled 2-3 days after the harvest of maize in the moist sub-surface layer. The moisture was adequate for good germination and to produce healthy seedlings without stress until 3 weeks. This practice saved the expenses for land preparation, water for 1 irrigation and advanced the time of sowing by about 2-3 weeks. The weeds are usually less due to the sub-merged conditions of puddle fields. Therefore, weed control with the herbicide sprays was very effective until the crop canopy covered the ground area and suppressed their growth. The herbicides mixture of Paraquat @ 5 ml and atrazine @ 4 g/litre was sprayed on the soil surface and stubbles of harvested rice within a couple of days. There was no rejuvenation of the rice stubbles and the newly emerging weeds were controlled. The light and medium soils were disturbed with the blade harrows after 3-4 weeks of sowing with bullock drawn harrows to break the capillaries for aeration and fertilizer is added. This is followed by irrigation. In heavy black soils fertilizers are dissolved in water and applied along the rows through the perforated poly vinyl chloride pipes.

Seed is also sown behind the plough by opening a furrow in light textured soils. This reduce expenditure on 25-30 labour for dibbling while 8-10 labour is sufficient to sow and fertilize behind the plough. This is not possible in heavy black soils.

In Nagarjuna Sagar commanded area, rice-rice is the predominant cropping systems. Maize, sunflower, groundnut and cluster bean are the alternate crops to rice in *rabi.* The water is released for cultivation in the 4th week of July to 1st week of August until the 4th week of April. Rice nurseries are raised dry with rain water in June-July. Seedlings are transplanted in August-September and the crop is harvested in November-December. Rice nurseries for the *rabi* are raised 2-3 weeks before the harvest of *kharif* crop. The land preparation is done within 10-15 days and the seedlings are transplanted during December-January.

Hail storms common in April-May coincide with the maturity stage and seriously damage the crop. This necessitated the farmers to search alternate crops. Maize was the best option. The seeds on the cobs are well protected with the layers of husk. Hence, the seeds are not in direct contact with the hail. It requires less water and the cultivation expenses are less. This was the turning point for the farmers during 2005 to grow maize by dibbling the seed between the rice stubbles with the help of ropes. Several demonstrations were conducted by the agriculturist of the Agricultural Research Station-Kampasagar to sow the seed with zero till fertilizer drill developed at Pantnagar. Currently the sowing methods followed by the farmers are hand dibbling, opening a small furrow with the country wooden plough. Zero till fertilizer drill, Zero till drum seeder and local devise developed by the farmers to move the wheel of a cycle embedded with spokes to make holes in the soil at desired spacing. The seeds are sown in the soft and moist sub-surface layer within 7-10 days after the harvest of rice in light textured soil in the low lying areas but within a week in the ridge areas where the moisture is lost quickly. In heavy clay soils, the seed is sown 10-15 days after the harvest of rice. The soil is irrigated immediately after sowing. Some farmers believe that the seed sown after a light irrigation is more beneficial. The germination is uniform and the plant population is well established. A pre-emergence herbicide spray of gramoxone @ 5 ml and atrazine @ 4 g/l water is the common practice to manage the weeds. The young seedlings develop a sickly appearance of nutrient starvation. Fertilizers are applied on the surface of the soil and irrigated during the 3rd week after sowing.

The farmers are convinced with the effective operation of sowing and fertilizer application with the zero till machine. Sowing is rapid. It covers 1 acre area/hour. The depth of seeding is uniform unlike in dibbling. Though many farmers cannot afford to purchase this drill at the cost of Rs. 35,000, there is a surge of interest to hire and sow the crop. However, farmers observed some limitations which need to be addressed in future. The height of 35-40 cm stubbles after the harvest of rice with combine machines comes in the way of swift movement of the machine. In the low lying patches, the soil stick to the tines and clog.

The seeds do not drop until cleaned and leave gaps. The plates which pick and drop the seed need refinement. Only 1 seed is dropped at a time, if the seed is bold and uniform in size. But, 1-3 seeds are dropped when they are of small size. This add extra burden of high seed rate on one hand and labour for thinning.

The zero till-cum-fertilizer drill fabricated by the ECF Agronomist, Dr. M. Govardhan at the Agricultural Research Station, Kampasagar, is less expensive and gaining popularity with the farmers. It costs < 10,000 rupees. As the drum moves, the spokes make holes for dibbling the seed and fertilizer application simultaneously adjacent to each other. This requires more labour for sowing and fertilizer application but less seed of 12-15 kg compared to 25 kg/ha by the zero till drill. There are no gaps and no need for thinning as there is no over seeding. One transparent set back is that the soil sticks to the spokes after making the holes. This needs cleaning. To overcome this problem, reduction in angle of the spokes on the roller is under consideration. The labour requirement and the seed rate of maize/ha for sowing by different methods are presented in Table 25.9.

TABLE 25.9: Relative seed and labour requirement for sowing maize with different methods after rice

Input	Method of sowing maize				
	Conventional	Zero till drill	Zero till drum	Dibbling	Behind the plough
Seed (kg/ha)	15-20	25	12-15	12-15	20
No. of labour/ha	10	2-3	15	15	6-8

The economics of maize cultivation due to the resource conservation technologies for uncommon operations are presented in Table 25.10.

TABLE 25.10: Cost of uncommon operations for zero till maize Cultivation (Rs/ha)

Operation/Input	Method of sowing maize				
	Conventional	Zero till drill	Zero till drum	Dibbling	Behind the plough
Land preparation	2,500				
Sowing	1,200	2,500	3,000	3,000	1,500
Weeding	3,000				
Harrowing	500				
Glyphosphate		875	875	875	875
Gramoxone		940	940	940	940
Atrazine		525	525	525	525
Seed	1,500	2,500	1,200	1,200	2,000
Total	8,700	7,340	6,540	6,540	5,840
Cost of machine	35,000	8,000			

"The ultimate object of our labour must always be to bring the phenomena into measurable form."

Heike Kamerlingh Onnes

Resource Recycling

Resource recycling clearly indicated that the interdependence of the different components of the total farming system to make the farmers self-sufficient in terms of ensuring the family members a balance diet for leading healthy life, increasing the standard of living through maximizing the total net return and providing more employment. The by-product of dairy, i.e. cow dung, forms a major raw material for biogas. Digested slurry of biogas forms a major part of feed of pisciculture for increasing plankton growth as well as supplying valuable manure to raise the productivity of crops. Similarly, byproduct of field crops like paddy straw forms a major ingredient of mushroom cultivation. Again straw used for mushroom production may be utilized for cattle feed and compost preparation. The by-product of poultry, i.e. poultry dropping, forms an important ingredient of pisciculture for increasing the plankton growth as well as increasing the fertility of land.

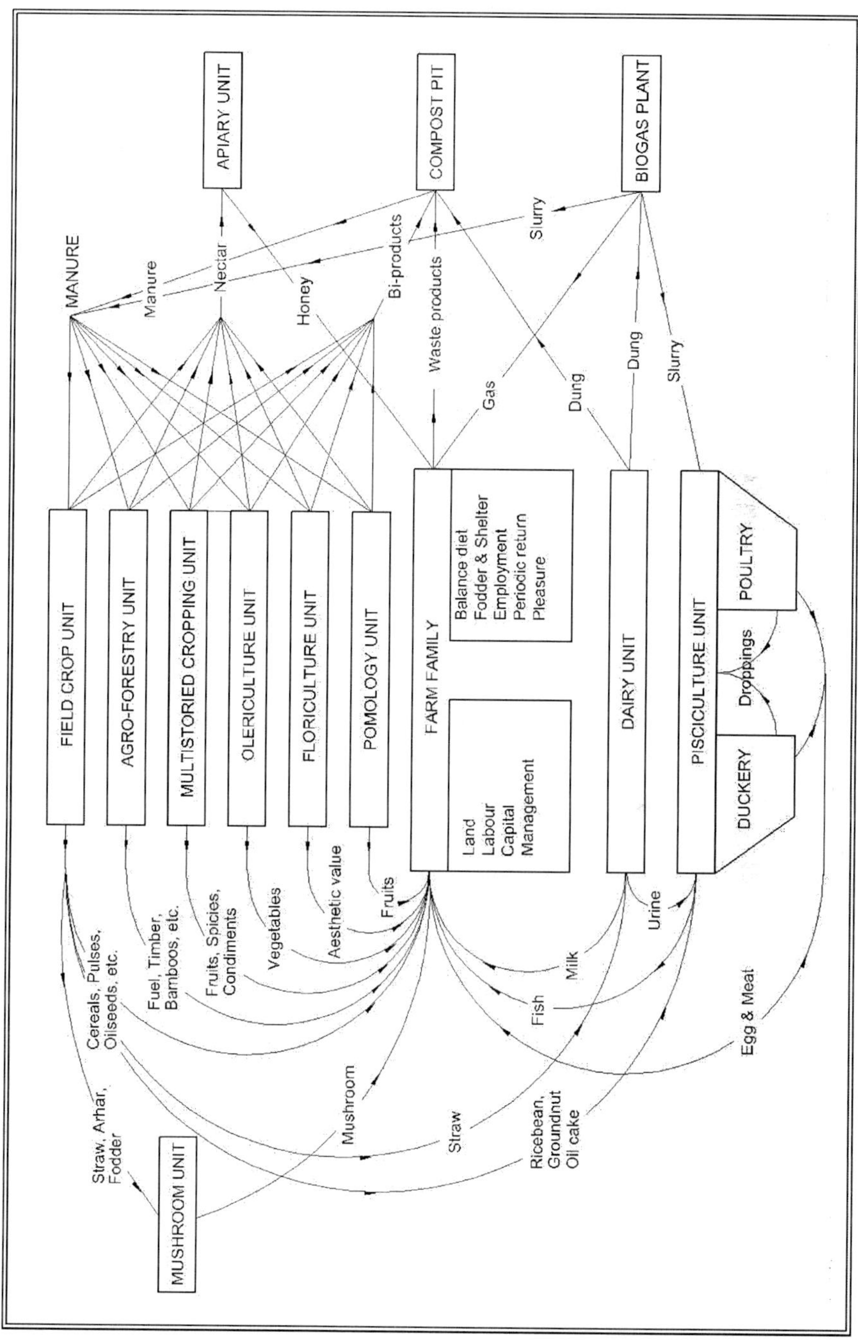

FIG. 25.1: Resource Recycling.

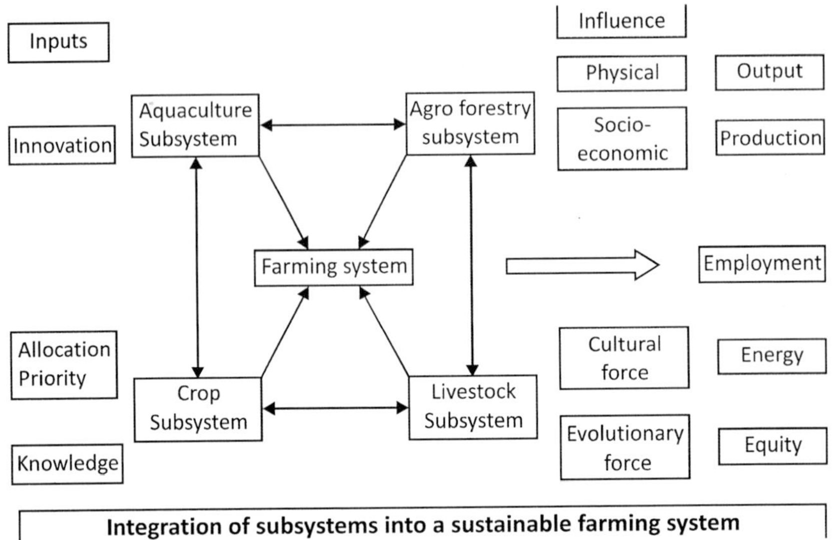

FIG. 25.2: Integration of Subsystems into a Sustainable Farming System.

Even an item like apiary has lot of indirect bearing on crop enterprise by increasing yield and quality though improvement in pollination, apart from giving a wholesome product like honey to the farmers. Farmers like to have flowers' which adds to the aesthetic sense of farm family and flowers also serve as foraging plants for honey bees. The entire philosophy of integrated farming system revolves around better utilization of time, money, resources and family labourers of farm families. The farm family gets scope for gainful employment round the year, thereby ensuring good income and better standards of living (Fig. 25.1).

BIOTECHNOLOGY: A SCIENCE OF SECURING FUTURE

Biotechnology

Biotechnology is one of the fastest growing sectors in India and elsewhere. This emerging field is going to take a leap forward in the next five years. As per the report of NASSCOM, by 2010, biotech sector may earn up to rupees five million dollar having job opportunities for 15 lakhs qualified biotechnologists in several fields such as **agriculture, medicine, environment, bioinformatics, industries, food processing units, pharmaceuticals, dairy and veterinary, biodiversity conservation, neuro-science, marine, informatics, teaching training and entrepreneurship development in both government and private sectors.** Modern science and technology along with research and discoveries are comfortable journeys from one revolution to the other, where the conventional technology blended with the new scientific discoveries and their fusion has been creating wonders. Biotechnology undoubtedly holds great promise for the future, Genetic engineering, Bio-Medical engineering, Bio-sensors and Bioinformatics etc. will be the hot topics decades to come. The importance of biotechnology was recognized after the establishment National Bio-technology Board in 1982. In 1986, the Government of India formed full-fledged department of Biotechnology (DBT) within the Ministry of Science and Technology. A number of Academic Institutions and Government Agencies such as the CSIR, ICAR, ICMR, DAE etc. are undertaking research and development activities related to it.

Stupendous strides have been made in making the earth a better place to live in, through human interference. Biotechnological innovations have been used as the means to relegate precarious predicaments. The firmament of Biotechnology has broadened to envelope the entire earthlings in its fold. Advances in Biotechnology have created a revolution in life sciences including genetics, and profoundly changes its face worldwide. Applications of biotechnology in human and animal health, agriculture, biomass and green cover, environmental protection and non-environmental energy sources are of great significance. The future of main kind depends on developing this issue and applied (bio) technology suited to the needs of humanity. Biotechnology has provided a great opportunity for direct manipulation of gene structure, function and expression to serve the humanity and meet the growing demand of the millions.

Several policy initiatives have been taken to create excellent human resource base, build up the requisite infrastructure, identity and set up centres of excellence, support R & D and manufacturing activities to meet the demands of future and growing population.

All basic economic needs can be fulfilled by application of Biotechnology as once imagined by the scientists. A wishful thinking has transformed itself to a glaring reality. Globalization is leading to intense competition with consequent higher stress levels in work places; healthcare sector is one such with a large share of global investment. The development of new generation drugs, their targeting, recombinant vaccines and newer diagnostic tools are some of the possibilities.

The development of vaccines for important diseases affecting livestock and diagnostic kits for farm animal healthcare are among several activities in this sector. All the antibiotics come under come under biotech banner. The chemicals we use are being replaced by biochemical or other alternative biotech products. It will provide new ways of influencing raw material processing. Most of the future medicines will be of biotech origin. Handling humanly impossible gene sequences have now become possible. Cloning and designer babies may become usual features in future. In the field of Seri-biotechnology, the DBT has initiated a programme in the genetic enhancement of mulberry and non- mulberry silkworm races and the host (mulberry) plants. The programme includes development of immune-diagnostics tools for detection of pathogens, bio-control measures for major pests and diseases of silkworm and new races of silkworm. Considerable efforts have been put in to ensure food and nutritional security through biotechnology R & D, under food biotech initiatives, technologies for scale-up of low cost nutrient foods have been standardized.

The **Tissue Culture industry** caters to horticulture and floriculture industries. The herbal medicine research has increased the level of interest in plants with potential in therapeutic benefits and commercial demands for herbal products in global markets. With increased interests in natural, biodegradable and/or recyclable material, development has focused research into biological (botanic) or organic materials. Biological and physical scientists come together and explore such exciting areas, a concept of systems Biology. **Agriculture Biotech** will give us plants that have disease resistance, improved yields and stress tolerance. Environmental biotechnology will help control pollution problem. In short, biotechnology is destined to touch anything and everything of human needs.

Biotechnology offers hope for solving many of the problems the world faces today. Most drugs are chemicals and have limitations in treatment, only helping managing the disease, except antibiotics. Understanding the pathways and protein chemistry at the molecular level will bring in better treatment. Understanding human genome would help us to predict the likelihood of diseases and improve disease diagnosis systems. With increased interest in natural, biodegradable and/or recyclable materials, developments have focused research into biological or organic materials. Biological, physical and other scientists come together and explore such exciting stress.

Biotechnology is inter-disciplinary, offering flexible eligibility criteria of the entry level. It encourages Systems Biology that provides that provides a common platform for applied areas of many fields of sciences such as chemistry, microbiology, biochemistry and chemical engineering among many other, having many branches of application in healthcare, agriculture, food processing, environmental and industrial facets as discussed earlier. Demand for a trained biotechnologist in these areas is high. To explore these, specialists in law, engineering, computer science and others as support system is highly desirable. With increased use of biotech products and genetically engineered organisms, bio-safety has been one of the main concerns, amongst the many social, ethical and legal issues. Guidelines in line with this are strictly being followed/imposed to prevent damage to human health and environment. A single window-processing cell has been created in the DBT for speedy processing of approvals for bio-safety.

The DBT has a strong focus now on the R&D activities. It interfaces with a number or other government departments, academic and R & D institutions, and other user agencies. It supports Post Graduate Programmes with focus on biotechnology in over 80 Universities, IIT and other premier institutions across the country, In collaboration with the university Grant Commission (UGC). Also, there are doctoral and post-doctoral courses as well as regular short term courses or training programmes for mid-career scientists. In order to encourage researchers for R & D towards products and processes, DBT has instituted number of awards, major thrust being in Genomics and Proteomics research.

In term of global industrial growth, biotech at present is referred as the sunrises sector. Now, the next likely concern is where to groom oneself to match to such global requirements?. Virtually, every Indian university is engaged in biotech research in some form or the other and various courses of biotech are offered at many centres of repute.

Today, there are over 150 modern biotech companies in India. Some of these have pursued a product based business model; others have adopted a service model in line with the IT sector. India's major strength in biotechnology is its highly trained workforce, a very good infrastructure base with a well-networked chain of laboratories and clusters of bio-resource-based industries. Added to these, the treasure of our biological diversity and the very rich traditional knowledge base (the concept of ITK, i.e., Indigenous Technical Knowledge) dating back to thousands of years. These put together are, by any standard, very formidable assets. The fruits of these efforts would hopefully become visible sooner or later.

Work of Biotechnologists

Any technique that uses living organisms or a part of living organism to make or modify products to improve plants, animals to process or to develop microorganisms for various uses and recent development in molecular biology and their commercial applications is known as **Biotechnology**. It includes micro-propagation, formation of eco-friendly products, cell and tissue culture technique as well as recombinant DNA technology which has opened avenues for existing possibilities and a new era research and development.

Few Important Functions of Biotechnology

1. Cloning of organisms.
2. Genome Sequencing.
3. To cure the hereditary disease through DNA Technology.
4. Formation and availability of gene bank, seed bank, cryo banks for future use.
5. Formation of new viral and bacterial vaccines.
6. Formation of organic fertilizers.
7. Development of disease resistant and hybrid plants and animals.
8. Drug designing.
9. Formation and production of alcohol, antibiotic, enzyme and secondary metabolites for welfare of humanity.
10. To find out alternate sources of energy.
11. Production of biogas, biodiesel and green energy.
12. Conservation and protection of forest.
13. Purification of environment.
14. Conservation of environment.
15. Minimizing pollution load through eco-friendly technology.
16. Waste management.
17. Production of genetically modified organisms.
18. Value addition in agricultural products.
19. Production of eco friendly pesticides, insecticides, herbicides and fertilizers.
20. Vermi composting.
21. Food processing industries.
22. Gene therapy, Stem cell culture, organ culture etc.
23. Production of designer plants and animals.
24. Tissue culture and cell culture.
25. Pharmaceutical research and generic medicines.

Biotechnology can be chosen as the area of specialization/study at graduation, post graduation or research level. The field offers late entrants the advantage of not necessarily having a prior degree in the subject. Some important prerequisites and entry/level biotech education available in India are:

Many state universities offer a B.Sc. Degree in biotech, like Bangalore University (Karnataka), CGS University (Uttar Pradesh), Kakatiya University (Andhra Pradesh), Madras University (Tamil Nadu), Osmania University (Andhra Pradesh), Patna and Magadh University (Bihar), and many others, for which the eligibility criterion is essentially 40% (minimum percentage may vary state) in PUC/10+2/intermediate/ equivalent with science subjects. The duration of the course is three years and admissions are either entrance exam by the University/institute or percentage-based.

B. Tech. / BE. Biotechnology

Some of the IITs offer a four-year B. Tech course in biotechnology. IIT, Kharagpur offers a B. Tech degree program in Biotechnology and Biochemical Engineering while IIT, Chennai and Guwahati give a B. Tech degree in Biotechnology. In 2004, IIT, Kanpur introduced a B. Tech degree in Biological Science and Bio-Engineering. Admission to these IIT undergraduate programs is through an all India competitive exam the Joint Entrance Examination (JEE).

Besides the IITs several other institutes and universities offer a BE/B. Tech Biotechnology course, e.g., the Anna University (Chennai) has been offering a B. Tech program in industrial Biotechnology since 1992. Some other universities offering B. Tech (Biotechnology) courses include Bharathidasan institute of Engineering and

Technology, Tiruchirappalli; Guru Gobind Singh University, Delhi, Guru Nanak Dev University, Amritsar (B.TechChemistry and Bio-Engineering); UP Technical University, Lucknow; Vellore Institute of Technology Vellore (Tamil Nadu).

Students are selected either through the all India Engineering Entrance Examination (AIEEE) or the common entrance exam conducted by respective State Government Organizations and Universities.

M. Sc. Biotechnology

Presently, there are 30 M. Sc. Courses in general Biotech, seven in agricultural biotech, one in medical biotech, two in marine biotech, one in industrial biotech and three in neuro-science running under various universities, supported by the DBT. Combined biotech entrance examination is conducted by Jawaharlal Nehru University (JNU), New Delhi (On behalf of DBI), in May every year.

M. Tech Biotechnology

IIT, Delhi and IIT, Kharagpur offer a five year integrated dual degree M. Tech course in Biochemical Engineering and Biotechnology with the support of DBT. IIT, Chennai offers a similar integrated M. Tech in Biotechnology. Students are eligible to enter these courses after completion of B. Sc with science or B. Tech.

M. Tech Biotedh program (2 years/4 semesters are offered by Anna University, Chennai and University of Technology, Kolkata and IIT, Kanpur with the support of the DBT, though many other academic institutions have also recently started offering the course. Candidates who have done B. Pharma or B. Tech/BE course in biotech or related disciplines or or M. Sc Life Science are eligible for this program. Admissions are either through the joint biotechnology entrance exam conducted by the JNU or through GATE. M. Tech degree in Bioprocess Technology offered by the University Institute of Chemical Technology, Mumbai. This course is also supported by the DBT. Students with Bachelor's Degree in Chemical Engineering, Pharmacy, Chemical technology with GATE are eligible and the selection is based on the GATE scores plus the ICT written test and interviews. The Birla Institute of Technology (BITS), Pilani offers a ME degree in Biotechnology.

MBA Biotechnology

The University of Pune pioneered a unique two-year fulltime MBA program in biotechnology in 2002. The syllabus is based on the requirements of the industry. About 60% of the syllabus covers the managerial aspect and the rest covers the nitty-gritty of biotechnology. Some other Universities, are also offering MBA Biotechnology course.

Bioinformatics

Bioinformatics, an offshoot of biotechnology, has been lately attracting a lot of attention. This has led to specialized B. Sc/M. Sc and B. Tech course in this branch Bharathiar University, Coimbatore and Pune University, which has a centre of excellence in Bioinformatics, offer a two year M. Sc course in University has been offering a B. Tech course in Bioinformatics, since 1999. DBT

DBT supports a one-year course (Advance diploma in Bioinformatics) in five Indian Universities viz., Calcutta University, Jawaharlal Nehru University, Madurai Kamaraj University, Pondicherry University and Pune University.

Apart from the DBT-supported courses, the Bangalore-based Institute of Bioinformatics and Applied Biotechnology (IBAB) offers a Ph.D. program in Bioinformatics.

Ph.D. Programmes

India boasts of a network of research institutes spread around the country established both by government and private sectors. Students wanting to pursue an active career in biotech research can enter any CSIP, ICAR, ICMR lab and medical colleges depending on their qualification and area of specialization.

CSIR

The Council for Scientific and Industrial Research holds national level joint CSIR-UGC Examination (NET) twice a year, in June and December. This examination determines the eligibility of Indian nationals for the award of Junior Research Fellowships (JRF) and appointment of lecturers in science and arts disciplines. Considering the emergence of new specialized disciplines and their need in drug R & D, CSIR has recently permitted fresh engineering and medical graduates and M. Pharmacy candidates with a valid GATE score to avail CSIR fellowships for pursuing a research career, in addition to CSIR lateral entry as Senior Research Fellow.

ICAR

He Indian Council of Agricultural Research (ICAR) promotes science and Technology programmes in agricultural research and education and carries out research directly through ICAR institutes and national research centres, project directorates and also in association with the State Agricultural Universities (SAUs) through All India Coordinated Research Project (AICRP) systems.

Agricultural Research Services (ARS)/National Eligibility Test (NET) examination is conducted by Agricultural Scientists Recruitment Board (ASRB), Krishi Anusandhan Bhavan, Pusa, New Delhi for filling up vacancies of scientists for the various ICAR Institutes and for requirement of Lecturers and assistant professors by the State Agricultural Universities (SAUs).

ICMR

The Indian Council of Medical Research (ICMR) formulates, coordinates and promotes biomedical research in India. Apart from the labs falling under CSIR/ICAR, almost all the IITs institutes and universities having a life/biological sciences department, further research in this frontier area.

Admission to these Ph.D. programs made on the basis of written test/interview conducted by the department/center concerned. The Indian Institute Science (IISc), Bangalore, a premier scientific research institute, conducts both Ph.D. and integrated Ph.D. programs (directly after B. Sc in biological sciences, which includes research in biotechnology also. Admission to IISc is through an entrance exam conducted by the institute or is based on the GATE score followed by an interview.

ENTREPRENEURSHIP POTENTIAL OF BIOTECHNOLOGIST

Professionally, trained biotechnologist can be successful entrepreneur by setting up industries/consultant house based on the biotechnology. There is great scope of starting biotechnology based business and industry in these fast changing times. Sustainable development of the globe is only possible with green scientific and technological breakthrough which can be brought by developing and adopting biotechnological tools and techniques. Food and nutritional security, industrial development, health care, climate change management are major areas which have the potential of establishing useful and profitable enterprises. Student of biotechnology need not to be a **"Job seeker"** but they can be **"Job provider"** by using their knowledge, skill and expertise for setting up biotechnology based enterprises. The research work going on in leading laboratories for breakthrough in biotechnological development can shape the new emerging world of sustainable development and properties for one and all. Indian students are sharp, intelligent and enterprising with biotechnology back ground they have the potential to make the nation developed and leader of the globe.

There are many reputed government organizations around our country, affiliated to one or the other technical and non-technical, state or central university offering biotechnology courses, as well.

"Some people dream of success while others wake up and work hard at it."

Ants work hard round the clock without sleeping. It makes us realize that we can achieve whatever we want only if we have the will to do it and work hard towards it.

"Countries that practice death control must also practice birth control or prepare now for a time when their people will have to living standing up because there won't be room to sit down or lie down".

Dr. C. Lement Market

"Poverty is the Greatest Polluter"

Prof. Galbraith

'If a population policy is not put in place, poverty will remain a curse'.

Biswadeb Chatterjee

26 References

Abrol, I. P. and Gupta, R.K. 1996. Sustaining high production systems: Research issues and approaches. Souvenir, 2nd International Crop Science Congress, Nov. 17-24, 1996, New Delhi, India

Abrol, I. P. and Palaniappan, S. P. 1987. Green manure crops in irrigated and rain fed lowland rice based cropping systems in South Asia. In: Green manure in rice farming. Proc. of a Symposium on "Sustainable Agriculture", Los Banos, Philippines, pp. 71.

Abrol, I.P. and Dhruva Narayana, V.V. 1990. Technologies for Wastelend Development., ICAR, New Delhi.

Adil, M.L. and Karte, R.K. 1992. Blue green algae in rice soil ecosystem. Proc. National seminar on "Organic Farming". Rai M. M. and LN. Verma (ed.), Ravi Printing Press, Jabalpur, India, pp. 200-210.

Agarwal, G. D., 1948. Economics of Mechanization of Agriculture in India, Indian J. Agril. Econ. Vol. IV No. 1.

Agarwal, M.C. and Khanna, S.S. 1983. Efficient soil and water management in Haryana., Haryana Agri. University, Hissar.

Agarwal, R. and Tilak, K.V.B.R. 1988. Establishment of *Azospirillum brasilense* in roots of finger millet. Microbio. 143: 539-544.

Agarwal, R.R., Jadev, J.S.P. and Gupta, R.N. 1982. Saline and alkaline soils of India., ICAR, New Delhi.

Ahmed, M. and Rab, A.B. 1992. Feasibility of adopting aquaculture to increase resource productivity in existing Bangladesh farming systems. NAGA, The ICLARM Quarterly, 15: 21-22.

Ahmed, S. 1990. Neem pest control. ILEIA Newsletter. 6: 7.

AICRPDA (All India Coordinated Research Project for Dry land Agriculture) 1982. A decade of dry land agricultural research in India - 1971-80. Hyderabad, Andhra Pradesh, India: AICRPDA.

Alam, A. 2002. Post harvest technology fetches better return. Indian Farming. 52 (8): 18-25.

Alan, R. Putnam and Chung Shih Tang. 1986. The Science of Allelopathy. John willey and sons. Canada

Albert, E.H., Norman, D. W. and Winch, F.E. 1980. Farming Systems Research. A critical appraisal. Michigan State University Rural Development-paper, 6, East Lansing Michigan, USA: Michigan State University.

Alden, S. Crafts and Wilfred W. Robbins, 1973. Weed Control – A text book and manual. Tata McGraw Hill Publishing Co. Ltd., New Delhi.

Alexander, M. 1977. Nitrogen fixation. In: Introduction to Soil Microbiology, 2nd ed. Willey, New York. pp. 305-330.

Allison, F.E. 1973. Soil organic matter and its role in crop production. Elsevier, Amsterdam, pp. 104-105.

Altieri, M.A. 1987. Agroecology: The scientific basis of alternative agriculture. Boulder/ London: West view press (ITP).

Altieri, M.A. and Liebman, M. Z. 1986. Insect, weed and plant disease management in multiple cropping systems. In: Francis, CA. (ed.), Multiple cropping systems, Mac Millan, New York, pp. 183-218.

Anandeshwari Awasthi. 1995. Indian Climatology., APH Publishing Corporation, New Delhi.

Andrews, D.J. and Kassam, A.H. 1979. The importance of multiple cropping in increasing world food supplies. *In:* Papendick, P.A., Sanchez, P.A. and Triplette,

G.B. (eds) Multiple cropping. American Society of Agronomy, Pub 27, Madison, Wisconson.

Annadurai, K., Palaniappan, S. P., Chinnusamy, C. and Jayanthi, C. 1994a. Integrated farming system, Kisan world 21 (8): 51.

Annadurai, K., Palaniappan, S.P., Chinnusamy, C. and Jayanthi, C. 1994b. Integrated farming system: Need of the hour. Kisan world 21 (9): 39.

Anonymous, 1982. Soil Conservation Problem, Approach and Progress in India. Soil Conservation Division, Dept. of Agric and Coopn, Ministry of Agric. Govt. of India, New Delhi.

Anonymous, 2008. Women's Empowerment. Yojana 2008 vol. 52.

Anuradha, Ch. 2000. Risks of transgenic plants. National Seminar on 'Transgenic crops and Foods' held on 7-9 November,2000 at Hyderabad. Pp 31-32.

Arakeri, H. R. and Roy Donahue. 1989. Principles of Soil Conservation and Water Management. Oxford & IBH Publishing Co. Pvt. Ltd., New Delhi.

Arakeri, H.R., Chalam, G.V., Satyanarayana, P. and Donahue, R.L. 1962. Soil Management in India. Asia Publishing House, London, 2nd Edition.

Arnon, I. 1972. Crop Production in dry regions. Vol. I. Leonard Hill London.

Arora, S.P. 1980. Fodder and feed resources requirement, availability and development. The Indian Dairyman. 32: 821-828.

ASCE, 1975. Sedimentation Engineering. Am. Soc. Of Civil Engrs. New York., 745 p.

Auld, B.A. and Kim, K.U. 1998. Weed Management in Rice. Oxford and IBH Publishing Co. Pvt. Ltd., New Delhi.

Aune, J.B. and Lal, R. 1994. The tropical soil productivity calculator: A model for assessing effect of soil management on productivity. In: Lal, R. and B.A. Stewart (eds.) Sustainable management of soils. Lewis Publishers, Chelsea, MI.

Awasthi, R.P. and Prasad, R.N. 1992. Farming Systems approach to manage resources for sustained productivity in Northeastern hill regions of India. Proceeding of the XII National symposium on "Resource Management for Sustained Crop Production" held at Rajasthan Agricultural University, Bikaner during Feb. 25-28. pp. 333-342.

Azzi Giromalo. 1968. Agricultural Ecology. Constable, London: 1-3

Baeumer, K. and Bakermans, W.A.P. 1973. Zero Tillage. Adv. Agron. 78-125.

Baguaraj, D.J., Manjunath, A. and Patil, R.C. 1979. Interaction between YAM and their effects on soybean in the field. New Phytol. 82: 141-145.

Bajwa, M. S., Singh, S. and Singh, P. 1993. Nitrate pollution of ground water under different systems of land management in Punjab (India): Proceedings of first agricultural Science Congress, National Academy of Agricultural Sciences. New Delhi- pp. 223-230.

Balagopal, K., 1994. Role of Chicken, Pigeon, Japanese quails and Duck rearing in sustainable IFS. In: Proceeding of Summer Institute on "Integrated Farming Systems Research and Development for Sustainable agriculture" organized by ICAR and Tamil Nadu Agric. Univ., Coimbatore, India during 6-15 June, 1994 pp. 235-240.

Balaguru, T. 1992. Gender Issue in Agriculture. National Academy of Agriculture Research Management, Hyderabad, India.

Balasubramanian, A., 1994. Farming system research and development in western zone of Tamil Nadu. In: Proc. of the Summer Institute on "Integrated Farming System Research and Management for Sustainable agriculture". Tamil Nadu Agric. Univ., Coimbatore. pp. 278-284.

Balasubramanian, T.N., 1988. Farming system research survey for new Moriones village in Philippines, p. 9. In: Abst. National seminar on "Farming Systems for Semiarid Tropics". Indian Council of Agricultural Research and Tamil Nadu Agric. Univ., Coimbatore.

Balasubramaniyan, P. and Palaniappan, S.P. 2001. Principles and Practices of Agronomy. Agrobios (India), Ludhiana.

Baldev, B., Ramanujam, S. and Jain, H.K. 1988. Pulse crops. Oxford and IBH, New Delhi.

Balls, W. L. 1917. Analysis of agriculture yield III. The influence of natural environmental factors upon the yield of Egyptian cotton. Phil. Trans. Roy. Soc. Ser. B. 208:157-233.

Balusamy, M. and Shanugasundaram, V.S. 1994. Integrated farming system studies at Agricultural Research Station, Bhavanisagar. In: Proc. of Summer institute on "Integrated Farming System Research and Management for Sustainable Agriculture". Tamil Nadu Agric. Univ. Coimbatore. pp. 270-277

Balusamy, M., 1996. Studies on nitrogen management in lowland rice-fishazolla integrated farming system. Ph.D. Thesis, Tamil Nadu Agric. Univ., Coimbatore.

Barry, R.G. and Chorley, R.J. 1976. Atmosphere, Weather and Climate. Third edition, English language Book Society and Methuen and Co., Ltd. Suffolk.

Barwale, B.R. 1993. Hybrid Rice. Food security in India. MacMillan India Ltd., Madras, India.

Battacharjee, J. P. Mechanization of Agriculture in India. Ibid, p. 121.

Beets, W.C. 1982. Multiple cropping and tropical farming systems. Westview Press, Boulder, Colorado.

Behera, U.K. and Mahapatra, I.C. 1999. Income and Employment Generation for Small and Marginal Farmers through Integrated Farming Systems. Indian J. Agron. 44(3): 431-439.

Beldev Sahai and Ajai. 1988. Application of Remote sensing techniques in Agriculture. Fert. News 33(4): 59-65.

Bennet, H.H., Bell, E.G. and Robinson, B.D. 1951. Raindrops and Erosion. Circular 895. USDA, FAO, 1979. Agriculture towards 2000, 20th Session of the FAO General Conferwence, Rome.

Berry, F.A. and Bollary, E. 1945. Hand book of Meteorolgy. McGraw Hill Book Company Inc., New York.

Bhandari, S.C. and Somani, L.L. 1990. Azotobactor In: Biofertilizers. Somani, L.L, S.C Bhandari, S.N. Saxena and KK Vyas (ed.), Scientific Publishers, Jodhpur, India. pp. 355-361.

Bhanot, K. K., Saful, L., Jena, J.K., Mohanty, S.N. and Ayyappan, S. 1999. Fishery technologies for women. Indian Farming 49(7): 46-48.

Bhardwaj, S.P., Khybri, M.L., Sewa Ram and Prasad, S.N. 1980. Effect of crop geometry on soil and water losses under maize in large sized runoff plot at 4 % slope. Rep. CSWCRTI, Dehradun: 125 p.

Bhatnagar, K.C., Khurana, C.L. and Leelawati, C.R. 1989. Statistical assessment of crop sequences in different agroclimatic zones in the country. Bulletin of the Indian Agricultural Statistics Research Institute, New Delhi, India: ICAR.

Bhatt, P.N., Gupta, O.P. and Tejwani, K.G. 1971. Influence of cropping and land use on plant nutrientlosses in Doon Valley. Intern. Symposium Soil Fertility Evaluation, Part. 1: 541 p held in Feb. 1971.

Bhusan, L.S. and Ghosh, S. 1981. Tillage and surface mulching for moisture conservation for double cropping under rainfed conditions. Dehradun Rep. CSWCRTI., Dehradun: 79 p.

Blackman, F. F. 1905. Optima and limiting factors. Ann. Bot. 19: 281-295.

Blackman, V. H. 1919. The compound interest law and plant growth. Ann. Bot. 33:253-360.

Boote, K. J. 1982. Growth stages of peanut (*Arachis hypoaea*). Peanut Sci. 9(1): 35-39.

Bourke Austin, P.M. 1968. The Aims of Agrometeorology. Agroclimatological Methods. Proceedings of the reading symposium UNESCO, Paris. 11-15

Bourke Austin, P.M. 1970. The Aims and Extent of Agricultural Meteorology. Agricultural Meteorology. Vol. 7.: 281-283.

Briggs, G. E., Kidd, F. and West, C. 1920. A quantitative analysis of plant growth. Ann. Appl. Biol. 7: 103-123, 202-223.

Briggs, G. E. 1925. Plant yield and the intensity of external factors. Ann. Bot. 155: 475-502.

Butler, W.L. Downs, R.J. 1970. Light and Plant Development. Plant Agriculture. Freeman, San Francisco: 78-85.

Byerlee, D. and Collinson, M. 1980. Planning technologies appropriate to farmers: Concepts and procedures, EI Batan, Mexico: CIMMYT.

Byerlee, D., Harrington, L. and Winkelmann, D.L. 1982. Farming systems research: Issues in research strategy and technology design. Amer J. Agric. Econ. 64: 897-904.

Campbell, CA., McConkey, B.G., Zentner, R.P., Dyck, F.B., Seller, F and Curtin, D. 1995. Carbon sequestration in a brown chernozen as affected by tillage and rotation. Can. J. Soil Sci. 75: 499-458.

Cannon, W.A. 1925. Physiological features of roots with special reference to the relation of roots to the aeration of soil. Carnegie Inst. Wash. Pub. 268: 1-168.

CGIAR, 1981. Second review of the CGIAR, Review report, Nov. 1981. Washington, D.C, CGIAR, Secretariat.

Chakraverty, A. 1995. Post harvest technology of cereals, pulses and oilseeds. Oxforrd anf IBH Publishing Co. Pvt. Ltd., New Delhi.

Chambers, R. 1980. The Small Farmer is a Professional. Cares. 13(2).

Chambers, R. 1990. Micro-nutrients unobserved. Gatekeeper Series 22. London 11ED.

Chandra, S. and Sharma, A.K. 1981. Runoff, soil and nutrient losses under different management practices on 2 % slope. Rep. CSWCRTI, Dehradun, 76 p.

Chandrasekharan, B. and Pandian, B. J. 2009. Rainwater harvesting and water saving technologies. Ind. Jour. Agronomy. 54 (1): 90-97.

Chang, J. 1968. Climate and Agriculture. Aldine Publishing Company, Chicago: 4-22.

Chatterjee, B.N. and Bhattacharya, K.K. 1986. Principles and Practices of grain legume production. Oxford and IBH Publishing Co. Pvt. Ltd., New Delhi.

Chatterjee, B.N. and Das, P.K. 1989. Forage crop production- principles and practices. Oxford and IBH Publishing Co. Pvt. Ltd., New Delhi.

Chatterjee, B.N. and Mandal, B.K. 1988. Rice based farming systems with integration of fish culture. In Abst. of National seminar on "Farming Systems for Semi-arid Tropics". Tamil Nadu Agric. Univ., Coimbatore. pp. 1-7.

Chatterjee, B.N, and Maiti, S. 1984. Cropping systems-theory and practice. Oxford and I.B.H. Publishing Co., Calcutta, India.

Cheema, S.S., Dhaliwal, B.K. and Sahota, T.S. 2001.Theory and Digest Agronomy. Kalyani Publishers, Ludhiana.

Chellaiah, S. and Chandrasekaran, B. 1989. Integrated farming system for Thanjavur district, Tamil Nadu. Thanjai Ulavar Mani, Dec. 88-Jan., 89 pp. 5-16.

Chen T.P. and Li, Y. 1980. Integrated agriculture aquaculture studies in Taiwan. In: Proc. of ICLARM-SEARCA conference on "Integrated Agriculture - Aquaculture Farming System". 4th Aug. 1979. Manila, Phillippines pp.: 239-241.

Chidda Singh, 1984. Modern techniques of raising field crops. Oxford and IBH publishing Co. Pvt. Ltd., New Delhi.

Chinnusamy, C. and Purushothantan, S. 1992. Integrated farming systems under garden land and their economics. In: Proc. National training on "Integrated Farming Systems". GOI and Tamil Nadu Agric Univ., Coimbatore. 10-17 Dec., 1992 pp. 46-48.

Chisci, G. and Zanchi, C. 1980. The influence of different tillage systems and different crops on soil losses on hilly silty clayey soil. Soil Conservation. Ed. Morgan John Willey & Sons.: 211-218.

CIMMYT and ISNAR, 1984. Issues in organization and management of research with a farming systems: Perspective aimed at technology generation (Ed. M. Joseph) The Hague, Netherlands: CIMMYT and ISNAR.

CIMMYT, 1974. Plan publication: Seven years of experience, 1967-1973. EL Batan, Mexico: CIMMYT.

Collinson, M.P. 1983. Farm management in peasant agriculture, 2nd edition, West View Press.

Collinson, M.P. 1987. FSR procedures for technology development. Expt. J. Agric. 23: 365-388.

Collinson, M.P. 1988. The development of African farming systems – Some personal views. Agric. Admin. and Extension 29: 7-22.

Collinson, M.P. 1982. Farming systems research in East Africa: The experience of CIMMYT and some National Agricultural Research Services, 19761981. Michigan State University International Development Paper, 7, East Lansing, Michigan, USA.

Concise Handbook for field official of Non-conventional energy sources, 1993. Published by Directorate of Rural Development, Government of Tamil Nadu, Chennai.

Cornick, T. 1983. The Social organization of production in Quimas, Ecuador: A case study of small farmer production in the highlands. Ph.D. Thesis, Cornep University, Ithaca, New York.

Cracknell, A. and hayes, L. 1987. Remote Sensing Year Book.: 33-73, Taylor and Francis, London.

Crafts, A.S. 1975. Modern Weed Control. University of California Press, California.

Crapman, S. R. and Carter, L. P. 1982. Crop Production – Principles and practices. Surjeet Publications, Delhi.

CRIDA (Central Research Institute for Dry land Agriculture), 1986. Fifteen years of dry land agricultural research in India. Hyderabad, Andhra Pradesh, India: Central Research Institute for Dry land Agriculture.

Critenfield, H.J. 1975. General Climatology. Third edition. Prentice-Hall of India, Private Ltd., New Delhi.

CSSRI, 1979. A decade of Research, Central Soil Salinity Research Institute (ICAR), Karnal, India.

Daniel Sundaraj, D. and Thulasidas, G. 1993. Botney of Field Crops. Mac Millan India Ltd., New Delhi.

DARE (Department of Agricultural Research and Education), 1990. Annual Report, New Delhi. India: Department of Agricultural

Research and Education of the Ministry of Agriculture, Government of India.

Dargan, K.S., Singh, O.P. and Gupta, I.C. 1982. Crop production in salt affected soils. Oxford and IBH Publishing Co. Pvt. Ltd., New Delhi.

Das, K.C., Panda, S.C., Sahu, S.K. and Rout, D. 1985. Effect of depth of submergence at different growth stagesof dwarf *indica* rice on the growth, yield and nutrient uptake by rice and mineral contents of soil. Oryza. 22(1): 69-83.

Das, L.K. and Panda, S.C. 2002. Economics of hybrid rice cultivation as influenced by Nitrogen and Potassium levels. Jour. Res. OUAT, 20(1):50-52.

Dash, A.N., Mahapatra, P.K. and Panda, S.C. 1991. Yield performance of rice under different irrigation schedules and soil fertility. SUJ. Sci. & Tech. X (6-13) Silver Jubilee Volume.

Datta, R.K. 1989. Weather Forcasting – Role of medium range techniques. In Survey of Indian Agriculture 1989. The Hindu, Madras.

Daubenmine, R.F. 1962. Plants and Environment. Willey, New York., 78-79.

Day, John A. 1967. The science of weather. Addison, London, 69-71.

Dayal, R., Singh, G. and Bhola, S.N. 1965. Investigation on the manuring under rainfed conditions in medium black soil. J. Indian Soc. Soil Sci. 13: 257-263.

De Datta, S.K. 1981. Principles and Practices of Rice production. John Wiley and Sons, New York.

De Datta, S.K., Bolton, F.R. and Lin, W.L. 1979. Prospects for Using Minimum Tillage in Tropical Low Land Rice., Weed Research. 19: 9-15.

De Gopal Chandra. 1989. Fundamentals of Agronomy. Oxford and IBH Publishing Co. Pvt. Ltd., New Delhi

Delmando, M.N. 1980. A review of Integrated livestock, bowls and fish farming systems. In: Proc. ICLARM - SEARCA Con£. on "Integrated agriculture - aquaculture farming systems", 4th Aug. 1979, Manila, Philippines.

Demnis, A.S. 1983. Augmentation of rainfall from summer cumulus clouds. Agric. Water Manage., 7:3-14.

Dev, G. and Sander, L.L. 1989. Maximum yield Research. Proceedings of symposium on Maximum yield research. Nov. 16-18, 1988, New Delhi. Potash and Phosphate Institute of Canada, India Programme, Potash Research Institute of India, Haryana, India.

Devadas, P., Rajammal, P. S. and Sithalakshmi, S. 1988. Appropriate home science technologies with their implications to farmwomen and farming. In: International Conference on "Appropriate Agricultural Technologies for Farm Women". Nov. 30 - Dec. 6, 1988.

Devasanapathy, P., Senthel Kumar and Shanmungam, P. M. 2009. Energy Management in crop production. Indian Jour. Agronomy: 54(1): 80-90.

Dhruvanarayana, V.V. 2002. Soil and Water Conservation Research in India. ICAR, New Delhi.

Dillon, J.L. and Anderson, J.R. 1984. Concept and practice in Farming Systems, Research. In: Proceedings of AGAR Consultation in Agricultural Research in E. Africa (Ed. J.V. Martin), Canberra, AGAR.

Dillon, J.L., Plucknett, D.L. and Vallaeys, G.J. 1978. Farming systems research in the International Agricultural Research Center, Rome Italy: Technical Advisory Committee to the Consultative Group on International Agricultural Research.

Djajadivedja, R., Jangkaru, Z. and Mohamed Junu, 1980. Freshwater aquaculture in Indonesia with special reference small-scale agriculture. In: Proc. ICLARM - SEARCA Conf. on "Integrated Agriculture Aquaculture Farming Systems", 4th Aug., 1979, Manila, Philippines pp. 143-165.

Donn, W.L. 1965. Meteorology, McGraw Hill. New York.

Donnel Hunt. 1997. Farm power and machinery management. Iowa State University Press, Iowa, USA.

Dora, B. 1983. Traditional agriculture in India: High yields and No waste. The Ecologist 13: 84-87.

Doughlass, G.K. (ed.), 1984. Agricultural sustainability is a changing. World order Boulder: West view.

Dunham, R.S. 1957. Introduction to Agronomy. Dryden Press, New York.

Dupre, G. (ed.) 1990. Farming knowledge and development: Proeedings on the 7th section on Science, Diffusion of technical knowledge and systems of local knowledge of the 7th World congress of rural sociology, June 1988, Bologna, Italy.

Dwivedi, A.P. 1992. Agroforestry – Principle and practices. Oxford and IBH Publishing Co. Ltd., New Delhi.

Easterbrook, L. F. 1945. Farming and Mechanized Agriculture, p 55.

Edwards, C.A. and Grove, T.C. 1991. Integrated nutrient management for crop production. Agric. Ecosystem and Environment.

Elliot, H. 1977. Farming systems research in Africa: Methods and results, Ford Foundation Farming Systems Seminar, Tunis, Tunisia, February 13, 1977.

Elsner, J. E., Wayne Smith, C. and Owen, D. F. 1979. Uniform stage descriptions in upland cotton. Crop Sci. 19: 361-363.

Enyi, B. A. C. 1962. Comparative growth stages of upland and swamp rice varieties. Ann. Bot. 26: 467-487.

Eugene, P.O. 1996. Fundamentals of Ecology. Natraj Publishers, Dehra Dun.

Evans, L. T. 1975. Crop Physiology. University Press, Cambridge.

Evans, L. T. and Wardlaw, I. F. 1976. Aspects of the comparative physiology of grain yield in cereals. Adv. Agronomy. 28: 301-359.

FAI. 1997. Fertilizer statistics 1995-96. The fertilizer association of India. New Delhi.

FAO, 1985. Production yearbook Vol. 37 FAO, Rome, Italy.

FAO, 1990. Comparative analysis of organic manure. Project field document. Vol. 24, FAO, Rome, Italy.

FAO. 1988. Sericulture manual Vol. I, II and III. Mulberry cultivation. Silk worm rearing and silk reeling. Oxford and IBH Publishing Co. Pvt. Ltd., New Delhi.

FAO. 1997. Year Book – Production Vol. 50. 1996. Food and Agricultural Organisation of the United Nations, Rome.

Fehr, W. E., Caviness, C. E., Burmood, D. T. and Pennigton, J. S. 1971. Stage of development description for soybeans, *Glycine max* (L). Merril, Crop Sci. 11: 929-931.

Francis, C.A. 1986. Multiple Cropping System. MacMillan Publishing Co., New York.

Francis, O. and Stem, W.R 1987. Cereal-legume intercropping systems. Advances in Agronomy, 41: 41-87

Franklyn, W. Cole. 1970. Introduction to Meteorology. John Wiley and Sons. Inc., New York.

Frasco. L. 1984. Comparing Anglophone and Francophone approaches to Farming systems research and extension. Paper prepared for the 4th annual conference on "Farming Systems Research" at Kansas State University, Manhattan, Kansas, USA, October, 1984.

Fresco, L.C. and Westphal, E. 1988. A hierarchical classification of farming system. Exptl. Agriculture 24: 399-419.

Gadkary, D.A. 1957. Mechanical Cultivation in India, p. 4.

Gallagher, E.J. 1984. Cereal production. Butter worths and Co. Ltd., London.

Ganesan, S., Chinnasamy, K.N., Chandrasekaran, B., Budhar, M.N. and Prince Jayaseelan, N.J. 1991. Duck-cum-fish culture in rice farming system in Cauvery delta region of Tamil Nadu. Indian Journal Agric. Economics 46(2): 180-185.

Gates, D.M. 1965. Radiant Energy, Its Receipt and Disposal. Agricultural Meteorology – Meteorological Monograph, Vol. 6 No. 28: 1-24.

Gaur, A.C. 1990. Principle and technology for composting urban wastes. Proc. of International Workshop on "Solid Waste Management and Resource Mobilization". Vol. B. Paper 3-12, Kathmandu.

Gaur, A.C. and Gaind. S. 1992. Role of P solubilizing microorganims and enriched organic manure in crop productivity. In: Proc. National seminar on "Organic Farming". Ravi Printing Press, Jabalpur, India. pp. 134-143.

Gaur, A.G., Neelakandan, S. and Dergan, K.S. 1984. Organic manures. ICAR, New Delhi.

Gautaman, K.C. and Sankaran, S. 1976. Effect of Minimum Tillage in Deccan Hybrid Maize. Madras Agri. Jour. 63: 445-448.

George, J. and Navalgund, R. 1991. Remote Sensing, Its Physical Basis and Evolution in India. Glimpse of Science in India. 357 p. National Academy of Science. Diamond Jubilee Commemoration Volume. Malhotra Pub. House, New Delhi

Ghadekar, S.R. 1995. Meteorology. 186 p Agromet Publishers, Nagpur.

Gilbert, EH., Narman. D.and Winsh, F.E. 1980. Farming systems research. A critical appraisal. Rural development paper No. 6, East Lansing Michigan State Univ., Michigan.

Gill, M. S. and Brar, L. S. 2005. Conservation Agriculture – Status and Prospects. Eds. Abrol, I. P., Gupta, R. K. and Malik, R. K. HAU, Hisar-Haryana: pp 64-71.

Gill, M. S., Pal, S. S. and Ahlawat, I. P. S. 2008. The approaches for sustainability of rice-wheat Cropping System in India – Gangetic plains of India – A review. Indian Journal of Agronomy 53: 81-96.

Gill, M.S. 2009. Farming System Approach: Its Principles and Role towards sustainability.

Resource Management in Crops and Cropping System under Changing Climate: 32-37.

Gips, T. 1986. What is sustainable agriculture? In: Allen, P. and D. Van Dusen (eds.) Global perspectives on Agroecology and sustainable agricultural systems: Proc. of the 6th International scientific conference on "International Federation of Organic Agricultural Movements" (Sanfa CVUZ: Agroecology Program), University of California, 1: 63-74.

Glenn T. Trewartha. 1968. An introduction to Climate. McGraw Hill Book Co. New York.

Gomez, A.A. 1990. Farming system research approach to identifying farmers' production problems. Farming system research approach (8): 63-70.

Gomez, A.A. and Gomeg, K.A. 1983. Multiple cropping in humid tropics of Asia. IDRC. Ottawa, Canada.

Gomez, K.A. and Gomez, A.A. 1984. Statistical procedures for agricultural research with emphasis on rice. John Wiley and Sons Inc., New York.

Gopalchari, N.C. 1984. Tobacco. ICAR. New Delhi

Gopalswamy, N. 1994. Agricultural Meteoroly. Rawat Publications, New Delhi.

Goswami, B.N. 1996. The Challenges of weather prediction: 1. The basic driving. Resonance, 1(2): 8-17.

Goswami, B.N. 1996. The Challenges of weather prediction: 2. Difficulties in predicting the weather. Resonance. 2(1): 8-15.

Goswami, B.N. 1997. The Challenges of weather prediction: 3. Old and modern ways of weather forecasting. Resonance, 2(3): 8-15.

Goud, J.V. 1988. Strategies for Food Production to meet the Challenges of 21st century. Agric. Situation in India. Sept.: 505-506.

Govina, L.D. 1990. Pig-Duck-Fish-Azolla integration in La union, Philippines. NAGA the ICLARM Quarterly, Apr. 18-20.

Govindan, R. 1988. Role of Popultry-cum-Fish culture on the economics of farming systems in Thanjavur district of Cauvery delta. M.Sc. (Ag.) Thesis, Tamil Nadu Agric. Univ. Coimbatore.

Govindan, R., Chinnasamy, K.N., Chandrasekaran, B., Budhar, M.N. and Price, J. 1990. Poultry - Fish culture in rice farming systems in Cauvery delta regions of Tamil Nadu, India. J. Agron. 35: 23-29.

Gregory, F. G. 1917. Annual Report. Experimental and Research Station, Chesnut.

Gulati, J.M.L., Panda, S.C. and Misra, B. 1988. Effect of water regimes on wheat. Ind. Jour. Agro. 33(4): 364-367.

Gulati, J.M.L., Panda, S.C., Misra, B., Taha, M., Acharya, N. Singh, N. and Dash, A.N. 1987. Effect of levels of nitrogen, spacing and planting direction on rice yield. Oryza. 24: 358-362.

Gupta, O.P. 1993. Weed Management – Principles and Practices. Agrobios (India), Jodhpur, India.

Gupta, O.P. 1999. Modern weed management. Agrobios (India), Jodhpur, India.

Gupta, P.K. 200. Soil, plant, water and fertilizer analysis. Agrobios (India), Jodhpur, India.

Gupta, P.K. 2000. Hand book of soil, fertilizers and manures. Agrobios (India) Jodhpur, India.

Gupta, R.S., Khybri, M.L. and Singh, B. 1963. Runoff plot studies with different grasses with special reference to conditions prevailing in Himalayas and Shiwalik region. Indian Forester. 89 (2): 128-133.

Gupta, S.K. and Ram Babu, 1977. Studies on efficiency of contour farming, channel terracing and channel terraces with graded furrows for erosion control. Soil Conservation Digest. 5(2): 29-32.

Gupta, U.S. 1982. Physiological aspects of dryland farming. Oxford and IBH Publishing Co. Pvt. Ltd., New Delhi.

Gurmel Singh, Arora, Y.K., Pratap Narain and Grewal, S.S. 1990. Agroferestry Research, Surya Publications, New Delhi.

Gurmel Singh, Venkataramanan, C., Sastry, G. and Joshi, B.P. 1990. Manual on soil and water conservation practices. Oxford and IBH Publishing Co. Pvt. Ltd., New Delhi.

Gururaj Hunsigi and Krish, K.R. 1998. Science of field crop production. Oxford and IBH Publishing Co. Pvt. Ltd., New Delhi.

Hake, R.E: 1982. Agroclimatic and dry-season maps of South, Southeast and East Asia Losbanos, Philippines: International Rice Research Institute.

Halmann, M. 1990. Synthetic plant growth regulators. Adv. Agronomy 43: 48-105.

Hardy, R. W. F. 1979. Chemical plant growth regulation in world agriculture. 'In plant regulation and world agriculture' Ed. Scott, T. K. Pleamum Press, New York.:

Harpal Singh, M. 1974. Agricultural Meteorology, Punjab Agricultural University, Ludhiana.

Harrington, I. and Erenstein, O. 2005. Conservation Agriculture and resource conserving technology – A global perspective. Conservation Agriculture – Status and Prospects. Eds. Abrol, I. P.: Gupta, R.K. and Malik, R. K. HAU, Hisar-Haryana, pp. 1-12.

Hart, R.D. 1982a. The ecological systems conceptual farmework for agricultural research and development. In: Readings in Farming Systems Research and Development, 44-59 (Eds. W. Shaner, P. Philip and W. Schmehl). Boulder, Colorado: West View Press.

Hart, R.D. 1982b. One farming system in Honduras: A case study in farming systems research. In: Readings in Farming Systems Research and Development, 59-73 (Eds. W. Shaner, P. Philip and W. Schemehl.) Boulder, Colorado: West View Press.

Harwood, R. 1977. Socio-economic considerations in multiple cropping systems. Invited Paper prepared for the round table discussion on "Agricultural Production System". XVI Annual reunion of the Board of Directors, Institute Intermericano de Ciencias Agricolas (IICA), San Domingo, Dominican Republic.

Harwood, R. 1979. Small-farm development: Understanding and improving farming systems in the Humid tropics. Boulder, Coforado: West View Press.

Harwood, R.R. 1987. Agroforestry and mixed farming systems. In: O.R Clark and RD. Child code). Ecological development in the humid tropics, Guidelines for planners. Morrilton, Arkanasas: Winrock International.

Heastie, H. 1978. A course in elementary meteorology. Her majesty's stationery office, London.

Heichel, G.H. 1987. Legume Nitrogen: Symbiotic fixation and recovery by subsequent crops. In: Z.R Helsel (ed.), Energy in plant nutrition and pest control. Elsevier scientific publishers, Amsterdam. pp 63-80.

Herdt, R W. 1993. Measuring sustainability using long-term experiment. Proc. Conf. held at Rothamstead experiment station, 29-30 April, 1993. Rothamstead, U.K.

Hickling, C.F. 1960. Observations on the Growth Rate of Chinese Grass Carp, *Ctenopharingodon idellus (c & v).* Malay Agric. J. 43(1): 43-53.

Hiebsch, C.K. and McCollum, R.E. 1987. Area x Time equivalency ratio: A method for evaluating the productivity of intercrops. Agron J. 79: 15-22.

Hildebrand, P. 1981. Motivating small-farmers to accept change. In: Readings in farming systems research and development, 100-109 (Ed. W. Shaner, P. Philip and W. Schmehl). Boulder, Colorado: West View Press.

Hiran, K.S., Jaspal Singh and Acharya, M.S. 1990. Irrigation scheduling. CBS Publishers and Distributors, New Dehi.

Horton, O. 1984. Social scientists in agricultural research: Lessons from the Mataro Valley Project, Peru, Ottawa, Canada: lore.

Horton, P. 1984. Technology evaluation: Five case studies from West Africa. In coming full circle: Farmers' participation in the development of technology, 95-118 (Ed. P. Matlon, R. Cantrell, D. King and M Benoitcattin). Ontario: IDRC.

Howard Critchfield, J. 1962. General climatology. Prentice hall of India Pvt. Ltd., New Delhi.

Hridai Ram Yadav. 1988. Dimensions of wasteland development. Proceedings of the National Seminar on Wasteland Development, New Delhi.

Hudson, N.W. 1957. Erosion control research-Progress report on experiments at Henderson Research Station, 1953-56. Rhod Agri. J. 54 (4): 297-323.

Hudson, N. 1971. Soil Conservation. B.T. Batsford Ltd. London.

IARI 1972. Recent research on multiple cropping. A bulletin. Indian Agricultural Research Institute, New Delhi.

IARI. 1977. Water requirement and irrigation management of crops in India. Water Technology Centre, IARI, New Delhi.

ICAR (Indian Council of Agricultural Research), 1972. A new strategy in agriculture: a collection of speeches of C. Subramanian, New Delhi: ICAR.

ICAR (Indian Council of Agricultural Research), 1983. Shifting cultivation in Northeast IndiJi. ICAR Research Complex for NE Hill region, Shillong. New Delhi, India: ICAR.

ICAR (Indian Council of Agricultural Research), 1989. Annual report 1987. 88. New Delhi, India: ICAR.

ICAR. 1960. Indigenous agricultural implements of India- An all India Survey. ICAR, New Delhi.

ICAR. 1977. Hand book of Animal Husbandry. ICAR, New Delhi.

ICAR. 1981. Indian Farming – Special number on Pulses Vol. 31 No. 5.

ICAR. 1982. A Decade of Dryland Agricultural Research in India 1971-80. All India Coordinated Research Project for Dry land Agriculture. Hyderabad, India.

ICAR. 1982. Indian Farming – special number on Oilseeds. Vol. 32 No. 8.

ICAR. 1985. Rice Research in India. ICAR, New Delhi.

ICAR. 1996 Indian Farming Special Issue. XXth World's poultry congress 2-8 Sept. 1996, New Delhi ICAR, New Delhi.

ICAR. 1997. Handbook of Agriculture (Revised). Indian Council of Agricultural Research, New Delhi.

ICRISAT, 1980. Proceedings of the International Workshop on the Agro-Climatological Research needs of the semi-arid Tropics, 22-24 November, 1978, Hyderabad, India.

ICRISAT. 1981. Proceedings of the International workshop on Intercropping, 10-13 Jan 1979. International Crops Research Institute for Semi-Arid Tropics. Hyderabad, India.

ICTC. 1960. Indian Tobacco. Indian Central Tobacco Committee, Ministry of Food and Agriculture, Government of India, Madras.

Indian Farming, 1989. Special issue on watershed management. ICAR, New Delhi, India.

Indian Institute of Tropical Meteorology, 1976. Proceedings of the symposium on Tropical Monsoon held at Pune-5, India, 8-10 September, 1976.

IOBC, 1993. International organization for biological and integrated control. Technical bulletin 16(1), 1993.

IRRI. Green manuring in rice farming. International Rice Research Institute, Manila, Philippines.

ISA. 1972. Multiple Cropping. Proceedings of a symposium. Oct. 7-8, 1972, Haryana Agricultural University, Hissar and Indian Society of Agronomy and ICAR, New Delhi.

ISA. 1986. Proceedings of the National Symposium on cropping system. 3-5 April 1985. Central Soil Salinity Research Institute, Karnal and Indian Society of Agronomy, New Delhi.

Ishizuka, 1971. Physiology of rice plant. Adv. Agronomy. 43: 241-315.

Isrealsen, O.W. and Hansen, V.E. 1962. Irrigation- Principles and practices. John Wiley and Sons Inc., New York.

Jain, B. K. S. 1965. Mechanization can lead to swift Changes. Productivity, Vol. VI, p. 499.

James, A. Taylor. 1972. Weather Forecasting for agriculture and industry. Davis and Charles (Publishers) Limited, Great Britain.

Jayanthi, C. 1995. Sustainable component linkage and resource recycling to lowland integrated farming system. Ph.D. Thesis, Tamil Nadu Agri. Univ., Coimbatore.

Jayanthi, C., Rangasamy, A., Venkitusamy, R. and Chinnusamy, C. 1994. Sustainable integrated farming systems research and management for sustainable agriculture. Summer Institute held at Tamil Nadu Agric. Univ., Coimbatore between 6-15th June, 1994, pp.: 97-109.

Jayaseelan, M.J.P., Venugopal, M.S. and Chelliah, S. 1988. Two innovative culture methods for increased integrated fish production. Farm forum 11: 23-26.

Jena, B.K., Patro, N. and Panda, S.C. 1994. Effect of enriched farm yard manure and fertilizer on finger millet in rainfed condition. Environment & Ecology. 15(1):136-138.

Jeremy, H., Davis C., Jonathan N, and Woolley, N. 1986. Multiple cropping with legumes and starchy roots. In: Francies, C.A. (eds.) Multiple cropping systems. Macmillan Publishing Co., New York

Jeswani, L.M. and Baldev, B. 1990, Advances in pulse production technology. ICAR., New Delhi.

Jha, L.K. 1995. Advances in Agroforestry, APH Publishing Co., New Delhi.

Jhingran, U.G. and Sharma, B.K. 1978. Operational Research Project on Aquaculture. Bull Cent. Inl. Fish Inst. 27.6.

Jhingran, U.G. and Sharma, B.K. 1980. Integrated livestock fish farming in India. In: Proc. ICLARM - SEARCA Conf. on "Integrated Agricultural -Aquaculture Farming Systems", 9th Aug. 1979, Manila Philippines, pp. 135-141.

John, K.C. 1991. Farming systems research in Eastern India: A new approach, New frontiers in rice research. In: Siddique, E.A. and K. Muralidharan (eds.), Directorate of Rice Research, Hyderabad.

John, Milton Poehlman, 1991. The mungbean. Oxford and IBH Publishing Co. Pvt. Ltd., New Delhi.

Johnson, H. and Clear, B. John. 1986. FSRE: Shifting the intersection between Research and Extension. Agric. Administration. 21: 61-93.

Josef D. Zimmerman. 1996. Irrigation. John Wiley and Sons Inc., New York.

Joseph, T. and Morrison, M. 2006. Nanotechnology in agriculture and food. A Nano forum report. www nanoforum.org.

Joshi, N.C. 1987. Manual of weed control. Researchco Publications, New Delhi.

Kakde, J.R. 1985. Agricultural Climatology, Metropolitan Book Co., New Delhi.

Kamaraj, S. 1993. Do's and Don'ts in Biogas plants. Published by Department of Bio energy, Tamil Nadu Agric. Univ., Coimbatore.

Kamaraj, S., Vijayaragavan, N.C. and Pugalendhi, S. 1994. Role of biogas in IFS. In: Proceedings of Summer Institute on "Integrated Farming Systems Research and Development for Sustainable Agriculture" organized by ICAR and Tamil Nadu Agric. Univ., Coimbatore, India during 6-15 June, 1994. pp. 203-212.

Kang, B.T. 1993. Changes in soil chemical properties and crop performance with continuous cropping on an entisol in the Humid tropics. In: Soil organic matter dynamics and sustainability of tropical agriculture. Mulongoy, K. and R. Mercisk (ed.). Wiley and Sons, U.K. pp. 297-306.

Kanitkar, N.V., Sirur, S. and Gokhale, D.H. 1960. Dry farming in India. New Delhi, India!: ICAR.

Kanwar, J.S. 1972. Cropping patterns, scope and concept. In: Proceedings of Cropping patterns in India, 11-38, New Delhi, India ICAR.

Kanwar, J.S. 1978. Soil fertility – Theory and practices. ICAR, New Delhi.

Kanwar, J.S., Virmani, S.M. and Das, S.K. 1992. Farming systems research in India: A Historical perspective. Expl. Agric. 28: 1-17.

Kar, S. 1995. Paper presented in group meeting on "Tillage requirement of rice-wheat system". Held at CSSRI, Karnal, 16-17 August, 1995.

Karuppiah, V. and Thangamuthu, G.S. 1986. Studies on the efficiency of phosphorus forms in combination with organic manures in lowland rice. Proc. Seminar on "Rock Phosphate in Agriculture". Tamil Nadu Agric. Univ., Coimbatore, India.

Katherkhan, S. 1994. Role of Dairy, Sheep and Goat, Piggery and Rabbit rearing in IFS development. In: Proceedings of Summer Institute on "Integrated Farming Systems Research and Development for Sustainable Agriculture" organized by ICAR and Tamil Nadu Agric. Univ., Coimbatore, India during 6-15 June, 1994. pp. 228-234.

Kathiresan, R. M. 2009. Integrated farm management for linking environment. Ind. Jour. Agronomy. 54(1): 9-14.

Katyal, J.C. 1997. Sustainability: Relevance of Integrated Concept. Survey of Indian Agriculture. The Hindu, Chennai, India

Katyayan, A. 2001. Fundamentals of Agriculture. Kushal Publications, Varanasi.

Kavimandan, S.K., Subba Rao, N.S. and Mohrir, A.V. 1978. Isolation of Spirilum lipoferum from the stems of wheat and N Fixation in enrichment cultures. Current science 47: 96-98.

Keshava Murthy, R. and Sankar Rao, M. 1992. The physics of monsoons. Allied publishers limited, New Delhi.

Khan, A.R. 1974. The Art and Science of Tillage. Published by Author, India.

Khan, S.A., Roy, A.S. and Chatterjee, B.N. 1981. Note on growing Winter Maize under Minimum Tillage after Transplanted Rice. Ind. Jour. Agri. Sci., 51: 54-55.

Khanda, C.M., Dixit, L. and Panda, S.C. 1997. Effect of zinc and graded levels of nitrogen on growth, yield and nutrient uptake of rice. Oryza. 34(1):43-46.

Khanna, S.S. 1989. The agro-climatic approach. In Survey of Indian Agriculture 1989. The Hindu, Madras.

Kirby, E. J. M. and Appleyar, M. 1984. Cereal development and its relation to crop management. In cereal production Ed. E. J. Gallaghar. Butterworths, London.

Klages, K.H. 1961. Ecological Crop Geography. Macmillan, New York: 266-80.

Kline, C.K., Green, D.A.G., Donahue, R.L. and Stout. 1969. Agricultural Mechanization in Equatorial Africa.

Klingman, G.C. 1981. Weed Science. Principles and practices. John Wiley and Sons Inc., New York.

Kondap, S. and Upadhyaya, U.C. 1985. A practical manual of weed control. Oxford and IBH Publishing Co., New Delhi.

Korwar, G.R., Osman, M., Tomar, D.S. and Singh, R.P. 1991. Dry land horticulture, CRIDA, Ext. bulletin 4: 33.

Kumar, D. and Rai, M. 1992. Advances in Oilseeds Research. Vol. I. Rapeseed and Mustard. Scientific publishers, Jodhpur, India.

Kumar, K and Singh, B.N. 1984. Preliminary observation on utilization of palletized poultry litter as feed for common carp. Int. J. Acad. Ichtyol: 151-154.

Kuppusamy, M. 1989. Feasibility of introducing sericulture as a component in the farming

System of Thanjavur district, Tamil Nadu. M.Sc. (Ag.) Thesis, Tamil Nadu Agric. Univ., Coimbatore, India.

Lal, R. 1993. Tillage effects on soil degradation, soil resilience, soil quality and sustainability. Soil and Tillage Res. 27: 1-8.

Lal, R. and Miller, F.P. 1990. Sustainable farming for tropics. In: Sustainable agriculture: Issues and prospective. Vol. 1. (Ed.) R.P. Singh. pp. 69-89, Indian Society of Agronomy, IARI., New Delhi.

Lal, R. and Miller, F.P. 1993. Soil quality and its management in humid subtropical and tropical environments. Proc. of XVI International Grass land Conference, Massey, New Zealand.

Lal, R., Kimble, J., Levine, E. and Stewart, B.A. (ed.), 1994. Soils and Global changes. Chelsea: Lewis Publishers.

Laird, R. D. Soviet Goals for 1965 and Problem of Agriculture. Productivity, Ibid. p. 282

Lal, R. 1991. Soil structure and sustainability, J. Sustainable Agric. 1: 67-92.

Lang, M. and Cantrell, R. 1984. Accenting the farmers' role: Pursue farming systems unit. In coming full circle: Farmers' participation in the development of technology, pp. 63-70 (Ed. P. Matlonn, R. Cantrell, D. King and M. Benoit-Cattin). Ottawa, Ontario: IDRC

Larson, K. J., Cassman, K.G. and Phillips, D.A. 1989. Yield, nitrogen fixation, and above ground nitrogen balance of irrigated white lupine i", a Mediterranean climate. Agron J. 81: 538-543.

Lawrence J. King. 1966. Weeds of the world – Biology and control. Interscience Publishers, Inc. New York.

Lawton, H.W. and Wilke, P.J. 1979. Ancient agricultural systems in dry regions. In: Hall, AC, G.H. Cannell and H.W. Lawton (ed.)., Agriculture in semi-arid environments. Springer Verlag, New York. pp. 1-44.

Leewrik, D.M., Panda, S.C. and Mahapatra, I.C. 1972. Multiple Cropping in Sambalpur. Bull. Ford Foundation. New Delhi.

Lenka, D. 1998. Climate, Weather and Crop in India. 490 p, Kalyani Publishers, Ludhiana.

Libuno, LP. 1990. Goat - fish integrated farming in the Philippines. Ambio. 19: 408-410.

Liebig, 1840. Chemistry in its Application to Agriculture and Physiology. Peterson, Philladelphia.

Lighfoot, C. and Minnick, D.R. 1990. Farmer - first qualitative methods: Farmers' diagram for improving methods of experimental design in integrated farming systems. A FSRE Symposium on "The Role of Farmers in FSRE and Sustainable Agriculture". 14-17 October 1990, Michigan State University, East Lansing, USA

Lightfoot, C., Bimbao, M.N., Dalsgaard, J.P.T. and Pulhin, R.S. 1992. Aquaculture and sustainability through integrated resource management. Quarterly Agric. 22: 143-150.

Little, D. and Muir, J. 1987. A guide to integrated warm water agriculture. Stirling. Scotland. Institute of Agriculture, University of Stirling.

Lock Wood, J.G. 1979. World Climatology-An Environmental Approach. The English Language Book Society and Edward Arnold (Publishers) Ltd., Bedform Square.

Loganathan, P. 1989. Mushroom cultivation as a component in mixed farming system in Thanjavur district, Tamil Nadu. M.Sc. (Ag.) Thesis, Tamil Nadu Agric. Univ. Coimbatore.

Lowry, W.P. 1967. Weather and Life. Academic, New York. 13-29.

M.K. Dowell, R. and Hidlebrand, P. 1980. Integrated crop and animal production: Making the most of resources available to small farms in developing countries. Working papers - The Rockfeller Foundation, New York: The Rockfeller Foundation.

Mahapatra, I.C. 1971. All India coordinated agronomic experiments scheme. In: National agricultural research systems in Asia, 226-232, (Ed. A.H. Mosenan). New York 10020, USA: Agriculture Development Council.

Mahapatra, I.C. 1994. Farming Systems Research: A Key to Sustainable Agriculture. Fertilizer News. 39(11): 13-25.

Mahapatra, I.C. and Panda, S.C. 1972. Uptake and utilization of nitrogen, phosphorus and potassium by dwarf indica rices. Rice News Letter. Xxi (3):1-19.

Mahapatra, I.C. and Panda, S.C. 1973. Light on fertilizing rice. Intensive Agri. XI (5): 9-11.

Mahapatra, I.C. and S.R. Bapat, 1992. Farming system research - challenges and opportunity. Proceeding of the XII National symposium on "Resource Management for Sustained Crop Production" held at Rajasthan Agricultural University, Bikaner during Feb. 25-28. pp. 382-390

Mahapatra, I.C., Bhatia, P.C., Singh, K.N., Dayanand and Panda, S.C. 1971. Agronomic management of upland rice in North and South Western India. Oryza 8(2) Supplement: 239-249.

Mahendra Pal, Jayanta Deka and Pai, R.K. 1996. Fundamentals of cereal crop production. Tata Mc Graw Hill Publishing Co Ltd., New Delhi.

Mahendra Pal, Singh, KA. and Ahalawat, I.P.S. 1986. Cropping systems research. 1. Concept, needs and directions. In: Proceedings of the National Symposium on "Cropping Systems", 1-20, New Delhi, India: Indian Society of Agronomy.

Maheswari, R. C., Srivastava, p. K., Bohra, C. R., Tomar, S. S. and Nema, B. R. 1981. Energy Census and Resource Assessment of village Islamnagar in the district of Bhopal, Central Institute of Agricultural Engineering, Bhopal.

Maiti, S., Hedge, M.R. and Chattopadhyay, S.B. 1988. Hand book on annual oilseed crops. Oxford and IBH publishing Co. Pvt. Ltd., New Delhi.

Make, R.E. 1982. Rice area by type of culture: South, Southeast, and East Asia, Losbanos, Philippines: International Rice Research Institute.

Malik, R. K.: Ashok Yadav and Sher Singh. 2005. Resource conservation technologies in rice-wheat cropping system of Indo-Gangetic plains. Conservation Agriculture – Status and Prospects. Eds. Abrol, I. P., Gupta, R. K. and Malik, R. K. HAU. Hisa-Haryana. Pp. 13-22.

Mandal, R.C. 1999. Weeds, weedicides and weed control. Principles and practices. Agro Botanica publishers, Bikaner. India.

Marlon, P., Cantrell, R., King, D. and Benoit-Cattin, M. (Ed), 1984. Coming full circle: Farmers' participation in the development of technology. Ottawa, Ontario: IDRC.

Marston, D. and Perrens, S. 1981. Effect of stubble on erosion of a black earth soil. South East Asian Regional Symposium on problems of soil erosion and sedimentation held at Asian Institute of Technology, Bangkok, Thailand: 289-300.

Martin, J. H.: Leonard, W. H. and Stamp, D. L. 1976. Principles of field crop production. Third Edition. Macmillan Publishing Co. New York.

Martsolf, J.D. and Decker, W.L. 1979. Microclimate Modification by Manipulation of Net Reduction. Agricultural Meteorology, 7: 197-216.

Mather, J.R. 1974. Climatology-Fundamentals and applications. McGraw Hill Book Co., New York.

Mavi, H.S. 1994. Introduction to Agricultural meteorology. Oxford & IBH Publishing Co. Pvt. Ltd, New Delhi.

Maxwell, S. 1984. Farming systems research: Hitting a moving target. Discussion paper No. 199. SUSSEX Institute of Development Studies. Mayer, H. 1979. Land Use: Ecology and agriculture in the Mantaro Valley of Peru with special reference to potatoes, Lima, Peru: Center for International Potato Research, Social Science Unit.

McInfire, J. and Cryseels, G. 1987. Crop-livestock interactions in sub-Saharan Africa and their implications for farming systems research. Expt. Agrl. 23: 235-243.

McIntosh, D.H. and Thom, A.S. 1969. Essentials of Meteorology. The Lykeham Publications (London) Ltd., London.

Meelu, O.P. and Morris, R.S. 1988. Green manure management in rice based cropping systems. In: "Green Manure in Rice Farming", International Rice Research Institute, Philippines. pp. 209-222.

Menon, P.A. 1989. Our weather. National Book Trust, New Delhi.

Metcalfe, D. S. and Elkins, M. M. 1980. Crop Production – Principles and Practices. Fourth Edition. Macmillan Publishing Co. New York.

Michael Haines, 1982. An introduction to farming system. Tien Mab lither Printing Co. Pvt. Ltd., London.

Michael, A.M. 1978. Irrigation – Theory and practice. Vikas Publishing House Pvt. Ltd., New Delhi.

Milthorpe, F. L. and Moorby, J. 1979. An Introduction to Crop Physiology. Second Edition. University Press, Cambridge.

Mishra, A. and Pasupalan, S. 1994. Integrated farming systems for rain fed lowland. The Hindu, 14th Sept. 1994.

Mishra, P. K. 2002. Indigenous Technical Knowledge (ITKs) on Soil and Water Conservation in semi-arid India. NATP, CRIDA, Hyderabad. P 151.

Mishra, R.D. and Ahmed, M. 1987. Manual on Irrigation Agronomy, Oxford and IBH Publishing Co. Pvt. Ltd., New Delhi.

Mittal, J. P., Sexena, R. P. and Singh, I. J. 1974. Mathematical expression of cost analysis of farm equipment. Ind. Jour. Agril. Economics. 19(1): 51-59.

Mittal, J. P.and Dhawan, K. C. 1988. Energy requirements in Agricultural Sector, PAU, Ludhiana.

Mittal, S.P. and Grewal, S.S. 1983. To evaluate the effect of land treatment measures on runoff and productivity and to evaluate norms of erosion for various crops. Rep. CSWCRTI, Dehradun: 62.

Mittal, S.P. and Singh, G. 1990. Resource conservation and management in watersheds of Shiwalik foot hills. In: proceeding of International Symposium on "Water Erosion". Sedimentation and resource conservation, 338-347. Dehradun, Uttar Pradesh, India: Central Soil and Water Conservation Research and Training Institute.

Mittal, V. K., Mittal, J. P. and Dhawan, K. C. 1985. Research Digest on Energy Requirements in Agriculture Sector, PAU, Ludhiana.

Mohamed Ali, A., Siddeswaran, K., Jegannathan, R., Muthusamy S. and Subramanian, S. 1998. Efficiency of mixed farming in the tribal areas of Sherveray Hills. Madras Agric. J. 71: 294-298.

Mohan, S. and Jeyarajan, R. 1994. Mushroom cultivation. In: Proceedings of Summer Institute on "Integrated Farming Systems Research and Development for Sustainable Agriculture" organized by ICAR and Tamil Nadu Agric. Univ., Coimbatore, India during 6-15 June, 1994. pp. 220227.

Mohandas, S. 1988. Field response of tomato inoculation with VAM Fungi and Rhizobium Phaseoli, Proc. Mycorrhiza Round Table. Varma, A.K. *et al.*, (ed.) pp. 411-416.

Moldenhauer, W.C., Wischmeier, W.H. and Parker, D.T. 1967. The influence of crop management on runoff, erosion and soil properties of Marshall silty clay loam. Pro. Soil Sci. Soc. Am. 31 (4): 541-546.

Molga, M. 1962. Agricultural Meteorology Part II Outline of Agrometeorogical Problems. Warsaw:5-8.

Montieth, J.L. 1990. How do crops respond to weather? Conference paper ICRISAT

Morachan, Y.B. 1980. Crop Production and Management, Oxford and IBH Publishing Co., New Delhi.

Munn, R.E. 1968. Descriptive Micrometeorology, Academic, New York. 8-22

Munn, R.E. 1970. Bio-meteorological Methods. Academic Press, New York

Murthy, P. S. S., Reddy, P. J. R. and Reddi, G. H. 1983. Variation in the Physiological parameters of popular groundnut varieties. Madras Agric J. 70 (9): 603-610.

Murty, V.V.N. 1969. Stone terracing of hill slopes. Jour. Soil and Water Conc. In India. 17(3 & 4): 35-38.

Nagaraja, B. 1989. Effect of Crop mixture on income and Employment, A case Study of Anantapur district.

Nagarajan, S., Gyanendra Singh and Tyagi, B.S. 1998. Wheat Research needs beyond 2000 A.D. Narosa Publishing House. New Delhi.

Nair, P.K.R. 1985. Agroforestry defined. In: Agroforestry systems in the Tropics. Nair, P.K.R. (Ed.) Kluswer Academic Publishers. Dordrecht, The Netherlands. pp. 13-18.

Nakra, C.P. 1980. Farm Machinery and Equipment. Dhanpati Rai and Sons, Delhi.

Nanda, S.S., Swain, K.C., Panda, S.C., Mohanty, A.K. and Alim, M.A. 1995. Effect of nitrogen and biofertilizers in fodder maize under rainfed condition of Orissa. Curr. Agri. Res. S (Suppl.): 45-47.

Nataraju, M.S. and Lovely, P.G. 1989 Role of farm women in farm and home activities. Indian Jour. Of Adult Education. 50(4): 64-67.

National Commission on Agriculture (NCA), 1976. Crop production during the pre-plan period. In Abridged report of the National Commission of Agriculture, 48-49. New Delhi, India: NCA.

National Research Council (NRC), 1993. Soil and Water quality: An agenda for agriculture. Committee on long range soil and water conservation board on agriculture. National Academy Press Washington, D.C.

Negi, M., Sachdev, M.S. and Tilak, K.V.B.R. 1991. Nitrogen uptake by Azospirillum brasilense inoculated barley as influenced by Nand P Fertilization. J. Nuclear Agric. Bio. 20: 25-32.

Nickell, L. G. 1981. Plant Growth Regulators – Agricultural Uses. Springer –Verlag. New York.

Noble, R.P. and Rshide, B. 1990. Aquaculture technology transfer to small holder farmers in Malawi, Southern Africa. NAGA, The ICLARM Quarterly. 16: 14-16.

Norman, D.W. 1974. Rationalizing mixed cropping under indigenous conditions, The example of Northern Nigeria. Journal of Development Studies 11: 3-21.

Norman, D.W. 1980. The farming system approach: Relevancy for the small farmers, Michigan State University Development Paper, 8, East Lansing, Michigan, USA: Michigan State University.

Norman, D.W. 1982. Farming systems approach to research. Paper prepared for the Farming Systems Research

Symposium, Kansas State University, Manhattan, Kansas, November 21-22, 1982.

Norman, D.W. and Collinson, M.P. 1985. Farming Systems Research: Theory and Practice. ACIAR Proc. No. 11: 16-30.

Norman, M.J.T. 1980. Your future depends on Agronomy, J. Aust. Inst. Agri. Sci. 46:105-111.

Norman, N. Potter and Joseph H. Hotchkiss 1996. Food Science. CBS Publishers and Distributors., New Delhi.

Nurhan Dunford 2006. Nanotechnology and opportunities for agriculture and food systems. Food technology. Fact sheet FAPC. 139.

Nyta, C. Brady. 1990. The nature and properties of soils (Tenth Edn) MacMillan Publishing Co. Inc., New York.

Okigbo, B.N. 1978. Cropping system and related research in Africa. AAASA occasional Pub. Series OT-I, Addis Abeba, Ethiopia. pp. 81.

Okigbo, B.N. and Lal, R. 1980. Residue mulches, intercropping and agri-silviculture potential in Tropical Africa. In: Basic techniques in ecological farming. Hill, S. (Ed.) IFOAM Conference, Montreal, Canada. pp. 54-69.

Oram, P. 1988. Building the agroecological framework. Environment pp. 14-34.

OTA, 1988. Enhancing Agriculture in Africa: A role for development. Assistance. DC Govt. Printing office, Washington, D.C

Owasu, D.Y. 1990. Experiences with agroforestry. ILEIA Newsletter 6(2): 9-10.

Page, J.B. and G.B. Bodman, G.B. 1952. The effect of soil physical properties on nutrient availability. Chapter 6, Mineral Nutrition of Plants. Ed. E. Truog, Wisconsin Univ. Press, Madison, Wis.

Palaniappan, S. P. and Annadurai, K. 2001. Diversification in farming system Paper presented in first International agronomy congress on "Agronomy, Environment and Food Security for 21st century held at IARI, New Delhi during Nov. 23-27, 1998.

Palaniappan, S. P. and Siddeswaran, K. 1990. Evaluation of the potential of green manures and grain legumes in rice farming system. National symposium of "Rice Farming System", Agri. College and Research Institute, Madurai, India.

Palaniappan, S.P. 1985. Cropping systems in tropics-Principles and practices. Willey Eastern Ltd., New Delhi.

Palaniappan, S.P. and Sivaraman, K. 1996. Cropping System in the Tropics. Principles and management. New Age International Pvt. Ltd. New Delhi.

Palaniappan. S.P. and Annadurai, K. 1999. Organic farming: Theory and Practice. Scientific publisher, Jodhpur, p. 257.

Palm, V., Desilva, A. and Guildford, M. 1987. Nutrient cycling in paddy of a traditional farming system. Tropical Agri. 87: 129-133.

Panda, S.C., 1967. The effect of number of splits in clonal multiplication on the growth and yield of rice at three levels of manuring. Ind. Jour, Agro. XII (4): 383-389.

Panda, S. C. 2007. Crop Production and Tillage. Agrobios (India), Jodhpur, India.

Panda, S. C. 2007. Soil Water Conservation and Dry Farming. Agrobios (India), Jodhpur, India.

Panda, S. C. 2008. Post Harvest Technology and Farm Mechanization. Agrobios (India), Jodhpur, India.

Panda, S. C. 2008. Soil Conservation and Fertility Management. Agrobios (India), Jodhpur, India.

Panda, S.C., Nanda, S.S. and Mishra, R.C. 1997. Response of *rabi* groundnut (*Arachis hypogaea*) in methods of sowing and sources of sulphur. Ind. Jour. Agro. 42(4):680-682.

Panda, S.C. 1972. Effect of different levels of nitrogen and phosphorus on the growth and yield of Pusa Baisakhi mung. Ind. Jour. Agro. XVII (3): 239-241.

Panda, S.C. 1972. Multiple Cropping in Sambalpur. Multiple Cropping: 148-151.

Panda, S.C. 1972. Performance of four potato varieties under Sambalpur condition. Ind. Jour. Agro. XVII (2): 110-111.

Panda, S.C. 1972. Performance of high yielding varieties of jowar under different levels of nitrogen. Ind. Jour. Agro. Xvii (1): 77-78.

Panda, S.C. 1974. Performance of soybean varieties under Sambalpur condition. Ind. Jour. Agro. 19(3):163-165.

Panda, S.C. 1977. Yield performance of some maize varieties. OUAT Jour. Res. 7 (1&2): 14-17.

Panda, S.C. 1978. Water Management of Wheat. Food Fmg. Agri. 10(4): 112-114.

Panda, S.C. 1980. Integrated nutrient supply for higher production of rice. Food Fmg. Agri. 13(1):18-20.

Panda, S.C. 1980. Water Management of Rice in Relation to Ground Water Table. Proc. Afro-Asian Reg. Conf. Subj. C. (9): 91-104.

Panda, S.C. 1981. Research on Water Management in Hirakud Command Area and its Achievements from 1967 to 1981

Panda, S.C. 1983. Water Management in Hirakud Command Area. 154 pp.

Panda, S.C. 1985. Research Base and Current Statues of Cropping System in Hirakud Command Area. 20 pages.

Panda, S.C. 1986. The Achievement of Research on Water Management from the inception of the Irrigation Research Centre at Chakuli/Chiplima for the period from 1967 to 1984.

Panda, S.C. 1995. Rice yield and physiochemical properties of soils as affected by integrated nitrogen management. National sympo. On sustainable agriculture in Sub-Humid Zone.

Panda, S.C. 2003. Principles and Practices of Water Management. Agrobios (India), Jodhpur. India.

Panda, S.C. 2004. Cropping and Farming Systems. Agrobios (India), Jodhpur. India.

Panda, S.C. 2004. Dry land Agriculture. Agrobios (India), Jodhpur, India.

Panda, S.C. 2005. Agronomy. Agrobios (India), Jodhpur, India.

Panda, S.C. 2006. Crop Management and Integrated Farming. Agrobios (India, Jodhpur, India.

Panda, S.C. 2006. Soil Management and Organic Farming. Agrobios (India), Jodhpur, India.

Panda, S.C. 2007. Farm Management and Agricultural Marketing, Kalyani Publishers, Ludhiana, India

Panda, S.C. 2008. Mechanization of Agriculture. Kalyani Publishers, Ludhiana, India.

Panda, S.C. 2010. Agro-meteorology and Contingent Crop Planning. Agrobios (India), Jodhpur, India.

Panda, S.C. 2010. Maize Crop Science. Agrobios (India), Jodhpur, India.

Panda, S.C. 2010. Rice Crop Science. Agrobios (India), Jodhpur, India.

Panda, S.C. 2010. Sustainable Food and Nutrition Security in National Economy. Agrobios (India), Jodhpur, India

Panda, S.C. and Das, K.C. 1975. Relative performance of high yielding varieties of rice at different levels of nitrogen. OUAT Jour. Res. 5(1 & 2):83-89.

Panda, S.C. and Das, K.C. 1978. Effect of age of seedlings and fertilization on the high yielding varieties of rice. Nonmonitary Inputs-Field Crop Production. 83-88.

Panda, S.C. and Das, K.C. 1979. Nitrogen for Rice. Ind. Jour. Agro. 23(2): 183-188.

Panda, S.C. and Leeuwrik, D.M. 1971. Effect of spacing on the growth and yield of high yielding varieties of rice. Oryza. 8(2): 30-46.

Panda, S.C. and Leeuwrik, D.M. 1972. Response of high yielding varieties of rice in high levels of nitrogen. Oryza. 9(2):23-27.

Panda, S.C. and Mahapatra, I.C. 1974. Growth and Development of Rice Plant-II. Flowering behaviour of individual tillers of different plant types. IL RISO. XXIII (3):259-262.

Panda, S.C. and Mahapatra, I.C. 1974. Growth and Development of Rice Plant- III. Analysis of Yield components. IL RISO. XXIII (3):262-276.

Panda, S.C. and Mahapatra, I.C. 1974. Growth and Development of Rice Plant-IV. Relation between grain yield and growth characters. IL RISO. XXIII (4):405-414.

Panda, S.C. and Mahapatra, I.C. 1974. Growth and Development of rice Plant- V. Correlation Studies. IL RISO. XXIII (4):414-419.

Panda, S.C. and Mahapatra, I.C. 1974. Growth and Development of Rice Plant-VI. Linear and Multiple Regression of grain yield per plant on primary, secondary and tertiary tillers for different characters. IL RISO. XXIII (4):420-438.

Panda, S.C. and Mahapatra, I.C. 1976. Relative performance of Taiwan indica, ponlai and local indica (different plant types) under varying levels of soil fertility, environmental condition and agronomic practices- I. Growth pattern of rice- Tillering behaviour and plant height. IL RISO. XXV (2):161-183.

Panda, S.C. and Mahapatra, I.C. 1976. Relative performance of Taiwan India, ponlai and local indica (different plant types) under varying levels of soil fertility, environmental condition and agronomic practices. II. Growth pattern of rice plant- Number of leaves per unit area, leaf area and leaf area index. IL RISO. XXVI (2): 157-178.

Panda, S.C. and Mahapatra, I.C. 1972. Growth and Development of Rice Plant- I. Growth Analysis. IL RISO. XXI (4): 315-336.

Panda, S.C. and Nayak, R.C. 1974. Effect of zinc on the growth and yield of rice. Ind. Jour. Agro. 19(1):9-13.

Panda, S.C. and Patro, H. 1998. Rice-Dairy Farming System: A boon to Marginal Farmers of Orissa. Intensive Agri. March-April: 24-31.

Panda, S.C. and Sahu, S.K. 1978. Harvest more from Usar soil. Food Fmg. Agric. 10(1):16-19.

Panda, S.C. and Sahu, S.K. 1978. Rhizobium for legume cultivation and fixation of atmospheric nitrogen. Food Fmg. Agric. 10(5):157-160.

Panda, S.C. and Verma, S.P. 1977. Comparison of puddling implements for rice cultivation. OUAT Jour. Res. 7(1 & 2): 82-88.

Panda, S.C., Sahu, S.K. and Misra, B. 1979. Leaching losses of Nitrate Nitrogen under different Water Tables and Water Management Practices in Rice Field. Oryza 16(2): 107-112

Panda, S.C., Acharya, N. and Misra, B. 1981. Effect of Irrigation Schedules on the Growth and Yield of Rice. Food Fmg. Agri. 13(9 & 10): 1282-185.

Panda, S.C., Das, K.C. and Sahu, S.K. 1986. Water table in relation to water management on rice. Proc. Sympo. Drainage of waterlogged and salt affected lands.

Panda, S.C., Das, K.C., Misra, B., Sahu, S.K. and Rout, D. 1980. Effect of depth of submergence of different growth stages of dwarf indica rice on the growth, yield and nutrient uptake of crop and mineral content of soil. I. Growth and yield of crop. Oryza. 17(2): 85-91.

Panda, S.C., Gulati, J.M.L. and Misra, B. 1988. Effect of Water Regimes and Nitrogen on Wheat. Indian J. Agron. 33(4): 364-367.

Panda, S.C., Leeuwrik, D.M. and Mahapatra, I.C. 1972. Optimum combination of calcium ammonim nitrate and sulphate of ammonium for efficient nitrogen utilization in rice as compared by grain yield. OUAT. Jour. Res. 2(1 & 2):1-4.

Panda, S.C., Leeuwrik, D.M. and Mahapatra, I.C. 1974. Performance of hybrid and composite varieties of maize under Sambalpur condition. Ind. Jour. Agro. 19(1):14-17.

Panda, S.C., Leewrik, D.M. and Mahapayra, I.C. 1972. Production Potential and Economic of Ten High Intensity one-year Crop Rotation in Sambalpur. Multiple Cropping: 102-108.

Panda, S.C., Misra, B and Singh, N. 1979. Water management of groundnut in relation to ground water tables. OUAT Jour. Res. 9(1 7 2): 110-114.

Panda, S.C., Misra, B. and Singh, N. 1978. Problems of waterlogging in the Hirakud Ayacut and its management. Food Fmg Agri. 9(12): 392-395.

Panda, S.C., Misra. B. and Rout, D. 1980. Effect of various irrigation schedules on potato under different depths of irrigation. Curr. Agri. 4 (4-4):153-163.

Panda, S.C., Mohanty, A., Sahoo, H.K. and Sahoo, B.C., 1999. Effect of manuring in Rice-Rice cropping system. Oryza. 36(2): 139-142.

Panda, S.C., Nayak, R.C. and Mahapatra, I.C. 1973. A note on the effect of cement dust on the soil fertility and crop growth around the factory area of Rajgangpur (Orissa). Ind. Jour. Agro. 14(3): 389-391.

Panda, S.C., Panda, P.C. and Nanda, S.S. 1995. Effect of levels of nitrogen and phosphates on yield and nutrient uptake of rice. Oryza. 32(1):146-148.

Panda, S.C., Patra, H., Panda, P.C. and Reddy, G.M.V. 1999. Effect of integrated nitrogen management on rice yield and physiological properties of soil. Crop Research. 18 (1): 25-28.

Panda, S.C., Pattanaik, A., Rath, B.S., Tripathy, R.K. and Behera, B. 1996. Response of wheat (*Triticum aestivum*) in crop geometry and weed management. Ind. Jour. Agro. 41(4):553-557.

Panda, S.C., Rath, B.S., Tripathy, R.K. and Satpathy, L. 1996. Contingent Planting of Rice for Aberrant Weather Conditions in Coastal Plains. Environment & Ecology 14(2): 455-458.

Panda, S.C., Rath, B.S., Tripathy, R.K. and Das, B. 1997. Effect of water management practices on yield and nutrient uptake in the dry season rice. Oryza. 34(1): 51-53.

Panda, S.C., Roul, P.K., Tripathy, R.K. and Chaudhury, Y.C. 1999. Effect of spacing and number of seedlings per hill on transplanted rice. Environment & Ecology. 17(3): 626-629.

Panda, S.C., Singh, N. and Misra, B. 1979. Performance of wheat varieties under different levels of soil salinity. Curr. Agri. 3 (1 & 2): 63-72.

Panda, S.C., Verma, S.P. and Patra, A. 1978. Maximize the production of rice through better puddling. Food Fmg. Agri. 10(2):31-34.

Pande, A. 1970. Modern Hygrometry, Somaiya Publications Private Ltd., Bombay.

Pandey, R.K., Sharma, S.K. and Hegde, D.M. 1992. Approach to farming system research in ICAR: Concepts and Methodology.

Project Directorate for Cropping Systems Research. Bulletin No.2. pp. 1-16.

Pandey, S.N. and Ajanta Chandha, 1996. Economic Botany. Vikas Pvt. Ltd., New Delhi.

Panse, V.G. and Sukhatme, P.V. 1967. Statistical methods for agricultural workers. ICAR., New Delhi.

Papadakis, J. 1970. Agricultural potentialities of World Climates. Buenos Aires, Argentina: 35-36.

Pati, M. 1992. An Economic Analysis Rice-based Farming System in Balasore District (Orissa). M.Sc (Ag) Thesis, OUAT, Bhubaneswar.

Patra, D.D. 1990. Cereal - legume associations. Effect on companion and succeeding cereals. Somani, LL, K.K. Vyas and S.N. Saxena (eds.) Biofertilizers. Scientific Publishers, Jodhpur, India, pp. 113-136.

Patterson, S. 1956. Weather Analysis and Forcasting. Sec. Edition, McGraw Hill Book Company, Inc. New York.

Paul, P. and Paddok, W. 1965. Famine.

Paulraj, S. 1984. Composite fish culture by pig-dung. Ceobios 11: 265-267.

Perrin, R., Winkelmann, D.L., Moscardi, E.R. and Anderson, J.R. 1976. From Agronomic data to farmer recommendations: An economic training manual. Information training bulletin, 21. El. Batan, Mexico: CIMMYT.

Pierce, EJ., Larson, W.E., Dowdy, R.H. and Graham, W.A.P. 1983. Productivity of Soils: Assessing long term changes due to erosion. J. Soil and Water Conservation 38: 39-44.

Plucknett, D.L., Dillon, J.L. and Vallaeys, G.J. 1987. Review of concepts of farming systems research: What, Why and How. In: Proc. of Workshop on "Farming Systems Research", ICRISAT, Hyderbad. pp. 2-9.

Prasad, C. 1989. Agricultural extension services. In: 40 years of Agricultural Research and Education in India, 234-267. New Delhi, India: ICAR.

Pratap Singh, Jaspal Singh and Mathur. 1995. Technology for Dryland Agriculture. Himanshu Publications, New Delhi.

Premsekhar, M. and Ranagsamy, A. 1994. Sustainable IFS research on lift irrigated garden lands. In: Proceeding of Summer Institute on "Integrated Farming Systems Research and Development for Sustainable Agriculture" organized by ICAR and Tamil Nadu Agric. Univ., Coimbatore, India during 6-15 June, 1994. pp. 106-114.

Prihar, S.S. and Sandhu, B.S. 1987. Irrigation of field crops – Principles and Practices. ICAR, New Delhi.

Purohit, S.S. and Rajan, R. 2003. Ecology, Environment and Pollution. Agrobios (India), Jodhpur, India.

Raddy, T. Y. and Reddy, G. H. S. 1992. Principles of Agronomy, Kalyani Publishers, Ludhiana.

Radha Vasiraju and Murty Krishna 1995. Practical Manual on Agricultural Meteorology. Kalyani Publishers, New Delhi.

Raheja, P.C. 1966. Soil Productivity and Crop Growth. Asia Publishing House, Bombay.

Rajendran, P. 1994. Integrated production technology under rainfed farming system. In: Proceeding of Summer Institute on "Integrated Farming System Research and Development for Sustainable Agriculture" organized by ICAR and Tamil Nadu Agric. Univ., Coimbatore, India during 6-15 June, 1994_ pp. 65-78.

Rajkumar, S. 1988. Feasibility of introducing dairy as complementary enterprise for small farmers in Thanjavur delta, Tamil Nadu, M.Sc. (Ag.) Thesis. Tamil Nadu Agric. Univ., Coimbatore, India.

Raju, R.A. 1996. Ecology of Aquatic weeds and their control. Kalyani Publishers, New Delhi.

Raju, R.A. 2000. Glimpses of rice technology. Agrobios (India), Jodhpur, India.

Ram Mohan Rao, M.S. and Chittranjan, S. 1983. Tillage during crop growth. Rep. CSWCRTI, Dehradun: 42 p.

Ram Mohan Rao, M.S., Agnihotri, R.C. and Patnaik, U.S. 1983. Influence of surface mulches in combination with and without gypsum on jowar yields. Rep. CSWCRTI., Dehradun, 110 p.

Rama Rao, M.S.V. 1962. Soil Conservation in India, ICAR, New Delhi

Rama Rao, M.S.V. 1974. Soil Conservation in India. ICAR., New Delhi.

Ramadas, L.A. 1964. Crop and Weather in India, ICAR, New Delhi.

Raman, K. V. 1998. FSR - its implication in future research strategies in India. FSR Newsletter 2 (1J): 3-5.

Raman, K. V. and Balaguru, T. 1991. Farming System Research in India: Strategies for Implementation. Proc. of National Workshop, NAARM, Hyderabad. pp. 1-39.

Ranagasamy, A. and Premsekhar, M. 1994. Sustainable integrated farming system research of lift irrigated garden lands.

Proceeding on "Integrated Farming System Research and Management for Sustainable Agriculture" held at Tamil Nadu Agric. Univ., Coimbatore, India, between 6-15, June, 1994. pp. 106-109.

Randhawa, M.S. 1980. A History of Agriculture in India, Vols. I & II, ICAR, New Delhi.

Randhawa, M.S. 1979. A history of the ICAR (1929-1979), ICAR, New Delhi, India: ICAR.

Rangasamy, A. 1992. Sustainable agriculture for rice based ecosystem. In: Proceeding of National training on "Integrated Farming System" held during 10-17th December, 1992. Gal and Tamil Nadu Agric. Univ., Coimbatore, India. pp. 5-12.

Rangasamy, A. 1994. Approaches to farming systems research. In: Proceeding of Summer institute on "Integrated Farming Systems Research and Management for Sustainable Agriculture". 6-15 June, 1994. Tamil Nadu Agric. Univ., Coimbatore, India. pp. 1-17.

Rangasamy, A. 1996. Organic farming system through integrated farming. system. Paper presented at the National Workshop on "Organic Farming for Sustainable Agriculture" held at WALAMTARI, Hyderabad, from 1820 January.

Rangasamy, A. and Jayanthi, C. 1994. Recycling of organic wastes in integrated farming systems. In: Proceeding National training on "Organic Farming". GOI and Tamil Nadu Agric. Univ. Coimbatore, India. pp. 114-118.

Rangasamy, A., Annadurai, K., Subbian, P. and Chinnusamy, J. 2002. Farming Systems in The Tropics. Kalyani Publishers. Ludhiana.

Rangasamy, A., Budhar, M.N., Venkidusamy, R. and Palaniappan, S.P. 1988. Integrated farming system for wetlands. In: Absts. of National seminar on "Farming Systems for Semi-arid Tropics". Tamil Nadu Agric. Univ., Coimbatore, India. pp. 1-6.

Rangasamy, A., Shanmugasundaram, V. S., Sankaran, S. and Subbian, P. 1990. Integrated Farming System Management: A viable approach. Research Bulletin. Tamil Nadu Agric. Univ. Coimbatore, India.

Rangasamy, A., Venkidusamy, R., Jayanthi, C., Purushothaman, S. and Palaniappan, S.P. 1995. Integrated farming system for rice based ecosystem. Madras Agric. J. 83: 287-290.

Rao, A. M. K. M. 2008. Emerging Rodent problems in South India and the strategies for their management. Souvenir – Regional Agricultural Fair on Sustainable Agriculture. Held at Hyderabad on 19-23 December 2008. pp. 89-96.

Rao, M.R. 1986. Cereals in multiple cropping. In: Francis, C.A. (eds.) Multiple cropping systems. Macmillan Publishing Co., New York

Rao, P. C. M. 2007. Water is Too Precious – Conserve it. Proceedings 12th National Water Convention at Podicherry from 1st to 3rd November, 2007.

Rao, P.Y. 1984. Mushroom farms in Himachal Pradesh. Intensive Agric. 22: 19-20.

Rao, V.M. 1991. Prospect of Poultry Development in India. Agric. Situation in India. June 121-126.

Rao, V.S. 2000. Principles of weed science (second edn.) Oxford and IBH Publishing Co. Pvt. Ltd., New Delhi.

Rath, R. 1989. Organic Fertilizer Augments Fish Production. Indian Farmers Digest. XXII (3): 31-33.

Rath, R. 1990. Integrated Fish Farming-a sound basis for the development of Integrated Rural Technology, National Seminar on Aquaculture in India-Problems and Prospects. 27-29 Nov. 1990. Thirubanantpuram Dept. of Aquatic Biology and Fisheries. Vhiv. Of Kerala p-18 Cochin.(Abstract P No. T.I.12)

Rathore, P.S. 2000. Techniques and management of Field crop production. Agrobios (India), Jodhpur, India.

Ratikanta Maiti. 1996. Sorghum Science. Oxford and IBH Publishing Co. Pvt. Ltd., New Delhi.

Raymond W. Miller and Roy L. Donahue. 1992. Soils. Prentice-Hall of India Pvt. Ltd., New Delhi.

Reddy, S.R. 2000. Principles of Agronomy. Kalyani Publishers, Ludhiana

Regnion, E. and Jahnke, R. 1990. Evolving strategies for managing weeds. In: Edwards, C.A., *et al.*, (ed.) Sustainable agricultural systems. Soil and Water (Conservation Society, Ankeny, IOWA pp. 174-202.

Reilly, M. L. 1984. Functional aspects of cereal structure. In: Cereal Production, ed, E. J. Gallagher, Butterworths, London, pp 137-160.

Rennie, R.J. and Kemp, G.A. 1984. ^{15}N-Determined time course for N_2 fixation in two cultivars of field beans. Agron J. 76: 146-154.

Rethinam, P. and Sreedhar, M. 2000. Transgenic crops and foods prospects and retrospects. National seminar on transgenic Crops and Foods. Herld at Hyderabad 7-8 November 2000. Pp 6-12.

Retner, E.T., Label, R., Teldhay, H. and Hartzook, A. 1979. Some characteristic of symbiotic nitrogen fixation, yield, protein and soil accumulation in irrigated peanuts. Plant Soil. 51: 373-376,

Rhoades, R. and Bootb, R. 1982. Farmer-Back to-Farmer. An model for generating acceptable technology. International Potato Research Center, Social Sciences Department, Working Paper 1982_1. Lima, Peru: International Potato Research Center. Research 1929-1979" ICAR, New Delhi, India: ICAR.

Richards, P. 1988. Experimenting 'farmers and agricultural, research. paper prepared for ILEIA Workshop on "Operational, Approaches for Participatory Technology Development in Sustainable Agriculture",11-12 April, 1,988, Leusden: The Netherlands.

Riehl, H. 1965. Introduction to the atmosphere, International student Edition, McGraw Hill Book Company, New York.

Rinaudo, G., Dreyfus, B. and Dommerges, Y. 1983. Sesbimia rostrata green manure and the nitrogen content of rice crop and soil. Soil Biochem. 15: 111-113.'

Ritchie, J.T., Singh, U., Goodwin, D.C. and Hunt, L. 1989. A users guide to CERES. Training Manual, IFDC, Muscle Shoais, AL,

Rorenberg, N.J. 1974. Microclimate – The biological environment. John Wiley and Sons, Inc., New York.

Rosenberg, N. J., Blad, B. L. and Verma, S. B. 1983. Micro-climate – the Biological Environment. John Wiley and Sons, New York.

Roy, B., Das, D.N. and Mukahopadhayay, P.K. 1990. Rice-fish - vegetables integrated farming: Towards a sustainable ecosystem. NAG A, The ICLARM Quarterly 13: 17-18.

Royal Commission on Agriculture (RCA), 1928. Report of the Royal Commission on Agriculture in India, Bombay, India.

Russell, E.W. and Keen, B.A. 1948. Studies in Soil Cultivation. VII. The Effect of Cultivation on Crop Yield. Jour. Agri. Science. 28: 272-298.

Russell, M.B. 1952. Soil Aeration and Plant Growth. Chapter 4.Soil Physical Conditions and Plant Growth. Agronomy Monograph., II. Ed. B.T. Shaw, Academic Press Inc., New York.

Sache, I. 1987. Towards a second green revolution. In: Glaeser, B. (ed.) The green revolution revised. Allen and Unwin, London, pp. 193-198.

Sadangi, B.N. 2004. Perception and management of drudgery in rice cultivation, proceedings. National seminar on drudgery reduction for farm women in rice farming. P 9-12.

Sahoo, B., Panda, S.C. and Mohanty, K. 1997. Effect of Rock phosphate under Rice-Green gram Cropping System in Acid Soils of Orissa. Oryza. 34(4): 358-361

Saint, S. K. and Sunitat T. Pandey 2009. Organic Farming: Development and Strategies in Indian Perspective. Ind. Jour. Agronomy. 54(2): 193-199.

Salathia, P.S., Bhagat, G.R., Khar, S.K. and Singh, S. 2004. Role of women in Agriculture and allied activities in hills. Intensive Agriculture. 42 (7-8): 7-10.

Samuel, L., Tisdale, warner. L. Nelson and Jemes D. Beaton. 1990. Soil fertility and fertilizers (fourth edition). MacMillan Publishing Co. Inc., New York.

Sankaran, S. and Rangasamy,.A. 1990. Farming system research. In: Agronomic research towards sustainable agriculture (Ed.) K.N. Singh and R. P. Singh, pp. 69-80. Indian Society of Agronomy, IARI, New DlehL

Sankaran, S. and Subbiah Mudaliar, V.T. 1992. Principles of Agronomy. The Bangalore and Publishing Co. Ltd., Bangalore, India.

Sankaran, S., Jayakumar, R. and Kampuchetty, N. 1993. Herbicide residues. Gandhi Book House, Coimbatore, India.

Santhanagopalkrishanan, 1988. Feasibility of rearing goat under deep litter system in the farming system of Thanjavur delta, Tamil Nadu M.Sc.(Ag.) Thesis, Tamil Nadu Agric. Univ., Coimbatore, India.

Sarkar, R.P. and Biswas, B.C. 1988. Weather forecasting and its application in Agriculture. Fert. News. 33(4):73-76.

Scroll, P. and Merrillsands, D. 1986. The Milpa and its alternatives. In: Farming systems (Eds. B.L. Turner and S. Brush).-New York: Guilford Press. Boulder, Colorado: West View Press.

Seetharam, A., Riley, K.W. and Harinarayaana, G. 1989. Small millets in Global Agriculture. Oxford and IBH Publishing Co. Pvt. Ltd., New Delhi.

Sellers, W.D. 1965. Physical Climatology. University of Chicago: 177-180.

Serman, D.W., Simmons, E.B. and Mays, H.M. 1982. Farming Systems in the Nigerian Savannah: Research Strategies for

Development. Boulder, Colorado: West View Press.

Shaik Mahammad 2009. Modern concepts of agricultural research: Ripples and options for conservation agriculture and resource use efficiency. Ind. Jour. Agronomy. 54(2): 149-158.

Shaner, W.W., Phillips, P.E. and Schmehl, W.R. 1982. Farming Systems Research and Development: A guideline for developing countries. Boulder, Colorado: West View Press.

Shanmugasundaram, V.S. and Balusamy, M. 1993. Rich-Fish - azolla integrated farming system in low lying wetlands. Farming Systems 9: 105-107.

Shanmugasundaram, V.S. and Ravi, K. 1992. Rice-fish - azolla integration. NAGA, The ICLARM Quarterly. 15: 27.

Shanmugasundaram, V.S., Balusamy, M. and Subramaniam, K.S. 1992. Integrated farming system in wetland ecosystem. In: Proceeding of National training on "Integrated Farming System", GOI and Tamil Nadu Agric. Univ., Coimbatore, India. pp. 18-20.

Sharma, B.C. 1991. Dryland Farming-Perspectives and prospects. Daya Publishing House, New Delhi.

Sharma, B.K. and Das, M.K. 1985. Integrated Fish Livestock crop Farming Systems in M. Mohan Joshep (ed.). The first Indian Fisheries Forum, Proceedings. Asian Fisheries Society, Indian Branch, Mangalore, pp: 27-30.

Sharma, B.K., Das, M.K. and Chakraborty, D.P. 1985. Integrated fish-cum-poultry farming. Agriculture extension manual news letter, Series. 5: 20-29.

Sharma, I., Kohitashw Kumar, P. and Kumar, P. 2004. Traditional and New Techniques of On-farm rainwater Harvesting in North – West Himalayan Region of India, International workshop on "Water harvesting and sustainable agriculture".

Sharma, L.R., Bhati, J.P. and Singh, R. 1991. Emerging Farming Systems in Himachal Pradesh: Key issues in sustainability. Indian J. Agri. Econ. 46(3): 421-427.

Sharma, P.N. and Kampen. 1977. Watershed based land and water management technology – Some tentative guidelines. ICRISAT. Hyderabad, India.

Sharma, S.K., Gangwar, K.S. and Singh, J.P. 1991. Farming Systems diagnosis, and analysis. Project report for Directorate of cropping system research on diagnostic team, Meerut, Uttar Pradesh, India. pp. 1-15.

Shende, S.T. 1987. Significance of Azotobacter inoculation for agricultural crops. Organic Manures and Fertilizers. Indian Agrl. Res. Institute, New Delhi. pp. 38-45.

Shippen, J.M., Ellin, C.R. and Clover, C.H. 1980. Basic farm machinery. Pergamon Press, Oxford, London

Shiv Kumar, M.V.K. 1990. Exploiting rainy season potential from onset of rains in Sahellan Zone of West Africa. Agricultural and Forest Meteorology 51: 321-32.

Siddeswaran, K. and Palaniappan, S.P. 1990. Role of integrated farming systems in minimizing the purchase input usage in small and medium farm holdings (Unpublished).

Simmods, N. 1985. The State of the Art of Farming Systems Research, Washington, D.C USA: World Bank.

Simmonds, N.W. 1985. Farming Systems Research - a review. Technical Paper No. 43. World Bank, Washington, D.C

Simmonds, N.W. 1986. A short review of Farming Systems Research in the Tropics Exptl. Agric. 22: 1-13.

Sin A.W.C. 1980. Integrated animal-fish husbandary systems in Hong Kong with case studies on duck-fish and goat-fish systems. In: Proceeding of KLARM - SEARCA Conf. on "Integrated Agriculture - Aquaculture Farming Systems." 4th Aug, 1979, Manila, Philippines.

Singh, B.P. 1990. Case of equity to farmers. Survey of Indian agriculture 1990. pp. 17-21. The Hindu, Chennai.

Singh, G. and Bhushan, L.S. 1978. Effect of surface mulches on yield of wheat. Rep., CSWCRTI, Dehradun: 118 p.

Singh, G.B. 1996. Integrated Nutrient Management: Strategy for balanced use. Survey of Indian Agriculture. The Hindu, Chennai, India.

Singh, G.B., Singh, K.N. and Singh, R.P. 1990. Sustainable land use systems in South Asia. In: Proceeding of International Workshop on "Sustainable Land use System Research". pp. 23-33. Oxford and IBH Publishing Co. (Pvt.) Ltd., New Delhi, India.

Singh, K.N. and Singh, R.P. 1990. Agronomic Research towards Sustainable Agriculture. Indian Society of Agronomy, IARI, New Delhi.

Singh, K.N. and Singh, R.P. 1992. Farming system research and development - A holistic approach for sustainable agriculture growth. In: Proceeding of the XII National symposium on "Resource Management for Sustained Crop Production" held at Rajasthan Agricultural University, Bikaner during Feb. 25-28. pp. 391-398.

Singh, N., Panda, S.C. and Gulati, J.M.L. 1989. Effect of moisture regimes of two stages of growth on groundnut. OUAT Jour. Agri. 2 93-4): 221-223.

Singh, P.K. 1989. Use of azolla in Asian Agriculture. Appl. Agric. Res. 4: 149161.

Singh, P.K. and Singh, D.P. 1992. Azolla and Rice Cultivation. In: Biofertilizers.

Singh, R.B. 1992. Integrated farming with magur fish is profitable to small farmers. Indian Fmg. (11): 5-12.

Singh, R.B. and Paroda, R.S. 1994. Sustainability and productivity of rice-wheat systems in the Asia-pacific regions. Research and technology development needs. pp. 1-35. In: Paroda, RS., T. Woodhead and RB (eds.), Sustainability of rice-wheat production systems in Asia, FAO of United Nations Regional Office for Asia and the Pacific, Bangkok, Thailand.

Singh, R.P. 1983. Dry land Agricultural Research in India: Thrusts in the eighties, 187-191 (Eds. RP. Singh and V. Balasubramanian). Hyderabad, Andhra Pradesh, India: AICRPDA.

Singh, R.P. 1991. Farm Ponds. All India Co-ordinated Research Project for Dryland Agriculture. Hyderabad, India.

Singh, R.P., Das, S.K., Bhaskara Rao, U.M. and Reddy, M.N. 1990. Towards sustainable dryland agriculture practices. Central Research Institute for Dryland Agriculture, Hyderabad, India.

Singh, R.P., Padmanabhan, M.V., Das, S.K. and Mishra, P.K. 1990. Field manual on watershed management. Central Research Institute for Dryland Agriculture, Hyderabad, India.

Singh, S. P. and Prasad, C. 1991. Technologies for maximizing risk in rainfed agriculture. Indian Society of Extension Education and ICAR, New Delhi.

Singh, S.N., Singh, K.P., Singh, N., Kadian, U.S. and Dahiya, S.S. 1988. Employment potentialities of different farming systems. In: Absts. of National Seminar on "Farming Systems for Semi-arid Tropics", Tamil Nadu Agric. Unlv., Coimbatore, India. p. 7.

Singh, S.P. 1987. Efficient Crop and Cropping Systems. In: Technological advances in dry land agriculture, 47-66. Hyderabad, India: CRIDA.

Singh, S.P. 1994. Hand book of agroforestry. Agrotech Publishing Academy. Udaipur. India.

Singh, S.S. 1990. Crop Management under Irrigated and Rainfed Conditions. Kalyani Publishers, Ludhiana.

Singh, S.S. 1996. Principles and practices of Agronomy. Kalyani Publishers, New Delhi.

Singh, S.S. 1996. Soil fertility and nutrient management. Kalyani Publishers, New Delhi.

Singh, S.S. 1997. Crop management under irrigated and rainfed conditions. Kalyani Publishers, New Delhi.

Singh, S.S., Gupta, P.and Gupta, A.K. 2004. Hand Book of Agricultural Science. Kalyani Publishers, Ludhiana.

Sinha, V.R.P. 1989. Fisheries Development in India, Fishing Chimes. 9(3): 23-29.

Sirohiet, A.S., Shanna, B.M., Singh, I. and Dayal, R. 1980. Role of dairy and poultry enterprises for increasing income and employment of farm in the Union Territory of Delhi. Indian J. Agrl. Econ. 35: 112-119.

Sivanappan, R.K. and Karai Gowder, K.R. 1977. Irrigation and drainage. Popular Book Depot., Madras, India.

Sivaraj, S. 1989. Economics of poultry-cum-fish culture as a component in the farming system under Thanjavur district, Tamil Nadu. M.Sc.(Ag.) Thesis, Tamil Nadu Agric. Univ., Coimbatore, India.

Sivasankaran, D. 1994. Integrated farming system research under rainfed lands. In: Proc. of Summer Institute on "IFS Research and Management for Sustainable Agriculture" held at Tamil Nadu Agric. Univ., Coimbatore, India, between 6-15th June, 1994. pp. 115-121.

Smith, L.P. 1975. Methods in Agricultural Meteorology, Elsevier Scientific Publishing Company, Amsterdam.

Somani, L.L. 1992. Dictionary of weed science. Agrotech Publishing Academy, Udaipur, India.

Somani, L.L. and Bhandari, S.C. 1990. Rhizobium In: Biofertilizers (Somani, L.L., S.C Bhandari, S.N. Saxena and KK Uyas (eds.), Scientific Publishers, Jodhpur, India. pp. 39-65.

Somani, L.L. *et al.*, (eds.). Scientific publishers, Jodhpur, India. pp. 39-65.

Somani, L.L., Vittal, K.P.R. and Venkateswarlu, 1992. Dryland Agriculture in India – State of Art of Research in India. Scientific Publishers, Jodhpur, India.

Soundara Rajan, M.S., Raddy, R.K., Venkateswarlu, M.S. and Sankar Raddy, G.H. 1981. Effect of Zero Tillage on weed control and yield of rainfed groundnut, Pesticides, 15(3): 17-18.

Spencer, D.S.C. 1991. Institutionalizing the farming systems perspective in multi-commodity research institutes. The role of systems based research groups. Exptl. Agric. 27: 1-9.

Sree Ramulu, V.S. 1995. Dry farming in India. Oxford and IBH Publishing Co. Pvt. Ltd., New Delhi.

Sreekumar, K.R. and Tilak, K.V.B.R. 1980. Nitrogen fixation in forage plants. Pasture and forage crops research: State of knowledge report. Singh P. (ed.) Proceeding of Third International Range Land Crops, New Delhi.

Srinivasan, V.M. 1994. Role of Agroforestry for sustainable agriculture. In: Proceedings of Summer Institute on "Integrated Farming Systems Research and Development for Sustainable Agriculture" organized by ICAR and Tamil Nadu Agric. Univ., Coimbatore, India during 6-15 June, 1994. pp. 168-173.

Srivastava, H.C., Bhaskaran, S., Menon, K.K.G., Ramanujam, S. and Rao, M.V. 1984. Pulse production constraints and opportunities. Oxford and IBH Publishing Co. Pvt. Ltd., New Delhi.

Srivastava, S.C., Johari, D.P. and Gill, P.S. 1988. Manual of Sugarcane production in India. ICAR, New Delhi.

Stephen, O. Duke. 1987. Weed physiology. Vol. II Herbicide physiology. CRC Press, Inc. Boca Raton, Florida.

Steppler, H. and Lundgren, B. 1988. Agroforestry: Now and in the Future. Outlook on Agriculture 17: 146-151.

Stinner, B. and Blair, J. 1990. Biological and agronomic characteristics of innovative cropping systems. In: Edwards, CA., *et al.*, (eds.). Sustainable Agricultural Systems. Soil and Water Conservation Society, Ankeng, Iowa. pp. 123-140.

Stocking, M. 1986. The cost of soil division in Zimbabwe in terms of the loss of three major nutrients. Consultant's working paper 3, Soil Conservation Programme, Land and Water Development division, FAa, Rome.

Stoop, W. 1984. Interdisciplinary and Intercommodity aspects in the organization of a farming systems research perspective. Paper presented at the INSORMIL workshop held at CIMMYT, EI Batan, Mexico, September 17-21, 1984.

Subba Rao, N.S. 1988. In Biological N fixation - Recent trends. Subba Rao, N. S (ed). Oxford and IBH, New Delhi. pp. 1-19.

Subbaian, P., Annadurai, K. and Palaniappan, S.P. 2000. Agriculture: Facts and Figures, Kalyani Publishers, Ludhiana, p. 359.

Subbian, P. and Palaniappan, S.P. 1992. Effect of integrated management practices on the yield and economics of crops under high intencity multiple cropping systems. India, J. Agron. 37: 1-5.

Sulton, O.G. 1953. Micrometeorology, McGraw Hill Company, Inc., New York.

Sundaresan, D. 1979. The role of improved buffaloes in rural development. Indian Dairyman. 31: 73-78.

Sunith, C.P. 1975. Methods in Agricultural Meteorology. Elsevier Scientific Publishing Co., Amstardam, India.

Surekha, K. 2007. Nitrogen release pattern from organic sources of different C : N ratios and lignin content and their contribution to irrigated rice. Indian Journal of Agronomy. 52: 220-224.

Swaminathan, K.R. 1992. Renewable energy technologies for rural women. Paper presented during Annual Biogas Seminar of AFPRO-CHFN Network, Nagercoil, October, 28, 1992.

Swaminathan, M. S. 2001. Science and sustainable food security. Indian Farming. 50: 4-6.

Swaminathan, M.S. 1975. Research advances in agriculture in relation to dairy development in India. Indian Dairyman. 27: 25-61.

Swaminathan, M.S. 1985. Imparting a rural women user perspective to agricultural research and development, IRRI, Phillippines.

Swaminathan, M.S. 1987. Inaugural address at the International symposium on "Sustainable Agriculture". The role of green manure crops in rice farming systems. 25-29 May, 1987. IRRI, Philippines.

TAC. 1989. Sustainable Agricultural Production: Implications for International Agricultural Research, CGIAR, Washington.

TAC/ICGIAR, 1988. Sustainable Agricultural Production: Implications for International Agricultural Research. FAO, Rome, Italy.

Tan, E.S.P. and Khoo, K.H. 1980. The integration of fish farming with agriculture in Malaysia. In: Proceeding of ICLARM -

SEARCA Conf. on "Integrated Agriculture - Aquaculture Farming Systems". 4th Aug., 1979., Manila, Philippines. pp. 175-187.

Taylore, J.A. 1967. (Ed) Weather and Agriculture. Pergamon, London: 1-10.

Tejwani, K.G., Gupta, S.K. and Mathur, H.N. 1975. Soil and Water Conservation Research 1956-71. New Delhi, India: ICAR.

Thakur, R.C., Bindra, A.D., Good, R.D. and Bhargava, M. 1995. Effect of Fertilizer application on wheat crop sequence. Indian J. Agron. 40: 4-13.

Thanh Hung, LE. 1992. Integration of aquaculture with coastal rice farming in Vietnam. NAG A, The ICLARM Quartly. 15 (2): 2-29.

Thiyagarajan, T. M. 2002. Experiments with a modified system of rice intensification in India. Research Report. India 1347-139.

Thomas, A. Blair and Robert C. Fite 1965. Weather Elements. Prentice-Hall, Inc. Englewood Cliff, n.j.

Thurlow, D.L and Hiltbott, A. 1985. Nitrogen fixation by soybeans in Alabama. Agron. J. 77: 432-436. "

Tilak, K.V.B.R. 1985. Interaction of VAM and N Fixers. In: Proceeding of National Symposium on "Current Trends in Soil Biology". Mishra, M.M. and K.K. Kapoor (eds.). HAU. Hisar, India, pp. 219:226.

Tilak, K.V.B.R. and Dwivedi, A. 1989. Field response of barley to *Azospirillum brasilense* inoculation. Indian J. Microbiol. 29: 61-64.

Tiuy, J. 1990. Agricultural Ecology. Longman Scientific and Technical, Essex, United Kingdom.

Tiwari, P.N. 1993. Integrated Farming Research for sustaining food production. J. Nuclear Agric. Biol. 20: 1-13.

Trewartha, A.T. 1964. An Introduction to climate, McGraw Hill Book Company, Inc. New York.

Tripathi, R.P. and Singh, H.P. 1993. Soil erosion and conservation. Wiley Eastern Ltd., New Delhi.

Troll, C. 1965. Seasonal climate of the earth. In: World maps of Climatology (Eds. E. Rodenwaldt and H. Jusatz). Berlin: Springer Verlag.

Ujah, J.E. and Adeoge, K.B. 1984. Effects of shelter belts in the Sudan Savanna Zone of Nigeria on Microclimate and yield of millet. Agric Forest Meteorol., 33:99-107.

Unger, P.W. and McCalla, T.M. 1980. Conservation Tillage, Adv. Agron. 33:2-58.

Vaidya, V.G. and Sahastrabudhe, K.R. 1970. Introduction to Agronomy and Soil and Water Management. Continental Prakasan, Poone.

Vangriensven, L.G. 1987. The cultivation of mushroom: Its present status and future development. Outlook on Agric. 6: 131-135.

Veerabadran, V. 1994. Farming systems research in southern parts of Tamil Nadu. In: Proceeding of Summer Institute on "Integrated Farming System Research and Management for Sustainable Agriculture". 6-15th June, 1994. Tamil Nadu Agric. Univ., Coimbatore. pp. 258-269.

Velayutham, M. 1996. Management of Natural Resources in India. Souvenir, 2nd International Crop Science Congress, Nov. 17-24, 1996. New Delhi, India.

Venkataraman, G.S. and Shanmugasundaram, S 1992. Algal bio-fertilizer technology for rice. Madurai Kamarajar University, Madurai, India. pp. 1-24.

Venkataraman, S. and Krishnan, A. 1992. Crops and Weather. ICAR, New Delhi.

Venkateswarlu, J. 1987. Efficient resource management system for Drylands of India. Advances in Soil Science. 7: 165-221.

Ventskevich, C.Z. 1961. Agrometeorology, Israel Program for Scientific Translations. Jerusalwm: 32-33.

Verma, B., Bhola, S.N., Prakash, C. and Prasad, S.N. 1983. Soil and water loss studies under different vegetative covers under 1 % slope in Kota clay soils. Rep. CSWCRTI, Dehradun: 96 p.

Verma, L.N. and Bhattacharyya, P. 1992. Production, distribution and production of biofertilizer. In: Fertilizers, Organic Manures, Recyclable Wastes and Biofertilizers. Tandon, H.L.S. (ed.), FDCD, New Delhi, pp. 132-147.

Vikraman Nair, R., Pillai, N.G. and Shaskanaman, 1979. Mixed farming in coconut gardens. Indian Farming. 26: 24-25.

Virmani, S.M. 1997. Rainfed Farming: Watershed based Approach. Survey of Indian Agriculture. The Hindu. Chennai, India.

Vishwa Nath and Chhotey Lal. 1995. Oilseeds in India – An overview. Westvill Publishing House, New Delhi.

Wali, M.K. (ed.), 1992. Ecosystem rehabilitation: Preamable to sustainable development. SPB Academic Publishing. The Hague, Netherlands.

Walker, T.S. 1982. The perceptions of farmers and the participation in the Taddanpalle watershed project in 1982-1983. ICRISAT economics program progress report, 44. Andhra Pradesh, India: ICRISAT.

Walker, T.S. and Ryan, J.G. 1990. Village and household economics in India's Semi-arid tropics. Baltimore, Maryland, USA: Johns Hopkins University Press.

Wang, J.Y. 1963. Agricultural Meteorology. Pace Maker. Milwaukee: 1-12.

Wani,. S.P. 1990. Inoculation with associative N fixing bacteria: Role of cereal grain production improvement. Indian J. Microbiol. 30: 363-393.

Wani,.S.P., Chandrapalaiah, I.S. and Dad, P.O.J. 1987. Nitrogen fixation potentialities of sorghum and millets. In: Biological Nitrogen Fixation: Recent Development. Subba Rao, N.S. (Ed.) Oxford and IBH, New Delhi. pp. 125-174.

Watson, D. J. 1947. Comparative physiological studies on the growth of field crops. I. Variation in net assimilation rates and leaf area between species and varieties, and within and between years. Ann. Bot. N.S. 11: 41-76.

Watson, D. J. 1952. The physiological basis of variation in yield. Adv. Agronomy. 4: 101-145.

WCED, 1987. Our common future. Oxford University Press.

Weiskel, T.C. 1989. The ecological sessions of the past: An Anthropology of environment decline. The Ecologist. 19: 98-103.

Weiss, E.A. 1983. Oilseed crops. Longman Group Ltd., London

Went, F.W. 1962. Weather and Indian Farmer, Government of India, Indian Meteorological Department, Pune: 4-22.

Went, F.W. 1970. Climate of Agriculture. Plant Agriculture. Reading from Scientific American. Freeman, San Francisco: 109-118.

Whyte, W.F. 1977. Towards a new strategy for research and development in agriculture; Helping small farmers in developing countries. Desarrollo Rural En Los Americas. 9: 51-62.

Whyte, W.F. 1981. Participatory approaches to agricultural research and development: A State of the Art Paper. Cornell University, Center for International Studies. Rural Development Committee, Special series on agricultural research and extension. 1, Ithaca, New York. Cornell. University.

Wild, A. 1988. Russel's soil conditions and plant growth. Longman group, London.

Wilken G.C. 1987. Good farmers: Traditional agricultural resource management in Mexico and Central America. Berkely: University of California Press.

Willey, R. W. 1979. Intercropping-Its importance and research needs 1. Competition and yield advantage and 2. Agronomy and research approaches. Field crop Abstracts, 32: 1-10, 73-85.

Willey, R.W. and Osiru, D.S.O. 1972. Studies on mixtures of maize and beans with particular reference to plant populations. J. Agric., Sci. 79: 579-529.

Williams, J.R., Jones, C.A. and Dyke, P.T. 1984. A modeling approach to determining the relationship between erosion and soil productivity. Trans. ASAE 27: 129-144.

Williams, R., Zangavil, A. and Karmidi, A. 1980. A portable evaporimeter for rapid measurement of the evaporation rate of water. Agric. Forest Meteorol. 32:217-224.

Winkelmann, D. 1984. Recent views in farming systems research. In: Issues in Organization and management of research with a farming systems perspective aimed at technology generation (Ed. M. Joseph Chang). The Hague, Netherlands: CIMMYT and ISNAR.

Wischmeier, W.H. 1955. Punch card record runoff and soil loss data. Agric. Engng. 36: 664-666.

World Resources, 1988-89. 1990-1991, 1992-1993. Reports by the World Resources Institute and International Institute for Environment and Development in Collaboration with the United Nations Environment Programme, Basic Book, New York.

Woyanarovich, E. 1980. Raising ducks on fishponds. In: Proceeding of ICLARM - SEARCA Conf. on "Integrated Agriculture-aquaculture Farming System", 4th Aug. 1979, Manila, Philippines.

Yadav, R.L. 1993. Agronomy of Sugarcane. Principles and practices. International Book Distributing Co. Lucknow, India.

Yellamada Reddi, T. and Sanka Reddi, G. H. 1992. Principles of Agronomy. Kalyani Publishers, New Delhi.

Yohida, S. 1981. Fundamentals of rice crop science. International Rice Research Institute, Manila, Philippines.

Young, A. 1989. Agroforestry for soil conservation. CAB International. Walling Ford, U.K.

Zandstra, H.G. 1979. Cropping systems research for Asian rice farmers. Agricultural Systems, 4: 135-155.

Zandstra, H.G. 1982. An overview of farming systems research. Keynote address for farming systems in the field "Farming Systems Research Symposium", Kansas State University, Manhattan, Kansas, November 20-21, 1982.

Zandstra, H.G., Pric, E.C., Litsinger, J.A. and Morris, R.A. 1981. A methodology for on-farm cropping systems research. Losbanos, Philippines: International Rice Research Institute.